Electronic Devices and Circuits

Dr. Sanjay Sharma

ELECTRONIC DEVICES AND CIRCUITS

Dr. SANJAY SHARMA
B.E., M.Tech, Ph.D, MIETE

Edited By
DEEKSHA SHARMA
&
V. SHARMA

S.K. KATARIA & SONS®

Publisher of Engineering Books
4885/109, Prakash Mahal, Dr. Subhash Bhargav Lane,
Opposite Delhi Medical Association, Daryaganj, New Delhi–110002 (INDIA)
Phone: +91-11-43551243, Mobile: +91-9871775858, 7678425209
e-mail: katariabook@gmail.com, katariabooks@yahoo.com
Website: www.skkatariaandsons.com
(e-books, subscriptions and bulk buying also available on our website)

Published by:

S.K. Kataria & Sons®

4885/109, Prakash Mahal, Dr. Subhash Bhargav Lane,
Opposite Delhi Medical Association, Daryaganj, New Delhi-110002 (INDIA)
Phone: +91-11-43551243; Mobile: +91-9871775858, 7678425209
e-mail: katariabook@gmail.com, katariabooks@yahoo.com
Website: www.skkatariaandsons.com

First Edition: July, 2009; Second Edition: 2012; Reprint: 2025

Laser Typesetting by : Plus Computers, Delhi

Preface to the Second Edition

I feel happy in presenting the second revised and enlarged edition of this book to the students. In fact, the favourable and warm reception which the first edition has earned has been a matter of great satisfaction for me. The author wishes to express his deep gratitude to the large number of readers who have used it and a particular to those of them, who sent valuable suggestions for the improvements of the book.

In this addition, the author has made sincere efforts to make the book up-to-date.

The present edition has been thoroughly revised and a lot of useful material has been added. To make it more useful, several solved examples and problems have been added from the Examination point of view. The mistakes and misprints, which had crept in last edition, have been eliminated in this edition. The author hopes that all these features will make the book even more useful and attractive to the readers

Any errors, misprints and suggestions for the improvements of the book, brought to my notice would be gratefully and thankfully acknowledged and incorporated in the next edition.

Dr. Sanjay Sharma
August, 2012

Preface to the First Edition

Electronic Devices and Circuits is a compulsory subject for the engineering students from all the disciplines. This book is primarily intended to serve as a textbook in accordance with the syllabus of Electronic Devices and Circuits offered by RGTU, Bhopal.

The objective of the book is to provide a clear explanation of the operation of all important electronic devices and circuits in general use today.

This book has been written in a very simple and lucid manner. Every effort has been made to make the treatments simple and comprehensive. Throughout the book, the stress has been given on fundamental concepts through illustrative examples. Neat and Clear diagrams have been used for explanation.

Summary of each chapter is provided to help the students to grasp the subject clearly. A large number of self explanatory accurate diagrams and tables have been used to supplement the text. Exercises and problems are also provided. These have been selected from examination papers of various Indian Universities. This will help the students to prepare accordingly.

This book is also intended as a textbook for the use in Electronic and Communication courses in polytechnics, IETE, AMIE and for various Engg. examinations. Thus, primarily designated as an undergraduate textbook for students of science and technology, this study will provide balanced coverage of devices, analog and digital circuits, with an emphasis on analog electronics. It also serves as a quick reference for practising professionals.

The whole text has been logically organized and spreaded into 13 chapters.

The author hopes that with these unique features the book will fulfill the genuine requirements of the student community. Lastly, much originality cannot be claimed in a book of this kind. The author takes this opportunity to place on record his indebtedness to the large number of books and journals that he has freely consulted in the preparation of this book.

Dr. Sanjay Sharma

July, 2009

Acknowledgement

I take this opportunity to express my gratitude and thanks to the respected Director General of KIET, **REAR ADMIRAL Dr. AJAY SHARMA** for his valuable technical suggestions and constant encouragement, without which this book would not have come into existence.

I am specially grateful to **Dr. S. NASEEM AHMAD**, Professor and Head, Department of Electronics and Communication Engg., Jamia Millia Islamia, New Delhi for his time-to-time, much needed valuable guidance.

I would like to express my sincere thanks to **Mr. Vibhav Sachan** (Associate Professor, E&C, KIET), and **Mr. Manish Gupta** (Lecturer, E&C, MIT, Modinagar) for constant inspiration and encouragement.

Special thanks to **Dr. B.P. Singh** (Former Professor, Electronics & Comm. Engg., MMMEC, Gorakhpur), **Dr. R.K. Chauhan** (Professor, Electronics & Comm. Engg., MMMEC, Gorakhpur), **Dr. G.S. Tripathi** (Electronics & Comm. Engg., MMMEC, Gorakhpur), **Dr. Arun Kumar** (Former Professor, Electronics & Comm. Engg., MMMEC, Gorakhpur) and **Dr. C.B. Tripathi** (Professor Electronics & Comm. Engg., MMMEC Gorakhpur) and **Dr. Vipin Kumar** (H.O.D. Physics, KIET), and **Mr. Manish Gupta** (Lecturer, MIT, Modinagar) for their guidance and valuable suggestions.

I am grateful to my younger brothers **Rajeev** and **Rakesh,** daughter **Ritika**, for inspiring me for this project.

I deeply express my heartful thanks to the publishers **S.K. Kataria & Sons** for publishing this book in such a beautiful get-up and well in time.

Lastly, I am also grateful to **Mr. Manoj Tyagi** and his staff from **Plus Computer's** for their hard work.

Dr. Sanjay Sharma

KIET, Ghaziabad

July, 2012

I DEDICATE THIS BOOK TO THE READERS OF THIS BOOK

CONTENTS

UNIT I

Semiconductor device thoery of P-N junction, temperature dependence and breakdown characteristics, junction capacitance, Zener diode, Varactor diode, PIN diode, LED, Photo diode Transistors BJT, FET, MOSFET, types, working principle, characteristics, and region of operation, load line biasing method, Transistor as an amplifier, gain, bandwidth, frequency response, h-parameters equivalent type of amplifier.

Above syllabus has been covered in following chapters :

Chapter 1

SEMICONDUCTOR DEVICES

Syllabus

Semiconductor device thoery of P-N junction, temperature dependence and breakdown characteristics, junction capacitance, Zener diode, Varactor diode, PIN diode, LED, Photo diode Transistors BJT, FET, MOSFET, types, working principle, characteristics, and region of operation, load line biasing method.

1.1. SEMICONDUCTORS

A semiconductor material is one whose electrical properties lie in between those of insulators and good conductors. At room temperature, such materials have conductivities considerably lower than those of conductors and much higher than those of insulators. Infact, the conductivity of a semiconductor lies in the range of 10^5 to 10^{-4} siemens per meter. An important property of a semiconductor is that its resistance decreases with the increase in temperature. This means that a semiconductor has a negative temperature coefficient of reistance. An even more important property of a semiconductor is that their resistivity or conductivity changes considerably when even a very small amount of certain other substances called the impurities, are added to them.

1.2. CLASSIFICATION OF SEMICONDUCTORS

Basically Semiconductors are of two Types: Elemental and compound. The examples of elemental semiconductors are germanium (Ge) and silicon (Si). Germanium was the mainly used material for the majority of the early solid state devices like transistors. Subsequently, germanium has been replaced by silicon in several applications because of the fact that silicon

* Distinguish between intrinsic and extrinsic semiconductor. What is the effect of temperature on conductivity of a semiconductor?

(RGTU, Bhopal, Sem. Exam. 2007-08)

has a better thermal stability, is readily available and has an advanced technology. The example of compound semiconductor is GaAs.

Semiconductors may also be classified in following way:

1. Intrinsic semiconductor
2. Extrinsic semiconductor

1.3. INTRINSIC SEMICONDUCTOR*

(MDU, Rohtak, Sem. Exam. 2005-06)

A semiconductor in an extremely pure form is known as **intrinsic semiconductor.** A semiconductor is not intrinsic unless its impurity content is less than one part impurity in 100 million parts of semiconductor.

The silicon and germanium are the two most widely used intrinsic semiconductors. Figure 1.1 shows the two dimensional crystal structure of silicon. There are four electrons in the outermost orbit or valence shell.

> **DO YOU KNOW?**
>
> Silicon as well as Germanium atoms contain 4-electrons in the valence shell. If there are 4-electrons in the outer-most orbit, the semiconductor material is called as pure or intrinsic semiconductor.

Each of the four valence electrons takes part in forming covalent bonds with the four neighbouring atoms. A covalent bond consists of two electrons, one from each adjacent atoms. Atoms bond together to form molecules because each atom attempts to get eight electrons in the valence shell to become a stable structure.

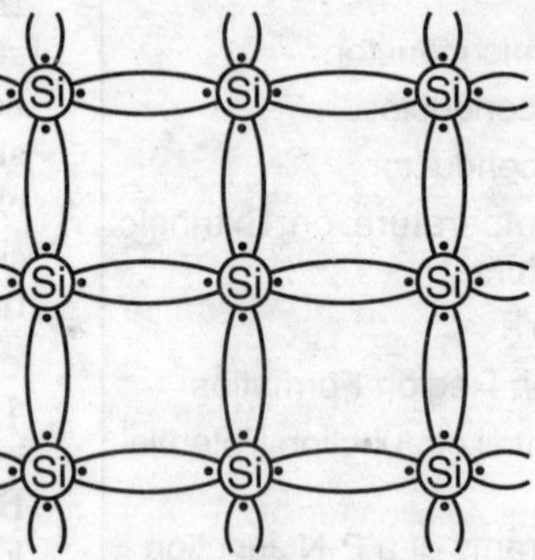

Fig. 1.1. Intrinsic semiconductor

At a low temperature such as absolute zero (0 K), all the valence electrons are tightly held by parent atoms and by covalent bonds with other atoms. Hence electrons cannot move through the crystal structure or in other words, no free electrons

---

* Define an intrinsic material, a negative temperature coefficient, and covalent bonding. List three materials that have negative temperature coefficient.

(JNTU, Hyderabad, Sem. Exam. 2006-07)

are available to conduct electricity. Thus, the intrinsic semiconductors, at absolute zero (0 K) behave as an **insulator**.

1.3.1. Crystalline Structure of Intrinsic Semiconductors

Some of the unique qualities of Silicon and Germanium are due to their special type of atomic structure. The atoms of both these materials form a very definite pattern which is periodic in nature, i.e., it continuously repeats itself. One such complete pattern is called as **crystal** and the periodic arrangement of atoms is called as lattice. For the Silicon and Germanium, the crystal has a three dimensional diamond shape as shown in figure 1.2.

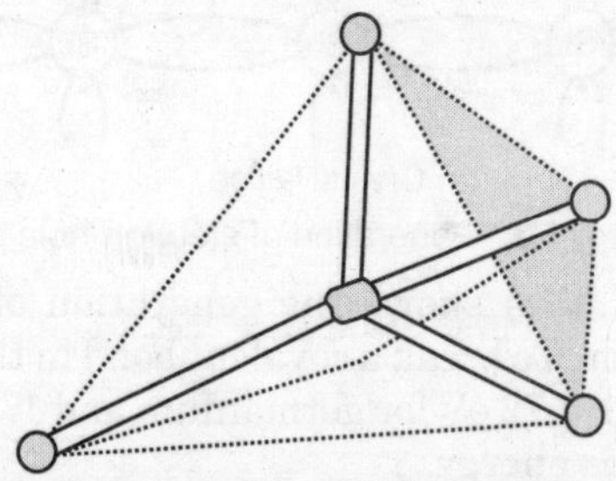

Fig. 1.2. Silicon and Germanium crystal structure

1.4. INTRINSIC SEMICONDUCTOR AT ROOM TEMPERATURE (300 K) OR THERMAL GENERATION OF ELECTRON-HOLE PAIRS

(RGTU, Bhopal, Sem. Exam. 2006-07)

At absolute zero, an intrinsic semiconductor behaves as a perfect insulator. But if the temperature is increased upto room temperature (300 K), some covalent bonds break. When a covalent bond breaks, the electron becomes free to move through the crystal. When an electron breaks a covalent bond and becomes free, a vacancy is also produced. This vacancy is known as a **hole**. Hence, when a free electron is produced, a hole is also produced simultaneously. This means that electrons and holes are produced in pairs.* In intrinsic semiconductors, the concentration of free electrons will always be equal to the concentration of holes. This type of simultaneous generation of electrons and holes due to temperature is also called **thermal generation**. Hence, at room temperature intrinsic semiconductor has some conductivity. Since, an electron is a negatively charged

* The librated electron and the hole left behind is called an electron-hole pair.

particle, the vacancy (or hole) created by this electron will be assumed as positively charged. These electrons and holes are also called **free charge carriers** because they are free to move throughout the crystal.

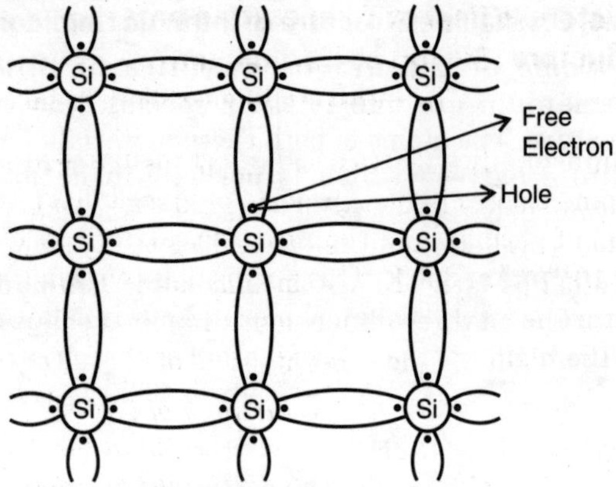

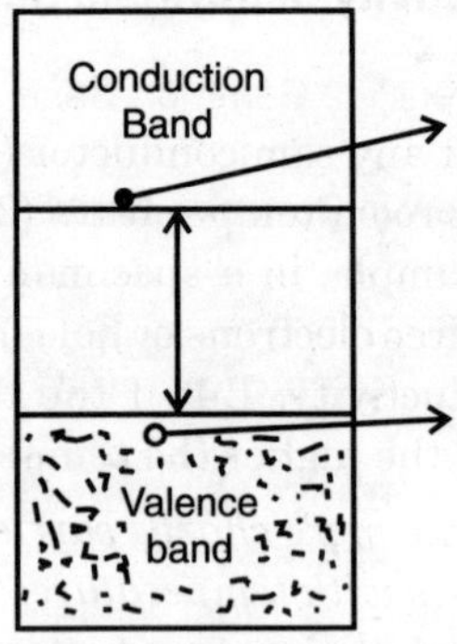

Fig. 1.3. Generation of electron-hole pair at room temperature in intrinsic semiconductor

Figure 1.3 (a) shows the generation of electron-hole pair at room temperature in intrinsic semiconductor. To break a covalent bond in the crystal lattice, a certain amount of energy is required. This energy is 0.72 eV for germanium and 1.12 eV for silicon. Room temperature (300 K) is sufficient to provide this energy.

Figure 1.3 (b) shows the energy band-diagram for electron-hole pair generation at room temperature in intrinsic semiconductor. Whenever an electron-hole pair generates, the hole remains in the valence band and electron lifts to the conduction band to take part in conduction of current. The energy to lift the electron from the valence band to conduction band is **0.72 eV for Ge** and **1.12 eV for Si**.

> **DO YOU KNOW?**
>
> A hole and an electron each possesses a charge of 1.6×10^{-19} C but of opposite polarity.

1.4.1. Why Intrinsic Semiconductors Behave Like Insulators at Low Temperatures?

(RGTU, Bhopal, Sem. Exam. 2008-09)

In each covalent bond, both the electrons involved are shared by both the parent atoms. Thus, with four covalent bonds associated with each atom, all the silicon atoms will effectively have eight electrons in their valence shells. Four of these electrons are their own and remaining four are due to the four covalent bonds. Due to this, the outermost shells of all the atoms are completely filled.

This makes the outermost shells of all atoms stable. Therefore, not a single free electron is available at the absolute zero temperature, i.e., at 0°K or – 273°C. Hence, the intrinsic semiconductor acts like an insulator at the absolute zero temperature.

1.4.2. Recombination

In case of an intrinsic semiconductor, electron-hole pairs are generated due to temperature or thermal agitation. These thermally generated electrons and holes move freely throughout the crystal. Due to this, there is a possibility of collision between electrons and holes. Whenever, there is a collision, an electron takes the position of holes and both of them disappear. This process is called **recombination**. Thus, in the process of recombination, both the free electron and hole disappear and energy* is released as a quantum heat or light. This quantum is absorbed by another electron to breakaway from its valence band and creates a new election-hole pair. At a given temperature, the rate of generation of electrons and holes is equal to the rate of recombination.

* Whenever an electron is freed from valence band, a certain amount of energy is absorbed. The same amount of energy is released when an electron falls into a hole.

Before recombination an electron or hole covers some distance or exists for a very small time. This average time of existence of an electron or hole is called ***mean lifetime*** *or lifetime. This lifetime falls in the range of 1μs to 10^3 μs.*

1.4.3. Conduction in Intrinsic Semiconductors: Effect of Temperature on Conductivity of Intrinsic Semiconductors (Silicon or Germanium)

(RGTU, Bhopal, Sem. Exam., 2005-06)

We know that any semiconductor (germanium or silicon) acts as a perfect insulator at absolute zero. However, at room temperature (300 K) some electron-hole pairs are produced due to thermal energy. As an example, in a specimen (sample) of germanium, the intrinsic carrier concentration (concentration of free electrons or holes) is $\mathbf{2.5 \times 10^{19}/m^3}$ at 300 K. This means that the semiconductor has a small conductivity. But if the temperature is raised further, more electron-hole pairs are produced. Hence, the higher the temperature, the higher is the concentration of charge carriers.

Therefore, since more charge carriers are made available, the conductivity of an intrinsic semiconductor increases with temperature. This means that the resistivity decreases with the increase in temperature. Hence, the semiconductors have ***negative temperature-coefficient*** *of resistance.*

1.4.4. Electron and Hole Current

(RGTU, Bhopal, Sem. Exam. 2008-09)

As we apply voltage across a piece of intrinsic semiconductor material, the thermally generated free electrons in the conduction band are attracted towards the positive end. The current constituted due to movement of free electrons is called as **electron current** as shown in figure 1.4.

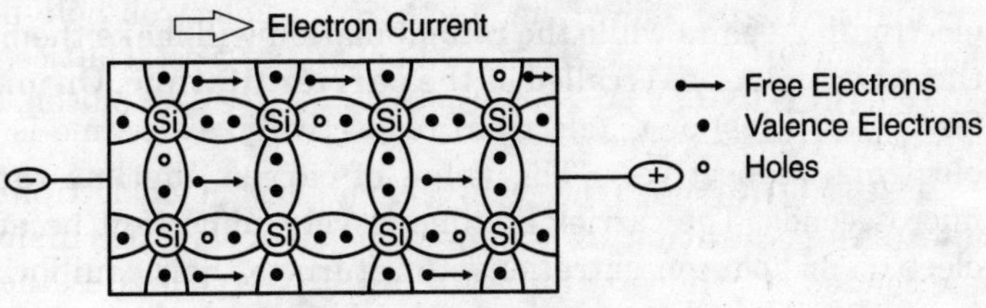

Fig. 1.4. Electron current

(i) Hole current

Figure 1.5 (a) shows three neigbouring silicon atoms A, B and C. The covalent bonds do exist but for convenience they have not been shown. A hole exists in a broken covalent bond with atom A as shown in figure 1.5(a). A valence electron may jump from another atom B to fill up this hole as shown in figure 1.5 (b). When it jumps, the electron leaves a hole behind it and thus the hole has moved in the direction opposite to that of the electron. Similarly the hole passes from atom B to atom C as shown in figure 1.5 (c). The current constituted due to this hole movement is called as the **hole current**. Thus, holes move in the opposite direction to that of electrons.

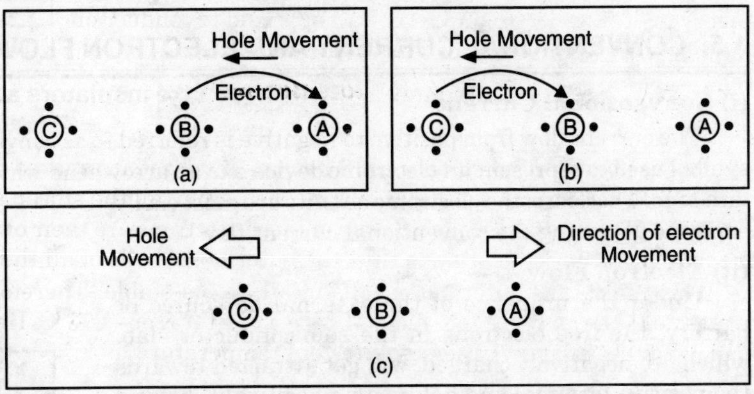

Fig. 1.5. Movement of electrons and holes.

Important Point: *Here, it may be noted that the hole current flows due to the valence electrons and not due to the free electrons. These electrons move due to the presence of holes. Therefore, holes are also considered as charge carriers like the electrons and the hole current always flows in opposite direction to that of the 'electron' current.*

(ii) Total current

The flow of an electric current is due to the movement of electrons in the conduction band and movement of holes in the valence band.

Thus, we can write

Total current = electron current + hole current

1.4.5. Semiconductors have a Negative Temperature Coefficient of Resistivity

Just now we have seen that as the temperature rises, more number of electrons will absorb the thermal energy to break the covalent bonds and contribute to the number of free carriers. The increased number of free electrons and holes will increase the conductivity of the semiconductor and results in a lower resistance level. Thus, resistance of the semiconductors decreases with increase in temperature. Therefore, semiconductors are said to have a *negative temperature coefficient of resistivity*.

1.4.6. Carrier Lifetime

In a pure semiconductor, the number of free electrons is equal to the number of holes. The thermal energy keeps producing new electron-hole pairs whereas some of the hole electron pairs will disappear due to the process of recombination. Thus, thermal agitation gives rise to production of electron-hole pairs while the recombination will make them disappear. The time duration between these two processes is called as the **carrier lifetime**. On an average, a hole or an electron will exist for τ_p (for holes) or τ_n (electron) respectively. This time is called as **mean lifetime** of holes and electrons respectively. The value of carrier lifetime ranges from a few nanoseconds to few microseconds. The carrier lifetime is very important because it indicates the time taken by the electron or hole concentrations, to return to their equilibrium value. The carrier lifetime can be precisely adjusted to a desired value by introducing gold as recombination agent into the silicon under a controlled condition.

1.5. CONVENTIONAL CURRENT AND ELECTRON FLOW

(i) Conventional Current

The current flow from positive to negative is referred to as conventional current. Every graphical symbol used to represent an electronic device has an arrowhead which indicates conventional current direction. Therefore, the electronic circuits are most easily explained by using the conventional current flow.

(ii) Electron Flow

Under the influence of the external dc source or battery, the free electrons in the semiconductor slab, which are negatively charged, will get attracted towards the positive terminal and holes being positively charged will be attracted towards the negative terminal of the external battery. Thus, as shown in figure 1.6, the electrons flow exactly in the opposite direction to that of the conventional current. The study of electron movement is essential in understanding the operation of any electronic device.

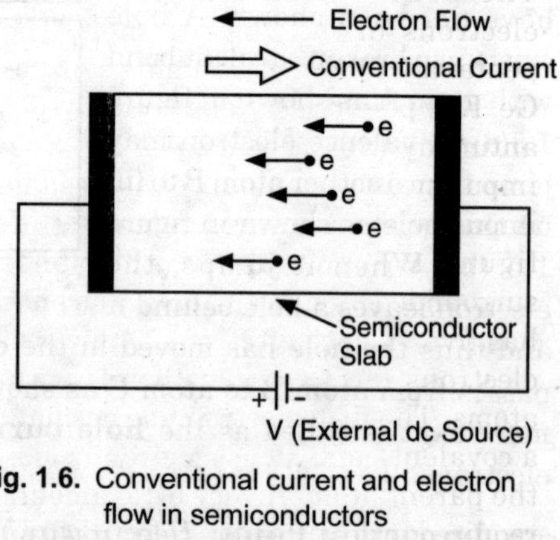

Fig. 1.6. Conventional current and electron flow in semiconductors

1.6. EXTRINSIC SEMICONDUCTORS

(RGTU, Bhopal, Sem. Exam. 2005-06)

As discussed earlier, intrinsic or pure semiconductors have small conductivity at room temperature. Therefore, they are not of much use. By adding some amount of impurity atoms to a pure or intrinsic semiconductor, we can change its conductivity or characteristics. This process of adding impurity to a pure semiconductor is called **doping.** Hence, the conductivity is increased by means of the doping process. Doping is done at a rate such that only one atom of impurity is added per 10^6 to 10^{10} semiconductor atoms.

> **DO YOU KNOW?**
>
> The extrinsic semiconductors have a better conductivity than the intrinsic semiconductors. Therefore, these are used in manufacturing of all the electronic components such as diodes, transistors etc.

We add impurities either to increase the number of electrons or holes in a semiconductor. The impurities are the materials used to dope the intrinsic semiconductor materials. These impurity materials can be of two types:

(i) Donor impurity
(ii) Acceptor impurity

(i) Donor impurity (Pentavalent impurity) **(RGTU, Bhopal, Sem., Exam., 2007-08)**

The material which is being used as impurity in the process of doping is called as *dopant*. When the dopant is a pentavalent atom, i.e., the atom containing five valence electrons then it is called as the *donor impurity* and the doping is called as *donor doping*. Donor doping is used to manufacture N-type extrinsic semiconductor. The examples of pentavalent or donor impurities are Arsenic, Phosphorous and Antimony.

(ii) Acceptor impurity (Trivalent impurity)

When the dopant is a trivalent atom, i.e., the atom consisting of only three valence electrons, then it is called as the *acceptor impurity* and the doping is called as *acceptor doping*. Acceptor doping is used to manufacture P-type extrinsic semiconductors. The examples of trivalent or acceptor impurities are Boron, Gallium, Aluminium and Indium.

A doped semiconductor is called **extrinsic semiconductor.** Depending upon the types of impurity added, there are two types of extrinsic semiconductors :

(a) N-type Semiconductor;
(b) P-type Semiconductor.

1.7. N-TYPE SEMICONDUCTOR*

If pentavalent impurity atoms are added to intrinsic semiconductor, N-type semiconductor is obtained. The pentavalent impurity atoms have five electrons in the valence shell. The size of the pentavalent atoms is roughly same as that of Si or Ge. Few pentavalent elements are phosphorus (P), antimony (Sb), arsenic (As) etc. Since amount of impurity is very small, it is assumed that each impurity atom is surrounded by silicon atoms. In figure 1.7, an impurity phosphorus atom is surrounded by silicon atoms. The phosphorus atom has five valence electrons. Four, out of these five electrons form covalent bonds with four silicon atoms. The fifth electron has no chance of forming a covalent bond. This electron is loosely bound to the parent atom. A very little amount of energy is required to de-attach this electron from the nucleus

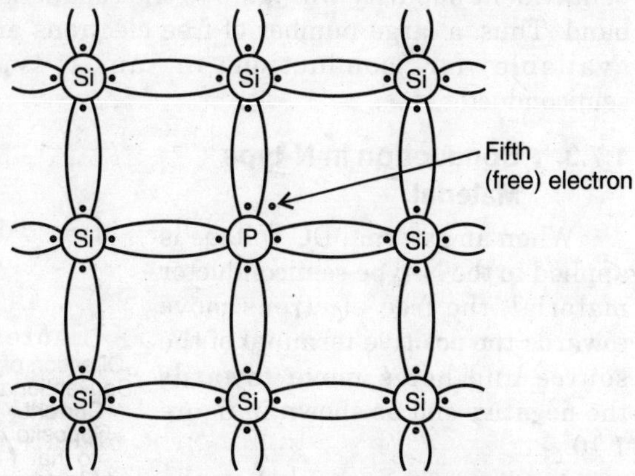

Fig. 1.7. N-Type semiconductor.

* Differentiate among conductor, insulator and semiconductor using energy band concept. Also differentiate between the n-type and p-type semiconductor materials. **(RGTU, Bhopal, Sem. Exam. 2006-07)**

of its parent atom. This energy is of the order of 0.05 eV for silicon and 0.01 eV for germanium. Practically, at room temperature, all such electrons become free.

In this way, each impurity atom donates one electron to the conduction band. Due to this reason, pentavalent impurity is also called **donor type impurity.** Besides a large number of donated electrons, there are also some thermally-generated, electron-hole pairs. A large number of electrons increases the rate of recombination of electrons with holes. Because of this, the net concentration of holes is much less than intrinsic value. Hence, in a **N-type semiconductor, electrons are majority carriers and holes are minority carriers.**

The pentavalent impurity atom donates one electron. After donating this electron, the impurity atom becomes a positively charged ion. This is also called **donor ion.** This donor ion is held tightly by four covalent bonds in crystal. This is unable to move and hence is called positively charged **immobile ion.**

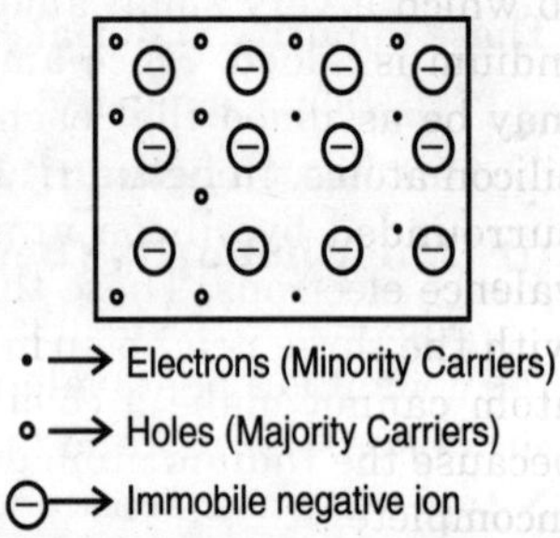

Fig. 1.8. Representation of N-type semiconductor

1.7.1. Representation of N-type Semiconductor

We know that in a N-type semiconductor, electrons are the majority carriers and holes are the minority carriers. Also, there are a large number of immobile positive ions. N-type semiconductor is not negatively charged but is electrically neutral. The total number of electrons is exactly equal to the total number of holes and immobile ions. Hence, negative charge is exactly balanced by the positive charge.

1.7.2. Changes in the Energy Band Diagram due to Doping

Addition of donor impurity to an intrinsic semiconductor is called *doping*. The process of doping will alter the energy band diagram to a great extent as shown in figure 1.9. The donor energy level (E_D) shown by the dotted lines is very close to the conduction band energy level (E_C). Therefore, the unbonded valence electrons of the impurity atoms can very easily jump into the conduction band and become free electrons. Thus, even at room temperature, almost all the extra electrons of pentavalent impurity will jump to the conduction band. Thus, a large number of free electrons are available for conduction in an N-type semiconductor.

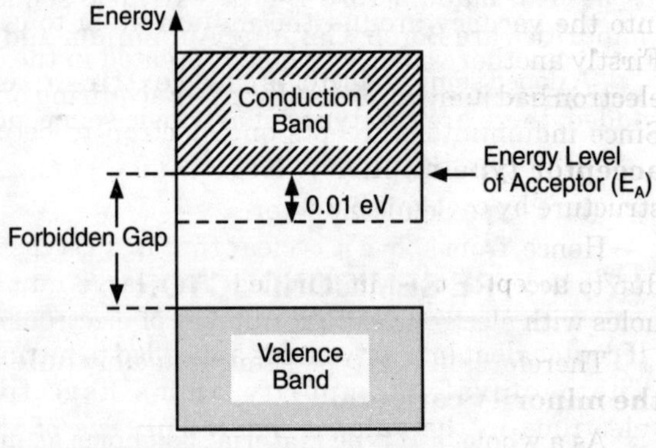

Fig. 1.9. Energy band diagram of N-type semiconductor

1.7.3. Conduction in N-type Material

When an external DC voltage is applied to the N-type semiconductor material, the free electrons move towards the positive terminal of the source and holes move towards the negative end as shown in figure 1.10.

As electrons outnumber the holes, the conduction in N-type material is mainly due to the majority charge carriers *electrons.*

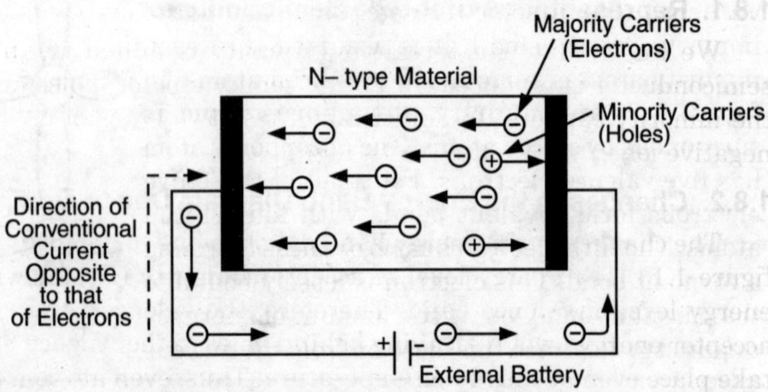

Fig. 1.10. Conduction in N-type material.

1.8. P-TYPE SEMICONDUCTOR

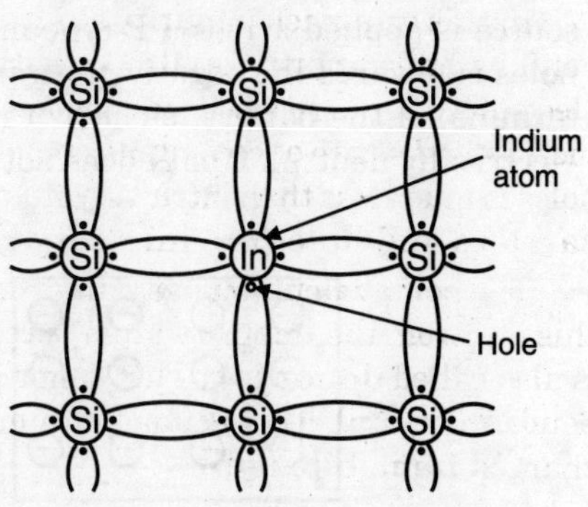

Fig. 1.11. P-type semiconductor.

If trivalent impurity atoms are added to intrinsic semiconductor, P-type semiconductor is obtained. The trivalent impurity atoms have three electrons in the valence shell. The size of the trivalent atoms is roughly same as that of Si or Ge. The trivalent impurities are boron, aluminium, gallium and indium. As an example, consider a sample of intrinsic silicon to which a very small amount of trivalent impurity such as indium is added. Since amount of impurity is very small, it may be assumed that each impurity atom is surrounded by silicon atoms. In figure 1.11, an impurity atom (Indium In) is surrounded by silicon atoms. The Indium atom has three valence electrons. These three electrons form covalent bonds with the three neighbouring silicon atoms. The fourth silicon atom cannot make a covalent bond with the Indium atom because the Indium atom does not have fourth valence electron. Hence, the fourth covalent bond is incomplete.

A vacancy that exists in the incomplete covalent bond constitutes a hole. This vacancy or hole have a tendency to complete the covalent bond from the neighbouring atoms. An electron from neighbouring atoms require some energy to jump into the vacancy.

At room temperature, this small amount of energy is provided by thermal energy. This energy is 0.05 eV for silicon and 0.01 eV for germanium. When an electron from the neighbouring atoms jump into the vacancy around the indium atom to complete the covalent bonds, the effect is two fold. Firstly another vacancy or hole is created in the covalent bond of surrounding atom from where the electron had jumped. Secondly, after acquiring an electron, the indium atom becomes a negative ion. Since indium atom accepts one electron to become negative ion, it is also called acceptor ion or **acceptor type impurity**. Negative ion is also immobile because it is held tightly in the crystal structure by covalent bonds.

Hence, from above it is clear that in a P-type semiconductor, a large number of holes are created due to acceptor type impurities. This large number of holes increases the rate of recombination of holes with electrons so that number of electrons is further reduced than intrinsic level.

Therefore, in a P-type semiconductor, **holes are the majority carriers and electrons are the minority carriers.**

As a whole, a P-type material has holes as majority carriers, electrons as minority carriers and negative immobile ions.

1.8.1. Representation of P-type Semiconductor

We know that if a trivalent impurity is added to an intrinsic semiconductor, a P-type semiconductor is produced. In P-type semiconductor, holes are the majority carriers, electrons are the minority carriers. In addition to these two types of charge carriers, there are also immobile negative ions.

1.8.2. Changes in the Energy Band Diagram Due to Doping

The changes in the energy band diagram due to doping with acceptor impurities is as shown in figure 1.13. The energy level of acceptor impurity (E_A) shown by dotted lines is very close to the energy level of valence band. Therefore, the valence band electrons can very easily jump to the acceptor energy level E_A leaving behind holes in the valence band. This process of hole creation can take place even at the room temperature. Thus, even at room temperature, a large number of holes are created inside a P-type semi-conductor. As holes are positively charged, the presence of a large number of holes will make the semiconductor a P-type semiconductor.

1.8.3. Conduction in P-type Semiconductors

The conduction inside a P-type semiconductor is as shown in figure 1.14. When an external dc source is applied across a P-type material, the holes move in the valence band. The movement of holes is towards the negative terminal of the supply. The free electrons move towards the positive terminal of the battery, as shown in figure 1.14. Note that the P and N-type semiconductors are electrically neutral. P or N does not mean that they are positively or negatively charged.

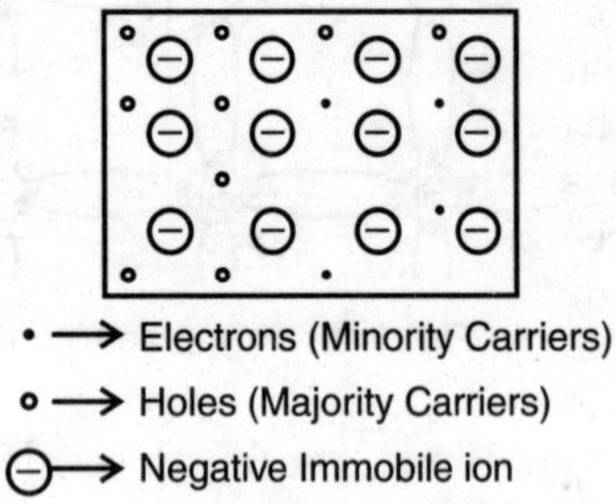

Fig. 1.12. Representation of P-type Semiconductor

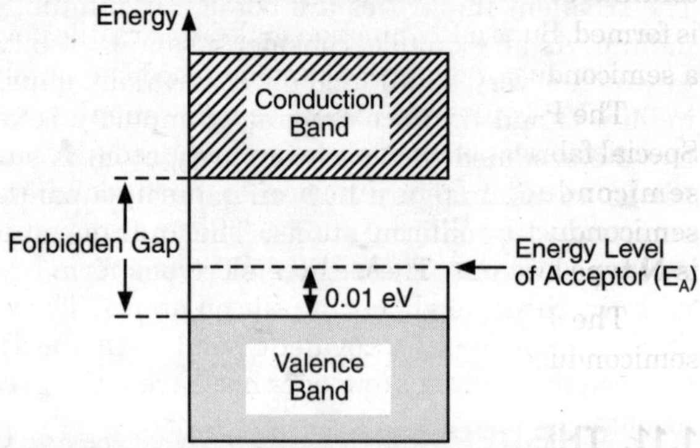

Fig. 1.13. Every and diagram of P-type semiconductor

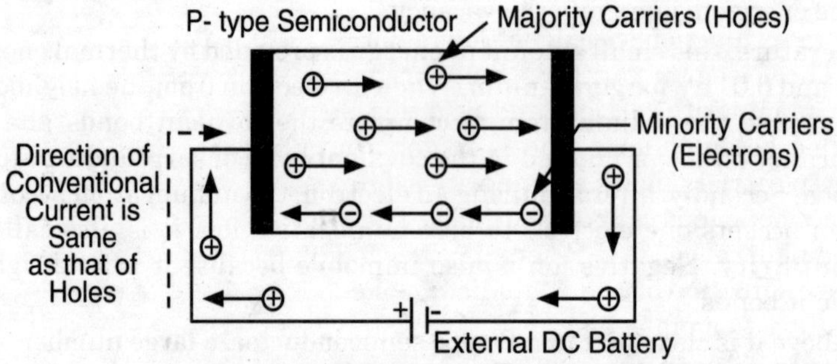

Fig. 1.14. Conduction in a P-type semiconductor

1.9. EFFECT OF TEMPERATURE ON EXTRINSIC SEMICONDUCTORS

(RGTU, Bhopal, Sem. Exam. 2007-08)

We know that addition of a small amount of donor or acceptor impurity generates a large number of charge-carriers in any extrinsic semiconductor. Actually, due to this large number of charge carriers, the conductivity of an extrinsic semiconductor is several times that of an intrinsic semiconductor at room temperature (300 K).

Now, let us observe what happens if we increase the temperature of extrinsic semiconductor (say N-type semiconductor) In this case because all the donor atoms have already donated their free electrons (at 300 K), the additional thermal energy due to increase in temperature only serves to increase the thermally produced carriers.

Due to this, the concentration of minority carriers increases. In this process a temperature is reached when the number of covalent bonds broken is quite large such that the number of holes is nearly equal to the number of electrons. Therefore, the extrinsic semiconductor now acts like an intrinsic semiconductor with higher conductivity. This critical temperature is 200°C for Si and 85°C for Ge.

1.10. P-N JUNCTION

We know about P-type semiconductors and N-type semiconductors. But P-type and N-type semiconductors, taken separately, are of a very little use. If a piece of P-type semiconductor is joined to a piece of N-type semiconductor in such a manner that the crystal structure remains continuous at the boundary, then a new structure called P-N junction is formed. Such a PN-junction makes a versatile device which is called a semiconductor diode, P-N junction diode or simply crystal diode.

The P-N junction cannot be made simply by joining two pieces. Special fabrication techniques are required to form a P-N junction. A semiconductor P-N junction is formed when a wafer of the semiconductor material such as silicon is doped so that one region is N-type and the other region is P-type.

The P-N junction is an important device because, all other semiconductor devices contain at least one P-N junction.

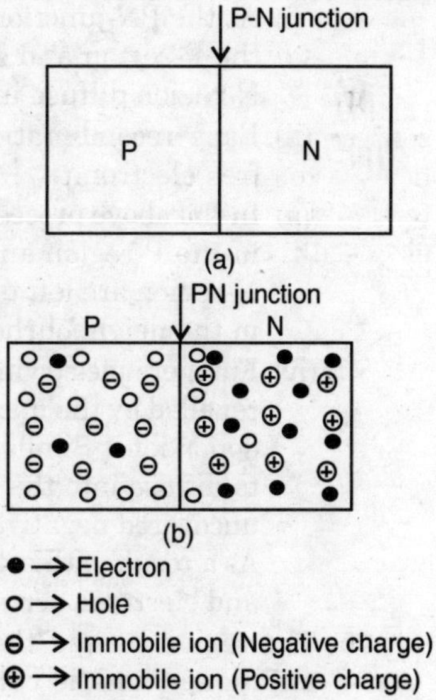

Fig. 1.15. PN-junction

1.11. THE DEPLETION REGION FORMATION

(RGTU, Bhopal, Sem. Exam., 2006-07)

> **DO YOU KNOW?**
>
> The depletion region is also known as space charge region. In the unbiased P-N junction, the depletion region gets formed very quickly after the formation of the junction.

Figure 1.16 shows a P-N junction with depletion region formed at the junction. The P-region contains holes as majority carriers, electrons as minority carriers and negative charged immobile ions called acceptor ions. The N-region contains electrons as majority carriers, holes as minority carriers and positive charged immobile ions called donor ions. The sample as a whole is the neutral so the P-region and N-region separately. In figure 1.16, no voltage is applied to the P-N junction. As soon as the PN-junction is formed, the following actions take place:

1. Holes from the P-region diffuse into the N-region. In the N-region, they combine with the electrons.
2. Electrons from the N-region diffuse into the P-region. In the P-region, they combine with the free holes.
3. The P-region has more number of holes and N-region has more number of electrons, therefore, there is a difference of concentrations in two regions. Due to this difference, the diffusion of holes and electrons takes place. A concentration gradient is produced due to the difference in concentration. Besides this, the electrons and holes move at random in all directions because of thermal energy. Some charge carriers cross the junction.

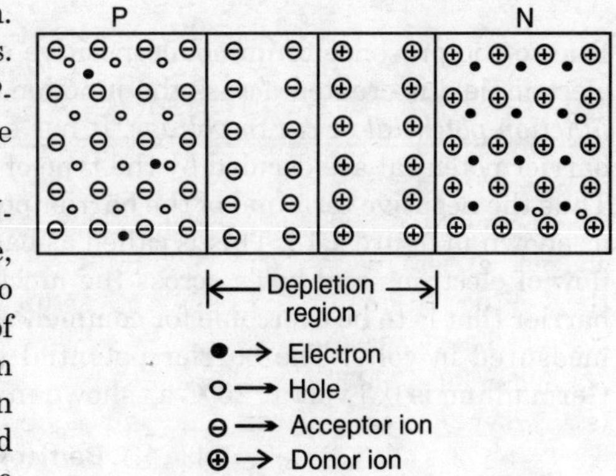

Fig. 1.16. Formation of depletion region in a P-N junction

4. The process of diffusion continues only for a short period of time. After a few recombination of holes and electrons, a restraining force is set up automatically in the neighbourhood of the junction. This restraining force is called barrier. As a result of this force, further diffusion of holes and electrons from one region to other is checked. The formation of this barrier may be given in following steps:

(i) As the PN-junction is formed, some of holes in the P-region and some of the electrons in the N-region diffuse in each other and recombine.

(ii) Each recombination eliminates a hole and a free electron.

(iii) In the above process, the negative acceptor ions in the P-region and positive donor ions in the N-region are left uncovered or uncompensated in the neighbourhood of the junction.

(iv) Further holes trying to diffuse into N-region are repelled by the uncovered positive charge of the donor ions. Similarly, further electrons trying to diffuse into the P-region are repelled by the uncovered negative charge of the acceptor ions. As a result of this, the further diffusion of holes and electrons across the junction is stopped.

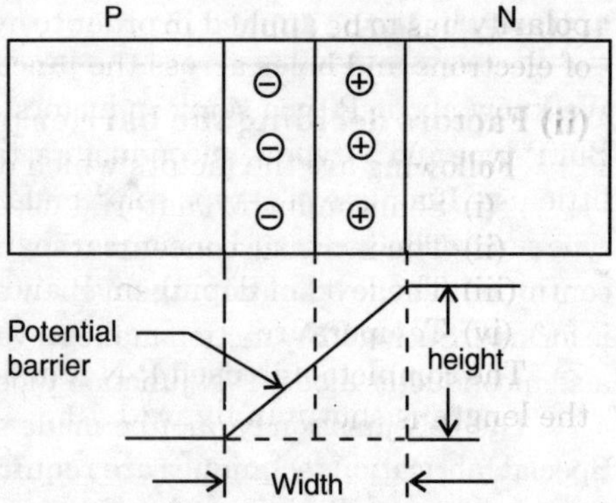

Fig. 1.17. Potential barrier

5. The region having the uncompensated accepted and donor ions is called depletion region. The reason for this is that there is a depletion (dirth) of mobile charge carriers in this region. This region contains immobile or fixed ions which are electrically charged. Therefore, this depletion region is also called as space-charge region.

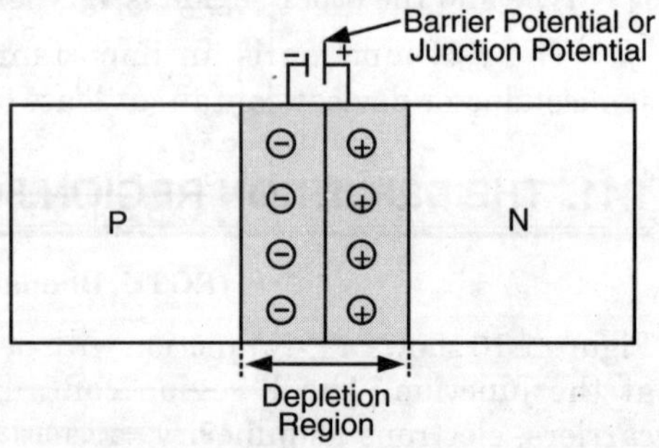

Fig. 1.18. Barrier potential

6. The width of depletion region depends upon the doping level of impurity in N-type or P-type semiconductor. The greater the doping level, the thinner will be the depletion region. This is due to the reason that a highly doped P-N junction contains a large number of electrons and holes.

1.12. BARRIER POTENTIAL OR JUNCTION POTENTIAL (V_j)

(RGTU, Bhopal, Sem. Exam. 2006-07)

Due to the presence of immobile positive and negative ions on opposite sides of the junction, an electric field is created across the junction. This electric field is known as the *barrier potential* or *junction potential* or *cut in voltage.* It has fixed polarities as shown in figure 1.17. The polarities of barrier potential are decided by the type of immobile ions present on the two sides of the junction. Thus the negative terminal of the barrier potential is on the P side and positive side is on the N-side as shown in figure 1.18. This is called as barrier potential because it acts as a barrier to oppose the flow of electrons and holes across the junction. The barrier potential represents the height of the barrier that is to be overcome for commencement of flow of electrons and holes. Barrier potential is measured in volts. The barrier potential for Silicon is about 0.7 volt whereas its value for the Germanium is 0.3 volt at 25°C as shown in Table 1.1.

Table 1.1. Barrier Potential for Different Materials

S.No.	Material	Barrier potential
1.	Silicon	0.7 volt
2.	Germanium	0.3 volt

(i) Importance of barrier potential

Barrier potential acts as a barrier (wall) which does not allow the electrons and holes to cross the junction. So, if we want them to cross the junction, then an external voltage of appropriate polarity has to be applied in order to overcome the opposition of the barrier potential. Then, the flow of electrons and holes across the junction can restart again.

(ii) Factors deciding the barrier potential value

Following are the factors which decide the value of the barrier potential:

(i) Semiconductor material used (Silicon or Germanium).

(ii) The intrinsic concentration of Si or Ge before doping.

(iii) The level of doping on P and N sides

(iv) Temperaure.

The complete unbiased P-N junction, immobile ions and the variation of barrier potential along the length is shown in figure 1.19.

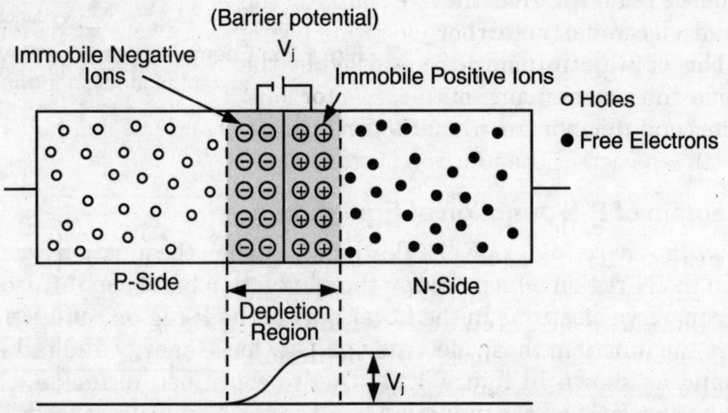

Fig. 1.19. Open circuited P-N junction and variation of junction potential

(iii) Penetration of depletion region

The penetration of the depletion region into P or N side depends on the doping levels of those sides. If both these sides are equally doped then the depletion region penetrates equally on both the sides as shown in figure 1.20 (a). But, if P-region is lightly doped as compared to the N-region then the penetration depletion region is more on the P-side as shown in figure 1.20 (b). Similarly, if N-side is lightly doped as compared to P-side then the depletion region extends more into the N-side as shown in figure 1.20 (c).

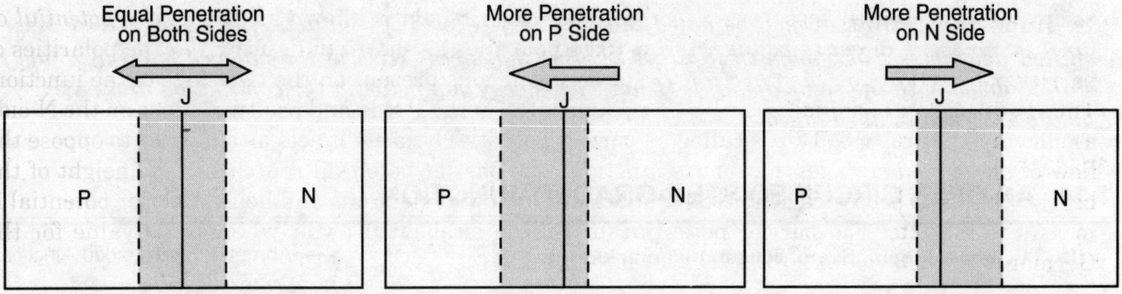

(a) Both sides equally doped (b) P side is lightly doped (c) N side is lightly doped

Fig. 1.20.

Thus, the depletion region always penetrates more on the side which is lightly doped as to the other.

1.13. ENERGY DIAGRAMS OF A P-N JUNCTION (RGTU, Bhopal, Sem. Exam., 2005-06)

The valence and conduction bands in an N- type material are at slightly lower energy levels than the valence and conduction bands in the P-type material. This is due to the difference in the atomic characteristics of the pentavalent (N-type) and trivalent (P-type) impurity atoms. The energy diagram for a P-N junction at the instant of formation is shown in figure 1.21. The valence and conduction bands on the N-side are at lower energy levels than those on the P-side. However, the conduction bands on both sides and valence bands on both sides overlap to a considerable extent.

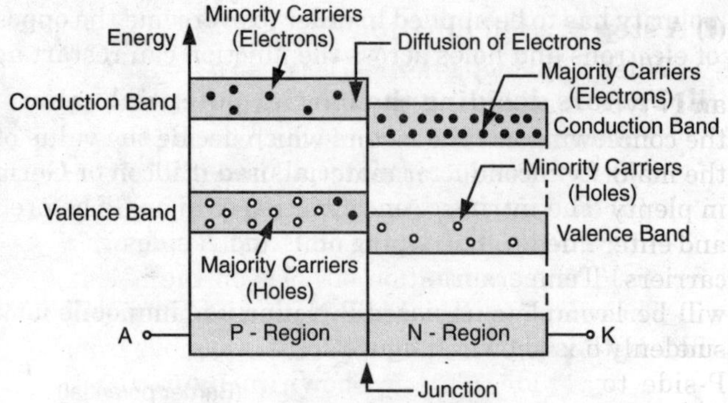

Fig. 1.21. Energy diagram of a P-N junction at the instant of junction formation

On P-side, the majority carriers, i.e., holes occupy the valence band whereas the thermally generated electrons (minority carriers) occupy the conduction band. Whereas on the N-side, the majority carriers (electrons) are in the conduction band whereas the minority carriers, i.e., holes occupy the valence band.

1.13.1. Energy Diagram of P-N Junction at Equilibrium

Figure 1.21 shows the energy diagram of a P-N junction when the junction reaches the equilibrium. The free electrons in the N-region which occupy the conduction band can diffuse across the junction and temporarily become free electrons in the lower part of the P-region conduction band. As soon as these electrons cross the junction, these electrons quickly loose energy and fall into the holes in the P-region valence band as shown in figure 1.21. Due to continued diffusion, the depletion region begins to form and the energy level of conduction band on the N-side decreases. This reduction in the energy level of the conduction band on N-side is because of the loss of higher energy electrons, which have diffused across the junction to the P-side. Very soon all the conduction band electrons on N-side having enough energy to cross the junction will cross the junction. So, no such electrons will be left on N-side.

At this point, the junction is at equilibrium and the formation of depletion region is complete and the diffusion process will stop. At equilibrium, there exists an energy gradient across the depletion region as shown in figure 1.21. This energy gradient acts as an energy hill. The conduction band electrons of N-side will have to climb this hill to go to the P-side.

Important Point: *It may be noted that the energy level of conduction band on N-side has shifted downwards as shown in figure 1.21. But the energy level of the valence band also has shifted down with the same amount. Hence, the energy gap between the conduction and valence bands will remain unchanged.*

1.14. AN OPEN CIRCUITED STEP-GRADED JUNCTION

(RGTU, Bhopal, Sem. Exam. 2001-02)(05 marks)

As we join the P and N-type semiconductors, at the boundary a P-N junction is formed. Let us consider a special case shown in figure 1.22 which shows a P-N junction formed between a P-type semiconductor with a constant concentration N_A and an N-type semicondctor with a uniform density N_D. The junction is a metallurgical (P-N) junction that separates the two sections of different concentrations. This type of doping, where the density changes abruptly from P to N type, is called as step grading. The junction is located at the plane where the concentration is zero.

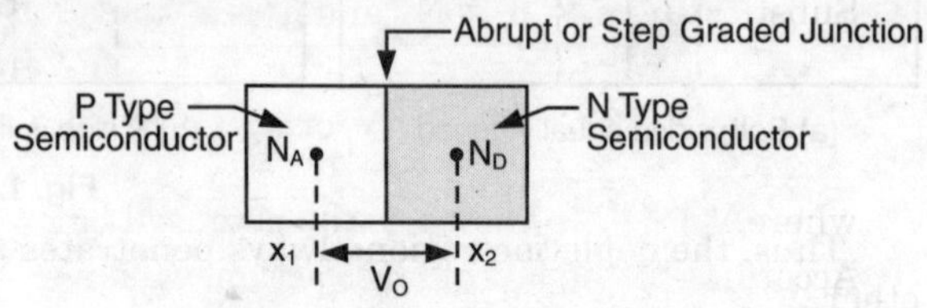

Fig. 1.22. An abrupt or step graded P-N junction

(i) A step graded junction

When a junction is formed between a P-type and an N-type material, there is an abrupt change in the concentration of holes and electrons. Holes are the majority carriers on P-side and hence available in plenty on that side. But as we cross the junction and enter into the N-side, the holes will be minority carriers. The concentration of holes on the N-side will be low. Thus, the concentration of holes will suddenly decrease at the junction as we move from P-side to N-side. This is shown in figure 1.23. Similarly, the concentration of electrons is high on the N-side and reduces drastically at the junction as they become minority carriers on the P-side.

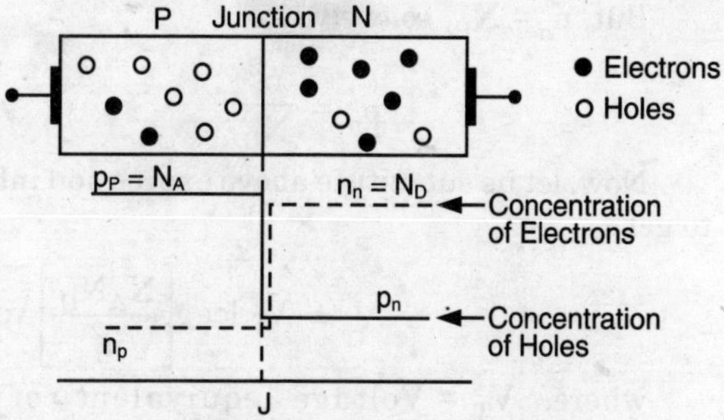

Fig. 1.23. A step graded P-N junction

Because there is a sudden change or step change in the concentrations of electrons and holes at the junction, this P-N junction is known as the step graded or an abrupt P-N junction.

The concentration of holes on the P-side is equal to the concentration of the acceptor impurity atoms.

Therefore, $p_p = N_A$ ...on P-side ...(1.1)

The concentration of electrons on the P-side is denoted by n_p which is very low as compared to p_p.

or $p_p >> n_p$ ...on P-side ...(1.2)

Similarly, the concentrations of electrons and holes on the N-side are denoted by n_n and p_n respectively and they are given by,

$n_n = N_D$ ..on N-side ...(1.3)

and $n_n >> p_n$ ..on N-side ...(1.4)

(ii) Contact difference potential or junction potential

In the previous section, we have proved that in a graded semiconductor, there exists an internal potential difference which is proportional to change in concentration. So, here due to the step or abrupt change in concentration at the junction, a potential gets developed which is called as contact difference potential or junction potential or barrier potential which is denoted by V_0 or V_j.

(iii) Expression for contact difference potential **(RGTU, Bhopal, Sem. Exam. 2005-06)**

Let us now obtain the expression for the barrier potential of the abrupt junction. We have

$$V_{21} = V_T \log_e \left[\frac{p_1}{p_2}\right] \quad ...(1.5)$$

At the junction, the hole concentration changes suddenly from p_p to p_n as we move from P to N side.

Substituting $p_1 = p_p = N_A$ and $p_2 = p_n$ in equation (1.5) we get,

$$V_j \text{ or } V_0 = V_T \log_e \left[\frac{N_A}{p_n}\right] \quad ...(1.6)$$

where V_j or V_0 is known as the potential barrier or junction potential or contact potential.

According to the law of mass action, we have

$$n_n \times p_n = n_i^2$$

Therefore, $$p_n = \frac{n_i^2}{n_n}$$

But, $n_n = N_D$, so, we write

$$p_n = \frac{n_i^2}{N_D}$$

Now, let us substitute above expression into equation (1.6) to get,

$$V_j = V_T \log_e \left[\frac{N_A N_D}{n_i^2}\right] \text{ volts} \quad ...(1.7)$$

where, V_T = Voltage equivalent of temperature

$$= \frac{T}{11600}$$

and n_i = Intrinsic concentration.

Equation (1.7) shows that the contact potential depends on the following four parameters

(i) Concentration of donor impurities N_D

(ii) Concentration of acceptor impurities N_A

(iii) Intrinsic concentration n_i.

(iv) V_T

PIONEER IN ELECTRONICS

The unit of measure for resistance, the ohm, was named for German physicist *George Simon Ohm (1787-1854)*. Ohm is also known for his development of Ohm's law: $I = \frac{V}{R}$

(iv) Conclusion

The important conclusions from equation (1.7) may be listed as under:

(i) The junction potential depends on the voltage equivalent of temperature V_T. This makes it dependent on temperature.

(ii) The junction potential is decided by the doping levels on the P and N sides (N_A and N_D).

(iii) The junction potential for Silicon diodes is 0.6 to 0.7 V while for Germanium diodes, it is 0.2 to 0.3 V.

EXAMPLE 1.1 Given a step graded Germanium P-N junction. It has $N_D = 10^3 N_A$ and N_A corresponds to 1 atom, per 10^8 Germanium atoms. Evaluate the junction potential. Assume: $n_i = 2.5 \times 10^{13}$ per cm^3. Atom density of Ge = 4.4×10^{22} atoms per cm^3.

Solution: Given that atom density of Ge = 4.4×10^{22} atoms/cm^3

N_A corresponds to 1 atom per 10^8 Ge atoms.

Hence, corresponding to 4.4×10^{22} atoms of Ge, the number of atoms of acceptor impurity will be,

$$N_A = \frac{4.4 \times 10^{22}}{10^8} = 4.4 \times 10^{14}/\text{cm}^3$$

or $\quad N_A = 4.4 \times 10^{14} \times 10^6/\text{m}^3 = 4.4 \times 10^{20}/\text{m}^3 \quad ...(i)$

It is given that, $\quad N_D = 10^3 N_A$

or $\quad N_D = 4.4 \times 10^{23}/\text{m}^3 \quad ...(ii)$

Intrinsic concentration, $\quad n_i = 2.5 \times 10^{13}/\text{cm}^3 = 2.5 \times 10^{19}/\text{m}^3 \quad ...(iii)$

The junction potential V_j is given by,

$$V_j = V_T \log_e \left[\frac{N_A N_D}{n_i^2}\right]$$

Assuming $\quad V_T = 26$ mV, we have

$$V_j = 26 \times 10^{-3} \log_e \left[\frac{4.4 \times 10^{20} \times 4.4 \times 10^{23}}{(2.5 \times 10^{19})^2}\right]$$

$$= 0.3287 \text{ Volts or } 328.7 \text{ mV}$$ **Ans.**

1.15. CURRENTS IN A P-N JUNCTION WITH NO EXTERNAL BATTERY

When a PN-junction is formed, a depletion region is established across the junction. Because of this a barrier potential exists across the junction which stops the further diffusion of holes and electrons across the junction. Few majority carriers have sufficiently high kinetic energy to balance the barrier potential and cross the junction. Hence due to majority carriers small current flows which is called majority carrier current.

> **DO YOU KNOW?**
>
> When the P-N junction is formed, the depletion region gets created and the movement of electrons and holes stops. Thus, the current flowing through an unbiased P-N junction is zero.

There is another current which flows across the junction. This current is due to the minority carriers. Since, there are some minority carriers in the P-region and in the N-region, they are accelerated across the junction by the barrier potential. This current due to minority carriers is called minority carrier current.

But the majority carrier current flows in a direction opposite to the direction of minority carrier current. Hence flow of minority carriers across the junction is balanced by the flow of some majority carriers. Therefore, there is no net flow of current across the junction when no external battery is applied.

1.16. BIASING OF A P-N JUNCTION

(RGTU, Bhopal, Sem, Exam. 2005-06)

When we apply a battery across the PN-junction, this process is called biasing of a PN-junction. By applying external voltage source across the PN-junction, the width of the depletion region can be controlled. By controlling the width of depletion region, P-N junction finds a large number of applications in the field of electronics.

The P-N junction can be biased in two ways:

1. Forward-Bias
2. Reverse-Bias

1.16.1. Forward-Biasing of P-N junction

(RGTU, Bhopal, Sem. Exam. 2005-06)

In forward-biasing of a P-N junction, positive terminal of the battery is connected to the P-side and the negative terminal to the N-side as shown in the figure 1.24. In this case, holes are repelled from the positive terminal of the battery and forced towards the junction. Similarly electrons are repelled from the negative terminal of the battery and forced towards the junction. Because of this increased energy, some holes and electrons enter the depletion region. This reduces the potential barrier. Therefore, width of depletion region also reduces. As a result of this, more majority carriers diffuse across the junction. Hence, this causes a large current to flow across the junction.

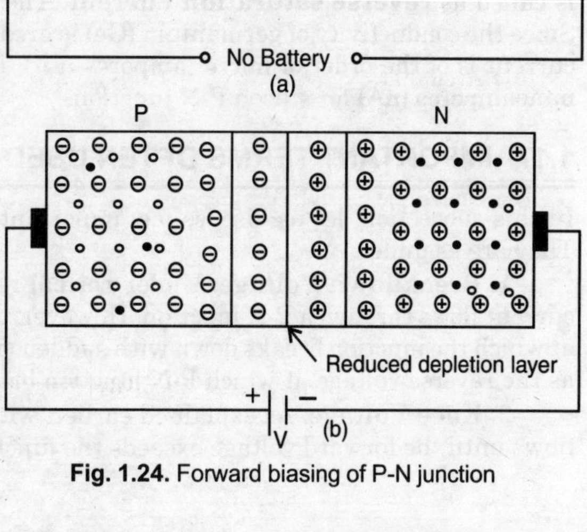

Fig. 1.24. Forward biasing of P-N junction

For each recombination of free electron and hole, an electron from the negative terminal of the battery enters the N-type region. This then moves towards the junction. Similarly, in the P-type region near the positive terminal of the battery, an electron breaks a bond in the crystal structure and enters the positive terminal of the battery. For each electron which breaks its bond, a hole is created. This hole moves towards the junction. Therefore, the current through the external circuit is due to the movement of electrons only. On the other hand, the current in the P-type region is due to the movement of holes and the current in the N-type region is due to the movement of electrons. Hence, there is a large current in the external circuit as long as the external battery is present in the circuit.

1.16.2. Reverse Biasing of P-N Junction (RGTU, Bhopal, Sem. Exam. 2005-06, 2006-07)

In reverse biasing of a PN-junction, a battery is connected across the P-N junction in such a way that its positive terminal is connected to the N-region and negative terminal is connected to the P-region.

In reverse bias, the holes in the P-region are attracted towards the negative terminal of the battery whereas the electrons are attracted towards the positive terminal of the battery. In this way majority charge carriers are drawn away from the PN-junction. Due to this, the depletion region becomes wider. This increases the barrier potential.

Due to the increased barrier potential, the majority charge carriers are not able to cross the junction. Hence, in reverse bias, there is no current due to the majority charge carriers.

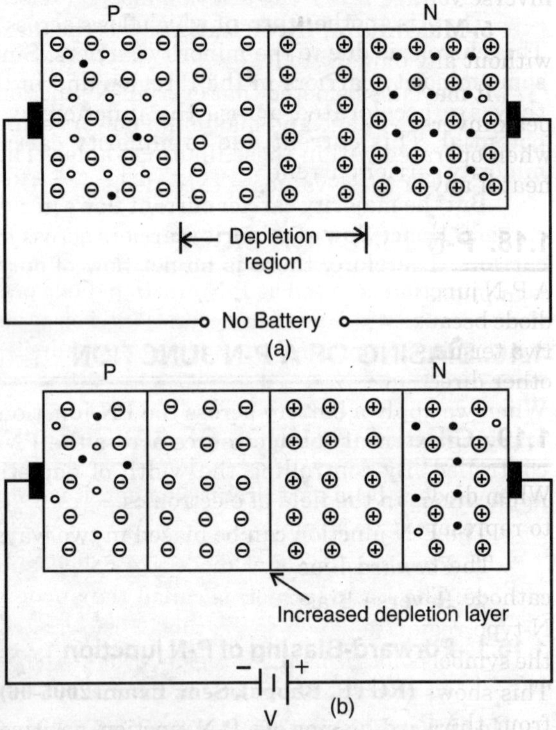

Fig. 1.25. Reverse biasing of PN-junction

But there are few thermally generated minority carriers in both the regions. The increased barrier potential enhances the flow of minority carriers across the junction. As a minority carrier is generated, it is drifted towards the P-N junction due to the barrier potenial. Therefore, a very small amount of current flows through the P-N junction. The generation of minority carriers depends upon the temperature and is independent of the reverse voltage applied. Hence if temperature is constant, the current due to the minority carriers also remains constant whether the reverse bias voltage is increased or decreased. Therefore, due to this reason, this current is called as **reverse saturation current**. The value of reverse saturation current is very small. Since the conductivity of germanium (Ge) is greater than silicon (Si), the value of reverse saturation current is of the order of micro- amperes (μA) for Germanium P-N junction and is of the order of nanoamperes (nA) for silicon P-N junction.

1.17. IMPORTANT TERMS OFTEN USED IN P-N JUNCTION

In this subsection, let us discuss few important terms which are generally used for P-N junction. They are as under:

1. Breakdown Voltage: Under normal reverse bias voltage, a very little amount of reverse current flows through a P-N junction. However, on increasing the reverse voltage, a point may reach at which the junction breaks down with sudden rise in reverse current. Breakdown voltage is defined as the reverse voltage at which P-N junction breaks down with sudden rise in reverse current.

2. Knee Voltage. As explained earlier, with forward bias to P-N junction a very little current flows until the forward voltage exceeds the junction barrier potential (0.3 V in case of Ge and 0.7 V in case of Si). The forward voltage, at which the current through the junction starts increasing rapidly, is known as the ***knee voltage*** or the ***cut-in-voltage.***

3. Maximum Forward Current. It is the highest instantaneous forward current that a P-N junction can conduct without damage to the junction. In case the forward current exceeds this rating, the junction will get destroyed due to over-heating.

4. Peak Inverse Voltage. It is the maximum reverse voltage that can be applied to the P-N junction without damage to the junction. If the reverse voltage across the junction exceeds its peak inverse voltage (PIV), the junction may get destroyed owing to excessive heat.

5. Maximum Power Rating. It is the maximum power that can be dissipated at the junction without any damage to it.

It may also be noted that every P-N junction has limiting values of maximum forward current, peak inverse voltage and maximum power rating. The P-N junction gives satisfactory performance when operated within these limiting values. The P-N junction may get destroyed due to excessive heat if any of these values is exceeded.

1.18. P-N JUNCTION DIODE

A P-N junction is called as P-N junction diode or a semiconductor diode. This is also called as crystal diode because it is formed in a crystal (Silicon or germanium). A P-N junction diode is an unidirectional two terminals semiconductor device. It conducts only in one direction and offers high resistance in other direction.

1.19. CIRCUIT SYMBOL OF A DIODE

When diode is used as a circuit element, it is convenient to represent it by a symbol.

The two terminals of diode are called anode and cathode. The P-type region is called the anode and the N-type region is called the cathode. The arrowhead in the symbol points from the P-region towards the N-region. This shows that the direction of conventional current is from the P-region to the N-region which is also the direction of flow of holes.

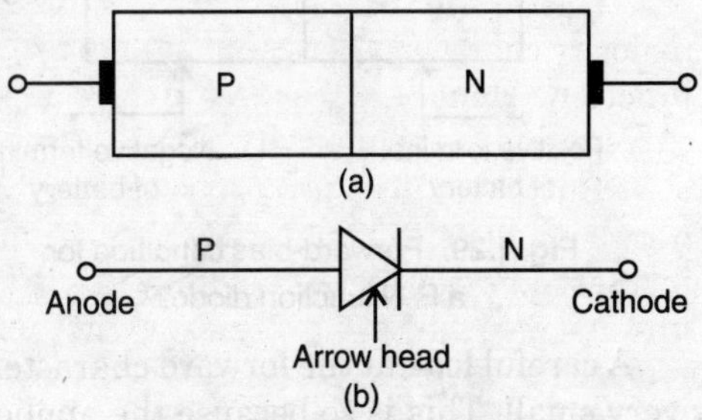

Fig. 1.26. Circuit symbol of diode

1.20. V-I CHARACTERISTICS OF P-N JUNCTION DIODE OR PRACTICAL DIODE*

If diode is used as a circuit element in an electrical circuit, there may be a requirement to know how it responds or behaves in electric circuit. This type of information may be obtained by a curve known as Volt-Ampere (V-I) characteristics of a practical diode.

The V-I characteristics of a diode is simply a curve or graph between the voltage applied across its terminals and the current that flows through the diode due to this applied voltage. The entire V-I characteristics may be divided into two parts namely:

1. Forward characteristics
2. Reverse characteristics

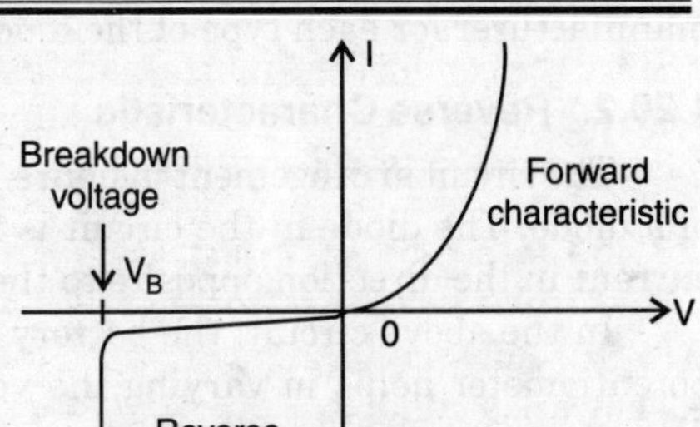

Fig. 1.27. Entire V-I characteristics of diode

* Explain the behaviour of p-n junction at no bias, reverse bias and forward bias. Sketch the V-I characteristics of p-n junction diode. **(RGTU, Bhopal, Sem. Exam. 2006-07)**

1.20.1. Forward Characteristic (RGTU, Bhopal, Sem. Exam. 2007-08)

The circuit arrangement shown in figure 1.28 is used in the laboratory to obtain forward characteristic of a diode. The diode in the circuit is in forward bias state because the battery V_{FF} is pushing the current in the direction of the arrowhead.

In the circuit, the battery V_{FF} is connected to the diode through potentiometer P. The potentiometer helps in varying the voltage across the diode. The load resistance R_L is used in the circuit to limit the current through the diode. A voltmeter is connected across the diode to measure the voltage. A milliammeter is also connected to measure the current in the circuit.

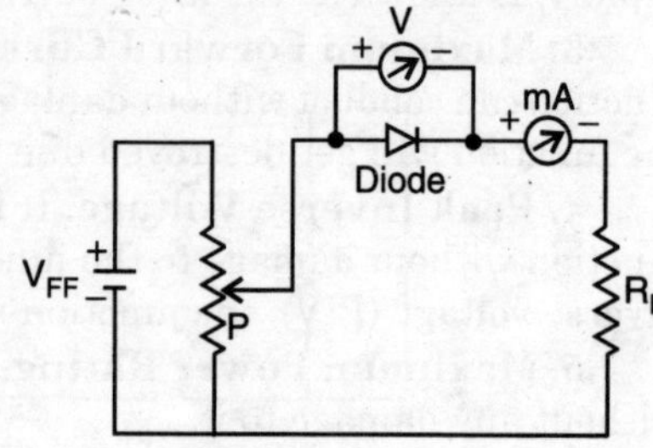

Fig. 1.28. Circuit arrangement for forward bias

To plot the characteristic, the voltage across the diode is increased gradually and the corresponding increase in the current is noted. The graph is plotted by taking voltage on the horizontal axis and current on the vertical axis.

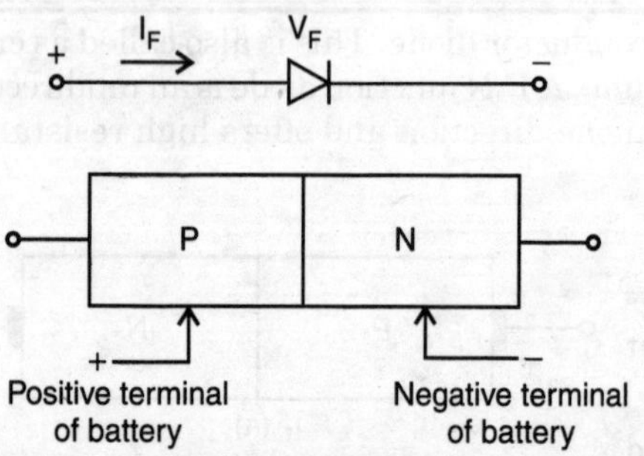

Fig. 1.29. Forward-bias condition for a P-N junction diode.

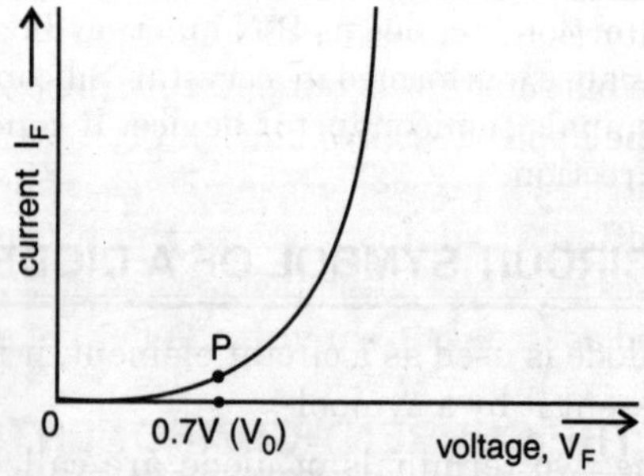

Fig. 1.30. Forward characteristics (Si diode)

A careful look at the forward characteristic reveals the fact that upto point P, the diode current is very small. This is so because the applied voltage has to overcome the barrier potential and the diode conducts poorly. Once the applied voltage is slightly greater than the barrier potential, the diode current increases rapidly and diode conducts heavily. This voltage at which the current starts increasing is called the **knee voltage (V_0)**. Its value is 0.7 V for a silicon diode and 0.3 V for a germanium diode. The applied voltage across the diode should not increase beyond a specified safe value otherwise the diode will burn out due to heat. This specified safe value is provided by the manufacturer for each type of the diode.

1.20.2. Reverse Characteristic (RGTU, Bhopal, Sem. Exam. 2007-08)

The circuit arrangement in figure 1.31 (a) is used in the laboratory to obtain reverse characteristic of a diode. The diode in the circuit is in reverse bias state because the battery V_{RR} is pushing the current in the direction opposite to the arrowhead.

In the above circuit, the battery V_{RR} is connected to the diode through potentiometer P. The potentiometer helps in varying the voltage across the diode. A voltmeter is connected across the diode to measure the voltage. A microammeter is also connected to measure the current in the circuit.

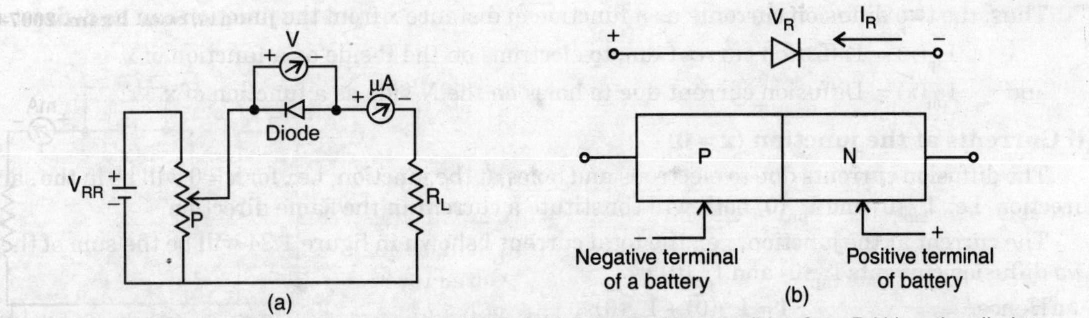

Fig. 1.31. (a) Circuit arrangement for Reverse bias (b) Reverse-bias condition for a P-N junction diode

A careful look at the reverse characteristic reveals the fact that below the breakdown voltage the diode current is very small and almost remains constant. This current is also called **reverse saturation current I_o.** It is of the order of μA for Germanium diode and nanoamperes for Silicon diode. It is due to the movement of minority carriers which are thermally generated. It means that reverse bias current is temperature dependent and does not depend upon the applied reverse bias voltage.

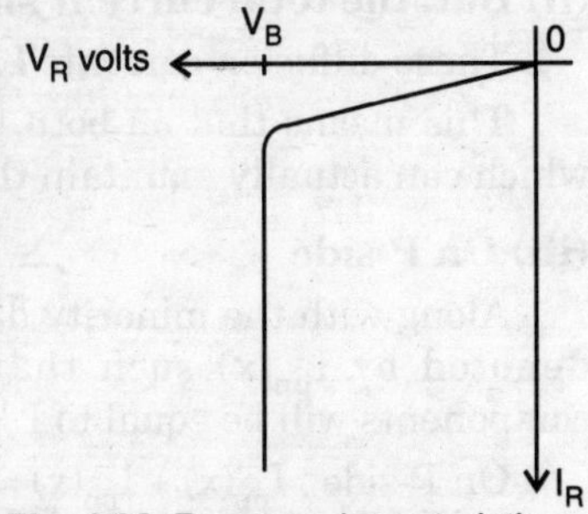

Fig. 1.32. Reverse characteristics

But, if a large reverse bias voltage is applied, then a process known as **junction breakdown** occurs. Due to this, the diode reverse current increases rapidly. The applied reverse bias voltage at which this occurs is called as **breakdown voltage (V_B)** of a diode.

1.21. THE CURRENT COMPONENTS IN A P-N JUNCTION DIODE*

(RGTU, Bhopal, Sem. Exam. 2006-07)

Figure 1.33 shows a forward biased P-N junction diode. The current I circulating in the circuit is the forward current of the diode. This current flows due to the majority carriers.

> **DO YOU KNOW?**
>
> The forward current through a P-N junction diode flows due to the majority carriers and its direction of flow (conventional) is always from anode to cathode.

Diffusion Currents on the P and N-side

Under the forward biased condition, the depletion region becomes extremely narrow and holes from P-side diffuse into the N-side and the electrons from the N-side diffuse into the P side as shown in figure 1.33. The current carried by these diffused electrons, which become minority carriers once they enter into the P-side, is denoted by I_{np} where n stands for electrons and p for the P-side.

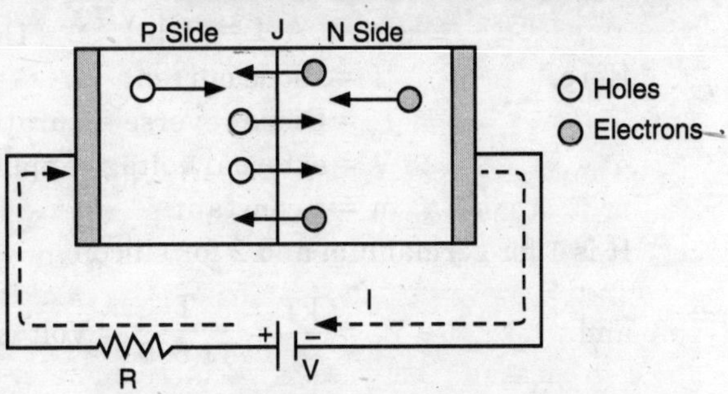

Fig. 1.33. A forward biased P-N junction

This current is the diffusion current on P-side and it decreases exponentially with increase in distance (x) from the junction. Similarly the current carried by the diffused holes which become minority carriers once they enter into the N-side is denoted by I_{pn} where p stands for holes and n for the N-side. This current is the diffusion current due to minority carriers on the N-side and it decreases exponentially with increase in distance (x) from the junction.

* Discuss the current flow mechanism in a p-n junction under:
(i) no bias (ii) forward bias (ii) reverse bias conditions **(RGTU, Bhopal, Sem. Exam. 2007-08)**

Thus, the two diffusion currents as a function of distance x from the junction can be defined as:

$I_{np}(x)$ = Diffusion current due to electrons on the P-side as a function of x.

and $I_{pn}(x)$ = Diffusion current due to holes on the N-side as a function of x.

(i) Currents at the junction (x = 0)

The diffusion currents due to electrons and holes at the junction, i.e., for x = 0 will be in the same direction, i.e., $I_{np}(0)$ and $I_{pn}(0)$ both will constitute a current in the same direction.

The current at the junction, i.e., the total current I shown in figure 1.34 will be the sum of these two diffusion currents $I_{np}(0)$ and $I_{pn}(0)$.

Hence, $$I = I_{np}(0) + I_{pn}(0) \quad \text{... (1.8)}$$

(ii) But, the total current should remain constant

These diffusion currents $I_{np}(x)$ and $I_{pn}(x)$ decrease exponentially with increase in x.

This means that on both the sides, there must be some other components of current present which can actually maintain the current I constant.

(iii) On P-side

Along with the minority diffusion current $I_{np}(x)$ there must be a majority current due to holes denoted by $I_{pp}(x)$ such that the sum of these two components will be equal to I.

On P-side : $I_{np}(x) + I_{pp}(x) = I$ ... (1.9)

(iv) On N-side

Similarly, on the N-side along with the minority diffusion current due to holes, i.e., $I_{pn}(x)$, there is a majority current component due to electrons denoted by $I_{nn}(x)$ such that the sum of these two components will be equal to I.

On N-side : $I_{pn}(x) + I_{nn}(x) = I$ ... (1.10)

All these current components have been plotted as a function of distance (x) from the junction in figure 1.34.

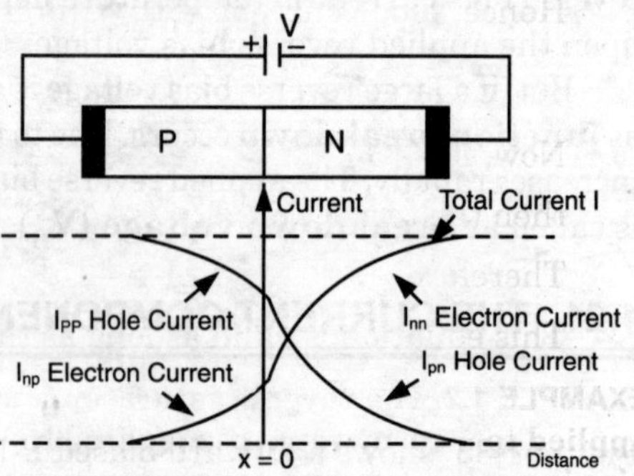

Fig. 1.34. Various current components

1.22. DIODE CURRENT EQUATION

With the help of solid state physics, the diode current equation, relating the voltage V and current I for the forward and reverse-bias regions, can be given by

$$I = I_o\left[e^{V/\eta V_T} - 1\right] \quad \text{...(1.11)}$$

Here, I = Diode current

I_o = Diode reverse saturation current at room temperature

V = external voltage applied to the diode

η = a constant

It is 1 for germanium and 2 for silicon

and $$V_T = \frac{kT}{q} = \frac{T}{11,600} = \text{Volt equivalent of temperature.} \quad \text{...(1.12)}$$

where k = Boltzmann's constant = 1.38066×10^{-23} J/Kelvin

q = Electronic Charge = 1.6×10^{-19} Coulomb

T = Diode junction temperature in (°K)

Now, at room temperature, (T = 300 °K), V_T = 26 mV

Putting these two values in current equation, we get

$$I = I_0[e^{40V/\eta} - 1] \quad ...(1.13)$$

For germanium diode, $\eta = 1$

Therefore, $$I = I_0[e^{40V} - 1] \quad ...(1.14)$$

and for silicon diode, $\eta = 2$

Therefore, $$I = I_0[e^{20V} - 1] \quad ...(1.15)$$

If the value of applied voltage is greater than unity then the diode current equation for germanium will be

$$I = I_0[e^{40V}]$$

and for silicon, $$I = I_0[e^{20V}] \quad ...(1.16)$$

If the diode is reverse biased, the current equation can be obtained by reversing the sign of the applied voltage V.

Hence, diode current with reverse bias is

$$I = I_0\,[e^{(-V/\eta V_T)} - 1] \quad ...(1.17)$$

Now, if $V >> V_T$

then the term $e^{(-V/\eta V_T)} << 1$

Therefore, $$I \cong -I_0 \quad ...(1.18)$$

This equation is valid as long as the external voltage is below the breakdown value.

PIONEERS IN ELECTRONICS

The unit of measure for capacitance, the farad (F), was named for Michael Faraday (1791-1867), an English chemist and physicist who discovered the principle of induction (1 F is the unit of capacitance that stores 1 coulomb (C) of charge when 1 volt (V) is applied).

EXAMPLE 1.2 The reverse saturation current at room temperature is 0.3 μA when a reverse bias is applied to a germanium diode. Find the value of current flowing in the diode when 0.15 V forward bias is applied at room temperature.

Solution: Given that $I_o = 0.3\ \mu A = 0.3 \times 10^{-6}$ A

and forward voltage, $V_F = 0.15$ Volt. The current flowing through the diode under forward bias is given by

$$I = I_o\left(e^{40V_F} - 1\right)$$

$$= 0.3 \times 10^{-6}\,(e^{40 \times 0.15} - 1)$$

or $$I = 120.73\ \mu A$$ **Ans.**

1.23. COMPLETE V-I CHARACTERISTICS OF SILICON AND GERMANIUM DIODES

The complete V-I characteristics is the combination of forward and reverse characteristics. The typical V-I characteristics of Silicon and Germanium diodes are as shown in figure 1.35.

Conclusions

We can draw following conclusions from V-I characteristics of figure 1.35:

(i) Cut-in voltages for Silicon and Germanium diodes are 0.6 and 0.2 V respectively.

(ii) Breakdown voltage of Silicon diode is higher than that of the Germanium diode. Therefore, Silicon diodes can withstand to a higher reverse voltage.

(iii) The reverse saturation current I_0 for a Germanium diode is few μA whereas that for a Silicon diode, it is in nA at room temperature.

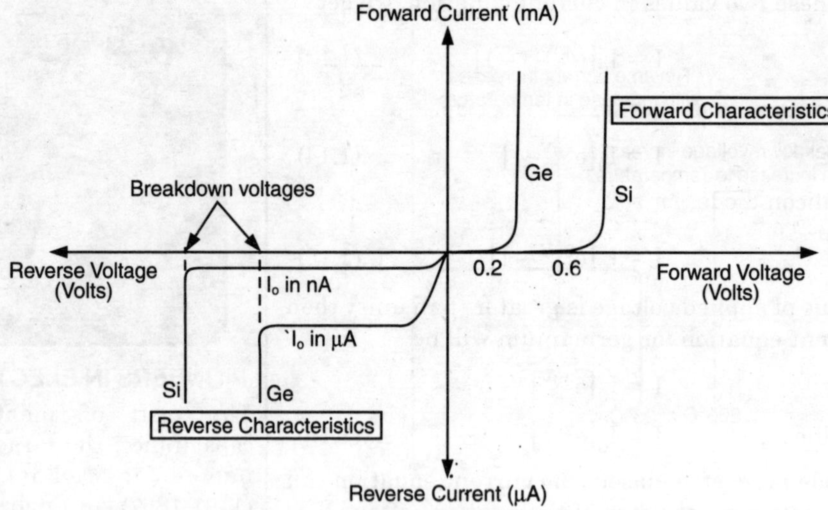

Fig. 1.35. V-I Characteristics of Silicon and Germanium diodes

1.24. COMPARISON OF SILICON AND GERMANIUM DIODES

S.No.	Parameter of comparison	Silicon Diode	Ge Diode
1.	Material used	Silicon	Germanium
2.	Cut-in voltage	0.6 V	0.2 V
3.	Reverse saturation current	In nanoamp	In microamp
4.	Effect of temperature	Less	More
5.	Breakdown voltage	Higher	Lower
6.	Applications	Rectifiers, clippers clampers, etc.	Low voltage Low temperature applications

1.25. EFFECT OF TEMPERATURE ON THE V-I CHARACTERISTICS

(RGTU, Bhopal, Sem. Exam. 2001-02)

We know that the expression for diode current is given by

$$I_D = I_0 \left[e^{V/\eta V_T} - 1 \right]$$

where I_0 = Reverse saturation current

$$V_T = \frac{T}{11600}$$

η = 1 for Ge diode = 2 for Si diode

V = diode voltage

> **DO YOU KNOW?**
>
> For any diode, the forward voltage, V_F, decreases as the temperature of the diode increases. As a rough approximation, V_F decreases by 2 mV for each degree Celsius rise in

The diode characteristics is mathematically expressed by the equation of I_D.

Two parameters I_0 and V_T are temperature dependent. Hence, the characteristics is dependent on the temperature. The effect of change in temperature on the V-I characteristics are as shown in figure 1.36.

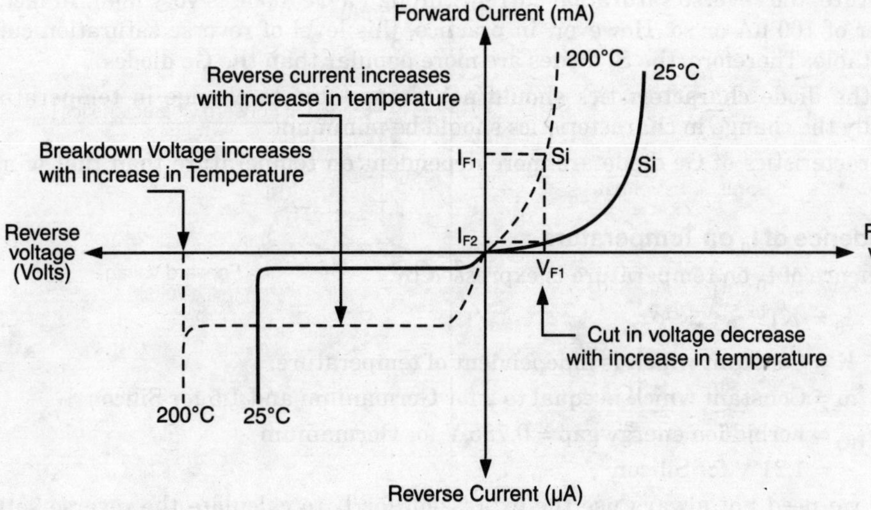

Fig. 1.36. Effects of temperature on V-I characteristics of a silicon diode

Conclusions

As observed from figure 1.36, the effects of temperature may be listed as under :

(i) Reduction in the cut-in voltage takes place with increase in temperature. Therefore, at the same forward voltage V_{F1}, a larger current I_{F1} flows through the diode at increased temperature.

(ii) The breakdown voltage increases with increase in temperature.

(iii) Reverse saturation current increases with increase in temperature.

Important Point: *In the expression for diode current, there are two factors which depend on the temperature. They are I_0 and V_T. Hence, the diode characteristics becomes the function of temperature.*

1.25.1. Germanium Diodes and Effect of Temperature

As a matter of fact, junction diodes fabricated from germanium are also available. The principle of operation is same as that of the silicon diodes and the expression for volt-ampere characteristic is also same as that of the silicon diodes. The value of $\eta = 1$ for germanium (Ge). The reverse saturation current of the Ge diodes is three to four times higher than the silicon diodes of same dimensions and doping densities. The cut-in voltage $V_\gamma = 0.2$ V for the Ge diodes.

1.25.2. Ge Diodes Produce Higher Reverse Saturation Current

We know that the four valence electrons of Ge are in the fourth shell whereas those of a Si atom are in the third shell. Hence, the force of attraction between the nucleus and valence electrons is weak in the generation atoms than that in the silicon atoms. The forbidden energy gap is smaller in Ge than Si atoms. Therefore, at the same temperature, more valence electrons will jump to the conduction band to produce higher reverse saturation current in case of Ge diode.

1.25.3. Si Diode is More Popular than Ge Diode **(V. Important)**

Following are the reasons for the popularity of silicon (Si) diode:

(i) The reverse saturation current for a Silicon diode is much lower than that of a Ge diode. Therefore, even with the two fold increase in I_0 after every 10°C rise in temperature, the reverse saturation current through Silicon diodes will still remain very low. But, at increased temperature, the reverse saturation current through a Ge diode is very high. In fact, it is of the order of 100 μA or so. However, in practice, this level of reverse saturation current is unacceptable. Therefore, the Si diodes are more popular than the Ge diodes.

(ii) Ideally the diode characteristics should not change due to change in temperature and practically the change in characteristics should be minimum.

(iii) The characteristics of Ge diode are more dependent on temperature than that of a silicon diode.

1.25.4. Dependence of I_0 on Temperature

The dependence of I_0 on temperature is expressed by

$$I_0 = KT^m e^{-V_{GO}/\eta V_T} \qquad \text{... (1.19)}$$

where K = Constant which is independent of temperature.

m = Constant which is equal to 2 for Germanium and 1.5 for Silicon

V_{GO} = Forbidden energy gap = 0.785 V for Germanium

= 1.21 V for Silicon

Practically we need not always use the expression for I_0 to calculate the reverse saturation current at a new temperature. As we know that there is a 7% rise in I_0 for every 1°C rise in temperature, we can write this as under:

$$I_{01} = (1.07)^{\Delta T} I_{02} \qquad \text{... (1.20)}$$

where I_{01} = Reverse saturation current at temperature T_1

I_{02} = Reverse saturation current at temperature T_2

and $\Delta T = |T_2 - T_1|$

1.25.5. Conclusions

Some important conclusions related to the reverse saturation current may be listed as under :

(i) The reverse saturation current increases at a rate of 7% for every 1°C rise in temperature. In other words, I_0 doubles its value for every 10°C rise in temperature.

(ii) Reverse saturation current for silicon diode is lower than that for a germanium diode.

EXAMPLE 1.3 The forward current through a Silicon diode is 10 mA at room temperature (27°C). The corresponding forward voltage is 0.75 volts. Calculate the reverse saturation current I_0.

Solution: Given that

$$I_F = 10 \text{ mA}, \; V_F = 0.75 \text{ Volt}, \; T = 27°C = 300°K, \; \eta = 2 \text{ for a Silicon diode}$$

We know that $V_T = \dfrac{T}{11{,}600} = \dfrac{300}{11{,}600}$ or $V_T = 26$ mV ... (i)

The forward current through a diode is given by

$$I_F = I_0 [e^{V_F/\eta V_T} - 1] \qquad \text{... (ii)}$$

Substituting all the values, we get

$10 \times 10^{-3} = I_0 |e^{0.75/2 \times 26 \times 10^{-3}} - 1|$ or $I_0 = 5.446$ nA. **Ans.**

The reverse saturation current for silicon diode = 5.446 nA.

EXAMPLE 1.4 Explain the effect of temperature on a P-N junction diode. Determine the forward current for a Germanium diode at room temperature when the voltage across it is 0.3 V. Assume room temperature is 27°C and reverse saturation current is 1 nA. Volt-equivalent of temperature is

$$V_T = \frac{T}{11{,}600}.$$

Solution: Given that $I_0 = 1 \text{ nA} = 1 \times 10^{-9}$ A at T = 27°C = 300°K

We know that $V_T = \frac{T}{11{,}600} = \frac{300}{11{,}600} = 25.86$ mV,

$V_F = 0.3$ V

For Ge diode $\eta = 1$

The forward current is given by

$$I_F = I_0\left[e^{(V_F/\eta V_T)} - 1\right] = 1 \times 10^{-9}\ [e^{300/25.86} - 1]$$

or $I_F = 1.091 \times 10^{-4}$ A or 0.1091 mA. **Ans.**

EXAMPLE 1.5 A diode operating at 300°K has V (forward) of 0.4 V across it when the current through it is 10 mA and 0.42 V when the current is twice as large. What values of I_0 and η allow the diode to be modelled by the diode equation ?

Solution: Given that T = 300°K, $V_1 = 0.4$ V, $V_2 = 0.42$ V, $I_1 = 10$ mA, $I_2 = 20$ mA

At T = 300°K, we have

$$V_T = \frac{T}{11{,}600} = \frac{300}{11{,}600} = 26 \text{ mV} \quad \text{... (i)}$$

First, let us calculate the value of η.

From the given data, we can write two equations of current for different values of V as under :

$$I_1 = I_0\left[e^{V_1/\eta V_T} - 1\right] \quad \text{and} \quad I_2 = I_0\left[e^{V_2/\eta V_T} - 1\right]$$

Substituting the given values, we get

$$10 \text{ mA} = I_0\left[e^{0.4/\eta \times 26 \times 10^{-3}} - 1\right] \quad \text{... (ii)}$$

$$20 \text{ mA} = I_0\left[e^{0.42/\eta \times 26 \times 10^{-3}} - 1\right] \quad \text{... (iii)}$$

Taking the ratio of equations (iii) and (ii), we obtain

$$\frac{20}{10} = \frac{I_0\left[e^{0.42/\eta \times 26 \times 10^{-3}} - 1\right]}{I_0\left[e^{0.4/\eta \times 26 \times 10^{-3}} - 1\right]}$$

Neglecting – 1 term, we have

$$2 = \frac{\left[e^{0.42/\eta \times 26 \times 10^{-3}}\right]}{\left[e^{0.4/\eta \times 26 \times 10^{-3}}\right]}$$

Taking log on both the sides, we obtain

$$0.693 = \frac{16.15}{\eta} - \frac{15.38}{\eta}$$

or $0.693\,\eta = 0.77$ or $\eta = 1.11$

Now, substituting this value of η into equation (ii), we get

$$10 \times 10^{-3} = I_0\left[e^{0.4/1.11 \times 26 \times 10^{-3}} - 1\right]$$

Solving the equation, we get

$I_0 = 9.69 \times 10^{-9}$ Amp or 9.69 nA. **Ans.**

1.26. DIODE RESISTANCE*

An ideal diode offers zero resistance when forward bias and infinite resistance when it is reverse biased. But practical or real diode does not behave ideally. It means that the forward resistance of a real diode is quite small and reverse resistance is very high.

1.26.1. Forward Resistance (RGTU, Bhopal, Sem. Exam., 2006-07)

The resistance offered by the diode in the forward bias is called forward resistance. This resistance is different for the flow of direct current (d.c.) and for alternating current (a.c.). Hence, this resistance is of two types:

(a) d.c. forward resistance
(b) a.c. forward resistance

1.26.2. D.C. Forward Resistance or Static Resistance** (RGTU, Bhopal, Sem. Exam. 2006-07)

When a dc voltage is applied to a diode, a dc current will flow through it and the operating point on the characteristic curve of the diode will not change its position with time. This is the resistance offered by the diode to dc in forward bias condition. It is also known as **static resistance.** It is given by the ratio of the d.c. forward voltage across the diode to the d.c. current flowing through it at a particular point or operating point on the forward characteristic curve.

Mathematically, the static forward resistance from the curve at point P

$$R_F = \frac{V_F}{I_F} = \frac{OA}{OB} \qquad ...(1.21)$$

Because of the non-linear shape of the diode characteristic, the value of R_F depends upon the exact location of the operating point on the curve. The typical value of forward static resistance R_F is between 10 Ω to 50 Ω.

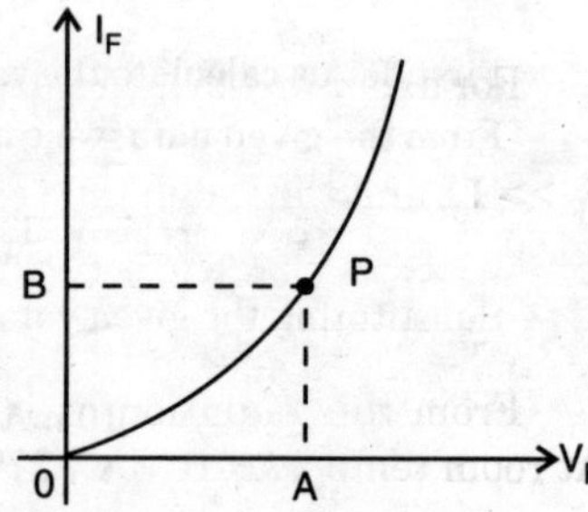

Fig. 1.37. (a). Static resistance of diode

1.26.3. A.C. Forward Resistance

It is the resistance offered by the diode to the a.c. in forward bias. It is also called **dynamic resistance.** It is given by the ratio of change in voltage across the diode to the resulting change in current through diode.

$$\text{a.c. forward resistance} = \frac{\text{Change in voltage across diode}}{\text{Resulting change in current through diode}}$$

From figure 1.37 (b), if P is operating point, then

change in forward voltage = CE
change in forward current = DF

$$\text{Therefore, a.c. forward resistance} = \frac{CE}{DF}$$

or

$$r_F = \frac{\Delta V_F}{\Delta I_F} \qquad ...(1.22)$$

Because of the non-linear shape of the forward characteristic, the value of a.c. resistance of a diode is in range of 1 to 25 Ω.

---

* What is static and dynamic resistance in the forward biasing of the diode? Explain with suitable diagram. (RGTU, Bhopal, Sem. Exam. 2005-06)

** Explain how you will determine the static resistance and dynamic resistance of p-n junction diode after defining both of them. (RGTU, Bhopal, Sem. Exam., 2006-07)

We know that the diode current equation is given as

$$I = I_0\left(e^{V/\eta V_T} - 1\right)$$

Let for forward-bias, the above equation be

$$I_F = I_0\left(e^{V_F/\eta V_T} - 1\right)$$

Differentiating, we get

$$\frac{dV_F}{dI_F} = \frac{\eta V_T}{I_0 . e^{V_F/\eta V_T}} = \frac{\eta V_T}{I + I_0}$$

Using equation (1.22), we get

$$\frac{dV_F}{dI_F} = r_F$$

Therefore, $$r_F = \frac{\eta V_T}{I + I_0}$$

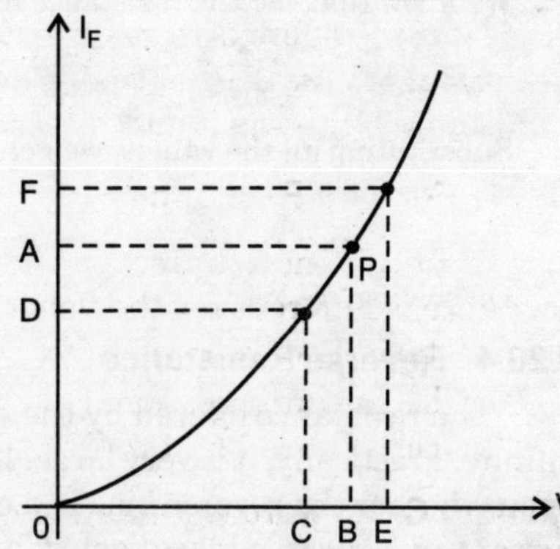

Fig. 1.37. (b) A.C. forward resistance

For a forward-bias greater than a few tenths of volt, $\left|\frac{V_F}{\eta V_T}\right| >> 1$ and according to equation (1.19)

$I_F >> I_0$, and thus, dynamic resistance r_F is given approximately by

$$r_F = \frac{\eta V_T}{I_F}$$

From this equation, it is clear that for forward-bias, r_F varies inversely with the current I_F. At room temperature (i.e., 27°C or 300 °K).

$$V_T = 0.026 \text{ volt or } 26 \text{ m Volt}$$

Thus, substituting $\eta = 1$ and $V_T = 26$ m Volt in the last equation, we get

$$r_F = \frac{26\,\text{mV}}{I_F\,(\text{mA})}$$

EXAMPLE 1.6 Determine the dynamic resistance of a P-N junction diode at a forward current of 2 mA. Assume that kT/*e* = 25 mV. **(A.M.I.E. Examinations, 1998)**

Solution: Given that

Forward current, $I = 2 \text{ mA} = 0.002 \text{ A}$

and volt-equivalent of temperature,

$$V_T = \frac{kT}{e} = 25 \text{ mV} = 0.025 \text{ V}$$

We know that the dynamic resistance is given approximately as

$$r_F = \frac{\eta V_T}{I_F}$$

Substituting, $\eta = 1$, $V_T = 0.025$ V and $I = 0.002$ A in above equation, we get

Dynamic resistance, $$r_F = \frac{0.025}{0.002} = 12.5 \text{ ohms}$$ **Ans.**

EXAMPLE 1.7 Find ac resistance for a semiconductor diode having a forward bias of 200 mV and reverse saturation current of 1 μA at room temperature.

Solution: At room temperature (i.e., 300°K)

$$V_T = 26 \text{ mV} = 0.026 \text{ V}$$

Applied forward voltage $$V_F = 200 \text{ mV} = 0.2 \text{ V}$$

Reverse saturation current $I_0 = 1\ \mu A = 1 \times 10^{-6}$ A

We know that the a.c. resistance for the diode is given as

$$r_F = \frac{\eta V_T}{I_0 e^{V_F/\eta V_T}}$$

Substituting all the values, we get

$$r_F = \frac{0.026}{1 \times 10^{-6} \times e^{0.2/0.026}}$$

On solving, we get $r_F = 11.86$ ohms **Ans.**

1.26.4. Reverse Resistance

The resistance offered by the diode in reverse bias is known as reverse resistance. Ideally it is infinite. Practically, it is very large. It can be d.c. reverse resistance or a.c. reverse resistance depending upon whether the reverse bias is d.c. or a.c. This reverse resistance is due to the existence of leakage current in a reverse bias diode. The reverse bias resistance is very large compared to the forward resistance.

1.27. DIODE CAPACITANCE (RGTU, Bhopal, Sem. Exam. 2005-06)

Diode offers two types of capacitance, one in forward bias and other in reverse bias. They are

(i) Diffusion or Storage Capacitance

(ii) Depletion or Transition Capacitance

1.27.1. Diffusion or Storage Capacitance* (RGTU, Bhopal, Sem. Exam., 2005-06, 2006-07)

The capacitance which exists in a forward-biased junction is called a diffusion or storage capacitance. It is called diffusion capacitance to account for the time delay in moving charges across the junction by diffusion process.

If in a forward-biased junction the applied voltage is suddenly reversed, then forward current I_F ceases at once but a lot of majority charge carriers are left in the depletion region. This charge represents a stored charge in the reverse bias condition and should be removed from the space charge region. This removal of charge takes a finite time. This effect is similar to the discharge of a capacitor. Therefore, the amount of stored charge represents the magnitude of diffusion capacitance. It may be expressed as

$$C_D = \frac{\tau . I_F}{\eta . V_T} \quad ...(1.23)$$

where C_D = Diffusion capacitance

η = Constant ($\eta = 2$ for Si, $\eta = 1$ for Ge)

V_T = Volt equivalent of temperature

τ = Mean life time of carriers

I_F = Forward current

Hence, the diffusion capacitance is directly proportional to the value of forward current I_F. The typical value of diffusion capacitance is about 0.02 μF. The effect of diffusion capacitance is negligible for a reverse-biased PN-junction.

1.27.2. Depletion or Transition Capacitance (RGTU, Bhopal, Sem. Exam. 2006-07)

When a PN-junction is formed, there exists a depletion region at the junction. This depletion region or layer consists of positive and negative immobile ions. This depletion layer is non-conductive

* Explain the transition capacitance and diffusiion capacitance of a P-N junction. (RGTU, Bhopal, Sem. Exam., 2005-06, 2006-07)

and hence acts as a dielectric medium between P-region and N-region. The P-region and N-region act as the two plates of a capacitor because they have a low resistance. These two plates are separated by a dielectric (depletion layer).

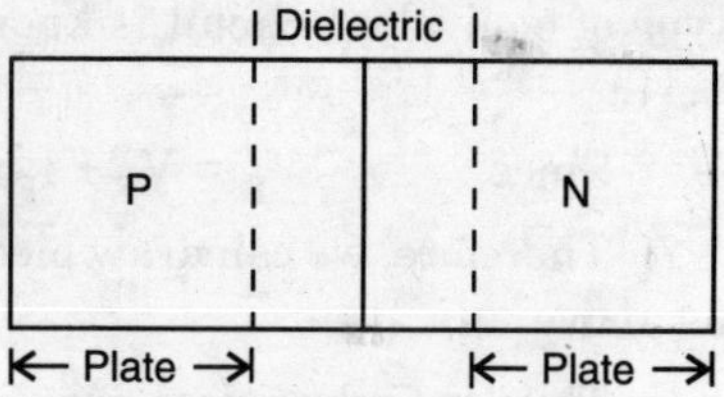

Fig. 1.38. Showing depletion or transition capacitance

The capacitance of a parallel plate capacitor is

$$C = \frac{\in .A}{d} \quad ...(1.24)$$

Here $\in$ is the permittivity of dielectric.

A is the area of plates and d is distance between plates.

Since the depletion layer width (d) increases with increase in reverse bias voltage, the depletion capacitance should decrease with the increase in reverse bias. In other words, the depletion capacitance dominates in reverse-bias.

Depletion capacitance is given as

$$C_T = \frac{K}{\sqrt{V}} \quad ...(1.25)$$

where K is a constant.

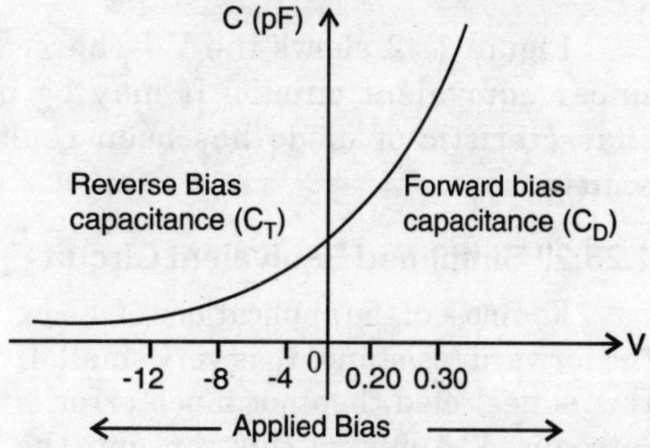

Fig. 1.39. Diffusion capacitance and depletion layer capacitance versus applied bias voltage for a diode

1.28. DIODE EQUIVALENT CIRCUIT

A system or device may be represented by its equivalent circuit. An equivalent circuit of a device consists of electric components or elements. The actual behaviour of the device is not affected by its equivalent circuit. With the help of equivalent circuit generally a type of network is obtained which may be solved for various parameters. We will now discuss the equivalent circuits of diode:

1.28.1. Piece-wise Linear Equivalent Circuit

(RGTU, Bhopal, Sem. Exam., 2007-08)

Figure 1.40 shows the forward characteristics of a diode.

Here, the forward voltage drop V_F has to overcome two voltage drops as under:

(i) Potential barrier V_T (0.7 V for Si and 0.3 V for Ge)

(ii) Forward resistance drop or internal drop $I_F r_F$*

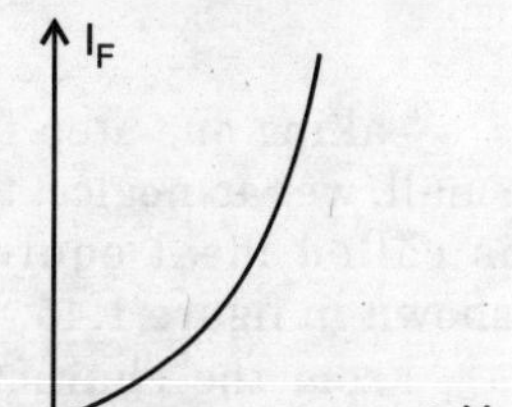

Fig. 1.40. Forward characteristics of diode

Therefore, $V_F = V_T + I_F r_F$ ...(1.26)

where r_F = forward resistance which is varying with point of operation because the V-I characteristic curve is non-linear.

Now, if the V-I characteristic is assumed as linear then forward resistance r_F is constant or it can be taken as average a.c. dynamic forward resistance. Hence due to this concept, we can replace the non-linear curve by straight-line segments. This

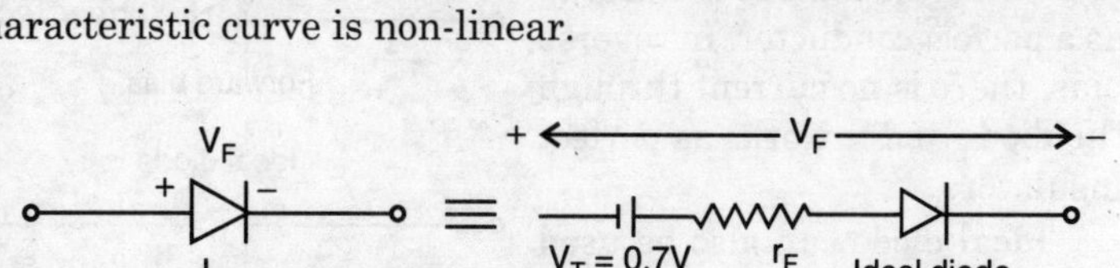

Fig. 1.41. Piece-wise linear equivalent circuit

type of equivalent circuit is known as **piece-wise-linear equivalent circuit**, as shown in figure 1.41.

Since, $V_F = V_T + I_F r_F$ ...(1.27)

Therefore, we can draw piece-wise linear equivalent circuit as above.

Thus, in finding piece-wise linear equivalent circuit, a barrier drop of 0.7 V is represented by a battery put in series with forward resistance r_F and an ideal diode.

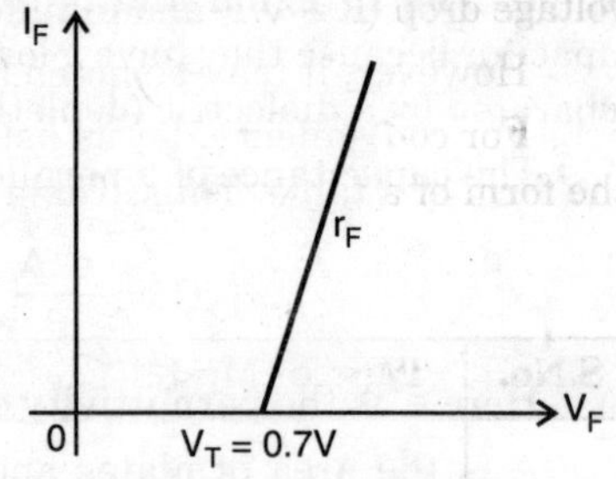

Fig. 1.42. V-I characteristics for piece-wise linear equivalent circuit

Figure 1.42 shows the V-I characteristics for piece-wise linear equivalent circuit. It may be noted that non-linear characteristic of diode has been replaced by straight-line segments.

1.28.2. Simplified Equivalent Circuit

For most of the applications of diode, the forward resistance r_F is very small. If this is neglected then not much error is introduced. Applying this concept, the piece-wise linear equivalent circuit is reduced to another equivalent circuit known as simplified equivalent circuit as shown in figure 1.43.

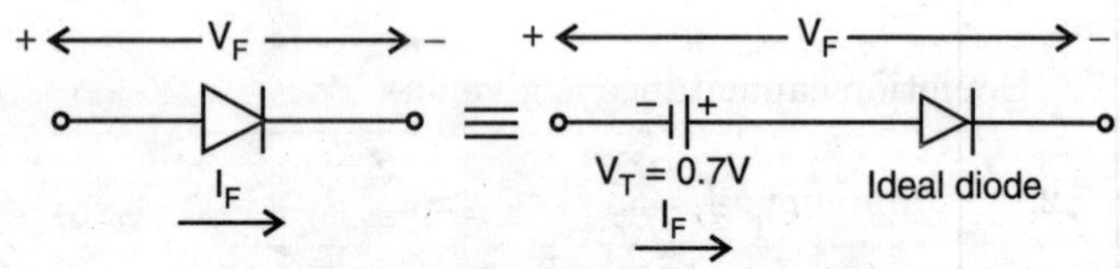

Fig. 1.43. Simplified equivalent circuit

Figure 1.44 shows the V-I characteristics of a silicon diode for simplified equivalent circuit. From figure 1.44, it may be noted that if forward resistance r_F is 0, then at V_T (i.e., knee voltage) the curve becomes a vertical straight line.

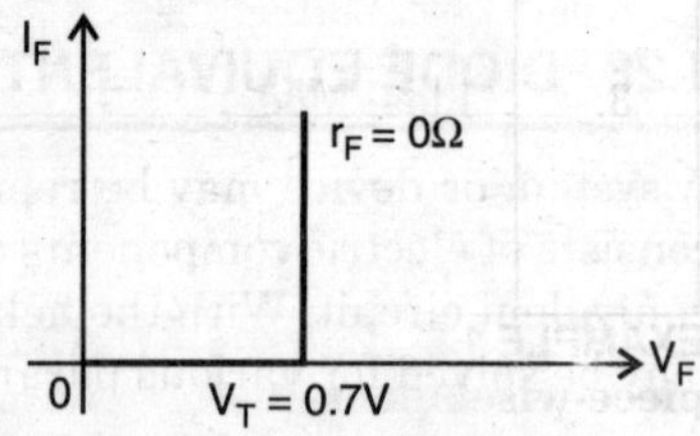

Fig. 1.44. V-I characteristics of a silicon diode for simplified equivalent

1.28.3. Ideal Diode: Ideal Equivalent Circuit

(RGTU, Bhopal, Sem. Exam. 2000-01, 2006-07)

Taking one step further, since barrier potential V_T is very small, we can neglect this also. In this case, the equivalent circuit is called ideal equivalent circuit as shown in figure 1.45.

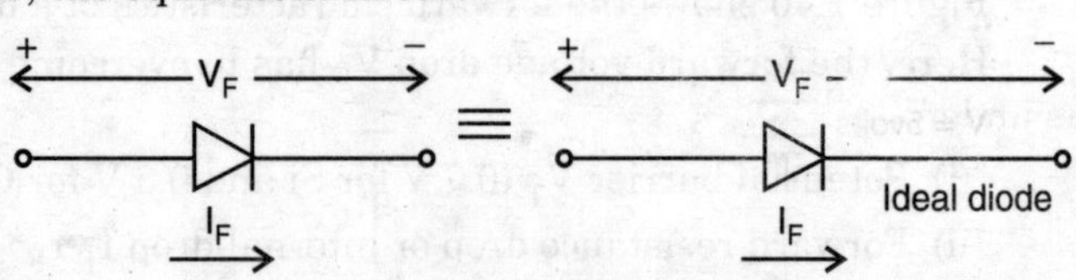

Fig. 1.45. Ideal equivalent circuit

From the characteristics of ideal diode, it may be observed that in forward bias, there is zero voltage drop across the diode and the current flowing through the diode is maximum. It means that in forward bias, diode is working as a perfect conductor. In reverse bias, there is no current through the diode and it works as perfect insulator.

Ideal diode can also be used as a perfect switch. The switch is closed if the diode is forward bias and is open if the diode is reverse bias as shown in figure 1.46.

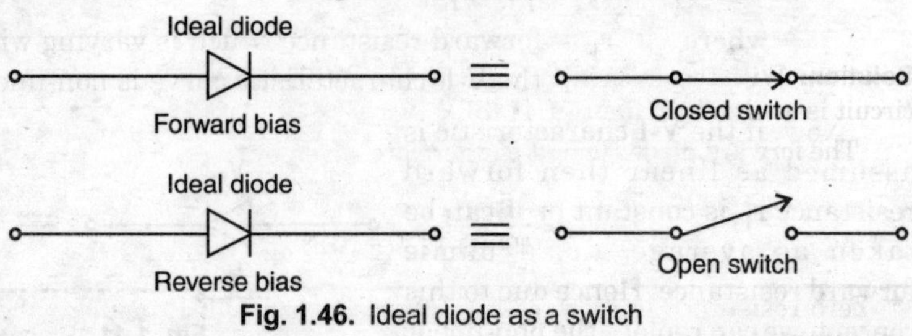

Fig. 1.46. Ideal diode as a switch

* Here r_F is the dynamic or a.c. resistance.

From above it is clear that an ideal diode offers zero resistance in forward bias* because there is zero voltage drop (R = V/I) and infinite resistance in reverse bias because there is zero current (R = V/I).**

However, it may be noted that an ideal diode cannot be manufactured in actual practice.

For convenience, let us list all the above diode equivalent circuits and their characteristics in the form of a table. Thus, Table 1.2 summarises the various diode models.

Table 1.2: The Various Diode Models

S.No.	Type of Model	Equivalent Circuit	Characteristics
1.	Piece-wise linear model	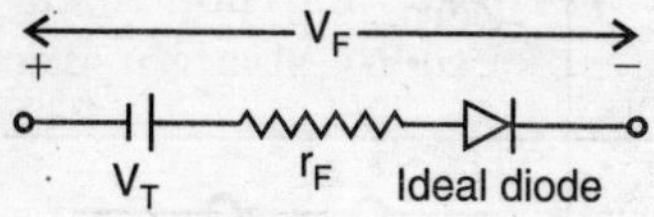	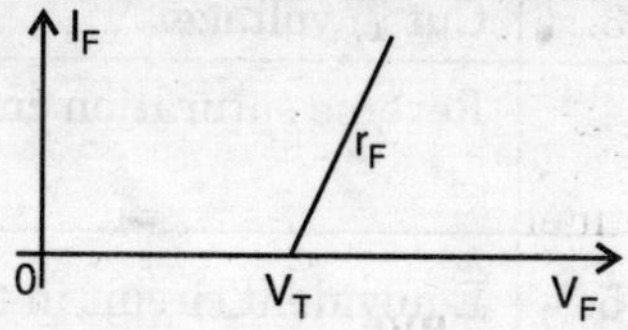
2.	Simplified Model	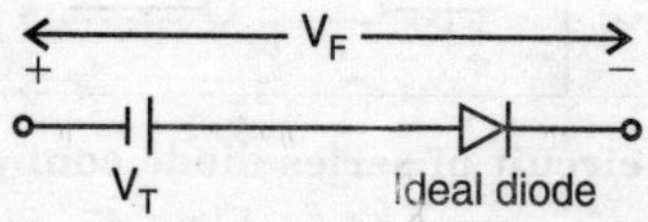	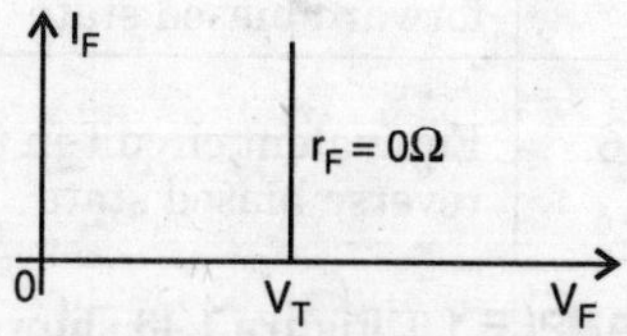
3.	Ideal Model	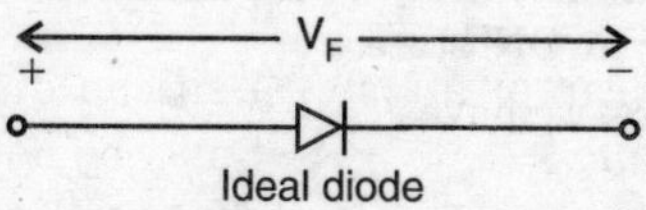	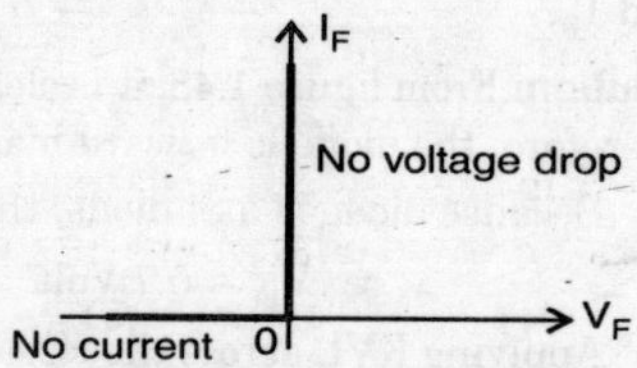

EXAMPLE 1.8 Calculate the diode voltage and current in the circuit shown in figure 1.47(a) using a piece-wise linear model. Assume the piece-wise linear diode parameters as

$$\mathbf{V_\gamma = 0.6\ V \quad and \quad r_F = 12\ \Omega.}$$

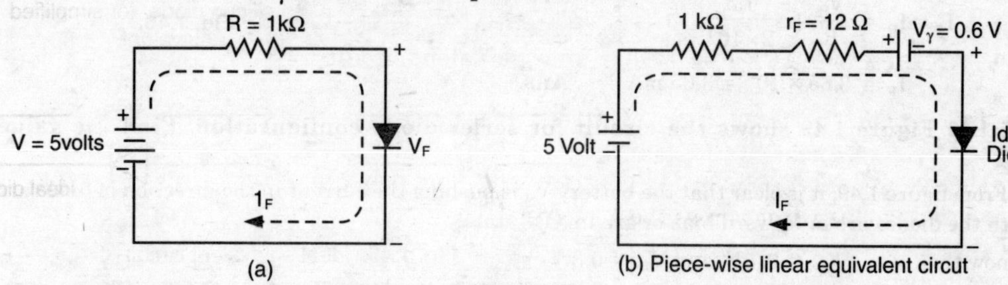

Fig. 1.47.

Solution: With the given input voltage polarity, the diode is forward biased. So, $I_F > 0$. Therefore, the equivalent circuit is as shown in figure 1.47 (b).

The forward diode current is, therefore, given by,

$$I_F = \frac{5 - V_\gamma}{1000 + 12} = \frac{5 - 0.6}{1012} = 4.3 \text{ mA} \qquad \textbf{Ans.}$$

* Zero resistance in forward-bias means that an ideal diode works as a short-circuit in forward-bias condition.

** Infinite resistance in reverse-bias means that an ideal diode works as an open-circuit in reverse-bias condition.

And the voltage across the diode is given by

$$V_F = V_\gamma + I_F r_F = 0.6 + (4.3 \times 10^{-3} \times 12) \quad \text{or} \quad V_F = 0.65 \text{ Volts.} \quad \textbf{Ans.}$$

1.28.4. Comparison of Ideal Diode and Practical Diode

S.No.	Parameter of comparison	Ideal Diode	Practical Diode
1.	Forward resistance	0 Ω	Few Ω
2.	Reverse resistance	∞	Few hundred kΩ
3.	Cut in voltage	0 V	0.6 V for Si and 0.2 V for Ge diode
4.	Reverse saturation current	Zero	Few nA for Si diode Few μA for Ge diode
5.	Equivalent circuit in the forward biased state		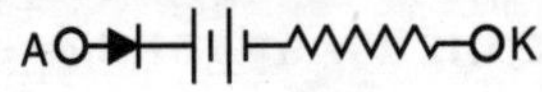
6.	Equivalent circuit in the reverse biased state	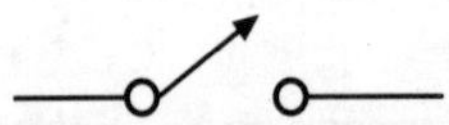	High resistance 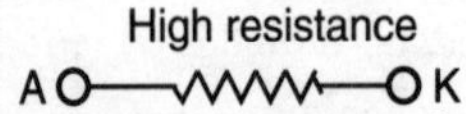

EXAMPLE 1.9 Figure 1.48 shows the circuit of series diode configuration. Find the value of V_F, V_R and I_F.

Solution: From figure 1.48, it is clear that the battery V_S is pushing the current in the direction of diode arrow. Therefore, the diode is forward biased or in 'ON' state.

Because diode is a Si diode, therefore, we have

$$V_F = 0.7 \text{ Volt}$$

Applying KVL across the circuit, we write

$$V_S = V_F + V_R$$

$$V_R = V_S - V_F = 8 - 0.7 = 7.3 \text{ Volt}$$

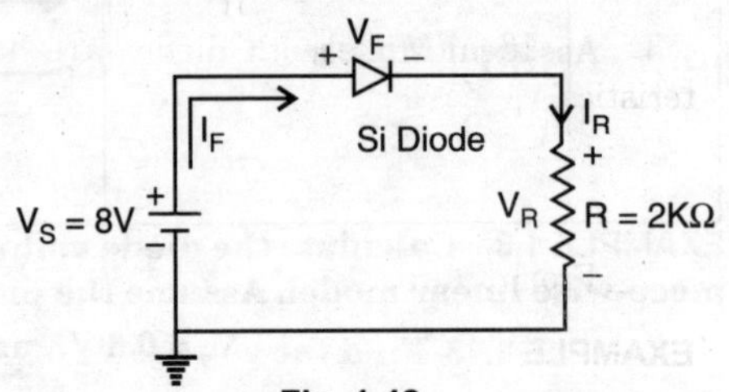

Fig. 1.48.

Also $I_F = I_R = \dfrac{V_R}{R} = \dfrac{7.3}{2 \times 10^3}$

$$I_F \cong 3.65 \times 10^{-3} \cong 3.65 \text{ mA} \quad \textbf{Ans.}$$

EXAMPLE 1.10 Figure 1.49 shows the circuit for series diode configuration. Find ıe value of V_R and I_F.

Solution: From figure 1.49, it is clear that the battery V_S is pushing the current in the direction of diode arrow. Hence, both the diodes are in forward bias or are in 'ON' state.

We know that $V_{F1} = 0.7$ V and $V_{F2} = 0.3$ V

Applying KVL across the circuit

$$V_S = V_{F1} + V_{F2} + V_R$$

$$12 = 0.7 + 0.3 + V_R$$

or $V_R = 12 - 1.0 = 11$ Volt

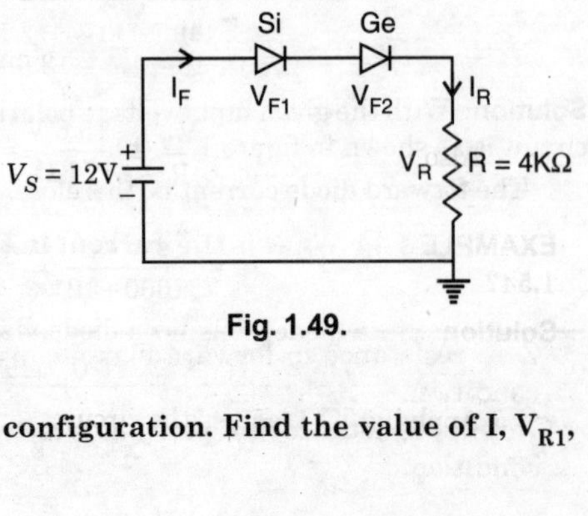

Fig. 1.49.

or $I_F = I_R = \dfrac{V_R}{R} = \dfrac{11}{4 \times 10^3}$

or $I_F = 2.75 \times 10^{-3} = 2.75$ mA **Ans.**

EXAMPLE 1.11 Figure 1.50 shows the circuit of series diode configuration. Find the value of I, V_{R1}, V_{R2} and V_0.

Solution: The polarities of batteries indicate that the diode is in forward bias. Since diode is a Si diode, $V_F = 0.7$ Volt.

Applying KVL across the circuit

$$V_{S1} + V_{S2} = V_F + V_{R1} + V_{R2}$$

or $$V_{R1} + V_{R2} = V_{S1} + V_{S2} - V_F$$

or $$IR_1 + IR_2 = V_{S1} + V_{S2} - V_F$$

$$I = \frac{V_{S1} + V_{S2} - V_F}{R_1 + R_2} = \frac{10 + 5 - 0.7}{(4+2)\times 10^3} = \frac{14.3}{6\times 10^3} = \frac{14.3\times 10^{-3}}{6}$$

$$I = 2.38 \times 10^{-3}\text{ A} = 2.38\text{ mA}$$

$$V_{R1} = IR_1 = 2.38 \times 10^{-3} \times 4 \times 10^3 = 9.52\text{ Volt}$$

and $$V_{R2} = IR_2 = 2.38 \times 10^{-3} \times 2 \times 10^3 = 4.76\text{ Volt}$$

Voltage V_0 is given by $V_0 = V_{R2} - V_{S2} = 4.76 - 5 = -0.24$ Volt **Ans.**

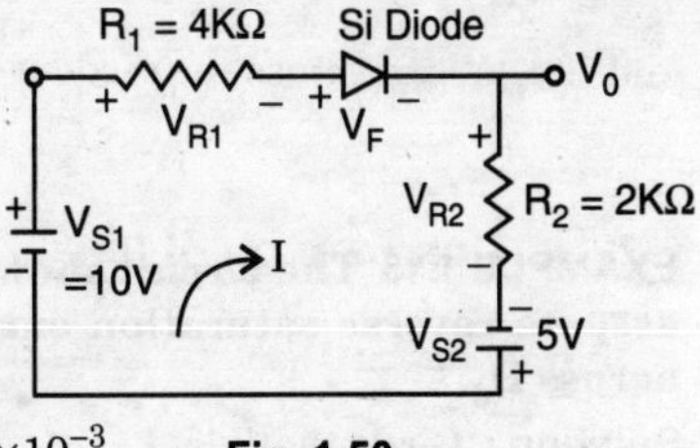

Fig. 1.50.

EXAMPLE 1.12 Figure 1.51 shows the circuit of parallel diode configuration. Find the values of V_0, I, I_{F1} and I_{F2}.

Solution: From figure 1.51, it is clear that the applied battery is pushing the currents in the directions of diode arrows. Hence, both diodes are in forward bias or are in 'ON' state.

Because each diode is a silicon diode, therefore, voltage drop across each diode is 0.7 Volt.

Hence, $V_{F1} = V_{F2} = 0.7$ Volt

The current *I* is given by

$$I = \frac{V_R}{R} = \frac{V_S - V_{F1}}{R} = \frac{10 - 0.7}{0.3\times 10^3} = \frac{9.30}{0.3\times 10^3} = 31\text{ mA}$$

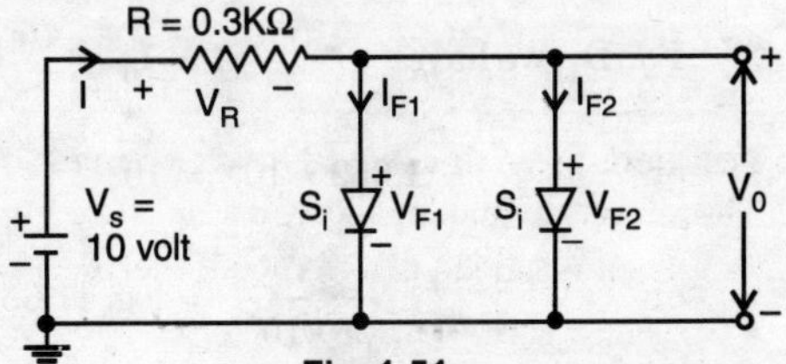

Fig. 1.51.

Assuming that both diodes are having similar characteristics, we have

$$I_{F1} = I_{F2} = \frac{I}{2} = \frac{31}{2} = 15.5\text{ mA}$$

From figure, $V_0 = V_{F2} = 0.7$ Volt **Ans.**

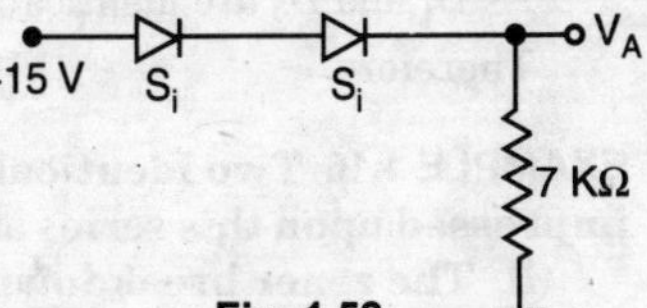

Fig. 1.52.

EXAMPLE 1.13 Find the voltage V_A of the current given in figure 1.52.

Solution: Let us redraw the given circuit.

Since, diodes are silicon diodes, hence

$$V_{F1} = V_{F2} = 0.7\text{ Volt}$$

From circuit of figure 1.53, both diodes are forward biased. Now, applying KVL across the circuit,

$$V_S = V_{F1} + V_{F2} + V_A$$

$$V_A = V_S - V_{F1} - V_{F2} = 15 - (0.7 + 0.7)$$

$$V_A = 15 - 1.4 = 13.6\text{ Volt}$$

Also $$I = \frac{V_A}{R} = \frac{13.6\text{ V}}{7\text{ K}\Omega} = 1.942\text{ mA}$$ **Ans.**

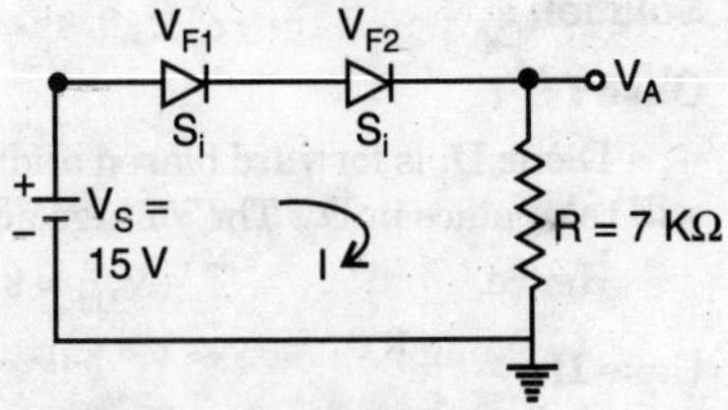

Fig. 1.53.

EXAMPLE 1.14 What is the current in the following circuit shown in figure 1.54?

Solution: Since diode is a silicon diode, hence

$$V_F = 0.7\text{ Volt}$$

Applying KVL across the circuit

$$5 = V_F + I.10 = 0.7 + 10\,I$$

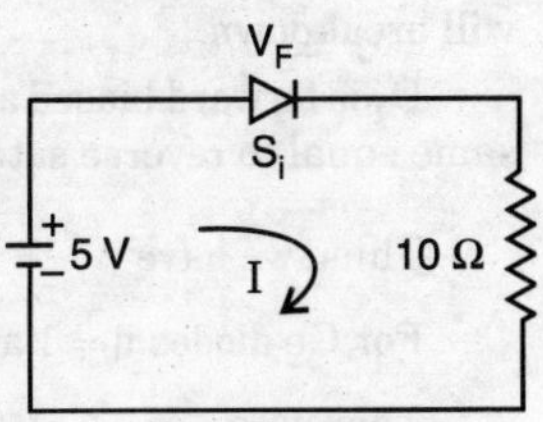

Fig. 1.54.

$$10\,I = 4.3$$

$$I = \frac{4.3}{10} = 0.43 \text{ A} \quad \textbf{Ans.}$$

EXAMPLE 1.15 The circuit shown in figure 1.55 uses identical diodes D_1, D_2 and D_3. For each diode, assume reverse saturation current of 10^{-3} A at T = 300°K and η = 1. Calculate V_{in} to get 600 mV across D_3.

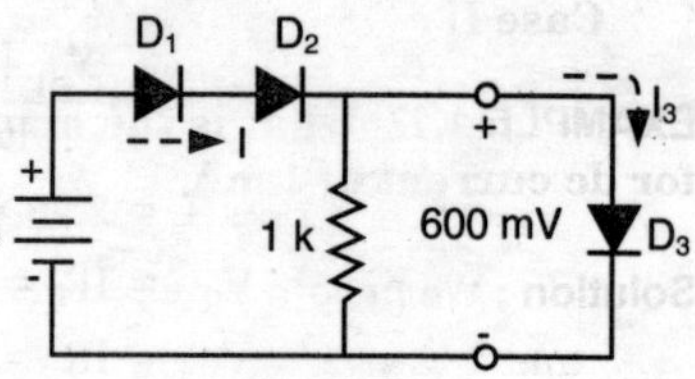

Fig. 1.55.

Solution : Given that $I_0 = 1 \times 10^{-3}$ A at T = 300° K,

$\eta = 1$, V = 0.6 V,

V_T at 300° K is 26 mV.

(i) Assume that current through $D_3 = I_3$.

Hence, we have

$$I_3 = I_0 \left[e^{V/\eta V_T} - 1\right]$$

$$= 10^{-3}\left[e^{0.6/26\times10^{-3}} - 1\right] = 1.052 \text{ mA} \quad \text{... (i)}$$

(ii) Current through 1 kΩ resistor $= \dfrac{600 \text{ mV}}{1 \text{ k}\Omega} = 0.6$ mA ... (ii)

(iii) Total current through D_1 and D_2 = 1.052 + 0.6 = 1.652 mA ... (iii)

(iv) Let us calculate voltage across diodes D_1 and D_2

For D_1, we have $I = I_0\left[e^{V/\eta V_T} - 1\right]$

or $$1.652 \times 10^{-3} = 10^{-13}\left[e^{V/26\times10^{-3}} - 1\right]$$

or $$\frac{V}{26\times10^{-3}} = \log_e\,[1.652 \times 10^{10}]$$

or $$V = 0.6117 \text{ Volts}$$

As D_1 and D_2 are identical diodes, the voltage drop across D_2 will be the same, i.e., 0.6117 Volt.

Therefore, V_{in} = 0.6117 + 0.6117 + 0.6 = 1.823 Volt. **Ans.**

EXAMPLE 1.16 Two Identical germanium diodes are connected in series opposition. A 10 V battery is impressed upon this series arrangement. Find the voltage across each diode at room temperature if:

(i) The zener breakdown voltage of each diode is 9.8 V.

(ii) The zener breakdown voltage of each diode is 10.2 V.

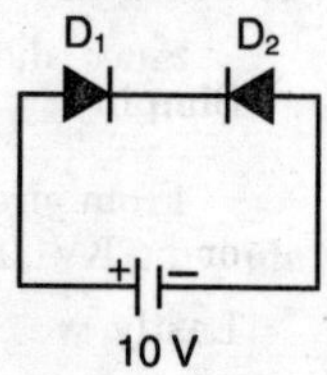

Fig. 1.56.

Solution :

Case I :

Diode D_1 is forward biased and D_2 is reverse biased. Hence, breakdown due to zener effect will take place in D_2. The voltage across it remains constant equal to V_Z = 9.8 V.

Hence, V_{D2} = 9.8 Volt and $V_{D1} = 10 - V_{D2}$ = 0.2 Volt. **Ans.**

Case II :

As the zener breakdown voltage V_Z = 10.2 volt, which is higher than the battery voltage, none of the diodes will breakdown.

D_1 is forward biased and D_2 is reverse biased. As they are connected in series, current through them remains same equal to reverse saturation current of D_2, which is I_0.

Thus, we have $$I = I_0\left[e^{V_1/\eta V_T} - 1\right] = I_0 \quad \text{...}(i)$$

For Ge diodes η = 1 and V_T = 26 mV at room temperature.

Therefore, $$e^{V_1/V_T} - 1 = 1$$

or $\quad e^{V_1/V_T} = 2 \quad$ or $\quad \frac{V_1}{V_T} = \log_e 2 = 0.693$

or $\quad V_1 = 0.693 \times 26 \times 10^{-3} = 0.018$ Volts

and $\quad V_2 = 10 - V_1 = 10 - 0.018 = 9.982$ Volts

Case I : $\quad V_{D1} = 0.2$ Volt and $V_{D2} = 9.8$ Volt

Case II : $\quad V_{D1} = 0.018$ Volt and $V_{D2} = 9.982$ Volt. **Ans.**

EXAMPLE 1.17 What is the magnitude of r for silicon P-N junction diode at room temperature and for dc current of 1 mA.

Solution : We have
$$r = \frac{\eta V_T}{I + I_0} \quad \text{...(i)}$$

Let us assume $\quad V_T = 26$ mV at room temperature

$I + I_0 = I$ since I_0 is negligible

$\eta = 2$ for silicon

Substituting all the values, we get
$$r = \frac{2 \times 26 \times 10^{-3}}{1 \times 10^{-3}} = 52\ \Omega. \quad \textbf{Ans.}$$

EXAMPLE 1.18 Each diode is having linearised volt ampere characteristics with internal resistance r and offset voltage V_γ. Diode D_1 is germanium with $V_\gamma = 0.2$ V and r = 20 Ω whereas diode D_2 is silicon with $V_\gamma = 0.6$ V and r = 15 Ω. Find diode current of the circuit shown in figure 1.57 (a) if R = 10 kΩ.

(RGTU, Bhopal, Sem. Exam. 2007-08)

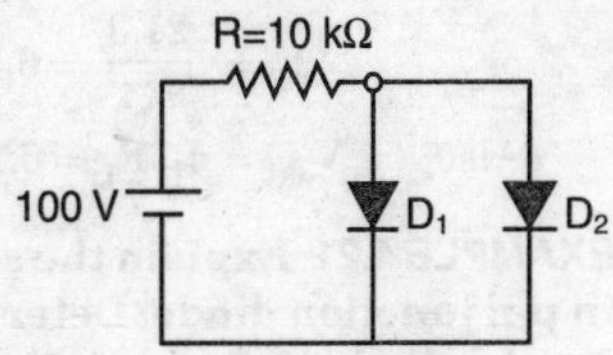

Fig. 1.57 (a)

Solution : First, let us replace the diodes by their equivalent circuits as shown in figure 1.57 (b)

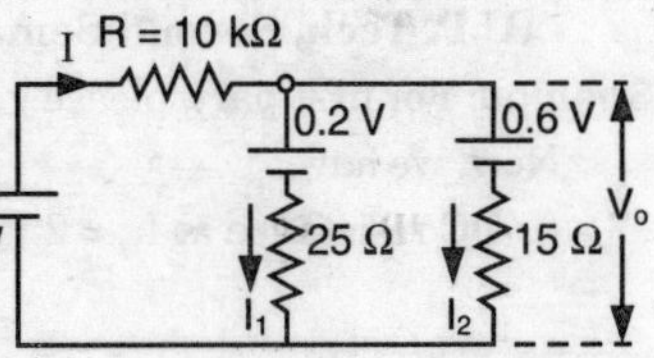

Fig. 1.57 (b)

Now, let us calculate V_D.

Here, we have $\quad V_D = 0.2 + 20\, I_1 = 0.6 + 15\, I_2 \quad$ ... (i)

Also $\quad I_1 + I_2 = I = \frac{100 - V_D}{10\ \text{k}\Omega} \quad$ ... (ii)

or $\quad \frac{V_D - 0.2}{20} + \frac{V_D - 0.6}{15} = \frac{100 - V_D}{10 \times 10^3}$

Simplifying, we get $\quad 10 \times 10^3[3V_D - 0.6 + 4V_D - 2.4] = 60\,[100 - V_D]$

$10 \times 10^3\,[7V_D - 3] = 6000 - 60V_D$

or $\quad 70 \times 10^3 V_D + 60V_D = (6000 + 30 \times 10^3) \quad$ or $\quad V_D = 0.5138$ volts

Lastly, we calculate diode currents as under:
$$I_1 = \frac{V_D - 0.2}{20} = \frac{0.5138 - 0.2}{20} = 0.01569 \text{ A} = 15.69 \text{ mA} \quad \textbf{Ans.}$$
$$I_2 = \frac{V_D - 0.6}{15} = \frac{0.5138 - 0.6}{15} = -5.7466 \text{ mA} \quad \textbf{Ans.}$$

EXAMPLE 1.19 A germanium diode carries a current of 1 mA of room temperature when a forward bias of 0.15 V is applied. Estimate the reverse saturation current at room temperature.

(RGTU, Bhopal, Sem. Exam., 2005-06)

Solution : Given that Ge diode, so, we have

$\eta = 1, \quad I = 1$ mA, $\quad V_F = 0.15$ Volts.

The expression for forward diode current is given by

$$I_F = I_0 \left[e^{V_F/\eta V_T} - 1\right]$$

Let V_T = 26 mV at room temperature

Substituting all the values, we get

$$1 \times 10^{-30} = I_0 \left[e^{0.15/1 \times 26 \times 10^{-3}} - 1\right]$$

or $1 \times 10^{-3} = I_0\,[403.43 - 1]$

or $I_0 = 2.4849 \times 10^{-6}$ A or 2.4849 μA. **Ans.**

EXAMPLE 1.20 Calculate V_{out} and I_D for the following figure 1.58.

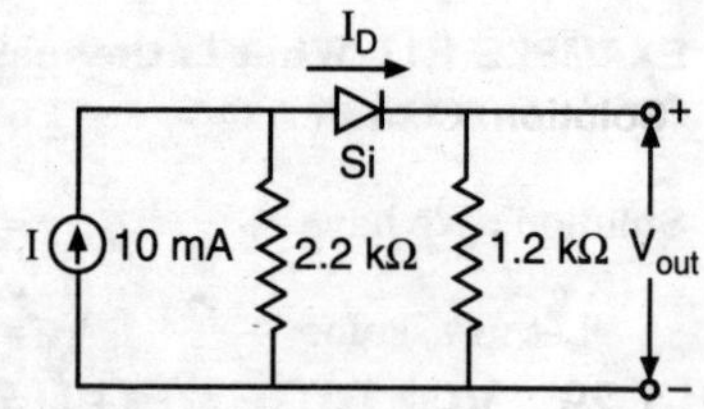

Fig. 1.58.

Solution: First, let us convert current source into voltage source in the given circuit. This has been shown in figure 1.59.

In figure 1.59, the value of V will be

$$V = I \times 2.2\text{ k}\Omega = 10 \times 10^{-3} \times 2.2 \times 10^3$$

or $V = 22$ Volt and $R_S = 2.2$ kΩ

Applying KVL, we have

$$V - 2.2\,I_D - 0.7 - (1.2 \times I_D) = 0$$

or $22 - 0.7 = 3.4\,I_D$

or $I_D = \dfrac{21.3}{3.4} = 6.26$ mA **Ans.**

Also, $V_{out} = I_D R_L = 6.26 \times 10^{-3} \times 1.2 \times 10^3 = 7.512$ Volt **Ans.**

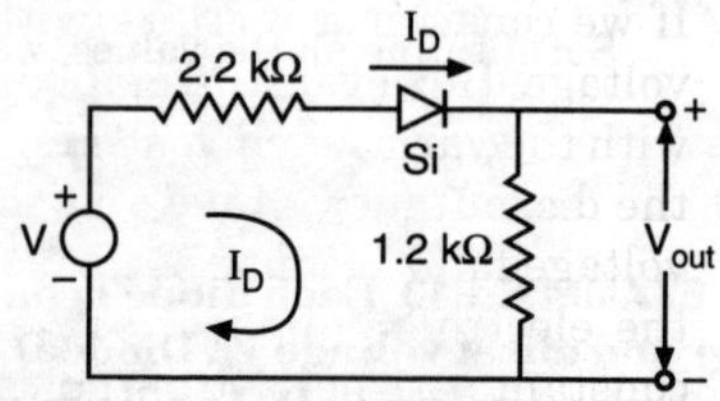

Fig. 1.59.

EXAMPLE 1.21 Explain the static and dynamic resistance in p-n junction diode. Determine the dc resistance levels for the diode in figure 1.60 at

(i) I_D = 2 mA; (ii) I_D = 20 mA; (iii) V_D = – 10 V

(U.P. Tech, Second Sem. Exam; 2006-07) (05 marks)

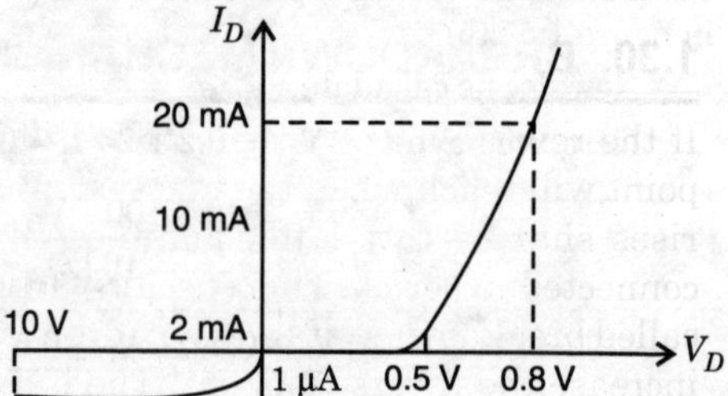

Fig. 1.60.

Solution: For first part, please go through article 1.26.

Now, we have

(i) DC resistance at I_D = 2 mA

$$r_{F1} = \frac{V_D}{I_D} = \frac{0.5\text{ V}}{2\text{ mA}} = 250\ \Omega \quad \textbf{Ans.}$$

(ii) DC resistance at I_D = 20 mA

$$r_{F2} = \frac{V_D}{I_D} = \frac{0.8\text{ V}}{20\text{ mV}} = 40\ \Omega \quad \textbf{Ans.}$$

(iii) DC resistance at V_D = – 10 V

$$r_R = \frac{10\text{ V}}{1\ \mu\text{A}} = 10\text{ M}\Omega \quad \textbf{Ans.}$$

EXAMPLE 1.22. Determine V_o and I for the following circuit shown in figure 1.61.

Solution: V_o = 10 V, I = 10 mA

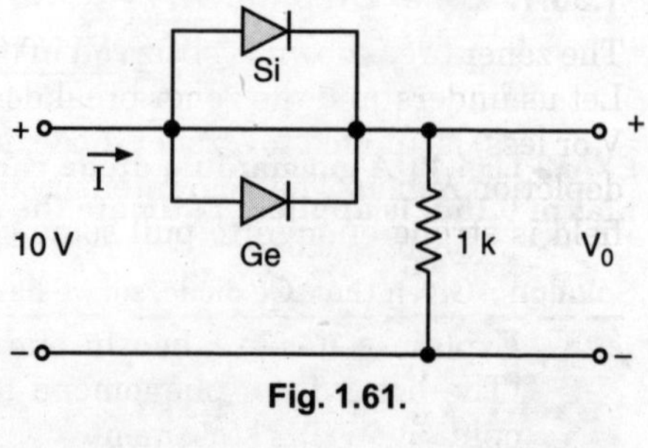

Fig. 1.61.

EXAMPLE 1.23. Calculate forward current I_F for the silicon diode with dynamic resistance r_d = 0.25 Ω used in the circuit of figure 1.62, V_γ = 0.6 V.

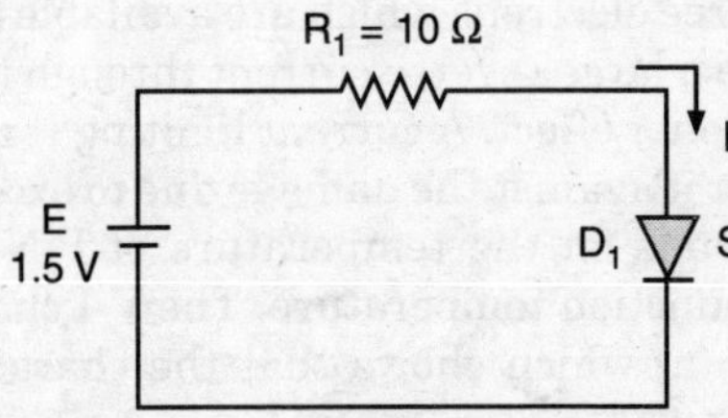

Fig. 1.62.

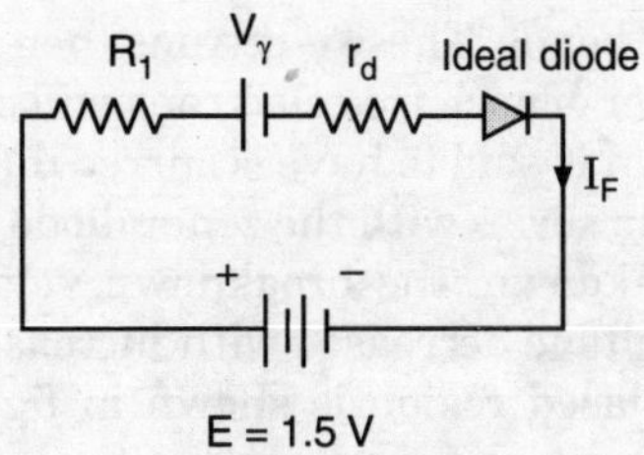

Fig. 1.63.

Solution: Considering figure 1.63, we get

$$I_F = \frac{E - v_\gamma}{(R_1 + r_d)} = \frac{1.5 - 0.6}{(10 + 0.25)} = 87.8 \text{ mA} \quad \textbf{Ans.}$$

1.29. VOLTAGE REGULATION OR STABILISATION

If we consider a rectifier with a suitable filter, then it will become a good source of d.c. output voltage. However, the main drawback of such a power supply is that the d.c. output voltage varies with the variation in load resistance or input voltage. Hence, if the applied input voltage increases, the d.c. output voltage also increases. In the same way, if the load-current increases, the output voltage falls because of the voltage drop in filter choke, rectifier element etc. However, in most of the electronic applications, it is desired that the d.c. output voltage of rectifier must remain constant irrespective of the variation in input a.c. voltage or load. Hence, to ensure this, a voltage stabilisation or regulation device known as voltage stabiliser is used. Although there are several voltage stabilising circuits which may be designed, but in this chapter, we shall discuss only zener diode as a voltage stabilizer or regulator.

1.30. BREAKDOWN MECHANISM **(RGTU, Bhopal, Sem. Exam; 2007-08)**

If the reverse-bias voltage applied to a P-N-junction is increased, a point will reach when the junction breaks down and reverse current rises sharply to a value limited only by the external resistance connected in series. This specific value of the reverse bias voltage is called breakdown voltage (V_Z). After breakdown, a very small further increase the reverse current. The breakdown voltage depends upon the width of depletion layer. This width of depletion layer depends upon the doping level.

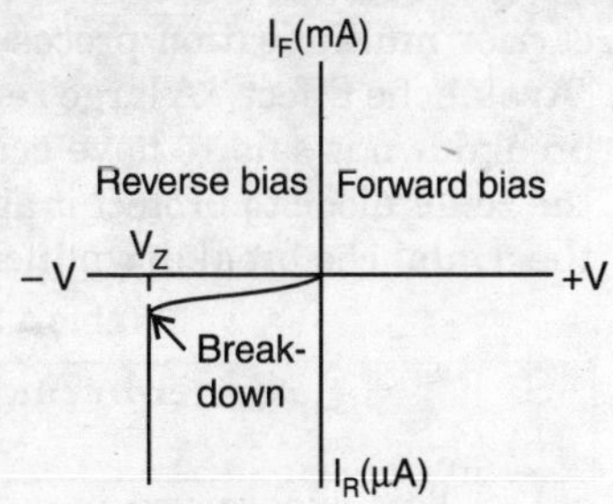

Fig. 1.64. Breakdown mechanism

The following two processes cause junction breakdown due to the increase in reverse bias voltage:

(i) Zener Breakdown

(ii) Avalanche Breakdown

1.30.1. Zener Breakdown*

The zener breakdown is observed in the zener diodes having V_z less than 5 V or between 5 to 8 Volts. Let us understand the zener breakdown mechanism for V_z less than 5 V. When a reverse voltage (5 V or less) is applied to a zener diode, it causes a very intense electric field to appear across a narrow depletion region. The field intensity is typically of the order of 3×10^5 V/cm. Such an intense electric field is strong enough to pull some of the valence electrons into the conduction band by breaking

* Explain zener breakdown and Avalanche multiplication phenomena in PN-junction diode. "The breakdown phenomena for voltage less than 6 V is zener breakdown". Explain your opinion on this statement. **(RGTU, Bhopal, Sem. Exam. 2005-06)**

their covalent bonds. These electrons then become free electrons which are available for conduction. A large number of such free electrons will constitute a large reverse current through the zener diode and breakdown is said to have occurred due to the zener effect. A current limiting resistance should be connected in series with the zener diode to protect it against the damage due to excessive heating. In zener breakdown, the breakdown voltage depends on the temperature of P-N junction. The breakdown voltage decreases with increase in the junction temperature. The V-I characteristics in the reverse biased region is shown in figure 1.65 (a) which shows that the characteristics after breakdown is very sharp, almost vertical.

1.30.2. Avalanche Breakdown in Zener Diodes*

(RGTU, Bhopal, Sem. Exam. 2005-06) (05 marks)

The avalanche breakdown is observed in the zener diodes having V_z higher than 8 V. Even though the mechanism of breakdown has changed, the device is still called as zener diode. Let us now understand the mechanism. In the reverse biased condition, the conduction will take place only due to the minority carriers. As we increase the reverse voltage applied to the zener diode, these minority carriers tend to accelerate. Therefore, the kinetic energy associated with them increases. While travelling, these accelerated minority carriers will collide with the stationary atoms and impart some of the kinetic energy to the valence electrons present in the covalent bonds. Due to this additionally acquired energy, these valence electrons will break their covalent bonds and jump into the conduction bond to become free for conduction. Now these newly generated free electrons will get accelerated. They will knock out some more valence electrons by means of collision. This phenomenon is called as carrier multiplication. In a very short time, a large number of free minority electrons and holes will be available for conduction, and the carrier multiplication process becomes self sustained. This self sustained multiplication is called "Avalanche Effect". A large reverse current starts flowing through the zener diode and the avalanche breakdown is said to have occurred. A current limiting resistor should be connected in series with the zener diode to protect it against the damage due to excessive heating. The breakdown voltage in the avalanche breakdown increases with increase in the junction temperature.

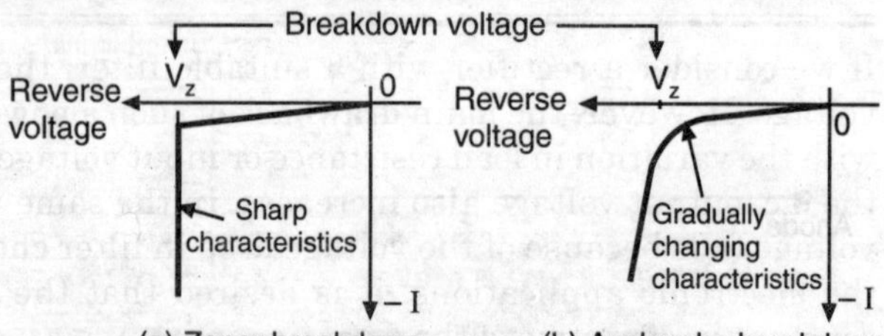

Fig. 1.65.

Table 1.3. Comparison of Zener and Avalanche Breakdown

S. No.	Zener breakdown	Avalanche breakdown
1.	This is observed in zener diodes having V_z between 5 to 8 volts.	This is observed in zener diodes having Vz greater than 8 volts.
2.	The valence electrons are pulled into conduction band due to very intense electric field appearing across the narrow depletion region.	The valence electrons are pushed into conduction band due to the energy imparted by colliding accelerated minority carriers.
3.	V-I characteristics with the zener breakdown is very sharp.	The V-I characteristics with the avalanche breakdown increases gradually. It is not as sharp as that with the zener breakdown.
4.	The breakdown voltage decreases with increase in temperature.	The breakdown voltage increases with increase in temperature.

* Explain 'Avalanche' breakdown. **(RGTU, Bhopal, Sem. Exam. 2005-06)(05 marks)**

1.31. ZENER DIODE

Zener diode is a reverse-biased heavily-doped PN-junction diode which operates in the breakdown region. The reverse breakdown of a PN-junction may occur either due to zener effect or avalanche effect. Zener effect dominates at reverse voltages less than 6 V whereas avalanche effect dominates above 6 V. Hence, first one should be called zener diode. But for simplicity, both types are called zener diodes.

The breakdown voltage of a zener diode can be set by controlling the doping level. For zener diodes, silicon is preferred to Ge because of its higher temperature and current capability.

1.32. CIRCUIT SYMBOL AND BIASING OF A ZENER DIODE

The circuit symbol of a zener diode is as shown in figure 1.66 (a). It is a two terminal device and the terminals are anode and cathode. The arrowhead in the symbol points towards the conventional direction of current through the zener diode, when it is forward biased.

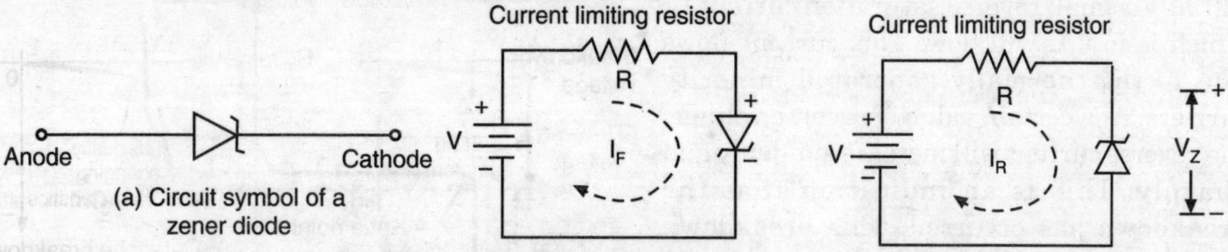

(a) Circuit symbol of a zener diode (b) Forward biasing of a zener diode (c) Reverse biasing of a zener diode

Fig. 1.66. Circuit symbol and biasing of a zener diode

(i) Forward biasing of zener diode

When the anode of the zener diode is connected to the positive terminal of the dc source and the cathode is connected to the negative terminal, the zener diode is said to be forward biased. The forward biased zener diode behaves identical to a forward biased diode. The forward biasing of a zener diode is shown in figure 1.66 (b). The zener diode is generally not used in the forward biased condition.

> **DO YOU KNOW?**
>
> Zener diodes with breakdown voltages near 7 V have the smallest zener impedance.

(ii) Reverse biasing of zener diode

When the cathode is connected to the positive terminal and anode is connected to the negative terminal of the dc source, the zener diode is said to be reverse biased. The operation of zener diode in the reverse biased condition is substantially different from that of a diode. The reverse biasing of a zener diode is shown in figure 1.66 (c). Zener diode in the reverse biased condition is used as a voltage regulator.

1.33. V-I CHARACTERISTICS OF A ZENER DIODE*

The V-I characteristics of a zener diode can be divided into two parts:

(i) Forward characteristics
(ii) Reverse characteristics.

(i) Forward characteristics

The forward characteristics of a zener diode is shown in figure 1.67 (a). It is almost identical to the forward characteristics of a P-N junction diode.

---

* Draw the V-I characteristics of zener diode and explain how does a zener regulate a voltage ?

(RGTU, Bhopal, Sem. Exam., 2005-06)

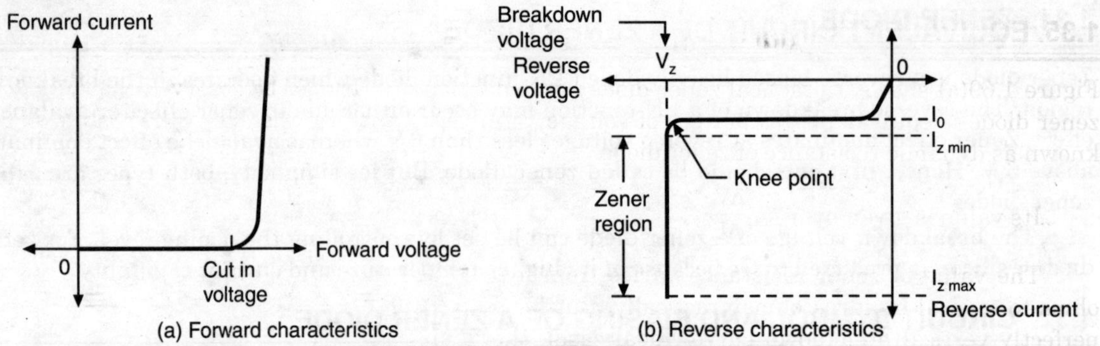

Fig. 1.67. V-I Characteristics of a zener diode

(ii) Reverse characteristics

As we increase the reverse voltage, initially a small reverse saturation current I_0, which is in μA, will flow. This current flows due to the thermally generated minority carriers. At a certain value of reverse voltage, the reverse current will increase suddenly and sharply. This is an indication that the breakdown has occurred. This breakdown voltage is called as **zener breakdown voltage or zener voltage and it is denoted by V_z**. The value of V_z can be precisely controlled by controlling the doping levels of P and N regions at the time of manufacturing a zener diode. After breakdown has occurred, the voltage across zener diode remains constant equal to V_z. Any increase in the source voltage will result in the increase in reverse zener current. The zener current after the reverse breakdown must be controlled by connecting a resistor R as shown in figure 1.66(b). This is essential to avoid any damage to the device due to excessive heating.

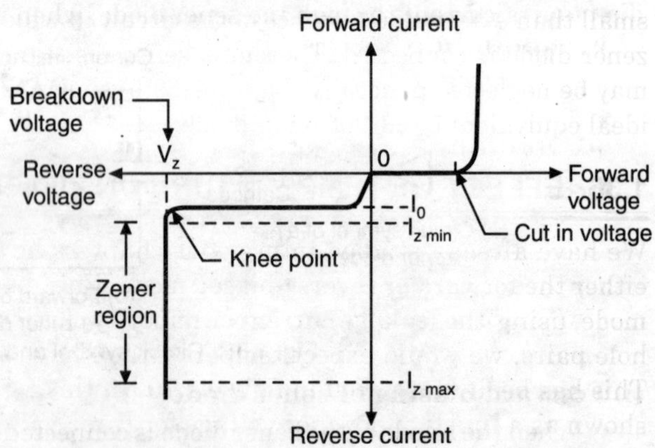

Fig. 1.68. Complete V-I characteristics of a zener diode

Zener Region and its importance

After reverse breakdown the zener diode operates in a region called zener region, as shown in figure 1.67 (b). In this region the voltage across zener diode remains constant but current changes depending on the supply voltage. Zener diode is operated in this region when it is being used as a voltage regulator. The complete V-I characteristics is as shown in figure 1.68.

1.34. APPLICATIONS OF ZENER DIODES* (RGTU, Bhopal, Sem. Exam., 2006-07)

Zener diodes have a large number of applications. Few of them are :

(i) Zener diode is used as a voltage regulator

(ii) Zener diode is used as a peak clipper in waveshaping circuits

(iii) Zener diode is used as a fixed reference voltage in transistor biasing circuits

(iv) Zener diode is used for meter protection against damage from accidental application of excessive voltage.

* Write the applications of zener diodes. **(RGTU, Bhopal, Sem. Exam., 2006-07)**

1.35. EQUIVALENT CIRCUIT OF A ZENER DIODE

Figure 1.69(a) shows a practical zener diode equivalent circuit. It may be observed that practical zener diode is equivalent to a battery of voltage V_z in series with a resistance r_z. Resistance r_z is known as dynamic resistance of zener diode.

Its value is given by $r_z = \dfrac{\Delta V_z}{\Delta I_z}$

The value of zener resistance varies from few ohms to several hundred ohms practically. But for a perfectly vertical breakdown curve (Ideal case), the value of zener resistance is zero.

Generally, the zener resistance value is quite small than external resistance of the circuit in which zener diode is connected. Therefore, zener resistance may be neglected generally. Figure 1.69(b) shows the ideal equivalent circuit of zener diode.

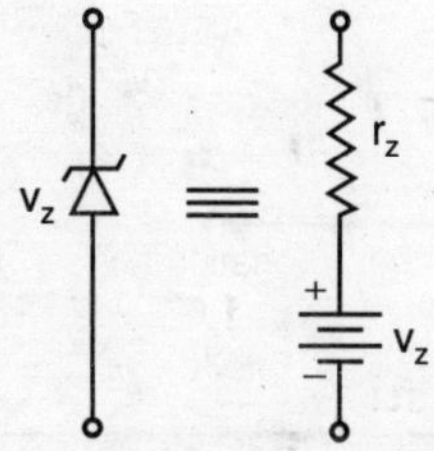

Fig. 1.69 (a). Practical equivalent circuit of zener diode

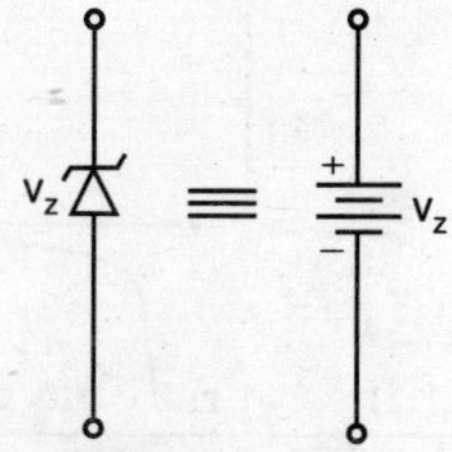

Fig. 1.69 (b). Ideal zener equivalent diode circuit

1.36. EFFECT OF TEMPERATURE ON ZENER DIODES

We have already studied in previous chapters how P-N junctions are temperature-sensitive in either the forward or reverse-biased mode. Now, because zener diodes operate only in the reverse mode, using the leakage current or minority carriers that are due to thermally generated electron hole pairs, we would expect temperature to have some profound effect on the zener voltage V_Z. This has been illustrated in figure 1.70. In this figure, the temperature coefficient α_Z in %/°C is shown as a function of zener voltage for three values of zener current.

This curve, typical for silicon voltage regulator diodes, may be interpretted as follows.

A 4 V Zener diode, for example, regardless of the level of current operation, will always be affected by temperature in a negative manner, decreasing zener voltage with an increase in temperature. On the other hand, zener diodes with V_Z in excess of 6 V will have an increase in breakdown voltage with an increase in temperature. If a diode is selected with a V_Z between 5 and 6 Volts and operated at the correct value of current, it will have very little change with temperature. The difference in sign of the temperature coefficient lies in the fact that reverse current is largely due to the true zener effect below 6 V, but is due to the avalanche effect above 6 V. Each of these mechanism is affected differently by temperature.

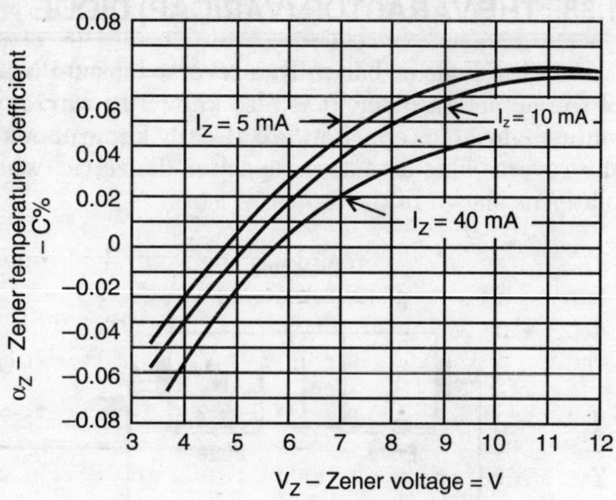

Fig. 1.70. Illustration of Graph of typical Zener temperature coefficients versus Zener voltage for silicon voltage regulator diodes

1.37. COMPARISON OF ZENER DIODE AND P-N JUNCTION DIODE

Zener diode	P-N-junction diode
1. Operates in reverse-bias condition	1. Operates in forward-bias condition
2. The characteristics lies in third quadrant 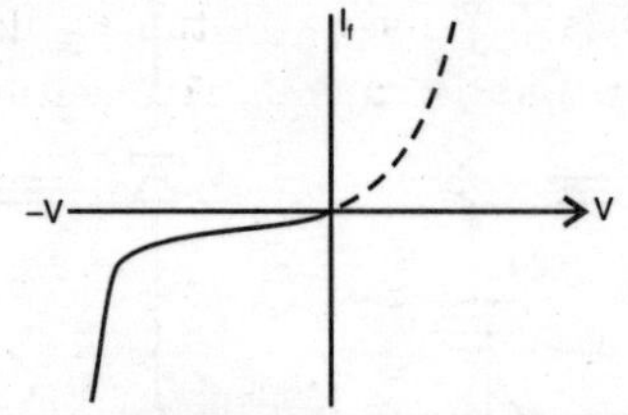	2. The characteristics lies in first quadrant 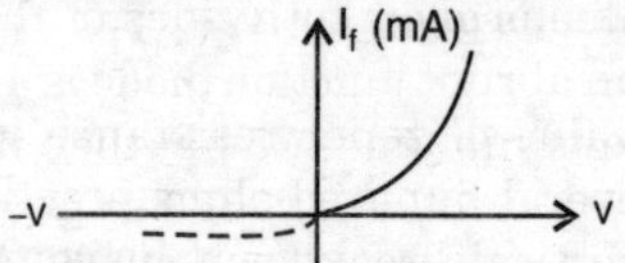
3. Dynamic Zener resistance is very small in reverse breakdown condition.	3. The diode resistance in reverse biased condition is very high
4. Zener diode symbol is 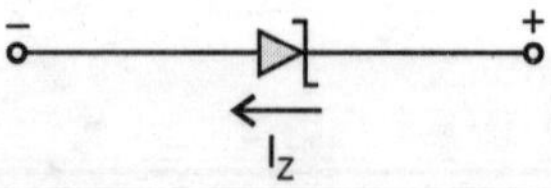	4. The P-N-junction diode symbol is 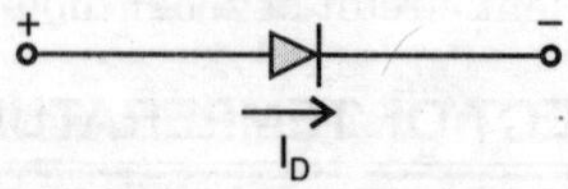
5. The conduction in zener is opposite to that of arrow in the symbol.	5. The conduction when forward biased is in same direction as that of arrow in the symbol.
6. Applications of Zener diode are voltage regula-tor, protection circuits, voltage limiters etc.	6. Applications of P-N-junction diode are rectifiers, voltage multipliers, clippers, clampers and many electronic devices.

1.38. THE VARACTOR (VARICAP) DIODE

A varacter diode is, basically, a reverse biased PN junction, which utilizes the inherent capacitance of the depletion layer. It is also known as **varicap, voltcap or tunning diode**. It is used as a voltage-variable capacitor. We already know about PN junction that the depletion layer created by the reverse bias acts as a capacitor dielectric, whereas the P-and N-regions act as the capacitor plates as shown in the figure 1.71 (a).

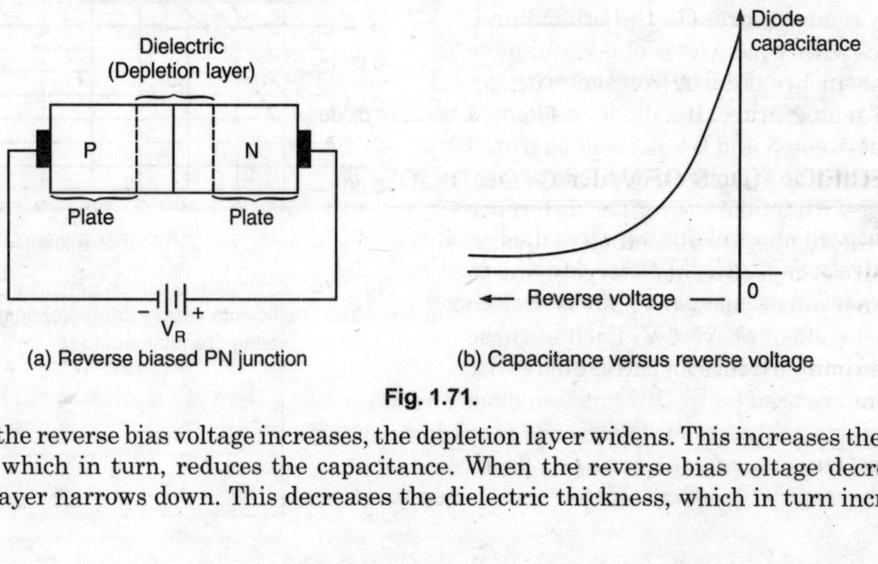

(a) Reverse biased PN junction (b) Capacitance versus reverse voltage

Fig. 1.71.

When the reverse bias voltage increases, the depletion layer widens. This increases the dielectric thickness, which in turn, reduces the capacitance. When the reverse bias voltage decreases, the depletion layer narrows down. This decreases the dielectric thickness, which in turn increases the capacitance. Figure 1.71 (b) shows the variation of capacitance with the reverse voltage. This indicates that the variation of capacitance is maximum when the reverse voltage is equal to zero. It reduces in a non-linear manner, as the value of reverse voltage is increased.

In a varactor diode, the capacitance parameters are controlled by the method of doping in the depletion layer, or the size and geometry of the diode construction. For example, figure 1.72 (a) shows the doping profile for an abrupt junction diode (i.e., the normal PN junction). In this type, the doping is uniform on both sides of the junction. The range of capacitance variation (called tuning range) of an abrupt junction diode is 4 : 1. It means that if its maximum transition capacitance is 100 pF and the minimum is 25 pF, then its tuning range is 4 : 1.

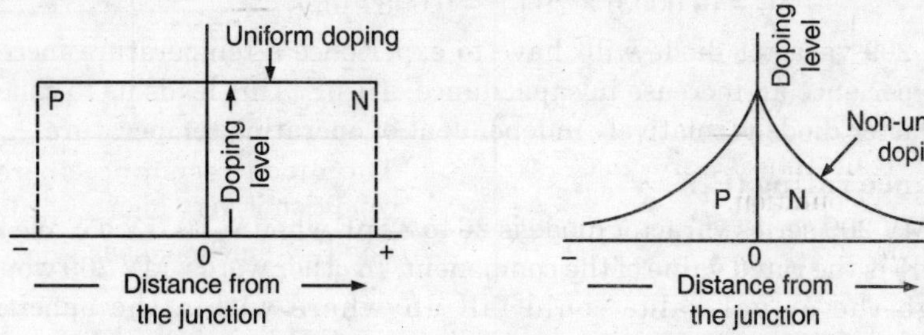

(a) Abrupt doping profile

(b) Hyper abrupt profile

Fig. 1.72.

It may be noted that the higher tuning range is obtained for varactors, which have hyper abrupt junction. The doping profile of such a junction will be as shown in figure 1.72(b). In this type, the doping level increases as we approach the junction. The heavy doping at the junction results in a narrower depletion layer and hence a larger capacitance. Moreover, any change in reverse voltage produces a larger variation in capacitance than that of abrupt junction.

A hyper abrupt varactor diode has a tuning range of 10 : 1. This range is quite enough to tune a broadcast receiver, over its frequency range of medium wave band (i.e., frequencies from 550 kHz to 1650 kHz).

Figure 1.73 (a) shows a sechematic symbol for a varactor diode and figure 1.73 (b) shows its equivalent circuit. Here, R_S is a reverse series resistance and C_T is a variable capacitance. The value of C_T ranges typically from 2 pF to 100 pF.

(a) Circuit symbol

(b) Equivalent circuit

Fig. 1.73. Varactor diode

1.39 SPECIFICATIONS OF VARACTOR DIODE

The specification sheet of any varactor diode will provide us with the information we need for any circuit analysis. Motorola MV 209 series varactor diodes are the most commonly used varactor diodes. Now, let us discuss birefly about various varactor diode parameter and characteristics.

> **DO YOU KNOW?**
>
> The tunning range of a varactor depends on the doping level. The tunning range of an abrupt-junction diode is between 3:1 and 4:1.

The maximum ratings for the varactor diode are the same as those that are used for the PN junction diode and the zener diode. The same holds true for the reverse breakdown voltage and reverse leakage current ratings which appear in the electrical characteristics section of the specification sheet. We already know about all these ratings for PN junction and zener diode. However, some of the additional ratings are given below:

(i) **The diode capacitance temperature coefficient (TC_C) rating**

This rating of a varactor diode provides the charge in the component's capacitance for each 1°C rise in temperature above 25°C. The TC_C rating for the MV 209 series varactor diodes is 300 ppm/°C. This means that varactor diode capacitance will increase by 300 parts per million for each 1°C rise in temperature above 25°C. For example, the normal capacitance rating of the MV 209 is typically 29 pF. Therefore, increase in capacitance for per 1°C rise in temperature will be,

$$\Delta C = (0.0003) \times 29 \text{ pF} = 0.0087 \text{ pF}$$

Hence, the MV 209 varactor diode will have to experience a temperature increase of nearly 115°C in order to experience an increase in capacitance of 1 pF. This leads us to conclude that the capacitance of a varacter diode is relatively independent of operating temperature.

(ii) **Diode capacitance rating (C_t)**

This rating for MV 209 series varactor diode is 26 to 32 pF when V_R is 3 V d.c. We know that the nominal value (29 pF) is the rated value of the component. In other words, MV 209 would be rated as 29 pF even though the actual value could fall anywhere within the specified range of V_R = 3 V dc.

(iii) **Capacitance ratio (C_R)**

This rating gives us the variation of the junction capacitance over the given range of voltages. For the MV 209 series, the C_R rating ranges from 5.0 to 6.5 for V_R = 3 to 25 V dc.

(iv) **Figure of merit (Q)**

It is the quality factor of a junction capacitance. We know that 'Q' of a capacitor is the ratio of energy stored in the capacitor to the energy lost through leakage current. In other words, it is the ratio of power returned to the circuit by the capacitor to the power dissipated by the capacitor. For the MV 209, Q rating is 200 (minimum).

DO YOU KNOW?

How is a varactor used? It is connected in parallel with an inductor to form a parallel resonant circuit. This circuit has only one frequency at which maximum impedance occurs. This frequency is called the resonant frequency. If the dc reverse voltage to the varactor is changed, the resonant frequency is also changed. This is the principle behind electronic tunning of a radio station, a TV channel and so on.

1.40. APPLICATIONS OF A VARACTOR DIODE

A major application of varactor diode is in tuning circuits. It is used in electronic tuners in radio, television and other commercial receivers. When a varactor is used in a tuning circuit, it allows a resonant frequency to be adjusted by a variable voltage level as shown in figure 1.74. In this circuit, two varactor diodes D_1 and D_2 provide the total variable capacitances in a parallel resonant circuit and V_C is a variable d.c. voltage, which controls the reverse bias and the capacitance of the diodes.

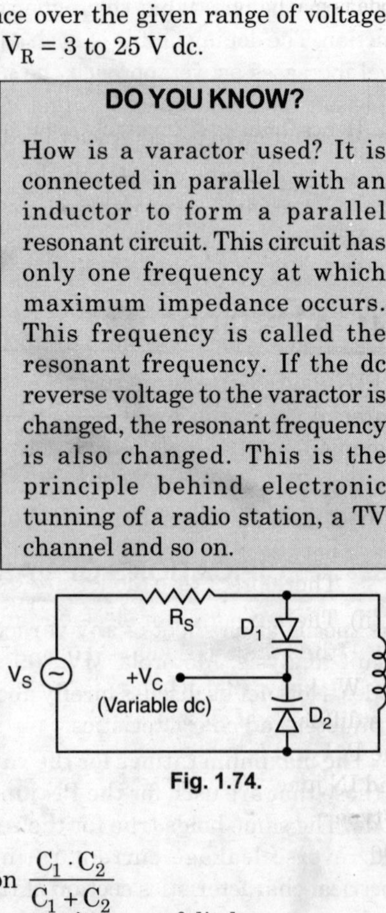

Fig. 1.74.

The resonant frequency of the tank circuit is given by the following relation,

$$f_0 = \frac{1}{2\pi\sqrt{LC}}$$

where L = value of the inductance,

C = value of total capacitance, which is given by the relation $\dfrac{C_1 \cdot C_2}{C_1 + C_2}$

where C_1 and C_2 are the maximum and minimum values of the capacitances of diode.

The varactor diode is also used in automatic frequency control device, adjustable band-pass filter and parametric amplifier.

EXAMPLE 1.24. The capacitance of certain varactor diode can be varied from 5 pF to 50 pF. The diode is used in a tuned circuit of a radio receiver as shown in figure 1.75.

Find the tuning range for circuit, if the inductance is 10 mH.

Solution: Given that C_1 (i.e., C_{min}) = 5 pF, C_2 (i.e., C_{max}) = 50 pF

and $L = 10\text{ mH} = 10 \times 10^{-3}\text{ H}.$

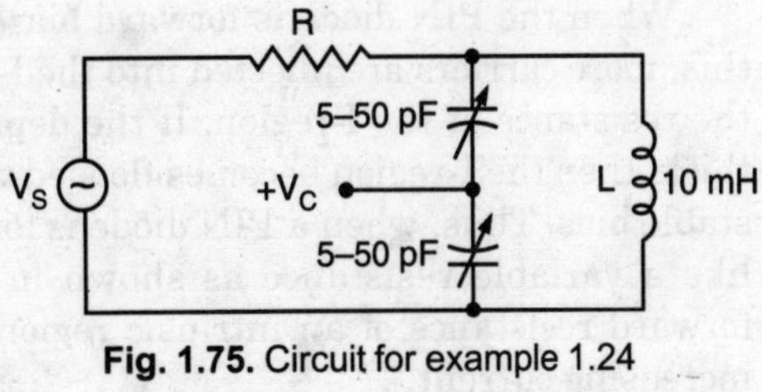

Fig. 1.75. Circuit for example 1.24

We know that the maximum resonant frequency is obtained, when the total capacitance of the varactor diode is minimum and its value is

$$C_{T\,(max)} = \frac{C_1 \cdot C_2}{C_1 + C_2} = \frac{5 \times 5}{5 + 5}$$

$$C_{T\,(max)} = 2.5 \times 10^{-12}\text{ F} = 2.5\text{ pF}$$

Hence, maximum resonant frequency will be

$$f_{o\,(max)} = \frac{1}{2\pi\sqrt{L \cdot C_{T(min)}}} = \frac{1}{2\pi\sqrt{(10 \times 10^{-3}) \times (2.5 \times 10^{-12})}}\text{ Hz.}$$

or $$f_{o(max)} = 1 \times 10^6\text{ Hz} = 1\text{ MHz} \quad \textbf{Ans.}$$

We also know that the minimum resonant frequency is obtained, when the capacitance of the varactor diode is maximum and its value is

$$C_{T\,(max)} = \frac{C_1 \cdot C_2}{C_1 + C_2} = \frac{50 \times 50}{50 + 50} = 25 \times 10^{-12}\text{ F} = 25\text{ pF}$$

Hence, minimum resonant frequency, will be

$$f_{o\,(min)} = \frac{1}{2\pi\sqrt{L \cdot C_{T(max)}}} = \frac{1}{2\pi\sqrt{(10 \times 10^{-3}) \times (25 \times 10^{-12})}}\text{ Hz}$$

or $$f_{o(min)} = 318 \times 10^3\text{ Hz} = 318\text{ kHz} \quad \textbf{Ans.}$$

1.41. A PIN DIODE **(RGTU, Bhopal, Sem. Exam., 2006-07)**

A PIN diode is made up of three semiconductor materials : two heavily doped P-and one N-type material separated by an intrinsic* semiconductor (I) as shown in figure 1.76 (a). The intrinsic region offers high resistance to the current flowing through it.

The PIN diode has the following two advantages over the normal PN junction diode:

(i) The capacitance between the P and N region decreases because of the increased separation between the P and N regions. This advantage allows the PIN diode to have fast response time. Therefore, these diodes are useful at very-high-frequencies (i.e., above 300 MHz).

(ii) There is a greater electron-hole pair generation because of the increased electric field between the P and N regions. This advantage allows the PIN diode to process even weak signals.

We know that when a PIN diode is unbiased (i.e., no voltage is applied across the diode), there is a diffusion of electron and holes across the junction due to the different concentration of atoms in the P, I and N regions. The diffusion of electrons and holes produce a depletion layer across the PI and IN junctions as shown in figure 1.76(b). The depletion layers penetrate to a little distance in the P-type and N-type semiconductor regions but to a larger distance in the I-region. Under such condition, the device has a high value of resistance.

* Undoped

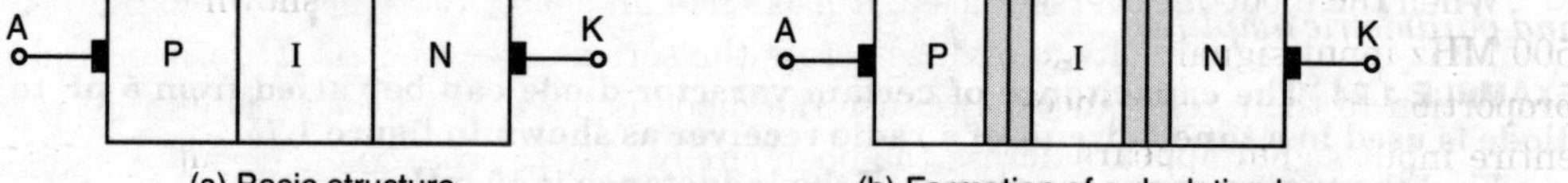

(a) Basic structure (b) Formation of a depletion layer

Fig. 1.76.

When the PIN diode is forward biased, the width of depletion layers decreases. As a result of this, more carriers are injected into the I-region. This reduces the resistance of the I-region. If the depletion layers are not thick, then the I-region becomes flooded with the carriers at a stable bias. Thus, when a PIN diode is forward biased, it acts like a variable resistance as shown in figure 1.77(a). The forward resistance of an intrinsic region decreases with the increasing current.

> **DO YOU KNOW?**
>
> When a PIN diode is forward-biased, it acts like a current-controlled resistance. When reverse-biased, the PIN diode acts like a fixed capacitor.

On the other hand when the PIN diode is reverse biased, the depletion layers become thicker. As the reverse bias is increased, the thickness of the depletion layer increases till the I-region becomes free of mobile carriers. The reverse bias, at which this happens, is called swept out voltage. At this stage, the PIN diode acts like an almost constant capacitance as shown in figure 1.77 (b).

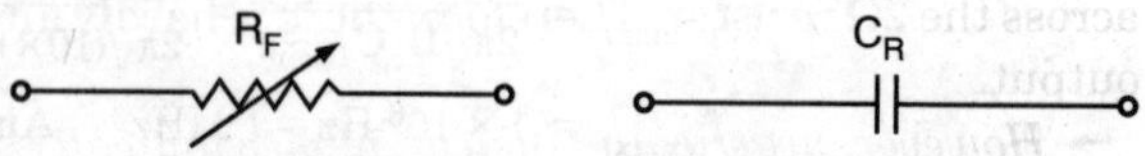

(a) Forward biased (b) Reverse bias

Fig. 1.77. Equivalent circuit of a PIN diode.

1.42. APPLICATIONS OF PIN DIODE

We know that when a PIN diode is forward biased, it acts like a variable resistance which decreases with the increasing current. Also when it is reverse biased, the PIN diode acts like an almost constant capacitance. These features make the PIN diode useful for a variety of applications as under:

(i) As a d.c. controlled microwave switch operated by rapid changes in bias or as a modulating device. This application takes advantage of the variable forward-resistance characteristic. Modulators are used to combine two signals of different frequencies into a single signal.

(ii) It is used in attenuator applications because its resistance can be controlled by the current.

(iii) As an AM modulator of very high frequency signals.

1.42.1. High Frequency Switching Operation

The use of PIN diode in electronic high frequency switching has been illustrated in figure 1.78.

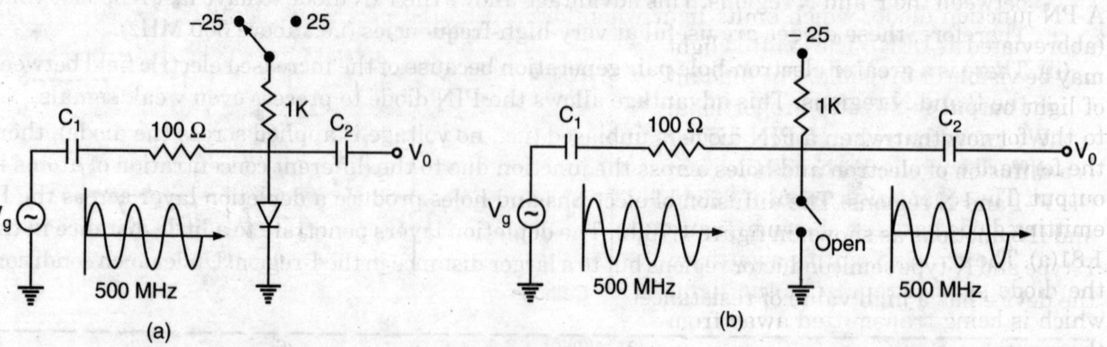

Fig. 1.78.

When the diode is reverse-biased, it looks like an open circuit as shown in figure 1.78 (b). The 500 MHz input signal voltage divides across the series-connected 100 Ω resistor and the diode in proportion to their resistances. Because the diode is open and hence has infinite resistance, the entire input signal appears across the open circuit. Therefore, the entire input signal passes out through coupling capacitance C_2 without any kind of loss or attenuation. Also, when the diode is forward-biased by a voltage source of +25 V dc, current I is given by

$$I = \frac{25}{1K} = 25 \text{ mA}$$

Because of this, the diode resistance becomes

$r_{ac} = \frac{50}{25} = 2\,\Omega$ as shown by its equivalent circuit shown in figure 1.79.

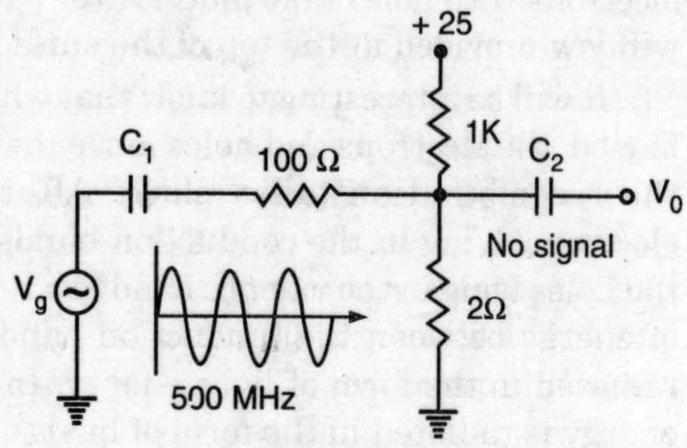

Fig. 1.79.

Now, in this case, almost all the input signal voltage drops across 100 Ω resistor and practically no voltage drops across the 2Ω resistor. Therefore, there is hardly any signal output.

However, practically, instead of mechanically switching the diode-biasing supply voltage from – 25V to + 25V, a transistor is used to do this switching operation. Hence, in this manner, we can turn a very high frequency signal (MHz range) OFF and ON with the speed of a transistor switching circuit.

1.42.2. PIN Diode Operation as an AM Modulator

Figure 1.80 shows the method in which a high frequency 500 MHz signal is modulated at 1 KHz rate. Here, a 1 KHz signal is fed into a PNP transistor where it varies its dc output current at the same rate. This varying dc current is applied as biasing current to the PIN diode as depicted in figure 1.80. It varies the diode ac resistance as seen by the 500 MHz signal. Therefore, the signal is modulated at 1 KHz rate as shown.

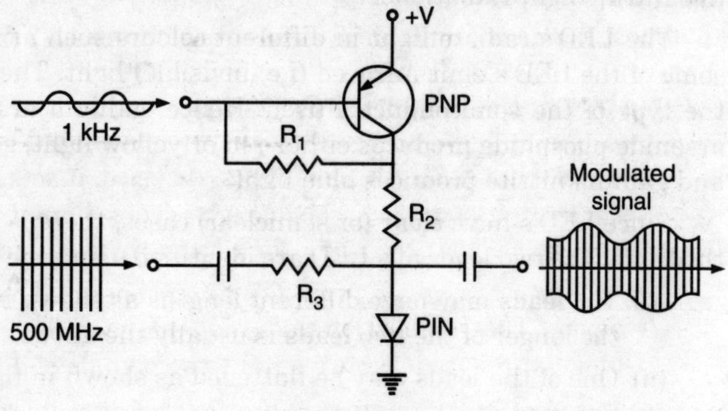

Fig. 1.80.

1.43. LIGHT EMITTING DIODE (LED) (Very Important)

A PN junction diode, which emits light when forward biased, is known as a light emitting diode (abbreviated as LED). The emitted light may be visible or invisible. The amount of light output is directly proportional to the forward current. Hence, higher the forward current, higher is the light output. The schematic symbol of a light emitting diode has been shown in figure 1.81(a). The arrows, pointing away from the diode symbol represent the light, which is being transmitted away from the junction.

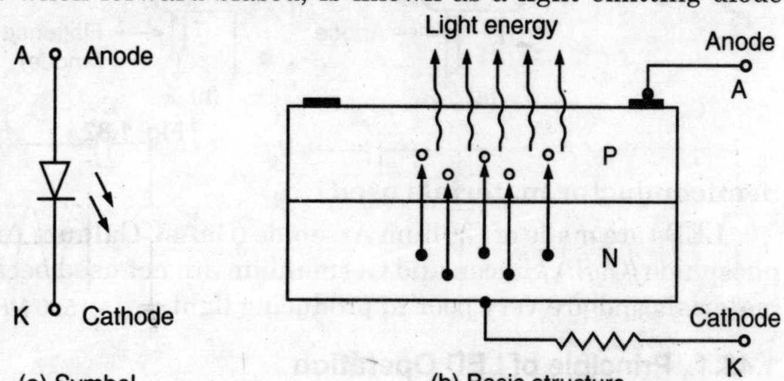

Fig. 1.81. Light emitting diode (LED)s

Basic Structure

Figure 1.81(b) shows the basic structure of a light emitting diode. Here, an N-type layer is grown on a P-type substrate (not indicated in figure) by a diffusion process. Then, a thin P-type layer is grown on the N-type layer. The metal connections to both the layers make anode and cathode terminals as indicated. The light energy is released at the junction, when the recombination of electrons with holes take place. After passing through the P-region, the light is emitted through the window provided at the top of the surface.

It will be interesting to know that when the LED is forward biased, the electrons and holes move towards the junction and the recombination takes place. After recombination, the electrons, lying in the conduction bands of N-region, fall into the holes lying in the valence band of a P-region. The difference of energy between the conduction band and valence band is radiated in the form of light energy. In ordinary diodes, this energy is radiated in the form of heat.

The semiconducting materials used for manufacturing light emitting diodes are gallium arsenide, gallium arsenide phosphide. The silicon and germanium are not used to manufacture light emitting diodes because these are heat producing materials. Moreover, these materials are very poor in emitting light radiations.

DO YOU KNOW?

The principal disadvantage of LEDs is that they draw considerable current in comparison to other types of visual displays. In many cases, LEDs are pulsed ON and OFF at a rapid rate, rather than being supplied with a steady drive current. LEDs appear to the eye to be on continuously, but they consume less power than if they were ON continuously.

The LED's radiate light in different colours such as red, green, yellow, blue, orange etc. Also, some of the LED's emit infrared (i.e. invisible) light. The colour of the emitted light depends upon the type of the semiconductor used. Hence, gallium arsenide emits infrared radiations, gallium arsenide phosphide produces either red or yellow light, gallium phosphide emits red or green light and gallium nitrite produces blue light.

Since LED's have clear (or semiclear) cases, there is normally no label on the cased to identify the leads. The two leads of a LED are identified using one of the several schemes as discussed below:

(i) The leads may have different lengths as shown in figure 1.82(a), when the scheme is used, the longer of the two leads is usually the anode.

(ii) One of the leads may be flattened as shown in figure 1.82(b). The flattened lead is usually the anode

(iii) One side of the case is flattened as shown in figure 1.82(c). The lead closest to the flattened side is usually the anode.

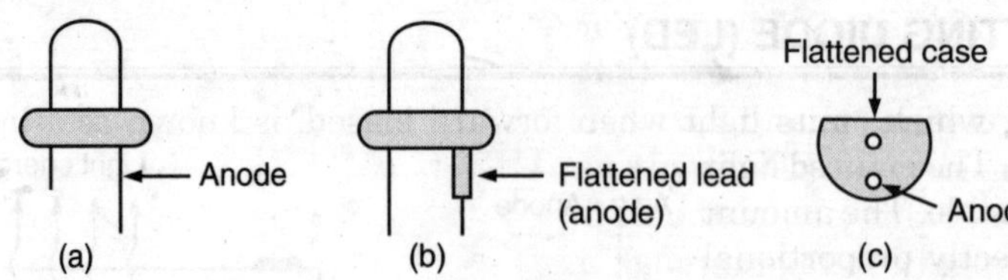

Fig. 1.82.

Semiconductor materials used

LEDs are made of Gallium Arsenide (GaAs), Gallium Arsenide phosphide (GaAsP) and Gallium phosphide (Ga P). Silicon and Germanium are not used because they are essentially heat producing materials and are very poor in producing light.

1.43.1. Principle of LED Operation

When the LED is forward biased, the electrons in the N-region will cross the junction and recombine with the holes in the P-type material. These free electrons reside in the conduction band and hence at a higher energy level than the holes in the valence band. When the recombination takes place, these electrons return back to the valence band which is at a lower energy level than the conduction band. While returning back, the recombining electrons give away the excess energy in the form of light. This is shown in figure 1.83. This process is called as **electro-luminescence**. In this way, an LED emits light. This is the principle of operation of LED.

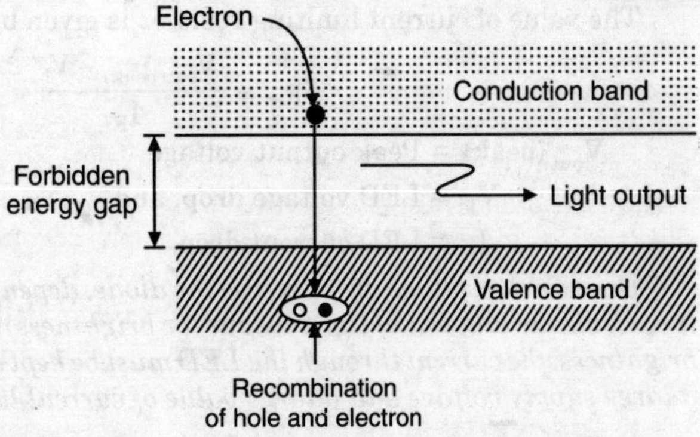

Fig. 1.83. Principle of operation of LED

Colour of the emitted light

The colour of the emitted light is decided by its wavelength. The wavelength of emitted light (and hence its colour) depends on the value of forbidden gap (E_G). Forbidden gap value depends on the material used. Hence the colour of emitted light is dependent on the material used. The colours associated with different materials are as under.

Material used	Colour of the emitted light
1. Gallium Arsenide (Ga As)	Infrared (IR)
2. Ga As P	Red or Yellow
3. Gallium phosphide (Ga P)	Red or Green

Silicon and Germanium are not used for LEDs because their forbidden gaps do not allow the light emission in visible spectrum. The commonly available LEDs emit green, red, yellow, orange and blue colours. Some LEDs are capable of emiting infrared light which is not visible to our eyes. Some LEDs are capable of emitting infrared light which is not visible to our eyes. Various impurities are added during the doping process to get precisely the required colour of the emitted light.

1.44. LED VOLTAGE DROP AND CURRENT

The LED's have also a certain forward voltage drop, when they are used in the circuit like an ordinarily diode (which has a voltage drop of 0.7 V). This voltage drop depends upon the LED current, colour of the emitted light etc. Usually, the voltage drop has a typical value that varies from 7.5 to 2.5 V for currents between 10 and 50 mA. It will be interesting to know that LED's are not able to withstand reverse bias of even every small voltages. For this reason, it is necessary to assure that reverse bias is never applied to an LED.

Now, let us consider a LED circuit as shown in figure 1.84. In this circuit, the LED is connected to the supply voltage through a current limiting resistor (R). The current through the LED is given by the following relation:

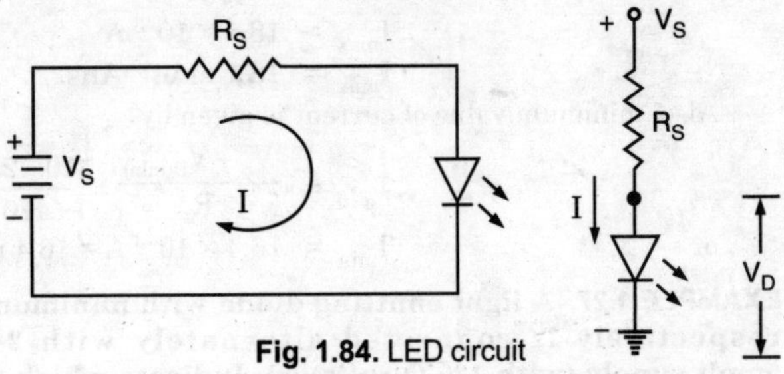

Fig. 1.84. LED circuit

$$I = \frac{V_S - V_D}{R_S}$$

where V_S = Supply voltage,
V_D = Voltage drop across LED, and
R_S = Current limiting resistor.

The value of current limiting resistor is given by the following expression:

$$R_S = \frac{V_{out(peak)} - V_F}{I_F}$$

V_{out} (peak) = Peak output voltage

V_F = LED voltage drop, and

I_F = LED current drop.

Because the voltage drop, across the diode, depends upon the value of current through it, therefore it controls the amount of light output (or brightness) of the LED. In order to avoid fluctuations in the brightness, the current through the LED must be kept constant. In actual practice, it is done by selecting a large supply voltage and a large value of current limiting resistor.

EXAMPLE 1.25. The driving circuit shown in figure 1.85(a) has a peak output voltage of 8 V. The LED has a rating of V_F = 1.8 to 2.0 V and $I_{F(max)}$ = 16 mA.

What will be the smallest standard resistor value that can be used as R_S.

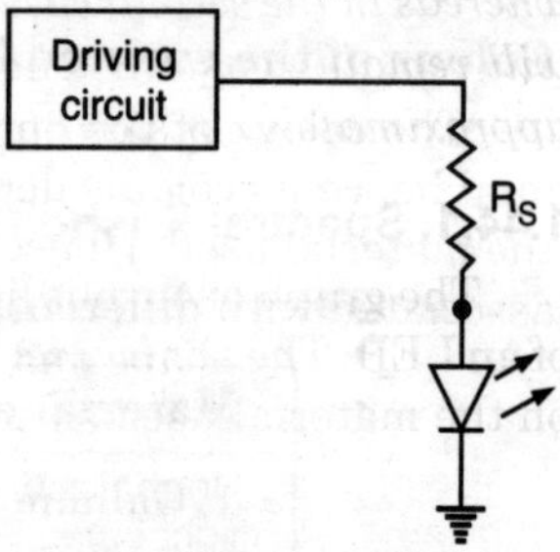

Fig. 1.85 (a). Circuit for example 1.25

Solution: Given that $V_{out(peak)}$ = 8 V; V_F = 1.8 to 2.0 V

and $I_{F(max)}$ = 16 mA = 16×10^{-3} A.

We know that current limiting resistor is given by

$$R_S = \frac{V_{out(pk)} - V_F}{I_F} = \frac{8 - 1.8}{16 \times 10^{-3}} = 387.5\ \Omega \quad \textbf{Ans.}$$

and if V_F = 2.0 V then R_S = 375 Ω. In either case, the smallest standard-value resistor that has a value greater than 387.5 Ω or 375 Ω is the 390 Ω resistor. This is the component which we have to use in the circuit of figure 1.85(a).

EXAMPLE 1.26. Figure 1.85(a) shows an LED circuit. The LED has a minimum drop of 1.5 V and a maximum drop of 2.3 V. If the supply voltage is 10 V and R_1 = 470 Ω. What are the minimum and maximum values of LED current?

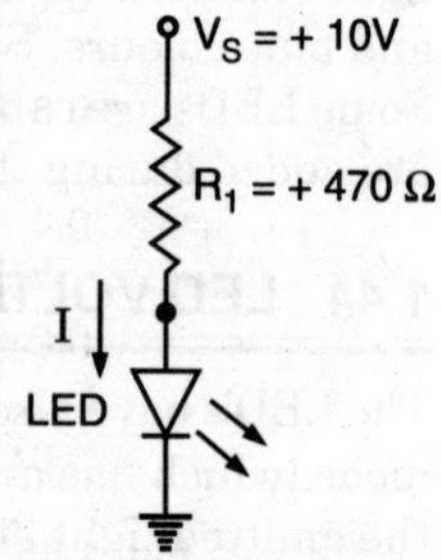

Fig. 1.85 (b).

Solution: Given that $V_{D(min)}$ = 1.5 volts;

$V_{D(max)}$ = 2.3 volts;

V_S = 10 volts

and R_1 = 470 Ω.

We know that the maximum value of current is given by

$$I_{max} = \frac{V_S - V_{D(min)}}{R_1}$$

$$I_{max} = \frac{10 - 1.5}{470}$$

$$I_{max} = 18.1 \times 10^{-3}\ \text{A}$$

$$I_{max} = 18.1\ \text{mA} \quad \textbf{Ans.}$$

Also, minimum value of current is given by

$$I_{min} = \frac{V_S - V_{D(min)}}{R_1} = \frac{10 - 2.3}{470}\ \text{A}$$

or $I_{min} = 16.4 \times 10^{-3}$ A = 16.4 mA **Ans.**

EXAMPLE 1.27. A light emitting diode with minimum and maximum voltage drops of 1.8V and 3 V respectively is connected alternately with 24 volt supply (with 820 Ω resistor), and 5 volt supply (with 120 Ω resistor). Indicate which supply voltage will keep the brightness of the diode as constant.

Solution: Given that $V_{D(min)}$ = 1.8 volts

and $V_{D(max)}$ = 3 volts

We know that minimum and maximum values of the current through the LED with 24 volt supply (V_S) and 820 Ω resistor (R_S) will be

$$I_{min} = \frac{V_S - V_{D(max)}}{R_S} = \frac{24-3}{820} = 25.6 \times 10^{-3}\ A = 25.6\ mA \qquad ...(i)$$

and

$$I_{max} = V_S - \frac{V_{D(min)}}{V_S} = \frac{24-1.8}{820} = 27.1 \times 10^{-3}\ A = 27.1\ mA \qquad ...(ii)$$

Also, minimum and maximum values of the current through the LED with 5 volt supply (V_S) and 120 Ω resistor (R_S) will be

$$I_{min} = \frac{V_S - V_{D(max)}}{R_S} = \frac{5-3}{120} = 16.7 \times 10^{-3}\ A = 16.7\ mA \qquad ...(iii)$$

and

$$I_{max} = \frac{V_S - V_{D(min)}}{R_S} = \frac{5-1.8}{120} = 26.7 \times 10^{-3}\ A = 26.7\ mA \qquad ...(iv)$$

It may be noted that the variation in current for the first case (27.1 – 25.6 = 1.5 mA) is small whereas in the second case it is (26.7 – 16.7 = 10.0 mA) large. Therefore, the brightness in the first case will remain constant whereas in the second case it will be changing. Therefore, in order to get an approximately constant brightness, we use as large a supply voltage as possible. **Ans.**

1.44.1. Spectral Response Curves of LED

The graph of output light (Y-axis) versus the wavelength is called as the spectral response curve of an LED. The shape and the wavelength at which maximum response is obtained varies depending on the material used as shown in figure 1.86.

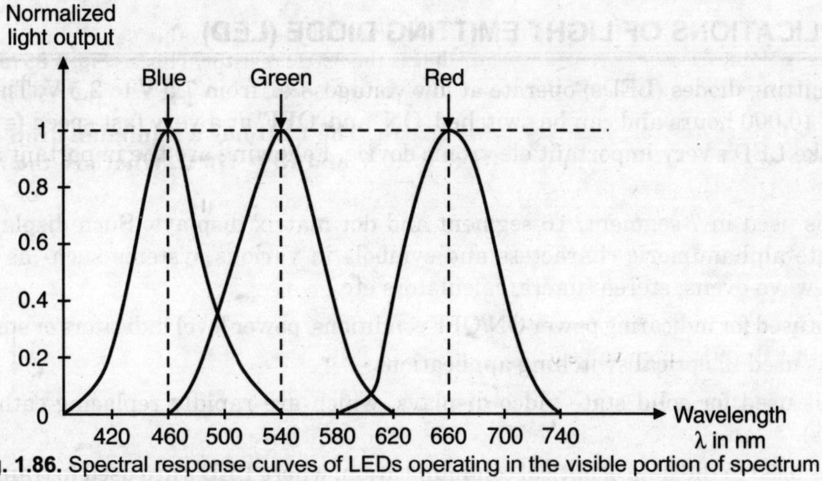

Fig. 1.86. Spectral response curves of LEDs operating in the visible portion of spectrum

1.44.2. Radiation Pattern of LED

As a matter fact, radiation pattern is a graph of light output plotted against angle as shown in figure 1.87. It gives us an idea about how directional the given light source is.

The LED is a highly directional source because it radiates only in a particular direction. Most of the radiated energy is confined within 20° of the direction of maximum light.

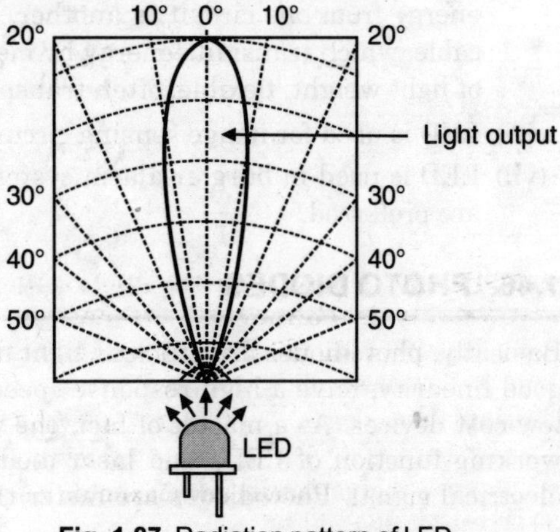

Fig. 1.87. Radiation pattern of LED.

1.44.3. Electrical and Optical Parameters of LED

Different electrical and optical parameters of LED are as follows:

- Reverse leakage current (I_R) = 50 nA.

- Reverse voltage (V_R) = 30 V
- Forward current (I_F) = 10 to 80 mA
- Forward voltage (V_F) = 1.2 V to 1.8 V
- Output power (P_0) = 200 to 500 μW.

1.44.4. Advantages of LED

(i) Small size and light weight.
(ii) Available in different spectral colours.
(iii) Longer life as compared to lamps.
(iv) Better linearity.
(v) Suitable for higher operating speeds.
(vi) Low cost.
(vii) Compatible with integrated circuits.

1.44.5. Disadvantages of LED

(i) Output power is affected by changes in temperature.
(ii) Overcurrent can damage it easily.
(iii) They need larger power for their operation.

1.45. APPLICATIONS OF LIGHT EMITTING DIODE (LED)

The light emitting diodes (LEDs) operate at low voltages i.e., from 7.5 V to 2.5 V. They have a long life of about 10,000 hours and can be switched 'ON' and 'OFF' at a very fast speed (≅1 n sec). These features make LED's very important electronic device. Following are the important applications of the LED's :

(i) LED is used in 7-segment, 16-segment and dot matrix displays. Such displays are used to indicate alphanumeric characters and symbols in various systems such as digital clocks, micro-wave-ovens, stereo tuners, calculators etc.

(ii) LED is used for indicating power ON/OFF conditions, power level indicators or stereo amplifiers.

(iii) LED is used in optical switching applications.

(iv) LED is used for solid state video displays, which are rapidly replacing cathode ray tubes (CRT's).

(v) LED is used in the field of optical communication, where LED's are used to transfer (or couple) energy from one circuit to another. They are also used to send light energy to fibre optical cable, which transmits energy by means of total internal reflection. The fiber optical cable is of light weight, flexible, often transparent and as small a 0.043 mm in diameter.

(vi) LED is used for image sensing circuits in picture phone.

(vii) LED is used in burglar alarm systems. In such applications, LED's radiating infrared light are preferred.

1.46. PHOTO DIODES (RGTU, Bhopal, Sem. Exam., 2002-03)

Basically, photodiodes are efficient light detectors. Photodiodes can be made in small sizes, have good linearity, have a high response speed, have simple biasing requirements and are relatively low-cost devices. As a matter of fact, the working function of a photodiode is just opposite to the working function of LEDs and laser diodes in the sense that photodiode converts light into an electrical signal. Photodiodes are faster than photoconductors since they have a P-N junction to collect the carriers, but they do not have the internal gain which photo-conductors have since the charge carriers are counted only once.

Further, the photodetection mechanism in photodiodes is similar to that in photoconductors in the sense that an electron-hole pair is created by a photon for which $h\nu \geq E_g$. However, it differs in the sense that only the photocarriers which diffuse to the depletion layer are counted and only the minority carriers are collected.

1.46.1. Types of Photodiodes

Usually, there are various types of photodiodes as under:

(i) P-N photodiodes

(ii) PIN photodiodes, and,

(iii) Avalanche photodiodes (APD)

1.47. P-N PHOTODIODE

As a matter of fact, photodiodes are made of semiconductor P-N junctions under reverse-bias condition. Whenever, a photon having energy $E_p = h\nu = \frac{hc}{\lambda} \geq E_g$ strikes the material, the photon is absorbed and its energy is acquired by an electron. Therefore, the electron is excited at the conduction band and is now able to move. Hence, electron-hole pairs are generated within the depletion region. When a photon strikes a depletion region, its energy separates an electron from its hole as depicted in figure 1.88(a). Now, the separated electron and hole are attracted by the positive and negative potentials of the depletion voltage respectively. Hence, a flow of charge carriers generates the current. Further, under the influence of reverse bias, the flow of electrons and holes is enhanced. Figure 1.88 (b) shows the electrical circuit of a photodiode.

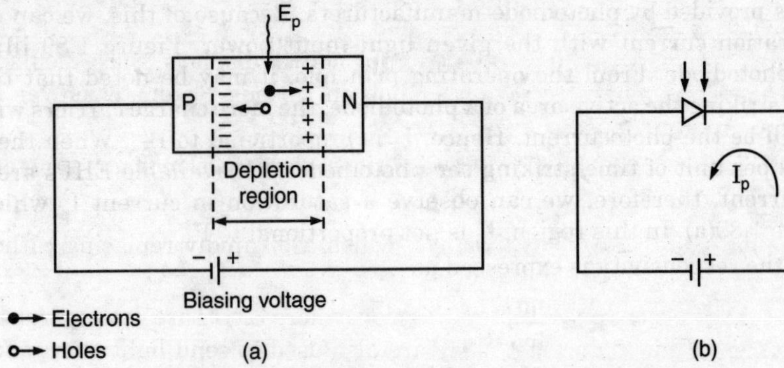

Fig. 1.88. Operating principle of a P-N photo diode
(a) P-N junction under reverse biasing , (b) Electrical circuit

1.47.1. Characteristics of a Photodetector

As a matter of fact, there are two important characteristics of a photodetector namely quantum efficiency and responsivity. In fact, these two characteristics or parameters depend upon the material band gap, the operating wavelength, the doping and the thickness of the P-N regions of the device.

(i) Quantum Efficiency

Quantum efficiency may be defined as the ratio of the number of electron-hole pairs (EHP) generated to the total number of incident photons and is expressed as

$$\eta = \frac{\text{Number of EHPs generated}}{\text{Total number of incident photons}} = \frac{I_p / q}{P_{op} / h\nu} = \frac{I_p h\nu}{qP_{op}} \quad ...(1.28)$$

where

I_p = the photogenerated current
P_{op} = the incident optical power
$h\nu$ = the photon energy
q = free charge carrier

Practically, 100 photons create 30 to 95 EHP, and hence giving η ranging from 30 to 95 percent. Also, to achieve a high quantum efficiency, the depletion layer must be thick enough to permit a large fraction of the incident light to be absorbed. Because of this, the photogenerated carriers have to drift a longer distance across the reverse-biased junction. Since, the carrier drift time determines the response speed of the photodiode, therefore, a compromise has to be made between quantum efficiency and response speed.

(ii) Responsivity

The responsivity may be defined as the ratio of photogenerated current to the incident optical power. It is expressed by the following expression:

$$R = \frac{I_p}{P_{op}} \quad ...(1.29)$$

where I_p = photogenerated current
P_{op} = the incident optical power

Further, the responsivity may be related to the quantum efficiency as under:

$$R = \frac{\eta q}{h\nu} \quad ...(1.30)$$

This parameter is very useful in the sense that it specifies the photocurrent generated per unit optical power.

Also, the typical values for responsivity R fall in the range from 0.5 A/W to 1.0 A/W. The value of responsivity R is provided by photodiode manufacturers. Because of this, we can evaluate the output photogeneration current with the given light-input power. Figure 1.89 illustrates the responsivity of a photodiode. From the operating principle, it may be noted that the more the number of photons striking the active area of a photodiode, the more charge carriers will be created i.e., the greater will be the photocurrent. Hence, I_p is proportional to P_{op}. When there are large number of photons per unit of time striking the photodiodes, all available EHPs are involved in producing photocurrent, therefore, we can observe a saturation in current I_p which has been illustrated in figure 1.89(a). In this region, I_p is not proportional to P_{op}.

We know that the responsivity is expressed as

$$R = \frac{\eta q}{h\nu}$$

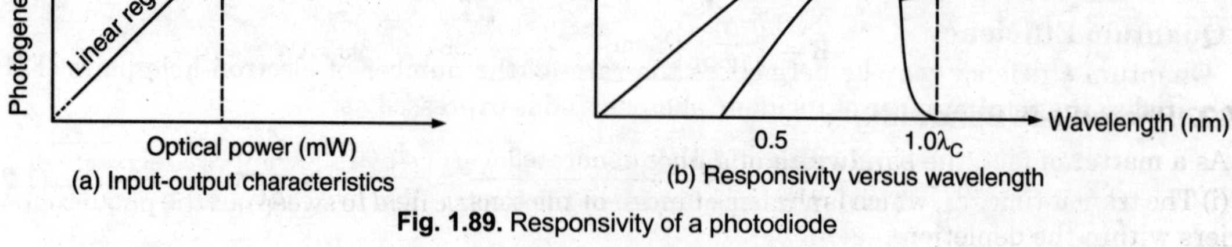

(a) Input-output characteristics

(b) Responsivity versus wavelength

Fig. 1.89. Responsivity of a photodiode

Above expression may be written as under:

$$R = \frac{\eta}{h\nu / q} \quad ...(1.31)$$

or

$$R(A/W) = \left(\frac{\eta}{1248}\right)\lambda\,(nm) \quad ...(1.32)$$

Hence, according to above expression, the graph between R and λ must be a straight line as depicted in figure (1.89 (b)) by the dotted straight line. However, practically, the curve for responsivity R as a function of wavelength has been shown in figure (1.89 (b)) as a solid line. In fact, this curve is very different from the expected straight line and also there exists a short and long cut-off waelengths.

DO YOU KNOW?

A photo diode has been optimized for its sensitivity to light. In this diode, a window lets light pass through the package to the junction. The incoming light produces free electrons and holes. The stronger the light, the greater the number of minority carriers and the larger the reverse current.

EXAMPLE 1.28. The responsivity of a photodiode is 0.85 A/W and the input power saturation is 1.5 mW. Calculate the photocurrent if the incident light power is

(i) 1 mW, and

(ii) 2 mW

Solution: We know that the Responsivity of a photodiode is given by

$$R = \frac{I_p}{P_{op}}$$

(i) Given that $R = 0.85$ A/W, $P_{op} = 1$ mW

$$I_p = RP_{op} = 0.85\left(\frac{A}{W}\right)\times 1\text{ mW} = 0.85\text{ mA} \quad \textbf{Ans.}$$

(ii) $P_{op} = 2$ mW but given that input-power saturation is 1.5 mW so I_p is not proportional to P_{op}, hence we cannot find the value of photocurrent. **Ans.**

EXAMPLE 1.29. In a 100 ns pulse, 6 × 10^6 photons at a wavelength of 1300 nm fall on an InGaAs phtodetector. On the average, 5.4 × 106 EHPs are generated. Calculate the quantum efficiency of photodetector.

Solution: We know that the quantum efficiency is expressed as

$$\eta = \frac{\text{Number of EHPs generated}}{\text{Number of incident photons}}$$

Substituting the given values, we get

$$\eta = \frac{5.4\times 10^6}{6\times 10^6} = 0.90$$

Hence, the quantum efficiency at 1300 nm is found to be 90%. **Ans.**

EXAMPLE 1.30. Determine the responsivity of an InGaAs photodiode if its quantum efficiency is equal to 70%. The energy gap of InGaAs is 0.75 eV.

Solution: We know that $E_g = \frac{hc}{\lambda}$

Substituting values, we get

$$\lambda = \frac{6.62\times 10^{-34}\times 3\times 10^8}{0.75\times 1.6\times 10^{-19}} = 1664\text{ nm}$$

Also, responsivity is given by

$$R = \left(\frac{\eta}{1248}\right)\lambda(nm) = \frac{0.70}{1248}\times 1664 = 0.933\text{ AW}^{-1}. \quad \textbf{Ans.}$$

1.47.2. Photodiode Bandwidth

As a matter of fact, the bandwidth of a photodiode is governed by following two factors:

(i) The transit time, τ_{tr}, which is the time needed or the electric field to sweep out the photoexcited carriers within the depletion region.

(ii) The RC time τ_{RC}, which is the time needed to discharge the junction capacitance (C_j) through a combination of internal and external resistances.

The transit time is expressed as

$$\tau_{tr} = \frac{W}{v_{sat}} \quad ...(1.33)$$

where v_{sat} = maximum drift velocity of charge carriers,

W = thickness of depletion region.

For silicon photodiodes, τ_{tr} may be estimated to a good approximation by using the expression

$$\tau_{tr} = \frac{\rho_n}{400} \text{ or } \frac{\rho_p}{1000} \text{ ns}$$

where ρ_n and ρ_p are the resistivities in ohm-cm of N-type or P-type materials.

The circuit defining the RC time constant of a silicon photodiode is shown in figure 1.90 in which R_S is the internal resistance, R_j is the junction resistance, R_L is the load resistance, C_j is the junction capacitance and I the photocurrent generated by the incident light.

R_S is comprised of three parts:

(i) The resistance of the undepleted region of the silicon chip,

(ii) The contact resistance,

(iii) Collection resistance associated with the resistivity of the front surface generated by diffusion.

Taking these factors into account, we get

$$\tau_{RC} = (R_S + R_L) C_j \cong R_L C_j$$

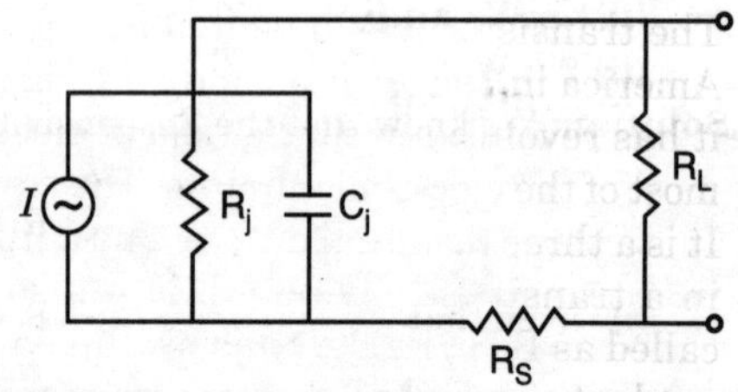

Fig. 1.90. Equivalent circuit of a photodiode.

C_j can be estimated as $C_j = \frac{\varepsilon A}{W}$, where ε is the permittivity of a semiconductor and A is the active area (the photosensitive area) of the photodiode.

Typical values of C_j are from 1 to 2 pF.

Hence, for $R_L = 50\ \Omega$, τ_{RC} is in the range of 50 to 100 ps.

Thus, the bandwidth of a photodiode is given by the following formula:

$$BW = \frac{1}{2\pi(\tau_{tr} + \tau_{RC})} = \frac{1}{2\pi\left[\left(\frac{W}{v_{sat}}\right) + R_L\left(\frac{\varepsilon A}{W}\right)\right]} \quad ...(1.34)$$

Hence, bandwidth of a photodiode is very much dependent upon the depletion region width (W). Becaue of this, P-N diode is an inefficient device for optical communication applications.

There is a third mechanism restricting the bandwidth of a P-N photodiode. As BW depends upon the depletion region width and since P and N regions are wider than the depletion region, many photons strike those regions, creating a diffusion photocurrent. But, diffusion flow is very slow compared with the drift flow of charge carriers. Therefore, the electric-output pulse of a P-N photodiode will be much wider than the optical input pulse and its tail will be determined by the diffusion photocurrent. This has been illustrated in figure 1.91.

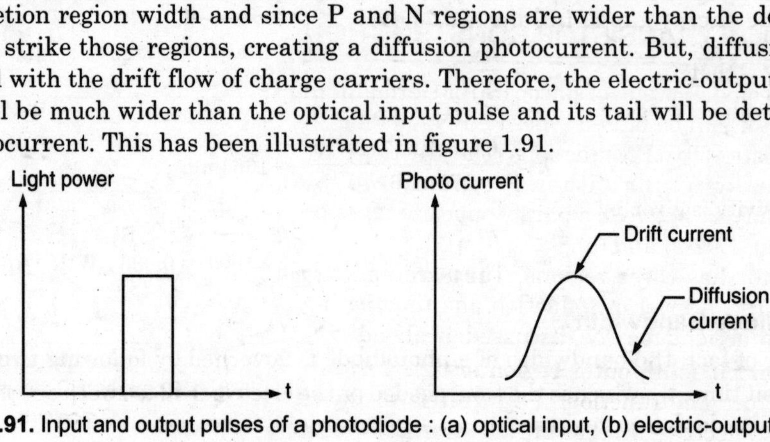

Fig. 1.91. Input and output pulses of a photodiode : (a) optical input, (b) electric-output pulse.

1.47.3. Comparison of LED and Photodiode

S.No.	LED	Photodiode
1.	It is a light emitting device.	It is a light detecting device.
2.	Electrical energy is converted into light.	Electric current proportional to light intensity is produced.
3.	LED is always forward biased.	It is always reverse biased.
4.	GaAs or GaP or GaAsP are the materials used.	Silicon is used.
5.	Due to recombination of electrons and holes, light is emitted.	Due to generation of electron hole pairs, the photocurrent will flow.

1.48. INTRODUCTION TO BJT

The transistor was invented by Dr. William Shockley and Dr. John Bardeen at Bell Laboratory in America in 1951. First time, in 1952 transistor was used in telephone switching circuits. Since then, it has revolutionized the field of electronics. The transistor has replaced the bulky vacuum tubes in most of the electronic circuits. The transistor is a basic building block of all modern electronic systems. It is a three terminal device. The output voltage, current or power are controlled by the input current in a transistor. Therefore, it is also called a current-controlled device. In short transistor is also called as BJT. BJT stands for Bipolar junction transistor because the transistor operation is carried out by two types of charge carriers: majority carriers and minority carriers.

A transistor has a very important property that it can raise the strength of an input weak signal. This property is called Amplification. Due to this quality, the transistor is one of the most widely used semiconductor devices.

> **DO YOU KNOW?**
>
> On December 23, 1947 Walter H. Brattain and John Bardeen demonstrated the amplifying action of the first transistor at the Bell telephone laboratories. The first transistor was called a point-contact transistor, which was the predecessor to the junction transistor invented by Shockley.

At present, transistor is used in almost every field of electronics.

Few applications are:

1. Transistor is used in control systems.
2. Transistor is used in satellites and mobile phones.
3. In digital computer electronics, the transistor is used as a high-speed electronic switch.
4. In communication systems, it is widely used as the primary component in the amplifier.

1.49. JUNCTION TRANSISTOR CONSTRUCTION

A transistor is a single crystal of silicon (Si) or Germanium (Ge). A transistor can be of two types : NPN or PNP. A transistor consists of two PN junctions. The junctions are formed by sandwitching either P-type or N-type semiconductor layers between a pair of opposite types as shown in figures 1.92(a) and (b).

A transistor has three regions. They are emitter, base and collector. A brief introduction and functioning of these three junctions may be discussed as ahead:

1. Emitter: It is an outer region situated in one side of transistor. The function of emitter is to inject charge carriers (electrons in case of NPN transistor and

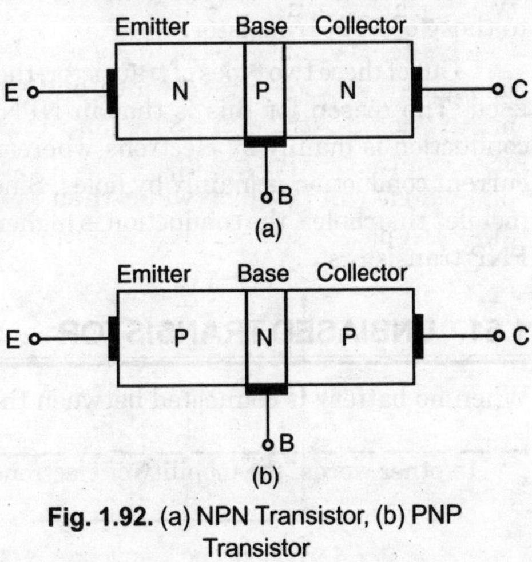

Fig. 1.92. (a) NPN Transistor, (b) PNP Transistor

holes in case of PNP transistors) into the base. Since emitter has to supply a large number of charge carriers, so it is heavily doped.

"More is the doping, more will be the available charge carriers."

2. **Base:** It is the middle region of the transistor. The base is a very thin and lightly doped region. The function of this region is to pass all the charge carriers (electrons or holes) onto the collector.

3. **Collector:** It is the other outer region situated in the other side of transistor. The doping of collector is between the heavy doping of emitter and light doping of Base. The function of collector is to collect charge carriers (electrons or holes). The collector region is physically larger than the emitter region. The reason for this is that the collector has to dissipate more heat. Hence, it is clear that although a BJT has two same type of outer regions, their function cannot be interchanged.

A transistor has two PN-junctions. One junction is formed between the emitter and the base. This is known as emitter-base junction. Another junction is formed between the base and the collector. This is known as collector-base junction.

1.50. TRANSISTOR SYMBOLS

(RGTU, Bhopal, Sem. Exam., 2005-06)

There are two types of transistors known as NPN or PNP. When the transistor is used as a circuit element in any electronic circuit, it is always represented by its symbol as shown in figures 1.93 (a) and 1.93(b).

DO YOU KNOW?

A bipolar junction transistor (BJT) is frequently used as a constant current source.

Figure 1.93(a) represents the symbol for NPN transistors. The emitter has an arrowhead. The arrowhead indicates the direction of a conventional current flow in a transistor. The arrowhead points from the P-region towards the N-region. Therefore, In a NPN transistor the conventional current will flow from the base to the emitter.

Figure 1.93(b) shows the symbol for PNP transistor. The arrowhead in the emitter points from the P-region towards the N-region. Therefore, In a PNP transistor the conventional current will flow from the emitter to the base.

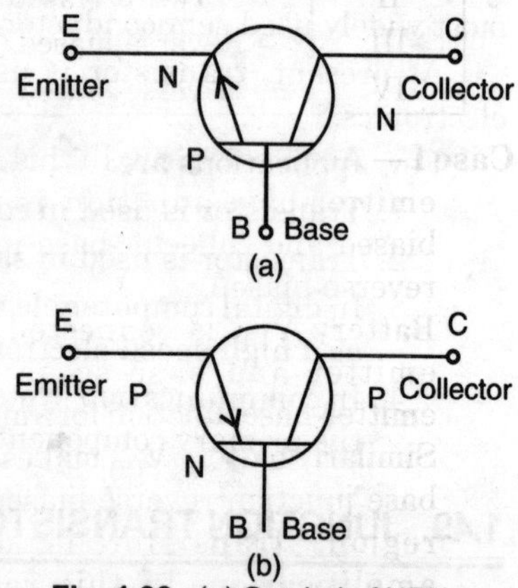

Fig. 1.93. (a) Symbol of NPN transistor, (b) Symbol of PNP transistor

The PNP transistor is a complement of the NPN transistor. Hence, in NPN transistor, the majority carriers are free electrons but in PNP transistor majority carriers are the holes. The currents and voltages involved in the action of PNP transistor are opposite to those of NPN transistor.

Out of these two types of transistor, the NPN transistor is mostly used. The reason for this is that in NPN transistors, the current conduction is mainly by electrons whereas in PNP transistors the current conduction is mainly by holes. Since the electrons are more mobile* than holes, the conduction is higher in NPN transistors than PNP transistors.

1.51. UNBIASED TRANSISTOR

When no battery is connected between the different terminals of a transistor then the transistor is

* In other words, the mobility of electrons is high compared to holes.

said to be in an unbiased state or in an open-circuit state. In this condition, the diffusion of charge carriers across the junction produces two depletion regions. The barrier potential for each of these two depletion layers is **0.7 V for Si** and **0.3 V for Ge.** But the three regions have different doping levels, so the depletion widths are not the same for two junctions. Since emitter is heavily doped than base and collector, the emitter-base depletion region width is smaller than that of collector-base depletion region width.

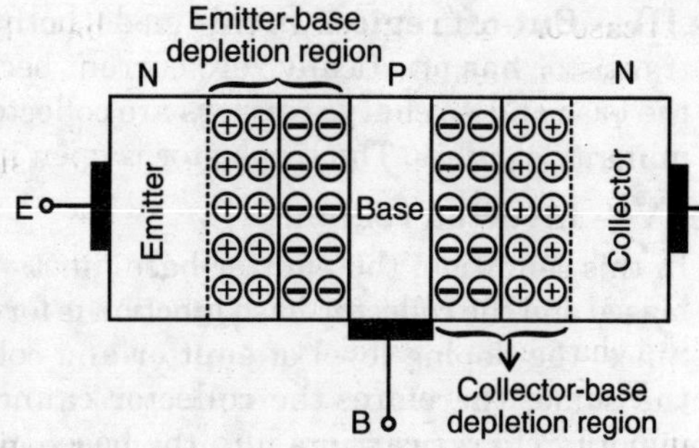

Fig. 1.94. Unbiased Transistor

1.52. BIASING OF A TRANSISTOR*

(RGTU, Bhopal, Sem. Exam., 2007-08)

When dc voltages are applied across the different terminals of a transistor, then this process is called biasing. Since, there are two junctions in a transistor namely emitter-base junction and collector-base junction, each of these two junctions may be forward-biased or reverse-biased. Therefore, there are four possible ways of biasing for these two junctions. These possible ways are also called **modes** of operation of a transistor. These different modes are listed in Table 1.4, given below:

> **DO YOU KNOW?**
>
> A bipolar transistor is frequently used as a constant current source.

Table 1.4

Cases	Emitter-base junction	Collector-base junction	Region of operation
I	Forward-biased	Reverse-biased	Active
II	Forward-biased	Forward-biased	Saturation
III	Reverse-biased	Reverse-biased	Cut-off
IV	Reverse-biased	Forward-biased	Inverted

Case I — Active region: In this mode, the emitter-base junction is forward-biased and collector-base junction is reverse-biased.

Battery V_{EB} is connected between emitter and base so as to make emitter-base junction forward-biased. Similarly battery V_{CB} makes collector-base junction reverse-biased. In this region, transistor is used for amplification. In this region, the collector current depends upon the base current (input current).

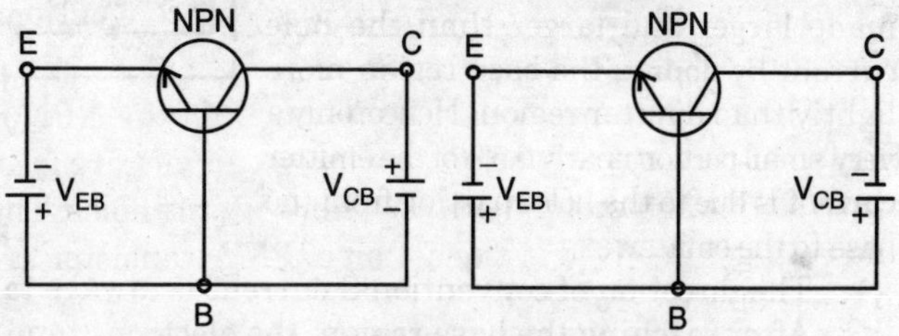

(a) Active region (b) Saturation region

Fig. 1.95.

Case II — Saturation region: In this region, both the junctions are forward-biased. The transistor works in saturation. In this region, the collector current becomes independent of the base current. Therefore, the transistor acts like a closed switch.

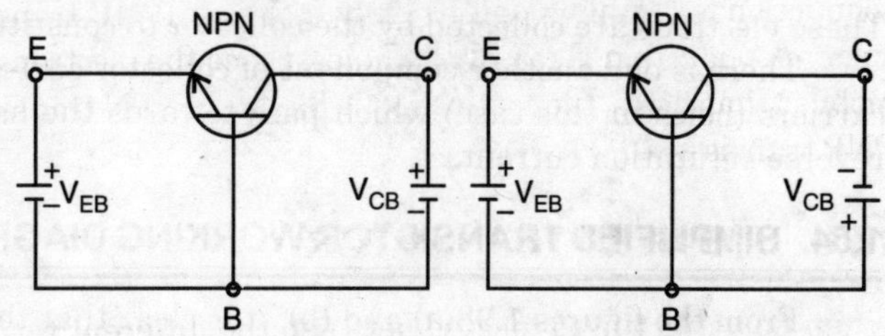

(c) Cut-off region (d) Inverted region

Fig. 1.96.

* In how many modes the BJT works ? Also explain the biasing pattern for each of them.
(RGTU, Bhopal, Sem. Exam., 2007-08)

Case III — Cut-off region: In this mode, both the junctions are reverse-biased. In this region, the transistor has practically zero current because the emitter does not emit charge carriers into the base and no charge carriers are collected by the collector except a few thermally generated minority carriers. The transistor is open in this mode when it is used as an open switch.

Case IV — Inverted region: **(RGTU, Bhopal, Sem. Exam., 2003-04)**

In this condition, the emitter-base junction is reverse-biased and the collector-base junction is forward-biased. Since, the doping level of emitter and collector is not the same, therefore, the collector cannot inject the majority charge carriers into the base. In this region, the action of transistor is poor. This region is of little importance.

DO YOU KNOW?

The term 'transistor' was derived from the words TRANSFER and RESISTOR. This term was adopted because it best describes the operation of a transistor, which is the transfer of an input signal current from a low resistance circuit to a high resistance circuit.

1.53. THE WORKING OF A TRANSISTOR

(RGTU, Bhopal, Sem. Exam., 2007-08)

To explain the working of a transistor, we consider an NPN transistor which is biased for active operation. That means, emitter-base junction is forward-biased by battery V_{EB} and collector-base junction is reverse-biased by battery V_{CB}.

The forward-biased voltage is small and reverse-biased voltage V_{CB} is quite large. The forward bias on the emitter-base junction pushes a large number of free electrons in the N-type emitter towards the base. This makes the emitter current (I_E). A very few holes also pass from the base region to the emitter region. This flow of electrons and holes constitute emitter current (I_E). Since, the electron current is useful in the action of transistor, it is made larger and larger than the hole current by doping the base region more lightly than emitter region. Hence, only a very small portion nearly 0.5% of the emitter current is due to the holes passing from the base to the emitter.

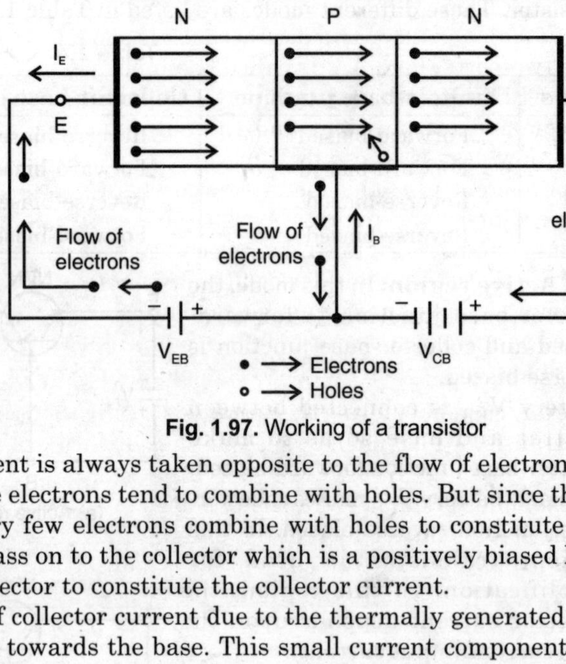

Fig. 1.97. Working of a transistor

The direction of conventional current is always taken opposite to the flow of electrons.

After reaching the base region, the electrons tend to combine with holes. But since the base is very thin and lightly doped, only a very few electrons combine with holes to constitute the base current (I_B). The remaining electrons pass on to the collector which is a positively biased N-region. These electrons are collected by the collector to constitute the collector current.

There is one another component of collector current due to the thermally generated minority carriers (holes in this case) which pass towards the base. This small current component is called reverse saturation current.

1.54. SIMPLIFIED TRANSISTOR WORKING DIAGRAM TO SHOW DIFFERENT CURRENTS

From the figures 1.98(a) and (b), it is clear that the emitter current is the sum of the collector and base currents.

Mathematically, $$I_E = I_C + I_B \quad ...(1.35)$$

* In how many modes the BJT works ? Also explain the biasing pattern for each of them.
(RGTU, Bhopal, Sem. Exam., 2007-08)

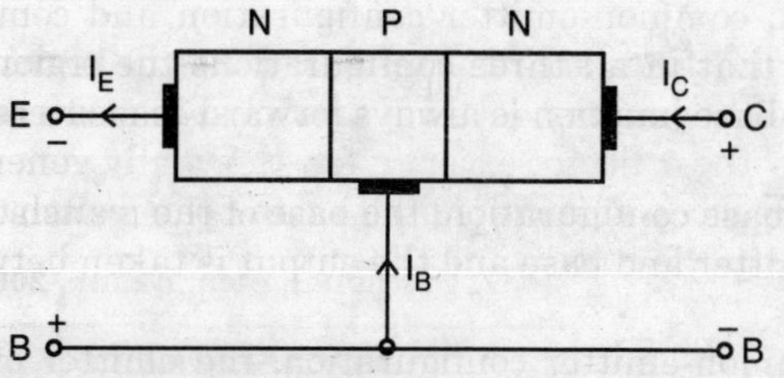

Fig. 1.98(a). Transistor currents

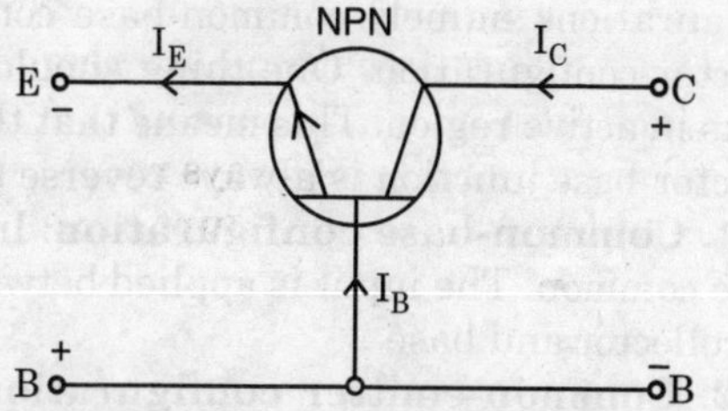

Fig. 1.98(b). Transistor currents with transistor symbol

The emitter current of a transistor consists of two components. These are base current and collector current. But the base current is only about 2% of the emitter current whereas the collector current is about 98% of the emitter current.

Therefore, from last equation (1.35), we have

$$I_E \cong I_C$$

Therefore, $$I_B \cong 0$$

1.55. TRANSISTOR AS A TWO PORT NETWORK

(RGTU, Bhopal, Sem. Exam., 2007-08)

In figure 1.99(a), a general two port network is shown. There are two ports. Port AA′ is called the input port and the port BB' is called the output port. One terminal is made common to the input and the output. The directions of voltages and currents shown are the standard or reference direction for any two port network. Hence, according to the figure 1.99, all the currents entering into the network are taken to be positive. A transistor is a three terminals device. If one terminal is made common to input and output, a transistor may be represented as a two port network. In the figure 1.99(b) an NPN transistor is taken and base is made common.

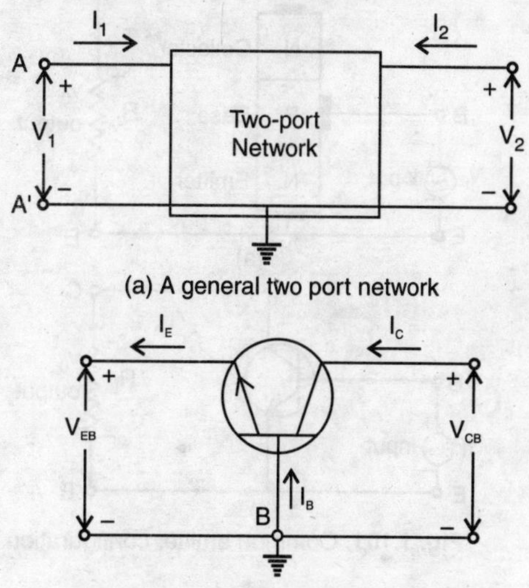

Fig. 1.99.

Since any current entering the network is positive and a current flowing out is negative, therefore, in case of transistor if actual current is in the outward direction, that will be taken with a negative sign.

For example, for an NPN transistor shown in figure 1.99 (b), the actual emitter current flows in the outward direction. Therefore, emitter current I_E is negative. Similarly I_B and I_C flow into the transistor. Hence they are positive.

1.56. TRANSISTOR CONFIGURATIONS

We know that a transistor may be represented as a two port network. A transistor has three terminals. Any of these three terminals can be made common to input and output. This common terminal is generally grounded or connected to the chasis. Hence, depending upon the common terminal, there are three transistor circuit

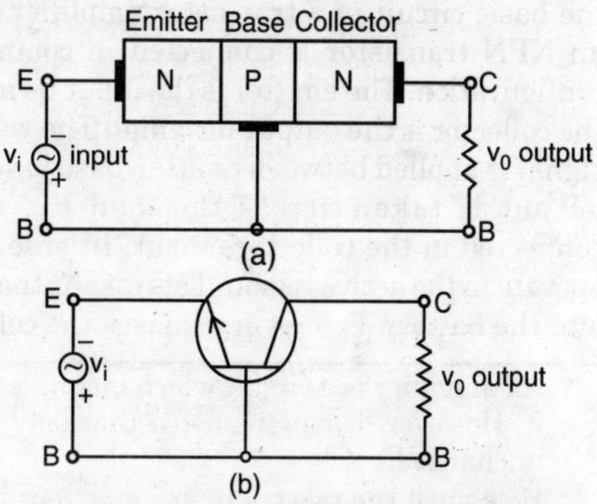

Fig. 1.100. Common base configurations

configurations namely common-base configuration, common-emitter configuration and common collector configuration. One thing should be noted that in all three configurations the transistor works in active region. This means that the emitter-base junction is always forward-biased and the collector-base junction is always reverse biased.

1. Common-base configuration: In common-base configuration, the base of the transistor is made common. The input is applied between the emitter and base and the output is taken between the collector and base.

2. Common-emitter configurations: In common-emitter configuration, the emitter of the transistor is made common. The input is applied between base and emitter and the output is taken between collector and emitter. This configuration is most widely used.

3. Common collector configuration: In common collector configuration, the collector is made common. The input is applied between base and collector and the output is taken between emitter and collector.

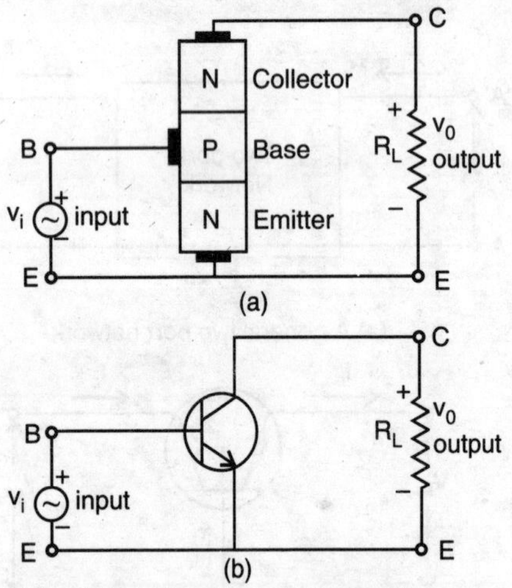

Fig. 1.101. Common emitter configuration

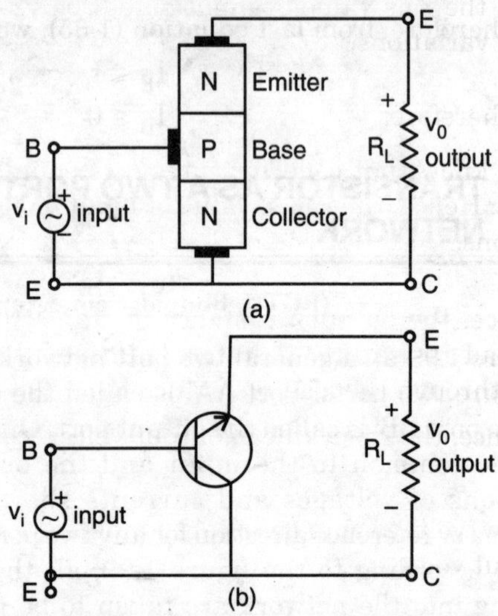

Fig. 1.102. Common collector configuration

1.57. TRANSISTOR AS AN AMPLIFIER*

A transister may be used to perform a number of functions. But mainly transistor is used in amplifying electrical signals.** Figure 1.103 shows the basic circuit of a transistor amplifier in which an NPN transistor is connected in common base configuration. The emitter is the input terminal and the collector is the output terminal. The weak input signal is applied between emitter-base junction and output is taken across the load R_L, which is connected in the collector circuit. In order to get faithful amplification, the transistor is biased to operate in the active region. This means that the battery V_{EE} forward-biases the emitter-base junction and the battery V_{CC} reverse-biases the collector-base junction.

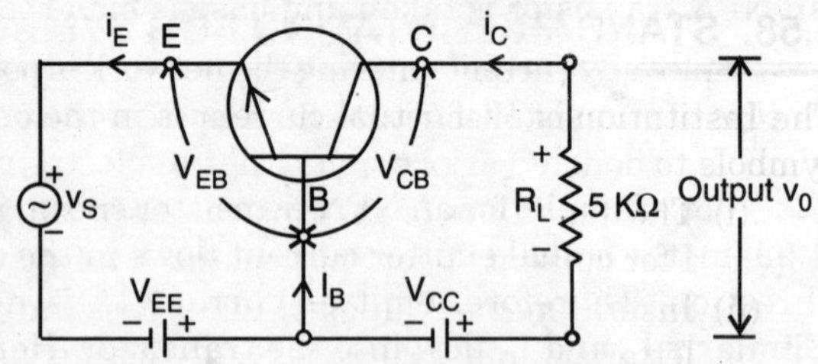

Fig. 1.103.

* An amplifier is a circuit which magnifies or increases the amplitude of a signal waveform applied to it. However, it may be noted that only amplitude is changed and shape of the waveform remains unchanged.

** How must the two transistor junctions be biased for proper transistor amplifier operation?

(RGTU, Bhopal, Sem. Exam. 2006-07)

Let us consider that an a.c. signal source v_s is connected in the input circuit and a load resistance R_L = 5 KΩ is also connected in the output circuit. An output voltage v_0 is obtained across this resistor R_L. When the input a.c. signal v_s is superimposed on the d.c. voltage V_{EE}, the emitter-base voltage (i.e., input voltage) V_{EB} varies with time. Due to this, the emitter current (i.e., input current) i_E also varies. We know that the collector current is a function of emitter current, therefore, similar variations occur in the collector current. This varying current passes through the load resistor R_L and so a varying voltage is developed across the load resistor R_L. This varying voltage is the output voltage v_0. The output voltage v_0 will be several times greater than the input signal v_s.

The emitter-base junction is forward-bias, so it offers very low resistance to the input signal v_s. In CB configuration, typical average value of input resistance is 40 ohms. The output junction or collector-base junction is reverse-bias and so output resistance is very high. Its typical value is 500 KΩ.

Let the rms value of input signal be v_s = 20 mV, then rms value of input current or emitter current variations,

$$I_E = \frac{v_s}{r_i} = \frac{20\text{ mV}}{40\ \Omega} = 0.5\text{ mA}$$

We know that collector current is equal to the emitter current. Hence, the r.m.s. value of input or collector current variations,

$$I_C \cong I_E = 0.5\text{ mA}$$

Since, the output resistance of transistor is very high than load resistance, therefore, almost all the current I_C passes through the load resistance R_L.

> **DO YOU KNOW?**
>
> The transistor can be used for two types of applications viz. amplification and switching. For the amplification applications, a transistor is biased in its active region. Whereas for switching applications, it is biased to operate in the saturation (fully ON) or cut off (fully OFF) regions.

Hence, the rms value of output voltage will be

$$v_0 = I_C.R_L = (0.5 \times 10^{-3}) \times (5 \times 10^3) = 2.5\text{ Volt}$$

Now, voltage gain or voltage amplification A of amplifier is given as ratio of output voltage v_0 to the input voltage v_s.

Therefore,
$$A = \frac{v_0}{v_s} = \frac{2.5}{20 \times 10^{-3}} = 125$$

Hence, amplitude of input voltage increases by 125 times.

1.58. STANDARD NOTATION FOR SYMBOLS (RGTU, Bhopal, Sem. Exam., 2007-08)

The Institution of Electrical and Electronics Engineers (IEEE) has adopted a standard notation of symbols to denote various quantities in electronics as under:

(i) The d.c. values or the rms values of a.c. are represented by upper case or capital letters like I for current, V for voltage and P for power.

(ii) Instantaneous values of quantities varying with time are denoted by lower case or small letters like i for current, v for voltage and p for powers.

(iii) Time-varying components or a.c. components are denoted by the small letter subscripts of the proper electrode symbol. As an example, the collector a.c. current is denoted by i_c and base a.c. voltage is denoted by v_b.

(iv) D.C. or average values and instantaneous values are denoted by the capital subscripts of the proper electrode symbol (E for emitter, C for collector etc.).
For example, d.c. collector current is denoted by I_C and instantaneous total value of emitter current is denoted by i_E.

(v) The reference direction of the current is denoted by an arrow. Also the voltage reference polarity is denoted by plus and minus signs or simply by an arrow which points from the negative to t' e positive terminal.

(vi) The current flow into an electrode from the external circuit is always taken as positive.

(vii) The d.c. supply magnitude is represented by using double subscripts of the proper electrode symbol.

For example, V_{CC} represents the magnitude of the d.c. supply in the collector circuit.

Table 1.5 shows some current and voltage components and their representations according to above standard notation.

Table 1.5. Standard Notations

S.No.	Name of Quantity	Instantaneous a.c.	d.c.	Total
1.	Collector current	i_c	I_C	i_C
2.	Emitter current	i_e	I_E	i_E
3.	Base current	i_b	I_B	i_B
4.	Collector-emitter voltage	v_{ce}	V_{CE}	v_{CE}
5.	Emitter-base voltage	v_{eb}	V_{EB}	v_{EB}

1.59. CURRENT GAIN OF A TRANSISTOR IN COMMON-BASE CONFIGURATION

Figure 1.104 shows the circuit of common-base configuration using NPN transistor and the actual directions of different currents in the circuit. In this configuration, emitter current is the input current and collector current is the output current. The current gain of a transistor is defined as the ratio of the transistor output current to the input current. Current gain may be of two types namely d.c. current gain and a.c. current gain depending upon whether the input and output currents are d.c. currents or a.c. currents.

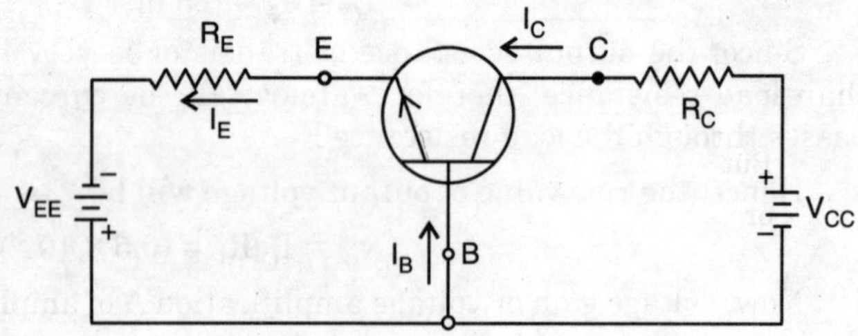

Fig. 1.104. Common-base configuration

(i) D.C. current gain in common-base configuration: D.C. current gain may be defined as the ratio of collector current I_C to emitter current I_E. It is denoted by α or α_{dc} or h_{FB}.

$$\alpha = \frac{I_C}{I_E} = h_{FB} \qquad ...(1.36)$$

But we know that the value of collector current is always less than the value of emitter current. Hence d.c. current gain in CB configuration is always less than unity.

(ii) A.C. current gain in common-base configuration: A.C. current gain may be defined as the ratio of change in collector current (ΔI_C) to the change in emitter current (ΔI_E) for a constant collector-to-base voltage (V_{CB}).

It is represented by α_0, α_{ac} or h_{fb}

$$\alpha_0 = \alpha_{ac} = \left.\frac{\Delta I_C}{\Delta I_E}\right|_{V_{CB} = \text{constant}} \qquad ...(1.37)$$

α_0 is also called **common-base short-circuit current gain.** The value of α_0 is also less than unity. α_0 is approximately same as α.

Therefore, for all practical purposes, we can write

$$\boldsymbol{\alpha = \alpha_0}$$

d.c. current gain = a.c. current gain ...(1.38)

EXAMPLE 1.31 **In a common base configuration, the value of collector current is 5.10 mA and the value of emitter current is 5.18 mA. Calcualte the common-base d.c. current gain.**

Solution: Common-base d.c. current gain is given as

$$\alpha_{dc} = \alpha = \frac{I_C}{I_E} = \frac{5.10}{5.18} = 0.9845 \quad \textbf{Ans.}$$

EXAMPLE 1.32 **Find the value of base current if the common-base d.c. current gain of a transistor is 0.987 and emitter current is 10 mA.**

Solution: Common-base d.c. current gain is given as

$$\alpha = \frac{I_C}{I_E} \Rightarrow I_C = \alpha I_E$$

$$I_C = 0.987 \times 10 = 9.87 \text{ mA}$$

Emitter current is expressed as

$$I_E = I_B + I_C$$

or

$$I_B = I_E - I_C$$

$$I_B = 10 - 9.87 = 0.13 \text{ mA} \quad \textbf{Ans.}$$

EXAMPLE 1.33 **Figure 1.105 shows a NPN transistor for which I_E = 10 mA, d.c. current gain α = 0.965. Find the values of I_C and I_B.**

Solution: Common-base d.c. current gain is

$$\alpha = \frac{I_C}{I_E}$$

$$I_C = \alpha I_E = 0.967 \times 10 = 9.67 \text{ mA}$$

But

$$I_E = I_B + I_C$$

or

$$I_B = I_E - I_C$$

$$I_B = 10 - 9.67 = 0.33 \text{ mA} \quad \textbf{Ans.}$$

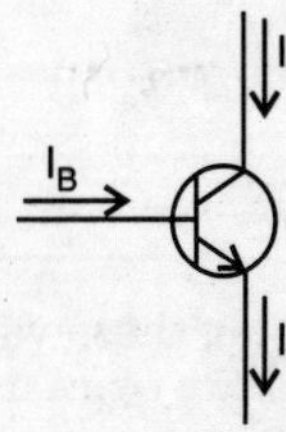

Fig. 1.105.

1.60. CURRENT GAIN OF A TRANSISTOR IN COMMON-EMITTER CONFIGURATION

Figure 1.106 shows the circuit of common-emitter configuration using NPN transistor and the actual directions of different currents in the circuit. In this configuration, base current is the input current and the collector current is the output current. In common-emitter configuration the current gain is defined as the ratio of collector current to the base current. Like CB configuration, current gain is of two types in common-emitter configuration.

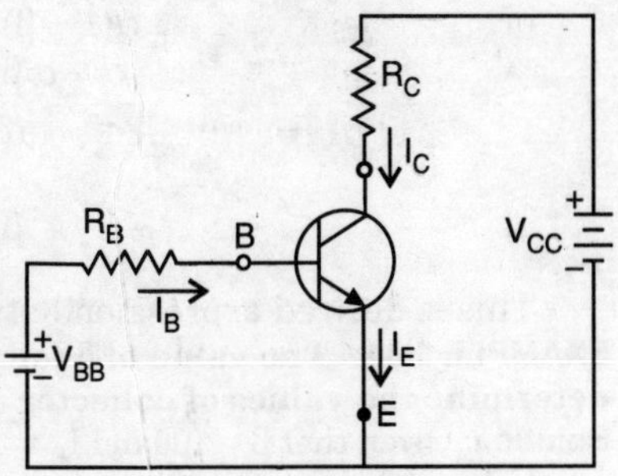

Fig. 1.106. Common emitter configuration

(i) Common-emitter d.c. current gain: D.C. current gain in common-emitter configuration may be defined as the ratio of collector current I_C to base current I_B. It is denoted by β or β_{dc} or h_{FE}.

$$\beta_{dc} = \beta = \frac{I_C}{I_B} \qquad \text{...(1.39)}$$

Since, the collector current is much larger than the base current, the value of β is much greater than unity. Typical range of β may be from 20 to 300.

(ii) A.C. current gain in common emitter configuration: A.C. current gain in common emitter configuration may be defined as the ratio of change in collector current ΔI_C to the change in base current ΔI_B keeping collector to emitter voltage V_{CE} constant. It is denoted by β_0, β_{ac} or h_{fe}.

$$\beta_0 = \left.\frac{\Delta I_C}{\Delta I_B}\right|_{V_{CE} = \text{constant}} \quad ...(1.40)$$

β_o is also known as **common-emitter short-circuit current gain.** The values of β_0 are approximately same as that of β.

For all practical purposes, we can write

$$\beta = \beta_0 \quad ...(1.41)$$

d.c. current gain = a.c. current gain*

1.61. RELATION BETWEEN CURRENT GAIN α AND β

(RGTU, Bhopal, Sem. Exam. 2007-08)

In this section, let us derive a relationship between the two current gains α and β.

We know that the emitter current I_E of a transistor is given as

$$I_E = I_B + I_C \quad ...(1.42)$$

Here I_B = Base Current

I_C = Collector Current

DO YOU KNOW?

The base current is very small by camparison, often less than 1 percent of the collector current.

Dividing above equation by I_C on both sides, we have

$$\frac{I_E}{I_C} = \frac{I_B}{I_C} + 1$$

But $\frac{I_C}{I_E} = \alpha$ and $\frac{I_C}{I_B} = \beta$

Putting these values in the last expression, we get

$$\frac{1}{\alpha} = \frac{1}{\beta} + 1 = \frac{1+\beta}{\beta}$$

or $$\alpha = \frac{\beta}{1+\beta}$$

or $$\alpha(1 + \beta) = \beta$$

$$\alpha + \alpha\beta = \beta$$

$$\alpha = \beta - \alpha\beta = \beta(1 - \alpha)$$

or $$\beta = \frac{\alpha}{1-\alpha} \quad ...(1.43)$$

This is desired expression between current gain α and β.

EXAMPLE 1.34 The value of β for a transistor is 100. If the value of emitter current is 10 mA, then determine the values of collector and base currents.

Solution: Given that β = 100and I_E = 10 mA

Value of β is given as

$$\beta = \frac{\alpha}{1-\alpha}$$

So that $$\alpha = \frac{\beta}{\beta+1} = \frac{100}{100+1} = \frac{100}{101} = 0.98$$ **Ans.**

Common-base current gain α is given as

$$\alpha = \frac{I_C}{I_E}$$

$$I_C = \alpha I_E = 0.98 \times 10 \times 10^{-3}$$

$$I_C = 0.98 \times 10^{-2} = 9.8 \text{ mA}$$ **Ans.**

* Define with respect to BJT the following:
(i) I_{CO} (ii) α (iii) β (iv) I_{CEO} (v) I_{CBO} **(RGTU, Bhopal, Sem. Exam. 2005-06)**

Value of emitter current is given as

$$I_E = I_B + I_C$$

$$I_B = I_E - I_C$$

$$I_B = 10 - 9.8 = 0.2 \text{ mA} \qquad \textbf{Ans.}$$

EXAMPLE 1.35 The value of α for a transistor is 0.950. Find the value of β. Also find the value of α if β changes to 100.

Solution: The relation between α and β is given as

$$\beta = \frac{\alpha}{1-\alpha} = \frac{0.950}{1-0.950} = \frac{0.950}{0.050} = 19$$

For second case

Therefore,
$$\beta = \frac{\alpha}{1-\alpha}$$

or
$$\alpha = \frac{\beta}{\beta+1} = \frac{19}{19+1} = \frac{19}{20} = 0.95 \qquad \textbf{Ans.}$$

EXAMPLE 1.36 Figure 1.107 shows a NPN transistor for which I_E = 10 mA and β = 100. Determine the values of I_C and I_B.

Solution:

We know that
$$\alpha = \frac{\beta}{1+\beta}$$

But
$$\alpha = \frac{I_C}{I_E}$$

$$\beta = \frac{I_C}{I_B}$$

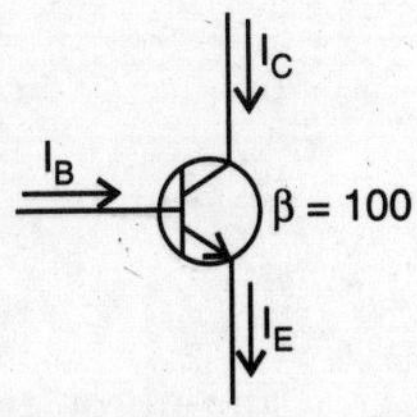

Fig. 1.107.

Putting the value of α and β, we get

$$\frac{I_C}{I_E} = \frac{I_C}{I_B}\cdot\left(\frac{1}{1+\beta}\right)$$

or
$$I_B = \frac{I_E}{1+\beta}$$

Given that β = 100, I_E = 10 mA

$$I_B = \frac{10}{1+100} = \frac{10}{101} = 0.098 \text{ mA}$$

Emitter current is given as

$$I_E = I_C + I_B$$

$$I_C = I_E - I_B = 10 - 0.098 = 9.902 \text{ mA} \qquad \textbf{Ans.}$$

1.62. LEAKAGE CURRENT IN A COMMON-BASE TRANSISTOR

Figure 1.108 shows the circuit of a common base NPN transistor. The polarity of battery V_{EE} is such that it makes the emitter-base junction forward biased. The battery V_{CC} sets the collector-base junction in reverse biased. Now, we consider two cases:

(i) Switch S is in closed position: In this case, electrons are injected into base region due to forward-bias. The most of these electrons are attracted by collector due to reverse-bias.

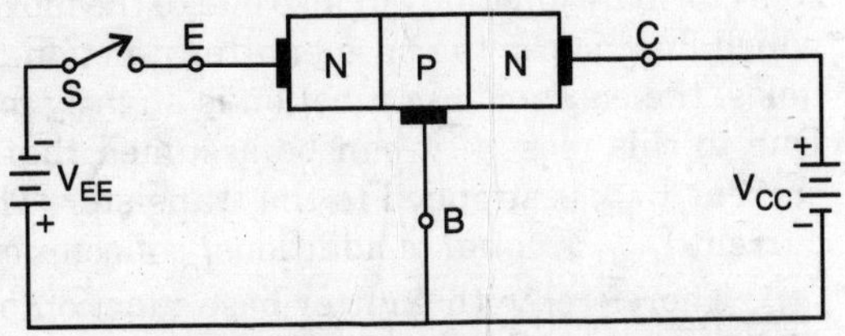

Fig. 1.108. Common-base transistor circuit

This accounts for collector current. Few electrons are combined with holes in base region and constitute base current. Therefore, total emitter current breaks up into two parts namely the collector current I_C and base current I_B.

$$I_E = I_B + I_C$$

(ii) Switch S is opened: Let us assume that the switch S is opened.

In this case, emitter-base junction of transistor is open-circuited. Therefore, there is no emitter current which means that there is no base current and collector current. But since the collector-base junction is reverse-biased, minority carriers diffuse across the collector base junction and this produce a small current. This small current is called leakage current and is denoted by I_{CBO} or I_{CO}. I_{CBO} means the leakage current between collector and base with emitter open. It is also known as reverse saturation current and it is like reverse saturation current in a PN-junction.

Hence, now it is clear that total collector current for a transistor contains following parts:

(a) The injected current due to majority carriers or a fraction of emitter current is equal to $\alpha.I_E$.

(b) The leakage current I_{CBO} or I_{CO} produced by thermally generated minority carriers.

Therefore,
$$I_C = \alpha I_E + I_{CBO} \quad \text{...(1.44)}$$

or
$$\alpha I_E = I_C - I_{CBO}$$

$$\alpha = \frac{I_C - I_{CBO}}{I_E} = \frac{I_C - I_{CBO}}{I_B + I_C} \qquad \text{(Since, } I_E = I_B + I_C\text{)}$$

or
$$\alpha I_B + \alpha.I_C + I_{CBO} = I_C$$

or
$$I_C(1-\alpha) = \alpha I_B + I_{CBO}$$

$$I_C = \frac{\alpha.I_B}{1-\alpha} + \frac{I_{CBO}}{1-\alpha} \quad \text{...(1.45)}$$

For silicon transistor, at room temperature, the value of I_{CBO} is of the order of nanoamperes and for germanium transistor, it is of the order of microamperes. For both Si and Ge, the value of I_{CBO} strongly depends on temperature. It becomes doubled for every 10°C rise in temperature. Since the value of I_{CBO} is less for Si than Ge, the silicon transistors can be used upto 200°C whereas Ge transistors can only be used upto 100°C. Therefore, silicon transistors are preferred over germanium transistors.

1.63. LEAKAGE CURRENT IN COMMON-EMITTER CONFIGURATION

Figure 1.109 shows a circuit of common-emitter configuration with NPN transistor. If the switch S is closed then the usual currents I_E, I_C and I_B flow. It means that emitter current is the sum of collector and base currents.

When the switch is open, the emitter-base junction is an open circuit and hence the value of input or base current is zero. But a leakage current I_{CEO} flows between collector and emitter. This leakage current I_{CEO} is not only due to the thermally generated minority carriers (I_{CBO}) across the collector-base junction but also due to the movement of holes which flow across the base-emitter junction. Due to flow of holes, the emitter base junction is slightly forward biased. Due to this reason, it can be assumed that a small base current I_{CBO} is supplied to the transistor. This small base current I_{CBO} is equal to additional collector current $\beta.I_{CBO}$.

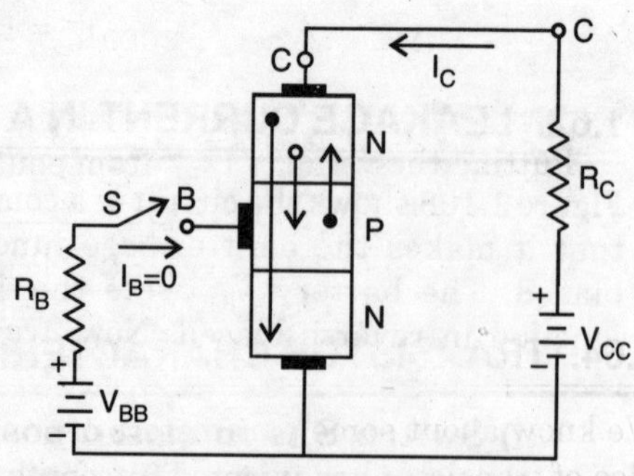

Fig. 1.109. Common emitter transistor circuit

Therefore, with emitter-base junction open, the total leakage current flowing through the transistor

$$I_{CEO} = I_{CBO} + \beta.I_{CBO} \quad ...(1.46)$$

$$I_{CEO} = I_{CBO}(1 + \beta) \quad ...(1.47)$$

Hence, the total collector current in transistor is the sum of injected current $\beta.I_B$ and I_{CEO}

$$\boxed{I_C = \beta.I_B + I_{CBO}(1 + \beta)} \quad ...(1.48)$$

The above relation can also be proved in the following manner:

We know that

$$I_E = I_C + I_B \quad ...(1.49)$$

For CB configuration $\quad I_C = \alpha I_E + I_{CBO} \quad ...(1.50)$

In CE configuration, I_B is the input current and I_C is the output current. Therefore, we have to find the relation between I_B and I_C.

Putting the value of I_E from equation (1.49) in equation (1.50), we get

$$I_C = \alpha(I_C + I_B) + I_{CBO} = \alpha I_C + \alpha I_B + I_{CBO}$$

$$I_C(1 - \alpha) = \alpha I_B + I_{CBO}$$

or

$$I_C = \frac{\alpha}{1-\alpha}.I_B + \frac{1}{1-\alpha}.I_{CBO}$$

To simplify the above equation, defining

$$\frac{\alpha}{1-\alpha} = \beta \quad ...(1.51)$$

and

$$\frac{I_{CBO}}{1-\alpha} = I_{CEO} \quad ...(1.52)$$

$$\boxed{I_C = \beta I_B + I_{CEO}} \quad ...(1.53)$$

where $\quad \beta \rightarrow$ common-emitter d.c. current gain

$I_{CEO} \rightarrow$ Leakage current in common emitter configuration.

Using equation (1.51), we have

$$\alpha = \frac{\beta}{1+\beta}$$

From equation (1.52), we have

$$I_{CEO} = \frac{I_{CBO}}{1-\alpha} = \frac{I_{CBO}}{1-\left(\frac{\beta}{1+\beta}\right)} = \frac{I_{CBO}(1+\beta)}{1+\beta-\beta} = (1+\beta)\, I_{CBO} \quad ...(1.54)$$

Putting the value of I_{CEO} from equation (1.54) in equation (1.53), we have

$$\boxed{I_C = \beta I_B + (1 + \beta)I_{CBO}} \quad ...(1.55)$$

1.64. TRANSISTOR CHARACTERISTICS

We know about some parameters of transistor such as α, β and leakage current. They give a small idea of transistor behaviour. The complete behaviour of transistor may be observed with the help of some types of curves known as characteristics curves. These characteristics curves relate the transistor currents and voltages. There are two types of characteristics curves:

(a) Input characteristics curves: Input characteristics curves relate the input current and input voltage for a given value of output voltage.

(b) Output characteristics curves: Output characteristics curves relate the output current and output voltage for a given value of input current.

With the help of these two types of characteristics curves the complete static operation of transistor can be explained. Hence, these curves are also known as *static characteristics curves.*

1.64.1. Transistor Characteristics in Common-Base Configuration*

(RGTU, Bhopal, Sem. Exam. 2005-06)

In common-base configuration, two types of characteristics are:

(a) Input characteristics: Input characteristics curves relate the input or emitter current I_E and input or emitter-to-base voltage V_{EB} keeping output or collector-to-base voltage V_{CB} constant.

(b) Output characteristics: Output characteristics curves relate the output or collector current I_C and output or collector-to-base voltage V_{CB} keeping the input or emitter current I_E constant.

Both these characteristics curves can be drawn with the same circuit arrangement shown in figure 1.110.

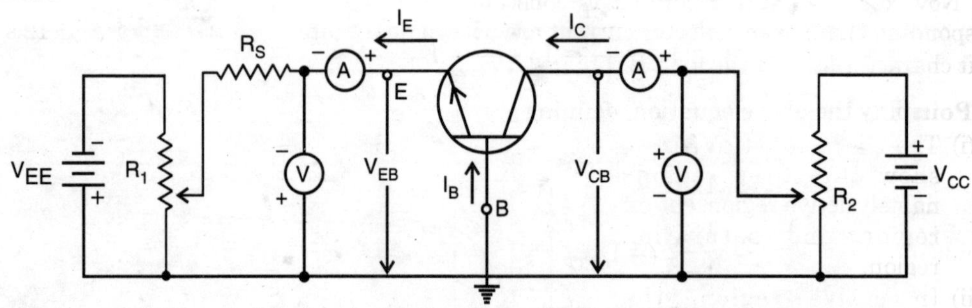

Fig. 1.110. Circuit arrangement for input and output characteristics curve in CB configuration

1.64.1.1. Transistor Input Characteristics in CB Configuration

Input characteristics curves are obtained with the help of circuit arrangement shown in figure 1.110. In this circuit NPN transistor is used. The series resistor R_s is connected to limit the emitter current. The value of voltage V_{EB} may be varied with the help of a potentiometer R_1. The collector voltage V_{CB} is varied by adjusting the potentiometer R_2.

The d.c. milliameters and d.c. voltmeters have been connected in the emitter and collector circuits to measure the voltages and currents.

The input characteristics in CB configuration are plotted between emitter current I_E and the emitter-base voltage V_{EB}, for different values of collector-base voltage V_{CB}. Input characteristics are shown in figure 1.111.

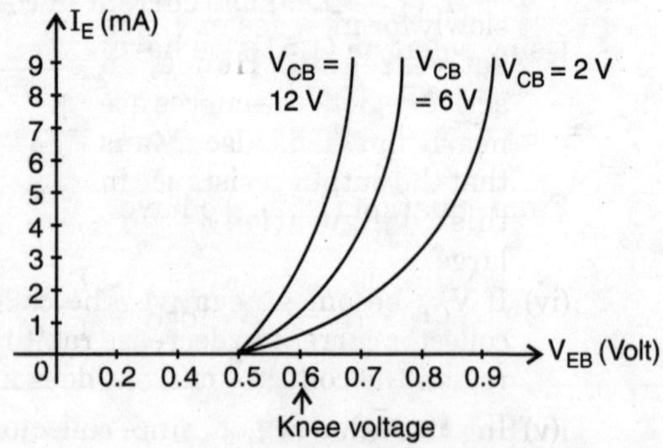

Fig. 1.111. Input characteristics in CB configuration

Few Points

(i) Below the cut-in voltage or knee voltage the emitter current is negligibly small. This value of cut-in-voltage is approximately 0.7 V for silicon and 0.3 V for germanium transistors.

(ii) After the knee voltage the emitter current I_E increases rapidly with a small increase in

* Draw and explain the input and output characteristics of common base configuration of BJT. Indicate all the region of operations. **(RGTU, Bhopal, Sem. Exam. 2005-06)**

emitter-to-base voltage V_{EB}. This indicates that input resistance in CB configuration is very small.

(iii) With the increase in V_{CB}, the curves shift upwards or become vertical. Hence, r_i decreases.

(iv) At any point on the curve, the value of a.c. or dynamic input resistance is given as

$$r_i = \left.\frac{\Delta V_{EB}}{\Delta I_E}\right| V_{CB} = \text{constant} \qquad ...(1.56)$$

where $\Delta V_{EB} \rightarrow$ Small change in emitter-to-base voltage V_{EB}.

$\Delta I_E \rightarrow$ Small change in emitter current I_E.

The dynamic input resistance r_i is very small nearly 25 to 100 Ω.

> **DO YOU KNOW?**
> A transistor is cut off when its collector current is zero.

1.64.1.2. Output Characteristics in CB configuration

The output characteristics curves in CB configuration can be obtained with the help of same circuit arrangement shown in figure 1.112. First, a suitable value of input current (emitter current) is fixed. Now keeping emitter current I_E constant. Collector-to-base voltage V_{CB} is increased and corresponding changes in collector current are noted and a graph is plotted. Figure 1.112 shows the output characteristics curve for CB configuration.

Few Points

(i) The characteristics curves show the three regions namely active region, cut-off region and saturation region.

(ii) In active region, the collector current is approximately equal to the emitter current I_E.

(iii) Emitter current increases slowly for increase in V_{CB} in active region. Hence, in active region, the curves are nearly flat. This also means that the output resistance in this configuration is very large.

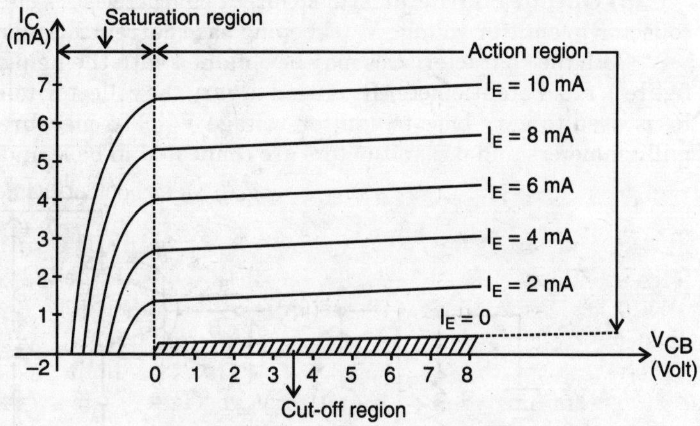

Fig. 1.112. Output characteristics in a CB configuration

(iv) If V_{CB} becomes negative, the collector-base junction is set in forward bias which causes collector current to decrease rapidly. This region is known as saturation region. In saturation region the collector current does not depend much on emitter current.

(v) In saturation region, since collector-base junction is also forward biased, a small increase in V_{CB} results in a large increase in collector current.

(vi) If emitter current I_E is made zero, then the collector current is not zero. It has a small value. This small value is leakage current I_{CBO}. The curve for $I_E = 0$ corresponds to cut-off region. In this region both the junctions are reverse-biased.

(vii) With the help of output characteristics, the dynamic output resistance r_0 can be found by the expression

$$r_0 = \left.\frac{\Delta V_{CB}}{\Delta I_C}\right|_{I_E = \text{constant}} \qquad ...(1.57)$$

Here ΔV_{CB} and ΔI_C are small changes in collector-to-emitter voltage and collector current for a given point.

(viii) With the help of output characteristics curve d.c. and a.c. current gains in CB configuration can also be determined.
D.C. current gain

$$\alpha_{dc} = \alpha = \frac{I_C}{I_E} \qquad \text{...(1.58)}$$

where I_C and I_E are the values of collector and emitter currents at any point on the curve.
a.c. current gain is

$$\alpha_0 = \left.\frac{\Delta I_C}{\Delta I_E}\right| V_{CB} = \text{constant} \qquad \text{...(1.59)}$$

where ΔI_C and ΔI_E are small changes in collector and emitter currents around a given point on the curve.

1.64.2. Transistor Characteristics in Common-Emitter Configuration*

In common-emitter configuration, a transistor has two types of characteristics namely input characteristics and output characteristics.

(a) Input characteristics: Input characteristics curves relate the base current I_B and the base-to-emitter voltage V_{BE} keeping collector-to-emitter voltage V_{CE} constant.

(b) Output characteristics: Output characteristics curves relate the collector current I_C and collector-to-emitter voltage V_{CE} keeping base current I_B constant.

Both the characteristics may be obtained with the help of single circuit arrangement shown in figure 1.113. Potentiometer R_2 is used to vary the collector-to-emitter voltage V_{CE} and potentiometer R_1 is used to vary base-to-emitter voltage V_{BE}. To measure different voltages and currents, d.c. milliammeters and d.c. voltmeters are connected in base and collector circuit of transistor.

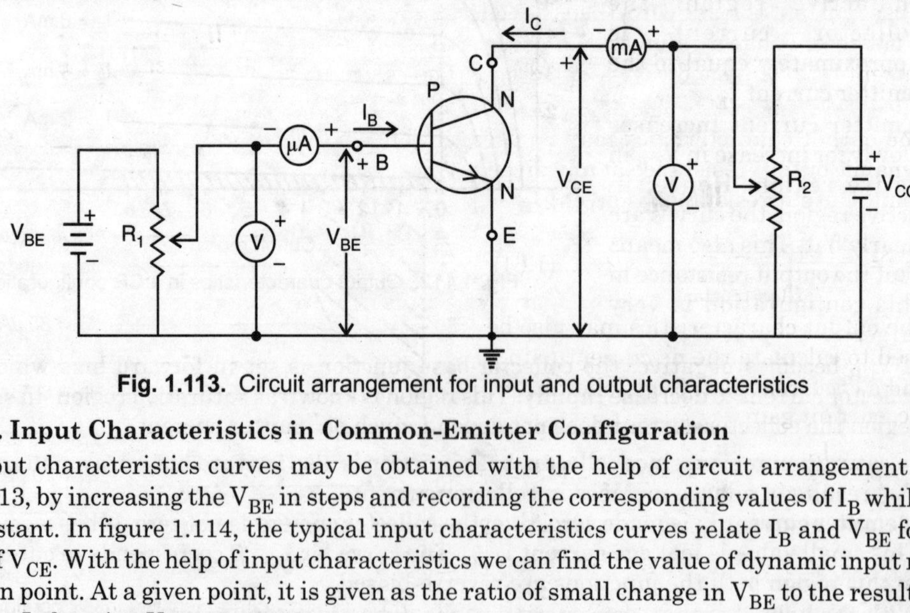

Fig. 1.113. Circuit arrangement for input and output characteristics

1.64.2.1. Input Characteristics in Common-Emitter Configuration

Input characteristics curves may be obtained with the help of circuit arrangement shown in figure 1.13, by increasing the V_{BE} in steps and recording the corresponding values of I_B while keeping V_{CE} constant. In figure 1.114, the typical input characteristics curves relate I_B and V_{BE} for various values of V_{CE}. With the help of input characteristics we can find the value of dynamic input resistance at a given point. At a given point, it is given as the ratio of small change in V_{BE} to the resulting small change in I_B keeping V_{CE} constant.

---

* Draw the circuit diagram of transistor in the CE configuration. Sketch the output characteristics and indicate the active, saturation and cut-off region.

$$r_i = \frac{\Delta V_{BE}}{\Delta I_B}\Bigg|_{V_{CE} = \text{constant.}} \quad ...(1.60)$$

1.64.2.2. Output Characteristics in Common-Emitter Configuration

Output characteristics in CE configuration can be plotted with the help of circuit arrangement shown in figure 1.113. The related output current I_C to the collector-to-emitter voltage for different values of input current I_B.

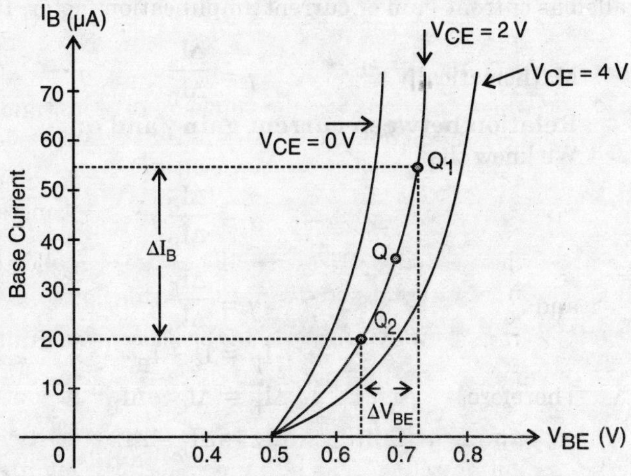

Fig. 1.114. Input characteristics in CE configuration

Figure 1.115 shows the typical output characteristics.

From the output characteristics, we can draw few points:

(i) In active region, the collector current increases slowly with the increase in V_{CE}. Output characteristics in CE configuration are less horizontal than those in CB configuration. This indicates that output resistance is small in CE configuration than CB configuration.

(ii) As the value of V_{CE} decreases below zero, I_C decreases rapidly. This occurs due to reason that when V_{CE} decreases below zero, collector-base junction becomes forward-biased. This means that now both the junctions are forward-biased. This is called saturation region. In saturation region, the collector current I_C does not depend upon the input current I_B.

(iii) If base current is made zero, then the collector current is not zero. It has a value known as reverse leakage current I_{CE0}. This condition is called the *cut-off region of transistor*.

DO YOU KNOW?

When a transistor is saturated, further increase in base current produce no further increase in collector current.

(iv) The output characteristics may be used to find the dynamic output resistance at any given point. Output resistance is given

$$r_0 = \frac{\Delta V_{CE}}{\Delta I_C}\Bigg|_{I_B = \text{constant}} \quad ...(1.61)$$

(v) The output characteristics may also be used to calculate the dc current gain β and a.c. current gain β_0 as follows:

d.c. current gain

$$\beta = \frac{I_C}{I_B} \quad ...(1.62)$$

a.c. current gain

$$\beta_0 = \frac{\Delta I_C}{\Delta I_B}\Bigg|_{V_{CE} = \text{constant}} \quad ...(1.63a)$$

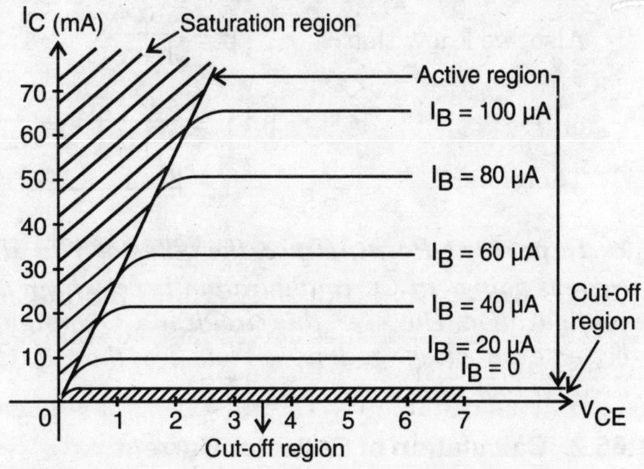

Fig. 1.115. Output characteristics in CE configuration

1.65. COMMON COLLECTOR (CC) CONFIGURATION

Figure 1.116 shows the Common Collector (CC) configuration for NPN transistor. In this configuration, the input is applied between base and collector whereas the output signal is taken across the emitter and collector. This means that the collector of the transistor is common to both input and output circuits and hence, the name common-collector configuration.

1.65.1. Current Gain or Current Amplification Factor in CC Configuration

In CC configuration, the ratio of change in emitter current to the change in base current is called as current-gain or current amplification factor. It is denoted by letter γ (Gamma).

Mathematically, $$\gamma = \frac{\Delta I_E}{\Delta I_B}$$

Relation between current gain γ and α:

We know that

$$\alpha = \frac{\Delta I_C}{\Delta I_E}$$

and $$\gamma = \frac{\Delta I_E}{\Delta I_B}$$

$$I_E = I_C + I_B$$

Therefore, $$\Delta I_E = \Delta I_C + \Delta I_B$$

or $$\Delta I_B = \Delta I_E - \Delta I_C$$

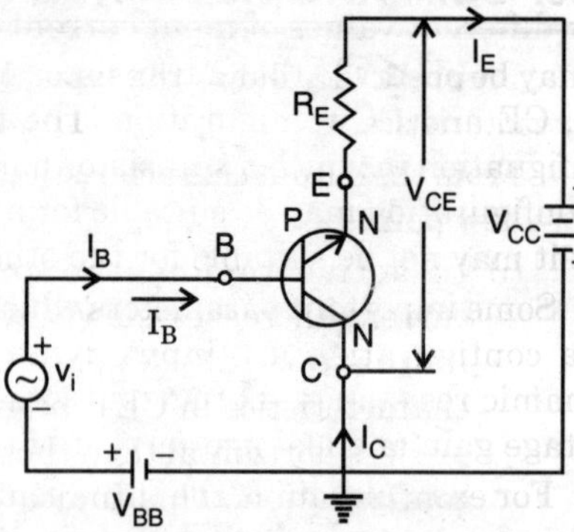

Fig. 1.116. Common-collector circuit for NPN transistor

Putting this value of ΔI_B in equation, $\gamma = \frac{\Delta I_E}{\Delta I_B}$, we obtain

$$\gamma = \frac{\Delta I_E}{\Delta I_E - \Delta I_C} = \frac{\Delta I_E / \Delta I_E}{\Delta I_E / \Delta I_E - \Delta I_C / \Delta I_E} = \frac{1}{1 - \frac{\Delta I_C}{\Delta I_E}}$$

But $$\frac{\Delta I_C}{\Delta I_E} = \alpha$$

Therefore, $$\gamma = \frac{1}{1 - \alpha}$$

Also, we know that $$\beta = \frac{\alpha}{1 - \alpha}$$

or $$\beta + 1 = \frac{\alpha}{1 - \alpha} + 1 = \frac{1}{\alpha - 1}$$

Therefore, $$\gamma = \beta + 1$$

***Important Point:** Hence, the value of γ is almost equal to β. This means that the value of current gain γ in CC configuration is very high and is comparable to the current gain β in CE configuration. However, this circuit in CC configuration is rarely used for amplification purpose because the voltage gain is very low (less than 1). Common-Collector (CC) configuration is mainly used for impedance-matching.*

1.65.2. Calculation of Collector Current

We know that $$I_C = \alpha I_E + I_{CBO}$$

Also, $$I_E = I_C + I_B = (\alpha I_E + I_{CBO}) + I_B$$

or $$I_E - \alpha I_E = I_B + I_{CBO}$$

or $$I_E (1 - \alpha) = I_B + I_{CBO}$$

or $$I_E = \frac{I_B}{1-\alpha} + \frac{I_{CBO}}{1-\alpha}$$

But $$\frac{1}{1-\alpha} = \beta + 1 = \gamma$$

Therefore, $$I_E = I_B(\beta + 1) + I_{CBO}(\beta + 1)$$

1.66. COMPARISON BETWEEN THREE TRANSISTOR CONFIGURATIONS

It may be observed that a transistor may be connected in any one of the three configurations namely CB, CE and CC configuration. The transistor behaves differently in different configuration. The configuation in which a transistor has to be connected depends upon the requirement of application. A configuration may be suitable for a particular application, but it may not be suitable for the other.

> **DO YOU KNOW?**
>
> All transistors have an active region and a breakdown region. The active region is the most important because amplification (enlargement) of signal is possible in the active region.

Some important parameters which govern the suitability of a configuration are input dynamic resistance, output dynamic resistance, d.c. current gain, a.c. current gain, a.c. voltage gain and leakage current of the transistor.

For example, out of the three configuration, the common collector configuration has very high input resistance but a small value of voltage gain. Hence, it is best suited for impedance matching. Therefore, for the application of a particular configuration, we must know its characteristics.

Table 1.6. shows the comparison of the characteristics of three configurations

Table 1.6. Comparison Between the Three Configurations

S.No.	Characteristics	CB configuration	CE configuration	CC configuration
1.	Input resistance	Very low (40 Ω)	Low (50 KΩ)	Very high (750 KΩ)
2.	Output resistance	Very high (1M Ω)	High (10 KΩ)	Low (50 Ω)
3.	Current gain	Less than unity (0.98)	High (100)	High (100)
4.	Voltage gain	Small (150)	High (500)	Less than unity
5.	Leakage current	Very small (5 μA for Ge and 1 μA for Si)	Very large (500 μA for Ge and 20 μA for Si)	Very large (500 μA for Ge and 20 μA for Si)
6.	Applications	For high-frequency applications	For audio freq. applications	For impedance matching

1.67. CE CONFIGURATION IS MOST WIDELY USED IN AMPLIFIER CIRCUITS

(RGTU, Bhopal, Sem. Exam., 2007-08)

For an ideal amplifier circuit, the input impedance should be as high as possible and the output impedance should be as low as possible. Although the input impedance for CC configuration is highest but gain for CC configuration is less than unity. So generally this is not preferred for amplifier circuits.

The input impedance of CE configuration is much higher than CB configuration since

$$r_i \text{ for CB} = \left.\frac{\Delta v_{EB}}{\Delta i_E}\right|_{V_{CB}=\text{constant}} \quad ...(1.64)$$

$$r_i \text{ for CE} = \left.\frac{\Delta v_{BE}}{\Delta i_B}\right|_{V_{CE}=\text{constant}} \quad ...(1.65)$$

The numerators are same in above two equations. But the denominator of the first is of the order of few mA, whereas that of the second is of the order of a few μA. Hence r_i is much higher for CE than CB. Similarly output impedance is much lower for CE than CB since the output characteristic curves are more horizontal for CB than CE.

Except this, the current gain of CB configuration is less than unity. It is generally 0.98. The current gain of CE configuration is quite high. It is generally 100 and may be as high as 250. Such high current gain in the CE configuration makes it possible to have quite high voltage gain as well as high power gain.

Hence, for the above reasons, CE configuration is most popular in amplifier circuits.

1.68. BASIC COMMON EMITTER AMPLIFIER AND D.C. LOAD LINE

Figure 1.117 shows an amplifier circuit in common emitter configuration using NPN transistor. The battery V_{BB} is used to forward bias the emitter-base junction and battery V_{CC} is used to reverse bias the collector-base junction. The resistor R_B is connected to limit the input or base current. The resistor R_C acts as a load resistance. The amplified a.c. output voltage appears across load resistor R_C.

DO YOU KNOW?

Centering the Q-point on a transistor load line is important because it allows for the maximum ac output voltage from the amplifier. Centering the Q-point on the dc load line is sometimes referred to as mid point bias.

The input a.c. signal to be amplified is v_i. This is applied to the base through capacitor C_1. The capacitors C_1 and C_2 are called coupling capacitors. The coupling capacitor C_1 blocks d.c. Hence, d.c. current flows only through R_B and is prevented from reaching the input signal source. The coupling capacitor C_2 also blocks d.c. from reaching the output terminals. Therefore, at the output terminals only amplified output a.c. signal voltage V_o appears.

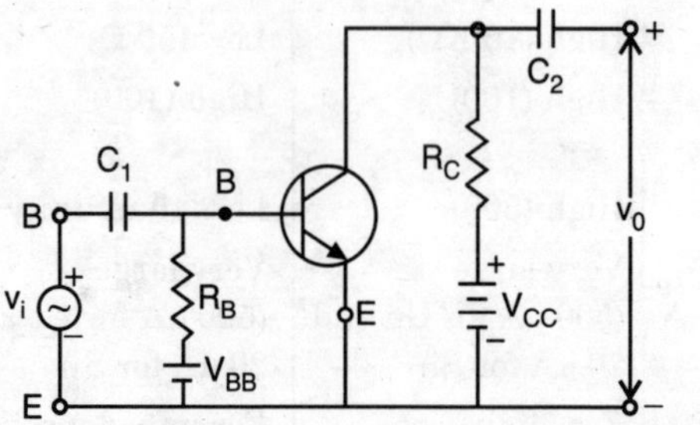

Fig. 1.117. Basic common emitter amplifier

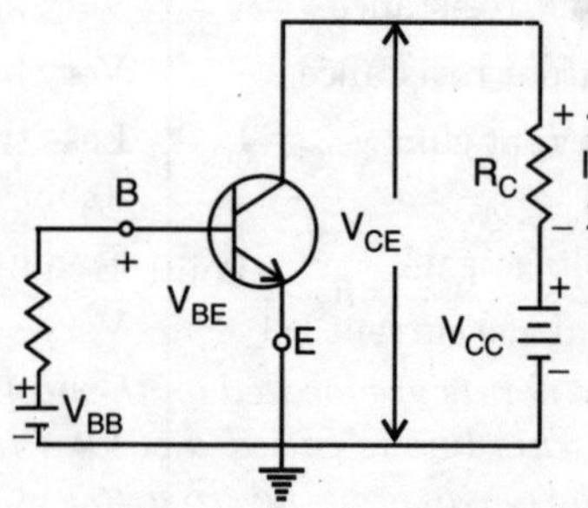

Fig. 1.118. Basic common emitter (D.C. equivalent circuit)

1.68.1. D.C. Load Line

(RGTU, Bhopal, Sem. Exam., 2007-08)

With the help of d.c. load line, we can easily analyse the performance of an amplifier circuit. In this section, we will learn how to draw a d.c. load line. For this, first we will find the d.c. equivalent circuit of the amplifier circuit shown in figure 1.118. This condition of having no input a.c. signal is also known as quiescent condition. Figure 1.118 shows the d.c. equivalent circuit of one shown in figure 1.117. This circuit is obtained by opening the two capacitors and make the value of signal source equal to zero. For d.c. the capacitor works as an open circuit.

Applying KVL to the collector portion of the circuit, we have

$$I_C R_C + V_{CE} = V_{CC}$$

or $$I_C R_C = -V_{CE} + V_{CC}$$

or $$I_C = \left(-\frac{1}{R_C}\right) V_{CE} + \frac{V_{CC}}{R_C} \quad ...(1.66)$$

The above equation is in the form of

$$Y = mx + c \quad ...(1.67)$$

The equation (1.67) is an equation of straight line.

Comparing equations (1.66) and (1.67) slope of straight line, we have

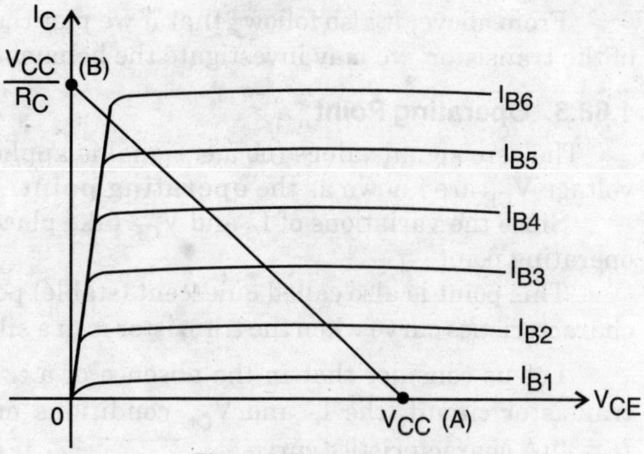

Fig. 1.119. dc load line

$$m = -\frac{1}{R_C}$$

Therefore, if equation (1.32) is plotted on the output characteristics curves, we will get a straight line having slope equal to $\left(-\frac{1}{R_C}\right)$ and intercept on vertical axis equal to $\left(\frac{V_{CC}}{R_C}\right)$.

This line plotted on the output characteristics is called d.c. load line.

To plot the d.c. load line, we use the concept that on V_{CE} axis, the I_C is zero and on I_C axis, the V_{CE} is zero.

Substituting these conditions in the equation (1.66), we get

$$V_{CC} = V_{CE} + I_C R_C$$

when $I_C = 0 \Rightarrow V_{CE} = V_{CC}$

Point A ≡ (V_{CC}, 0)

when $V_{CE} = 0 \Rightarrow V_{CC} = I_C R_C$

or $$I_C = \frac{V_{CC}}{R_C}, \text{ Point B} \equiv \left(0, \frac{V_{CC}}{R_C}\right)$$

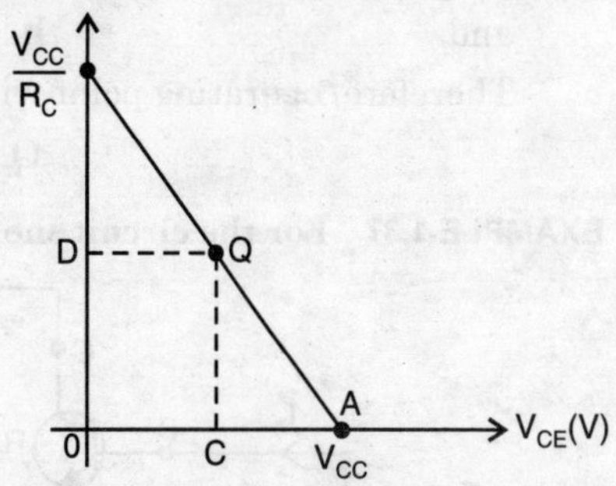

Fig. 1.120.

If two points are located on the output characteristics and joined by a straight line, we get a d.c. load line. This line is called d.c. load line because its slope depends upon the value of resistor R_C which is the d.c. load of the circuit.

1.68.2. Importance of d.c. Load Line

(RGTU, Bhopal, Sem. Exam., 2006-07)

The collector or output current I_C and collector to emitter voltage V_{CE} conditions in the transistor circuit are represented by a point on the output characteristics curve. This same information may be obtained from the d.c. load line.

Hence, when I_C is maximum $\left(I_C = \frac{V_{CC}}{R_C}\right)$ then V_{CE} will be zero ($V_{CE} = 0$) and when $I_C = 0$, then V_{CE} is maximum ($V_{CE} = V_{CC}$) as shown in figure 1.120.

For some other value of output or collector current I_C (say OD), the collector-to-emitter voltage V_{CE} will be equal to OD. It means that d.c. load line provides a more convenient and direct solution to the problem.

From above, it also follows that if we plot the d.c. load line on the output characteristics curve of the transistor, we may investigate the behaviour of transistor amplifier.

1.68.3. Operating Point

The zero signal values (no a.c. signal is applied) of collector current I_C and collector-to-emitter voltage V_{CE} are known as the **operating point.**

Since the variations of I_C and V_{CE} take place about this point, therefore, this is known as the operating point.

This point is also called quiescent (stable) point or simply Q-point since it is a point on output characteristics curve when the transistor is in a silent condition (i.e., the absence of a.c. input signal).

Let us consider that in the absence of a.c. input signal, the base current is 4 μA. Then, in transistor circuit, the I_C and V_{CE} conditions must be represented by a point on I_B curve i.e., $I_B = 4\mu A$ characteristics curve.

But at the same time we know that I_C and V_{CE} conditions in transistor circuit must also be represented by a point on the d.c. load line AB. Therefore, it follows that the point Q where the d.c. load line and the characteristics curve intersect will be the only point which satisfies both of these conditions.

Hence, the point Q describes the actual state of circuit in the zero signal conditions and is known as the operating point.

According to figure 1.121, for $I_B = 4$ μA, the zero signal values are

$$V_{CE} = \text{OC volts} = 10 \text{ Volts}$$

and $$I_C = \text{OD mA} = 6 \text{ mA}$$

Therefore, operating point will be

$$Q \equiv (V_{CE}, I_C) \equiv (10 \text{ V}, 6 \text{ mA})$$

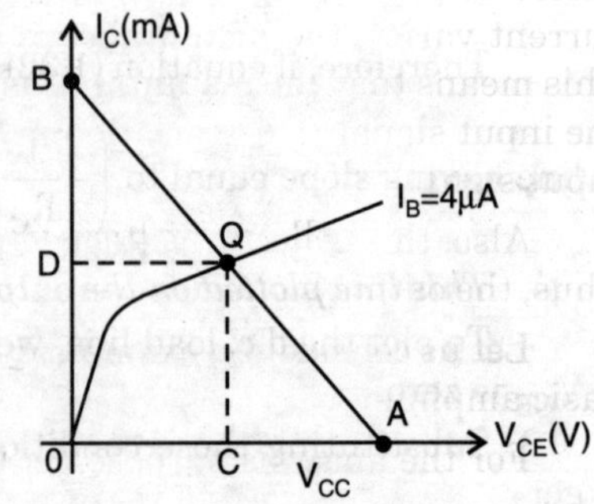

Fig. 1.121.

EXAMPLE 1.37 For the circuit shown in figure 1.122, draw the d.c. load line.

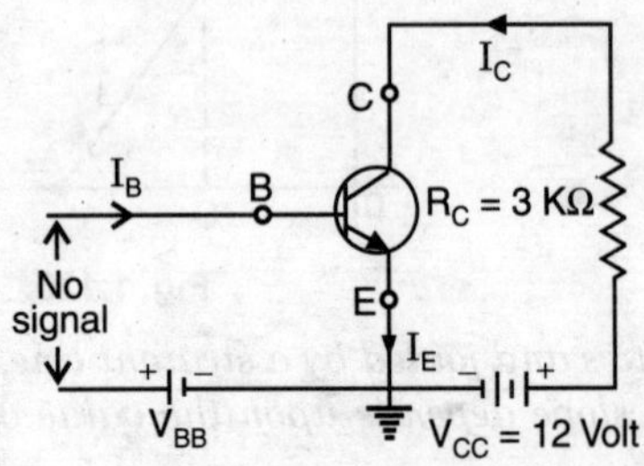

Fig. 1.122.

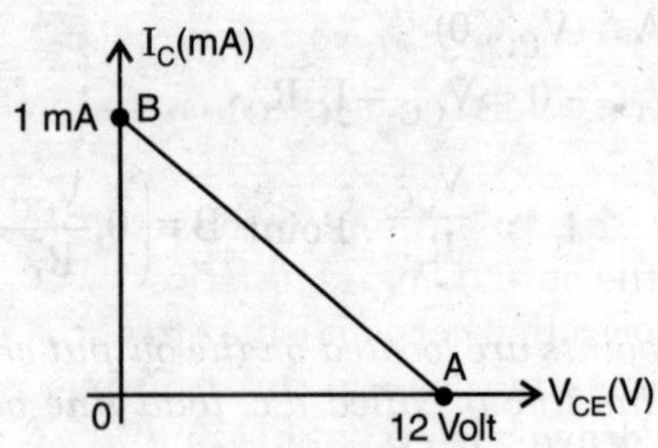

Fig. 1.123.

Solution: Applying KVL in the output portion, we get

$$V_{CC} = I_C R_C + V_{CE}$$

or $$V_{CE} = V_{CC} - I_C \cdot R_C \quad \text{...(i)}$$

Now, when $I_C = 0$

From equation (i) $$V_{CE} = V_{CC} - 0 = V_{CC}$$

$$V_{CE} = 12 \text{ Volt}$$

This locates the point A of the d.c. load line on the collector-emitter voltage (V_{CE}) axis.

Similarly, when $V_{CE} = 0$.

From equation (i) $$0 = V_{CC} - I_C \cdot R_C$$

or $$I_C R_C = V_{CC}$$

or $$I_C = \frac{V_{CC}}{R_C} = \frac{12\,V}{3\,K\Omega} = 4\ mA$$

This locates the point B of the d.c. load line on the collector current (I_C) axis.

Now, by joining these two points A and B, we get the d.c. load line AB as shown in figure 1.123.

1.68.4. Analysis of Amplifier Using d.c. Load Line

As discussed earlier, a transistor may be used to amplify ac signals after fixing its operating point suitably.

In last article, we learnt how to fix the Q-point on the output characteristics of a transistor. In fact, the Q-point must lie in the middle portion on the output characteristics of a transistor. This helps the transistor to amplify a.c signals faithfully, i.e., without producing distortion in waveshape of input signal.

In quiescent condition, the base current is fixed at a constant d.c. value. It is found from the Q-point. After this, when we apply an a.c signal at the input of the amplifier circuit, the base voltage varies as per the signal voltage v_s. Due to this, the base current will also vary. Now, as the base current varies, the instantaneous operating point of the transistor moves along the d.c. load line. This means that the instantaneous values of the collector current and voltage also vary according to the input signal. This variation in collector voltage is several times larger than the variation of the input signal.

Also, this collector-voltage variation reaches the output terminals through coupling capacitor. Thus, the output signal is several times larger than the input signal.

Let us consider an example. Figure 1.124 shows the circuit of a basic amplifier.

For the analysis purpose, let us assume that

$$V_{CC} = 12\ V,\ R_C = 1\ k\Omega,$$
$$V_{BB} = 10.7\ V \text{ and } R_B = 200\ k\Omega$$

Here the d.c. base current is found to be 50 μA and the collector d.c. current and collector d.c. voltage are 6 mA and 6 V respectively.

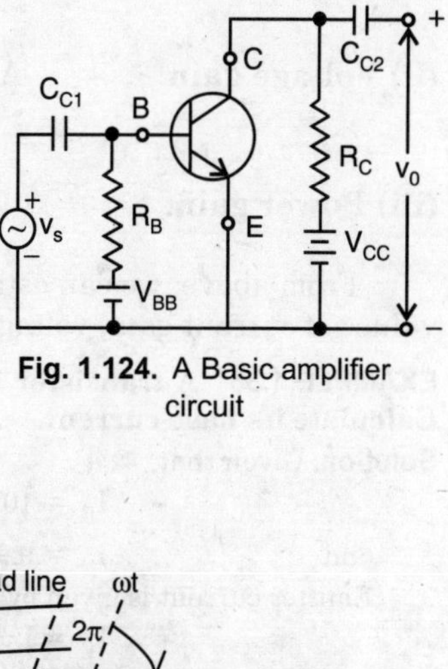

Fig. 1.124. A Basic amplifier circuit

Now, let us apply a small a.c. signal voltage of 7 mV at the input.

To show the variations in base current I_B on the output characteristics of the transistor, a line perpendicular to the d.c. load line and passing through the Q-point is drawn. We take this line as ωt-axis and then variations in I_B are plotted.

As the instantaneous operating point moves along the d.c. load line between the points A and B, both the collector-current and collector voltage would vary.

The current i_C varies between the points A_1 (I_C = 7.3 mA) and B_1 (I_C = 4.8 mA). These variations are shown on the left side of the output characteristics. The voltage v_{CE} varies between points A_2 (v_{CE} = 4.9 V) and B_2 (v_{CE} = 7.1 V).

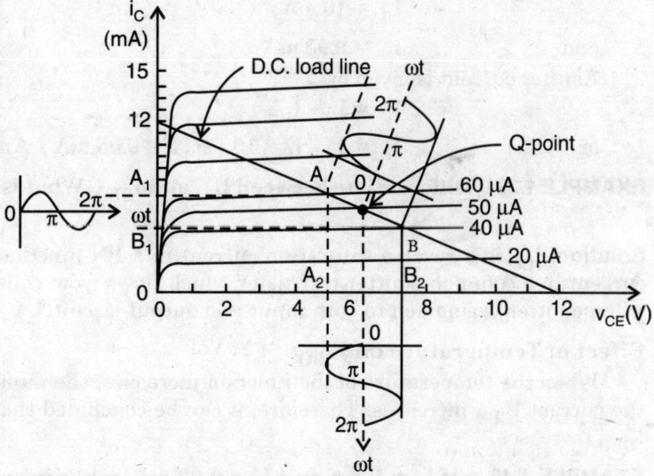

Fig. 1.125. Illustration of variation in base current which produces variation in collector current and voltage in a CE amplifier.

Also the collector-voltage variations are shown at the bottom of the output characteristics.

This voltage would have about 20 mV peak-to-peak variations. Now if the input dynamic resistance r_i of the transistor is taken to be 1 kΩ, then the input voltage variation would produce a peak-to-peak variation of 20 μA in base current, since

$$\Delta i_B = \frac{20\text{ mV}}{1\text{ K}\Omega}$$

$$\Delta i_B = 20\ \mu\text{A}$$

This variation in base current exists around its quiescent value (i.e., operating point) of 50 μA. Now as the base current varies, the instantaneous operating point moves along the d.c. load line between the points A(I_B = 60 μA) and B(I_B = 40 μA).

Now we can compute the current gain, voltage gain and the power gain of the amplifier on the basis of peak-to-peak variation as under:

(i) Current gain:

$$A_i = \frac{\Delta i_C}{\Delta i_B}$$

Substituting all the values, we get

or $$A_i = \frac{(7.3-4.8)\times 10^{-3}}{(60-40)\times 10^{-6}} = 125$$

(ii) Voltage gain: $$A_v = \frac{\Delta v_{CE}}{\Delta v_{BE}} = \frac{7.1-4.9}{20\times 10^{-3}} = 110$$

(iii) Power gain: $$Q = \frac{\text{Output ac power}}{\text{Input ac power}} = \frac{i_c.v_{ce}}{i_b.v_{be}} = A_i \times A_V = 125\times 110 = 13750$$

From above, we can establish the fact that a common emitter amplifier has sufficiently large values of current gain, voltage gain and the power gain.

EXAMPLE 1.38 A transistor has an emitter current of 10 mA and a collector current of 9.95 mA. Calculate its base current.

Solution: Given that

$$I_E = 10\text{ mA}$$

and $$I_C = 9.95\text{ mA}$$

Emitter current is given by

$$I_E = I_C + I_B$$

or $$I_B = I_E - I_C = 10 - 9.95 = 0.05\text{ mA}$$ **Ans.**

EXAMPLE 1.39 Differentiate between I_{CO} and I_{CBO}. What is the effect of temperature on I_{CBO}?

(RGTU, Bhopal, Sem. Exam., 2008-09)

Solution: I_{CO} is a reverse saturation current in a PN-junction when it is reverse biased. I_{CBO} is a leakage current in a common emitter transistor which flows from collector to base due to minority charge carriers, when emitter is common to both input and output circuit.

Effect of Temperature on I_{CBO}

When the temperature of the junction increases, the number of minority carriers increases and hence, the current I_{CBO} increases. Therefore, it can be concluded that on increasing the temperature, the current I_{CBO} increases.

EXAMPLE 1.40. If I_C = 5 mA and I_B = 0.02 mA, what is current gain?

(RGTU, Bhopal, Sem. Exam., 2007-08)

Solution: Given $I_C = 5$ mA

and $I_B = 0.02$ mA

Current gain $\beta = \frac{I_C}{I_B} = \frac{5 \times 10^{-3}}{0.02 \times 10^{-3}} = 250$ **Ans.**

1.69. INTRODUCTION TO FET

We have already discussed that a bipolar junction transistor (BJT) is a current controlled device. Due to this reason, the output characteristics of BJT are controlled by the base current and not the base voltage. Another type of transistor, known as the field-effect transistor (FET), is a voltage-controlled device. This means that the output characteristics of FET are controlled by the applied input voltage and not by the input current. In addition to this, FET is an unipolar device. This means that the current flowing through it is only due to one type of charge particles, i.e., holes or electrons. BJT, on the other hand, is a bipolar device because holes and electrons both contribute to the flow of current.

In a broad sense, there are two main types of field-effect transistors as under:

(i) Junction Field-Effect Transistors (JFET).

(ii) Metal Oxide Semiconductor Field-Effect Transistor (MOSFET) or Insulated Gate Field-Effect Transistor (IGFET).

Both of these types of FETs may be further sub-divided as under:

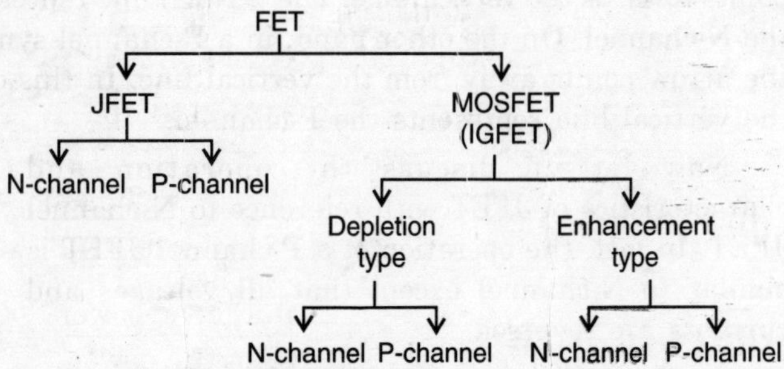

Fig. 1.126. Different types of FETs.

Now, let us discuss the construction, operation and characteristics of all the above-mentioned FET in the sections to follow.

1.70. JUNCTION FIELD-EFFECT TRANSISTOR (JFET)

As shown in figure 1.126, the junction field-effect transistor (JFET) may be divided depending on their structure into the following two categories:

(i) N-channel JFET.

(ii) P-channel JFET.

Figure 1.127 shows the basic construction of an N-channel JFET. It simply consists of an N-type semiconductor bar with two P-type heavily doped regions diffused on opposite sides of its middle part. The P-type regions form two PN-junctions. The space between the junctions (i.e., N-type regions) is known as a **channel**. Both the P-type regions are connected internally and a single wire is taken out in the form of a terminal known as the gate (G). The electrical connections (known as ohmic contacts) are made to both ends of the N-type semiconductor and are taken out in the form of two terminals known as drain (D) and source (S). The drain (D) is a terminal through which electrons leave the semiconductor bar and source (S) is a terminal through which the electrons enter the semiconductor.

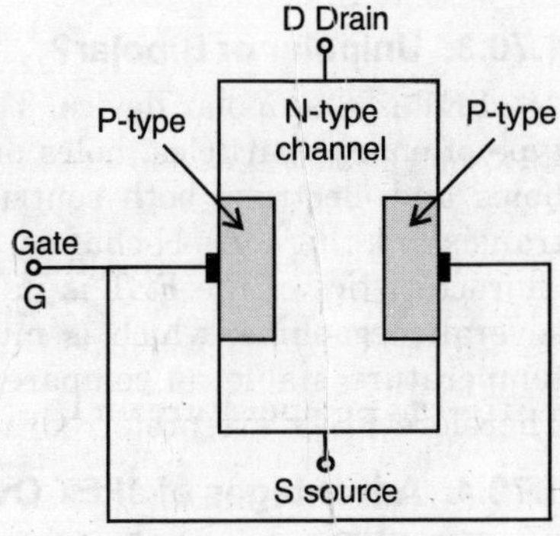

Fig. 1.127. A N-channel JFET

Whenever a voltage is applied across the drain and source terminals, a current flows through the N-channel. The current consists of only one type of carriers (i.e., electrons). Therefore, the field-effect transistor (FET) is known as an unipolar device. In fact, this quality distinguishes an FET from a BJT (i.e., a bipolar Junction transistor) where the current consists of the flow of both the electrons and holes.

Figure 1.128 shows a P-channel JFET. Its construction is similar to that of N-channel JFET, except that it consists of a P-channel and N-type junctions. The current carriers in JFET are the holes which flow through the P-type channel.

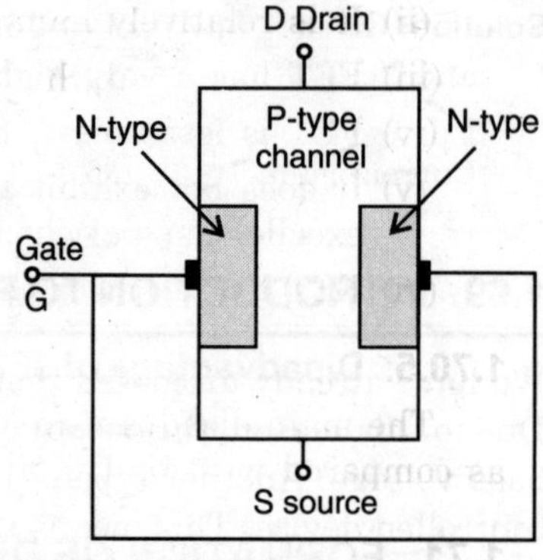

Fig. 1.128. A P-channel JFET

1.70.1. Symbol for JFET

Figure 1.129(a) shows the schematic symbol for an N-channel JFET. Similarly, figure 1.129(b) shows the schematic symbol for a P-channel JFET.

It may be noted that in an N-channel JFET, the arrow points towards the vertical line. The vertical line represents the N-channel. On the other hand, in a P-channel symbol, the arrow points away from the vertical line. In this case, the vertical line represents the P-channel.

> **DO YOU KNOW?**
>
> Because of the extremely high input impedance of a JFET, the input current is generally assumed to be 0 μA, and the current gain of a JFET amplifier is an undefined quantity.

Now let us discuss the operation and characteristics of JFET with reference to N-channel JFET. In fact, the operation of a P-channel JFET is similar to N-channel except that all voltages and currents are reversed.

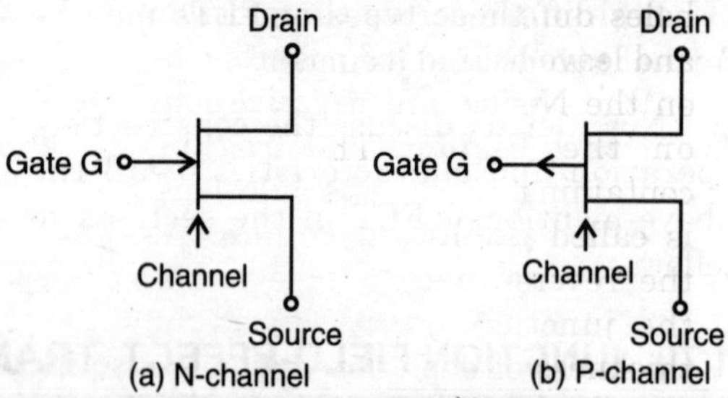

Fig. 1.129. Symbols for JFET

1.70.2. Voltage Controlled or Current Controlled?

The voltage applied between gate and source (V_{GS}) controls the drain current I_D. Therefore, FET is a voltage controlled device. The transistor on the contrary is a current controlled device because the output current I_C is controlled by varying the base current I_B. The name field effect is derived from the fact that the current flow in FET is controlled by an electric field set by an external voltage.

1.70.3. Unipolar or Bipolar?

(RGTU, Bhopal, Sem. Exam., 2006-07)

FET is a unipolar device. This means that the current flowing through it is only due to one type of charge particles, holes or electrons. Transistor, on the other hand, is a bipolar device as holes and electrons both contribute to the flow of current. Just as there are NPN and PNP transistors, there are N-channel and P-channel field effect transistors. One of the most important characteristics of the FET is its very high input impedance. Typically, it is in the range of 1 to several megaohms, which is much higher than the input resistance of a BJT. FETs are more temperature stable as compared to the BJT and it requires less space than that for a BJT. Therefore FETs are preferred in integrated circuits.

1.70.4. Advantages of JFET Over BJT

The FET enjoys several advantages over the conventional bipolar junction transistor (BJT). Some of them may be listed as under :

(i) It is a unipolar device so its operation depends on the flow of majority carriers only.

(ii) It is relatively immune to radiation.

(iii) FET has a very high input resistance (typically few megaohms)

(iv) FET is less noisy

(v) It does not exhibit any offset voltage at zero drain current hence it can be used as an excellent signal chopper.

(vi) FET has a better thermal stability.

1.70.5. Disadvantage of JFET

The main disadvantage of the JFET is that it has a relatively small gain-bandwidth product as compared with that of a conventional transistor (i.e., BJT).

1.71. FORMATION OF DEPLETION REGION IN JFET (RGTU, Bhopal, Sem. Exam., 2006-07)

Let us consider an N-channel JFET as shown in figure 1.130 (a). Here the P-type gate and N-type channel constitute a PN-junction. This PN-junction is always reverse-biased in JFET operation. The reverse bias is applied by a voltage V_{GG} connected between the gate and source terminals as shown in figure 1.130 (b). Here, it can be noted that the positive terminal of the voltage V_{GG} is connected to source (S) and negative terminal to gate (G).

Whenever a PN-junction is reverse-biased, the electrons and holes diffuse across the junction and leave behind the positive ions on the N-side and negative ions on the P-side. The region, containing these immobile ions, is called **depletion region.** As the reverse bias voltage across the junction is increased, the thickness of depletion region is also increased. If both the P-side and N-side of the junctions are equally doped, the depletion region will extend equally in both the regions.

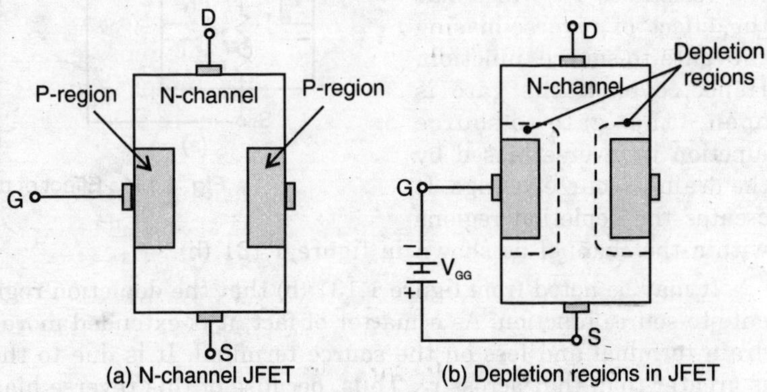

Fig. 1.130.

However, if one side of the junction is heavily doped as compared to the other side, the depletion region extends more into the region of lower doping.

A similar action takes place in JFET. Because the P-region of an N-channel JFET is heavily doped as compared to the N-channel, the depletion region extends less into the P-region and deeper into the N-channel as depicted in figure 1.130 (b). Incidentally, when there is no applied voltage between the drain and the source, the depletion region is symmetrical around the junction. The conductivity of depletion region is zero since there are no mobile charge carriers in this region. Therefore, the effective width of the N-channel is reduced. It further reduces with the increased reverse bias voltage applied across the gate and source terminals of the JFET.

The reverse-bias across the gate-source junction of a

> **DO YOU KNOW?**
>
> In a transistor, the emitter-base depletion layer is narrower than the collector-base depletion layer. The reason can be attributed to the different doping levels of the emitter and collector regions. With much heavier doping in the emitter region, the penetration into the n material is minimal because of the availability of many more free electrons. However, on the collector side, fewer free electrons are available and the depletion layer must penetrate more deeply in order to set-up the barrier potential.

JFET can also be achieved simply by applying a voltage across the drain and source terminals as shown in figure 1.131 (a)

It may be noted that the drain (D) is connected to the positive terminal of the d.c. supply (V_{DD}) and source (S) is connected to the negative terminal.

Now, let us discuss as to how the drain-to-source voltage produces a reverse-bias across the gate-source junction in the same way as the gate-to-source voltage. In order to do so, we represent the channel resistances r_a and r_b as shown in figure 1.131 (a). These resistances have been shown variable since their values depend on the magnitude of the drain-to-source voltage (V_{DS}) and gate-to-source voltage (V_{GS}). In the presence of positive supply voltage V_{DD} (with gate open) the electrons flow from source to drain through the N-channel and constitutes a current known as **drain current (I_D)**. The conventional direction of the drain current is indicated from drain-to-source through the device. The drain current causes a voltage drop across the resistance r_b, which has the effect of reverse-biasing the gate-to-source junction. Hence, even if the gate is open, the gate-to-source junction is reverse biased by the drain-to-source voltage. It creates the depletion regions within the channel as shown in figure 1.131 (b).

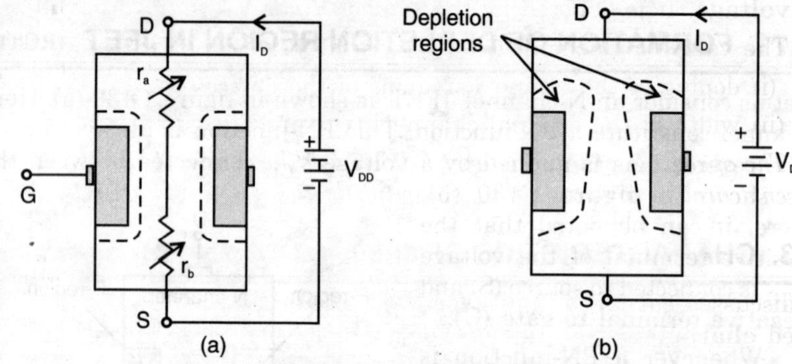

Fig. 1.131. Effect of drain-to-source supply

It may be noted from figure 1.131 (b) that the depletion region is not symmetrical around the gate-to-source junction. As a matter of fact, it is extended more deeper into the channel near the drain terminal and less on the source terminal. It is due to the fact that voltage drop across r_a is greater than that across r_b. Thus, because of this reverse bias voltage is higher near the drain end of the channel as compared to the source end.

1.72. OPERATION OF JFET*

(RGTU, Bhopal, Sem. Exam., 2006-07)

In last article we have discussed the application of negative gate voltage or positive drain voltage, with respect to source, reverse biases the gate-source junction of an N-channel JFET. The effect of reverse-bias voltage is to form the depletion regions within the channel. When a voltage is applied between the drain and source with a dc supply voltage (V_{DD}), the electrons tend to flow from source to drain side through the narrow channel existing between the depletion regions. This constitutes the drain current (I_D) and its conventional direction is indicated from drain-to-source. The value of this drain

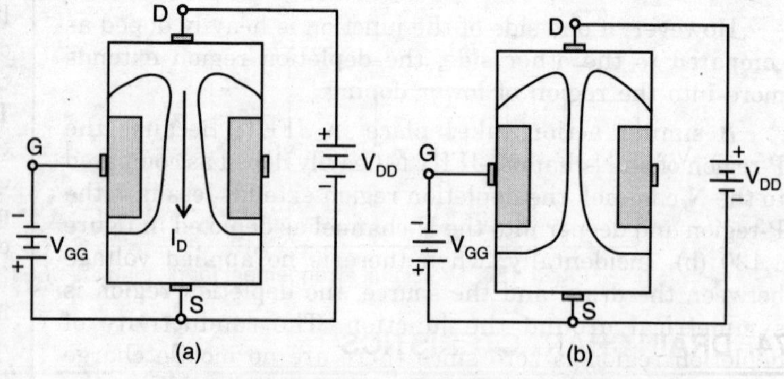

Fig. 1.132. Illustration of operation of JFET

* Explain the working of a JFET along with its construction with neat sketch. Also draw its transfer characteristic curve with explanation. **(RGTU, Bhopal, Sem. Exam., 2006-07)**

current is maximum when no external voltage is applied between the gate and the source and is designated by the symbol I_{DSS}.

When the gate-to-source voltage (V_{GS}), is applied by a dc supply (V_{GG}) and increased above zero as depicted in figure 1.132 (a), the reverse-bias voltage across the gate-source junction is increased. Because of this, the depletion regions are widened. This reduces the effective width of the channel and hence controls the flow of drain current through the channel.

When the gate-to-source voltage (V_{GS}) is further increased, a stage is reached at which two depletion regions touch each other as shown in figure 1.132 (b). At this gate-to-source voltage, the channel is completely blocked or pinched off and drain current is reduced to zero. The gate-to-source voltage (V_{GS}) at which the drain current is zero (or completely cut-off) is known as **pinch off voltage**. It is denoted by the symbol V_P or V_{GS} (off).

The value of pinch-off voltage V_P is negative for N-channel JFET. It depends upon

(i) doping of the N and P regions of the device, and

(ii) width of the original channel structure.

The operation of the P-channel JFET is exactly similar to the N-channel JFET except that the current carriers are holes and polarities of the dc supply voltage V_{GG} and V_{DD} are reversed.

1.73. CHARACTERISTICS OF JFET

(RGTU, Bhopal, Sem. Exam., 2005-06)

As discussed earlier, a family (or a set) of curves which relate device currents and voltages are called **characteristic curves**. For a JFET, there are two important characteristics as under:

(i) V-I or drain characteristics

These curves give relationship between the drain current (I_D) and drain-to-source voltage (V_{DS}) for different values of gate-to-source voltage (V_{GS}).

(ii) Transfer characteristics

These curves give relationship between drain current (I_D) and gate-to-source voltage (V_{GS}) for different values of drain-to-source voltage (V_{DS}).

The drain and transfer characteristics of JFET can be obtained by using an N-channel JFET connected in the common source mode as shown in figure 1.133. Here, the potentiometers R_1 and R_2 are used to vary the voltages V_{GS} and V_{DS} respectively. The voltages V_{DS} and V_{GS} can be measured by the voltmeters connected across the JFET terminals. The drain current (I_D) can be measured by the milli-ammeter connected in series with the JFET and the supply voltage V_{DD}.

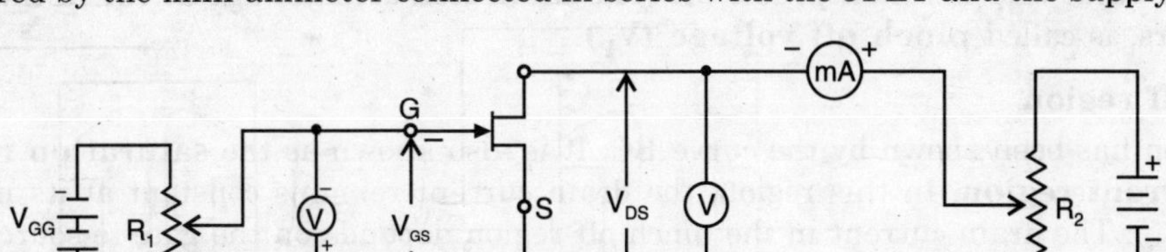

Fig. 1.133. The circuit arrangement for plotting the JFET characteristics

1.74. DRAIN CHARACTERISTICS

The drain characteristics curves may be obtained by using the circuit arrangement shown in figure 1.133. Firstly, we adjust the gate-to-source voltage (V_{GS}) to zero volt. After this, we increase the drain-to-source voltage (V_{DS}) in small suitable steps and note the corresponding values of drain current (I_D) at each step. Now, if we plot a graph taking drain-to-source voltage along the horizontal axis and drain current along the vertical axis, then we shall obtain a curve marked V_{GS}

= 0 as shown in figure 1.134. A similar procedure can be followed to obtain curves for different values of gate-to source voltage i.e., V_{GS} = 1, 2, 3 and 4 volt etc.

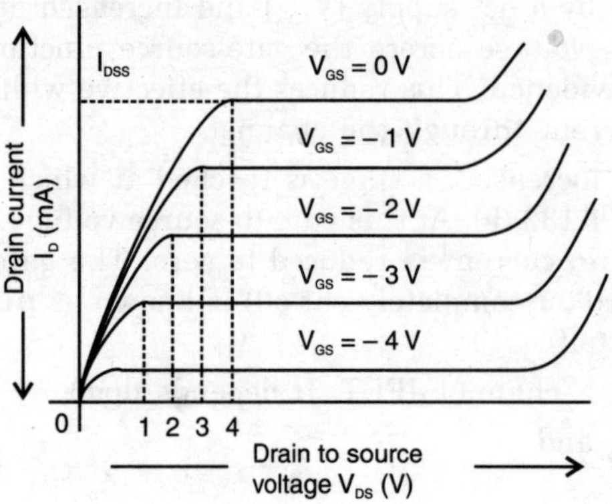

Fig. 1.134. The drain characteristics of JFET

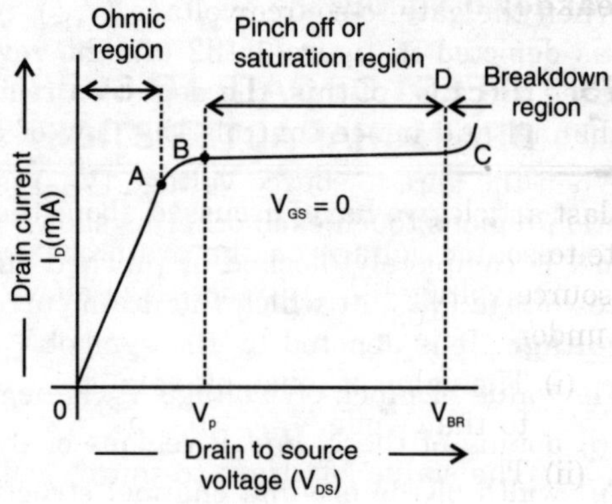

Fig. 1.135. The drain characteristics with V_{GS} = 0 Volt

Now, in order to explain the typical shape of drain characteristics, let us select the curve with V_{GS} = 0 volt as shown in figure 1.135. The curve can be sub-divided into the following regions:

(i) Ohmic region

Ohmic region has been shown as a curve OA in figure 1.135. In this region, the drain current increases linearly with the increase in drain-to-source voltage obeying Ohm's law. The linear increase in drain current is due to the fact that N-type semiconductor bar acts like a simple resistance.

(ii) Curve AB

In this region, the drain current increases at the reverse square law rate with the increase in drain-to-source voltage. It means that drain current increases slowly as compared to that in the ohmic region. It is due to the fact that with the increase in drain-to-source voltage, the drain current increases. This, in turn, increases the reverse-bias voltage across the gate-source junction. As a result of this, the depletion region grows in size thereby reducing the effective width of the channel. At the drain-to-source voltage, corresponding to point B, the channel width is reduced to a minimum value and is called **pinch off**. The drain-to-source voltage at which the channel pinch-off occurs, is called **pinch off voltage (V_P)**.

(iii) Pinch-off region

This region has been shown by the curve BC. It is also known as the **saturation region** or **constant current region**. In this region, the drain current remains constant at its maximum value (i.e., I_{DSS}). The drain current in the pinch off region depends on the gate-to-source voltage and is given by the following relation:

$$I_D = I_{DSS}\left(1 - \frac{V_{GS}}{V_P}\right)^2 \quad ...(1.68)$$

This relation is known as ***Shockley's equation****. The pinch-off region is the normal operating region of JFET, when used as an amplifier.*

(iv) Breakdown region

This region has been shown by the curve CD. In this region the drain current increases rapidly as the drain-to-

DO YOU KNOW?

The pinch off voltage. V_p is the point at which further increases in V_{DS} are offset by a proportional increase in the channel's resistance. This means that if the channel resistance is increasing in direct proportion to V_{DS} above V_P, I_D must remain the same above V_P.

source voltage is also increased. It happens because of the breakdown of gate-to-source junction due to the avalanche effect. The drain-to-source voltage corresponding to point C is known as the **breakdown voltage.**

1.75. EFFECT OF GATE-TO-SOURCE VOLTAGE ON DRAIN CHARACTERISTICS

(RGTU, Bhopal, Sem. Exam., 2006-07)

In last article, we have discussed about the drain characteristics of JFET. The effect of increasing gate-to-source voltage on the drain characteristics can be noted from figure 1.134. As the gate-to-source voltage (V_{GS}) is increased above zero to the 1, 2, 3, and 4 volts, two effects take place as under:

(i) The value of pinch-off voltage is reached at a smaller value of drain current as compared to that when V_{GS} = 0, and

(ii) The value of drain-to-source voltage (V_{DS}) is decreased as compared to that when V_{GS} = 0.

Now, the first effect can be understood from the fact that reverse bias voltage across the gate-to-source junction of JFET is the sum of gate-to-source voltage (V_{GS}) and drain-to-source voltage (V_{DS}). Hence, for fixed value of V_{DS}, the increased value of voltage V_{GS} would increase the reverse-bias voltage across the junction. For example, if V_{GS} of –1 volt is applied between the gate and the source of JFET, then the gate-to-source junction becomes reverse-biased even when the drain current is zero. This causes the depletion regions to be formed within the channel. As the voltage V_{DS} is increased above zero, the drain current increases. This produces voltage drop across channel resistance, which in turn, produces reverse-bias voltage across the gate-source junction. The amount of reverse-bias voltage required to be produced by drain current will obviously be decreased by one volt. In other words, a smaller voltage drop along the channel (i.e., smaller than that when V_{GS} = 0) will increase the depletion regions to the point where they will pinch off the drain current. As a result of this the pinch-off voltage (V_P) is reached at a lower value of drain current than that when V_{GS} = 0.

Also the second effect is caused due to the fact that the increased value of gate-to-source voltage (V_{GS}) keeps adding to the reverse-bias voltage at the junction produced by the drain current. Hence, its lower value is required to reach the breakdown voltage.

It can be noted that with zero gate-to-source voltage the drain current saturates at I_{DSS} value and the characteristic indicates the pinch-off voltage (V_P) equal to 4 volt. However, if V_{GS} of – 1 volt is applied, the channel still requires –4 volt to achieve pinch-off condition. It means that now the channel requires only 3 volt drop (instead of previously 4 volt) from the drain-to-source voltage. This 3 volt drop can be achieved with the lower value of drain current. Similarly when V_{GS} is –2 and –3 volts, pinch-off is achieved with 2 volt and 1 volt drops respectively along the channel. These voltage drops are obtained with further reduced values of the drain current. It can be noted that when V_{GS} is –4 volt (i.e., equal to pinch-off voltage), no channel drop is required and thus the drain current is zero.

1.76. TRANSFER CHARACTERISTICS

(V. Important)

Transfer characteristics are also known as transconductance curves which give us the relationship between drain current (I_D) and gate-to-source voltage (V_{GS}) for a constant value of drain-to-source voltage (V_{DS}). The transfer characteristics can be obtained by using the circuit arrangement shown in figure 1.133. Firstly, we adjust the drain-to-source voltage to some suitable value and increase the gate-to-source voltage in small suitable steps. After this, note the corresponding values of drain current at each step. If we plot a graph with gate-to-source voltage (V_{GS}) along the horizontal axis and the drain current (I_D) along the vertical axis, we shall obtain a curve as shown in figure 1.136. A similar procedure can be used to obtain curves at different values of the gate-to-source voltage (V_{GS}).

It may be observed that the upper end of the curve has been shown by the drain current value equal to I_{DSS}, while the lower end is indicated by a voltage equal to $V_{GS(off)}$ or V_P. It can be noted that the curve is part of a parabola and hence can be expressed by the following expression

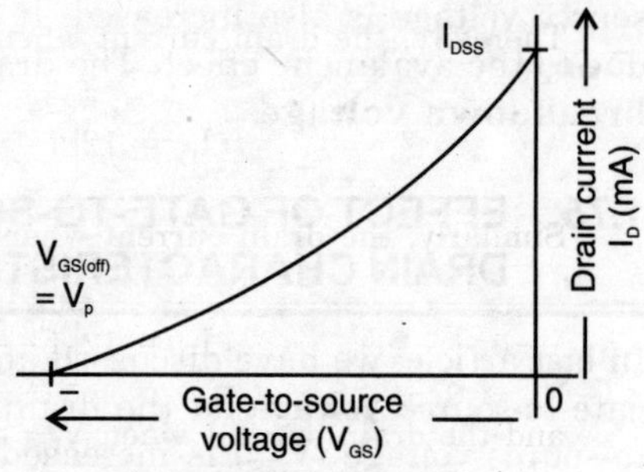

Fig. 1.136. The transfer characteristics of N-channel.

$$I_D = I_{DSS}\left[1-\frac{V_{GS}}{V_{GS(off)}}\right]^2 = I_{DSS}\left(1-\frac{V_{GS}}{V_P}\right)^2$$

There is one important point which must be mentioned at this stage. Most JFET specifications sheet will list more than one value of V_{GS} (off) and I_{DSS}.

DO YOU KNOW?

The transconductance curve of a JFET is unaffected by the circuit or configuration in which the JFET is used.

For example, the specifications sheet for the JFET 2N 5457 lists the following:

$V_{GS\ (off)}$ (minimum) = – 0.5 volt

and I_{DSS} (min) = 1 mA

$V_{GS(off)}$ (maximum) = – 6 volt

and I_{DSS} (max) = 5 mA

In such a situation, we must use the two minimum values to plot one curve and the two maximum values to plot another curve on the same graph.

EXAMPLE 1.41 The data sheet of a certain JFET indicates that I_{DSS} is equal to 15 mA and $V_{GS(off)}$ is equal to –5 volt. Find the drain current for V_{GS} equal to 0 volt, –1 volt and – 4 volt.

Solution: Given that I_{DSS} = 15 mA and, $V_{GS(off)}$ = –5 volt

We know that the drain current is given by

$$I_D = I_{DSS}\left[1-\frac{V_{GS}}{V_{GS(off)}}\right]^2$$

Therefore, drain current when (V_{GS} = 0), will be

$$I_D = 15\left(1-\frac{0}{-5}\right)^2 = 15 \text{ mA} \quad \textbf{Ans.}$$

Similarly, drain current when V_{GS} – 1 V, will be

$$I_D = 15\left(1-\frac{-1}{-5}\right)^2 = 9.6 \text{ mA} \quad \textbf{Ans.}$$

and the drain current when V_{GS} = –4 V, will be

$$I_D = 15\left(1-\frac{-4}{-5}\right)^2 = 0.6 \text{ mA} \quad \textbf{Ans.}$$

EXAMPLE 1.42 A JFET has parameters of $V_{GS(off)}$ equal to –20 volt and I_{DSS} equal to 12 mA. Plot the transconductance curve for the device using V_{GS} values of 0 V, –5 V, –10 V, –15 V, and –20 V.

Solution: Firstly, let us find the value of drain currents for the given values of V_{GS} (0 V , –5 V, –10 V, –15 V and –20 V) as discussed below:

Given that $V_{GS(off)}$ = –20 volt and, I_{DSS} = 12 mA

We know that the drain current is given by

$$I_D = I_{DSS}\left[1-\frac{V_{GS}}{V_{GS(off)}}\right]^2$$

Therefore, the drain current when $V_{GS} = -5$ V, will be

$$I_D = 12\left(1 - \frac{-5}{-20}\right)^2 = 6.75 \text{ mA}$$

Similarly, the drain current when $V_{GS} = -10$ V, will be

$$I_D = 12\left(1 - \frac{-10}{-20}\right)^2 = 3 \text{ mA}$$

and the drain current when $V_{GS} = -15$ V, will be

$$I_D = 12\left(1 - \frac{-15}{-20}\right)^2 = 0.75 \text{ mA}$$

The value of V_{GS} and the corresponding values of I_D obtained above have been shown in Table 1.7.

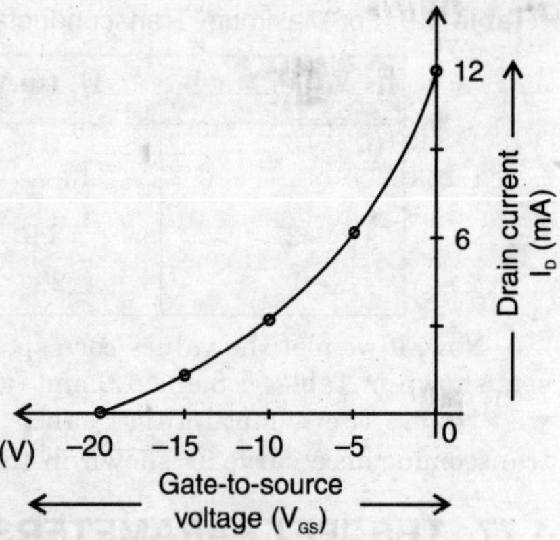

Fig. 1.137.

Table 1.7.

V_{GS} (V)	I_D (mA)
–20	0
–15	0.75
–10	3
–5	6.75
0	12

These points are now plotted and joined together to form a curve shown in figure 1.137.

EXAMPLE 1.43 A 2N 5486 JFET has values of $V_{GS(off)}$ equal to –2 volt to –6 volt and I_{DSS} is equal to 8 mA to 20 mA. Plot the minimum and maximum transconductance curves for the device.

Solution: Given that

$V_{GS\ (off)} = -2$ volt to -6 volt and $I_{DSS} = 8$ mA to 20 mA

Firstly, let us find the values of drain currents for the given values of V_{GS}.

We know that the drain current is given by

$$I_D = I_{DSS}\left[1 - \frac{V_{GS}}{V_{GS(off)}}\right]^2 \quad \text{...(i)}$$

In this equation, substituting the values of $V_{GS} = -0.5$, -1.0 and -1.5 with $V_{GS(off)} = -2.0$ volt, we get the corresponding values of $I_D = 4.5$ mA, 2 mA and 0.5 mA.

With these values, we have the set of V_{GS} versus I_D value as shown in Table 1.8.

Table 1.8. For Minimum Transconductance Curve

V_{GS} (V)	I_D (mA)
– 2.0	0
– 1.5	0.5
– 1.0	2.0
– 0.5	4.5
0	8

Similarly, using the maximum values of $V_{GS\ (off)} = -6$ Volt and $I_{DSS} = 20$ mA, we can use equation (i) again for $V_{GS} = -4$ V and -2 V to obtain the values of I_D in between the end points. The end points of the transcondutance curve are (–6 V, 0 mA) and (0 V, 20 mA). The values of I_D for in between points are 2.2 mA and 9.9 mA. The values obtained have been shown in Table 1.9.

Table 1.9. For Maximum Transconductance Curve

V_{GS}(V)	I_D (mA)
–6	0
–4	2.2
–2	3.9
0	20

Now, if we plot the values corresponding to the sets shown in Tables 1.8 and 1.9 and joining them, we get the maximum and minimum transconductance curve as shown in figure 1.138.

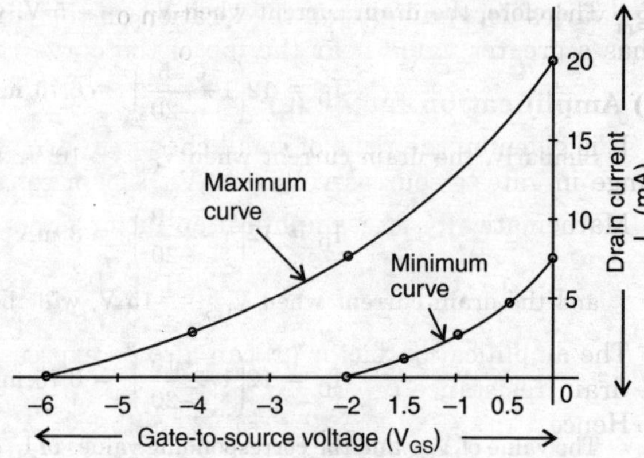

Fig. 1.138.

1.77. THE JFET PARAMETERS

The electrical behaviour of the JFET can be described in terms of certain parameters known as JFET parameters. Such parameters can be obtained from the characteristic curves. The important parameters of JFET have been discussed below:

(i) D.C. drain resistance (R_{DS})

It is also known as the static or ohmic resistance of the channel and is given by the ratio of voltage (V_{DS}) to the drain current (I_D). Mathematically, the d.c. drain resistance is expressed as

$$R_{DS} = \frac{V_{DS}}{I_D}$$

(ii) A.C. drain resistance (r_d)

It is also known as the dynamic drain resistance and is the a.c. resistance between the drain and the source terminal when the JFET is operating in the pinch-off or saturation region. It is given by the ratio of small change in drain-to-source voltage (ΔV_{DS}) to the corresponding change in drain current (ΔI_D) for a constant gate-to-source voltage (V_{GS}).

Mathematically, the a.c. drain resistance is expressed as

$$r_d = \frac{\Delta V_{DS}}{\Delta I_D}$$

It is, sometimes, written as r_{ds}, which indicates that it is the resistance from drain-to-source terminals. Since r_d is the output resistance of JFET, it can also be expressed as an output admittance y_{DS} (i.e., equal to $1/r_d$).

(iii) Transconductance (g_m) (RGTU, Bhopal, Sem. Exam. 2005-06)

It is also known as the forward transconductance or forward transmittance. It is given by the ratio of small change in drain current (ΔI_D) to the corresponding change in gate-to-source voltage (ΔV_{GS}) for a constant drain-to-source voltage (V_{DS}).

Mathematically, the transconductance is expressed as

$$g_m = \frac{\Delta I_D}{\Delta V_{GS}}$$

The value of transconductance (g_m) is expressed in siemens (S) or mhos. We know that transfer characteristic of a JFET is a part of a parabola i.e., it is non-linear. Therefore, the value of g_m varies depending upon the location on the curve as set by the gate-to-source voltage (V_{GS}). It has a greater value near the top of the curve than near the bottom.

(iv) Amplification factor (μ)

It is given by the ratio of small change in drain-to-source voltage (ΔV_{DS}) to the corresponding change in gate-to-sources voltage (ΔV_{GS}) for a constant drain current (I_D).

Mathematically, the amplification factor is expressed as

$$\mu = \frac{\Delta V_{DS}}{\Delta V_{GS}}$$

The amplification factor (μ) can also be expressed as a product of transconductance (g_m) and a.c. drain resistance (r_d).

Hence,

$$\mu = \frac{\Delta V_{DS}}{\Delta V_{GS}} = \frac{\Delta V_{DS}}{\Delta I_D} \times \frac{\Delta I_D}{\Delta V_{GS}} = r_d \times g_m$$

(v) Input resistance (R_i)

We know that a JFET operation with its gate-to-source junction reverse biased. As the reverse current of a junction is, usually, very small, therefore, gate-reverse current of a JFET is also very small.

Because the gate-source junction of JFET is used as an input, therefore, the input resistance of a JFET is very high. This high input resistance is an advantage of JFET over the bipolar transistor.

Mathematically the input resistance is given by

$$R_i = \frac{V_{GS}}{I_{GSS}}$$

where V_{GS} = Gate-to-source voltage

and I_{GSS} = Gate reverse current

It is of the order of nano-amperes.

1.78. THE MATHEMATICAL EXPRESSION FOR TRANSCONDUCTANCE

(RGTU, Bhopal, Sem. Exam., 2007-08)

As discussed earlier, the drain current is expressed by the relation

$$I_D = I_{DSS}\left(1 - \frac{V_{GS}}{V_P}\right)^2$$

Differentiating the above equation on both sides with respect to V_{GS}, we have

$$\frac{dI_D}{dV_{GS}} = 2I_{DSS}\left(1 - \frac{V_{GS}}{V_P}\right) \times \left(\frac{0-1}{V_P}\right)$$

or

$$g_m = -\frac{2I_{DSS}}{V_P}\left(1 - \frac{V_{GS}}{V_P}\right) \quad \text{...(i)}$$

Substituting $V_{GS} = 0$ in the above expression, we get

$$g_m = -\frac{2I_{DSS}}{V_P} = g_{mo}$$

The value of g_m at $V_{GS} = 0$ is the maximum and is equal to $-\frac{2I_{DSS}}{V_P}$. This value of g_m is designated as g_{mo}. The equation (i) can now be rewritten as

$$g_m = g_{mo}\left(1 - \frac{V_{GS}}{V_P}\right)$$

EXAMPLE 1.44 The following information is included on the data sheet for an N-channel JFET:

$$I_{DSS} = 20 \text{ mA} \qquad V_P = -8 \text{ Volt} \quad \text{and,} \quad g_{mo} = 5000 \ \mu s$$

Find the values of the drain current and transconductance at V_{GS} = –4 Volts.

Solution: Given that: I_{DSS} = 20 mA; V_P = –8 Volts; g_{mo}= 5000 μs; and V_{GS} = – 4 Volts

Now, we know that the drain current is given by

$$I_D = I_{DSS}\left(1 - \frac{V_{GS}}{V_P}\right)^2$$

Substituting values, we get

or $$I_D = 20\left(1 - \frac{-4}{-8}\right)^2 = 5 \text{ mA} \quad \textbf{Ans.}$$

Also the transconductance is given by

$$g_m = g_{mo}\left(1 - \frac{V_{GS}}{V_P}\right) = 5000\left(1 - \frac{-4}{-8}\right) = 2500 \ \mu s \quad \textbf{Ans.}$$

EXAMPLE 1.45 The data sheet for a certain enhancement-type MOSFET reveals that $I_{D(on)}$ = 10 mA at V_{GS} = – 12 V and $V_{GS(th)}$ = – 3 V. Is this device P-channel or N-channel? Find the value of I_D, when V_{GS} = – 6 V.

Solution: Given that: $I_{D(on)}$ = 10 mA; V_{GS} = – 12 volts and $V_{GS(th)}$ = – 3 Volts.

Since the value of V_{GS} is negative for the enhancement-type MOSFET, this indicated that the device is P-channel.

We know that the drain current is given by

$$I_D = K\,(V_{GS} - V_{GS(th)})^2 \qquad \text{...(i)}$$

or $$I_{D(on)} = K\,[(V_{GS} - V_{GS(th)}]^2$$

$$10 = K\,[-12 - (-3)]^2 = 81\ K$$

$$K = 10/81 = 0.12 \text{ mA/V}$$

Substituting this value of K and V_{GS} (equal to – 6) in the equation (i),

$$I_D = 0.12\,[-6 - (-3)]^2 = 1.08 \text{ mA} \quad \textbf{Ans.}$$

EXAMPLE 1.46 Show that if $|V_{GS}| << |V_P|$, then drain current can be approximated as

$$I_{DS} = I_{DSS} - g_{mo} V_{GS}$$

Solution: We know that

$$I_{DS} = I_{DSS}\left[1 - \frac{V_{GS}}{V_P}\right]^2$$

Simplifying, we get $$I_{DS} = I_{DSS}\left[1 - 2\frac{V_{GS}}{V_P} + \left(\frac{V_{GS}}{V_P}\right)^2\right]$$

Neglecting the square term, we get

$$I_{DS} = I_{DSS} - \frac{2 I_{DSS} V_{GS}}{V_P}$$

Putting $g_{mo} = \frac{2I_{DSS}}{V_P}$, we get

$$I_{DS} = I_{DSS} + g_{mo} V_{GS} \quad \textbf{Hence Proved.}$$

EXAMPLE 1.47 An N-channel JFET has I_{DSS} = 8 mA and V_P = – 5 V. Find the minimum value of V_{DS} for pinch-off region and the drain current I_{DS}, for V_{GS} = – 2 V in the pinch-off region.

Solution: The minimum value of V_{DS} for pinch-off to occur for V_{GS} = – 2 V is

$$V_{DS(min)} = V_{GS} - V_P = -2 - (-5) = 3 \text{ V}$$

$$I_{DS} = I_{DSS}\left[1 - \frac{V_{GS}}{V_P}\right]^2$$

Substituting values, we get

$$I_{DS} = 8 \times 10^{-3} [1 - (-2)/(-5)]^2$$

Solving, we get I_{DS} = 2.88 mA **Ans.**

EXAMPLE 1.48 V_P = – 2 V, I_{DSS} = 1.55 mA for the circuit in figure 1.139. It is desired to bias the circuit at I_D = 0.8 mA, V_{DD} = 24 V. Find: (i) V_{GS} (ii) g_m.

Solution: (i) We know that

$$V_{GS} = V_P\left[1 - \sqrt{\frac{I_D}{I_{DSS}}}\right]$$

Substituting values, we get

$$V_{GS} = -2\left[1 - \sqrt{\frac{0.8 \times 10^{-3}}{1.65 \times 10^{-3}}}\right]$$

Simplifying, we get V_{GS} = – 0.6074 V

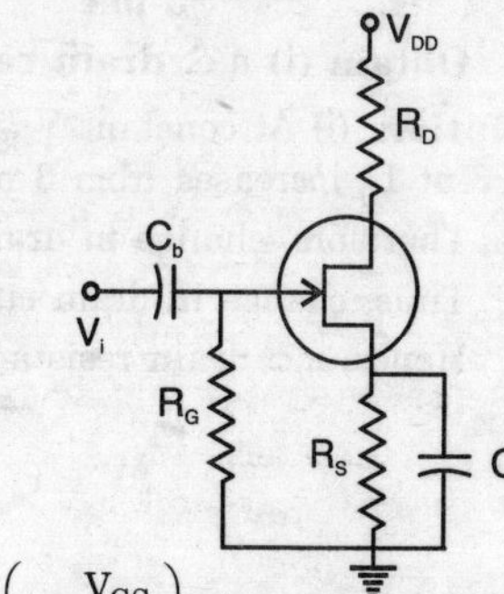

Fig. 1.139.

(ii) $$g_{mo} = -\frac{2I_{DSS}}{V_P} = -\frac{2 \times 1.65}{-2} = 1.65 \text{ ms} = g_{mo}\left(1 - \frac{V_{GS}}{V_P}\right)$$

or $$g_m = 1.65\left(1 - \frac{0.6074}{2}\right) \times 10^{-3} = 1.65 \times 10^{-3} \times 0.6963 = 1.15 \text{ ms} \quad \textbf{Ans.}$$

EXAMPLE 1.49 The pinch-down voltage of a P-channel junction FET is V_P = 5 V and the drain-to-source saturation current I_{DSS} = – 40 mA. The value of drain-source voltage V_{DS} is such that the transistor is operating in the saturated region. The drain current is given as I_D = – 15 mA. Determine the gate-source voltage V_{GS}. (GATE Examination, 2001)

Solution: It is found experimentally that a square-law characteristic closely approximate the drain current in saturation:

$$I_{D(sat)} \approx I_{DSS}\left(1 + \frac{V_{GS}}{V_P}\right)^2$$

Givan that : V_P = 5 V; I_{DSS} = – 40 mA; I_D = – 15 mA

$$-15 \text{ mA} = -40 \text{ mA}\left(1 + \frac{V_{GS}}{5}\right)^2$$

or $$\sqrt{\frac{15}{40}} = 1 + \frac{V_{GS}}{5}$$

Therefore, $$1 + \frac{V_{GS}}{5} = 0.612$$

or $$\frac{V_{GS}}{5} = 0.612 - 1 = -0.3876$$

or $$V_{GS} = 1.938 \text{ volt.} \quad \textbf{Ans.}$$

EXAMPLE 1.50 Determine the value of transconductance of a FET, when the drain current changes from 1 mA to 1.9 mA with a change in gate-source voltage from –3.3 V to –3 V.

Solution: Change in drain current,

$$\Delta I_D = 1.9 - 1.0 = 0.9 \text{ mA}$$

Change in gate-source voltage,

$$\Delta V_{GS} = 3.3 - 3.0 = 0.3 \text{ V}$$

Now, transconductance is given by

$$g_m = \frac{\Delta I_D}{\Delta V_{GS}} = \frac{0.9 \text{ mA}}{0.3 \text{V}} = 3 \text{ mA/V}$$

$$g_m = 3000 \ \mu \text{ mhos} \quad \textbf{Ans.}$$

EXAMPLE 1.51 The following readings were obtained experimentally from a FET:

V_{GS}	– 0.1 V	– 0.1 V	– 0.4 V
V_{DS}	5 V	14 V	14 V
I_D	3 mA	3.3 mA	0.71 mA

Obtain (i) a.c. drain resistance, (ii) transconductance and (iii) amplification factor.

Solution: (i) At constant V_{GS} (i.e., at V_{GS} = – 0.1 V), when V_{DS} is increased from 5 V to 14 V the drain current I_D increases from 3 mA to 3.3 mA.

Therefore, change in drain-source voltage, $\Delta V_{DS} = 14 - 5 = 9$ V

Thus, change in drain current, $\Delta I_D = 3.3 - 3 = 0.3$ mA

Hence, a.c. drain resistance will be

$$r_d = \frac{\Delta V_{DS}}{\Delta I_D} \text{ at constant } V_{GS}$$

$$r_d = \frac{9 \text{ V}}{0.3 \text{ mA}} = 30 \text{ K ohms} \quad \textbf{ANs.}$$

(ii) At constant V_{DS} (i.e., at V_{DS} = 14 V), when V_{GS} is changed from – 0.1 V to – 0.4 V, the drain current I_D changes from 3.3 mA to 0.71 mA.

The change in gate-source voltage will be

$$\Delta V_{GS} = 0.4 - 0.1 = 0.3 \text{ V}$$

The change in drain current will be

$$\Delta I_D = 3.3 - 0.71 = 2.59 \text{ mA}$$

Therefore, transconductance will be

$$g_m = \frac{\Delta I_D}{\Delta V_{GS}} \text{ at constant } V_{DS}$$

or
$$g_m = \frac{2.59 \text{ mA}}{0.3 \text{V}} = 8.63 \text{ mA/V} = 8630 \ \mu\text{mhos}$$

(iii) Amplification factor is given by

$$\mu = r_d \times g_m = (30 \times 10^3) \times (8630 \times 10^{-6}) = 258.9 \quad \textbf{Ans.}$$

1.79. COMPARISON BETWEEN FET AND BJT (RGTU, Bhopal, Sem. Exam., 2005-06)

Table 1.10 shows some of the important points of comparison between the field effect transistor (FET) and bipolar junction transistor (BJT).

Table 1.10. Comparison Between Field-effect Transistor (FET) and Bipolar Junction Transistor (BJT) **(Important)**

S.No.	Field-Effect Transistor (FET)	Bipolar junction transistor (BJT)
1.	It is a unipolar device i.e., current in the device is carried either by electrons or holes.	It is bipolar device, i.e., current in the device is carried by both electrons and holes.
2.	It is a voltage-controlled device, i.e., voltage at the gate (or drain) terminal controls the amount of current flowing through the device.	It is a current-controlled device, i.e., the base current controls the amount of collector current.
3.	Its input resistance is very high and is of the order of several megaohms.	Its input resistance is very low as compared to FET and is of the order of few kilo-ohms.
4.	It has a negative temperature coefficient at high current levels. It means that current decreases as the temperature increases. This characteristic prevents the FET from thermal breakdown.	It has a positive temperature coefficient at high current levels. It means that collector current increases with the increase in temperature. This characteristic leads the BJT to thermal breakdown.
5.	It does not suffer from minority-carrier storage effects and therefore, has higher switching speeds and cut-off frequencies.	It suffers from minority carrier storage effects and therefore, has lower switching speed and cut-off frequencies than that of FET's.
6.	It is less noisy than a BJT or vacuum tube and is thus more suitable as an input amplifier for low-level signals. It is used extensively in high fidelity frequency modulated receivers.	It is comparatively more noisy than a field-effect transistor.
7.	It is much simpler to fabricate as an integrated circuit (IC) and occupies a less space on IC chip than that of BJT.	It is comparatively difficult to fabricate as an integrated circuit (IC) and occupies more space on IC chip than that of FET.

1.80. FET AS VOLTAGE VARIABLE RESISTANCE (VVR)

In almost all the linear applications of FET, it is operated in the constant current portion of its output characteristics (i.e., in the saturation region).

Now, let us observe the operation of JFET in the region just before pinch off. In this region, V_{DS} is small but I_D is large. This region is marked as ohmic region in the output characteristics of a JFET. In this region, the FET can be used as a voltage variable resistor (VVR) or voltage dependent resistor (VDR). The drain to source resistance r_{DS} in this region is dependent on the gate to source voltage. As V_{GS} becomes less negative, more drain current flows and r_{DS} decreases. For higher negative gate voltage, I_D decreases and r_{DS} will increase. Thus, the JFET acts as a voltage variable resistor (VVR).

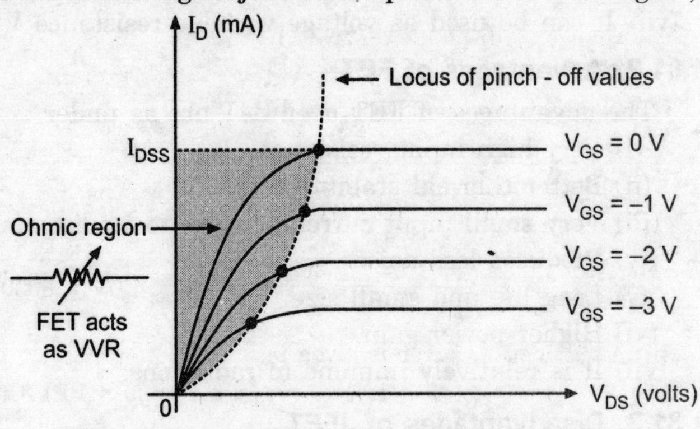

Fig. 1.140. Ohmic region in which FET acts as VVR

1.80.1. Application of VVR

(i) The FET used as VVR finds applications in many areas where this property can be used.

(ii) The VVR, for example, can be used to vary the voltage gain of a multistage amplifier A as the signal level is increased. This is called as **automatic gain control**.

(iii) A typical arrangement for controlling the gain of an amplifier automatically is shown in figure 1.141.

Working Operation

The gain of the common emitter amplifier Q_1 is controlled. The output signal v_0 is rectified and filtered and the dc voltage proportional to v_0 is applied to the gate of FET operating in the ohmic region. Depending on the dc gate voltage, the resistance r_{DS} of FET will change, this will change the emitter resistance of Q_1 since R_E is in parallel with r_{DS}.

The gain of CE amplifier is approximately equal to $-\frac{R_C}{R_E}$. Hence, change in R_E will change the gain automatically. If v_0 increases, then dc voltage at the rectifier filter output will increase. Hence, V_{GS} becomes more negative. This decreases I_D or increases r_{DS} of FET. Hence, the parallel combination ($R_E \| r_{DS}$) will increase and the gain of Q_1 will increase. Exactly opposite action takes place to increase gain of Q_1 if v_0 decreases. This is how the automatic gain control is achieved.

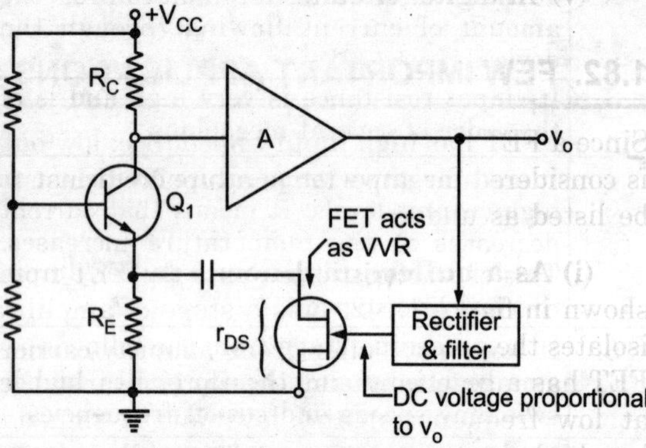

Fig. 1.141. AGC amplifier using FET as VVR

1.81. SUMMARY OF VARIOUS FEATURES OF JFET

1.81.1. Peculiarities

(i) It is a voltage controlled device. (V_{GS} controls I_D).
(ii) It is equivalent to a voltage controlled current source.
(iii) The gate-source junction is always reverse biased.
(iv) Very small gate current ($I_G = 0$).
(v) High input resistance and input capacitance.
(vi) Can be used as switch (operated in ohmic region) or as an amplifier (operated in saturation region).
(vii) It can be used as voltage variable resistance VVR.

1.81.2. Advantages of FET

The advantages of FET over BJT are as under :

(i) Very high input resistance.
(ii) Better thermal stability.
(iii) Very small input current, hence, to loading on the input source.
(iv) Produces less noise.
(v) Long life and small size.
(vi) Higher power gain.
(vii) It is relatively immune to radiations.

1.81.3. Disadvantages of JFET

(i) It can not be used as a current amplifier.
(ii) The value of transconductance g_m is small.
(iii) The voltage gain of JFET amplifier is lower than that of the transistor amplifier.
(iv) Special handling precautions are required to be taken.

1.81.4. Applications of JFET

(i) JFET can be used as an amplifier.

(ii) JFET can be used as a switch.

(iii) It can be used as analog switch in circuits like sample and hold, amplitude modulation, ADC/DAC (analog to digital or digital to analog) converters.

(iv) As a voltage variable resistor (VVR).

(v) In digital circuits.

1.82. FEW IMPORTANT APPLICATIONS OF FET (RGTU, Bhopal, Sem. Exam., 2005-06)

Since a FET has high input impedance, low output impedance and low noise level, therefore, FET is considered far superior to a bipolar transistor. Some of the important applications of FET can be listed as under:

(i) As a buffer amplifier: A buffer amplifier stage is shown in figure 1.142. It is a stage of amplification which isolates the preceding stage from the following stage. Since, FET has a high input impedance (well into the megaohms at low frequencies) and low output impedance (a few hundred ohms), it can act as an excellent buffer amplifier. The high input impedance means high load on the preceding stage. This permits almost the entire output from stage A to appear at the buffer input. The low output impedance means that the buffer (FET) can drive heavy loads. This ensures that the output from buffer reaches at the input of stage B.

DO YOU KNOW?

In general, JFETs are more temperature stable than bipolar transistors. Furthermore, JFETs are typically much smaller than bipolar transistors. This size difference makes them particularly suitable for use in ICs, where the size of each component is very critical.

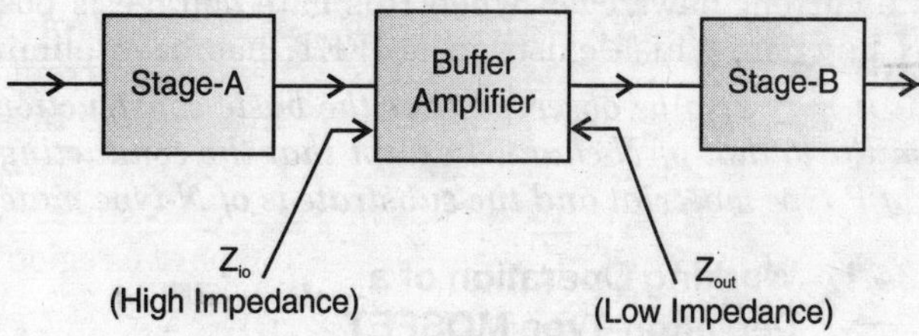

Fig. 1.142. The buffer amplifier stages

(ii) As a low-noise amplifier: Noise is an unwanted disturbance superimposed upon the useful signal. The useful signal contains the required information. As the noise level increases the information becomes inferior.

For instance, the noise in radio receivers produces hissing; severe noise may completely mask the voice or music. Similarly, the noise in television receivers produces white spots (snow) on the picture; a heavy noise may wipe off the picture. Noise is independent of the signal strength because it exists even when the signal is off.

Since FET is an outstanding low-noise device, therefore, it is preferred to apply it near the front end of the receivers and other electronic equipment. Hence, when we use FET amplifier at the front end, we get less amplified noise (disturbance) at the final output.

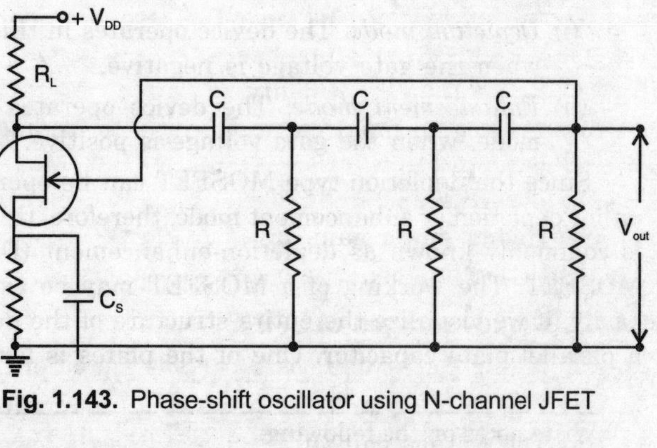

Fig. 1.143. Phase-shift oscillator using N-channel JFET

(iii) As phase-shift oscillator: Although all the oscillators can also work with FETs, however, the high input impedance of FET is especially valuable in a phase-shift oscillator to minimise loading. The circuit diagram of a phase-shift oscillator using N-channel JFET is shown in figure 1.143.

> **DO YOU KNOW?**
>
> In many cases, bipolar transistors and MOSFETs are used in the same electronic circuit.

1.83. MOSFETs

The MOSFET is an abbreviation for metal-oxide semiconductor field-effect transistor Like JFET, it has a source, gate and drain. However, unlike JFET, the gate of a MOSFET is insulated from the channel. Due to this, the MOSFET is sometimes called an IGFET which stands for insulated-gate field effect transistor.

Basically, the MOSFETs are of two types namely Depletion type MOSFET and Enhancement type MOSFET.

1.84. DEPLETION-TYPE MOSFETs*

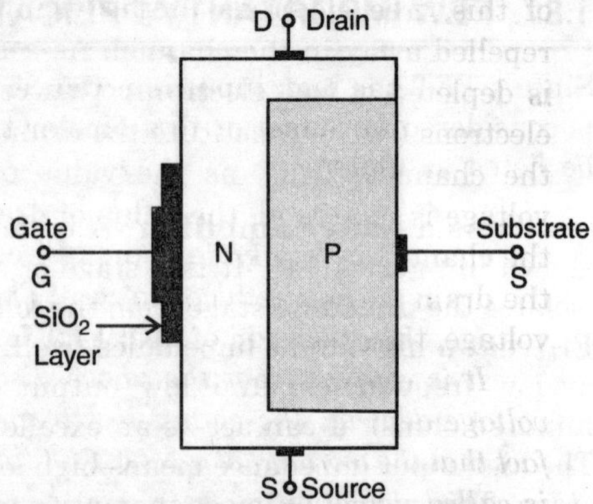

Fig. 1.144. Basic structure of an N-channel depletion-type MOSFET

Figure 1.144 illustrates the basic structure of an N-channel depletion type MOSFET. It simply consists of a conducting bar of N-type material with an insulated gate on the left and P-region on the right. Free electrons can flow from source to drain through the N-type material. The P-region is called substrate (or body). It physically reduces the conducting path a narrow channel. A thin layer of silicon dioxide SiO_2 is deposited on the left side of the channel. This layer insulates the gate from the channel. Because of this, a negligible gate current flows even when this gate voltage is positive. It will be interesting to know that a PN junction, which exists in a JFET, has been eliminated in the MOSFET.

It may also be observed that the basic construction of a depletion-type P-channel MOSFET is similar to that of N-channel except that the conducting bar is of P-type material and the substrate is of N-type material.

> **DO YOU KNOW?**
>
> Like a JFET, a depletion-mode MOSFET is considered a normally on device. This is because both devices have drain current when $V_{GS} = 0$ V. Recall that for a JFET, I_{DSS} is the maximum possible drain current. With a depletion-mode MOSFET, the drain current can exceed I_{DSS} if the gate voltage is of the correct polarity to increase the number of charge carriers in the channel. For an n-channel D-MOSFET, I_D is greater than I_{DSS} when V_{GS} is positive.

1.84.1. Working Operation of a Depletion-Type MOSFET

The depletion-type MOSFET can be operated in two different modes as given below:

(i) *Depletion mode.* The device operates in this mode, when the gate voltage is negative.

(ii) *Enhancement mode.* The device operates in this mode, when the gate voltage is positive.

Since the depletion-type MOSFET can be operated in either depletion or enhancement mode, therefore, this device is commonly known as depletion-enhancement (DE) type MOSFET. The working of a MOSFET may be explained easily, if we visualize the entire structure of the device as a parallel plate capacitor. One of the plates is formed by

* Write notes on the following:
(i) depletion type MOSFET (ii) pinch-off voltage (iii) transfer characteristics of JFET
(iv) Transconductance **(RGTU, Bhopal, Sem. Exam. 2005-06)**

the gate and the other by the semiconductor channel. The plates are separated by a dielectric (SiO_2) layer. We know that if one plate of a capacitor is made negative, it induces a positive charge on the opposite plate and vice versa. This principle is used below in explaining the working of MOSFETs in the depletion and enhancement modes.

(i) Depletion mode: Figure 1.145 (a) shows a MOSFET with a negative gate-to-source voltage. The negative voltage, on the gate, induces a positive charge in the channel. Because of this, free electrons in the vicinity of positive charge are repelled away in the channel. As a result of this, the channel is depleted of free electrons. This reduces the number of free electrons (which constitute the drain current) passing through the channel. Thus, as the value of negative gate-to-source voltage is increased, the value of drain voltage, called $V_{GS\,(off)}$, the channel is totally depleted of free electrons and, therefore, the drain current reduces to zero. Thus, with the negative gate voltage, the operation of MOSFET is similar to that of a JFET.

It is evident from the above discussion that negative gate voltage depletes the channel of free electrons. It is due to this fact that the working of a MOSFET, with a negative gate voltage, is called depletion mode.

(ii) Enhancement mode: Figure 1.145 (b) shows a MOSFET with a positive gate-to-source voltage. The positive gate voltage increases the number of free electrons passing through the channel. The greater the gate voltage, greater is the number of free electrons passing through the channel. This increases *i.e.*, enhances the conducting of the channel. Because of this fact, positive gate operation is called **enhancement mode.**

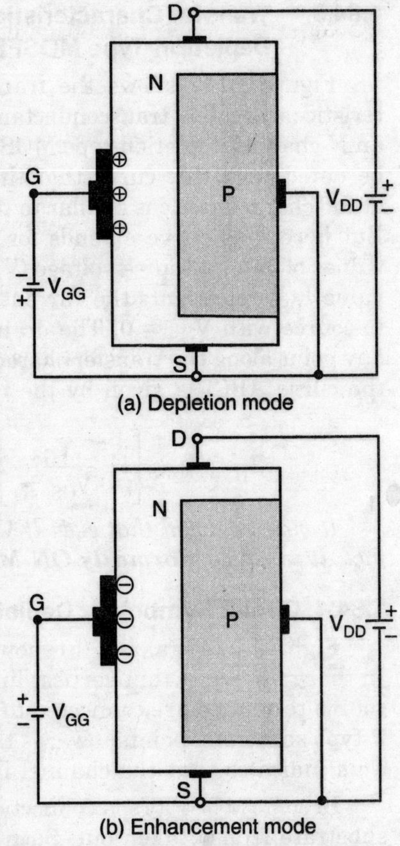

Fig. 1.145. Working of a depletion-type MOSFET

It will be interesting to know that depletion-type MOSFET can conduct even if the gate-to-source voltage (V_{GS}) is zero. Because of this, it is commonly known as ***Normally-ON MOSFET.***

1.84.2. Drain Characteristic of Depletion-Type MOSFET

Figure 1.146 shows the drain characteristic for the N-channel depletion-type MOSFET in the common source configuration. These curves are plotted for both negative and positive values of gate-to-source voltage (V_{GS}). The curves shown above the curve for $V_{GS} = 0$ have a positive zero whereas those below it have a negative value of V_{GS}. When V_{GS} is zero and negative, the MOSFET operates in the depletion-mode. On the other hand, if V_{GS} is zero and positive, the MOSFET operates in the enhancement-mode.

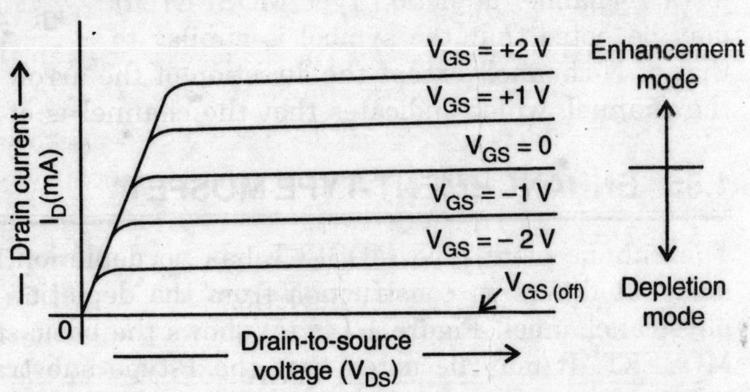

Fig. 1.146. Drain characteristic of N-channel depletion-type MOSFET

It may be noted that the drain characteristics of depletion-type MOSFETs are similar to that of JFET. The only difference is that JFET does not operate for positive values of gate-to-source voltage (V_{GS}).

1.84.3. Transfer Characteristic of Depletion-Type MOSFET

Figure 1.147 shows the transfer characteristic (also called transconductance curve) for an N-channel depletion-type MOSFET. It may be noted from this curve that the region AB of the characteristic is similar to that of JFET. But here, this curve extends for the positive values of gate-to-source voltage (V_{GS}) also. The value I_{DSS} represents the current from drain-to-source with $V_{GS} = 0$. The drain current at any point along the transfer characteristic (i.e., the curve ABC) is given by the relation,

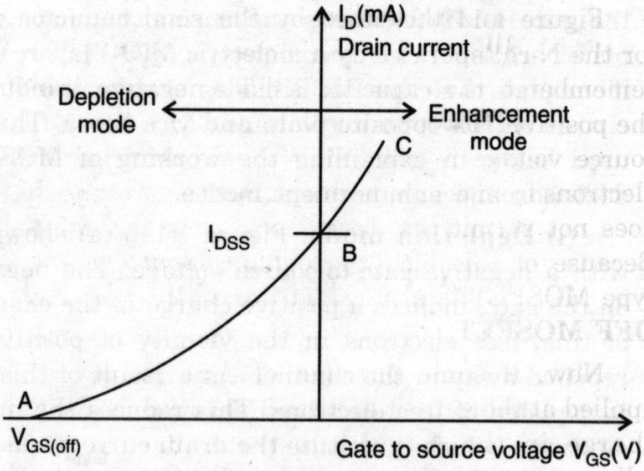

Fig. 1.147. Transfer characteristics of a N-channel depletion-type MOSFET

$$I_D = I_{DSS}\left[1 - \frac{V_{GS}}{V_{GS(off)}}\right]^2$$

It may be noted that even if $V_{GS} = 0$, the device has a drain current equal to I_{DSS}. Due to this fact, it is called ***normally-ON MOSFET.***

1.84.4. Circuit Symbol for Depletion-Type MOSFET

Figures 1.148 (a) and (b) show the circuit symbols for the N-channel depletion-type MOSFET. In these figures, a thin vertical line (just right to the gate) represents the channel. The drain and source terminals are connected to the top and bottom of the channel as shown. The arrow, on the P-type substrate, points towards the channel. This indicates that the channel is N-type.

In some MOSFETs, a connection from the substrate is also taken out. Such MOSFETs have 4-terminals as indicated in the figure 1.148 (a). However, in most of the MOSFETs, the substrate is internally connected to the source. This results in a three terminal device, whose circuit symbol is as shown in figure 1.148 (b).

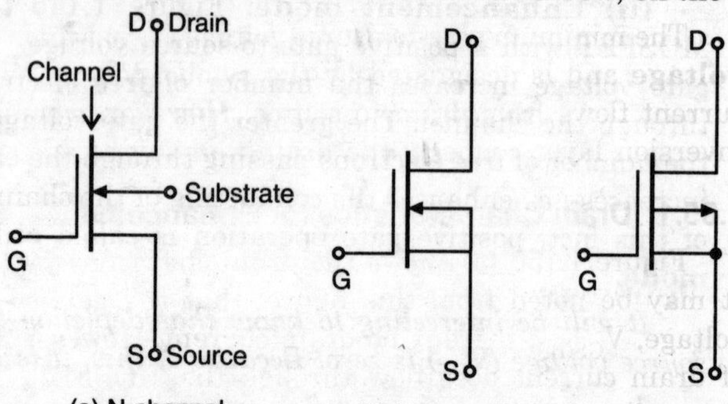

Fig. 1.148. Circuit symbols for depletion type MOSFETs

Figure 1.148 (c) shows the circuit symbol for a P-channel depletion type MOSFET. It may be noted that the symbol is similar to that of N-channel, except the direction of the arrow on the substrate. Its direction is away from the channel, which indicates that the channel is of P-type material.

1.85. ENHANCEMENT-TYPE MOSFET*

The enhancement-type MOSFET has no depletion mode and it operates only in enhancement mode. It differs in construction from the depletion-type MOSFET in the sense that it has no physical channel. Figure 1.149 (a) shows the basic structure of the N-channel enhancement-type MOSFET. It may be noted that the P-type substrate extends the silicon dioxide (SiO_2) layer completely.

* What is the significant difference between the construction of an enhancement type MOSFET and depletion type MOSFET ? Explain with suitable diagram ?

(RGTU, Bhopal, Sem. Exam. 2005-06)

Figure 1.149 (b) shows the normal biasing polarities for the N-channel enhancement-type MOSFET. It must be remembered that this MOSFET is always operated with the positive gate-to-source voltage (V_{GS}). When the gate-to-source voltage is zero, the V_{DD} supply tries to force free electrons from source-to-drain. But the presence of P-region does not permit the electrons to pass through it. Thus, there is no drain current for $V_{GS} = 0$. Because of this fact, the enhancement type MOSFET is also called **normally-OFF MOSFET**.

DO YOU KNOW?

With the E-MOSFET, V_{GS} has to be greater than $V_{GS(th)}$ to get any drain current at all.

Now, if some positive voltage is applied at the gate, it induces a negative charge in the P-type substrate just adjacent to the silicon dioxide (SiO_2) layer. The induced negative charge is produced by attracting the free electrons from the source. When the gate is positive enough, it can attract a number of free electrons. This forms a thin layer of electrons, which stretches from source to drain. This effect is equivalent to producing a thin layer of N-type channel in the P-type substrate. This layer of free electrons is called **N-type inversion layer.**

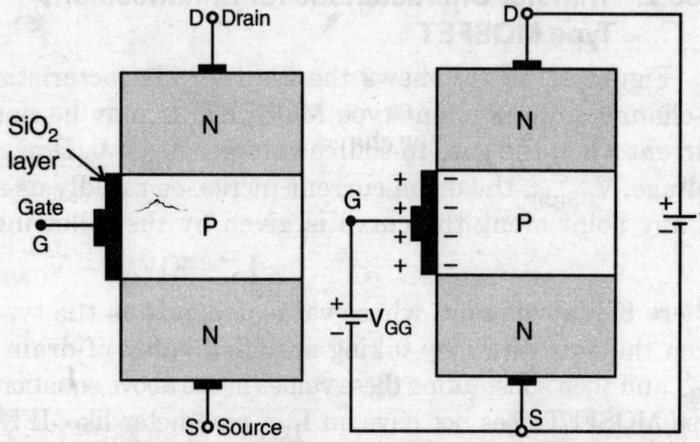

(a) Basic construction, (b) Formation of inversion layer.

Fig. 1.149. Enhancement type MOSFET

The minimum gate-to-source voltage (V_{GS}), which produces inversion layer, is called **threshold voltage** and is designated by the symbol $V_{GS(th)}$. When the voltage V_{GS} is less than $V_{GS(th)}$, no current flows from drain to source. However, when the voltage V_{GS} is greater than $V_{GS(th)}$, the inversion layer connects the drain and source and we get a significant value of current.

1.85.1. Drain Characteristics for Enhancement-Type MOSFET

Figure 1.150 (a) shows the drain characteristics for N-channel enhancement-type MOSFET. It may be noted from this figure, that the gate-to-source voltage (V_{GS}) is less than threshold voltage, $V_{GS(th)}$, there is no drain current. However, in actual practice, an extremely small value of drain current does flow through the MOSFET. This current flow is due to the pressure of thermally generated electrons in the P-type substrate. When the value of V_{GS} is kept above $V_{GS(th)}$, a significant amount of drain current flows.

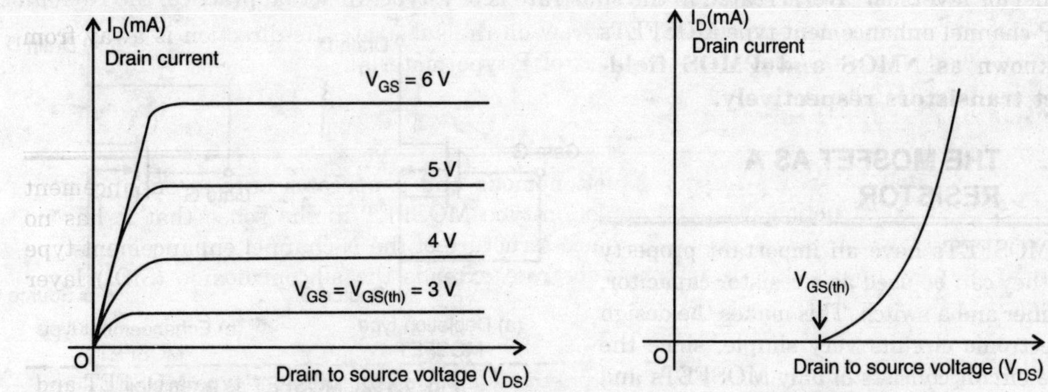

(a) Drain characteristic (b) Transfer characteristic

Fig. 1.150. Characteristics for N-channel enhancement-type MOSFET

The values of drain current increases with the increase in gate-to-source voltage. It is because of the fact that the width of inversion layer widens for increased values of V_{GS} and, therefore, allows more number of free electrons to pass through it. The drain current reaches its saturation value above a certain value of drain-to-source voltage (V_{DS}).

DO YOU KNOW?

The depletion-mode MOSFET was part of the evolution toward the enhancement-mode MOSFET, abbreviated E-MOSFET. Without the E-MOSFET, the personal computers that are now so widespread would not exist.

1.85.2. Transfer Characteristic for Enhancement-Type MOSFET

Figure 1.150 (b) shows the transfer characteristic for N-channel enhancement type MOSFET. It may be noted from this figure that there is no drain current when the gate-to-source voltage, $V_{GS} = 0$. However, if V_{GS} is increased above the threshold voltage, $V_{GS(th)}$, the drain current increases rapidly as shown in figure 1.150(b). The drain current at any point along the curve is given by the following relation:

$$I_D = K[V_{GS} - V_{GS(th)}]^2$$

where K is a constant, whose value depends on the type of MOSFET. Its value can be determined from the data sheet by taking specified value of drain current called $I_{D(ON)}$ at the given value of V_{GS} and then substituting these values in the above equation. Incidently, it may be noted that enhancement-type MOSFET does not have an I_{DSS} parameter like JFET and depletion-type MOSFET.

1.85.3. Circuit Symbol for Enhancement-Type MOSFET

Figure 1.151 (a) shows the circuit symbol for N-channel enhancement-type MOSFET. In this figure, the broken line indicates that there is no conducting channel between drain and source, when $V_{GS} = 0$. Because of this fact, this device is also known as **"Normally-OFF MOSFET"**. The drain and source terminals are shown at the top and bottom end of the broken line. The substrate is internally connected to the source as shown. The arrow points in the direction of channel (or inversion layer), which is created when V_{GS} is increased above $V_{GS(th)}$.

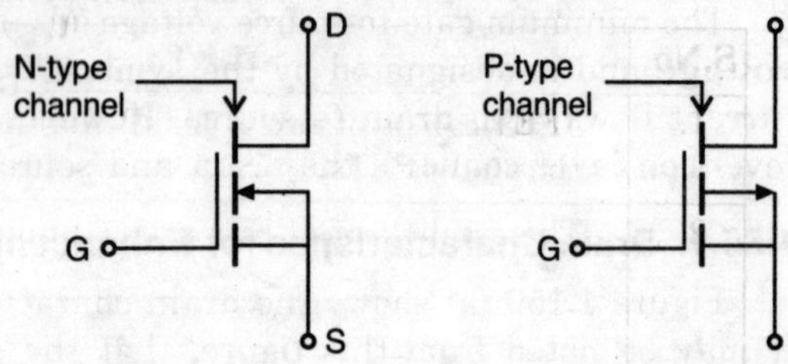

Fig. 1.151. Circuit symbols for enhancement-type MOSFET

Figure 1.151 (b) shows the circuit symbol for the P-channel enhancement-type MOSFET. This symbol is similar to N-channel except that the arrow points outwards. This indicates that the channel (or inversion layer) created in the substrate is of P-type. In actual practice, the N-channel and P-channel enhancement-type MOSFETs are known as **NMOS and PMOS field-effect transistors respectively.**

1.86. THE MOSFET AS A RESISTOR

The MOSFETs have an important property that they can be used as a resistor capacitor, amplifier and a switch. This makes the design of electronic circuits very simple, since the entire circuit consists of only MOSFETs and no other component. The examples of such circuits are microprocessors and memory circuits.

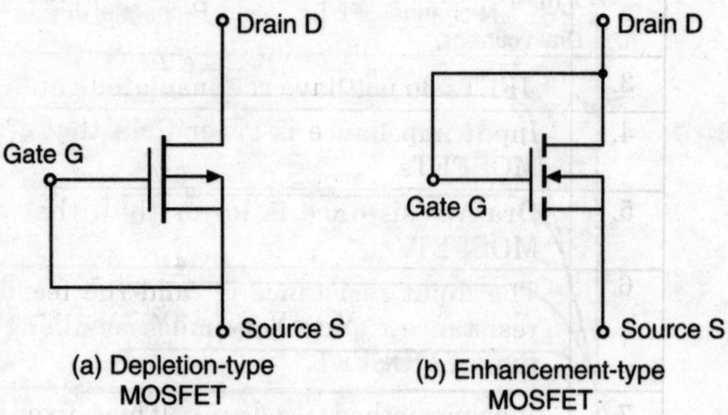

Fig. 1.152. MOSFET as a resistor

The depletion type MOSFET may be connected as a resistor, if its gate is connected to the source so that the gate-to-source voltage (V_{GS}) is zero, as shown in figure 1.152 (a).

The enhancement-type MOSFET may be connected as a resistor, if its gate is connected to drain so that $V_{GS} = V_{DS}$ as shown in figure 1.152 (b).

1.87. ADVANTAGES OF N-CHANNEL MOSFETs OVER P-CHANNEL

These days N-channel MOSFETs have become much popular than P-channel for fabricating MOS circuits and systems. The P-channel MOSFETs find their use only in complementary metal-oxide semiconductor (CMOS) integrated circuits. The main advantage of N-channel MOSFET, over P-channel, is due to the fact that charge carriers in N-channel devices are the electrons, which have a mobility of about 1300 cm^2/volt sec. On the other hand, the charge carriers in P-channel devices are the holes which have a mobility of about 500 cm^2/volt sec. Since the current, in a semiconductor, is directly proportional to the mobility, therefore, the current in the N-channel MOSFET is more than two times that of P-channel for the same dimension.

The ON resistance of the N-channel MOSFET is one-third of that for the P-channel. It means that in order to achieve the same value of current and ON resistance, the P-channel MOSFET requires three times the area of an equivalent N-channel MOSFET. Thus, electronic currents using N-channel MOSFETs are much smaller in size than those of P-channel. This results in higher packing density.

1.88. COMPARISON OF JFET AND MOSFET

Table 1.11. Comparison of JFET and MOSFET

S.No.	JFET	MOSFET
1.	JFET can be of two types, P-channel and N-channel JFETs.	MOSFETs can be of depletion type or enhancement type.
2.	The symbols of JFETs are as under : 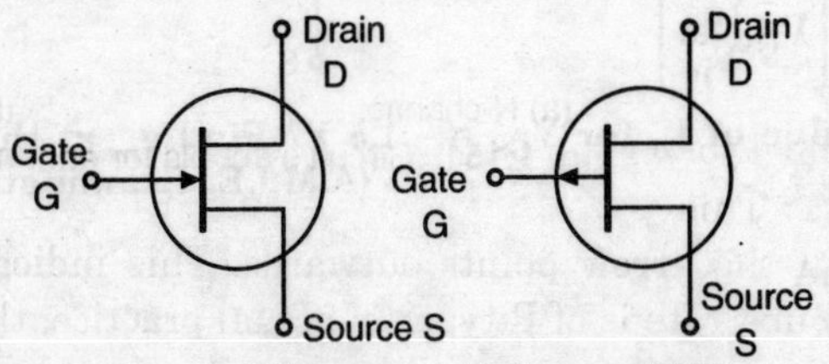 N-channel JFET, P-channel JFET	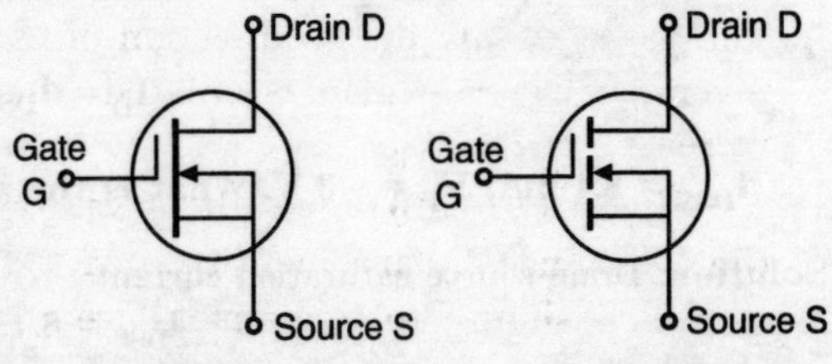 N-channel depletion MOSFET, N-channel enhancement MOSFET
3.	JFETs do not have the insulated gate.	MOSFETs have the insulated gate structure.
4.	Input impedance is lower than that of the MOSFETs.	Input impedance is higher than that of JFET due to the insulated gate structure.
5.	Drain resistance is lower than that of a MOSFET.	Drain resistance is higher than that of a JFET.
6.	The input resistance r_{gs} and the feedback resistance r_{gd} are very much smaller than that of MOSFET.	r_{gs} and r_{gd} are much larger than that of a JFET.
7.	Biasing methods used are self bias, fixed bias and voltage divider bias.	Biasing circuits of depletion MOSFET are same as those used for JFET but biasing circuits for enhancement MOSFET are, feedback biasing and voltage divider biasing.

1.89. COMPARISON OF JFET AND DMOSFET

In this section, let us compare JFET and DMOSFET in the form of a Table 1.12.

Table 1.12.

S.No.	JFET	DMOSFET
1.	Symbol of N-channel JFET. 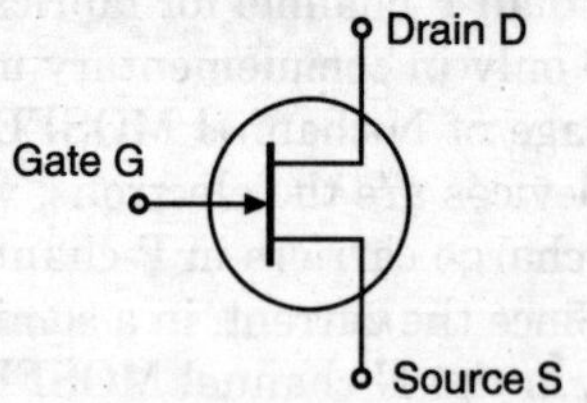	Symbol of N-channel depletion-type MOSFET. 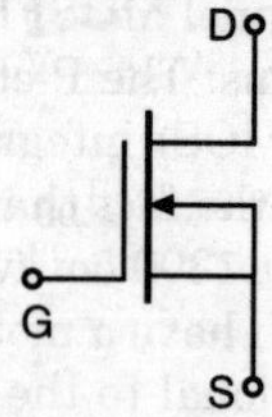
2.	The oxide layer is absent hence gate is not isolated from the channel.	The insulating layer of SiO_2 is present between the gate and channel.
3.	Drain current decreases with increase in the negative V_{GS} voltage.	Drain current decreases as negative V_{GS} is increased.
4.	Reduction in drain current is due to the narrowing of the channel width with increase in $-V_{GS}$.	Reduction in I_D is due to the recombination process taking place under the influence of negative V_{GS}.
5.	Drain characteristics are drawn only for $V_{GS} \leq 0$ volts.	Drain characteristics are same as those of JFET except for the part corresponding to positive V_{GS}.
6.	Input impedance is high	Input impedance is still higher due to presence of SiO_2 layer.

EXAMPLE 1.52 **A FET follows the following relation:**

$$I_D = I_{DSS}\left[1 - \frac{V_{GS}}{V_P}\right]^2$$

I_{DSS} = 8.4 mA, V_P = – 3 V. What is the value of I_D for V_{GS} = – 1.5 V? Find g_m at this point.

(A.M.I.E. Examination, 1993)

Solution: Drain-source saturation current,

$$I_{DSS} = 8.4 \text{ mA}$$

$$\text{Pinch-off voltage, } V_P = -3 \text{ V}$$

$$\text{Gate-source voltage, } V_{GS} = -1.5 \text{ V}$$

We know that the drain current is given by

$$I_D = I_{DSS}\left[1 - \frac{V_{GS}}{V_P}\right]^2$$

Substituting all the values,

$$I_D = 8.4\left(1 - \frac{-1.5}{-3}\right)^2 = 2.1 \text{ mA} \quad \textbf{Ans.}$$

Transconductance for $V_{GS} = 0$ is given by

$$g_{mo} = \frac{-2I_{DSS}}{V_P} = \frac{-2 \times 8.4}{-3} = 5.6 \text{ mA/V or } 5.6 \text{ mS}$$

Transconductance,

$$g_m = g_{mo}\left(1 - \frac{V_{GS}}{V_P}\right) = 5.6\left(1 - \frac{-1.5}{-3}\right) = 2.8 \text{ mS} \quad \textbf{Ans.}$$

EXMAPLE 1.53 An N-channel JFET has a pinch off voltage of – 4.5 V and I_{DSS} = 9 mA. At what value of V_{GS} will I_{DS} be equal to 3 mA? What is its g_m at this I_{DS}?

Solution: Pinch-off voltage, $V_P = -4.5$ V

Drain-source saturation current, $I_{DSS} = 9 \text{ mA} = 9 \times 10^{-3}$ A

Drain-source current, $I_{DS} = 3 \text{ mA} = 3 \times 10^{-3}$ A

From Shockley's equation, we have

$$V_{GS} = V_P\left[1-\sqrt{\frac{I_{DS}}{I_{DSS}}}\right] = -4.5\times\left[1-\sqrt{\frac{3\times10^{-3}}{9\times10^{-3}}}\right] = 1.902 \text{ V} \quad \textbf{Ans.}$$

Transconductance g_m for I_{DS} = 3 mA for which V_{GS} = – 1.902 V

$$g_m = \frac{-2I_{DSS}}{V_P}\left[1-\frac{V_{GS}}{V_P}\right] = \frac{-2\times9\times10^{-3}}{-4.5}\left(1-\frac{-1.902}{-4.5}\right) = 2.31 \text{ mA/V} = 2.31 \text{ mS} \quad \textbf{Ans.}$$

EXAMPLE 1.54 When drain-source voltage is changed by1.5 volts, the change in drain current is of 120 μA, the gate-source voltage remaining unchanged. Determine the a.c. drain resistance of the JFET.

Solution: Change in drain-source voltage,

$$\Delta V_{DS} = 1.5 \text{ V}$$

Change in drain-current, $\Delta I_o = 120 \times 10^{-6}$ A

$$\text{AC drain resistance of the JFET, } r_d = \frac{\Delta V_{DS}}{\Delta I_D} = \frac{1.5}{120\times10^{-6}} = 12.5 \text{ k}\Omega \quad \textbf{Ans.}$$

EXAMPLE 1.55 The readings obtained experimentally from a JFET are as follows:

Drain-source voltage, V_{DS} in volts	**5**	**12**	**12**
Gate-source voltage V_{GS} in volts	**0**	**0**	**– 0.25**
Drain current, I_D in mA	**8**	**8.2**	**7.5**

Determine (i) a.c. drain resistance (ii) transconductance and (iii) amplification factor.

Solution: (i) The drain current increases from 8.0 mA to 8.2 mA when drain-source voltage is increased from 5 V to 12 V, keeping gate-source voltage constant at 0.V.

$$\text{Therefore, A.C. drain resistance, } r_d = \frac{\Delta V_{DS}}{\Delta I_D} = \frac{(12-5)\text{V}}{(8.2-8)\text{mA}} = \frac{7}{0.2\times10^{-3}} = 35 \text{ k}\Omega \quad \textbf{Ans.}$$

(ii) The drain current changes from 8.2 mA to 7.5 mA when gate-source voltage is changed from 0 to – 0.25 V, keeping drain-source voltage constant at 12 V.

$$\text{Hence, Transconductance,} \quad g_m = \frac{\Delta I_D}{\Delta V_{GS}} = \frac{(7.5-8.2)\text{mA}}{(-0.25-0)\text{V}} = 2.8 \text{ mA/V} = 2.8 \text{ mS} \quad \textbf{Ans.}$$

(iii) Amplification factor, $\mu = r_d \times g_m = 35 \times 1,000 \times 2.8 \times 10^{-3} = 98$ **Ans.**

EXAMPLE 1.56 A JFET has V_P = – 4.5 V, I_{DSS} = 10 mA and I_{DS} = 2.5 mA. Determine the transconductance. **(A.M.I.E. Examination, 1996)**

Solution: Drain-source saturation current,

$$I_{DSS} = 10 \text{ mA}$$

Pinch-off voltage, $V_P = -4.5$ V

Drain-source current, $I_{DS} = 2.5$ mA

From Shockley's equation, drain-source current,

$$I_{DS} = I_{DSS}\left[1-\frac{V_{GS}}{V_P}\right]^2 \quad \text{...(i)}$$

or

$$V_{GS} = V_P\left[1-\sqrt{\frac{I_{DS}}{I_{DSS}}}\right] = -4.5\left[1-\sqrt{\frac{2.5}{10}}\right] = -2.25 \text{ V}$$

From equation (i), transconductance will be

$$g_m = \frac{-2I_{DSS}}{V_P}\left[1-\frac{V_{GS}}{V_P}\right] = \frac{-2\times10\times10^{-3}}{-4.5}\left[1-\frac{-2.25}{-4.5}\right] = 2.22 \text{ m A/V} \quad \textbf{Ans.}$$

EXAMPLE 1.57 A JFET has g_m = 10 mS, I_{DSS} = 10 mA. Calculate $V_{GS(OFF)}$.

Solution: Transconductance, $g_m = 10 \text{ mS} = 10 \times 10^{-3}$ S

Drain-source saturation current, $I_{DSS} = 10 \text{ mA} = 10 \times 10^{-6}$ A

$$V_{GS(OFF)} \rightarrow V_{GS} = V_P$$

$$g_m = g_{mo} = \frac{-2I_{DSS}}{V_P} = \frac{-2I_{DSS}}{V_{GS(OFF)}}$$

or $$10 \times 10^{-3} = \frac{-2\times10\times10^{-6}}{V_{GS(OFF)}}$$

or $$V_{GS(OFF)} = \frac{-2\times10\times10^{-6}}{10\times10^{-3}} = 2 \text{ mV} \quad \textbf{Ans.}$$

EXAMPLE 1.58 An N-channel JFET has I_{DSS} = 10 mA and V_P = – 4 V. Determine the minimum value of V_{DS} for pinch-off region and drain current I_D for V_{GS} = – 2 V in pinch-off region.

Solution: Pinch-off voltage, $V_P = -4$ V

Gate-source voltage, $V_{GS} = -2$ V

Drain-source saturation current, $I_{DSS} = 10 \text{ mA} = 10 \times 10^{-3}$ A

$$\text{Drain current, } I_D = I_{DSS}\left[1-\frac{V_{GS}}{V_P}\right]^2 = 10\times10^{-3}\left[1-\frac{-2}{-4}\right]^2 = 2.5 \text{ mA} \quad \textbf{Ans.}$$

The minimum value of V_{DS} for pinch-off region is equal to V_P. Thus the minimum value of V_{DS}

$$V_{DS(min)} = V_P = -4 \text{ V} \quad \textbf{Ans.}$$

EXAMLE 1.59 For an N-channel JFET

$$\mathbf{I_{DSS} = 8.7 \text{ mA}}$$
$$\mathbf{V_P = -3 \text{ V}}$$
$$\mathbf{V_{GS} = -1 \text{ V.}}$$

Find the values of (i) I_D (ii) g_{mo} (iii) g_m.

Solution: Drain-source saturation current,

$$I_{DSS} = 8.7 \text{ mA} = 0.0087 \text{ A}$$

Pinch-off voltage, $V_{GS} = -1$ V

(i) $$\text{Drain current, } I_D = I_{DSS}\left(1-\frac{V_{GS}}{V_P}\right)^2 = 0.0087\left(1-\frac{-1}{-3}\right)^2 = 3.8667 \text{ mA} \quad \textbf{Ans.}$$

(ii) Transconductance for V_{GS} = 0, is given by

$$g_{mo} = \frac{-2I_{DSS}}{V_P} = \frac{-2\times 8.7 \text{ mA}}{-3} = 5.8 \text{ mA/V} = 5.8 \text{ mS} \quad \textbf{Ans.}$$

(iii) Now, transconductance, $g_m = g_{mo}\left(1-\frac{V_{GS}}{V_P}\right) = 5.8\left(1-\frac{-1}{-3}\right) = 3.867 \text{ mA/V} = 3.867 \text{ mS}$ **Ans.**

EXAMPLE 1.60 The FET shown in figure 1.153 has the following parameters: I_{DSS} = 5.6 mA, V_P = 4 Volt. If V and V_0 are constant, then

determine: (i) V_0 when V_i = 0

(ii) V_0 when V_i = 10 Volts

(iii) V_i when V_0 = 0

Solution: Applying KVL to the gate source loop in figure 1.153, we have

$$12 = I_D \times 10 - V_{GS}$$

or $$V_{GS} = 10 \times I_D - 12$$

or $$I_D = \frac{V_{GS}+12}{10} \quad ...(i)$$

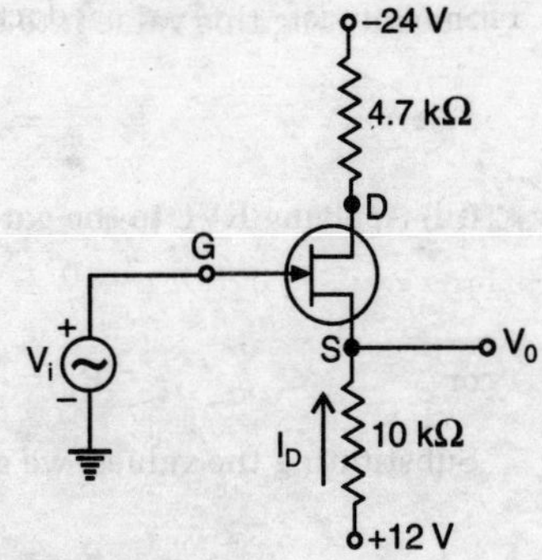

Fig. 1.153. Circuit for Example 1.60.

Since, we do not know the value of I_D, therefore, let us find the value of I_D

$$I_D = I_{DSS}\left[1-\frac{V_{GS}}{V_P}\right]^2 = 5.6\left[1-\frac{V_{GS}}{4}\right]^2 \quad ...(ii)$$

Now substituting V_{GS} from equation (i) into equation (ii), we get

$$I_D = 5.6\times10^{-3}\left[1-\frac{10I_D-12}{4}\right]^2$$

$$= 5.6\times10^{-3} - \frac{56}{4}I_D + 3\times5.6\times10^{-3}$$

Therefore, $15\ I_D = 22.4 \times 10^{-3}$ or $I_D = 1.49$ mA ...(iii)

Substituting the value of I_D into equation (i), we get

$$V_{GS} = (10 \times 1.49) - 12 = 2.9 \text{ Volts}$$

The output voltage will be

$$V_o = 12 - I_D R_S$$

Substituting values, we get

$$V_o = 12 - (1.49 \times 10) = -29 \text{ Volts} \quad \textbf{Ans.}$$

EXAMPLE 1.61 The JFET used in figure 1.154 has $V_P = -2$ V, $I_{DSS} = 1.65$ mA. It is desired to bias the circuit at $I_D = 0.8$ mA. If $V_{DD} = 24$ V and $r_d >> R_D$.

Find: (i) V_{GS}, (ii) g_m, (iii) R_S.

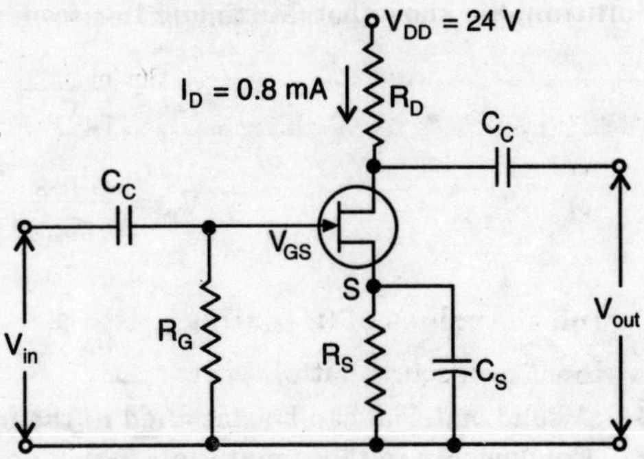

Fig. 1.154.

Solution: (i) We know that

$$I_D = I_{DSS}\left[1-\frac{V_{GS}}{V_P}\right]^2 \quad ...(i)$$

or $$V_{GS} = V_P\left[1-\sqrt{\frac{I_D}{I_{DSS}}}\right]$$

Substituting the values, we get

$$V_{GS} = -2\left[1-\sqrt{\frac{0.8}{1.65}}\right] = -2\left[1-\frac{1}{\sqrt{2}}\right]$$

$$= +0.6074 \text{ Volts} \quad \textbf{Ans.}$$

(ii) We know that $$g_m = g_{mo}\left[1-\frac{V_{GS}}{V_P}\right] \quad ...(ii)$$

But we do not know the value of g_{mo} so let us find it first as under:

$$g_{mo} = \frac{-2I_{DSS}}{V_P}$$

Substituting the values of I_{DSS} and V_P, we get

$$g_{mo} = \frac{-2\times1.65\times10^{-3}}{-2} = 1.65 \times 10^{-3} \quad ...(iii)$$

Substituting this value into equation (iii), we get

$$g_m = 1.65\times10^{-3}\left[1-\frac{0.6074}{2}\right] = 1.15\times10^{-3}\text{ mho}\quad \textbf{Ans.}$$

(iii) Applying KVL to the gate to source circuit of figure 1.154, we get

$$0 = V_{GS} + I_D R_S$$

or $$R_S = \frac{-V_{GS}}{I_D}$$

Substituting the values, we get

$$R_S = \frac{0.6074}{0.8\times10^{-3}} = 759.25\ \Omega \quad \textbf{Ans.}$$

EXAMPLE 1.62 The maximum transconductance of a certain N-channel JFET is 9.4 mS. If I_{DSS} is 20 mA, what is the pinch off voltage ?

Solution: The maximum transconductance is given by

$$g_{mo} = \frac{-2I_{DSS}}{V_P} \quad \text{or} \quad V_P = \frac{-2I_{DSS}}{g_{mo}}$$

Substituting the values, we get

$$V_P = \frac{-2\times20\times10^{-3}}{9.4\times10^{-3}}$$

or Pinch off voltage $V_P = -4.255$ Volts **Ans.**

EXAMPLE 1.63 The maximum transconductance of a certain N-channel JFET is 3.8×10^{-3} S. If I_{DSS} = 18 mA. What would be the pinch off voltage?

Solution: We know that, Maximum transconductance,

$$g_{mo} = \left|\frac{2I_{DSS}}{V_P}\right|$$

or $$V_P = \frac{2I_{DSS}}{g_{mo}} = \frac{2\times18\text{mA}}{9.8\times10^{-3}} = 3.6\text{ Volts} \quad \textbf{Ans.}$$

SUMMARY

- A solid material can be classified in the form of conductors, insulators and semiconductors.
- Conductors are those materials which are good conductors of electricity.
- Insulators are those materials which are bad conductors of electricity.
- Semiconductors are those materials whose conductivities lie between conductors and insulators.
- In a solid, the closely placed energy levels, which differ slightly in energy, form an energy band.
- The range of energies possessed by electrons of the same orbit in a solid, is called energy band.
- The energy band which possesses the free electrons is called valence band.
- The energy band which possesses the free electrons after exciting is called conduction band.
- Electrons present in the conduction band take part in conduction.
- Between the valence band and the conduction band there is an energy gap. This energy gap is called forbidden energy gap and is denoted by E_G.
- Insulators, metals and semiconductors can also be defined on the basis of band gap.
- In insulators, there is generally no electron in the conduction band and the valence band is filled.
- In metals, the forbidden energy gap between the valence band and conduction band is zero. Infact, the valence and conduction bands overlap each other.
- In case of semiconductors, forbidden energy gap is not very wide.
- Semiconductors can be classified in two forms: intrinsic semiconductors and extrinsic semiconductors.

- A semiconductor in an extremely pure form is known as intrinsic semiconductor.
- At absolute zero (i.e., 0 K) the intrinsic semiconductor behaves as an insulator.
- At room temperature the intrinsic semiconductor has some conductivity.
- In intrinsic semiconductors, the concentration of free electrons will always be equal to the concentration of holes. This type of simultaneous generation of electrons and holes due to temperature is called thermal generation.
- At a given temperature, the rate of generation of electrons and holes is equal to the rate of recombination.
- Before recombination, an electron or hole covers some distance or exists for a very small time. This average time of existence of an electron or hole is called mean life time.
- If a piece of P-type semiconductor is joined to a piece of N-type semiconductor in such a manner that the crystal structure remains continuous at the boundary, then a P-N junction is formed.
- When a battery is connected across the P-N junction, then this process is called biasing of a P-N junction.
- The P-N junction can be biased in two ways as under:
  (i) Forward-bias, (ii) Reverse-bias.
- In forward-biasing of a P-N junction, positive terminal of the battery is connected to the P-side and the negative terminal to the N-side.
- In reverse biasing of a P-N junction a battery is connected across the P-N junction in such a way that its positive terminal is connected to the N-region and negative terminal is connected to the P-region.
- A PN-junction is known as a P-N junction diode or a semiconductor diode.
- The V-I characteristics of a diode is simply a curve or graph between the voltage applied across its terminals and the current that flows through the diode due to this applied voltage.
- An ideal diode offers zero resistance when forward-biased and infinite resistance when it is reverse-biased.
- Practical or real diode does not behave ideally. Its forward resistance is quite small and reverse resistance is very high.
- A diode offers two types of capacitance as under:
  (i) Diffusion or storage capacitance. (ii) Depletion or transition capacitance.
- The breakdown voltage depends upon the width of depletion layer.
- This width of depletion layer depends upon the doping level.
- The following two processes cause junction breakdown due to the increase in reverse bias voltage :
  (i) Zener Breakdown, (ii) Avalanche Break-down.
- The depletion layer is narrow for heavily doped junctions.
- This form of breakdown occurs in junctions which are lightly doped. Lighly-doped junctions have wide depletion layers where the electric field is not strong enough to cause Zener breakdown.
- Zener diode is a reverse-biased heavily-doped PN-junction diode which operates in the breakdown region.
- The reverse breakdown of a PN-junction may occur either due to zener effect or avalanche effect.
- Zener effect dominates at reverse voltages less than 6 V whereas avalanche effect dominates above 6 V.
- The breakdown voltage of a Zener diode can be set by controlling the doping level.
- For zener diodes, silicon is preferred to Ge because of its higher temperature and current capability.
- Zener diodes have a large number of applications. Few of them are:
  (i) Zener diode is used as a voltage regulator.
  (ii) Zener diode is used as a peak clipper in wave shaping circuits.
  (iii) Zener diode is used as a fixed reference voltage in transistor biasing circuits.
  (iv) Zener diode is used for meter protection against damage from accidental application of excessive voltage.
- A varacter diode is, basically, a reverse biased PN junction, which utilizes the inherent capacitance of the depiction layer. It is also known as varicap, voltcap or tunning diode. It is used as a voltage-variable capacitor.
- The maximum ratings for the varactor diode are the same as those that are used for the PN junction diode and the zener diode.
- A major application of varactor diode is in tuning circuits. It is used in electronic tuners in radio, television and other commercial receivers. When a varactor is used in a tuning circuit, it allows a resonant frequency to be adjusted by a variable voltage level.

- The varactor diode is also used in automatic frequency control device, adjustable band-pass filter and parametric amplifier.
- A PIN diode is made up of three semiconductor materials : two heavily doped P-and one N-type material separated by an intrinsic (i.e., undoped) semiconductor (I).
- A PN junction diode, which emits light when forward biased, is known as a light emitting diode (abbreviated as LED). The emitted light may be visible or invisible. The amount of light output is directly proportional to the forward current. Hence, higher the forward current, higher is the light output.
- The LED's which emit one colour of light when forward biased and another when reverse biased, are called multicolored LED's.
- Photodiode is a two terminal PN junction device, which operates in a reverse bias. It has a small transparent window, which allows light to strike the PN junction.
- A transistor is a three terminal device. The output voltage, current or power are controlled by the input current in a transistor. Therefore, it is also called a current-controlled device.
- A transistor is also called a BJT. BJT stands for Bipolar junction transistor.
- A transistor has a very important property that it can raise the strength of an input weak signal.
- A transistor has three regions namely emitter, base and collector.
- When no battery is connected across the different terminals of a transistor then the transistor is said to be in an unbiased state or in an open-circuit state.
- When dc voltages are applied across the different terminals of a transistor, then this process is called biasing.
- There are two junctions in a transistor namely emitter-base junction and collector-base junction.
- There are four possible ways of biasing a transistor. They are as under:
  (i) Active region, (ii) Saturation region
  (iii) Cut-off region, (iv) Inverted region.
- A transistor can be treated as a two-port network.
- There are three transistor configurations namely Common-Base (CB) configuration, Common-Emitter (CE) configuration and Common-Collector (CC) configuration.
- The biasing deals with setting a fixed level of current which should flow through the FET with a desired fixed voltage drop across the FET junctions.
- The commonly used biasing circuits for FET are as under:
  (i) Fixed-bias circuit. (ii) Self-bias circuit.
  (iii) Voltage divider bias circuit.
- The Q-point (i.e., quiescent or operating point or bias point) for a self-biased JFET is established by determining the value of drain current (I_D) for a desired value of gate-to-source voltage (V_{GS}) or vice-versa.
- In analytical method, we note down the values of the maximum drain current (I_{DSS}) and the gate-to-source cut-off voltage [$V_{GS(off)}$] from the data sheets of JFET supplied by the manufacturer.
- In graphical method, we draw a self-bias line such that it intersects the transfer characteristic curve near its mid-point.
- The Q-point for the self-biased JFET circuit can also be obtained graphically by using d.c. load line.
- However, it has been observed that the JFET parameters like maximum drain current (I_{DSS}) and the gate-to-source cut-off [$V_{GS(off)}$] vary with the temperature.
- The transfer characteristics of the enhancement type MOSFET are completely different from that of a JFET.
- The other most popular biasing arrangement for the enhancement type MOSFET is voltage divider biasing circuit.
- The field-effect transistor (FET), is a voltage-controlled device.
- The output characteristics of FET are controlled by the applied input voltage and not by the input current.
- FET is an unipolar device.
- The current flowing through it is only due to one type of charge carriers, i.e., holes or electrons.

SHORT QUESTIONS WITH ANSWERS

Q.1. What is doping?

Ans. The process of adding impurity to a pure semiconductor is called doping. Doping is done at a rate such that only one atom of impurity is added per 10^6 to 10^{10} semiconductor atoms.

Q.2. What are the types of extrinsic semiconductor ?

Ans. Depending upon the types of impurity added, there are two types of extrinsic semiconductors:

(a) N-type semiconductor; (b) P-type semiconductor.

Q.3. What is the energy gap?

Ans. The energy gap between the valence band and conduction band is known as forbidden energy gap. Forbidden energy gap is a region in which no electron can stay as there is no allowed energy state. The greater the forbidden energy gap more tightly the valence electrons are bound to the nucleus.

Q.4. Define diode resistance.

Ans. An ideal diode offers zero resistance when forward bias and infinite resistance when it is reverse biased. But practical or real diode does not behave ideally. It means that the forward resistance of a real diode is quite small and reverse resistance is very high.

Q.5. What is forward resistance? Explain.

Ans. The resistance offered by the diode in the forward bias is called forward resistance. This resistance is different for the flow of direct current (d.c.) and for alternating current (a.c.). Hence, this resistance is of two types:

(a) d.c. forward resistance (b) a.c. forward resistance.

Q.6. What is a P-N junction?

Ans. The contact surface between the layers of P-type and N-type semiconductor pieces placed together so as to form a P-N junction is called the P-N junction.

Q.7. What is the effect of reverse bias on the width of a P-N junction?

Ans. When the P-N junction is reverse biased, the width of depletion region is increased.

Q.8. What types of carriers are present in space charge region?

Ans. No mobile carrier is present in the space charge region.

Q.9. What is cut-in voltage in semiconductors?

Ans. The forward voltage, at which the current through the P-N junction starts increasing rapidly, is called the cut-in voltage.

Q.10. What do you mean by breakdown mechanism? Explain.

Ans. If the reverse-bias applied to a P-N-junction is increased, a point will reach when the junction breaks down and reverse current rises sharply to a value limited only by the external resistance connected in series. This specific value of the reverse bias voltage is called breakdown voltage (V_B). The breakdown voltage depends upon the width of depletion layer. This width of depletion layer depends upon the doping level.

The following two processes cause junction breakdown due to the increase in reverse bias voltage:

(i) Zener Breakdown (ii) Avalanche Breakdown.

Q.11. What is Zener breakdown? Explain.

Ans. This form of breakdown occurs in junctions that are heavily doped. The depletion layer is narrow for heavily doped junctions. When the reverse bias voltage is increased, electric field also increases across the narrow depletion layer. This strong electric field (about 10^8 V/m) causes covalent bonds to break in the crystal structure. Due to this reason a large number of carriers are generated. This causes a large current to flow.

Q.12. What is zener voltage ?

Ans. The voltage at which the zener diode breaks down is called the zener voltage.

Q.13. What happens to the series current, load current and zener current when the dc input voltage of a zener regulator increases ?

Ans. Zener current and series current increase while the load current remains unchanged.

Q.14. What is a zener diode ?

Ans. Zener diode is a P-N junction diode specially designed for operation in its breakdown region.

Q.15. What do you mean by the Varactor Diode?

Ans. A varacter diode is, basically, a reverse biased PN junction, which utilizes the inherent capacitance of the depletion layer. It is also known as varicap, voltcap or tunning diode. It is used as a voltage-variable capacitor.

Q.16. Write few applications of BJT.

Ans.
1. Transistor is used in control systems.
2. Transistor is used in satellites and mobile phones.
3. In digital computer electronics, the transistor is used as a high-speed electronic switch.
4. In communication systems, it is widely used as the primary component in the amplifier.

Q.17. What do you mean by operating point? Explain.

Ans. The zero signal values (no a.c. signal is applied) of collector current I_C and collector-to-emitter voltage V_{CE} are known as the **operating point.**

Since the variations of I_C and V_{CE} take place about this point, therefore, this is known as the operating point.

This point is also called quiescent (stable) point or simply Q-point since it is a point on output characteristics curve when the transistor is in a silent condition (i.e., the absence of a.c. input signal).

Q.18. Explain why an ordinary transistor is called bipolar?

Ans. Because the transistor operation is carried out by two types of charge carriers (majority and minority carriers), an ordinary transistor is called bipolar.

Q.19. What is done to the base region of a transistor to improve its operation ?

Ans. Base is made thin and very lightly doped in comparison to either emitter or collector so that it may pass most of the injected charge carriers to the collector.

Q.20. Why transistor is called current controlled device ?

Ans. The output voltage, current or power is controlled by the input current in a transistor so it is called the current controlled device.

Q.21. Why silicon type transistor are more often used than germanium type ?

Ans. Because, silicon has smaller cut-off current I_{CBO}, small variations in I_{CBO} due to variations in temperature as compared to those in case of germanium.

Q.22. Why the width of the base region of a transistor is kept very small compared to other regions?

(A.M.I.E. Examination, 1994)

Ans. Base region of a transistor is kept very small and very lightly doped so as to pass most of the injected charge carriers to the collector.

Q.23. Why is there a maximum limit of collector supply voltage for a transistor ?

Ans. Although collector current is practically independent of collector supply voltage over the transistor operating range, but if V_{CB} is increased beyond a certain value, collector current I_C eventually increases rapidly and possibly destroys the device.

Q.24. Why CE configuration is most popular in amplifier circuits ?

Ans. CE configuration is mainly used because its current, voltage and power gains are quite high and the ratio of output impedance and input impedance are quite moderate.

Q.25. Why is CC configuration seldom used ?

Ans. CC configuration is seldom used because its voltage gain is always less than unity.

Q.26. What is meant by D.C. load line?

Ans. With the help of d.c. load line, we can easily analyse the performance of an amplifier circuit.

Q.27. What meant by Operating point?

Ans. The zero signal values (no a.c. signal is applied) of collector current I_C and collector-to-emitter voltage V_{CE} are known as the **operating point.**

Q.28. What do you mean by field-effect transistor amplifier?

Ans. The field-effect transistor (FET) has a capability to amplify a.c. signals like a bipolar junction transistor (BJT). Depending upon the types of configuration, the FET amplifiers may be classified under the following three heads:

(i) Common source amplifier (ii) Common drain amplifier

(iii) Common gate amplifier.

Q.29. Why are field-effect transistors called unipolar transistors ?

Ans. In field-effect transistors, current conduction is by only one of type of majority carriers (either by electrons or by holes) and, therefore are called unipolar transistors.

Q.30. Why the channel of a JFET is never completely closed at the drain end ?

Ans. If the channel is completely closed in JFET, then there will be no drain current, so there will be no voltage drop along the channel length and amount of reverse bias will become uniform and wedge-shaped depletion region will become rectangular one.

Q.31. How is drain current controlled in a JFET?

Ans. In a JFET, drain current is controlled by controlling the reverse bias given to its gate (i.e., V_{GS}).

Q.32. What is meant by drain characteristics of FETs ?

Ans. The curve drawn between drain current and drain-source voltage with gate-to-source voltage as the parameter is called the drain characteristic.

Q.33. What is meant by transfer characteristic of FETs ?

Ans. The curve drawn between drain current and gate-source voltage for a given value of drain-source voltage is called the transfer characteristic.

Q.34. What is pinch-off voltage in a JFET ?

Ans. The value of drain-source voltage at which channel is pinched-off (i.e., all the free charges from the channel get removed) is called the pinch-off voltage.

Q.35. What is meant by saturation region ?

Ans. The region of drain characteristic of a FET in which drain current remains fairly constant is called the saturation or pinch-off region.

Q.36. Why is input impedance of the FET very high ?

Ans. FET has very high input impedance because its input circuit (gate-to-source) is reverse biased and the input gate current is very small (of the order of few nano-amperes).

Q.37. Why E-MOSFET is called sometimes normally-off MOSFET ?

Ans. E-MOSFET operates with large positive gate voltages only and does not conduct when gate-source voltage $V_{GS} = 0$, so it is called normally-off MOSFET.

Q.38. Why MOSFETs are never connected or disconnected in the circuit when power is ON ?

Ans. If a MOSFET is connected or disconnected in circuit while the power is ON, transient voltages caused by inductive kickback and other effects may exceed $V_{GS(max)}$ and thus wipe out the MOSFET.

Q.39. Name the factors which make the JFET superior to BJT?

Ans. The high input impedance, low output impedance and low noise level make JFET far superior to the BJT.

Q.40. Write Disadvantage of JFET.

Ans. The main disadvantage of the JFET is that it has a relatively small gain-bandwidth product as compared with that of a conventional transistor (i.e., BJT).

REVIEW QUESTIONS

1. Describe the difference between P-type and N-type semiconductor materials. Sketch the atomic structure of silicon and insert the impurity of arsenic in silicon.
2. Explain clearly the difference between acceptor and donor impurities and state what type of carriers are contributed by them.
3. Define diffusion current in a semiconductor.
4. Compare insulators, conductors and semicon ductors on the basis of energy band diagram.
5. Write note on P-type and N-type semiconductors.
6. Define mobility. Give its dimensions. A semiconductor is doped with both donors and acceptors of concentration N_D and N_A. Derive the equation to determine electron and hole concentration.
7. What is a PN-junction ? How is it formed.
8. Explain the formation of depletion region in a PN-junction.
9. Explain how a barrier potential is developed at the PN-junction.
10. Explain the action of PN-junction diode under forward bias and reverse bias.
11. Draw and explain the V-I characteristics of a PN junction diode.

12. What is meant by biasing a P-N junction diode? Why is it necessary ?
13. What is meant by the term 'Barrier potential'? What are the values for Silicon and Germanium diodes ?
14. Define reverse saturation current of a diode. What are the factors that decide the magnitude of reverse saturation current ?
15. Define α and β for a transistor.
16. Derive the relationship between α and β.
17. What is meant by Q-point?
18. What is the need for biasing a transistor?
19. Explain the requirements of a biasing circuit.
20. What are the factors responsible for the instability of operating point.
21. Draw a fixed bias circuit and obtain the value of d.c. Voltage and currents in the circuit.
22. Define, stability factor. Derive general expression for stability factor of a biasing circuit.
23. Draw a self-bias circuit and derive an expression for stability factor.
24. State various methods of improving stability.
25. Define Stability factor w.r.t. transistor biasing.
26. Explain the function of emitter in the operation of a junction transistor.
27. Illustrate with neat circuit diagrams the technique of biasing an enhancement MOSFET.
28. Explain giving necessary diagrams how JFET can be biased suitable to overcome the effect of unit to unit variation.
29. Explain briefly the technique of biasing a JFET for zero current drift with temperature.
30. Write notes on: Biasing techniques of MOSFET.
31. Explain any one method of biasing a single stage JFET amplifier.
32. With the help of transfer curve and bias line, explain the biasing techniques of N-channel FET.

NUMERICAL PROBLEMS

1. **Determine the forward bias voltage applied to a silicon diode to produce a forward current of 10 mA and reverse saturation current $I_0 = 25 \times 10^{-7}$ A at room temperature.** **[Ans.** 0.4 Volt]
2. **A germanium diode draws 50 mA with a forward bias of 0.27 V. The junction is at room temperature of 27°C. Determine the reverse saturation current of the diode.**
   [Ans. 1.545 μA]
3. **Determine ac resistance for a semiconductor diode with a forward bias of 0.25 V. Reverse saturation current at room temperature is of 1.2 μA.**
   [Ans. 1.445 Ω]

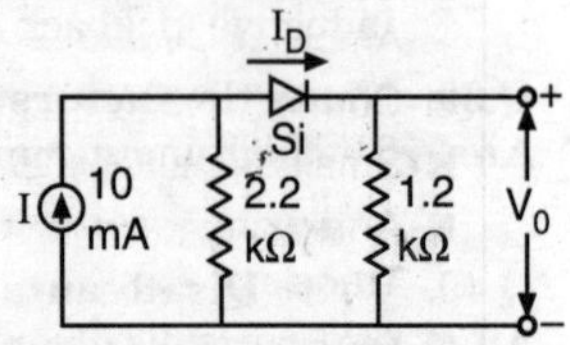

Fig. 1.155.

4. **Determine the current flowing in the curcuit shown in figure 1.155.**
5. **A light emitting diode (LED) is connected across a voltage source + 10 volts through a series resistance of 820 Ω. Calculate the LED current. Assume the voltage drop across an LED of 1.5 volt.**
6. **An n-channel depletion type MOSFET is operated in the pinch off region. If I_{DSS} = 15 mA and V_P = – 4 V, calculate I_D when: (a) V_{GS} = – 3 V and (b) V_{GS} = + 2.5 V.** **[Ans. I_D = 3.75 mA, 3961 mA]**
7. **An N-channel JFET having a pinch off voltage of – 3 volts has a saturation current of 2.5 mA; when V_{GS} = – 1 V. What is its saturation current when,**
   (a) V_{GS} = 0 Volts
   (b) V_{GS} = – 2 Volts
   [Ans. (i) I_{DSS} = 5.625 mA at V_{GS} = 0 V,
   (ii) I_{DSS} = 0.625 mA at V_{GS} = –2 V]

Chapter 2

TRANSISTOR AS AN AMPLIFIER

Syllabus

Transistor as an amplifier, gain, bandwidth, frequency response, h-parameters equivalent type of amplifier.

Chapter Outline

- Introduction
- Analog Signals and Linear Amplifiers
- The Bipolar Linear Amplifier
- Amplifier Analysis
- Small Signal Analysis
- Types of BJT Amplifiers
- Common Emitter (CE) Amplifier
- Common Collector or Emitter Follower Amplifier Circuit
- Common Base Amplifier
- Small Signal Operation
- Analyzing an Amplifier: Analysis of Transistor Amplifier
- AC Equivalent Circuit
- AC Emitter Resistance
- AC Beta (β_{ac})
- Various Models used For AC Analysis
- The R_e Model of Transistor.
- Analysis of CE Amplifier using R_e Model
- Hybrid Parameters (h-parameters)
- h-Parameter Representation for Transistor.
- General Hybrid Equivalent Circuit

2.1. INTRODUCTION

Earlier, we have discussed the operating principle and biasing circuits for the BJT. In this chapter, let us understand how to use it as a linear amplifier. The linear amplifiers are meant to amplify the analog signals. The magnitude of analog signal can take on any value and the signal can vary continuously with time.

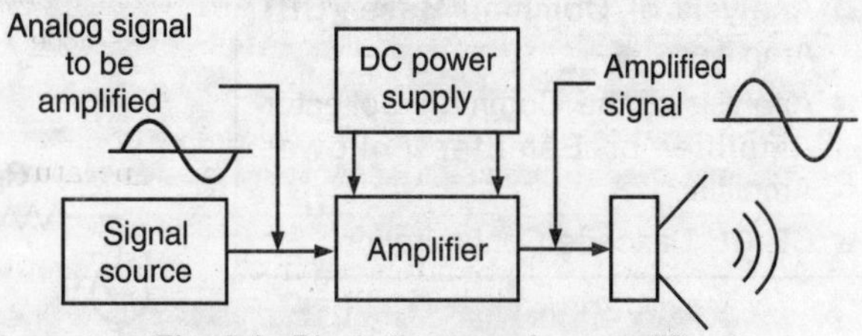

Fig. 2.1. Block diagram of an amplifier

A linear amplifier is then suppose to multiply the input signal by a constant to produce the output. This multiplying factor is greater than 1 and called as gain of the amplifier.

2.2. ANALOG SIGNALS AND LINEAR AMPLIFIERS

Let us consider the block diagram of figure 2.1. The analog signal applied at the amplifier input is an information signal. It may be derived from a CD player, a microphone or from some kind of a transducer. The magnitude of this analog signal can take on any value and may vary continuously with time. Electronic

- Hybrid Equivalent Circuit for Common Emitter Configuration
- Hybrid Equivalent Circuit for Common Base Configuration
- Hybrid Equivalent Circuit for Common Collector Configuration
- Analysis of a Transistor Amplifier Using h-Parameters
- h-Parameter (Hybrid) Expression for Common-Emitter Amplifier
- h-Parameter (Hybrid) Expressions for Common Base Amplifier
- h-Parameter (Hybrid) Expression for Common Collector Amplifier
- Simplified or Approximate Model
- The Common Emitter (CE) Amplifier With other type of Bias
- Common Collector (CC) Amplifier With Outer Type of Bias
- Common Base (CB) Amplifier With Other Type of Bias
- Base Biased Amplifiers
- CE Amplifier with Unbypassed Emitter Resistor.
- Analysis of CE Amplifier with Voltage-Divider Bias (VDB)
- Analysis of Common Base (CB) Amplifier
- Analysis of the Common Collector Amplifier or Eemitter Follower Amplifier
- CE-CC Cascade Configuration

circuits which process the analog signals are called as analog circuits. A Linear Amplifier magnifies the input signal and produces an output signal which is larger in size and has same shape as that of the input. The dc power supply in figure 2.1 supplies power to the amplifier which contains a transistor acting as an amplifying device.

2.2.1. Why is a linear amplifier required ?

We want the output signal to be linearly proportional to the input signal because only then the amplified signal will exactly have the same shape as that of the input. If the amplifier is non-linear, then the shape of the output signal will be different from that of the input and the output signal is said to be distorted. Hence, the amplifier must be linear.

2.3. THE BIPOLAR LINEAR AMPLIFIER

As a matter of fact, the transistor is heart of an amplifier. Let us consider the transistor circuit shown in figure 2.2 (a). The voltage transfer characteristics of this circuit is shown in figure 2.2 (b). To operate the transistor as an amplifier, we have to bias it at the Q point shown in figure 2.2 (b) so that it is biased in the forward active region.

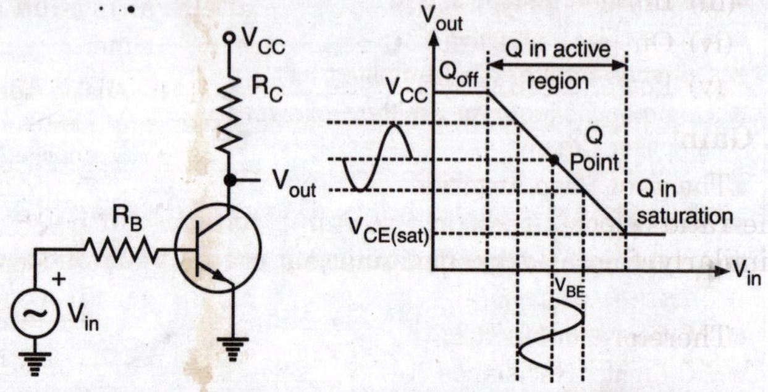

(a) A transistor circuit (b) Transfer characteristics

Fig. 2.2.

Working Operation

The d.c. voltage at the base is say V_{BB}. If a time varying signal say a sine wave is superimposed on the d.c. voltage V_{BB}, then the output voltage will change along the transfer curve. This will produce a time varying output voltage. If the time varying output voltage is proportional to and larger than the time varying input signal, then the circuit is a linear amplifier. If the transistor is not biased in its forward active mode, then its output will not vary proportional to the input. Hence, the circuit will no more act as a linear amplifier.

2.3.1 Block Diagram of an Amplifier

The block diagram of an amplifier has been shown in figure 2.3.

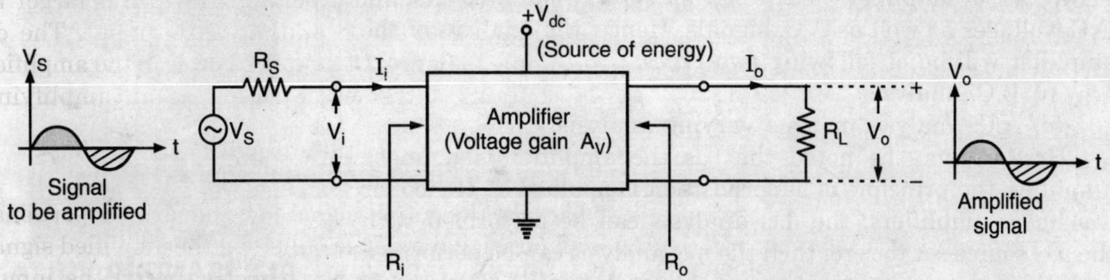

Fig. 2.3 Block diagram of an amplifier showing important parameters

In order to magnify the input signal V_s, all the amplifiers need a source of energy. This may be provided by a battery or a dc power supply. The dc power supply is also essential for biasing the BJT used in the amplifier circuit. The amplifier should contain atleast one active device such as a transistor or field effect transistor (FET) or an operational amplifier (op-amp). If a transistor is used, then it should be biased in the active region. In figure 2.3, the source V_S represents the ac input signal to be amplified. The amplified signal is applied to the load resistance R_L.

2.3.2 Amplifier Characteristics

For comparing the amplifier performance, we have to define certain important characteristics. Some of them are:

(i) The voltage gain A_V
(ii) The current gain A_I
(iii) Input resistance R_i
(iv) Output resistance R_o
(v) Power gain A_p

1. Gain

The gain of an amplifier is defined as the ratio of output quantity to the input quantity. Hence, the ratio of output voltage to input voltage will be called as **voltage gain A_V** of the amplifier. Similarly, the ratio of output current to input current is called as **current gain A_I** of the amplifier.

Therefore, current gain, $$A_I = \frac{I_o}{I_i} \quad ...(2.1)$$

and voltage gain, $$A_V = \frac{V_o}{V_i} \quad ...(2.2)$$

The voltage and current gains of an amplifier should be as large as possible.

2. Input resistance (R_i)

It is resistance seen looking into the input terminals of an amplifier. More generally, an input impedance Z_i is defined, however Z_i is resistive for the transistorised amplifiers, therefore, it is replaced by R_i.

3. Output resistance (R_o)

It is the resistance seen looking into the output terminals of an amplifier when the input signal source V_S is short circuited ($V_i = 0$) and output terminals are open circuited.

Thus, current gain, voltage gain, input resistance and output resistance are the four important parameters of an amplifier.

2.4. AMPLIFIER ANALYSIS

Figure 2.1 of amplifier block diagram shows that it consists of D.C. voltages as well as A.C. signals. Hence, the analysis of the amplifier will be of following two types:

(i) D.C. analysis

(ii) A.C. analysis or time varying analysis

Here, it may be noted that as the amplifier is a linear amplifier, the principle of superposition is applicable. Hence for the linear amplifiers, the d.c. analysis can be performed with the a.c. source set to zero, then the a.c. analysis can be performed with the dc source set to zero and the total response is the sum of the d.c. and a.c. responses.

ALESSANDRO VOLTA (1745-1827)

In 1796 Italian physicist *Alessandro Volta (1745-1827)* invented the first chemical battery, which provided the first practical source of electricity.

D.C. analysis

The D.C. analysis includes, obtaining the Q-point co-ordinates (V_{CEQ}, I_{CQ}) and plotting of D.C. load line. We have already done this earlier. Therefore, let us see how to perform the a.c. analysis.

2.5. SMALL SIGNAL ANALYSIS

In this chapter, let us perform the a.c. analysis of bipolar transistor amplifiers. For A.C. analysis, we shall use following two techniques:

(i) Graphical analysis, and

(ii) Mathematical analysis using a small signal equivalent model.

Notations used

The notations used in this chapter have been summarised in table 2.1.

Table 2.1. Summary of notations

S.No.	Variable	Meaning
1.	i_B, v_{BE}, i_C, etc	Instantaneous values
2.	I_B, I_C, V_{BE}, V_{CE} etc	D.C. values
3.	I_b, I_c, V_{be}, V_{ce} etc	A.C. values

2.6. TYPES OF BJT AMPLIFIERS

The transistor amplifiers may be classified into following three categories:

(i) Common emitter (CE) amplifier

(ii) Common base (CB) amplifier

(iii) Common collector (CC) amplifier.

Let us discuss them one by one.

2.7. COMMON EMITTER (CE) AMPLIFIER

1. Basic Concepts

Let us consider a practical transistor amplifier circuit, discuss its operation and then analyze the behaviour of this circuit using the dc load line. A single stage RC coupled amplifier using transistor as an active device has been shown in figure 2.4.

2. Description

The capacitors C_1 and C_2 are called as the coupling capacitors. As the load resistor R_L is coupled to the amplifier through the coupling capacitor, this amplifier is called as RC coupled amplifier. The transistor is connected in the common emitter (CE) amplifier configuration. Therefore, this amplifier is called CE amplifier.

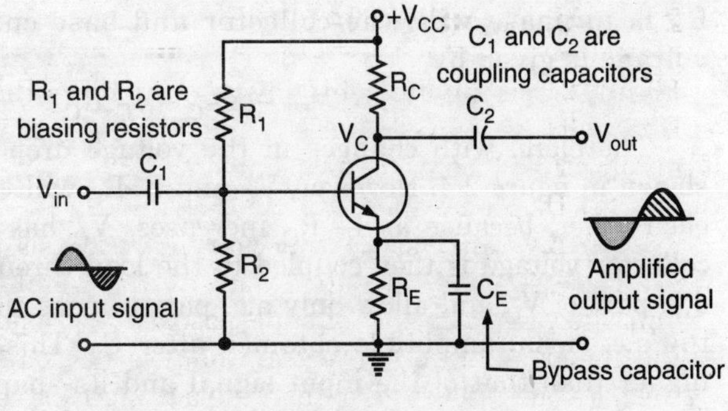

Fig. 2.4. Single stage RC coupled CE amplifier

(i) Circuit Components and their functions

Let us now know about the function of each component in the RC coupled amplifier. Resistors R_1, R_2 and R_E are used for biasing the transistor in the active region because for operating the transistor as an amplifier, it is necessary to bias it in the active region. The type of **biasing circuit used here is voltage divider bias or self bias.** R_C is the collector resistor used for controlling the collector current.

(ii) Input coupling capacitor C_1

The input coupling capacitor C_1 is used for coupling the ac input voltage V_{in} to the base of the transistor. As a capacitor blocks dc, this capacitor helps to block any dc component present in V_{in} and couples only the ac component of the input signal.

(iii) Bypass capacitor C_E

The capacitor connected in parallel with the emitter resistor R_E is called as the emitter bypass capacitor. This capacitor offers a low reactance to the amplified ac signal. Therefore, the emitter resistor R_E gets bypassed through C_E for only the a.c. signals. This will increase the voltage gain of the amplifier. Moreover, as C_E acts as an open circuit for d.c. voltages, it does not bypass R_E for d.c. conditions. Thus, presence of C_E does not alter the d.c. biasing conditions.

(iv) Output coupling capacitor C_2

This capacitor couples the amplifier output to the load resistance or to the next stage of the amplifier. It is used for blocking the d.c. part and passing only the a.c. part of the amplified signal to the load.

(v) Working Operation of the RC Coupled Amplifier

When the a.c. input signal is absent, the value of d.c. base current of the transistor is I_{BQ} which is called as the quiescent point base current. Corresponding to I_{BQ}, a quiescent d.c. collector current I_{CQ} also flows through the transistor ($I_{CQ} = \beta_{dc} I_{BQ}$) and the collector emitter voltage is V_{CEQ}.

All this happens because we have biased the transistor in its active region using the biasing components. When small a.c. sinusoidal signal is applied at the input of the amplifier, an alternating base current, starts flowing in the circuit. This base current varies above and below the Q point value of the base current (I_{BQ}) as shown in figure 2.4.

Thus, we can say that the a.c. signal is superimposed on the dc current I_{BQ}. Due to these variations in the base current, proportional variations take place in the collector current, because $I_C = \beta I_B$. As I_B increases, I_C also increases and with decrease in I_B. Also, I_C will also decrease as shown in figure 2.4. Thus, I_B and I_C are in phase but I_C is a magnified version of I_B, I_C varies above and below its Q point value I_{CQ}. This varying I_C passes through the collector resistor R_C to produce a varying voltage drop I_C R_C across it. This voltage drop I_C

R_C is in phase with the collector and base currents as shown in figure 2.5. The collector voltage is given by,

$$V_C = V_{CC} - I_C R_C \quad ...(2.3)$$

Therefore, with changes in the voltage drop $I_C R_C$, the collector voltage also will vary as shown in figure 2.4. However, V_C and $I_C R_C$ will vary in exactly opposite manner with respect to each other, because as $I_C R_C$ increases, V_C has to decrease according to equation (2.3). This collector voltage is then coupled to the load through the coupling capacitor C_2. It will block the d.c. part of V_C and allow only a.c. part to pass through as shown in figure 2.5. Let us consider the a.c. signal amplitude obtained after C_2. This is the output voltage. Its magnitude is much higher than that of the input signal and its shape is exactly same as that of the input signal. Thus, the input a.c. signal has been successfully amplified.

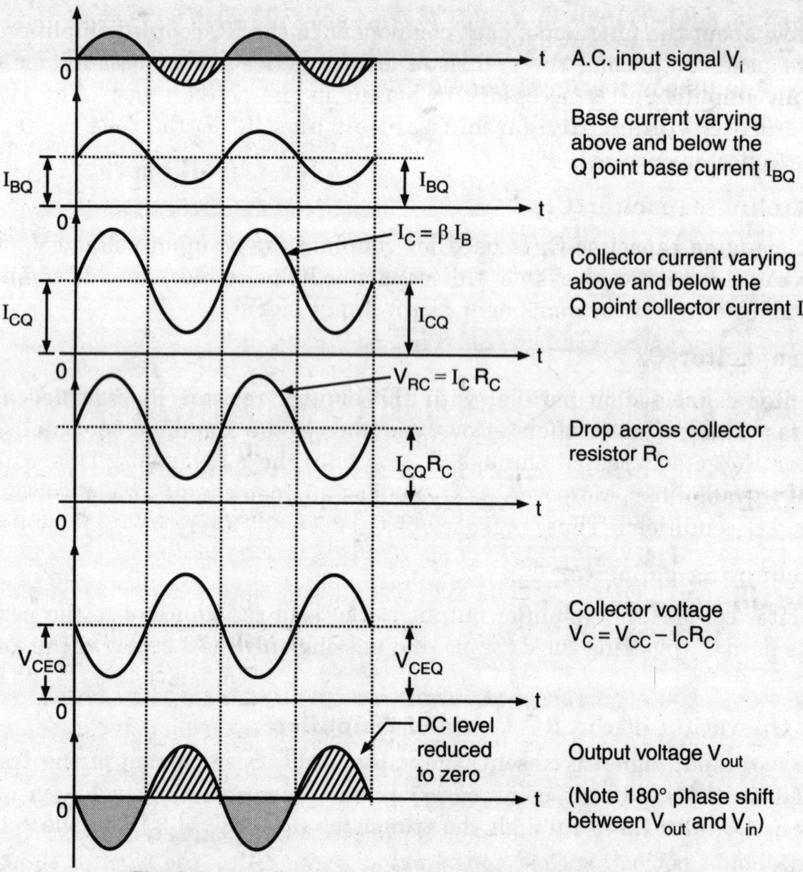

Fig. 2.5. Waveforms showing the process of amplification

3. Summary of working operation of a RC coupled CE amplifier

(i) In the absence of ac input signal, $I_B = I_{BQ}$, $I_C = I_{CQ}$ and $V_{CE} = V_{CEQ}$. The Q-point is selected to be in the active region of transistor.

(ii) As V_{in} is applied, the base current varies above and below I_{BQ}.

(iii) Hence, $I_C = \beta I_B$ varies above and below I_{CQ}. Variation in I_C is large.

(iv) Therefore, voltage across R_C varies because $V_{RC} = I_C \times R_C$.

(v) Hence, collector voltage V_C varies above and below V_{CEQ} as $V_C = V_{CC} - I_C R_C$.

(vi) Through C_2, only the ac part of V_C is coupled to the load. V_{out} is of same shape as V_{in}, but of larger size. Thus, amplification has taken place.

4. Phase relationship between input and output

As observed from figure 2.5, there is a 180° phase shift between the output and input or the output is said to be an inverted version of input. Thus, for a CE amplifier, there is a phase shift of 180° between V_{in} and V_{out}.

5. Role of coupling capacitor C_2

The coupling capacitor C_2 couples on the A.C. part of the signal to the output and blocks the D.C. part as shown in figure 2.4. Therefore, the signal obtained after the coupling capacitor C_2 has a zero D.C. value.

6. Does the frequency change in the amplification process?

Comparing the waveforms of V_{in} and V_{out}, we conclude that the frequency of the amplified output signal V_{out} is same as that of the input voltage. Thus in the amplification process, the frequency of the signal remains unchanged.

2.7.1. Graphical Analysis of the Amplifier with the help of DC load Line

The amplifier operation discussed earlier can also be explained with the help of DC load line. Let us consider figure 2.5 which illustrates the amplification process graphically.

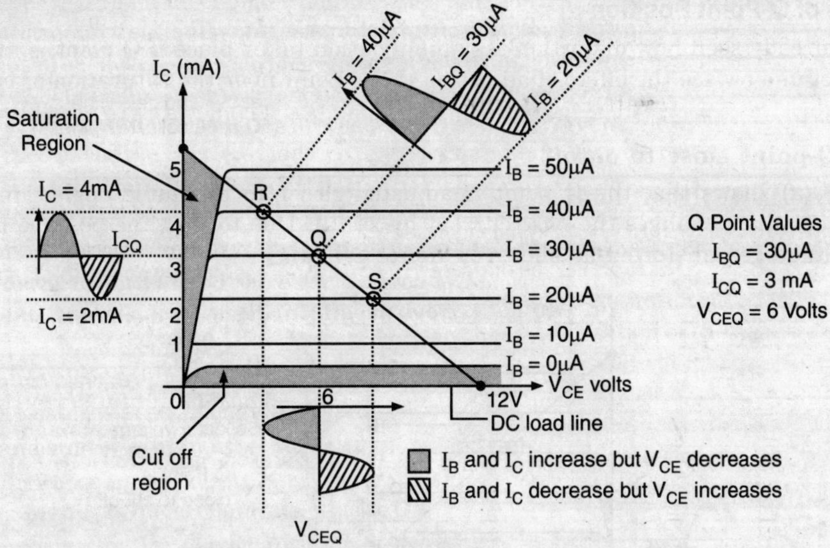

Fig. 2.6. Graphical representation of amplification process

1. Selection of Q-Point

The criteria for selecting quiescent point to operate the transistor as an amplifier is as under:

The Q point should not be too close to the cutoff region. Also, the Q point should not be too close to the saturation region. The above two conditions should be satisfied in order to avoid any waveform distortion in the amplified output signal. The Q point should be located at the centre of the d.c. load line so that the variation in the amplified voltage is equal corresponding to the positive and negative half cycles of the input signal. This will ensure that the amplified signal will be an exact replica of the input signal. In figure 2.5, the component values are so adjusted that the Q point is situated exactly at the centre of the dc load line. The co-ordinates of the Q point are as under:

$$\text{Q point} = (V_{CEQ}, I_{CQ}) = (6\text{V}, 3 \text{ mA}) \qquad ...(2.4)$$

The value of quiescent base current is i.e., $I_{BQ} = 30 \ \mu A$.

2. Amplification Process

As shown in figure 2.6, due to application of the input signal, the base current varies sinusoidally above and below I_{BQ}. The change in base current i.e.,

$$\Delta I_B = I_{B(max)} - I_{B(min)}$$

or $$\Delta I_B = 40\ \mu A - 20\ \mu A = 20\ \mu A$$

The operating point will move along the load line from Q to R and then from R to S corresponding to change in I_B. Due to this, the collector current I_C will change above and below I_{CQ} as shown in figure 2.6. Here, it may be noted that $\Delta I_C = \beta\, \Delta I_B = 2$ mA and I_C and I_B are in phase. The collector to emitter voltage V_{CE} (which is same as V_{out}) will vary above and below its Q point value V_{CEQ}. This is the amplified output voltage. Here, it may be noted that V_{CE} is 180° out of phase with I_C and I_B.

Important Point: Here, it may be noted that the shape of the output voltage is exactly same as that of the input. This means that no waveform distortion has taken place. Thus, when Q point is at the centre of the load line, no waveform distortion will take place.

2.7.2. Effects of Q-Point Position

We have already seen how distortionless amplification takes place if Q point is at the centre of load line. Let us now see the effect of adjusting the Q point near the saturation or cutoff region of the transistor.

1. Effect of Q-point close to cutoff

Figure 2.7 (a) shows that the Q point is adjusted closer to the cutoff region and an input signal is applied which changes the base current by 20 μA. Due to this, the positive half cycle of the output voltage gets distorted as shown in figure 2.7 (a).

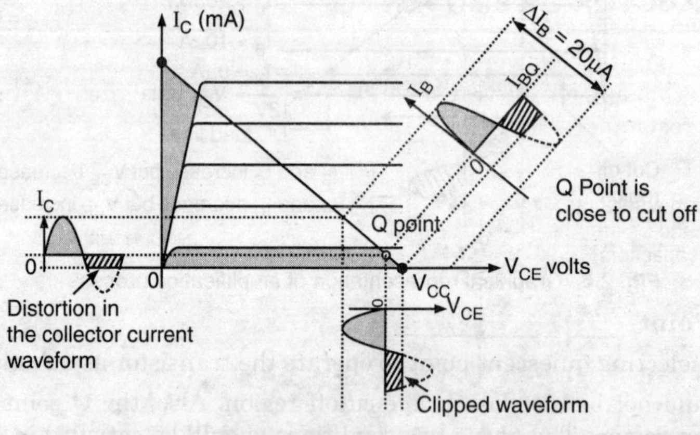

Fig. 2.7 (a) Effect of Q point close to cut-off region

2. Effect of Q point close to saturation

Figure 2.7 (b) shows the effect of Q point adjusted close to the saturation region. In the positive half cycle of I_B, the transistor goes into saturation. Hence, the collector current remains constant at $I_{C(sat)}$. Hence, the negative half cycle of the output voltage and positive half cycle of I_C gets distorted as shown in figure 2.7 (b).

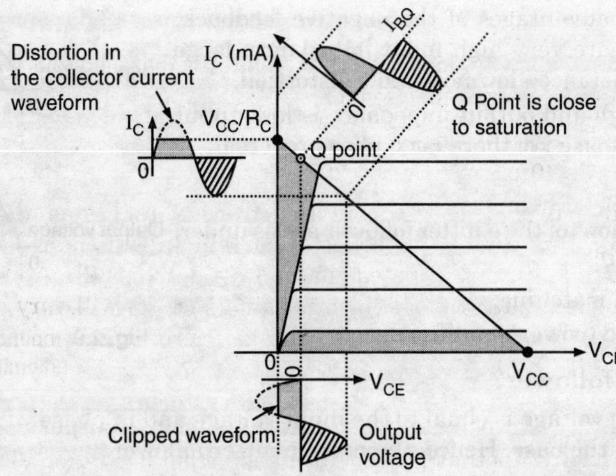

Fig. 2.7 (b) Effect of Q point close to saturation region

2.7.3. Salient Features of CE amplifier

(i) Input signal is applied at the base and amplified output is obtained at the collector.

(ii) Voltage gain is high.

(iii) Current gain is high

(iv) Input impedance is moderately high.

(v) Output impedance is moderately high.

(vi) There is a phase shift of 180° between the input and output.

2.8. COMMON COLLECTOR OR EMITTER FOLLOWER AMPLIFIER CIRCUIT

The common collector or emitter follower configuration has been shown in figure 2.8. The resistors R_1, R_2 and R_E provide dc biasing for the transistor. Here, C_1 and C_2 are the coupling capacitors and R_L is the external load resistance.

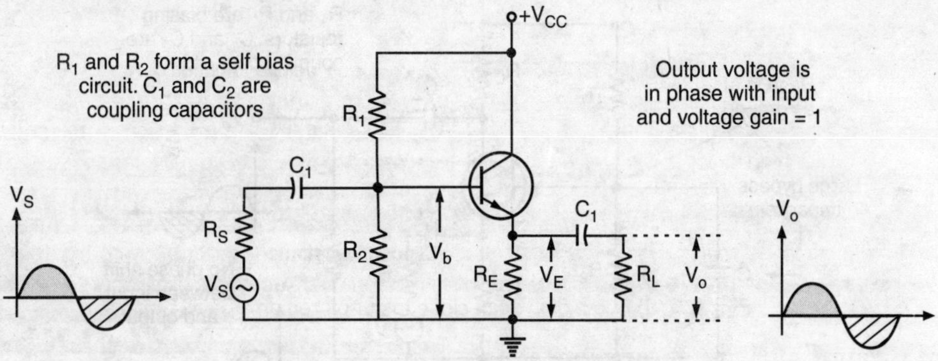

Fig. 2.8 Common collector amplifier circuit.

Waveforms

The input voltage is applied at the base with respect to ground and the output of amplifier is taken from the emitter with respect to ground. As the emitter voltage follows the base voltage, the gain of this amplifier is approximately is equal to 1. As shown in figure 2.8, the input and output voltage waveforms are in phase with each other. Also observe that the output voltage is exactly equal to the input voltage. Due to the voltage series negative feedback present in the emitter follower circuit, it posseses all the advantages of the negative feedback. Some of these advantages are very high input impedance, large bandwidth, low output resistance., low noise, low distortion.

Input impedance is high and output impedance is low. Input and output signals are in phase i.e. there is no phase reversal.

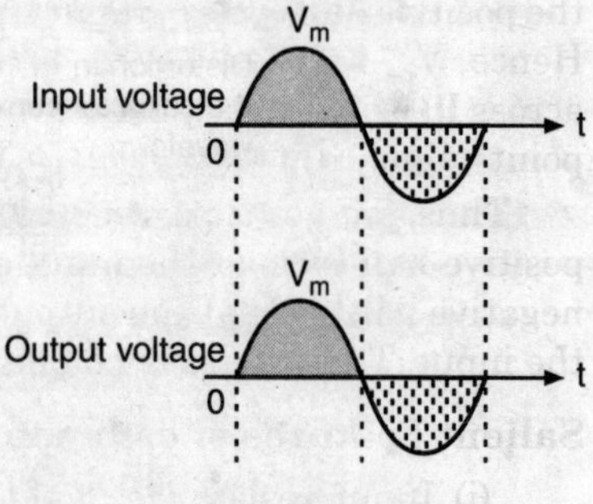

Fig. 2.9 Input output waveforms of an emitter follower

Applications

The important applications of the emitter follower are as under:

(i) As buffer amplifier.
(ii) For the impedance matching.
(iii) As the output stage (power amplifier).

Why is it called emitter follower?

As the output (emitter) voltage is equal to the input voltage and in phase with input voltage, it is said that emitter follows the base. Hence, the name emitter follower.

Salient Features of CC amplifier

(i) Input signal is applied at base and output is obtained at emitter.
(ii) Voltage gain is less than 1, but very close to 1 and current gain is high equal to $(1 + \beta)$.
(iii) Input impedance is high and output impedance is low.
(iv) Input and output signals are in phase i.e. there is no phase reversal.

Applications

(i) The applications of common collector amplifier are as buffer stages or as an output stage that can increase the current sourcing capacity.
(ii) It is also used for impedance matching.

2.9. COMMON BASE AMPLIFIER

The common base amplifier configuration has been shown in figure 2.10. The resistors R_1, R_2 and R_E provide the voltage divider biasing for the transistor.

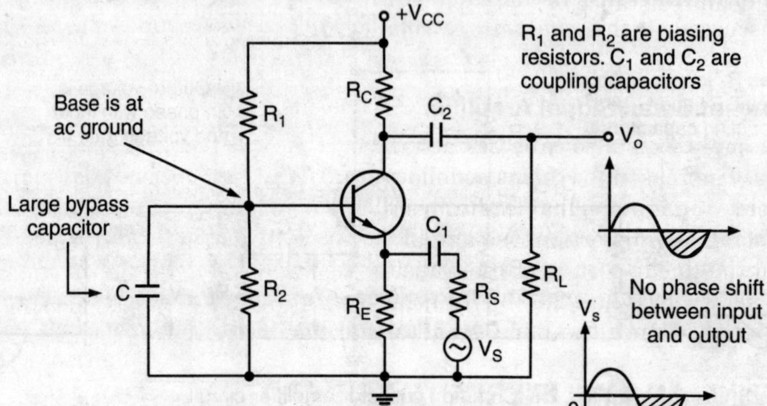

Fig. 2.10 A Common base amplifier

Here, C_1 and C_2 are the coupling capacitors and R_L is the external load.

Working Operation

The input is applied to the emitter while output is taken from the collector. The base is connected to ground (for ac signals only) via a large capacitor C which acts as a short circuit for ac signals. In the positive half cycle of input signal, the emitter voltage varies sinusoidally above its Q-point value. Hence, V_{be} reduces, hence, I_b and I_C will reduce. Thus, the voltage drop across R_C will reduce. Hence, the collector voltage will increase above its Q-point value.

Thus, we get a positive half cycle at the output corresponding to the positive half cycle at the input as shown in figure 2.11. Similarly, we get a negative half cycle at the output corresponding to the negative half cycle at the input. Thus, there is no phase shift between the input and output.

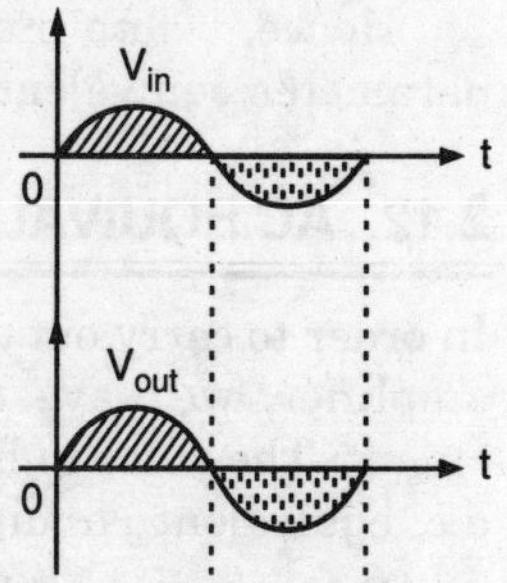

Fig. 2.11.

Salient Features of common base amplifier

(i) Input resistance of this configuration is very low, hence, source loading can take place.

(ii) Output resistance is very large.

(iii) Current gain is approximately equal to 1. So, there is no current amplification.

(iv) Voltage gain is high.

Applications

The common base amplifier are used for the applications where impedance matching is to be achieved. They are also used as **wideband amplifiers** i.e. the amplifiers having a large bandwidth.

2.10. SMALL SIGNAL OPERATION

If the amplitude of the input ac signal being applied to the amplifier is small (few mV), then the amplifier is called as a **small signal amplifier**, and the operation of the amplifier is called as **small signal operation**. As the input signal is small, the transistor will operate on the linear region of its transfer characteristics and produces a distortionless output as shown in figure 2.12.

This is why the small signal amplifiers are also called as **linear amplifiers**. All the amplifiers (CE, CB and CC) discussed in this chapter are small signal amplifiers.

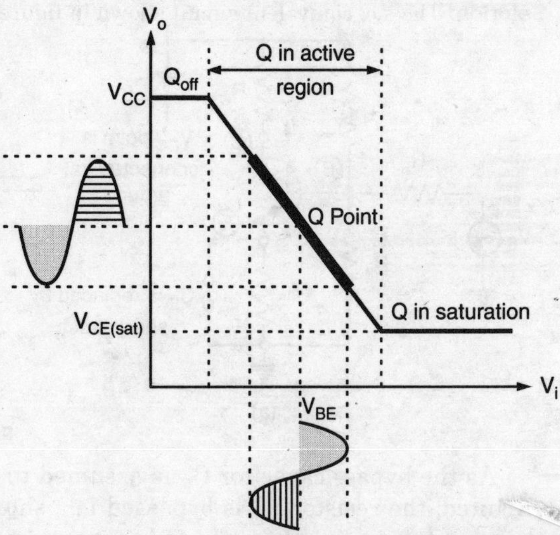

Fig. 2.12 Transfer characteristics.

2.10.1. Meaning of Small Signal Amplifier

The linear amplifier is also called as a small signal amplifier. The reason for this is as follows. In order to amplify the input signal faithfully i.e., without introducing any distortion, the amplifier must operate in the linear portion of the transfer characteristics shown in figure 2.6 (b). To achieve this, the amplitude of the input signal has to be sufficiently small. Hence the name small signal amplifiers.

2.11. ANALYZING AN AMPLIFIER: ANALYSIS OF TRANSISTOR AMPLIFIER

In this chapter, let us analyze various transistor amplifier configurations. The procedure to be followed for such an analysis is as under :

Procedure for analysis

(i) First, we draw a.c. equivalent circuit of the given amplifier.

(ii) Then, we replace the transistor by its h-parameter equivalent circuit.

(iii) Lastly, we calculate voltage gain, current gain, input impedance and output impedance.

Hence, before proceeding further, we have to learn about the a.c. equivalent circuit and h-parameter equivalent circuit.

2.12. AC EQUIVALENT CIRCUIT

In order to carry out only the a.c. analysis of the amplifier, we have to use its a.c. equivalent circuit. The rules to be followed for drawing the a.c. equivalent circuit are as under :

Rules to be followed to draw the a.c. equivalent circuit

(i) First, we replace all the d.c. sources by a short circuit. The d.c. voltages are important only in deciding the Q-point of the amplifier.

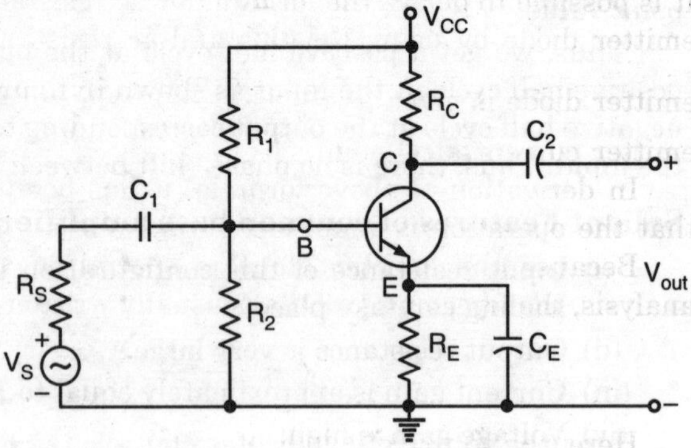

Fig. 2.13. CE single stage transistor amplifier

(ii) Then, the coupling capacitors C_1 and C_2 and the bypass capacitor C_E are to be replaced by short circuit.

This is because the reactance of these capacitors is very small at the frequency of application.

EXAMPLE 2.1. Draw the a.c. equivalent circuit of the amplifier shown in figure 2.13.

Solution: The a.c. equivalent circuit shown in figure 2.14 is drawn from figure 2.13.

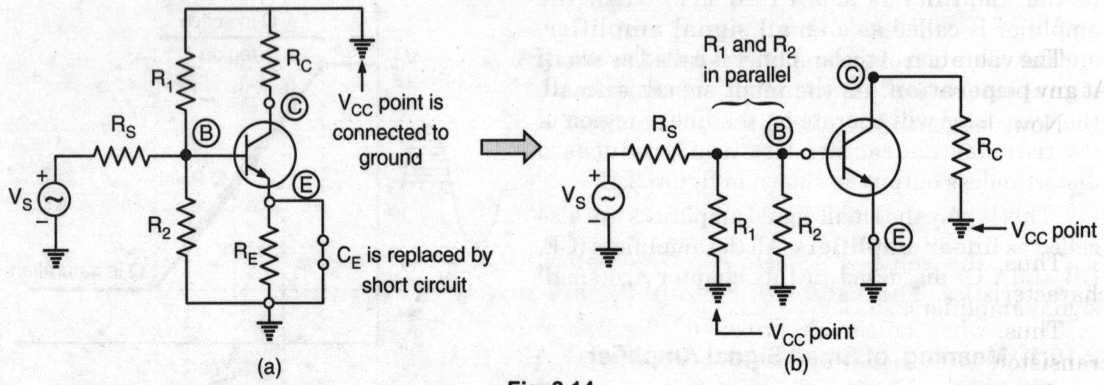

Fig. 2.14

As the bypass capacitor C_E is assumed to be short circuited, the resistor R_E is bypassed i.e., shorted out, thus, emitter is directly connected to ground point.

As the dc supplies are to be replaced by short circuits, the + V_{CC} point is connected to ground. Therefore, R_C appears between collector and ground and R_1 appears in parallel with R_2 as shown in figure 2.14 (b). The ac equivalent circuit has been shown in figure 2.14 (a). The final a.c. equivalent circuit has been shown in figure 2.14 (c).

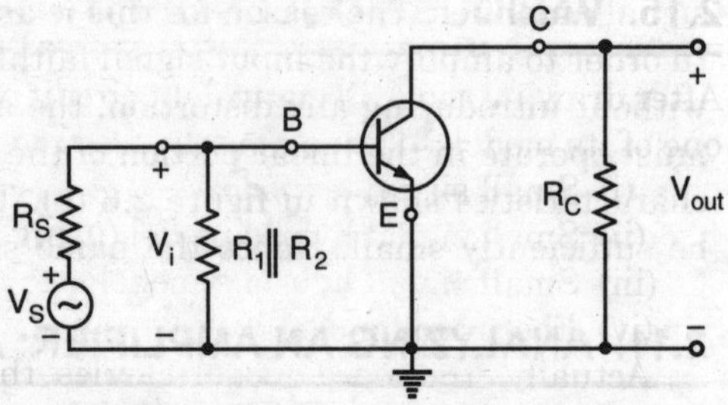

Fig. 2.14 (c) AC equivalent circuit of the amplifier shown in figure 7.13

2.13. AC EMITTER RESISTANCE

The ac or dynamic resistance of the emitter-base junction (or simply emitter) diode is called the **ac emitter resistance**. It is given as change in base-emitter voltage divided by the corresponding change in emitter current *i.e.*,

$$r_{ac} = \frac{\Delta V_{BE}}{\Delta I_E} \text{ (Figure 2.15)} \quad ...(2.5)$$

It is possible to derive the formula for ac resistance of the emitter diode by using calculus and ac resistance of the emitter diode is given as $r_{ac} = \frac{25 \text{ mV}}{I_E}$, where I_E is the dc emitter current at Q-point.

In derivation of above formula, it has been assumed that the operation is small signal one.

Because the emitter diode is special in transistor analysis, the foregoing formula is usually written as below:

$$r'_e = \frac{25 \text{ mV}}{I_E} \quad ...(2.6)$$

Here, the subscript *e* indicates emitter, the lower case *r* is used to indicate ac or dynamic resistance and the prime,', indicates the internal resistance.

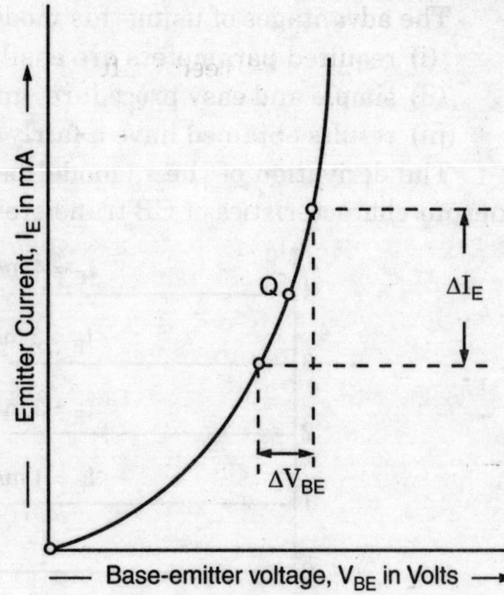

Fig. 2.15

2.14. AC BETA (β_{ac})

We know about β_{dc} as under:

$$\beta_{dc} = \frac{I_C}{I_B} \quad ...(2.7)$$

The value of β_{dc} can be obtained from the output characteristics of transistor in CE configuration. At any point on the characteristics, we can calculate β_{dc} by taking the ratio of I_C and I_B at that point.

Now, let us define AC beta of a transistor as under:

$$\beta_{ac} = \left.\frac{\Delta I_C}{\Delta I_B}\right|_{V_{CE} \text{ constant}} \quad ...(2.8)$$

Thus, the value of ac beta can be obtained at a constant value of V_{CE} from the output characteristics. **The values of β_{dc} and β_{ac} are nearly the same.**

Thus, when no signal is applied, then the ratio of I_C to I_B is called the **dc beta (β_{dc})** of a transistor. With signal applied, the ratio of ΔI_C to ΔI_B is called as the **ac beta (β_{ac})**.

2.15. VARIOUS MODELS USED FOR A.C. ANALYSIS

After drawing the A.C. equivalent circuit of an amplifier, the next step is to replace the transistor by one of its models. The commonly used models for the small signal a.c. analysis are as under:

(i) Small signal r_e model
(ii) Small signal hybrid model (using h-parameters)
(iii) Small signal hybrid π model
(iv) Ebers Moll model.

Actually, the word model carries the meaning of equivalent circuit. Let us discuss these equivalent circuits one by one.

2.16. THE r_e MODEL OF TRANSISTOR.

As the h-parameter values vary considerably with individual transistors of the same type and the accuracy of the analysis depends on the operating conditions, an alternate model known as r_e model which employs transistor β and resistance values is gaining importance now-a-days.

The advantages of using this model are as under :

(i) required parameters are easily available

(ii) simple and easy procedure, and

(iii) results obtained have a fairly good accuracy for the study of amplifier circuits.

The derivation of the r_e model begins with a close examination of approximated input and output characteristics of CB transistor configuration shown in figures 2.16 (a) and (b).

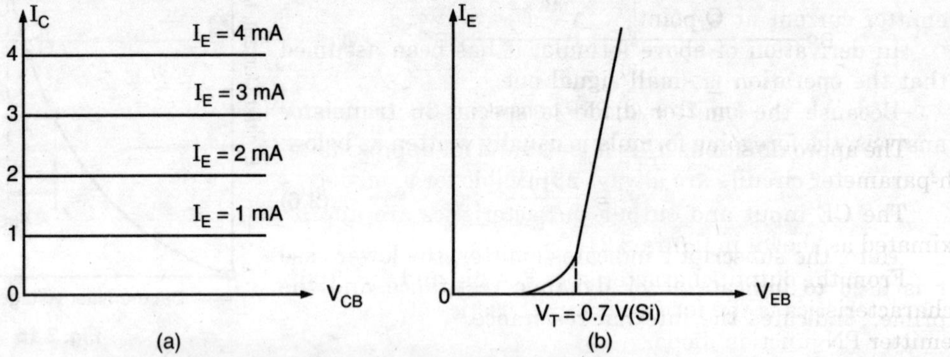

Fig. 2.16. (a) Output characteristic of CB configuration.
(b) Input characteristic of CB configuration

From the output characteristics $I_C = I_E$ and the input characteristics is the forward characteristic of the emitter-base PN junction diode. Thus, the equivalent circuit of the PNP transistor for d.c. conditions is shown in figure 2.17.

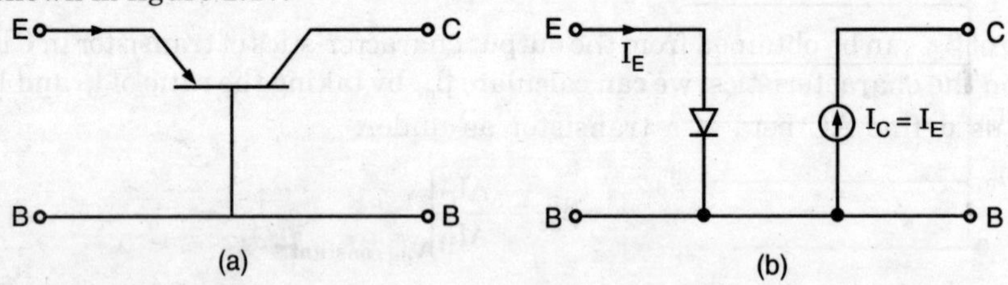

Fig. 2.17. (a) PNP transistor in CB configuration (b) DC equivalent circuit.

For an NPN transistor, the orientation of the diode and the direction of current flow are reversed as shown in figure 2.18.

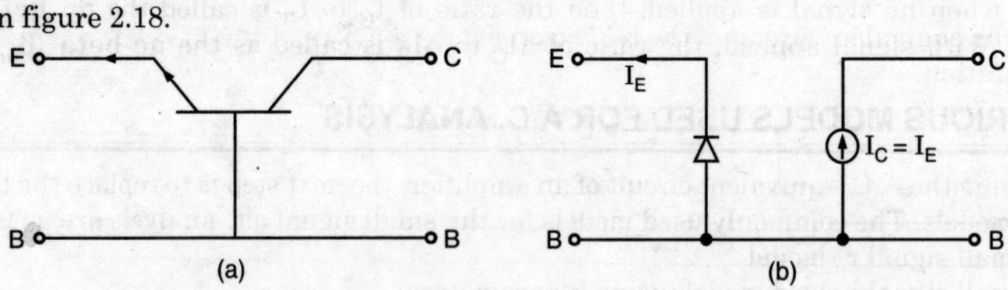

Fig. 2.18. NPN transistor in CB configuration (b) DC equivalent circuit

For a.c conditions, i.e., when small input a.c. signals are applied, the emitter node does not rectify, instead it offers a resistance called a.c resistance. Since the diode current is the emitter current, the notation for diode resistance is r_e and its value is determined by

$$r_{a.c} \cong r_e = \frac{26\text{mV}}{I_E}\,\Omega \qquad ...(2.9)$$

Substituting r_e in figure 2.17 gives the r_e equivalent circuit for CB configuration shown in figure 2.19. The approximate h-parameter equivalent circuit of a transistor in CB configuration is shown in figure 2.20 for comparison. Comparing figures 2.19 and 2.20, we have

$$h_{ib} = r_e \quad ...(2.10)$$
$$h_{fb} = -1 \quad ...(2.11)$$

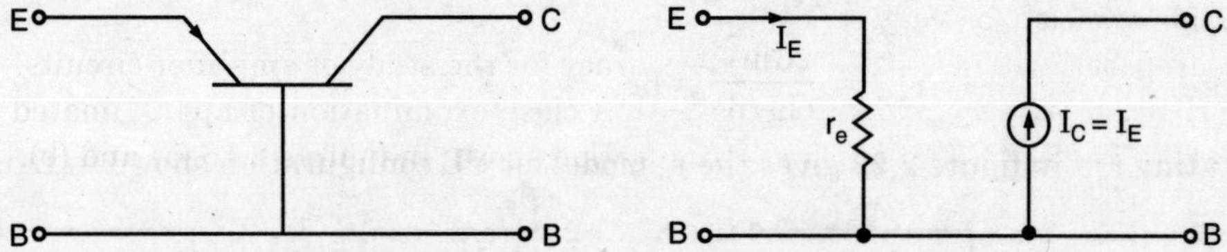

(a) (b)

Fig. 2.19. (a) PNP transistor in CB configuration (b) r_e equivalent circuit.

The approximations $h_{rb} = h_{ob} = 0$ valid for approximate h-parameter circuits are always applicable for r_e model.

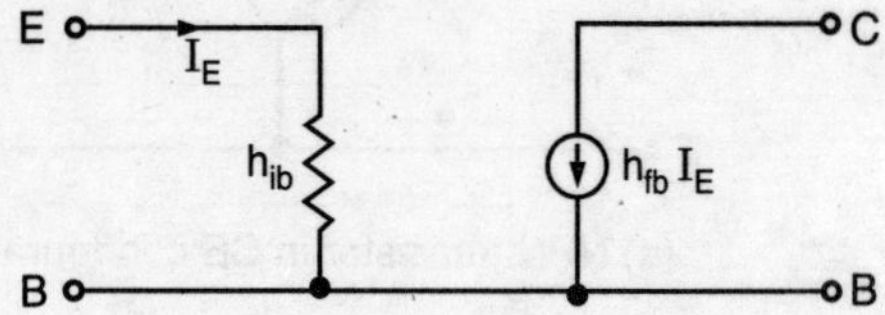

Fig. 2.20. Approximate h-parameter equivalent circuit for CB configuration.

The CE input and output characteristics are approximated as shown in figure 2.21.

From the output characteristics $I_C = \beta I_B$ and the input characteristics is the forward characteristics of the base-emitter PN junction diode.

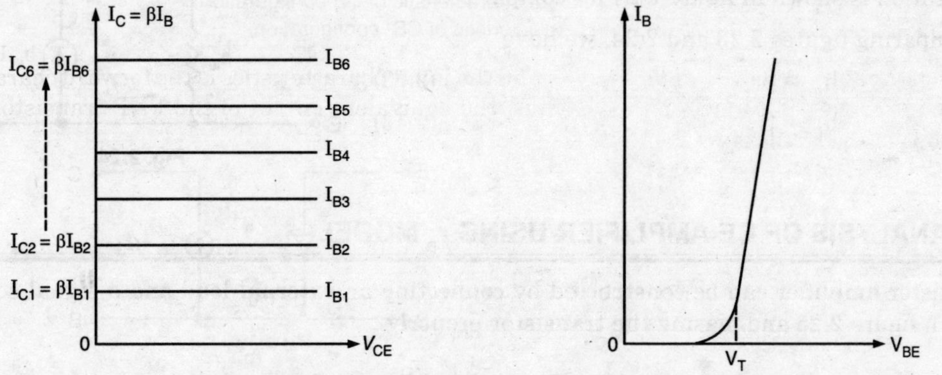

(a) Output characteristics of CE configuration (b) Input characteristics of CE configuration

Fig. 2.21.

Thus, the equivalent circuit of the NPN transistor for d.c. conditions is shown in figure 2.22 for CE configuration.

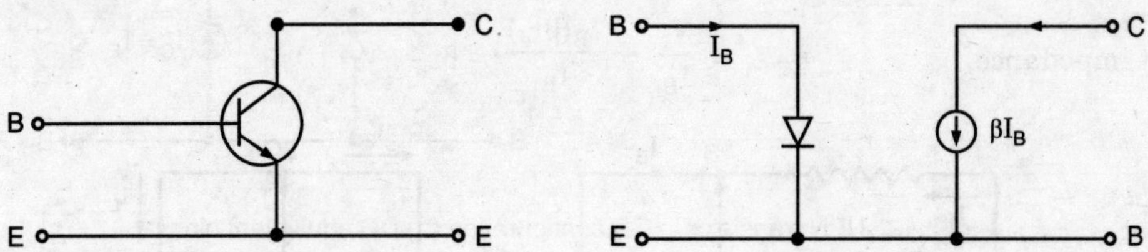

(a) NPN transistor in CE configuration (b) DC equivalent circuit.

Fig. 2.22.

For a.c conditions, the resistance offered by the base-emitter junction is given by

$$r_{a.c.} = \frac{26\,\text{mV}}{I_B}$$

But $$I_B = \frac{I_C}{\beta} = \frac{I_E}{\beta}$$

Therefore, $$r_{a.c.} = \frac{26\,mV}{I_E/\beta} = \beta r_e$$

Substituting $r_{a.c.}$ in figure 2.22 gives the r_e model for CE configuration shown in figure 2.23.

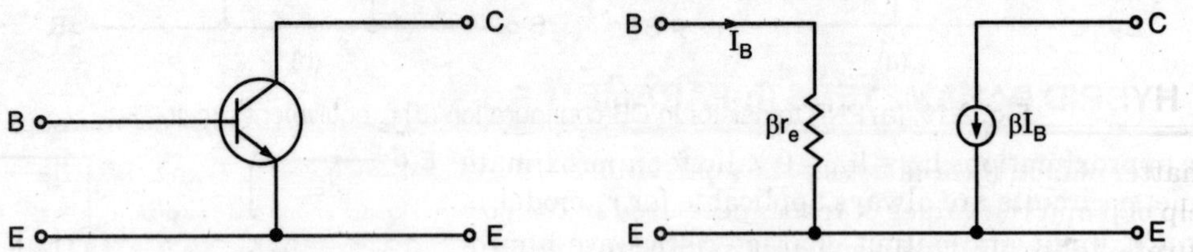

(a) NPN transistor in CE configuration (b) r_e equivalent circuit.

Fig. 2.23.

The approximate hybrid equivalent circuit for the CE configuration is shown in figure 2.24 for comparison.

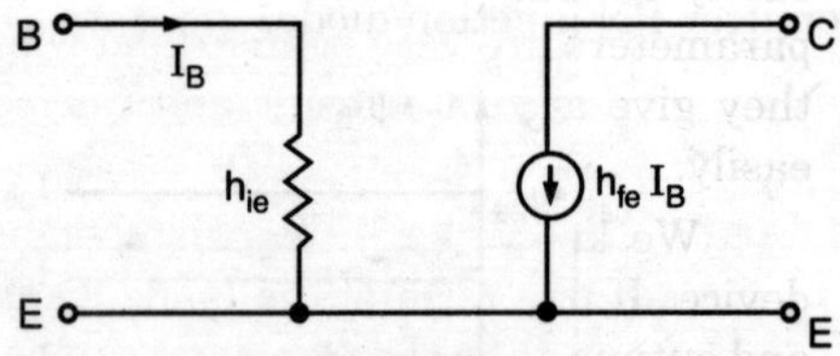

Fig. 2.24.

Comparing figures 2.23 and 2.24, we have

$$h_{ie} = \beta r_e \quad \text{...(2.12)}$$

$$h_{fe} = \beta \quad \text{...(2.13)}$$

2.17. ANALYSIS OF CE AMPLIFIER USING r_e MODEL

A transistor amplifier can be constructed by connecting an external load and a signal source as shown in figure 2.25 and biasing the transistor properly.

Voltage gain, $$A_v = \frac{V_2}{V_1} = \frac{-\beta\, I_B\, R_L}{I_B\, \beta r_e} = \frac{-R_L}{r_e} \quad \text{...(2.14)}$$

Current, gain, $$A_i = \frac{I_L}{I_i} = \frac{-\beta I_B}{I_B} = -\beta \quad \text{...(2.15)}$$

Input impedance, $$Z_i = \frac{V_1}{I_B} = \frac{I_B(\beta r_e)}{I_B} = \beta r_e \quad \text{...(2.16)}$$

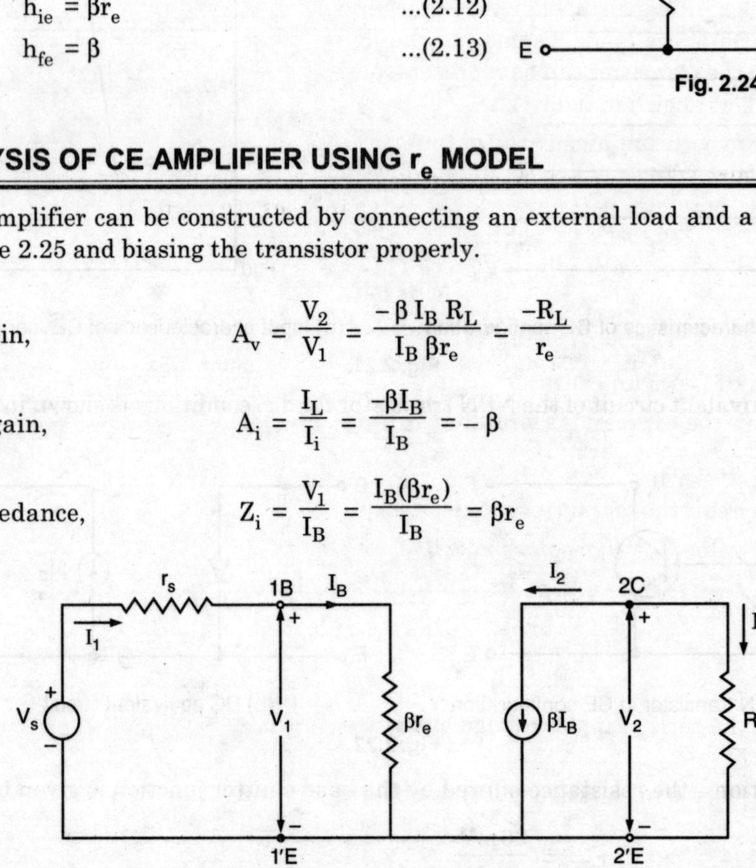

Fig. 2.25. CE amplifier using r_e model.

Output admittance $Y_0 = \frac{I_2}{V_2}$ with $V_S = 0$ and $R_L = \infty$

with $V_S = 0$, $I_B = 0$ and $I_2 = 0$

Hence, $Y_0 = 0$

Output impedance, $Z_0 = \frac{1}{Y_0} = \infty$

2.18. HYBRID PARAMETERS (h-PARAMETERS)

As a matter of fact, if we carry out the analysis of a transistor amplifier, to study its behaviour, with the help of β and the values of resistances used in the circuit, then this type of analysis is not found so accurate. In fact, the more accurate method to analyse a transistor amplifier is hybrid parameters or h-parameters method. We can study h-parameters at low-frequency and at high frequency.

Hybrid parameters are also called as h-parameters. Characteristics of amplifier can be specified easily in terms of h-parameters. The hybrid or h-parameters are very popular and widely used because they give accurate result and can be measured very easily.

We know that a transistor is a three terminals device. If one terminal is made common to the input and output, then the transistor can be represented as a two-port network as shown in figure 2.26.

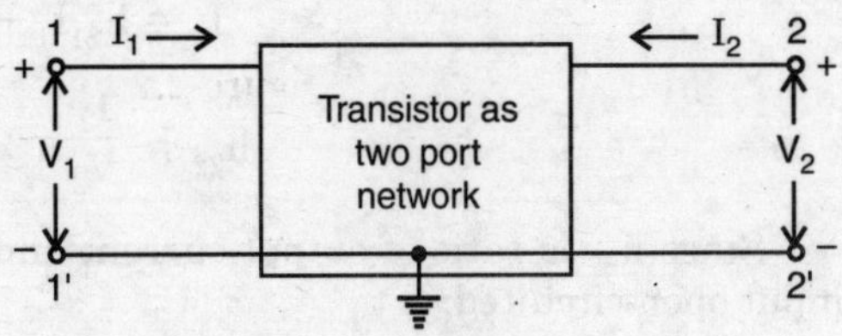

Fig. 2.26. Transistor as a two port network

There are two ports, input port and output port. For each port, there are two variables, currents and voltages. Figure 2.26 shows the standard directions of currents and polarities of voltages for a two-port network. For a two-port network, the four variables are related in terms of h-parameters as

$$V_1 = h_{11}I_1 + h_{12}V_2 \quad ...(2.17)$$

$$I_2 = h_{21}I_1 + h_{22}V_2 \quad ...(2.18)$$

The parameters h_{11}, h_{12}, h_{21} and h_{22} are called hybrid parameters or h-parameters.

Determination of h-parameters

We can derive the expressions for different h-parameters as under :

(i) Value of h_{11} and h_{21}

If we short-circuit the output terminals, then $V_2 = 0$

Putting this value $V_2 = 0$ in equation (2.17), we get

$$V_1 = h_{11}I_1 + h_{12}V_2 = h_{11}I_1 + 0$$

or

$$h_{11} = \frac{V_1}{I_1}$$

Since h_{11} is the ratio of input voltage and current, it is called **input impedance** with output short-circuit.

Now putting $V_2 = 0$ in equation (2.18), we obtain

$$I_2 = h_{21}I_1 + h_{22}V_2 = h_{21}I_1 + 0$$

$$h_{21} = \frac{I_2}{I_1}$$

Since, h_{21} is ratio of output and input currents, it is called **forward current gain** with output short-circuit.

(ii) Value of h_{12} and h_{22}

If we open the input terminals, then $I_1 = 0$

putting $I_1 = 0$ in equation (2.17), we get

$$V_1 = h_{11}I_1 + h_{12}V_2 = 0 + h_{12}V_2$$

$$h_{12} = \frac{V_1}{V_2}$$

Since h_{12} is the ratio of input and output voltages, it is called **reverse voltage gain** with input open- circuited.

Now, putting $I_1 = 0$ in equation (2.18), we get

$$I_2 = h_{21}I_1 + h_{22}V_2 = 0 + h_{22}V_2$$

$$h_{22} = \frac{I_2}{V_2}$$

Since h_{22} is ratio of output current and output voltage, it is called **output admittance** with input open-circuited.

In short, h-parameters can be summarized as

1. $h_{11} = \frac{V_1}{I_1}$ = input impedance with output shorted and unit is ohm.

   This is also represented as h_i

   $h_{11} = h_i$

2. $h_{21} = \frac{I_2}{I_1}$ = forward current gain with output shorted and unit is none.

   This is also represented as h_f.

   $h_{21} = h_f$

3. $h_{12} = \frac{V_1}{V_2}$ = reverse voltage gain with input open circuit and has no unit.

   This is also represented as h_r

   $h_{12} = h_r$

4. $h_{22} = \frac{I_2}{V_2}$ = output admittance with input open circuit and has unit ohm^{-1} or mho.

   This is also represented as h_o.

   $h_{22} = h_o$

Important Point: *From above, it is clear that all the four parameters have different unit and thus the parameters are hybrid (mixed) in nature. Therefore, they are called **hybrid or h-parameters.***

Table 2.2. h-parameters and their meaning

S.No.	h-Parameter	Meaning	Condition
1.	h_{11}	Input impedance	Output is shorted
2.	h_{21}	Forward current gain	Output is shorted
3.	h_{12}	Reverse voltage gain	Input open
4.	h_{22}	Output admittance	Input open

2.19. h-PARAMETER REPRESENTATION FOR TRANSISTOR

The h-parameter representation for transistor depends upon the type of configuration (CE, CB, or CC). Therefore, each h-parameter is assigned a second subscript letter e, b or c. depending upon the type of configuration. The letter 'e' is used when the configuration is common emitter, the letter 'b' is used if the configuration is common base and letter 'c' is used if the configuration is common collector.

Table 2.3. Shows the h-parameter representation for three types of configuration

S.No.	h-Parameter	Types of Configuration		
		Common emitter	Common base	Common collector
1.	h_{11}	h_{ie}	h_{ib}	h_{ic}
2.	h_{21}	h_{fe}	h_{fb}	h_{fc}
3.	h_{12}	h_{re}	h_{rb}	h_{rc}
4.	h_{22}	h_{oe}	h_{ob}	h_{oc}

2.20. GENERAL HYBRID EQUIVALENT CIRCUIT

There are three transistor configurations CB, CE, and CC. Each configuration will have its separate hybrid or h-parameter equivalent circuit. But, first we will develop a general h-parameter equivalent circuit which can be applied for any of three configurations. The equivalent circuit thus obtained for any type of configuration is also called as small-signal, low-frequency hybrid model of a transistor.

DO YOU KNOW?

The symbol h_{FE} represents the forward current transfer ratio in the common-emitter configuration. The symbol h_{FE} is a hybrid (h)-parameter symbol. The h-parameter system is the most common system in use today for specifying transistor parameters.

Thus, h-parameter equations are

$$V_1 = h_{11}I_1 + h_{12}V_2$$

$$I_2 = h_{21}I_1 + h_{22}V_2$$

or

$$V_1 = h_i.I_1 + h_r.V_2 \quad \text{...(2.19)}$$

$$I_2 = h_f.I_1 + h_o.V_2 \quad \text{...(2.20)}$$

Since, each term of equation (2.19) has the units of volts, we can use Kirchoff's voltage law to find a circuit that fits this equation. Figure 2.27 shows this circuit.

Similarly, each term of equation (2.20) has the units of currents, we can use Kirchoff's current law to find a circuit that fits this equation. Figure 2.28 shows this circuit.

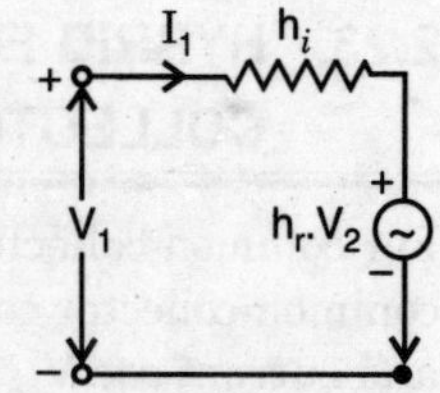

Fig. 2.27.

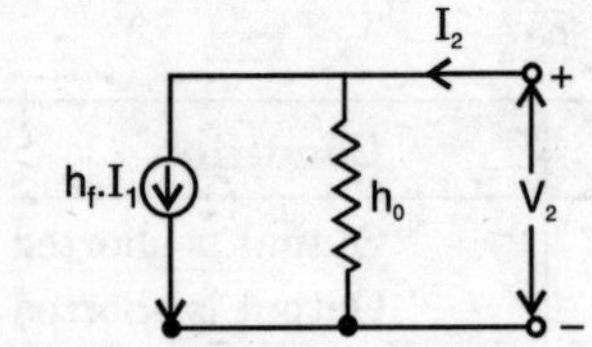

Fig. 2.28.

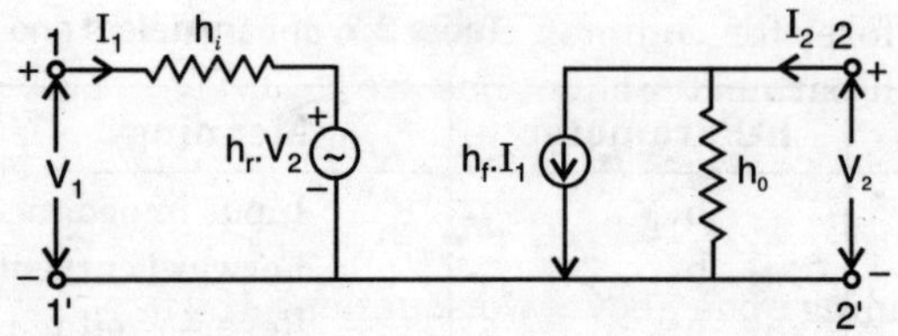

Fig. 2.29. General h-parameter equivalent circuit.

Combining these two figures, we get the general hybrid (h-parameter) equivalent circuit as shown in figure 2.29.

2.21. HYBRID EQUIVALENT CIRCUIT FOR COMMON EMITTER CONFIGURATION

We have already discussed that for common emitter configuration, an additional suffix 'e' is added to the h-parameters. Also, for common-emitter configuration, input voltage and current are V_{be} (base emitter voltage) and I_b (base current). The output voltage and current are V_{ce} and I_c.

Threfore, for common emitter configuration, the h-parameter equations are

$$V_{be} = h_{ie}.I_b + h_{re}.V_{ce} \qquad ...(2.21)$$

$$I_c = h_{fe}.I_b + V_{ce}.h_{oe} \qquad ...(2.22)$$

Depending upon above two equations, the h-parameter equivalent circuit for common-emitter configuration is shown in figure 2.30.

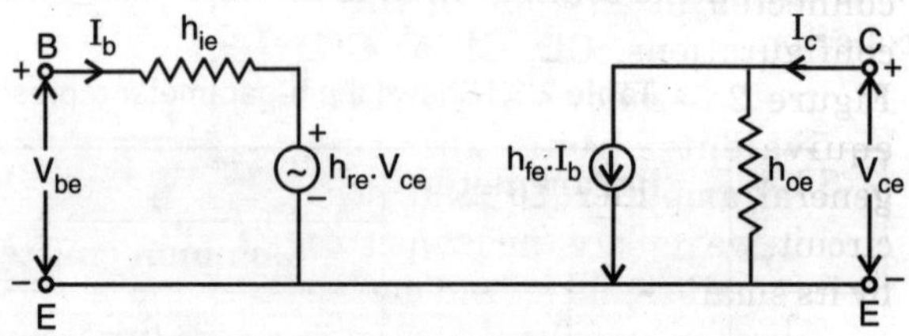

Fig. 2.30. Hybrid equivalent circuit for common-emitter configuration

2.22. HYBRID EQUIVALENT CIRCUIT FOR COMMON BASE CONFIGURATION

For common-base configuration, an additional suffix 'b' is added to the h-parameters. Also, for common-base configuration, input voltage and input current are I_e and V_{eb}. The output voltage and curent are V_{cb} and I_c.

Therefore, for common-base configuration, the h-parameter equations are

$$V_{eb} = h_{ib}.I_e + h_{rb}.V_{cb}$$

$$I_c = h_{fb}.I_e + V_{cb}.h_{ob}$$

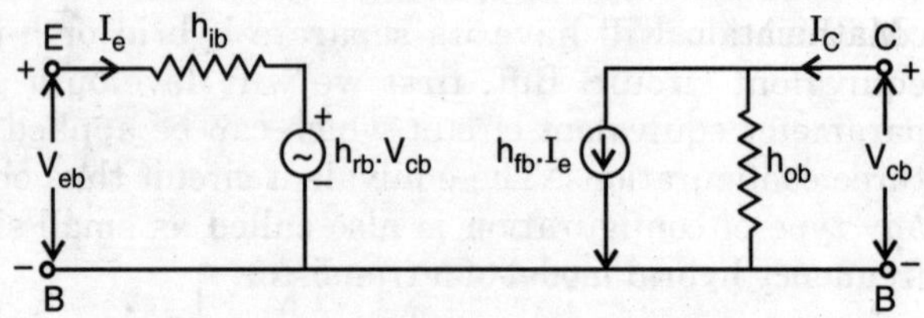

Fig. 2.31. Hybrid equivalent circuit for common-base configuration

Depending upon above two equations, the h-parameter equivalent circuit for common-base configuration is shown in figure 2.31.

2.23. HYBRID EQUIVALENT CIRCUIT FOR COMMON COLLECTOR CONFIGURATION

For common-collector configuration an additional suffix 'c' is added to the h-parameters. Also, for common-collector configuration, input voltage and input current are V_{bc} and I_b. The output voltage and current are V_{ec} and I_e.

Therefore, for common-collector configuration, the h-parameter equations are

$$V_{bc} = h_{ic}.I_b + h_{rc}.V_{ec}$$
$$I_e = h_{fc}.I_b + V_{ec}.h_{oc}$$

Depending upon above two equations, the h-parameter equivalent circuit for common-collector configuration is shown in figure 2.32.

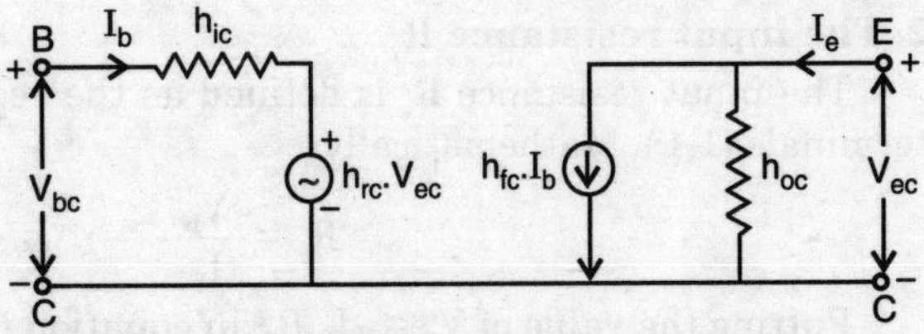

Fig. 2.32. Hybrid equivalent circuit for common-collector configuration

2.24. ANALYSIS OF A TRANSISTOR AMPLIFIER USING h-PARAMETERS.

Figure 2.33 shows the circuit of a general amplifier. In this circuit, the transistor may be connected in any one of three configurations, CE, CB or CC. Figure 2.34 shows the hybrid equivalent circuit for this general, amplifier. To obtain this circuit, we replace the transistor by its small-signal hybrid model.

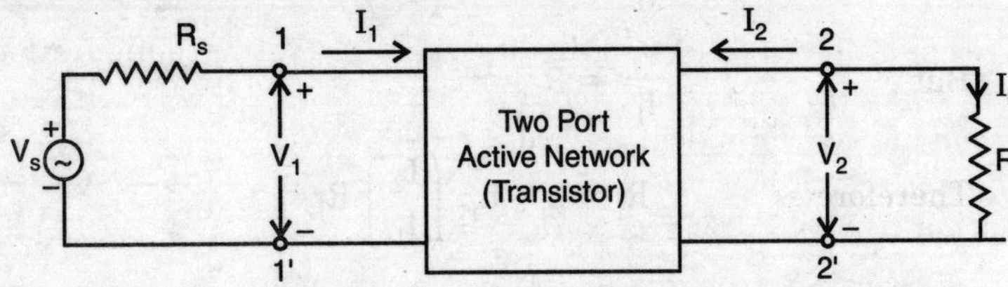

Fig. 2.33. General Amplifier Circuit

We shall derive expressions for current gain, voltage gain, input resistance, output resistance for amplifier in terms of h-parameters.

First writing the h-parameter equations as under:

$$V_1 = h_i.I_1 + h_r.V_2 \quad ...(2.23)$$
$$I_2 = h_f.I_1 + h_o.V_2 \quad ...(2.24)$$

1. The Current Gain

The current gain A_I for transistor amplifier is defined as the ratio of output to input currents. Mathematically,

$$A_I = \frac{I_L}{I_1} = -\frac{I_2}{I_1} \qquad \text{(Since, } I_L = -I_2\text{)}$$

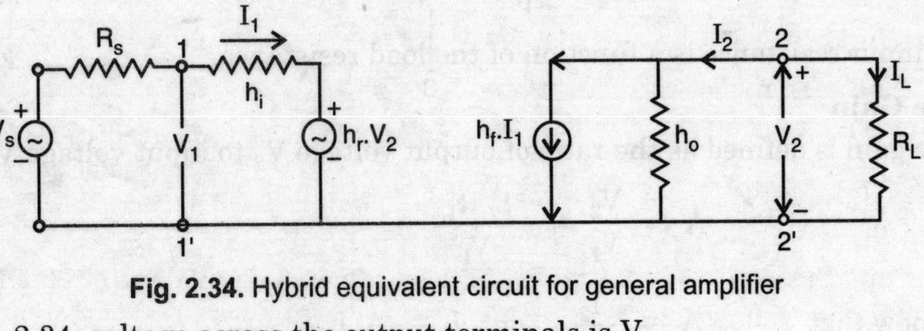

Fig. 2.34. Hybrid equivalent circuit for general amplifier

From figure 2.34, voltage across the output terminals is V_2

$$V_2 = I_L.R_L = -I_2R_L$$

Putting value of V_2 in equation (2.24), we get

$$I_2 = h_f.I_1 - h_0 I_2 R_L$$

$$I_2 (1+h_o.R_L) = h_f.I_1$$

$$\frac{I_2}{I_1} = \frac{h_f}{1+h_o.R_L} \quad \text{or} \quad A_I = \frac{-I_2}{I_1} = \frac{-h_f}{1+h_o.R_L} \quad ...(2.25)$$

2. The input resistance R_i

The input resistance R_i is defined as the resistance we see looking into the amplifier input terminals (1,1'). Mathematically.

$$R_i = \frac{V_1}{I_1}$$

Putting the value of $V_2 = -I_2.R_L$ in equation (2.23), we get

$$V_1 = h_i.I_1 + h_r.V_2 = h_i.I_1 + h_r.(-I_2R_L) = h_i.I_1 - h_r.I_2.R_L$$

Dividing the above equation on the both sides by I_1, we have

$$\frac{V_1}{I_1} = h_i - h_r.\left(\frac{I_2}{I_1}\right)R_L$$

But $\quad \dfrac{V_1}{I_1} = R_i$

Therefore $\quad R_i = h_i - h_r.\left(\dfrac{I_2}{I_1}\right).R_L$

We know that $\quad A_I = \dfrac{-I_2}{I_1}$

Thus, $\quad R_i = h_i + h_r.A_i.R_L$

Putting the value of $A_i = \dfrac{-h_f}{1+h_o.R_L}$ in last equation, we have

$$R_i = h_i + h_r.\left(\frac{-h_f}{1+h_o.R_L}\right).R_L$$

$$R_i = h_i - \frac{h_r.h_f}{h_o + \frac{1}{R_L}} = h_i - \frac{h_r.h_f}{h_o + Y_L} \qquad \text{...(2.26)}$$

where $\dfrac{1}{R_L} = Y_L$ = Load admittance.

Therefore, input resistance is a function of the load resistance.

3. The Voltage Gain

The voltage gain is defined as the ratio of output voltage V_2 to input voltage V_1.

$$A_v = \frac{V_2}{V_1} = -\frac{I_2.R_L}{V_1} \qquad \text{...(2.27)}$$

Since we know that $\quad A_i = -\dfrac{I_2}{I_1}$ or $I_2 = -A_i.I_1$

Therefore, $\quad A_v = \dfrac{A_i.I_1.R_L}{V_1} = A_i.R_L\left(\dfrac{I_1}{V_1}\right)$

But $\quad \dfrac{V_1}{I_1} = R_i$

$$A_v = \frac{A_iR_L}{R_i} \qquad \text{...(2.28)}$$

Again, since $$A_i = \frac{-h_f}{1+h_o.R_L}$$

and $$R_i = h_i - \frac{h_r \cdot h_f}{h_o + \frac{1}{R_L}}$$

Substituting the values of A_i and R_i in equation (2.28) and rearranging the terms, we have

$$A_v = -\frac{h_f.R_L}{h_i + (h_i.h_o. - h_r.h_f)R_L}$$

Replacing $$\Delta h = h_i h_o - h_r.h_f$$

Hence, $$A_v = -\frac{h_f.R_L}{h_i + \Delta h.R_L} \quad ...(2.29)$$

4. Output Resistance

The output resistance may be calculated by reducing the source voltage V_s to zero and load resistance R_L to infinity and driving the output terminals by a voltage generator V_2. This arrage-ment is shown in figure 2.35.

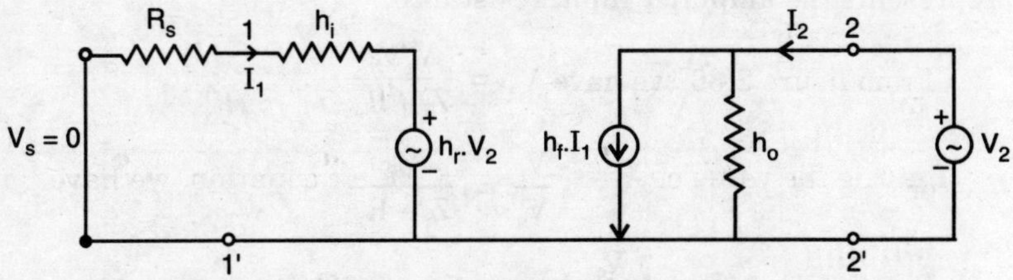

Fig. 2.35. Hybrid equivalent circuit to calculate output resistance

From figure 2.35, output resistance R_o is the ratio of voltage V_2 and the current drawn from the voltage source, I_2.

Mathematically, $$R_0 = \frac{V_2}{I_2}$$

But $$I_2 = h_f.I_1 + h_o.V_2$$

Hence $$R_0 = \frac{V_2}{h_f.I_1 + h_o.V_2} \quad ...(2.30)$$

Applying KVL to the input side of the circuit, we have

$$R_s.I_1 + h_r.V_2 + h_i.I_1 = 0$$

or $$I_1 (R_s + h_i) = -h_r.V_2$$

or $$I_1 = -\frac{h_r.V_2}{R_s + h_i} \quad ...(2.31)$$

Now, putting the value of I_1 in equation (2.30), we get

$$R_0 = \frac{V_2}{h_f.\left(\frac{-h_r.V_2}{R_s + h_i}\right) + h_o.V_2} = \frac{V_2}{\frac{-h_f.h_r.V_2}{R_s + h_i} + h_o.V_2}$$

Rearranging, we get $$R_0 = \frac{R_s + h_i}{h_o.(R_s + h_i) - h_f.h_r} = \frac{R_s + h_i}{R_s.h_o + (h_i h_o - h_f.h_r)} = \frac{R_s + h_i}{R_s.h_o + \Delta h}$$

where $$\Delta h = h_i h_o - h_f h_r$$

If source resistance R_s is zero, then output resistance will be

$$R_0 = \frac{h_i}{\Delta h} \quad ...(2.32)$$

5. The overall voltage gain

The overall Voltage gain of the transistor amplifier is the ratio of output voltage V_2 to the source voltage V_s. It is denoted by A_{vs}. Hence, A_{vs} is the voltage amplification factor taking into account the source resistance R_s.

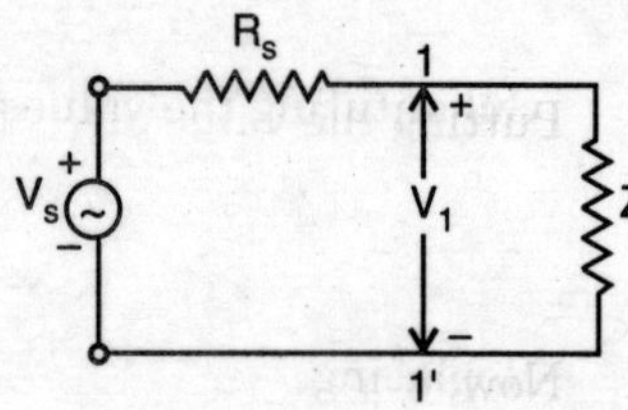

Fig. 2.36. Equivalent input circuit

Mathematically, overall voltage gain

$$A_{vs} = \frac{V_2}{V_s} = \frac{V_2}{V_1}.\frac{V_1}{V_s}$$

But $$A_v = \frac{V_2}{V_1}$$

Therefore, $$A_{vs} = A_v.\frac{V_1}{V_s} \quad ...(2.33)$$

To determine A_{vs}, we take help of the equivalent input circuit shown in figure 2.36. Here, Z_i represents the amplifier input resistance.

From figure 2.36, we have $$V_1 = \frac{V_s.Z_i}{Z_i + R_s}$$

or $$\frac{V_1}{V_s} = \frac{Z_i}{Z_i + R_s}$$

Putting the value of $\frac{V_1}{V_s}$ in equation (2.33), we get

$$A_{vs} = A_v \frac{Z_i}{Z_i + R_s} \quad ...(2.34)$$

Note that if $R_S = 0$.

Then $A_{vs} = A_v$

Therefore A_v is the voltage gain with an ideal voltage source ($R_s = 0$).

6. Overall current gain

Overall current gain of the transistor amplifier is the ratio of output current (I_L) to the current delivered by the source (I_s).

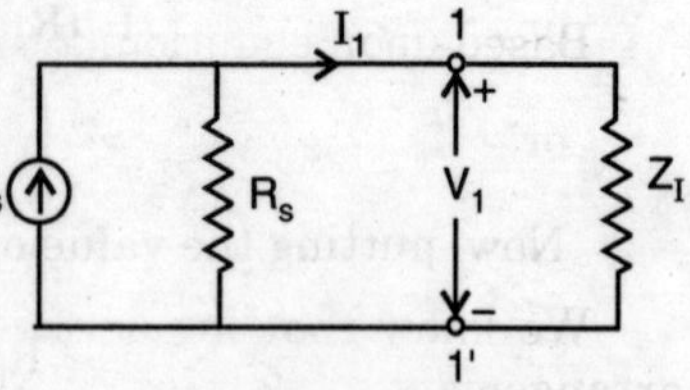

Fig. 2.37. Modified input equivalent circuit

Overall current gain

$$A_{is} = \frac{I_L}{I_s} = \frac{-I_2}{I_s}$$

Modifying the above expression, we have

$$A_{is} = \frac{-I_2}{I_1} \times \frac{I_1}{I_s}$$

But, $$\frac{I_2}{I_1} = A_i$$

$$A_{is} = -A_i \times \frac{I_1}{I_s} \quad ...(2.35)$$

To determine A_{is}, we take help of modified input equivalent circuit shown in figure 2.37. Here Z_i represent the amplifier input resistance.

From figure 2.37, we have

$$I_1 = I_s \times \frac{R_s}{R_s + Z_i}$$

or $$\frac{I_1}{I_s} = \frac{R_s}{R_s + Z_i}$$

Putting the value of $\frac{I_1}{I_s}$ in equation (2.35), we get

$$A_{is} = -A_i \times \frac{R_s}{R_s + Z_i} \quad ...(2.36)$$

Now, if $R_s = \infty$

then $A_{is} = A_i$

Therefore, A_i is the current gain with an ideal current source (infinite resistance)

2.25. h-PARAMETER (HYBRID) EXPRESSION FOR COMMON-EMITTER AMPLIFIER

The h-parameter expressions for CE amplifier will be obtained with the help of general h-parameter formulas discussed in last article.

1. Current Gain : The general expression for current gain is

$$A_i = -\frac{h_f}{1 + h_o.R_L} \quad ...(2.37)$$

For common-emitter amplifier, we add an additional subscript 'e'.

Hence, current gain for CE Amplifier is given by

$$A_i = -\frac{h_{fe}}{1 + h_{oe}.R_L} \quad ...(2.38)$$

Here R_L is the a.c. load resistance

2. Input resistance : The resistance looking into the amplifier input terminals (base of a transistor) is

General expression $$R_i = h_i - \frac{h_r.h_f}{h_o + \frac{1}{R_L}} \quad ...(2.39)$$

Based upon equation (2.39), expression for common emitter amplifier will be given by

$$R_i = h_{ie} - \frac{h_{re}.h_{fe}}{h_{oe} + \frac{1}{R_L}} \quad ...(2.40)$$

We know that input resistance of the amplifier stage, Z_i depends upon the type of biasing arrangement

As an example, for a fixed bias circuit, we have

$$Z_i = R_i \| R_B$$

If there is no biasing resistance

then $Z_i = R_i$ ...(2.41)

3. Voltage Gain : General Expression is expressed as

$$A_v = -\frac{h_f.R_L}{h_i + \Delta h.R_L} \quad ...(2.42)$$

Hence, expression for common emitter amplifier will be given by

$$A_v = -\frac{h_{fe}.R_L}{h_{ie} + \Delta h.R_L} \quad ...(2.43)$$

Here, $\Delta h = h_{ie}.h_{oe} - h_{re}.h_{fe}$

4. Output resistance : The resistance looking into the amplifier output terminals is general expression

$$R_0 = \frac{R_s + h_i}{R_s.h_o + \Delta h} \quad ...(2.44)$$

R_S = Source Resistance

$\Delta h = h_{ie}.h_{oe} - h_{re}.h_{fe}$

The output resistance of the amplifier stage will be

$$Z_0 = R_o \| R_L \quad ...(2.45)$$

5. Overall Voltage Gain :
General expression is given by

$$A_{vs} = A_v.\frac{Z_i}{Z_i + R_s} \quad ...(2.46)$$

Hence, expression for common emitter amplifier is also same

Thus, $$A_{vs} = \frac{A_v.Z_i}{Z_i + R_s} \quad ...(2.47)$$

6. Overall current Gain

$$A_{is} = -\frac{A_i.R_s}{R_s + Z_i} \quad ...(2.48)$$

EXAMPLE 2.2. A common emitter amplifier uses a voltage source having internal resistance R_s = 800Ω and the load resistance is R_L = 1000Ω. The h-parameter are h_{ie} = 1 KΩ, $h_{re} = 2\times10^{-4}$, h_{fe} = 50 and h_{oe} = 25 μA/volt. Calculate the current gain A_I, input resistance R_I, Voltage gain A_V.

Solution : Given that $R_s = 800\Omega \quad R_L = 1000\Omega$

$h_{ie} = 1K\Omega \quad h_{re} = 2\times10^{-4}$

$h_{fe} = 50 \quad h_{oe} = 25\ \mu A/volt = 25\times10^{-6}$ A/Volt

(i) **Current Gain**

$$A_i = -\frac{h_{fe}}{1 + h_{oe}.R_L}$$

Putting the given values, we get

$$A_i = -\frac{50}{1 + 25\times10^{-6}\times10^3} = -48.78 \quad \textbf{Ans.}$$

(ii) **Input resistance**

$$R_i = h_{ie} - \frac{h_{re}.h_{fe}}{h_{oe} + \frac{1}{R_L}} = 1000 - \frac{50\times2\times10^{-4}}{25\times10^{-6} + \frac{1}{1000}} = 990.24\Omega \quad \textbf{Ans.}$$

(iii) **Voltage gain** $$A_v = A_i.\frac{R_L}{R_i} = -48.78\times\frac{1000}{990.24} = -49.26 \quad \textbf{Ans.}$$

2.26. h-PARAMETER (HYBRID) EXPRESSIONS FOR COMMON BASE AMPLIFIER

(i) Current Gain

General Expression $$A_i = -\frac{h_f}{1 + h_o.R_L} \quad ...(2.49)$$

Hence, expression for common base amplifier will be

$$A_i = \frac{-h_{fb}}{1 + h_{ob}.R_L} \quad ...(2.50)$$

(ii) Input resistance:

The resistance looking into the input terminals is

general Expression $$R_i = h_i - \frac{h_r.h_f}{h_o + \frac{1}{R_L}} \quad ...(2.51)$$

Hence, expression for common base amplifier will be

$$R_i = h_{ib} - \frac{h_{rb}.h_{fb}}{h_{ob} + \frac{1}{R_L}} \quad ...(2.52)$$

The input resistance of the amplifier stage depends upon the biasing arrangement. If there is no biasing resistance, then

$$Z_i = R_i \quad ...(2.53)$$

(iii) Voltage gain

General Expression $$A_v = -\frac{h_f.R_L}{h_i + \Delta h.R_L} \quad ...(2.54)$$

Hence, expression for common base amplifier will be

$$A_v = \frac{h_{fb}.R_L}{h_{ib} + \Delta h.R_L} \quad ...(2.55)$$

(iv) Output resistance

The resistance looking into amplifier output terminals is

general expression $$R_o = \frac{R_s + h_i}{R_s.h_o + \Delta h} \quad ...(2.56)$$

Hence, expression for common-base Amplifier will be

$$R_o = \frac{R_s + h_{ib}}{R_s.h_{ob} + \Delta h} \quad ...(2.57)$$

The output resistance of the amplifier stage will be

$$Z_o = R_o \| R_L \quad ...(2.58)$$

(v) Overall voltage gain

$$A_{vs} = \frac{A_v.Z_I}{Z_I + R_s} \quad ...(2.59)$$

(vi) Overall current gain

$$A_{is} = -\frac{A_i.R_s}{R_s + Z_i} \quad ...(2.60)$$

EXAMPLE 2.3. For a common base amplifier, a transistor has the following values of h-parameters. $h_{ib} = 28\ \Omega$, $h_{fb} = -0.98$, $h_{rb} = 5\times10^{-4}$ & $h_{ob} = 0.34 \times 10^{-6}$S. Find the values of current gain, input resistance, voltage gain and output resistance. The load resistance is 1.2 kΩ. The source resistance is zero.

Solution: (i) **The current gain**

$$A_i = -\frac{h_{fb}}{1+h_{ob}.R_L} = -\frac{(-0.98)}{1+0.34\times10^{-6}\times1200} = 0.98 \quad \textbf{Ans.}$$

(ii) **The input resistance**

$$R_i = h_{ib} + h_{rb}.A_i.R_L = 28 + 5 \times 10^{-4} \times 0.98 \times 1200$$

$$R_i = 28 + 0.6 = 28.6\ \Omega \quad \textbf{Ans.}$$

(iii) **The voltage gain**

$$A_v = \frac{A_i.R_L}{R_i} = \frac{0.98\times1200}{28.6} = 41 \quad \textbf{Ans.}$$

(iv) **Output resistance**

$$R_o = \frac{R_s + h_{ib}}{R_s h_{ob} + \Delta h} \cong \frac{h_{ib}}{\Delta h}$$

where $\Delta h = h_{ib}.h_{ob} - h_{rb}.h_{fb}$

Substituting all the values, we have

$$\Delta h = 28 \times 0.34 \times 10^{-6} - 5 \times 10^{-4} \times (-0.98) = 5 \times 10^{-4}$$

Hence, $$R_o = \frac{h_{ib}}{\Delta h} = \frac{28}{5\times10^{-4}} = 5.6 \times 10^4 \text{ ohm} = 56\ \text{K}\Omega \quad \textbf{Ans.}$$

2.27. h-PARAMETER (HYBRID) EXPRESSION FOR COMMON COLLECTOR AMPLIFIER

(i) Current gain

General Expression $$A_i = -\frac{h_f}{1+h_o.R_L} \quad ...(2.61)$$

Therefore, expression for common collector amplifier will be

$$A_i = -\frac{h_{fc}}{1+h_{oc}.R_L} \quad ...(2.62)$$

(ii) Input resistance

The input resistance looking into the input terminals is

$$R_i = h_{ic} - \frac{h_{rc}.h_{fc}}{h_{oc} + \frac{1}{R_L}} \quad ...(2.63)$$

The input resistance of the amplifier stage depends upon the biasing arrangement. If there is no biasing resistance, then

$$Z_i = R_i \quad ...(2.64)$$

(iii) Voltage Gain $$A_v = \frac{A_i.R_L}{R_i} \quad ...(2.65)$$

$$A_v = -\frac{h_{fc}.R_L}{h_{ic} + \Delta h.R_L} \quad ...(2.66)$$

(iv) Output resistance $$R_o = \frac{R_s + h_{ic}}{R_s.h_{oc} + \Delta h} \quad ...(2.67)$$

(v) Overall voltage gain $$A_{vs} = \frac{A_v.Z_i}{R_s + Z_i} \quad ...(2.68)$$

(vi) Overall current gain $A_{is} = \dfrac{-A_i . R_s}{R_s + Z_i}$ ...(2.69)

2.28. SIMPLIFIED OR APPROXIMATE MODEL

As a matter of fact, the h-parameters are widely used in amplifier analysis. To simplify the analysis, the exact h-parameters expressions are converted into approximated expressions. The complete behaviour of the transistor circuit can be obtained by using this simplified hybrid model*.

Figure 2.38 shows the h-parameter equivalent circuit of transistor in CE configuration.**

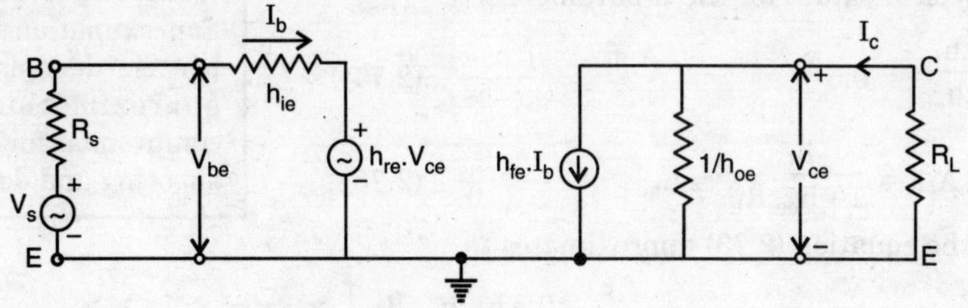

Fig. 2.38. Exact CE hybrid Model

In exact CE hybrid model, $1/h_{oe}$ and R_L are in parallel. The parallel combination of these two is approximately equal to lower value, i.e. R_L. Hence, If $1/h_{oe} >> R_L$, then the term h_{oe} may be neglected provided that $h_{oe} R_L << 1$.

Further, if h_{oe} is omitted, the collector current I_c is given by $I_c = h_{fe}.I_b$

Under this condition, the magnitude of voltage of generator in the emitter circuit.

$$h_{re}|V_{ce}| = h_{re}.I_c R_L = h_{re}.h_{fe}.I_b.R_L$$

But, since, $h_{re} \cdot h_{fe} \cong 0.01$

Therefore, the voltage $h_{re}.V_{ce}$ may be neglected in comparison to the voltage drop across h_{ie} = h_{fe}. I_b provided that R_L is not too large.

To summarize, if the load resistance R_L is small, it is possible to neglect the parameter h_{re} and h_{oe} and obtain the approximate equivalent circuit as shown in figure 2.39. It may also be shown that if $h_{oe}R_L \leq 0.1$, the error in calculating A_i, A_v, R_i and R_o is less than 10%.

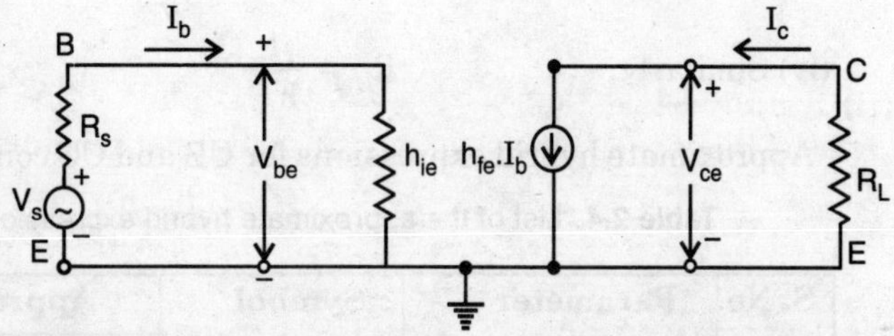

Fig. 2.39. Approximate common emitter (CE) hybrid model

Approximate hybrid expressions may be obtained in following way :

(i) Current Gain : Exact expression is given by

$$A_i = -\frac{h_{fe}}{1 + h_{oe}R_L} \qquad ...(2.70)$$

---

* Since the h-parameters themselves vary widely for the same type of transistor, it is justified to make approximations and simplify the expressions for A_i, A_v, R_i and R_o. In addition, a better understanding of the behaviour of the transistor circuit can be obtained by using the simplified hybrid model.

** Since CE configuration is more useful and general, it is taken for consideration.

If $h_{oe}.R_L < 0.1$ ...(2.71)

then $A_i = -h_{fe}$

(ii) Input resistance : Exact expression is given by

$$R_i = h_{ie} + h_{re}A_i.R_L \quad ...(2.72)$$

It can be written as

$$R_i = h_{ie}\left[1 - \frac{h_{re}\,h_{fe}|A_i|\,h_{oe}\,R_L}{h_{ie}\,h_{oe}\,h_{fe}}\right] \quad ...(2.73)$$

Using the typical values for the h-parameters

$$\frac{h_{re}.h_{fe}}{h_{ie}.h_{oe}} \cong 0.5 \quad ...(2.74)$$

Further, $$|A_i| = \frac{h_{fe}}{1 + h_{oe}.R_L} \cong h_{fe} \quad ...(2.75)$$

Therefore, the equation (2.73) approximates to

$$R_i \simeq h_{ie}\left[1 - \frac{0.5\,h_{fe}.h_{oe}.R_L}{h_{fe}}\right]$$

If $h_{oe}.R_L < 0.1$

then $$R_i \cong h_{ie} = \frac{V_b}{I_b} * \quad ...(2.76)$$

(iii) **Voltage gain :** We know that the voltage gain is given by

$$A_v = A_i.\frac{R_L}{R_i}$$

But $A_i = -h_{fe}$ and $R_i = h_{ie}$

Therefore, we have $$A_v = -\frac{h_{fe}.R_L}{h_{ie}}$$

(iv) Similarly $$R_o = \frac{1}{h_{oe}}$$

Approximate hybrid expressions for CB and CC configurations may be found in a similar way.

DO YOU KNOW?

Exact answers are a waste of time in most electronics work. Almost everybody uses approximations because the answers are adequate for most applications. The ideal transistor is useful for basic trouble-shooting. The third approximation is needed for precise design. The second approximation is a good compromise for both trouble-shooting and design.

Table 2.4. List of the approximate hybrid expressions for CE, CB and CC configuration

S. No.	Parameter	Symbol	Approximate hybrid expressions for		
			Common Emitter	Common Base	Common Collector
1.	Current gain	A_i	$-h_{fe}$	$-h_{fb}$	$-h_{fc}$
2.	Input resistance	R_i	h_{ie}	h_{ib}	$h_{ic}+h_{fc}.R_L$
3.	voltage gain	A_v	$-\frac{h_{fe}.R_L}{h_{ie}}$	$-\frac{h_{fb}.R_L}{h_{ib}}$	$-\frac{h_{fc}R_L}{h_{ic}}$
4.	output resistance	R_o	$\frac{1}{h_{oe}}$	$\frac{1}{h_{ob}}$	$\frac{1}{h_{oc}}\ \frac{R_s+h_{ie}}{h_{fe}}$

* This over estimates the input resistance by less than 5%

Table 2.5. Typical h-parameter values for transistor

S.No.	Parameter	Values of parameter for		
		CE	CB	CC
1.	h_i	1100Ω	1100Ω	22Ω
2.	h_r	2.5×10^{-4}	1	3×10^{-4}
3.	h_f	50	–51	–0.98
4.	h_o	25μA/V	25 μA/V	0.49 μA/V

Table 2.6. Conversion Expression (Formulae) for h-parameters

S. No.	Types of Configuration	
	CB	CC
1.	$h_{ib} = \frac{h_{ie}}{1+h_{fe}}$	$h_{ic} = h_{ie}$
2.	$h_{rb} = \frac{h_{ie}\ h_{oe}}{1+h_{fe}} - h_{re}$	$h_{rc} = 1$
3.	$h_{fb} = \frac{-h_{fe}}{1+h_{fe}}$	$h_{fc} = -(1+h_{fe})$
4.	$h_{ob} = \frac{h_{oe}}{1+h_{fe}}$	$h_{oc} = h_{oe}$

2.29. THE COMMON EMITTER (CE) AMPLIFIER WITH OTHER TYPE OF BIAS

Figure 2.40 shows the circuit of a single stage CE amplifier which makes use of a NPN transistor. Here, the emitter-base junction is forward biased by power supply V_{BB} and collector-base junction is reverse biased by power supply V_{CC}, so that the transistor remains in the active region throughout its operation. The operating point Q is determined by V_{CC}, R_B and R_C. Input signal is applied to the base-emitter circuit and the amplified output signal is taken from the C-E circuit. C_1 and C_2 are coupling capacitors to provide d.c. isolation at the input and output of the amplifier.

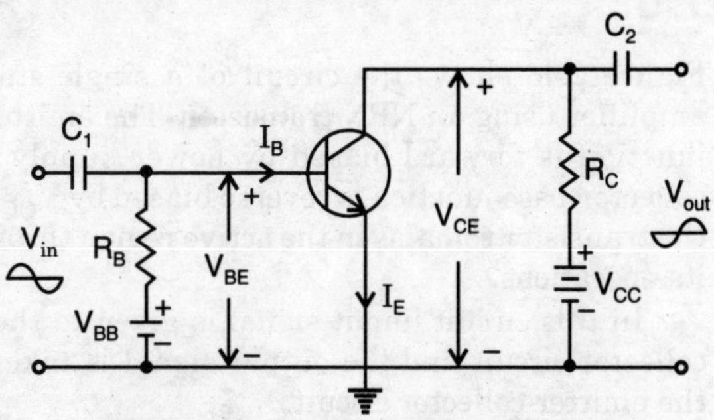

Fig. 2.40. A common emitter (CE) amplifier

However, instead of using two power supplies, the same bias conditions can be set up in the CE amplifier circuit with a single power supply as shown in figure 2.41.

Now referring to figure 2.41, when no a.c. signal input is given, i.e., under d.c. conditions, we have

$$I_B = \frac{V_{CC} - V_{BE}}{R_B} \cong \frac{V_{CC}}{R_B} \quad ...(2.77)$$

$$I_C = \beta\, I_B \quad ...(2.78)$$

$$V_{CE} = V_{CC} - I_C R_C. \quad \text{...(2.79)}$$

When a sinusoidal a.c. signal is applied at the input terminals of the circuit, during the positive half cycle, the forward bias of the base-emitter junction V_{BE} is increased, resulting in an increase in I_B. The collector current I_C is increased by β times the increase in I_B. From equation (2.79), V_{CE} is correspondingly decreased, i.e., the output voltage gets decreased as shown in figure 2.42.

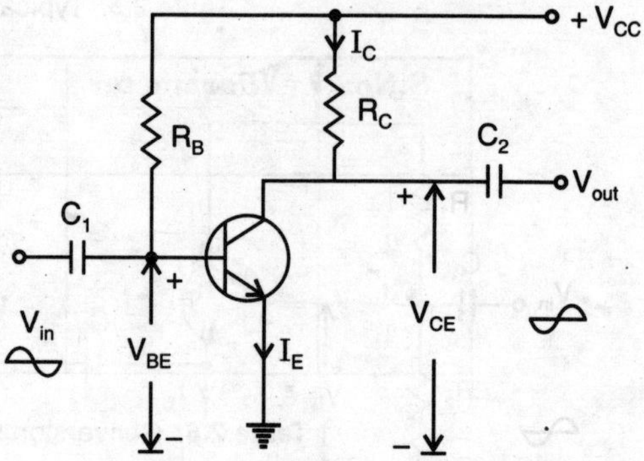

Fig. 2.41. CE amplifier with a single power supply

Hence, in a CE amplifier a positive going signal is converted into a negative going output signal, i.e., a 180° phase shift is introduced between the output and input signal and further the output signal is an amplified version of the input signal.

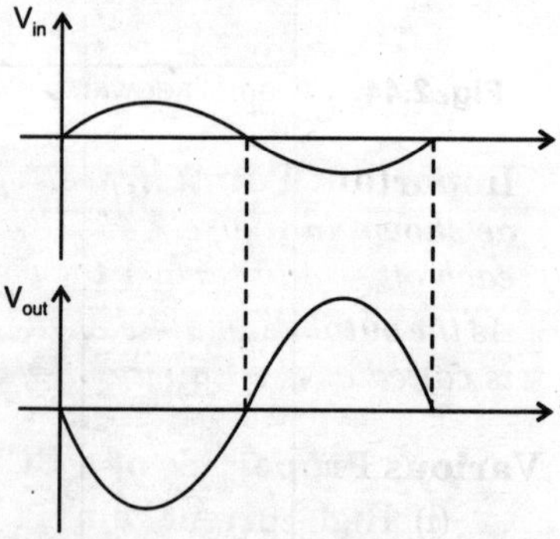

Fig. 2.42. Input and output voltage of CE amplifier

Various Properties of a CE Amplifier

(i) Large current gain (A_I)

(ii) Large voltage gain (A_V)

(iii) Large power gain ($A_P = A_I.A_V$)

(iv) Voltage phase shift of 180°.

(v) Moderate input impedance

(vi) Moderate output impedance

2.30. COMMON COLLECTOR (CC) AMPLIFIER WITH OTHER TYPE OF BIAS

(RGTU, Bhopal, Sem. Exam; 2008-2009)

Figure 2.43 shows the circuit of a single-stage CC amplifier using an NPN transistor. The emitter-base junction is forward biased by power supply V and collector-base junction is reverse biased by V_{CC}. Here, the transistor remains in the active region throughout its operations.

In this circuit, input signal is given to the base-collector circuit and the output signal is taken from the emitter-collector circuit.

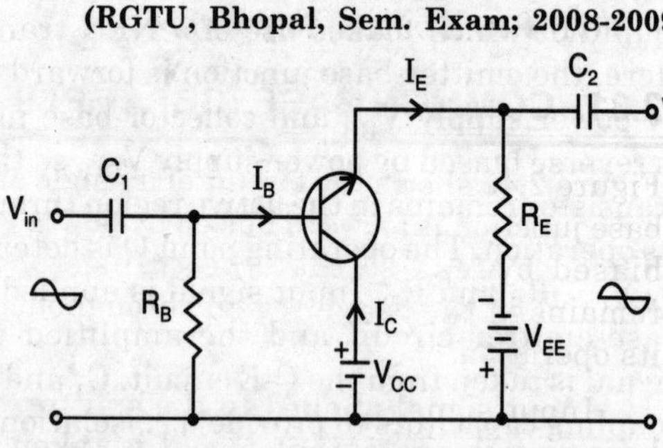

Fig. 2.43. The common collector (CC) amplifier

Instead of using two power supplies, the required bias condition can be set up in a CC amplifier circuit with a single power supply as shown in figure 2.44.

Here, C_1 and C_2 are coupling capacitors to provide d.c. isolation at the input and output of the amplifier.

From figure 2.44, we have

Output voltage $\quad V_{out} = I_E R_E = \beta I_B R_E$

When a sinusoidal a.c. signal is applied at the input, during the positive half cycle of the signal applied, the base potential V_B increases thereby increasing the base current I_B. Hence, emitter current $I_E = I_C + I_B$ increases and voltage drop across R_E, i.e., the output voltage increases.

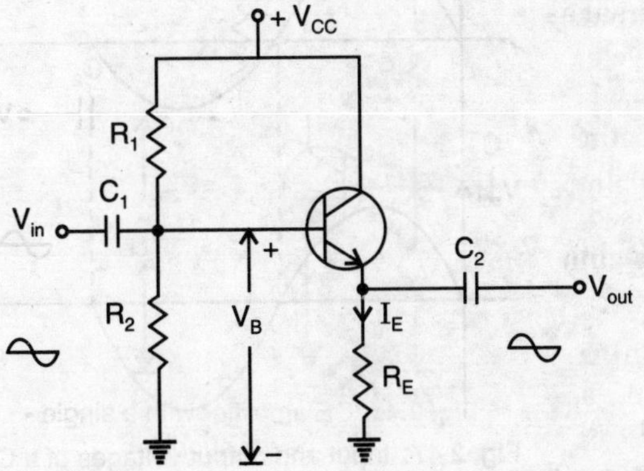

Fig. 2.44. CC amplifier with a single power supply

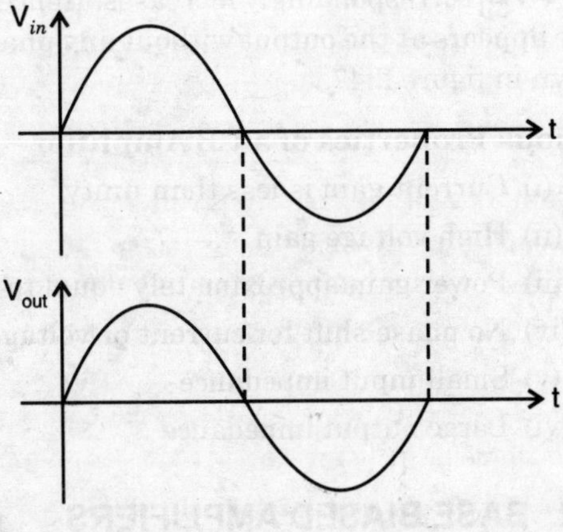

Fig. 2.45. Input and output voltages of CC amplifier

Important Point: *Hence, a positive going input signal results in a positive going output signal as shown in figure 2.45. Hence, in a CC amplifier, input and output signals are in phase with each other and further the voltage gain is approximately unity.*

As the output signal taken at the emitter terminal almost follows the input signal, the CC amplifier is called as the ***Emitter Follower***.

Various Properties of a CC amplifier

(i) High current gain

(ii) Voltage gain of approximately unity

(iii) Power gain approximately equal to current gain

(iv) No current or voltage phase shift

(v) Large input impedance

(vi) Small output impedance

2.31. COMMON BASE (CB) AMPLIFIER WITH OTHER TYPE OF BIAS

Figure 2.46 shows the circuit of a single stage CB amplifier using an NPN transistor. The emitter-base junction is forward biased by power supply V_{EE}, whereas the collector-base junction is reverse biased by V_{CC} so that the transistor remains in the active region throughout its operation.

Input signal applied to the emitter-base circuit and output signal is taken from collector-base circuit. The output voltage (= V_{CB}) is given by the equation:

$$V_{out} = V_{CC} - I_C R_C \quad ...(2.80)$$

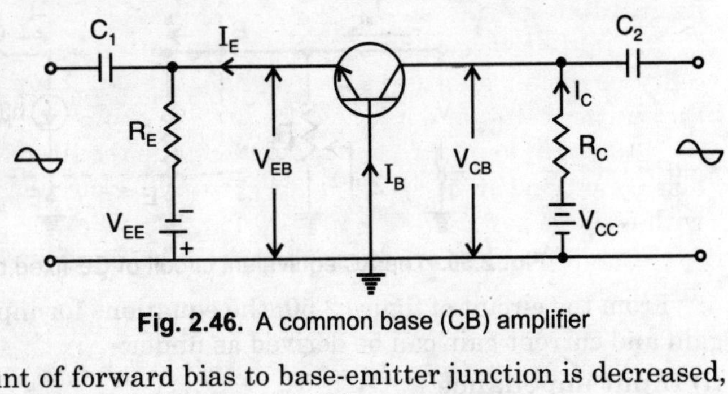

Fig. 2.46. A common base (CB) amplifier

When a sinusoidal a.c. signal is applied at the input, during the positive half cycle of the applied signal, the amount of forward bias to base-emitter junction is decreased, resulting in a decreases in I_B. As a result I_E (= βI_B) and hence, I_C also decreases.

From equation (2.80), the drop $I_C R_C$ decreases, hence $V_{out} = V_{CB}$ correspondingly increases. Hence, positive half cycle appears at the output without any phase reversal as shown in figure 2.47.

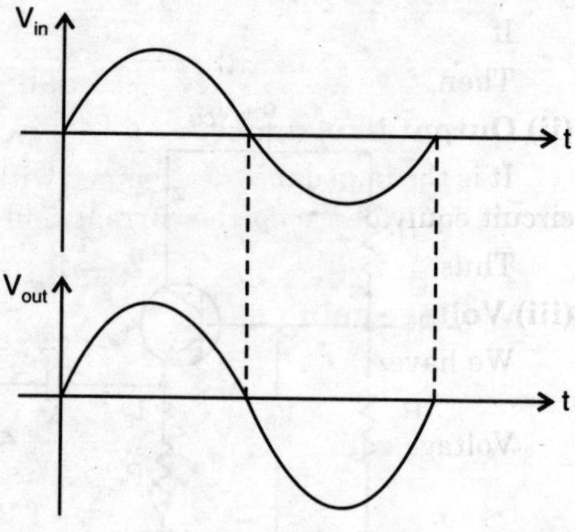

Fig. 2.47. Input and output voltages of a CB amplifier

Various Properties of a CB Amplifier

(i) Current gain is less than unity
(ii) High voltage gain
(iii) Power gain approximately equal to voltage gain
(iv) No phase shift for current or voltage
(v) Small input impedance
(vi) Large output impedance

2.32. BASE BIASED AMPLIFIERS

The circuit in figure 2.48 shows a CE amplifier in fixed bias configuration. The a.c. equivalent circuit of the amplifier can be drawn by the steps mentioned as follows:

(i) Remove the d.c. effects of power supply (V_{CC}) by grounding them.
(ii) Replace the capacitors (C_1 and C_2) by short circuits.

Thus, the circuit of figure 2.48 reduces to the a.c. equivalent circuit shown in figure 2.49.

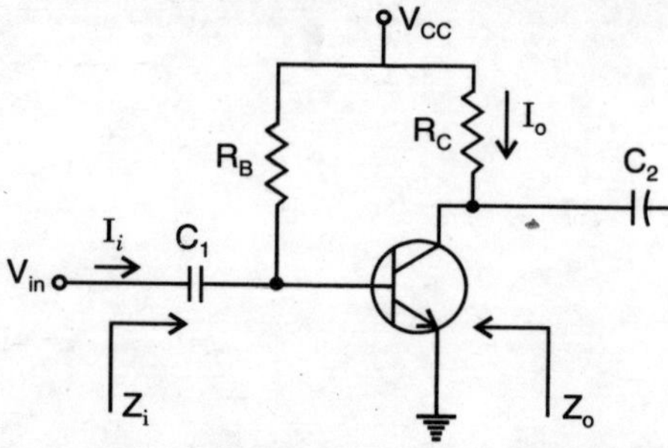

Fig. 2.48. CE fixed bias configuration

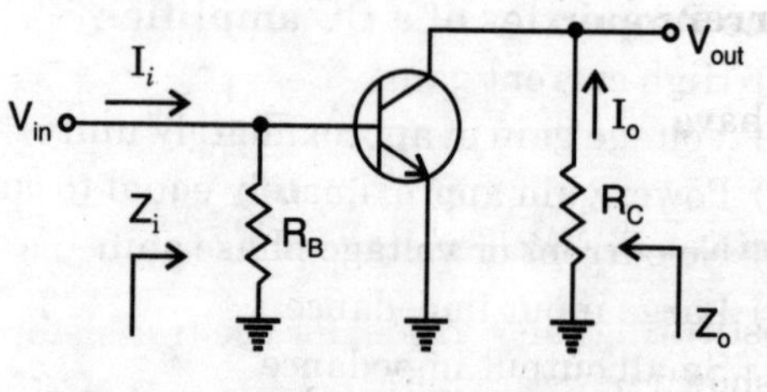

Fig. 2.49. AC equivalent circuit of CE fixed bias amplifier

Now, let us substitute the approximate hybrid model for the transistor in figure 2.49 to get the circuit shown in figure 2.50.

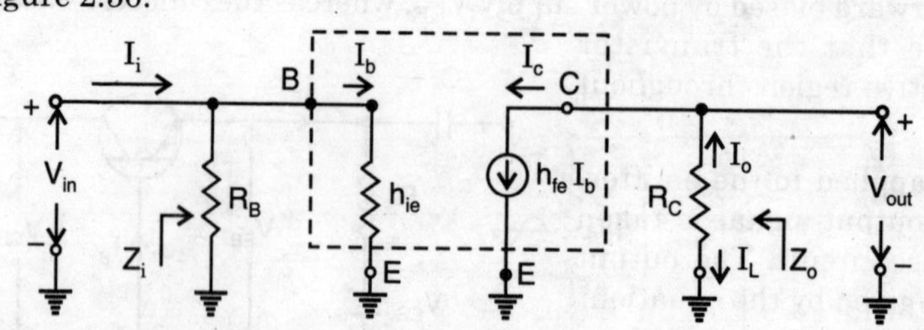

Fig. 2.50. The AC equivalent circuit of CE fixed bias amplifier using hybrid model

From the circuit of figure 2.50, the equations for input impedance, output impedance, voltage gain and current gain can be derived as under:

(i) Input impedance

We have

$$Z_i = R_B \,||\, h_{ie} \qquad ...(2.81)$$

If $R_B >> h_{ie}$

Then, $Z_i \cong h_{ie}$ ...(2.82)

(ii) Output impedance

It is the impedance determined with $V_{in} = 0$. With $V_i = 0$, $I_b = 0$ and $h_{fe} I_b = 0$ indicating an open circuit equivalence for the current source.

Thus, $Z_o = R_L (= R_C)$

(iii) Voltage gain

We have

Voltage gain, $A_V = \dfrac{V_{out}}{V_{in}}$

$V_{out} = -I_o R_C$

Substituting $I_o = h_{fe} I_b$

$V_{out} = -h_{fe} I_b R_C$

Assuming that $R_B >> h_{ie}$,

$I_i = I_b$ and $V_{in} = I_b h_{ie}$

Hence
$$A_V = \frac{-h_{fe} I_b R_C}{I_b h_{ie}} = \frac{-h_{fe} R_C}{h_{ie}} \quad ...(2.83)$$

As h_{fe} and h_{ie} are positive, A_V is negative. The negative sign indicates a 180° phase shift between input and output signals.

(iv) Current gain

We have
$$A_I = \frac{I_L}{I_i} = \frac{-I_o}{I_i} \cong \frac{-h_{fe} I_b}{I_b} = -h_{fe} \quad ...(2.84)$$

Important Point: The sign for A_I will be positive if A_I is defined as the ratio of I_o to I_i.

Substituting the r_e model for the transistor in CE connection, the circuit of figure 2.50 reduces to that shown in figure 2.51.

(i) Input impedance

$$Z_i = R_B \mid\mid \beta r_e \quad ...(2.85)$$

If $R_B >> \beta r_e$

$$Z_i = \beta r_e \quad ...(2.86)$$

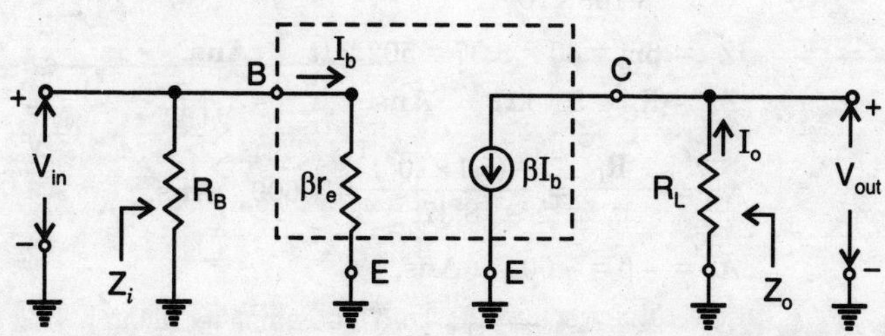

Fig. 2.51. The AC equivalent circuit of CE fixed bias amplifier using r_e model.

(ii) Output impedance

We have $Z_o = R_C$ ...(2.87)

(iii) Voltage gain $A_V = \dfrac{V_{out}}{V_{in}}$

$$V_{out} = -I_o R_C$$

Substituting $I_o = \beta I_b$

$$V_{out} = -(+\beta I_b)R_C = -\beta I_b R_C$$

Assuming $R_B >> \beta r_e$,

$$V_{in} \cong I_b \beta r_e$$

or $$A_V = \frac{-\beta I_b R_C}{I_b \beta r_e} = \frac{-R_C}{r_e} \quad ...(2.88)$$

(iv) Current gain

We have $$A_I = \frac{-I_o}{I_i} = \frac{-I_o}{I_b} = \frac{-\beta I_b}{I_b} = -\beta \quad ...(2.89)$$

EXAMPLE 2.4. Determine the input impedance, output impedance, voltage gain and current gain for the CE amplifier of figure 2.52. The h-parameters of the transistor are h_{fe} = 60, h_{ie} = 500 Ω at I_C = 3 mA.

Solution: $R_B = 220 \text{ k}\Omega >> h_{ie} = 500\ \Omega$

From h-parameter model, we have

$$Z_i = h_{ie} = 500\ \Omega$$

$$Z_o = R_C = 7.1 \text{ k}\Omega$$

We have $$A_V = \frac{-h_{fe}R_C}{h_{ie}}$$

$$= \frac{-60(5.1\times10^3)}{500} = -612$$

Also, $A_I = -h_{fe} = -60$

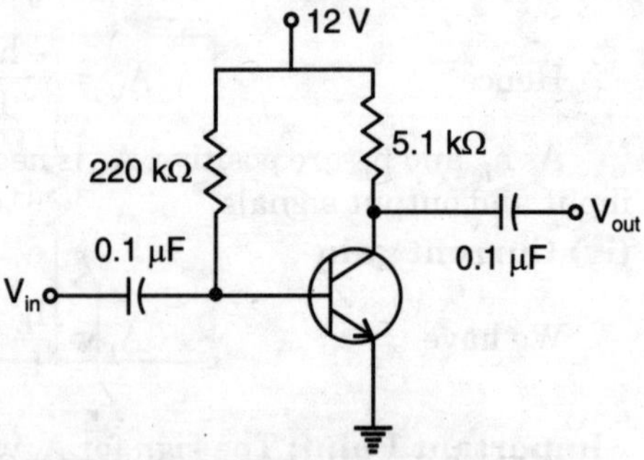

Fig. 2.52.

From r_e model, we have $Z_i = \beta \cdot r_e$ where $r_e = \frac{26 \text{ mV}}{I_e}$

From the circuit, $$I_b = \frac{V_{CC} - V_{BE}}{R_B} = \frac{12 - 0.6}{220\times10^3} = 51.8\ \mu A$$

$$I_e \cong I_c = \beta I_b = 60 \times 51.8 \times 10^{-6} = 3.108 \text{ mA}$$

Also, $$r_e = \frac{26\times10^{-3}}{3.108\times10^{-3}} = 8.37\ \Omega$$

Further, $Z_i = \beta r_e = 60 \times 8.37 = 502.2\ \Omega$ **Ans.**

$Z_o = R_C = 5.1 \text{ k}\Omega$ **Ans.**

$$A_V = \frac{-R_C}{r_e} = \frac{-(5.1\times10^3)}{8.37} = -609 \quad \textbf{Ans.}$$

$A_I = -\beta = -60$ **Ans.**

Important Point: Small differences in the calculation of amplifier parameters between the h-parameter and r_e models might arise due to the approximation made in the circuit design.

2.33. CE AMPLIFIER WITH UNBYPASSED EMITTER RESISTOR.

Figure 2.53 shows the CE amplifier with unbypassed emitter resistor. The a.c. equivalent circuit for this amplifier shown in figure 2.55 can be drawn by following the steps mentioned earlier.

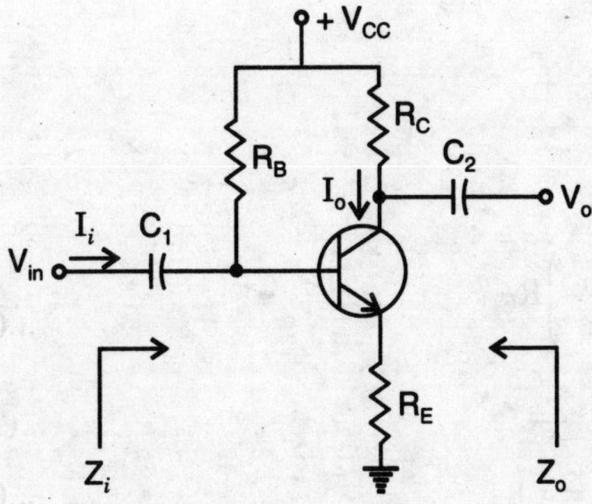

Fig. 2.53. CE amplifier with unbypassed

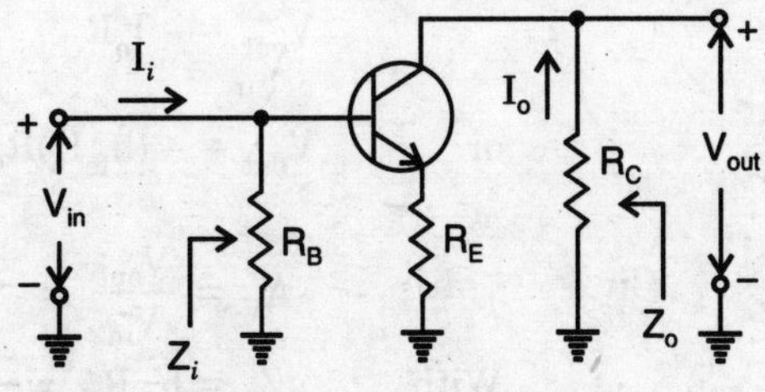

Fig. 2.54. Hybrid equivalent circuit

By substituting the hybrid equivalent circuit for the transistor of figure 2.54, the circuit reduces to that shown in figure 2.55.

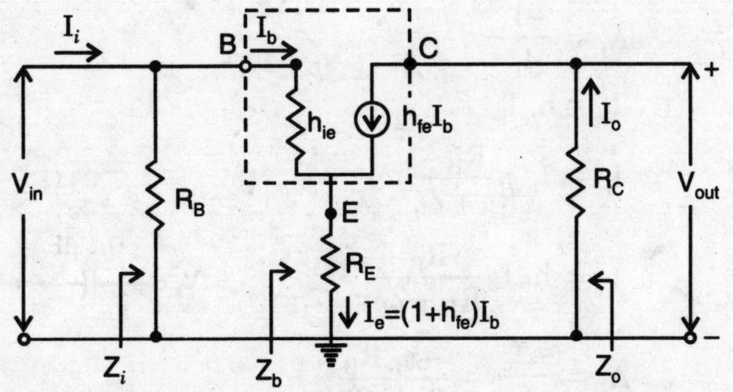

Fig. 2.55. AC equivalent circuit for CE amplifier with unbypassed emitter resistor using hybrid model

(i) Input impedance

Current through the emitter resistor R_E is

$$I_e = I_b + h_{fe}I_b = (1 + h_{fe})I_b$$

From figure 2.55, we have

$$V_{in} = I_b h_{ie} + (1 + h_{fe})\, I_b R_E \quad \text{...(2.90)}$$

$$Z_b = \frac{V_{in}}{I_b} = h_{ie} + (1 + h_{fe})R_E$$

As $h_{fe} >> 1$, we have ...(2.91)

$$Z_b \cong h_{ie} + h_{fe}R_E$$

Normally, $h_{fe}R_E >> h_{ie}$ leading to

$$Z_b \cong h_{fe}R_E \quad \text{...(2.92)}$$

From the circuit of figure 2.55, we get

$$Z_i = R_B || Z_b \quad \text{...(2.93)}$$

(ii) Output impedance

With $V_{in} = 0$, $I_b = 0$, $h_{fe}I_b = 0$ indicating open circuit for current source.

Hence, $$Z_o = R_C \quad \text{...(2.94)}$$

(iii) Voltage gain

We have $$A_V = \frac{V_{out}}{V_{in}}$$

From the circuit of figure 2.55, we have

$$V_{out} = -I_o R_C$$

or $$V_{out} = -(h_{fe} I_b) R_C = -h_{fe}\left(\frac{V_{in}}{Z_b}\right) R_C$$

$$A_V = \frac{V_{out}}{V_{in}} = \frac{-h_{fe} R_C}{Z_b} \quad \text{...(2.95)}$$

With $Z_b = h_{fe} R_E$, we obtain

$$A_V = \frac{-h_{fe} R_C}{h_{fe} R_E} = -\frac{R_C}{R_E} \quad \text{...(2.96)}$$

(iv) Current gain

We have $$A_I = \frac{-I_o}{I_i}$$

$$I_o = h_{fe} I_b$$

where, $$I_b = I_i \frac{R_B}{R_B + Z_b}$$

Therefore, $$I_o = h_{fe} I_i \frac{R_B}{R_B + Z_b}$$

$$A_I = \frac{-I_o}{I_i} = \frac{-h_{fe} R_B}{R_B + Z_b} \quad \text{...(2.97)}$$

Substituting the r_e model for the transistor of figure 2.55, the circuit reduces to that shown in figure 2.56.

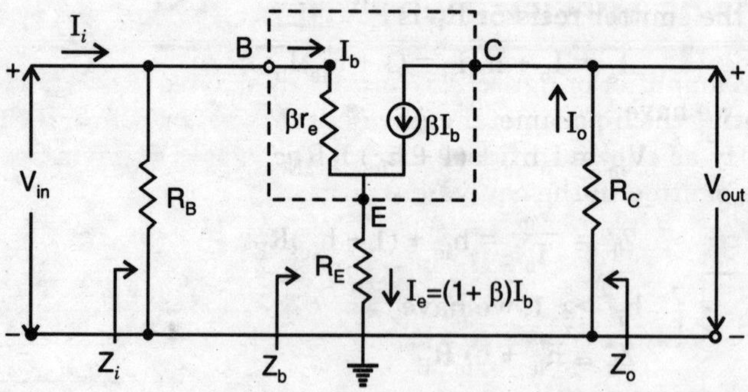

Fig. 2.56. AC equivalent circuit of CE amplifier with unbypassed emitter resistor using r_e model.

(i) Input impedance

$$V_{in} = I_b(\beta r_e) + (1 + \beta) I_b R_E$$

$$Z_b = \frac{V_{in}}{I_b} = \beta r_e + (1 + \beta) R_E$$

As, $\beta >> 1$, we have

$$Z_b \cong \beta(r_e + R_E) \cong \beta R_E \quad \text{...(2.98)}$$

$$Z_i = R_B \,||\, Z_b \qquad ...(2.99)$$

(ii) Output impedance

For similar reasons, we have

$$Z_o = R_C \qquad ...(2.100)$$

(iii) Voltage gain

We have, $A_V = \dfrac{V_{out}}{V_{in}}$

$$V_{out} = -I_oR_C = -\beta I_bR_C = -\beta\frac{V_{in}}{Z_b}R_C$$

Therefore, $A_V = \dfrac{V_{out}}{V_{in}} = \dfrac{-\beta R_C}{Z_b}$

Substituting $Z_b \cong \beta R_E$,, we get

$$A_V = \frac{-R_C}{R_E} \qquad ...(2.101)$$

(iv) Current gain

$$A_I = -\frac{I_o}{I_i}$$

$$I_o = \beta I_b$$

Where $I_b = I_i\dfrac{R_B}{R_B + Z_b}$

Therefore, $I_o = \beta I_i\dfrac{R_B}{R_B + Z_b}$

$$A_I = \frac{-I_o}{I_i} = -\frac{\beta R_B}{R_B + Z_b} \qquad ...(2.102)$$

2.34. ANALYSIS OF CE AMPLIFIER WITH VOLTAGE-DIVIDER BIAS (VDB)

Let us consider the CE amplifier of figure 2.57 in which the base bias is obtained by two resistors R_1 and R_2. By substituting the h-parameter equivalent for the transistor, the a.c. equivalent circuit can be obtained directly as shown in figure 2.58. As capacitor C_E bypasses R_E in the operating frequency range, R_E is omitted in the equivalent circuit.

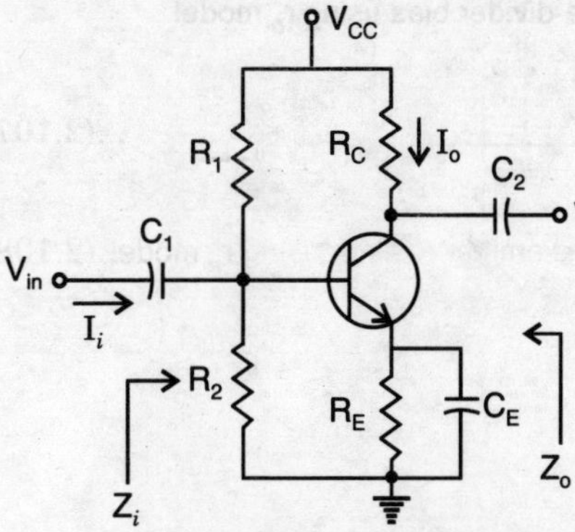

Fig. 2.57. CE amplifier with voltage-divider bias

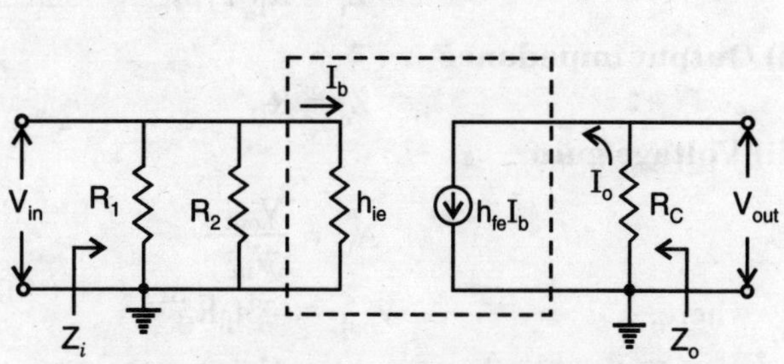

Fig. 2.58. AC equivalent circuit of CE amplifier with voltage-divider bias using hybrid model

(i) Input impedance

Let $$R_B = R_1 \,||\, R_2$$

or $$R_B = \frac{R_1R_2}{R_1 + R_2}$$

$$Z_i = R_B \,||\, h_{ie} \quad ...(2.103)$$

(ii) Output impedance

$$Z_o = R_C \quad ...(2.104)$$

(iii) Voltage gain $$A_V = \frac{V_{out}}{V_{in}}$$

Substituting $V_{out} = -h_{fe}I_bR_C$ and

$V_{in} = I_bh_{ie}$

$$A_V = \frac{-h_{fe}I_bR_C}{I_bh_{ie}} = \frac{-h_{fe}R_C}{h_{ie}} \quad ...(2.105)$$

(iv) Current gain $$A_I = -\frac{I_o}{I_i}$$

Substituting $I_o = -h_{fe}I_b$

where $$I_b = I_i \frac{R_B}{R_B + h_{ie}}$$

Therefore, $$I_o = -h_{fe}\, I_i \frac{R_B}{R_B + h_{ie}}$$

$$A_I = -h_{fe} \frac{R_B}{R_B + h_{ie}} \quad ...(2.106)$$

Substituting the r_e model for the transistor, the a.c. equivalent circuit is as given in figure 2.59.

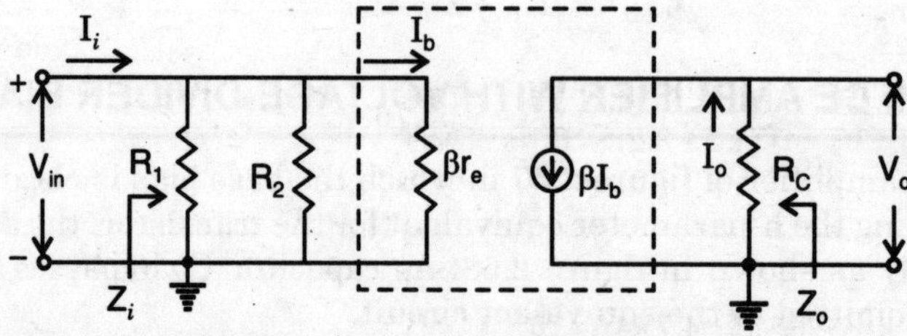

Fig. 2.59. AC equivalent circuit of CE amplifier with voltage-divider bias using r_e model.

(i) Input impedance

$$Z_i = R_B \,||\, \beta r_e \quad ...(2.107)$$

(ii) Output impedance

$$Z_o = R_C \quad ...(2.108)$$

(iii) Voltage gain

$$A_V = \frac{V_{out}}{V_{in}}$$

where, $$V_{out} = -\beta I_bR_C$$

$$I_b = \frac{V_{in}}{\beta r_e}$$

Therefore, $$V_{out} = -\beta\left(\frac{V_{in}}{\beta r_e}\right)R_C$$

$$A_V = \frac{V_{out}}{V_{in}} = \frac{-R_C}{r_e} \qquad ...(2.109)$$

(iv) Current gain

$$A_I = \frac{-I_o}{I_i}$$

From the circuit of figure 2.59, $I_o = -\beta I_b$

where $$I_b = I_i \frac{R_B}{R_B + \beta r_e}$$

Therefore, $$I_o = -\beta I_i \frac{R_B}{R_B + \beta r_e}$$

$$A_I = \frac{I_o}{I_i} = -\beta \frac{R_B}{R_B + \beta r_e} \qquad ...(2.110)$$

EXAMPLE 2.5. Determine the input impedance, output impedance, voltage gain and current gain of the CE amplifier using h-parameter and r_e model equivalents for the transistor with h_{ie} = 3.2 kΩ and h_{fe} = 100 at the operating conditions.

Solution: h-Parameter analysis

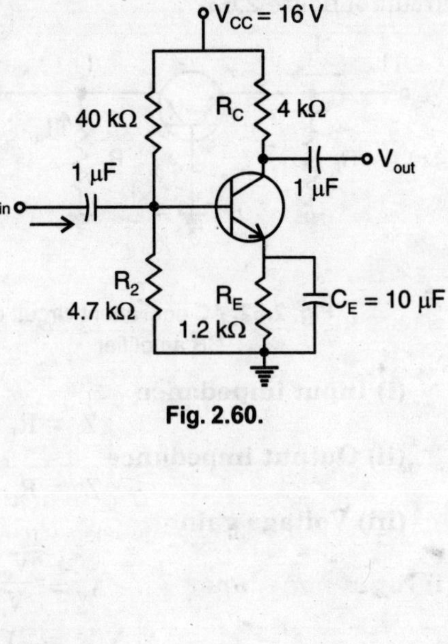

Fig. 2.60.

$$Z_i = R_B \,||\, h_{ie}$$

$$R_B = R_1 \,||\, R_2 = 40\text{ k}\Omega \,||\, 4.7\text{ k}\Omega = 4.2\text{ k}\Omega$$

$$Z_i = 4.2\text{ k}\Omega \,||\, 3.2\text{ k}\Omega = 1.82\text{ k}\Omega$$

$$Z_o = R_C = 4\text{ k}\Omega$$

$$A_V = \frac{-h_{fe}R_C}{h_{ie}} = \frac{-100\times 4\times 10^3}{3.2\times 10^3} = -125$$

$$A_I = \frac{-h_{fe}R_B}{R_B + h_{ie}} = \frac{-100\times 4.2\times 10^3}{(4.2+3.2)\times 10^3} = -56.76$$

Using r_e Model

To find I_B, we have

$$V_B = \frac{R_2 V_{CC}}{R_1 + R_2} = \frac{4.7\times 16\times 10^3}{44.7\times 10^3} = 1.682\text{ V}$$

Using Thevenin's equivalent for input portion, we have

$$I_B = \frac{V_B - V_{BE}}{R_B + (1+\beta)R_E} = \frac{1.682 - 0.6}{4.2\times 10^3 + (1+100)(1.2\times 10^{-3})}$$

or $$I_B = \frac{1.082}{125.4\times 10^3} = 8.628\ \mu\text{A}$$

Also, $$I_C = \beta I_B = 100(8.628\times 10^{-6}) = 0.86\text{ mA}$$

$$I_E = I_C = 0.86\text{ mA}$$

$$r_e = \frac{26\text{ mV}}{I_E} = \frac{26\times 10^{-3}}{0.86\times 10^{-3}} = 30.23\ \Omega$$

We have, $Z_i = R_B \,||\, \beta r_e = (4.2 \times 10^3) \,||\, (100 \times 30.23)$

or $Z_i = 4200 \,||\, 3023 = 1.76\ k\Omega$

$Z_o = R_C = 4\ k\Omega$

$$A_V = \frac{-R_C}{r_e} = \frac{-4000}{30.23} = -132.3$$

Further, we know that

$$A_I = \frac{-\beta R_B}{R_B + \beta r_e} = \frac{-100\,(4.2\times10^3)}{(4.2\times10^3)+(100\times30.23)} = -58.147 \quad \textbf{Ans.}$$

2.35. ANALYSIS OF COMMON BASE (CB) AMPLIFIER

Figure 2.61 shows the circuit of a common base amplifier

By applying the steps mentioned earlier to draw the a.c. equivalent circuit, the circuit of figure 2.61 reduces to the circuit of figure 2.62. Substituting the approximate hybrid model for the transistor in CB connection, the circuit of figure 2.62 reduces to the circuit of figure 2.63.

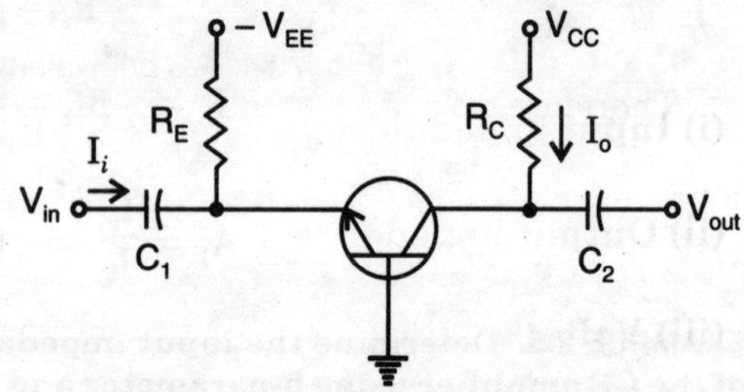

Fig. 2.61. Common base (CB) amplifier

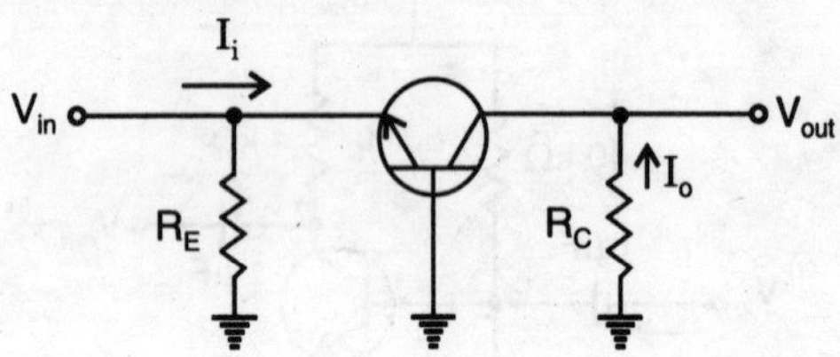

Fig. 2.62. AC equivalent circuit of CB amplifier

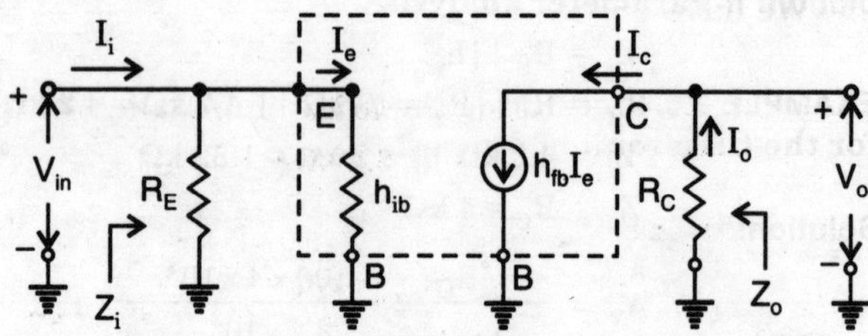

Fig. 2.63. AC equivalent circuit of CB amplifier using hybrid model

(i) Input impedance

$$Z_i = R_E \,||\, h_{ib} \quad ...(2.111)$$

(ii) Output impedance

$$Z_o = R_C \quad ...(2.112)$$

(iii) Voltage gain

$$A_V = \frac{V_{out}}{V_{in}}$$

$$V_{out} = -I_o R_C = -h_{fb} I_e R_C$$

Assuming that $R_E >> h_{ib}$, we get

$$I_e \cong I_i \text{ and } V_{in} = I_e h_{ib}$$

$$A_V = \frac{V_{out}}{V_{in}} = \frac{-h_{fb} R_C}{h_{ib}} \quad ...(2.113)$$

(iv) Current gain

We have $A_I = \dfrac{-I_o}{I_i} \cong \dfrac{-h_{fb}\, I_e}{I_e}$

or $A_I = -h_{fb}$ ...(2.114)

Important Point : As h_{fb} is negative, A_V and A_I are positive for CB configuration.

For substituting the r_e model for the transistor, let us substitute $h_{ib} = r_e$ and $h_{fb} = -1$. Hence, the orientation of current source is reversed. This results in the a.c. equivalent circuit for the CB amplifier using r_e model shown in figure 2.64.

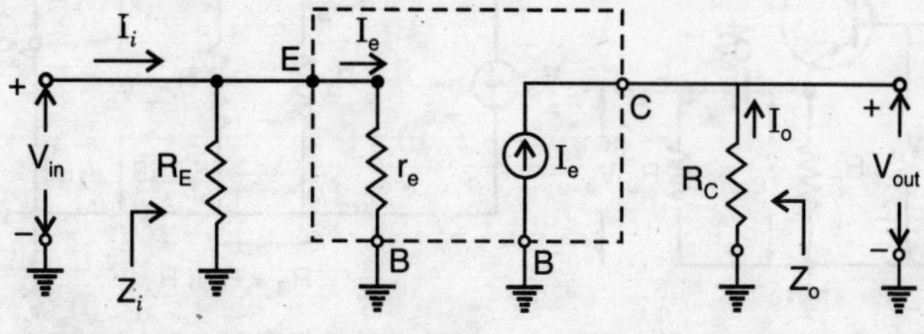

Fig. 2.64. AC equivalent circuit for the CB amplifier using r_e model

(i) Input impedance

$$Z_i = R_E \,||\, r_e \qquad ...(2.115)$$

(ii) Output impedance $\quad Z = R_C \qquad ...(2.116)$

(iii) Voltage gain

$$A_V = \frac{V_{out}}{V_{in}} = \frac{(-I_o)R_C}{I_e r_e} = \frac{(-I_e)R_C}{I_e r_e} = \frac{R_C}{r_e} \qquad ...(2.117)$$

(iv) Current gain

We have

$$A_I = \frac{-I_o}{I_i} \cong \frac{-(-I_E)}{I_E} = 1 \qquad ...(2.118)$$

EXAMPLE 2.6. Determine the input impedance, output impedance, voltage gain and current gain for the CB circuit of figure 2.65.

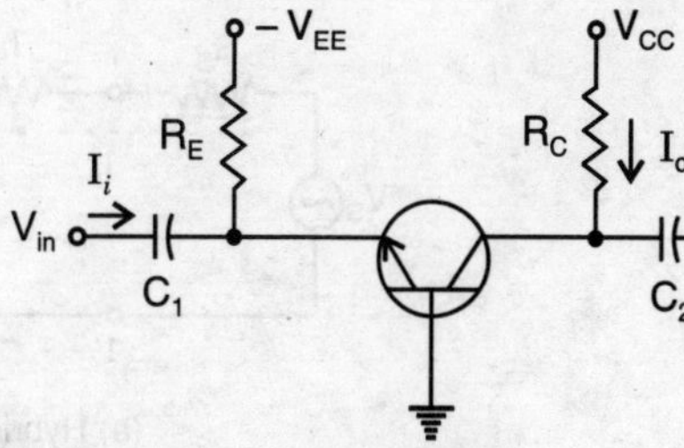

Fig. 2.65.

Solution: We have

$$|I_E| = \frac{V_{EE} - V_{BE}}{R_E}$$

or

$$|I_E| = \frac{8 - 0.6}{4000} = 1.85 \text{ mA}$$

We know that

$$r_e = \frac{26 \text{ mA}}{I_E} = \frac{26}{1.85} = 14.05\ \Omega$$

Also,

$$Z_i = R_E \,||\, r_e = 4000 \,||\, 14.05 = 14\ \Omega$$

$$Z_o = R_C = 3\ k\Omega$$

$$A_V = \frac{R_C}{r_e} = \frac{3000}{14.05} = 213.52$$

or

$$A_I = 1 \quad \textbf{Ans.}$$

2.36. ANALYSIS OF THE COMMON COLLECTOR AMPLIFIER OR EMITTER FOLLOWER

The common collector or emitter follower configuration has been shown in figure 2.66. The resistance R_1, R_2 and R_E provide dc biasing for the transistor. Here, C_1 and C_2 are the coupling capacitors and R_L is the external load resistance.

2.36.1 AC Equivalent Circuit

The ac equivalent circuit of the common collector configuration has been shown in figure 2.67.

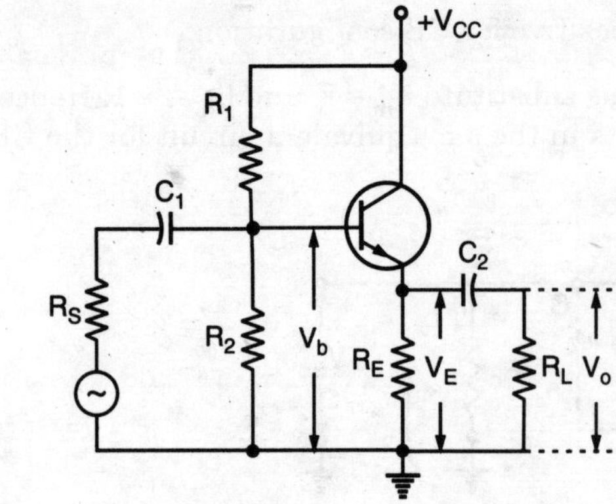

Fig. 2.66. Common collector circuit

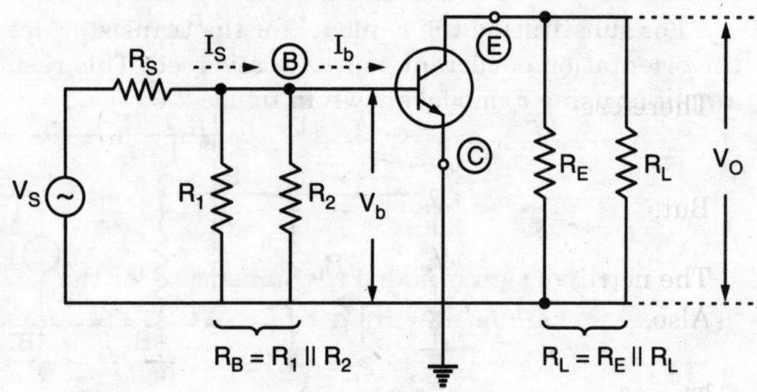

Fig. 2.67. AC equivalent circuit of a CC configuration

2.36.2 Why is this Circuit Called Common Collector Circuit?

Let us consider figure 2.67 which clearly shows that in the ac equivalent circuit, collector acts as a common terminal for input and output ports. Hence the name is common collector configuration. Base side acts as input side while emitter side acts as output side. All ac voltages are specified with respect to collector.

2.36.3 Analysis

For analysis, we shall use the standard results derived earlier. In order to do so, the hybrid equivalent of this circuit has to match with that of the standard circuit. Figure 2.68 (a) shows the standard circuit and figure 2.68 (b) shows the hybrid equivalent of CC amplifier drawn on the similar lines.

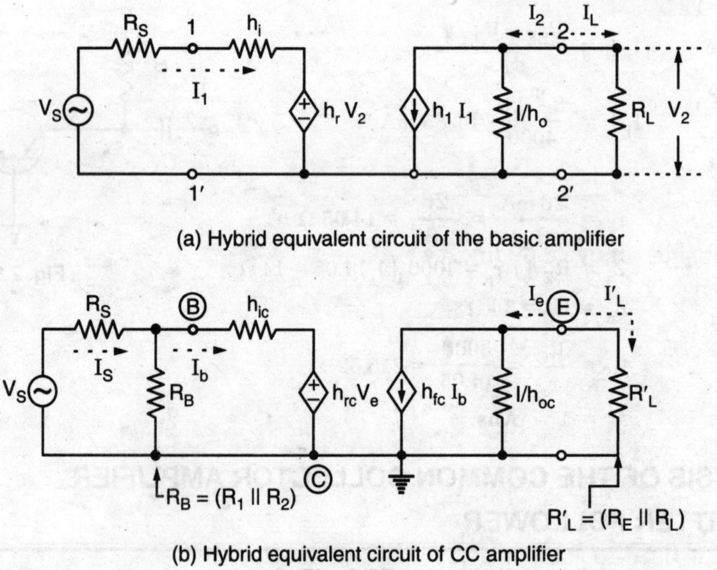

Fig. 2.68.

1. Current gain (A_I)

We have, $$A_I = \frac{I_L}{I_1} = \frac{-h_f}{1 + h_o R_L}$$ (for basic amplifier)

Comparing circuits in figures 2.68 (a) and (b), get

$$A_I = \frac{I_L'}{I_b} = \frac{-h_{fc}}{1 + h_{oc} R_L'} \quad \text{(for CC amplifier)}$$

Therefore, $$A_I = \frac{-h_{fc}}{1 + h_{oc}(R_E \parallel R_L)} \quad \text{Since, } R_L' = (R_E \parallel R_L)$$

But, $$h_{fc} = -\frac{I_e}{I_b} = -(1 + h_{fe})$$

The negative sign is added to compensate for the incorrect direction of I_e in figure 2.68 (b).

Also, $h_{oc} \cong h_{oe}$

Therefore, $$A_I = \frac{-[-(1 + h_{fe})]}{1 + h_{oe}(R_E \parallel R_L)} \quad ...(2.119)$$

As $h_{oc} \cong h_{oe}$,

It is small and hence can be neglected.

Therefore, $$A_I = (1 + h_{fe}) \quad ...(2.120)$$

This expression actually represents the current gain of the transistor.

2. Current gain of the overall circuit (A_{IS})

We have, $$A_{IS} = \frac{I_L}{I_S} = \frac{I_L}{I_e} \times \frac{I_e}{I_b} \times \frac{I_b}{I_S} \quad ...(2.121)$$

Let us consider figures 2.69 (a) and (b) for further simplification.

Let us consider figure 2.69(a) to write,

$$I_b = \frac{R_B}{R_B + R_i} \times I_S$$

where, $R_B = R_1 \parallel R_2$

Therefore, $$\frac{I_b}{I_S} = \frac{R_B}{R_B + R_i} \quad ...(2.122)$$

Now, let us consider figure 2.69 (b) to write

$$I_L = \frac{-R_E}{(R_L + R_E)} \times I_e$$

or $$\frac{I_L}{I_e} = \frac{-R_E}{R_L + R_E} \quad ...(2.123)$$

and $$\frac{I_e}{I_b} = -(1 + h_{fe}) \quad ...(2.124)$$

Negative sign is used to compensate for wrong direction of I_e.

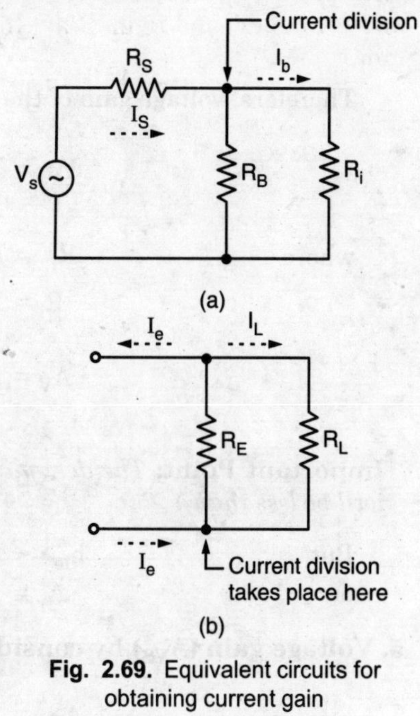

Fig. 2.69. Equivalent circuits for obtaining current gain

Substituting equations (2.122), (2.123) and (2.124) into equation (2.121), we get,

$$A_{IS} = \frac{-R_E}{(R_L + R_E)} \times -(1 + h_{fe}) \times \frac{-R_B}{(R_B + R_i)}$$

or, $$A_{IS} = \frac{(1+h_{fe})R_B R_E}{(R_L+R_E)(R_B+R_i)} \quad ...(2.125)$$

3. Input resistance R_i and R'_i

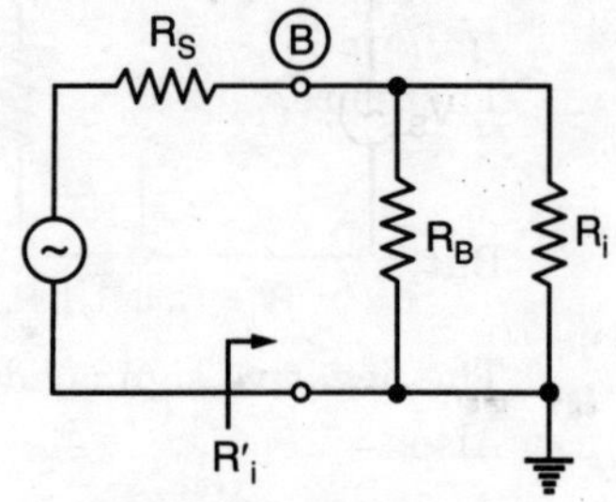

Fig. 2.70. Equivalent circuit to calculate R'_i

For the basic amplifier, we have

$$R_i = h_i + h_r A_I R_L$$

For CC amplifier, we have

$$R_i = h_{ic} + h_{rc} A_I R'_L$$

where $h_{ic} \cong h_{ie}$, $h_{rc} = 1$

$$R_i = h_{ie} + A_I (R_E \parallel R_L)$$

Since $R'_L = R_E \parallel R_L$

But $A_I = (1 + h_{fe})$

Therefore, $R_i = h_{ie} + [(1 + h_{fe}) (R_E \parallel R_L)]$ ...(2.126)

Hete, it may be noted that the input resistance R_i of the CC configuration is much higher than that of a CE configuration. The input resistance seen by the source V_S is given by,

$$R'_i = R_B \parallel R_i = (R_1 \parallel R_2) \parallel R_i$$

or $$R'_i = [h_{ie} + (1 + h_{fe}) (R_E \parallel R_L)] \parallel (R_1 \parallel R_2) \quad ...(2.127)$$

4. Voltage gain (A_V)

The voltage gain of the basic amplifier is given by,

$$A_V = A_I . \frac{R_L}{R_i}$$

Therefore, voltage gain of the CC amplifier is given by,

$$A_V = A_I \frac{R'_L}{R_i}$$

where $R'_L = (R_L \parallel R_E)$

$$R_i = h_{ie} + (1 + h_{fe}) (R_E \parallel R_L)$$

$$A_V = \frac{(1+h_{fe})(R_L \parallel R_E)}{h_{ie} + [(1+h_{fe})(R_L \parallel R_E)]} \quad ...(2.128)$$

Important Point: *The denominator of this expression is larger than the nuerator. Hence, A_V will be less than* 1.

But $h_{ie} << [(1 + h_{fe}) (R_L \parallel R_E)]$ ...(2.129)

or $A_v \cong 1$

5. Voltage gain (A_{VS}) by considering the source resistance

$$A_{VS} = \frac{V_o}{V_S} = \frac{V_o}{V_b} \times \frac{V_b}{V_S}$$

or, $$A_{VS} = A_V . \frac{V_b}{V_s} \quad ...(2.130)$$

From figure 2.71 (a), we can write

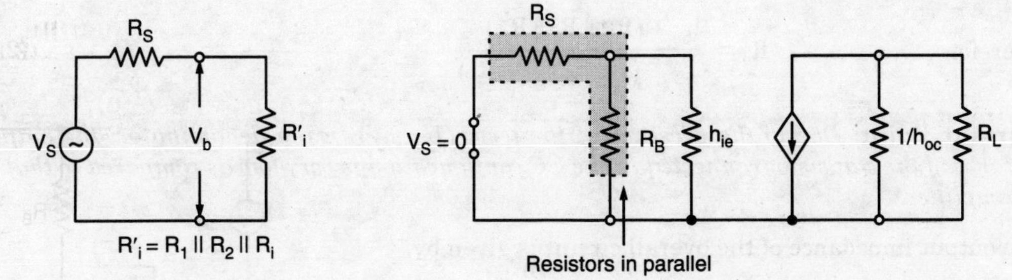

(b) Equivalent circuit for the measurement of R_o

Fig. 2.71.

$$V_b = \frac{R_i'}{(R_S + R_i')} V_S$$

$$\frac{V_b}{V_S} = \frac{R_i'}{(R_S + R_i')}$$

Substituting this equation in equation (2.130), we get

$$A_{VS} = A_V \cdot \frac{R_i'}{(R_S + R_i')}$$

Now, substituting the expression for A_V from equation (2.71) , we get,

$$A_{VS} = \frac{(1 + h_{fe})(R_E \parallel R_L)}{R_i} \times \frac{R_i'}{(R_S + R_i')} \quad \text{...(2.131)}$$

6. Output resistance (R_o and R'_o)

We know that the expression for output admittance of the basic amplifier is given by,

$$Y_o = h_o - \frac{h_f \, h_r}{h_{ie} + R_S} \quad \text{...(2.132)}$$

To apply this expression to the given CC amplifier, let us consider figure 2.71 (b). The source voltage V_S has been reduced to 0. Hence, R_S and R_B appear in parallel with each other.

Let $\qquad R_S' = (R_S \parallel R_B) = (R_S \parallel R_1 \parallel R_2)$

Output admittance of CC amplifier is given by,

$$Y_o = h_{oc} - \frac{h_{rc} \, h_{fc}}{h_{ie} + (R_S \parallel R_1 \parallel R_2)} \quad \text{...(2.133)}$$

In equation (2.133), we can neglect h_{oc} and the value of $h_{rc} = 1$.

We have $\qquad Y_o = \dfrac{-h_{fc}}{h_{ie} + (R_S \parallel R_1 \parallel R_2)} \quad \text{...(2.134)}$

But, $\qquad R_o = \dfrac{1}{Y_o}$

Therefore, $\qquad R_o = \dfrac{h_{ie} + (R_S \parallel R_1 \parallel R_2)}{-h_{fc}}$

But, $\qquad -h_{fc} = (1 + h_{fe})$

Therefore, $$R_o = \frac{h_{ie} + (R_S \parallel R_1 \parallel R_2)}{(1 + h_{fe})} \qquad ...(2.135)$$

Important Point: *Due to the presence of the term $(1 + h_{fe})$ in the denominator, the output impedance of the transistor connected in the CC configuration is very low as compared to that of a CE amplifier.*

The output impedance of the overall circuit is given by,

$$R'_o = R_o \parallel (R_E \parallel R_L) \qquad ...(2.136)$$

As R_E and R_L have values much higher than R_o, therefore,

$$R'_o \cong R_o \qquad ...(2.137)$$

Summary of analysis of CC amplifier

Table 2.7 Summary of the approximate analysis of a CC amplifier

S. No.	Parameter	Expression
1.	Current gain of transistor	$A_I = (1 + h_{fe})$
2.	Current gain of the overall circuit	$A_{IS} = \frac{(1+h_{fe}) R_B R_E}{(R_E + R_L)(R_B + R_i)}$
3.	Input resistance	$R_i = h_{ie} + (1 + h_{fe}) (R_E \parallel R_L)$
4.	Input resistance (Overall)	$R'_i = R_i \parallel (R_1 \parallel R_2)$
5.	Voltage gain	$A_V \cong 1$, practically $A_V < 1$
6.	Votage gain (Overall)	$A_{VS} \cong \frac{R'_i}{(R_S + R'_i)}$
7.	Output resistance	$R_o = \frac{h_{ie} + (R_S \parallel R_1 \parallel R_2)}{(1 + h_{fe})}$
8.	Output resistance of overall circuit	$R'_o = R_o$

EXAMPLE 2.7. The amplifier circuit has been shown in figure 2.72 (a). It uses a transistor with h_{fe} = 100 and h_{ie} = 3.37 kΩ. The other two parameters are negligible.

Calculate (i) $\frac{V_o}{V_i}$ (ii) $\frac{I_o}{I_i}$ (iii) R'_i

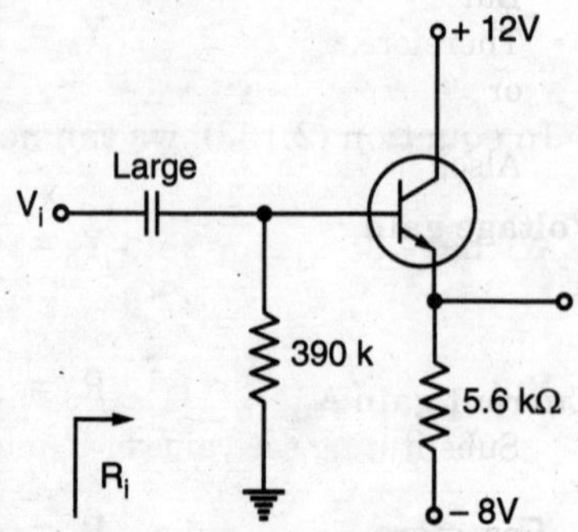

Fig. 2.72. (a)

Solution: (i) The ac equivalent circuit

The ac equivalent circuit has been shown in figure 2.72(b). This circuit shows that the given amplifier is a common collector amplifier. Hence, let us use the standard expression to find the required quantities.

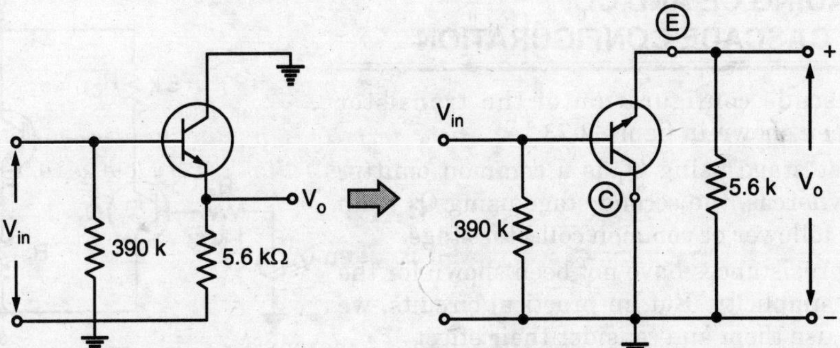

(b) AC equivalent circuit (c) Simplified AC equivalent circuit

Fig. 2.72.

(ii) The hybrid equivalent circuit.

The hybrid equivalent circuit is shown in figure 2.72 (d).

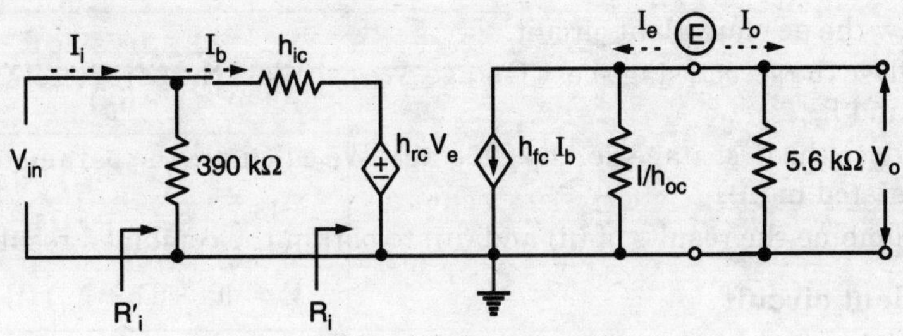

Fig. 2.72. (d) Hybrid equivalent circuit

Current gain

Given

$$h_{fc} = -(1 + h_{fe}) = -101$$
$$h_{ic} = h_{ie} = 3.37\ K\Omega$$

$$A_I = \frac{I_o}{I_b} = \frac{-I_e}{I_b} = -h_{ic}$$

$$A_I = +(1 + h_{fe}) = +101$$

$$A_{IS} = \frac{I_o}{I_b} \times \frac{I_b}{I_i} = 101 \times \frac{390\ k}{390\ k + R_i} \qquad \ldots(i)$$

Input Resistance

$$R_i = h_{ic} + h_{rc} A_I R_L$$

But $h_{rc} \cong 1.$

Therefore, $R_i = h_{ic} + A_I R_L = 3.37\ k + (101 \times 5.6\ k)$

or $R_i = 568.97\ k\Omega$ **Ans.**

Also, $R_i' = R_i \parallel 390\ k = 568.97\ k \parallel 390\ k = 231.4\ k\Omega$ **Ans.**

Voltage gain

$$A_V = \frac{A_I R_L}{R_i} = \frac{(101) \times 5.6}{568.97} = \frac{V_o}{V_i} = 0.994$$ **Ans.**

Current gain A_{IS}

Substituting the value of R_i into equation (i), we get

or, $$A_{IS} = 101 \times \frac{390\ k}{390\ k + 568.97\ k} = 41.075$$ **Ans.**

2.37. CASCADING CE AND CC: CE-CC CASCADE CONFIGURATION

The CE-CC cascade configuration of the transistor amplifier has been shown in figure 2.73.

(i) The first stage using Q_1 is a common emitter stage, whereas, the second stage using Q_2 is an emitter follower or common collector stage.

(ii) Biasing resistances have not been shown for the sake of simplicity. But, in practical circuits, we have to use them and consider their effect.

(iii) Input is applied to the CE stage and output is obtained from the CC stage.

Let us analyze the CE-CC amplifier.

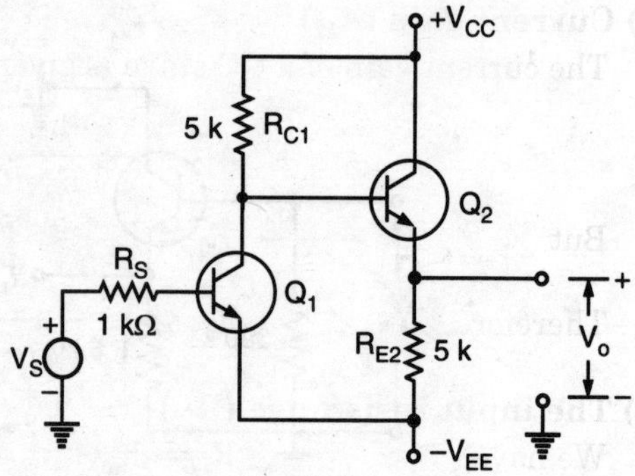

Fig. 2.73. A CE-CC amplifier

Steps to be followed:

(i) First, we draw the ac equivalent circuit.

(ii) Then, we analyze the second stage i.e. CC stage. We calculate the parameters in the sequence of A_I, R_i, A_V and R_o.

(iii) Then, we analyze the first stage i.e. the CE stage. We calculate the parameters in the same sequence as stated in (ii).

(iv) Finally, we combine the results of (ii) and (iii) to obtain the combined results.

(i) The ac equivalent circuit

The ac equivalent circuit has been shown in figure 2.74.

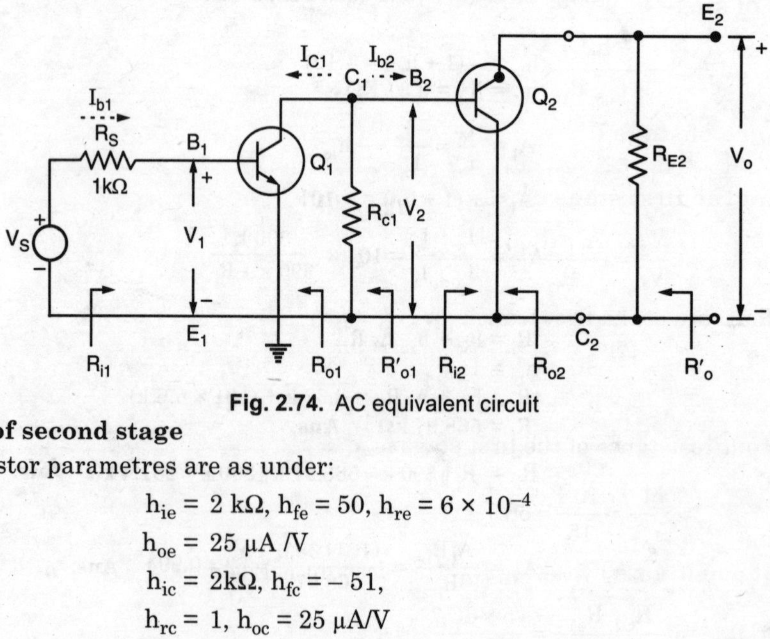

Fig. 2.74. AC equivalent circuit

(ii) Analysis of second stage

The transistor parametres are as under:

$$h_{ie} = 2\ k\Omega,\ h_{fe} = 50,\ h_{re} = 6 \times 10^{-4}$$

$$h_{oe} = 25\ \mu A/V$$

$$h_{ic} = 2k\Omega,\ h_{fc} = -51,$$

$$h_{rc} = 1,\ h_{oc} = 25\ \mu A/V$$

Approximate or exact analysis?

Here, $h_{oc} R_L = 25 \times 10^{-6} \times 5 \times 10^3 = 0.125$. Since $h_{oc} R_L$ is greater than 0.1, we will carry out the exact analysis for the second stage. Therefore, here, we shall use the expressions derived for the basic amplifier.

(a) Current gain (A_{I2})

The current gain of a CC stage is given by,

$$A_{I2} = \frac{-h_{fc}}{1 + h_{oc} R_L} \quad ...(2.138)$$

But $R_L = R_{E2} = 5\ k\Omega$

Therefore,

$$A_{I2} = \frac{51}{1 + \left(25 \times 10^{-6} \times 5 \times 10^3\right)} = 45.3 \quad ...(2.139)$$

(b) The input resistance (R_{i2})

We have, $R_{i2} = h_{ic} + h_{rc} A_{I2} \cdot R_{E2} = 2k + (1 \times 45.3 \times 5) = 228.5\ k\Omega$ ...(2.140)

Note that the input resistance of CC stage is very high.

(c) Voltage gain (A_{V2})

We have,

$$A_{V2} = \frac{V_o}{V_2} = A_{I2} \frac{R_{E2}}{R_{I2}} = \frac{45.3 \times 5}{228.5} = 0.99 \quad ...(2.141)$$

Note that the voltage gain of the second stage is very close to 1.

(iii) Analysis of first stage

The net load resistance R_{L1} of the first stage is the parallel combination of R_{C1} and R_{i2}.

Thus,

$$R_{L1} = R_{C1} \parallel R_{i2} = \frac{R_{C1}\ R_{i2}}{(R_{C1} + R_{i2})} = \frac{5 \times 228.5}{(5 + 228.5)}\ 4.9\ k\Omega \quad ...(2.142)$$

Approximate or exact analysis?

Here, $h_{oe} R_{L1} = 25 \times 10^{-6} \times 4.9 \times 10^3 = 0.1225$. Since, $h_{oe} R_{L1}$ is higher than 0.1, we have to perform the exact analysis even for the first stage.

(a) Current gain of stage I

We have,

$$A_{I1} = \frac{-h_{fe}}{1 + h_{oe} R_{L1}} = \frac{-50}{1 + \left(25 \times 10^{-6} \times 4.9 \times 10^3\right)} = -44.5 \quad ...(2.143)$$

(b) Input resistance of stage I

$$R_{i1} = h_{ie} + h_{re} A_{I1} R_{L1} = 2 - (6 \times 10^{-4} \times 44.5 \times 4.9) = 1.87\ k\Omega \quad ...(2.144)$$

(c) Voltage gain of the first stage

$$A_{V1} = \frac{V_2}{V_1} = \frac{A_{I1}\ R_{L1}}{R_{i1}} = \frac{-44.5 \times 4.9}{1.87} = -116.6 \quad ...(2.145)$$

The output admitance of the first transistor is given by,

$$Y_{o1} = h_{oe} - \frac{h_{re}\ h_{fe}}{h_{ie} + R_S} = 25 \times 10^{-6} - \frac{50 \times 6 \times 10^{-4}}{2 \times 10^3 + 1 \times 10^3} = 15 \times 10^{-6}\ \text{mho} \quad ...(2.146)$$

Hence, the output resistance of the first stage is given by,

$$R_{o1} = \frac{1}{Y_{o1}} = \frac{10^6}{15} = 66.7\ k\Omega \quad ...(2.147)$$

And the output admittance by taking R_{C1} into account is given by,

$$R'_{o1} = \frac{R_{C1}\ R_{o1}}{R_{C1} + R_{o1}} = \frac{5 \times 66.7}{(5 + 66.7)} = 4.65\ k\Omega \quad ...(2.148)$$

(d) The output resistance of second stage

The output admittance of the second stage will be,

$$Y_{o2} = h_{oe} - \frac{h_{fc}\, h_{rc}}{h_{ic} + R'_{S2}}$$

where $R'_{S2} = R'_{o1} = 4.65\ k\Omega$

Therefore, $$Y_{o2} = 25 \times 10^{-6} - \frac{(-51)(1)}{(2 \times 10^3 + 4.65 \times 10^3)} = 7.7 \times 10^{-3}\ A/V \quad ...(2.149)$$

or, $$R_{o2} = \frac{1}{Y_{o2}} = \frac{10^3}{7.7} = 130\ \Omega$$

This is the output resistance of transistor Q_2. The output resistance of second stage considering R_{E2}, is given by,

$$R'_{o2} = R_{o2} \parallel R_{E2} = \frac{R_{o2}\, R_{E2}}{(R_{o2} + R_{E2})} = \frac{130 \times 5000}{(130 + 5000)} = 127\Omega \quad ...(2.150)$$

(iv) Overall voltage and current gain

(a) Total current gain of both stages

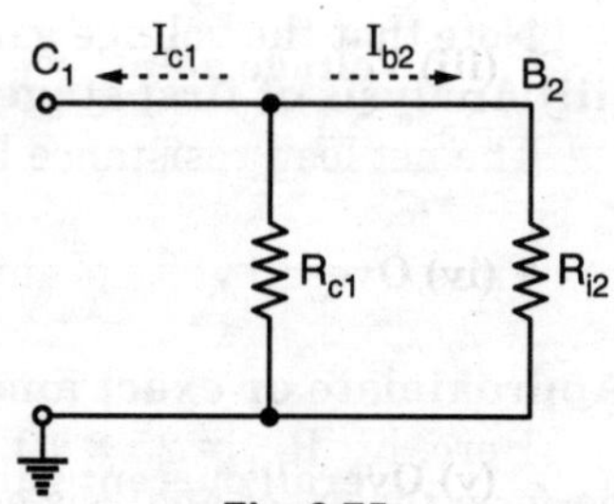

Fig. 2.75.

We have, $$A_I = \frac{-I_{e2}}{I_{b1}} = \frac{-I_{e2}}{I_{b2}} \times \frac{I_{b2}}{I_{c1}} \times \frac{I_{c1}}{I_{b1}}$$

or $$A_I = -A_{I2} \times \frac{I_{b2}}{I_{c1}} \times A_{I1} \quad ...(2.151)$$

From figure 2.75, we can write that,

$$\frac{I_{b2}}{I_{c1}} = \frac{-R_{c1}}{R_{i2} + R_{c1}} \quad ...(2.152)$$

Substituting this value into equation (2.151), we get,

$$A_I = A_{I1}\, A_{I2} \cdot \frac{R_{c1}}{R_{c1} + R_{c1}}$$

Hence, $$A_I = 45.3 \times (-44.5) \times \frac{5}{(5 + 228.5)} = -43.2 \quad ...(2.153)$$

(b) Overall voltage gain

We have $$A_V = A_{V1}\, A_{V2} = -0.99 \times -116.6 = -115 \quad ...(2.154)$$

And the overall voltage gain including the effect of R_s is given by,

$$A_{VS} = \frac{V_o}{V_S} = A_V \frac{R_{i1}}{R_{i1} + R_S} = \frac{-115 \times 1.87}{(1.87 + 1)} \quad ...(2.155)$$

or $$A_{VS} = -75.3 \quad ...(2.156)$$

Salient Features of CE-CC cascade

(i) The CE-CC cascade has high current gain.

(ii) The input impedance of CE-CC cascade is low.

(iii) The output impedance of CE-CC cascade is very low.

(iv) The voltage gain of CE-CC cascade is very high.

IMPORTANT SOLVED EXAMPLES

EXAMPLE 2.8. A CB transistor amplifier uses a voltage source of internal resistance R_s = 1200 Ω and the load resistance is R_L = 1200 Ω. The h-parameters are h_{ib} = 22 Ω, h_{rb} = 3×10⁻⁴, h_{fb} = –0.98 and h_{ob} = 0.5 μA/V. Calculate the following:

(i) Current gain
(ii) Input impedance
(iii) voltage gain
(iv) overall voltage gain
(v) overall current gain.

Solution:

(i) Current gain,

$$A_i = \frac{-h_{fb}}{1+h_{ob}.R_L} = -\frac{-(0.98)}{1+0.5\times10^{-6}\times1000} = 0.98 \quad \textbf{Ans.}$$

(ii) Input resistance,

$$R_i = h_{ib} + h_{rb}.A_iR_L$$

$$R_i = 22 + (3 \times 10^{-4}) \times 0.98 \times 1000 = 22.3\ \Omega. \quad \textbf{Ans.}$$

(iii) Voltage gain

$$A_v = \frac{A_i.R_L}{R_i} = \frac{0.98\times1000}{22.3} = 43.946 \quad \textbf{Ans.}$$

(iv) Overall voltage gain

$$A_{vs} = \frac{A_v.R_i}{R_i+R_s} = \frac{43.946\times22.3}{22.3+1200} = 0.802 \quad \textbf{Ans.}$$

(v) Overall current gain

$$A_{is} = \frac{A_i.R_s}{R_i+R_s} = \frac{0.98\times1200}{22.3+1200} = -0.962 \quad \textbf{Ans.}$$

EXAMPLE 2.9. Figure 2.76 shows the circuit of common collector amplifier or emitter follower. The h-parameters are h_{ic} = 2kΩ, h_{fc} = –51, h_{rc} = 1 and h_{oc} = 25×10⁻⁶ mhos. Determine the following:

(i) Current gain
(ii) Input resistance
(iii) Voltage gain
(iv) Output resistance.

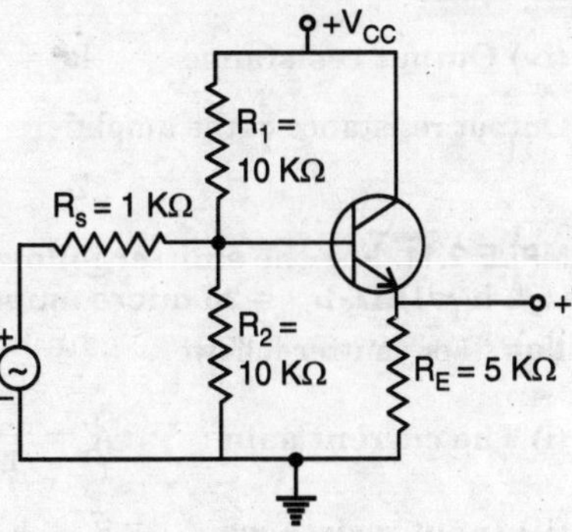

Fig. 2.76.

Solution :

(i) Current gain

$$A_i = \frac{h_{fc}}{1+h_{oc}.R_L}$$

$$A_i = \frac{-51}{1+25\times10^{-6}\times5\times10^3}$$

$$A_i = -45.3 \quad \textbf{Ans.}$$

(ii) Input resistance

$$R_i = h_{ic} + h_{rc}.A_i\,R_L$$

$$R_i = 2000 + 1 \times 45.3 \times 5 \times 10^3$$

or

$$R_i = 228\ k\Omega. \quad \textbf{Ans.}$$

Input resistance of amplifier stage

$$Z_i = R_i \| (R_1 \| R_2)$$

or

$$Z_i = 228 \| (10 \| 10) = 4.9\ k\Omega \quad \textbf{Ans.}$$

(iii) Voltage gain $A_v = \frac{A_i.R_L}{R_i} = \frac{45.3\times5\times10^3}{228\times10^3} = 1$ **Ans.**

(iv) Output Resistance

$$R_o = -\frac{R_s + h_{ic}}{h_{fc}} = -\frac{1000+2000}{-51} = 58.8\,\Omega \quad \textbf{Ans.}$$

Output resistance of the amplifier stage

$$Z_o = R_o \| R_E = 58.8 \| 5\times10^3 = 581.1\Omega \quad \textbf{Ans.}$$

EXAMPLE 2.10. Figure 2.77 shows the circuit of a single stage CE amplifier. The values of h-parameters are, h_{ie} = 1.5 kΩ, $h_{re} = 5\times10^{-3}$, h_{fe} = 50, $h_{oe} = 2\times10^{-5}$ μA/V.
Determine the following : (Use Approximate Analysis)

(i) Current gain
(ii) Input resistance
(iii) Voltage gain
(iv) Output resistance.

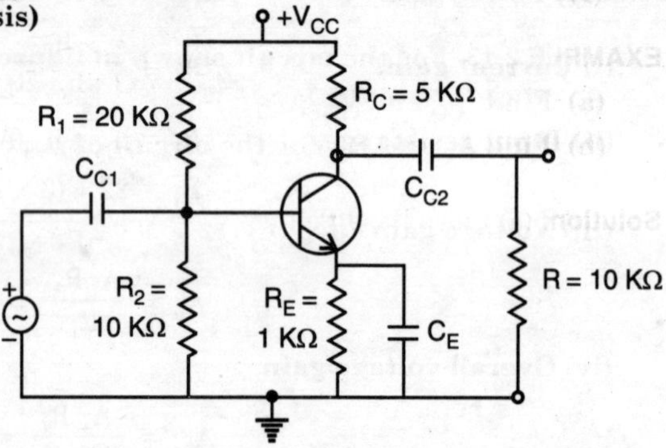

Fig. 2.77.

Solution: In this case, value of a.c. load resistance

$$R_L = R_C \| R = 5 \| 10 = 3.3k\Omega$$

(i) Current gain $A_i = -h_{fe} = -50$ **Ans.**

(ii) Input resistance $R_i = h_{ie} = 1.5$ kΩ **Ans.**

Input resistance of amplifier stage

$$Z_i = R_i \| (R_1 \| R_2)$$

$$= 1.5 \| (20 \| 10) = 1.22 \text{ k}\Omega \quad \textbf{Ans.}$$

(iii) Voltage gain $A_v = \frac{-h_{fe}.R_L}{h_{ie}} = \frac{(-50)\times(3.3\times10^3)}{1500} = -110$ **Ans.**

Overall voltage gain $A_{vs} = A_v = -110$ **Ans.**

(iv) Output resistance $R_o = \frac{1}{h_{oe}} = \frac{1}{20\times10^{-6}} = 0.05\times10^6$ Ohm = 50 kΩ.

Output resistance of the amplifier

$$Z_o = R_o \| R_L = 50 \| 3.3 = 3.1 \text{ k}\Omega \quad \textbf{Ans.}$$

EXAMPLE 2.11. For the emitter follower with R_s = 0.5KΩ and R_L = 5KΩ, calculate A_i, R_i, A_v. Assume h_{fe} = 50, h_{ie}=1KΩ, h_{oe} = 25 micro amp/volt. (U.P.Tech – Tutorial Question Bank)

Solution : For emitter follower,

(i) The current gain : $A_i = \frac{1+h_{fe}}{1+h_{oe}.R_L} = \frac{1+50}{1+25\times10^{-6}\times5\times10^3} = \frac{51}{1.125} = 45.33$ **Ans.**

(ii) Input resistance: $R_i = h_{ic} + h_{rc}.A_i.R_L = h_{ie} + 1.A_i.R_L = h_{ie} + A_i.R_L$

Putting all the values

$$R_i = 1\times10^3 + 45.33\times5\times10^3 = (1+226.6)10^3 = 227.6 \text{ k}\Omega \quad \textbf{Ans.}$$

(iii) Voltage gain: $A_v = \frac{V_{out}}{V_{in}} = \frac{A_i.R_L}{R_i} = \frac{45.33\times5}{227.6} = \frac{226.6}{227.6} = 0.9958$ **Ans.**

EXAMPLE 2.12. Find the values of voltage gain, current gain, input resistance and power gain for a common emitter transistor amplifier with R_L=1600 ohm and R_s=1KΩ. The transistor has h_{ie}=1100 ohm, $h_{fe} = 2.5\times10^{-4}$, h_{oe} = 25 micro amp/V. (U.P.Tech – Tutorial Question Bank)

Solution: For common emitter transistor amplifier

(i) current gain $A_i = -\dfrac{h_{fe}}{1+h_{oe}.R_L} = -\dfrac{2.5\times10^{-4}}{1+25\times10^{-6}\times1600} = -\dfrac{2.5\times10^{-4}}{1.04}$

$A_i = -2.4\times10^{-4}$ **Ans.**

(ii) Input resistance

$R_i = h_{ie} + h_{re}\,A_i.R_L \cong h_{ie}$ (neglecting the factor $h_{re}A_i.R_L$)

$R_i \equiv 1100$ ohm **Ans.**

(iii) Voltage gain $A_v = A_i\dfrac{R_L}{R_i} = -2.4\times10^{-4}\dfrac{1600}{1100} = -\dfrac{38.4\times10^{-4}}{11} = -3.49\times10^{-4}$ **Ans.**

(iv) Power gain $= A_i.A_v = (-2.4\times10^{-4})\times(-3.49\times10^{-4}) = 8.37\times10^{-8}$ **Ans.**

EXAMPLE 2.13. For the circuit shown in figure 2.78, given $\beta_{dc} = h_{FE} = 130$.

(a) Find I_{CQ} and V_{CEQ}

(b) Find A_v and R_{in} for the circuit of part (a) if $h_{fe} = 50$, $h_{ie} = 1K\Omega$, $h_{re} = 0$ and $h_{oe} = 0$.

Solution: (a) Since, Base Voltage $V_B = V_{CC}\times\dfrac{R_2}{R_1+R_2}$

$$V_B = 18\times\frac{510}{510+510}$$

$$V_B = \frac{18}{2} = 9 \text{ Volt}$$

Value of I_C is given by $I_C = \dfrac{V_B - V_{BE}}{R_E}$

$$I_C = \frac{V_B - 0}{R_E} \text{ (neglecting } V_{BE})$$

$$I_C = \frac{V_B}{R_E} = \frac{9}{7.5\times10^3} = \frac{9\times10^{-3}}{7.5} = 1.2 \text{ mA}$$

or $I_{CQ} \equiv 1.2$ mA **Ans.**

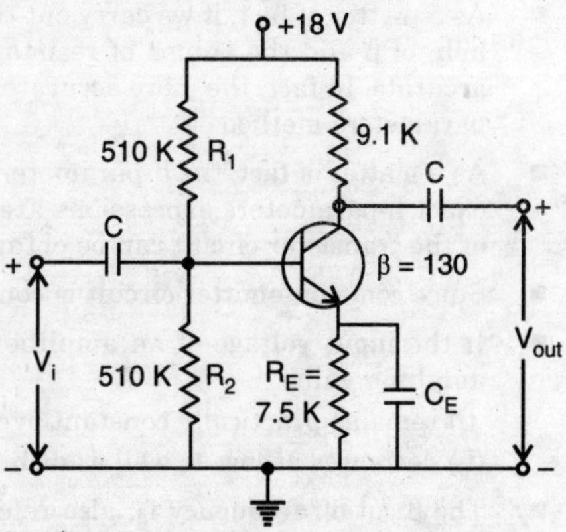

Fig. 2.78.

Now, $V_{CE} = V_{CC} - I_C.(R_C + R_E) = 18 - 1.2\times10^{-3}(9.1+5.5)\times10^3$

$V_{CE} = 18 - 1.2\times16.6$

or $V_{CE} = 18 - 19.9 = -1.9 \text{ Volt} \equiv -1.9$ Volt **Ans.**

(b) Input resistance $R_i \cong h_{ie} \cong 1K\Omega$ **Ans.**

Input resistance of amplifier stage

$$R_{in} = R_i \,\|\, (R_1 \,\|\, R_2) = 1 \,\|\, (510 \,\|\, 510)$$

$$R_{in} = 1 \,\|\, 255 = 0.99\ K\Omega \quad \textbf{Ans.}$$

Current gain $A_i \cong -h_{fe} = -50$ **Ans.**

Voltage gain $A_v = \dfrac{A_i.R_L}{R_i} = \dfrac{A_i.R_C}{R_i} = \dfrac{-50\times9.1}{1} = -455$ **Ans.**

SUMMARY

- We want the output signal to be linearly proportional to the input signal because only then the amplified signal will exactly have the same shape as that of the input. If the amplifier is non-linear, then the shape of the output signal will be different from that of the input and the output signal is said to be distorted. Hence, the amplifier must be linear.

- The linear amplifier is also called as a small signal amplifier. In order to amplify the input signal faithfully i.e., without introducing any distortion, the amplifier must operate in the linear portion of the transfer characteristics. To achieve this, the amplitude of the input signal has to be sufficiently small. Hence the name small signal amplifiers.
- The commonly used models for the small signal ac analysis are as under:
  (i) Small signal r_e model
  (ii) Small signal hybrid model (using h-parameters)
  (iii) Small signal hybrid π model
  (iv) Ebers Moll model.
- As the h-parameter values vary considerably with individual transistors of the same type and the accuracy of the analysis depends on the operating conditions, an alternate model known as r_e model which employs transistor β and resistance values is gaining importance now-a-days.
- The advantages of using this model are as under:
  (i) required parameters are easily available
  (ii) simple and easy procedure, and
  (iii) results obtained have a fairly good accuracy for the study of amplifier circuits.
- As a matter of fact, if we carry out the analysis of a transistor amplifier, to study its behaviour, with the help of β and the values of resistances used in the circuit, then this type of analysis is not found so accurate. In fact, the more accurate method to analyse a transistor amplifier is hybrid parameters or h-parameters method.
- As a matter of fact, the h-parameters are widely used in amplifier analysis. To simplify the analysis, the exact h-parameters expressions are converted into approximated expressions. The complete behaviour of the transistor circuit can be obtained by using this simplified hybrid model
- Since common-emitter circuit is considerable the most important practical configuration.
- If the input voltage of an amplifier is kept constant but its frequency is varied, it is found that the amplifier gain:
  (i) remains practically constant over a sizable range of mid-frequencies.
  (ii) decreases at low as well as at high frequencies.
- The β cut-off frequency f_β, also referred to as h_{fe} is the CE short-circuit small signal forward-current-transfer-ratio cut off frequency f_β is the frequency at which a transistor's CE short-circuit current gain drops 3-dB from its value at lower (mid) frequencies. Further, f_β represents the maximum attainable bandwidth for the current gain of a CE amplifier with a given transistor.
- A CB amplifier has a much higher 3 dB frequency than a CE amplifier. The short-circuit current gain for CB amplifier which can be derived from the approximate high frequency circuit of the CB amplifier with output shorted
- Various Properties of a CE Amplifier
  (i) Large current gain (A_I)
  (ii) Large voltage gain (A_V)
  (iii) Large power gain ($A_P = A_I.A_V$)
  (iv) Voltage phase shift of 180°.
  (v) Moderate input impedance
  (vi) Moderate output impedance

SHORT QUESTIONS WITH ANSWERS

Q. 1. What are the Hybrid Parameters (h-parameters)? Explain.

Ans. Hybrid parameters are also called as h-parameters. Characteristics of amplifier can be specified easily in terms of h-parameters. The hybrid or h-parameters are very popular and widely used because they give accurate result and can be measured very easily.

We know that a transistor is a three terminals device. If one terminal is made common to the input and output, then the transistor can be represented as a two-port network.

Q.2. What do you mean by cascade amplifier?

Ans. Cascade amplifier is a composite amplifier pair with a large bandwidth used for RF applications and as a video amplifier. It consists of a CE stage followed by a CB stage directly to each other and combines some of the features of both the amplifiers.

For high frequency applications, CB configuration has the most desirable characteristics. However, it suffers from low input impedance ($Z_i \approx h_{ib}$). The cascade configuration is designed to have the input impedance essentially that of CE amplifier, the current gain that of CE amplifier, the voltage gain that of CB amplifier and good isolation between the input and output.

Q.3. What is an amplifier ?

Ans. The device that amplifies the amplitude of the input signal is called the amplifier. An amplifier may be defined as a device that increases the current, voltage or power of an input signal with the help of a transistor by furnishing the additional power from a separate source of supply.

Q.4. What is meant by small signal amplifier?

Ans. When the input signal is quite weak and produces small fluctuations in the output current in comparison to its quiescent value, the amplifier is called the small signal or voltage amplifier.

Q.5. What is meant by phase reversal?

Ans. In a CE configuration, the output voltage increases in the negative direction when the input signal voltage increases in the positive direction and vice-versa. This is called the phase reversal and causes a phase difference of 180° between the input signal voltage and output voltage.

Q.6. What is an ac emitter resistance?

Ans. The dynamic resitance of the emitter-base junction diode is called the ac emitter resistance. It is given as $\frac{25\text{ mV}}{I_E}$, where I_E is the dc emitter current at Q-point.

Q.7. What is the effect of source resistance on voltage gain of a common base transistor amplifier?

Ans. The voltage gain of a CB transistor amplifier will decrease if source resistance is considered because in such a case, there will be a voltage drop across the source resistance and output voltage will decrease.

Q.8. What is the effect of removal of emitter bypass capacitor in a CE amplifier circuit?

Ans. Removal of bypass capacitor in a CE amplifier circuit causes excessive degeneration in the amplifier circuit.

Q.9. Why common-collector circuit is known as an emitter follower ?

Ans. The CC circuit amplifier is called an emitter follower because in this circuit the output voltage at the emitter terminal follows the input signal applied to the base terminal.

Q.10. What are the main purposes for which a common-collector amplifier may be used?

Ans. For a common collector amplifier, current gain is as high as for CE amplifier, voltage gain is less than unity, input resistance is the highest and the output resistance is the lowest of all the three (CE, CC and CB) configurations. This circuit finds wide application as a buffer amplifier between a high impedance source and a low impedance load.

Q.11. What is meant by frequency response of an amplifier ?

Ans. The curve drawn between the voltage gain and signal frequency of an amplifier is known as the frequency response of an amplifier.

Q.12. Define f_T of a BJT.

Ans. The f_T of a BJT is defined as the frequency at which the short-circuit common emitter current gain β falls to unity. It is equal to $\beta\, f_\beta$.

Q.13. What is Miller effect ?

Ans. According to Miller effect, when viewed from input base terminal of a CE connected transistor the collector-base capacitance C_{bc} appears to be $(1 + A_v)\, C_{bc}$, i.e., the C_{bc} is amplified by a factor $(1 + A_v)$.

Q.14. What is meant by distortion in an amplifier ?

Ans. Change in amplified output signal waveform is called the distortion.

Q.15. Why distortion in amplifier is considered harmful ?

Ans. Because distortion may change the intelligence (useful information) carried by the signal.

REVIEW QUESTIONS

1. Draw a simple circuit of a common-collector amplifier and explain its operation. Obtain an expression for its voltage gain.
2. Classify the amplifiers on the basis of type, frequency, coupling and mode of operation.
3. Write short notes on the following :
   (a) r'_e model of a common base transistor amplifier.
   (b) Analysis of a common-emitter transistor amplifier using h-parameters.
   (c) Emitter follower.
   (d) Comparison of transistor amplifier configurations.
4. What is AC equivalent circuit? How is it drawn?
5. Draw the h-parameter equivalent of a transistor amplifier in CE, CB and CC mode.
6. How will you calculate the h-parameters from the input and output characteristics?
7. Explain the operation of CE transistor amplifier.
8. State and explain the necessity of each component in the CE amplifier.
9. Why is the bypass capacitor C_E used?
10. Draw and explain the common base transistor amplifier.
11. Draw and explain the common base transistor amplifier for an emitter follower.
12. State the features and applications of CB, CE and CC amplifiers.
13. What is the effect of unbypassed R_E on the input resistance of an amplifier?
14. Why are h-parameters called small signal parameters?
15. Show graphically the effect of biasing a transistor near the saturation and cut off regions.
16. Why are the CC, CB, CE amplifiers called small signal amplifiers?
17. Explain the working of BJT as a switch **(U.P. Tech., Sem. Exam., 2006-07) (Marks 5)**
18. Draw the circuit schematic of a common emitter BJT amplifier. Sketch and explain the small signal high frequency model of the amplifier and hence obtain the expression for the input impedance. **(U.P. Tech., Sem. Exam., 2008-09) (Marks 10)**

NUMERICAL PROBLEMS

1. A parallel resonant circuit has a bandwidth of 15 kHz and Q-factor of 120. What is the resonant frequency of the circuit ? [**Ans.** 1800 KHz]
2. A junction transistor has $r_e = 50\ \Omega$, $r_b = 1$ K, $r_e = 1$ M and $\alpha = 0.96$. It is used in common base-circuit with a load resistance of 8 K. Determine input and output resistances and current, voltage and power gains. [**Ans.** $R_{in} = 90\ \Omega$; $R_{out} = 85.7$ K; $A_i = 0.96$; $A_v = 85.33$; $A_p = 82$]
3. A PNP junction transistor is used as voltage amplifier in the grounded-base connection, the load resistance being 300 K and the internal resistance of the original source being 200 Ω.
   Derive the expression for the voltage gain of the amplifier and calculate its magnitude if the transistor T-network parameters are :
   $r_e = 18\ \Omega$; $r_b = 700\ \Omega$; $r_c = 1$ M and $R_{in} = 976$ K. [**Ans.** 1,250]

1. Ebers-Model of a transistor represents two diodes
   (a) in series (b) in parallel (c) back-to-back (d) none of the above
2. I_{CBO} in a transistor can be reduced by reducing
   (a) I_B (b) V_{CC} (c) I_E (d) temperature
3. If $\alpha = 0.98$, $I_{CO} = 6\ \mu A$ and $I_B = 100\ \mu A$ for a transistor then the value of I_C will be
   (a) 2.3 mA (b) 3.1 mA (c) 4.6 mA (d) 5.2 mA
4. For the transistor to work as an amplifier
   (a) the base emitter junction should be forward-biased
   (b) reverse biased (c) none of these (d) forward biased
5. The transistor configuration producing highest output resistance is
   (a) CC (b) CB (c) CE (d) none of the above
6. The input resistance of a common emitter transistor is of the order of
   (a) 1 MΩ (b) 1 KΩ (c) 0.01 Ω (d) 0.001 Ω

Answers

1. (c) 2. (c) 3. (a) 4. (a) 5. (a) 6. (c)

UNIT II

Feedback amplifier, negative feedback, voltage-series, voltage shunt, current series and current shunt feedback, Sinusoidal oscillators, L-C (Hartley-Colpitts) oscillators, RC phase shift, Wien bridge and crystal oscillators, Power amplifiers, Class A, Class B, Class C amplifiers, their efficiency and power dissipation.

Above syllabus has been covered in following chapters :

CHAPTER 3

Feedback

CHAPTER 4

Sinusoidal Oscillators

CHAPTER 5

Power Amplifiers

Chapter 3

FEEDBACK

Syllabus

Feedback: Feedback amplifier, negative feedback, voltage-series, voltage shunt, current series and current shunt feedback.

Chapter Outline

3.1 INTRODUCTION

As discussed earlier, an ideal linear amplifier in the mid frequency region, provides an output signal that is an exact replica of the applied input signal. But, the practical amplifiers depart from the ideal one for several reasons like non-linearity of transistor characteristics, parameter variation, temperature effects etc. As a matter of fact, some of these factors can be minimized by improving the involved basic devices. However, there is a limit, beyond which this approach does not work. The best approach is to use the principle of feedback to achieve a desired degree of improvement. The feedback may be defined as a process of injecting some energy from the output and then return it back to the input. In other words, when a part or fraction of output is combined to the input, feedback is said to exist.

Thus, the process of combining a fraction of the output energy (i.e., voltage or current) back to the input is called the feedback. The amplifiers, which use the feedback principle, are known as feedback amplifiers.

A feedback amplifier, basically, consists of two parts namely an amplifier circuit and feedback circuit (or network). The function of the feedback network is to return a fraction of the output energy (i.e., voltage or current) to the input of the amplifier. The feedback network is usually in the form of a passive two port network and may be formed by resistors, inductors and capacitors. However, feedback network mostly uses only resistors.

3.2 CLASSIFICATION OF FEEDBACK

DO YOU KNOW?

Special low-noise pre-amplifiers are used in several applications such as high-quality stereo systems to exploit the benefit of feedback and improve the signal to noise ratio.

Depending upon whether the feedback signal increases or decreases the input signal, there are two basic types of feedback in amplifiers :

(i) Positive Feedback, and
(ii) Negative Feedback

(i) Postive Feedback

If the feedback signal (i.e., voltage or current) is applied in such a way that it is in phase with the input signal and thus increases it, then it is called a positive feedback. Some-times, it is also known as regenerative feedback or direct feedback. Figure 3.1 illustrates the concept of positive feedback. Here, both amplifier and feedback network introduce a phase shift of 180°. The result is a 360° phase shift around the loop and thus causing a 360° phase shift around the loop and thus causing the feedback voltage V_f to be in phase with the input signal V_{in}.

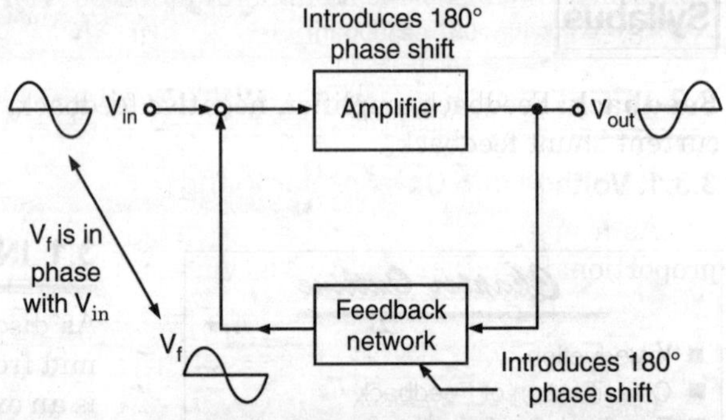

Fig. 3.1. Concept of positive feedback

Since positive feedback causes excessive distortion and instability, it is seldom used in amplifiers. However, because of its capability of increasing the power of the original signal, it is used in oscillator circuits.

(ii) Negative Feedback

If the feedback signal (i.e., voltage or current) is applied in such a way that it is out of phase with the input signal and thus decreases it, then it is called a negative feedback. Sometimes, it is also called as degenerative feedback or inverse feedback. Figure 3.2 illustrates the concept of negative feedback. Here, the amplifier introduces a phase shift of 180° into the circuit while the feedback network is so designed that it introduces no phase shift (i.e., 0° phase shift). Because of this, the feedback voltage V_f is 180° out of phase with the input signal V_{in}.

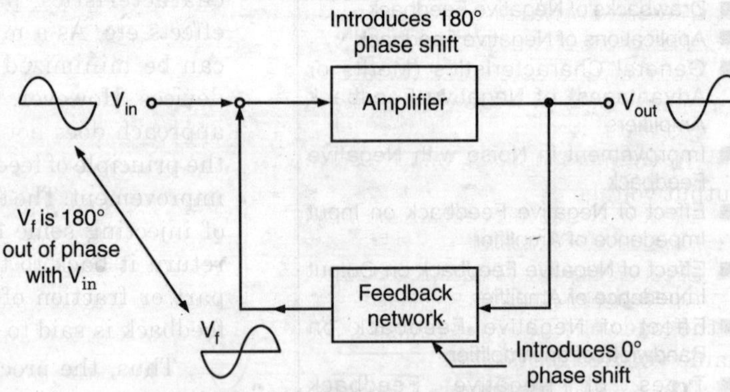

Fig. 3.2. Concept of negative feedback

Note that negative feedback reduces the amplifier gain but it has a numerous advantages such as gain stability, reduction in non-linear distortion, reduction in noise, increase in bandwidth etc. Therefore, negative feedback is most widely used in amplifier circuits.

The feedback may also be classified as voltage feedback and current feedback. In voltage feedback, the energy fed back to the input terminals is proportional to the output voltage whereas in current feedback, the energy fed back to the input terminals is proportional to the current through the load.

3.2.1. Performance Comparison of Positive and Negative Feedback

S.No.	Parameter of comparison	Positive feedback	Negative feedback
1.	Overall phase shift	0 or 360°	180°
2.	Feedback signal and input signal	are in phase	are out of phase
3.	Input voltage	Increases due to feedback	Decreases due to feedback
4.	Output voltage	Increases due to feedback	Decreases due to feedback
5.	Voltage gain	Increases due to feedback	Decreases due to feedback
6.	Stability	Becomes poor as feedback increases	Becomes better as feedback increases
7.	Application	Oscillators, schmitt trigger	Amplifier
8.	Noise	Increases with feedback	Decreases with feedback

3.3 FEEDBACK CONFIGURATIONS

3.3.1. Voltage and Current Sampling

As a matter of fact, the sample supplied by the β network (i.e., feedback network) may be proportional to either load voltage or load current, as shown in figure 3.3(a) and figure 3.3(b).

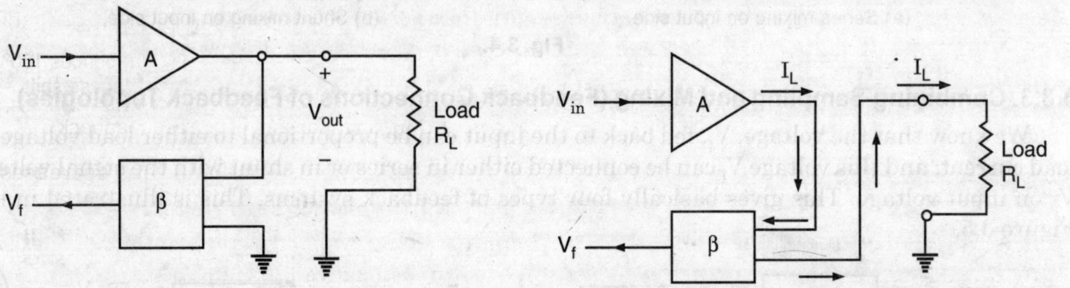

Fig. 3.3. (a) Voltage sampling voltage feedback (b) Current sampling current feedback.

In voltage feedback, output voltage is sampled, and voltage fed back, V_f, is proportional to the output voltage V_{out}, i.e.,

$$V_f \propto V_{out}$$
$$V_f = \beta V_{out}$$

Note that V_{out} is also the voltage across the load, i.e., V_{out} is the load voltage and V_f is part of load voltage and hence the name **voltage feedback.**

The current sampling has been illustrated in figure 3.3 (b).

In this case, the current flowing through output of amplifier, input of β network, and load resistance R_L is the same viz. load current I_L. Then, the voltage V_f which is fed back is made proportional to output current or load current. Hence, this type of feedback is known as **current feedback.**

> **DO YOU KNOW?**
>
> A feedback amplifier is sometimes known as a closed-loop amplfier because the feedback forms a closed-loop between the input and the output. It essentially consists of two blocks: an amplifier and a feedback circuit.

Note that in both voltage feedback and current feedback, a voltage V_f is fed back. In voltage feedback system, V_f is proportional to output voltage (load voltage), while in current feedback, V_f is proportional to the output or load current.

The general feedback equation is given by

$$A_f = \frac{A}{1 + A\beta}$$

is applicable to both voltage and current feedbacks, with proper interpretation of β. In both systems, negative feedback reduces the gain of the amplifier. This is the disadvantage of the negative feedback. However the negative feedback has lot of advantages. Hence, the negative feedback, although it reduces the gain, is almost invariably used in the amplifier.

3.3.2. Series and Shunt Mixing

The voltage V_f fed back to the input side can be mixed with the original signal input voltage V_s in two ways, as shown in figure 3.4.

Figure 3.4 (a) shows that the two voltages V_s and V_f are connected in series opposition. Thus, it is series mixing.

Figure 3.4 (b) indicates that the two voltages V_s and V_f are connected in parallel. Thus, it is shunt mixing.

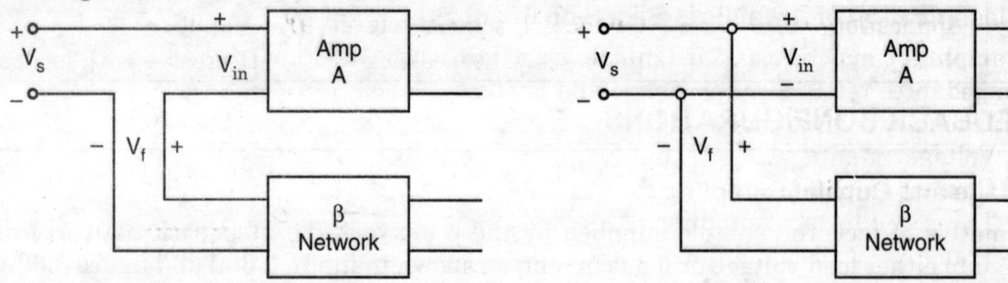

(a) Series mixing on input side

(b) Shunt mixing on input side.

Fig. 3.4.

3.3.3. Combining Sampling and Mixing (Feedback Connections of Feedback Topologies)

We know that the voltage, V_f, fed back to the input can be proportional to either load voltage or load current; and this voltage V_f can be connected either in series or in shunt with the signal voltage V_s on input voltage. This gives basically four types of feedback systems. This is illustrated in the Figure 3.5.

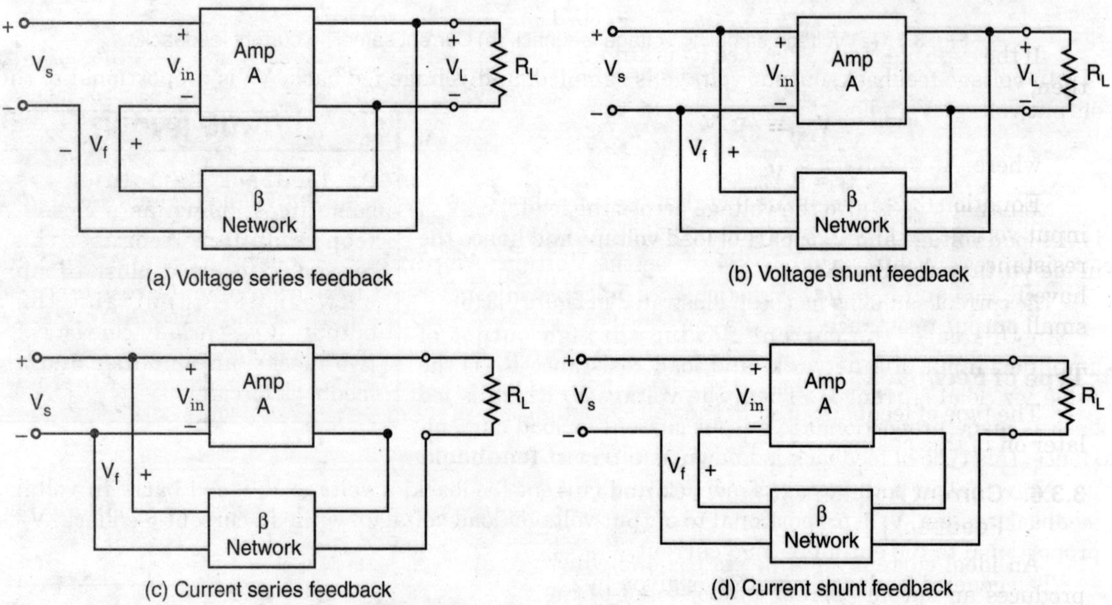

(a) Voltage series feedback

(b) Voltage shunt feedback

(c) Current series feedback

(d) Current shunt feedback

Fig. 3.5. Basic Feedback connections of feedback topologies

This means that both voltage and current can be fedback to the input either in series on in parallel which results in four basic feedback connections as under :

(i) Voltage–series feedback
(ii) Voltage-shunt feedback
(iii) Current–series feedback
(iv) Current–shunt feedback

Here, in describing all these connections, the voltage refers to connecting the output voltage as input to the feedback network and current refers to tapping off some output current through the feedback network. Also, series refers to connecting the feedback signal in series with input signal voltage and shunt refers to connecting the feedback signal in parallel (i.e., shunt) with an input current source.

3.3.4. Classification of Amplifiers Based on Feedback Topology

The amplifiers may be classified into four categories based on the magnitudes of the input and output impedances relative to the source and load impedances respectively. These categories may be listed as follows :

(i) Voltage amplifier
(ii) Current amplifier
(iii) Transconductance amplifier
(iv) Transresistance amplifier

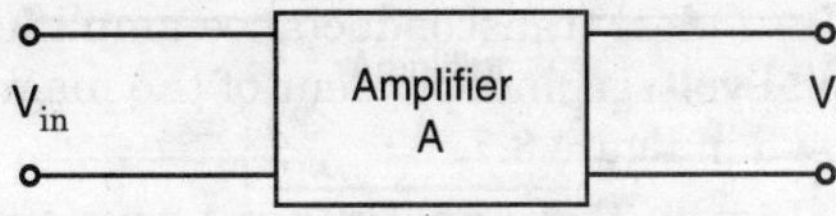

Fig. 3.6 (a).

Let us discuss these four types one by one in the following subsections.

3.3.5. Voltage Amplifiers (Series–Shunt Feedback)

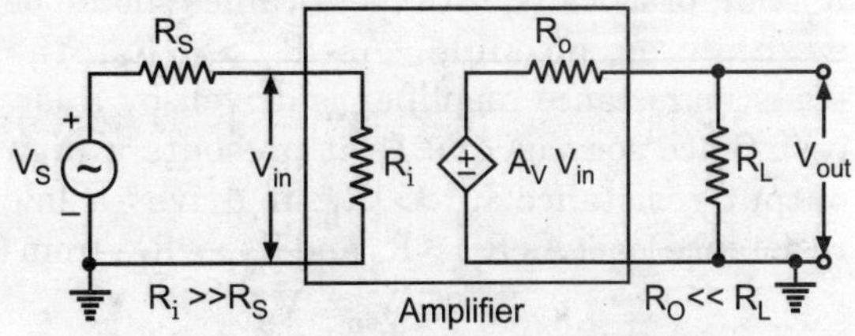

Fig. 3.6. (b) Thevenin's equivalent circuit of an amplifier

Figure 3.6 (b) shows a Thevenin's equivalent circuit of a two port network which represents an amplifier. If the amplifier input resistance R_i is large as compared to the source resistance R_S.

Then, we have

$$V_{in} \cong V_s \quad \text{... (3.1)}$$

If the external load resistance R_L is large as compared to the output resistance R_o of the amplifier then,

$$V_{out} = A_V V_{in} \cong A_V V_S \quad \text{... (3.2)}$$

where, A_V = Voltage gain

Equation (3.2) reveals the fact that the amplifier produces an output voltage proportional to the input voltage and the proportionality factor does not depend on the magnitudes of the source load resistances. Such a circuit is known as the **Voltage Amplifier**. An ideal voltage amplifier must have $R_i = \infty$ and $R_o = 0\ \Omega$. Practically, a voltage amplifier must have a large input resistance and small output resistance.

Type of Feedback

The type of feedback used in the voltage amplifiers is series-shunt feedback as will be discussed later on in this chapter.

3.3.6. Current Amplifier (Shunt - Series Feedback)

An ideal current amplifier is the one which produces an output current proportional to the signal current and the proportionality factor does not depend on R_S or R_L. Figure 3.6 (c) shows the Norton's equivalent circuit of a current amplifier.

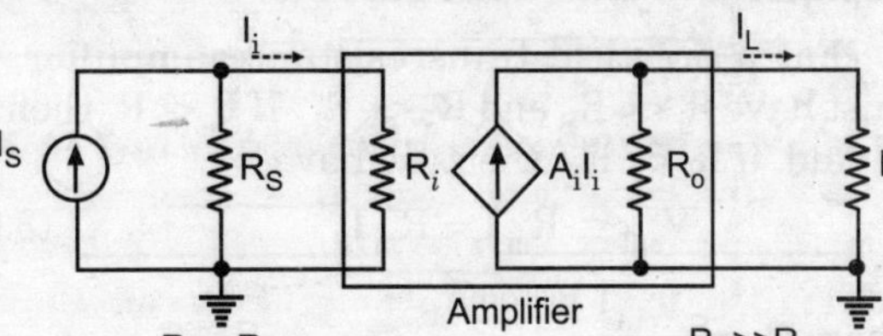

Fig. 3.6. (c) Norton's equivalent of a current amplifier

An ideal current amplifier must have $R_i = 0$ and $R_o = \infty$. Practically, the input resistance should be as small as possible and the output resistance must be as large as possible. As $R_S >> R_i$, the current amplifier is being driven by a high resistance source and as $R_o >> R_L$, the current amplifier drives a low resistance load.

The current gain, $A_i = \dfrac{I_L}{I_i} = \dfrac{I_L}{I_S}$ because $R_i << R_S$

Also as $R_o >> R_L$, we have

$$I_L = A_i\, I_i = A_i\, I_S \qquad \text{... (3.3)}$$

Thus, the output current is proportional to the signal current.

Type of Feedback

The type of feedback used in the current amplifier is shunt series feedback as will be discussed later on in this chapter.

3.3.7. Transconductance Amplifier (Series - Series Feedback)

An ideal transconductance amplifier produces an output current which is proportional to the signal voltage, independent of the magnitudes of R_S and R_L. The transconductance amplifier is as shown in figure 3.7.

Ideally, this amplifier must have an infinite input resistance R_i and infinite output resistance R_o. But, practically, both these values should be as high as possible. As $R_i >> R_S$, the transconductance amplifier is driven by a low resistance source. And as it presents a high output resistance $R_o >> R_L$, it drives a low resistance load. As $R_i >> R_S$ and $R_o >> R_L$, from the equivalent circuit of figure 3.7. we can write that

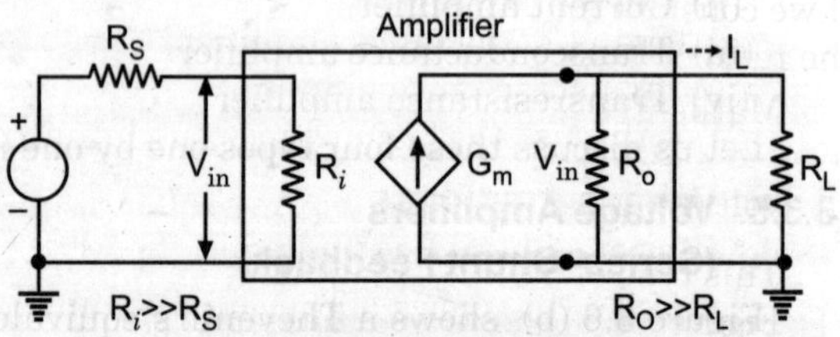

Fig. 3.7. A transconductance amplifier

$$V_{in} \cong V_S \qquad \text{... (3.4)}$$

Also, if $R_o >> R_L$, then we have

$$I_L \cong G_m\, V_{in} \cong G_m\, V_S \qquad \text{... (3.5)}$$

Hence, the output current is proportional to the input signal voltage. The proportionality factor is defined as $G_m = \dfrac{I_L}{V_{in}}$ at $R_L = 0$ and G_m represents the short-circuit mutual or transfer conductance.

Type of Feedback

The type of feedback used in the transconductance amplifier is series-series feedback as will be discussed later on in the chapter.

3.3.8. Transresistance Amplifier (Shunt-Shunt Feedback)

Transresistance amplifier is one which provides output voltage V_{out} in proportion with the signal current I_S, independent of the value of R_S and R_L. The Norton's equivalent circuit of transresistance amplifier is as shown in figure 3.8.

For a practical transresistance amplifier, we must have $R_i << R_S$ and $R_o << R_L$. If $R_i << R_S$ then $I_i \cong I_S$. And, if $R_o << R_L$, then, we have

$$V_{out} \cong R_m\, I_i = R_m\, I_S \qquad \text{... (3.6)}$$

where $R_m = \left.\dfrac{V_{out}}{I_i}\right|_{R_L = \infty}$

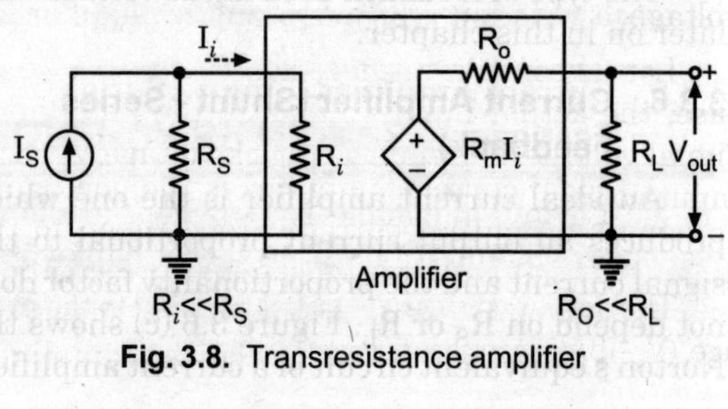

Fig. 3.8. Transresistance amplifier

In other words, R_m is the open circuit mutual or transfer resistance.

The common emitter circuit may be considered as transresistance amplifier if $R_L >> R_0$. The ideal amplifier characteristics are as listed in table 3.1.

Table 3.1. Ideal amplifier characteristics

S.No.	Parameter	Amplifier type			
		Voltage	Current	Transconductance	Transresistance
1.	Input resistance (R_i)	∞	0	∞	0
2.	Output resistance (R_o)	0	∞	∞	0
3.	Transfer characteristics	$V_o = A_V V_S$	$I_L = A_i I_S$	$I_L = G_m V_S$	$V_{out} = R_m I_S$

The type of feedback used for the transresistance amplifier is the shunt-shunt feedback as will be explained later on in this chapter.

3.4 PRINCIPLE OF FEEDBACK IN AMPLIFIERS

If we consider an ordinary amplifier as shown in the figure 3.9 (a), the voltage gain is expressed as the ratio of output voltage to the input voltage.

Mathematically,

voltage gain, $$A = \frac{V_{out}}{V_{in}}$$

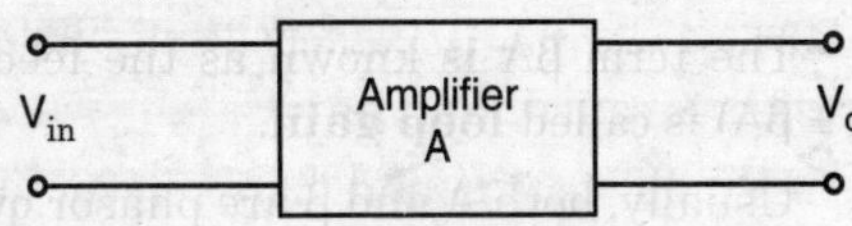

Fig. 3.9 (a)

This type of gain is, sometimes, known as the open-loop gain.

Figure 3.9(b) shows an amplifier with the concept of feedback. Here, it may be noted that amplifier, with two input terminals and one output terminal, has been shown in triangular representation. Let a fraction (say β) of the output voltage V_{out} be supplied back to the input and also let A be the open-loop gain where

$$A = \frac{V_{out}}{V_{in}}$$

Now, from figure 3.9(b), the input voltage can be written as

$$V_{in} = V_s + V_f$$

For positive feedback we have

$$V_{in} = V_s + \beta.\ V_{out} \quad ...(3.7)$$

and $$V_{in} = V_s - V_f$$

For negative feedback, we have

$$V_{in} = V_s - \beta.\ V_{out} \quad ...(3.8)$$

This means that $$V_{in} = V_s \pm \beta.\ V_{out}$$

depending upon whether the feedback voltage is in phase with or in phase opposition to the input signal voltage.

Let us consider, a negative feedback case. In this case, the actual input voltage to the amplifier is the signal voltage V_s minus feedback voltage V_f i.e., actual input voltage to the amplifier is given by

$$V_{in} = V_s - \beta.V_{out}$$

The output voltage V_{out} must be equal to the input voltage, $(V_s - \beta.V_{out})$ multiplied by the amplifier open loop gain A.

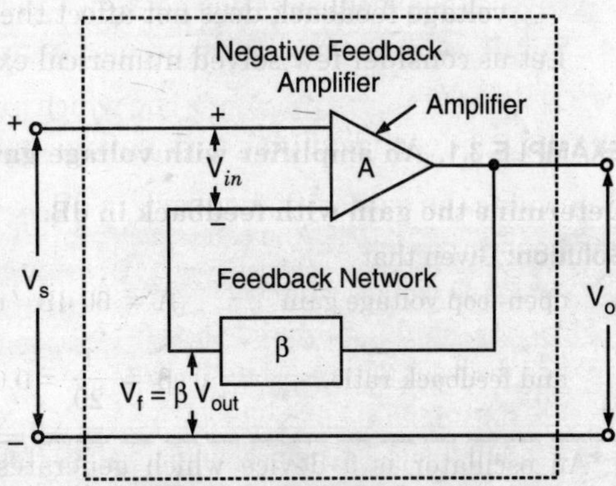

Fig. 3.9 (b). An Amplifier with feedback

Therefore,

The output voltage, $V_{out} = A\,(V_s - \beta.V_{out})$

or $(1 + \beta A)V_{out} = A.V_s$

or $$\frac{V_{out}}{V_s} = \frac{A}{1+\beta A} \quad ...(3.9)$$

DO YOU KNOW?

Depending on the desired application, the feedback circuit can be made using passive components or active components or a combination of both.

But, $\frac{A}{1+\beta A}$ is known as the voltage gain of the amplifier with feedback and is denoted by A_f. This is also **known as the closed-loop gain.**

Hence, the voltage gain with negative feedback will be

$$A_f = \frac{V_{out}}{V_s} = \frac{A}{1+\beta A} \quad ...(3.10)$$

Similarly, we can compute the voltage gain with positive feedback i.e.,

$$A_f = \frac{A}{1-\beta A} \quad ...(3.11)$$

The term βA is known as the feedback factor whereas β is called the **feedback ratio** and (1 ± βA) is called **loop gain.**

Usually, both A and β are phasor quantities (i.e., they have both magnitude as well as phase). Each amplifier stage introduces a phase shift of 180°. Hence for a 3-stage amplifier total phase shift between the output and input would be 3 × 180° = 540° or 180°. β is usually arranged to have a phase value of 0° or 180° (i.e., either in phase with input signal voltage or in phase opposition to it).

Important Observations : Because (1 – βA) is a complex quantity, we may note the following points:

(i) If (1 – βA) is less than unity in equation (3.11) then A_f exceeds A. This condition corresponds to positive feedback because voltage feedback adds to inputs signal voltage and hence increases input voltage V_{in}. Although, positive feedback increases the gain but it reduces the stability and increases the distortion and therefore it is generally avoided.

(ii) If (1 – βA) is equal to zero then the gain A_f becomes infinite. This is only possible when input is zero. Hence, the amplifier is then capable of giving output voltage even with zero signal. Under such conditions, the circuit operates as an oscillator.*

(iii) If (1 – βA) is greater than unity then A_f is smaller than A. This means that the feedback voltage V_{in} becomes smaller than input signal voltage V_S. This corresponds to negative feedback in an amplifier. Although, negative feedback reduces the gain of the amplifier but it improves performance of an amplifier in several aspects. It can be noted that the negative voltage feedback does not affect the current of the circuit.

Let us consider few solved numerical examples to illustrate the concept of feedback.

EXAMPLE 3.1. An amplifier with voltage gain of 60 dB uses $\frac{1}{20}$ of its output in negative feedback. Determine the gain with feedback in dB.

Solution: Given that

open-loop voltage gain $A = 60$ dB or $A = 1000$

and feedback ratio, $\beta = \frac{1}{20} = 0.05$

*An oscillator is a device which generates a.c. signals of desired frequency. Oscillators will be discussed later on in next chapter.

Now, gain with feedback is given by

$$A_f = \frac{A}{1+\beta A}$$

Substituting given values, we get

$$A_f = \frac{1,000}{1+0.05\times 1,000}$$

Solving we get A_f = 19.6 or 20 $\log_{10}$ 19.6 dB = 25.85 dB **Ans.**

EXAMPLE 3.2. The voltage gain of an amplifier without feedback is 60 dB. It decreases to 40 dB with feedback. Determine the feedback factor.

Solution: Given that

Voltage gain of amplifier without feedback A = 60 dB or A = 1,000

Voltage gain with feedback will be

$$A_f = 40 \text{ dB or } 100$$

Therefore, gain with feedback is given by

$$A_f = \frac{A}{1+\beta A}$$

Now, feedback factor, $\beta A = 9$ **Ans.**

EXAMPLE 3.3. A single stage transistor amplifier has a voltage gain of 600 without feedback and 50 with feedback. Find the percentage of output which is fedback to the input side.

Solution: Given that

Voltage gain without feedback, A = 600

Voltage gain with feedback,

$$A_f = 50$$

We know that

$$A_f = \frac{A}{1+\beta A}$$

Substituting values, we get

$$50 = \frac{600}{1+600\beta}$$

Solving, we get $\beta = 0.01833$

Now, the percentage of output voltage which is fedback to the input side, is given as

$$= \frac{V_f}{V_{out}}\times 100 = \beta\times 100 = 0.01833\times 100 = 1.833\%$$ **Ans.**

EXAMPLE 3.4. An amplifier with a negative feedback provides an output voltage of 5 V with an input voltage of 0.2 V. On removing feedback, it requires only 0.1 V input to provide the same output. Calculate: (i) gain without feedback, (ii) gain with feedback, and (iii) feedback ratio.

Solution: We know that gain without feedback is given by

$$A = \frac{\text{Output voltage}}{\text{Input voltage}} \text{ without feedback}$$

Substituting values, we get

Gain without feedback $A = \frac{5}{0.1} = 50$ **Ans.**

Gain with feedback is given by

$$A_f = \frac{\text{Output voltage}}{\text{Input voltage}} \text{ with feedback} = \frac{5}{0.2} = 25$$ **Ans.**

We know that

$$A_f = \frac{A}{1+\beta A}$$

Substituting values, we get

$$25 = \frac{50}{1+50\beta}$$

Solving, we get

$$\text{feedback ratio} = \frac{\frac{50}{25}-1}{50} = 0.02 \quad \textbf{Ans.}$$

EXAMPLE 3.5. A negative feedback of β = 0.002 is applied to an amplifier of gain 1,000. Calculate the change in overall gain of the feedback amplifier if the internal amplifier is subjected to a gain reduction of 15%.

Solution: Given that the voltage gain without feedback is

$$A = 1,000$$

and $\beta = 0.002$

Voltage gain with feedback is given by

We know that the $$A_f = \frac{A}{1+\beta A}$$

Substituting all the values we get

$$A_f = \frac{1,000}{1+0.002\times 1,000} = 333.33$$

When open-loop gain is reduced by 15%, then we have

$$A' = (1-0.15)\times 1,000 = 850$$

Voltage gain with feedback is given by

$$A'_f = \frac{850}{1+0.002\times 850} = 314.8$$

Percentage change in overall gain will be

$$= \frac{A_f - A'_f}{A_f}\times 100 = \frac{333.333-314.8}{333.333}\times 100 = 5.6\% \quad \textbf{Ans.}$$

3.5 MERITS OF NEGATIVE FEEDBACK

(RGTU, Bhopal, Sem. Exam., 2007-08)

As discussed earlier, there are a number of advantages if we use negative feedback in amplifier which outweigh its only drawback of reduction in gain. Some of the important advantages can be listed as under :

(i) Stabilization of Gain (Desensitivity of Gain)

The voltage gain of an amplifier with negative feedback is given by $A_f = \frac{A}{1+A\beta}$. Now, if $A\beta >> 1$, then the expression becomes $A_f = \frac{1}{\beta}$ i.e., overall gain of feedback amplifier A_f is independent of internal gain and depends only on feedback ratio β and β, in turn, depends on the passive elements such as resistors. We know that the resistors remain fairly constant and therefore the gain is stabilized.

(ii) Reduction of Non-linear Distortion

As a matter of fact, a large signal stage has non-linear distortion which is reduced by a factor $(1 + A\beta)$ when negative feedback is employed.

(iii) Reduction of Noise

There is always a noise voltage in the amplifier which is reduced by a factor (1 + A β) when negative feedback is employed.

(iv) Increase in Bandwidth (Improved Frequency Response)

As a matter of fact, the bandwidth (BW) of an amplifier without feedback is equal to the separation between the 3-dB frequencies f_1 and f_2. If A is gain then gain-bandwidth products is A × BW. With the negative feedback the amplifier gain is reduced and since gain bandwidth product has to remain constant in both cases, therefore, the bandwidth would increase to compensate for the reduction in gain.

DO YOU KNOW?

The designer has to choose a suitable scheme and to strike a balance between gain loss and noise reduction by the negative feedback technique.

(v) Increase in Input Impedance

The input impedance of the amplifier with negative feedback is increased by a factor (1 + A β).

(vi) Reduction in Output Impedance

The output impedance of the amplifier with negative feedback is reduced by factor (1 + A β).

Thus from above discussion, it is clear that using negative feedback in amplifiers provides a number of advantages.

Now, let us discuss all these advantages of negative feedback in detail, in the sections to follow.

3.6 DRAWBACKS OF NEGATIVE FEEDBACK

Some of the disadvantages of negative feedback may be listed as follows:

(i) Reduction in gain.
(ii) Reduction in input resistance in case of voltage shunt and current shunt feedback amplifiers.
(iii) Increase in output resistance in case of current shunt and current series feedback amplifiers.
(iv) More number of amplifier stages are required to be used, in order to obtain the required gain.

3.7 APPLICATIONS OF NEGATIVE FEEDBACK

(i) In almost all the electronic amplifiers.
(ii) In the regulated power supplies.
(iii) In wideband amplifiers (amplifiers having a large bandwidth)
(iv) In special circuits like bootstrapping.

3.8 GENERAL CHARACTERISTICS (MERITS OR ADVANTAGES) OF NEGATIVE FEEDBACK AMPLIFIERS

3.8.1. Expression for Transfer Gain with Feedback (A_f) or Ideal Closed Loop Gain

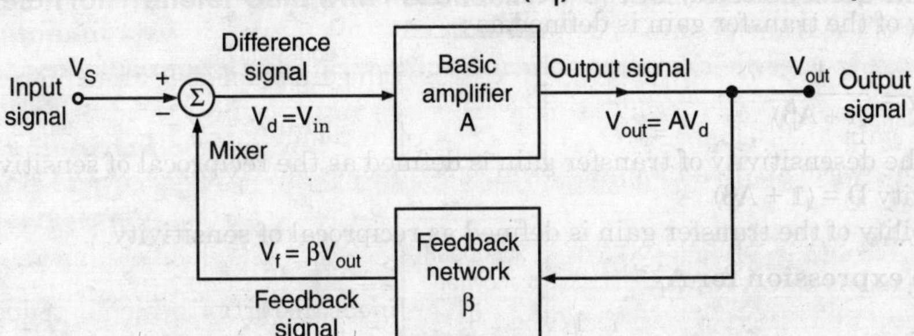

Fig. 3.10. Block diagram of a negative feedback amplifier

Figure 3.10 shows the schematic diagram of a negative feedback amplifier.

Here, A = Transfer gain of the basic amplifier without feedback and

A_f = Transfer gain of the basic amplifier with feedback.

Thus, $$A = \frac{V_{out}}{V_{in}} \text{ and } A_f = \frac{V_{out}}{V_S} \quad ...(3.12)$$

where, V_{out} = Output signal (Voltage or current)

V_{in} = Input signal (Voltage or current)

V_S = Source signal (Voltage or current)

From figure 3.10, we can write that,

$$V_{in} = V_S - V_f \quad ...(3.13)$$

where, V_f = Feedback signal (Voltage or current). Substituting the value of V_S from equation (3.13) in equation (3.12), we obtain

$$A_f = \frac{V_{out}}{V_{in} + V_f}$$

Dividing numerator and denominator by V_{in}, we obtain

$$A_f = \frac{V_{out} / V_{in}}{1 + (V_f / V_{in})} \quad ...(3.14)$$

But, $\frac{V_{out}}{V_{in}} = A$,

Therefore, $$A_f = \frac{A}{1 + (V_f / V_{in})}$$

or $$A_f = \frac{A}{1 + \left[\frac{V_f}{V_{out}} \cdot \frac{V_{out}}{V_{in}}\right]} = \frac{A}{1 + \beta A}$$

Because $\beta = \frac{V_f}{V_{out}}$, therefore, β is called as feedback factor or feedback ratio. It is always between 0 and 1.

Transfer gain with feedback $$A_f = \frac{A}{1 + \beta A} \quad ...(3.15)$$

Conclusions

From equation (3.15), it is obvious that the transfer gain with feedback A_f will always be less than that without feedback (A), i.e., $A_f < A$.

As, we increase the value of feedback factor β, the value of A_f will decrease.

Sensitivity and Desensitivity of the transfer gain

Sensitivity of the transfer gain is defined as :

$$\text{Sensitivity} = \frac{1}{(1 + A\beta)} \quad ...(3.16)$$

Whereas the desensitivity of transfer gain is defined as the reciprocal of sensitivity.

$$\text{Desensitivity } D = (1 + A\beta) \quad ...(3.17)$$

Thus, stability of the transfer gain is defined as reciprocal of sensitivity.

Approximate expression for A_f

We have $$A_f = \frac{1}{(1 + A\beta)}$$

If $A\beta >> 1$, then, $1 + A\beta \cong A\beta$

so $$A_f \cong \frac{A}{A\beta} \cong \frac{1}{\beta} \quad ...(3.18)$$

This equation shows that A_f is dependent only on the feedback network. By adjusting the value of β (feedback factor), it is possible to adjust the value of A_f.

This also shows that the amplifier gain A_f is independent of the active device amplifier gain.

3.8.2. The Gain Stability with Negative Feedback

As a matter of fact, the gain of an amplifier reduces due to many reasons. Some of these reasons are variations in temperature, power supply voltages, variations in transistor parameters with replacement. However, the overall gain of the amplifier can be made independent of these variations if we use negative feedback in amplifiers. This is an important advantage of negative feedback.

We know that the voltage gain with negative feedback is expressed as

$$A_f = \frac{A}{1 + A\beta}$$

if β is made much larger than unity

then, we have $$A_f \cong \frac{A}{A\beta} \cong \frac{1}{\beta} \quad ...(3.19)$$

Hence, the gain is independent of internal gain of the amplifier and depends only on the passive elements such as resistors. The values of resistors remain quite constant because they can be chosen very precisely with almost zero temperature coefficient or resistance. Therefore the gain becomes stabilized.

Further, even if open-loop gain A is not very large, some improvement in gain stability can be achieved.

Differentiating equation (3.15) with respect to A, we get

$$\frac{dA_f}{dA} = \frac{(1 + A\beta) - A.\beta}{(1 + A\beta)^2}$$

Simplifying, we get $$\frac{dA_f}{dA} = \frac{1}{(1 + A\beta)^2}$$

$$d\,A_f = \frac{dA}{(1 + A\beta)^2} \quad ...(3.20)$$

Dividing equation (3.20) by equation (3.15), we get

$$\frac{dA_f}{A_f} = \frac{dA}{A} \times \frac{1 + A\beta}{(1 + A\beta)^2}$$

$$\frac{dA_f}{A_f} = \frac{1}{(1 + A\beta)} \frac{dA}{A} \quad ...(3.21)$$

DO YOU KNOW?

The gain stability of practical amplifiers is not very high because of device parameter variations due to changes in ambient temperature and manufacturing spread. This problem is more acute in the case of BJT. The gain stability can be improved appreciably by the technique of negative feedback wherein a portion of the output signal is fedback to the input in phase opposition to the input.

Important Point: *In negative feedback $(1 + A\beta) >> 1$, the percentage change in gain with negative feedback is less than the percentage change in gain without feedback. This means that negative feedback improves the gain stability of the amplifier.*

EXAMPLE 3.6. It is required to design a feedback amplifier with a closed-loop voltage gain of 100. The closed loop gain should not vary by more than 1% despite 10% variation in the open-loop voltage gain. Find the open-loop voltage gain and the feedback factor for the amplifier.

Solution : Given that variation in closed-gain,

$$\frac{d A_f}{A_f} = \frac{1}{100} = 0.01 \qquad \text{...(i)}$$

Also, variation in open-loop gain,

$$\frac{dA}{A} = \frac{10}{100} = 0.1 \qquad \text{...(ii)}$$

We know that $\frac{d A_f}{A_f} = \frac{1}{(1+A\beta)} \frac{dA}{A}$

Using equations (i) and (ii), we get

or $0.01 = \frac{1}{1+A\beta} \times 0.1$

Solving, we get

Feedback factor, $A\beta = \frac{0.1}{0.01} - 1 = 9$ **Ans.**

Also, since closed-loop gain is given as

$$A_f = \frac{A}{1+A\beta}$$

therefore $100 = \frac{A}{1+9}$

Thus, open-loop voltage gain will be

$$A = 100 \times 10 = 1000 \text{ Ans.}$$

EXAMPLE 3.7. An amplifier gain changes by ± 10%. Using negative feedback, the amplifier is to be modified to provide a gain of 100 with ± 0.1% variation. Obtain the required open-loop gain of the amplifier and the amount of negative feedback.

Solution : Given that variation in open-loop gain

$$\frac{d A}{A} = \frac{10}{100} = 0.1 \qquad \text{...(i)}$$

Variation in closed-loop gain

$$\frac{d A_f}{A_f} = \frac{0.1}{100} = 0.001 \qquad \text{...(ii)}$$

We know that $\frac{d A_f}{A_f} = \frac{1}{(1+A\beta)} \cdot \frac{d A}{A}$

Using equations (i) and (ii), we get

or $0.001 = \frac{1}{1+A\beta} \times 0.1$

or Feedback factor, $A\beta = \frac{0.1}{0.001} - 1 = 99$

Since, the closed-loop gain is expressed as

$$A_f = \frac{A}{1+A\beta}$$

therefore open-loop voltage gain

$$A = A_f(1 + A\beta) = 100\,(1 + 99) = 10{,}000 \text{ Ans.}$$

and the amount of negative feedback $= \frac{A_f}{A} = \frac{100}{10.000} = \frac{1}{100}$ **Ans.**

3.8.3. Reduction in Frequency Distortion with Negative Feedback

For a negative feedback amplifier, having $\beta A >> 1$, the gain with feedback is given by $A_f \cong \frac{1}{\beta}$.

From this expression, it follows that if the feedback network is purely resistive then the gain with feedback becomes independent of frequency even though the basic amplifier is frequency dependent. In practice, the frequency distortion arising because of varying amplifier gain with frequency is considerably reduced when negative feedback is employed.

DO YOU KNOW?

The frequency distortion, because of varying amplifier gain with frequency, is considerably reduced in a negative voltage feedback amplifier circuit.

3.8.4. Reduction in Non-Linear Distortion with Negative Feedback (Important)

As discussed earlier, harmonic distortion occurs when a transistor or other device is driven beyond its range of linear operation. If negative feedback is used then the strength of the input signal increased by the same factor by which the gain is reduced. Hence, the magnitude of output signal remains unaltered and, therefore there is a reduction in harmonic or non-linear distortion. It can be explained as under :

Let us consider an amplifier having open-loop gain A which produces a distortion D in the output signal without feedback. Now, if feedback is introduced the output as well as distortion is feedback to the input side.

Let us consider that the gain with feedback is A_f and distortion in the output is D_f. Now, a Part βD_f of this distortion is fedback to the input side and its gets amplified by factor A and becomes $\beta A D_f$.

Therefore, net distortion will be

$$D_f = D - A\beta D_f$$

or

$$D_f = \frac{D}{1 + A\beta} \quad ...(3.22)$$

Hence, the distortion is reduced by the same factor as the gain.

Important Point : However, it can be noted that improvement in distortion is possible only when the distortion is introduced by the amplifier itself, not when it is already present in the input signal.

DO YOU KNOW?

Non-linear distortion is introduced at the power amplification stage because of large signal swing required for high power efficiency. Distortion thus caused introduces second and higher harmonic signals at the output stage.

EXAMPLE 3.8. An amplifier has an input of 10 m V and a gain of 200 without feedback. The distortion produced at the output of the amplifier is 10%. It is required to reduce the distortion to 1% by employing negative feedback. Find the gain, input voltage and output voltage with feedback.

Solution : Given that : Distortion without feedback

$$D = 10\% = 0.1 \quad ...(i)$$

and distortion with feedback

$$D_f = 1\% = 0.01 \quad ...(ii)$$

Also, amplifier gain without feedback

$$A = 200$$

We know that

$$D_f = \frac{D}{1 + A\beta}$$

using equations (i) and (ii), we get

or

$$0.01 = \frac{0.1}{1 + 200\beta}$$

or

$$\beta = \left(\frac{0.1}{0.01} - 1\right) \Big/ 200$$

Solving, we get $\beta = 0.045$ or 4.5%

Now, gain with feedback is given by

$$A_f = \frac{A}{1 + \beta A}$$

Substituting given values, we get

$$A_f = \frac{200}{1+0.045\times 200} = 20 \quad \textbf{Ans.}$$

New output voltage, $V_o = A_f\, V_s = 20 \times 10 \text{ mV} = 200 \text{ mV or } 0.2\text{V}$ **Ans.**

New input voltage $V_i = 0.01\text{ V} + (-0.045 \times 0.2) = 0.001\text{V or } 1\text{mV}$ **Ans.**

EXAMPLE 3.9. An amplifier with an open-loop voltage gain of 1,000 delivers 10 W output power at 10% second harmonic distortion when the input signal is 10 mV. If 40 dB negative voltage series feedback is applied and the output power is to remain at 10 W, find: (i) the required input signal, (ii) percentage second harmonic distortion, and (iii) closed loop voltage gain.

Solution : Given that

Open-loop gain, $A = 1{,}000$

Distortion without feedback $D = 10\%$ or 0.1

Input signal voltage, $V_S = 10$ mV

and $A\beta = 40 \text{ dB} = 100$

Because with feedback the output is reduced by $\frac{1}{1+A\beta}$, therefore, the input must be increased by a factor $(1 + A\,\beta)$ so as to deliver the same output voltage.

Hence

(i) Input signal voltage

$$V'_s = V_s \times (1 + A\,\beta) = 10 \text{ mV} \times (1 + 100) = 1.01 \quad \textbf{Ans.}$$

(ii) Percentage second harmonic distortion will be

$$D_f = \frac{D}{1+A\beta} = \frac{10}{1+100}\% = 0.1\% \quad \textbf{Ans.}$$

(iii) Closed-loop voltage gain will be

$$A_f = \frac{A}{1+A\beta} = \frac{1{,}000}{1+100} = 9.9 \quad \textbf{Ans.}$$

3.9 IMPROVEMENT IN NOISE WITH NEGATIVE FEEDBACK

As a matter of fact, noise in an amplifier is reduced or improved in the same manner as the distortion. However, distortion is usually introduced in the last stage of the amplifier and therefore is always within the feedback loop whereas the noise in an amplifier may be generated or introduced from outside source at any stage that may not be within the feedback loop. Therefore, the noise can be reduced by negative feedback only if its source lies within the feedback loop after the first stage.

As non-linear distortion generated within the amplifier is reduced with negative feedback by a factor $(1 + \beta\, A)$, the noise introduced within amplifier also gets reduced by the same factor $(1 + \beta\, A)$ on using negative feedback and hence the performance of the amplifier is very much improved.

3.10 EFFECT OF NEGATIVE FEEDBACK ON INPUT IMPEDANCE OF AMPLIFIER

As discussed earlier, the high input impedance is always desirable for an amplifier because it will not load the preceding stage or the input voltage source. Such a desirable characteristic can be achieved with the help of negative feedback. The effect of negative feedback on the input impedance of an amplifier can be explained as under.

Figure 3.11 illustrates the detailed circuit diagram of a voltage series feedback amplifier. The input impedance can be determined as under :

We know that

$$I_i = \frac{V_{in}}{Z_i} = \frac{V_s - V_f}{Z_i}$$

Now, substituting $V_f = \beta. V_{out}$, we get

$$I_i = \frac{V_s - \beta V_{out}}{Z_i}$$

Also, since $V_{out} = A.V_{in}$

Therefore $$I_i = \frac{V_s - A\beta\, V_N}{Z_i}$$

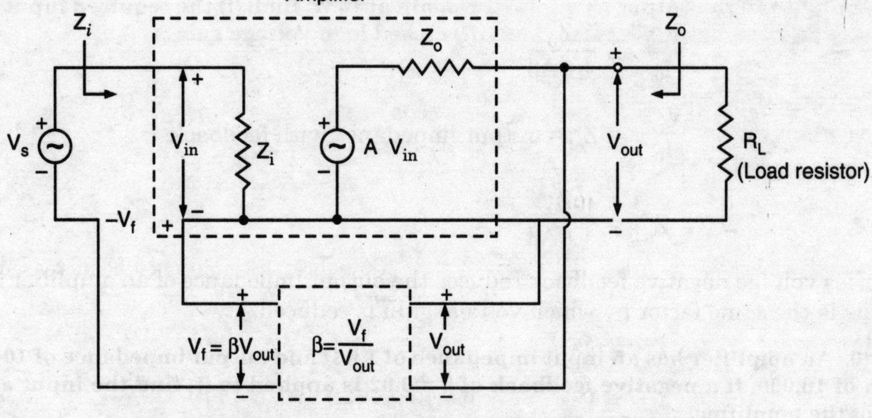

Fig. 3.11. Illustration of Voltage-Series Feedback Connections

or $$I_i Z_i = V_s - A\,\beta\, V_{in}$$

or $$V_S = I_i Z_i + A\,\beta\, V_{in}$$

or $$V_S = I_i Z_i + A\,\beta\, I_i Z_i = I_i (Z_i + A.\beta\, Z_i)$$

or $$\frac{V_s}{I_i} = Z_i + A\,\beta\, Z_i$$

Now, since $= Z_{if}$ (input impedance with feedback)

therefore $$Z_{if} = \frac{V_s}{I_i} = Z_i + (A\,\beta)\, Z_i = Z_i(1 + A\beta) \quad ...(3.23)$$

Thus, we have $$Z_{if} = Z_i\,(1 + A\beta)$$

Therefore, series voltage negative feedback increases the input impedance of an amplifier by factor $(1 + A\,\beta)$. This is the same factor by which voltage gain is reduced.

3.11 EFFECT OF NEGATIVE FEEDBACK ON OUTPUT IMPEDANCE OF AMPLIFIER

Like high input impedance, the low output impedance is also a desirable feature for an amplifier. In fact, an amplifier with low output impedance, is better suited to drive a low impedance load. This desirable characteristic can be achieved with the use of the negative feedback. The effect of negative feedback on the output impedance of an amplifier can be explained as under :

We shall use the voltage-series feedback circuit shown in figure 3.11 to determine output impedance with feedback. Here, the input terminals are short-circuited so that $-V_f$ is now the only input voltage to the amplifier.

Next, a voltage source V_{out} is applied at the output terminals so that current I_o is drawn from the applied source.

Now, from circuit, $$V_{out} = I_o Z_o + AV_{in}$$

But, since $V_{in} = -V_f$

therefore $V_{out} = I_o Z_o - AV_f$

Now, since $V_f = \beta.V_{out}$

Therefore $V_{out} = I_o Z_o - A(\beta V_{out})$

or $V_{out} + A\beta V_{out} = I_o Z_o$

or $V_{out}(1 + A\beta) = I_o Z_o$

or $$\frac{V_{out}}{I_0} = \frac{Z_0}{1+A\beta} \quad ...(3.24)$$

But $\frac{V_{out}}{I_0} = Z_{of}$ = o tput impedance with feedback.

therefore, $$Z_{of} = \frac{Z_0}{1+A\beta}$$

Here, series voltage negative feedback reduces the output impedance of an amplifier by a factor $(1 + A\beta)$. This is the same factor by which voltage gain is reduced.

EXAMPLE 3.10. An amplifier has an input impedance of 1 k Ω and output impedance of 10 k Ω and a voltage gain of 10,000. If a negative feedback of β = 0.02 is applied to it, find the input and output impedance of the amplifier.

Solution : Given that open-loop gain of amplifier,

$$A = 10{,}000$$

Feedback ratio, $\beta = 0.02$

Input impedance without feedback,

$$Z_i = 1\text{ k}\Omega$$

Output impedance without feedback,

$$Z_o = 10\text{ k}\Omega$$

Input impedance with feedback is given by,

$$Z_{if} = (1 + A\beta)\, Z_i = (1 + 0.02 \times 10{,}000) \times 1\text{ k}\Omega = 201\text{ k}\Omega \quad \textbf{Ans.}$$

Also, output impedance with feedback is given by,

$$Z_{of} = \frac{Z_0}{1+A\beta}$$

Substituting given values, we get

$$Z_{of} = \frac{10\text{ k}\Omega}{1+0.02\times 10{,}000} = 49.75\ \Omega \textbf{ Ans.}$$

EXAMPLE 3.11. An amplifier with a gain of 60 dB has an output impedance of 10 k Ω. It is required to modify its output impedance to 1 k Ω. What type of feedback has to be applied ? Find the feedback factor. Also find the percentage change in the overall gain, for a 10% change in the open-loop gain of the amplifier.

Solution : Given that

$$Z_i = 1\text{ k}\Omega;$$

$$Z_o = 10\text{ k}\Omega;$$

and $A = 1{,}000$

Open-loop voltage gain $A = 60\text{ dB}$ or antilog $\frac{60}{20} = 1{,}000$

We know that,
the output impedance with feedback is given by,

$$Z_{of} = \frac{Z_0}{1 + A\beta}$$

Substituting given values, we get

$$1 \times 10^3 = \frac{10 \times 10^3}{1 + \beta \times 1,000}$$

Solving, we get $\beta = 0.009$

Now feedback factor, $A\,\beta = 0.009 \times 1,000 = 9$ **Ans.**

For 10% decrease in open-loop gain, we have new open-loop gain given by

$$A' = \left(1 - \frac{10}{100}\right)A = 0.9\,A = 0.9 \times 1,000 = 900$$

New gain with feedback will be,

$$A'_f = \frac{A'}{1 + A'\beta}$$

Substituting the values we get

$$A'_f = \frac{900}{1 + 0.009 \times 900} = 98.9$$

Percentage change in overall gain will be

$$= \frac{A_f - A'_f}{A_f} \times 100 = \frac{100 - 98.9}{100} \times 100\ {}^{*} = 1.1\%\ \textbf{Ans.}$$

3.12 EFFECT OF NEGATIVE FEEDBACK ON BANDWIDTH OF AN AMPLIFIER

In this section, let us consider the effect of negative feedback on bandwidth of amplifier. Voltage gain with feedback is given by

$$A_{Vf} = \frac{A_V}{1 + A_V\beta} \qquad \text{...(3.25)}$$

Now, if $A_V\beta >> 1$

Then, $A_{Vf} \cong \frac{1}{\beta}$

Thus, from this expression, we conclude that the voltage gain can be made to depend entirely on the feedback network. However, it may be noted that even if parameter β is constant the voltage gain A_V is not constant since it depends on frequency. This means that at certain low or high frequencies $\beta\,A_V$ will not be much larger than unity, and, therefore equation (3.25) will not be valid.

DO YOU KNOW?

Negative feedback in amplifiers has indeed the effect of maintaining the gain-bandwidth product which means as the gain gets reduced, the bandwidth is increased. This property of feedback is used in op-amps wherein feedback adjusts the gain to a desired (low) stable value, while the bandwidth is considerably increased.

When negative feedback is applied in an amplifier, cut-off frequencies are also affected. This means that lower cut-off frequency f_L is lowered by the same factor $(1 + A\,\beta)$ whereas upper cut-off frequency f_H is raised by the same factor $(1 + A\,\beta)$.

It is possible to obtain the desired bandwidth by selecting a proper feedback factor β. The increase in bandwidth due to the negative feedback has been shown in figure 3.12.

---

*Since $A_f = \frac{A}{1 + \beta A} = \frac{1,000}{1 + 0.009 \times 1,000} = 100$

We may also note the reduction in gain due to negative feedback.

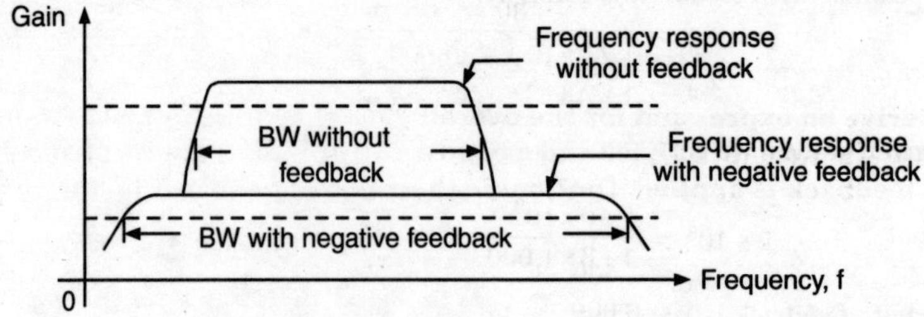

Fig. 3.12. Effect of negative feedback on bandwidth of an amplifier.

Thus, we conclude that bandwidth of an amplifier with negative feedback is increased by the same factor by which voltage gain is reduced and gain bandwidth product remains the same. However, since the amplifier with feedback has lower gain, the net operation is to trade bandwidth for gain.

3.12.1. Effect of Negative Feedback on Amplifier Characteristics

Table 3.2 summarizes the effect of negative feedback on various amplifier characteristics.

Table 3.2. Effect of negative feedback on amplifier characteristics

S. No.	Parameter/Characteristics	Type of feedback			
		Series shunt	**Series series**	**Shunt shunt**	**Shunt series**
1.	Gain (A_f)	Reduces	Reduces	Reduces	Reduces
2.	Gain stability	Improved	Improved	Improved	Improved
3.	Input resistance	Increases	Increases	Decreases	Decreases
4.	Output resistance	Decreases	Increases	Decreases	Increases
5.	Non-linear distortion	Decreases	Decreases	Decreases	Decreases
6.	Noise	Decreases	Decreases	Decreases	Decreases
7.	Frequency distortion	Decreases	Decreases	Decreases	Decreases
8.	Bandwidth	Increases	Increases	Increases	Increases

EXAMPLE 3.12. An R-C coupled amplifier has $A_V = 50{,}000$, $f_H = 20$ kHz $f_L = 30$ Hz. A resistive voltage negative feedback is added such that $\beta = 5 \times 10^{-5}$. Determine A_{vf}, f_{Hf} and f_{Lf}. Derive the relations to be used for the calculations.

Solution : Given that open-loop gain, $A = 50{,}000$

and feedback ratio, $\beta = 5 \times 10^{-5}$

We know that the gain with feedback is given by

$$A_{Vf} = \frac{A_v}{1 + A_v \cdot \beta}$$

Substituting, the values, we get

$$A_{Vf} = \frac{50{,}000}{1 + 5 \times 10^{-5} \times 50{,}000} = 14{,}285.7 \quad \textbf{Ans.}$$

The upper cut-off frequencies with feedback is given by

$$f_{Hf} = f_H(1 + A_V \beta)$$

Substituting the values, we get

$$f_{Hf} = (20 \times 10^3)(1 + 5 \times 10^{-5} \times 50{,}000) = 70 \text{ kHz} \quad \textbf{Ans.}$$

Lower cut-off frequency with feedback is given by,

$$f_{Lf} = \frac{f_L}{1 + A_V \beta}$$

Substituting all the given values, we get

$$f_{Lf} = \frac{30}{1+5\times10^{-5}\times50,000} = 8.57 \text{ Hz} \quad \textbf{Ans.}$$

EXAMPLE 3.13. Derive an expression for the overall gain of a voltage series feedback amplifier. An amplifier has the midband gain of 1,500 and a bandwidth of 4 MHz. The midband gain reduces to 150 when a negative feedback is applied. Determine the value of feedback factor and the bandwidth.

Solution : Given that

Midband gain $A_V = 1,500$

Midband gain with feedback is given by

$$A_{Vf} = \frac{A_V}{1+A_V\beta} \quad \text{...(i)}$$

Substituting the value of

$$A_{Vf} = 150$$

and $A_V = 1,500$

in equation (i), we get

$$150 = \frac{1,500}{1+A_V\beta}$$

or feedback factor, $A_V\beta = \frac{1,500}{150} - 1 = 9$ **Ans.**

Bandwidth with feedback will be

$$BW_f = (1 + A_V\beta) \times (\text{ Bandwidth without feedback}) = (1 + 9) \times 4$$

or $BW_f = 40$ MHz **Ans.**

3.13 TYPES OF NEGATIVE FEEDBACK CONNECTIONS

As discussed earlier, there are four different combinations in which negative feedback may be accomplished as under:

(i) Voltage-series feedback

(ii) Voltage-shunt feedback

(iii) Current-series feedback

(iv) Current-shunt feedback

The series feedback connections tend to increases the input impedance whereas the shunt feedback connections tend to decrease the input impedance. In addition to this, the voltage feedback tends to decrease the output impedance whereas the current feedback tends to increase the output impedance. In electronic circuits, it is desirable the most of the cascade amplifiers are required to have higher input impedance and lower output impedance. The voltage-series type of feeedback has the high input impedance and low output impedance. However, the drawback of voltage-series feedback is that it suffers the highest decrease in voltage gain. On the other hand, the current-shunt feedback connection has the least desirable effects in the sense that it decreases input impedance and, sametimes, increases output impedance.

The input signal V_S, output signal V_{out}, feedback signal V_f and the difference signal V_d alongwith gain and feedback factor β have been listed in table 3.3.

Table 3.3. Various signals in feedback amplifiers.

Signal or ratio	Type of feedback			
	Voltage series (series shunt)	Current series (series series)	Current shunt (shunt series)	Voltage shunt (shunt shunt)
V_{out}	voltage	current	current	voltage
V_S, V_f, V_d	voltage	voltage	current	current
A	A_V	G_M	A_I	R_M
β	V_f/V_{out}	V_f/I_0	I_f/I_0	I_f/V_{out}

Now, let us discuss all the four feedback connections in detail in the subsections to follow:

3.14 VOLTAGE-SERIES FEEDBACK CONNECTION

Figure 3.13 shows a block diagram of a voltage series feedback connection. In this block diagram, the input to the feedback network is in parallel with the output of the amplifier. A fraction of the output voltage is applied in series with the input voltage of the amplifier through the feedback network. Here, it may be noted that the shunt connection at the output reduces the output resistance R_o, whereas, the series connection at the input increases the input resistance. Further, the worthnoting point is that a voltage-series feedback amplifier is a true voltage amplifier. From figure 3.13, it is clear that the voltage feedback factor is given by $\beta = V_f/V_{out}$.

DO YOU KNOW?

The amplifier and feedback circuits are connected in series-shunt. The inputs of amplifier and feedback circuits are in series but the outputs are in parallel. In practice, this circuit is widely used.

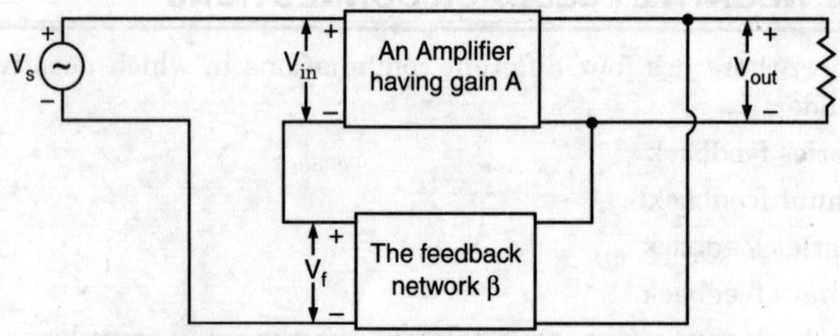

Fig. 3.13. Block diagram of the voltage-series feedback

3.14.1. Analysis of a Voltage-series Feedback Circuit

(RGTU, Bhopal, Sem. Exam., 2006-07)

Figure 3.14 shows the voltage-series feedback circuit. We shall use this circuit to calculate input and output resistances.

From figure 3.14, we have

$$V_s = V_{in} + V_f = I_iR_i + V_f = I_iR_i + \beta V_{out}$$

But since $V_{out} = AI_iR_i$

therefore, $$V_s = I_iR_i + \beta\, AI_i\, R_i \quad ...(3.26)$$

Now, $$R_{if} = \frac{V_s}{I_i}$$

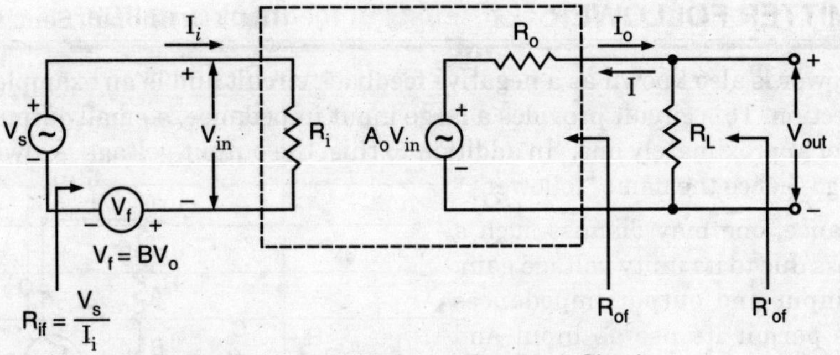

Fig. 3.14. Voltage-Series Feedback Circuit

Using equation (3.26), we have

$$R_{if} = \frac{V_s}{I_i} = R_i + A\beta.R_i = R_i\,(1 + A\beta)$$

Thus, we have $R_{if} = R_i(1 + A\beta)$

Hence, the input-resistance of a voltage-series feedback amplifier is given by

$$R_{if} = R_i\,(1 + A\beta) \qquad ...(3.27)$$

where R_i is the input resistance of the amplifier without feedback.

After this, for measuring the output resistance, the load resistance R_L is disconnected and voltage source V_s is set to zero. Then an external voltage V is applied across the output terminals and the current I delivered by V is calculated.

Then, $R_{of} = \frac{V}{I}$

Because of feedback, the input voltage V_f reduces the output voltage AV_{in} which opposes V. Therefore, we write

$$I = \frac{V - AV_{in}}{R_0} = \frac{V + A\beta V}{R_0}$$

Hence, we have $R_{of} = \frac{V}{I} = \frac{R_0}{1 + A\beta}$

Thus, the output resistance of a voltage-series feedback amplifier is given by

$$R_{of} = \frac{R_0}{1 + A\beta} \qquad ...(3.28)$$

where R_o is the output resistance of the amplifier without feedback.

EXAMPLE 3.14. A voltage-series negative feedback amplifier has a voltage gain without feedback A = 500, input resistance R_i = 3kΩ, output resistance R_o = 20 kΩ and feedback ratio β = 0.01. Find the voltage gain A_f, input resistance R_{if} and the output resistance R_{of} of the amplifier with feedback.

Solution: Given that A = 500, R_i = 3 kΩ, R_o = 20 kΩ and β = 0.01.

We know that for voltage-series feedback amplifier, the voltage gain is given by

$$A_f = \frac{A}{1 + A\beta} = \frac{500}{1 + 500 \times 0.01} = \frac{500}{6} = 83.33 \quad \textbf{Ans.}$$

Input resistance with feedback is given by

$$R_{if} = R_i(1 + A\beta) = 3 \times 10^3 \times (1 + 500 \times 0.01)$$

or $R_{if} = 18\ k\Omega$ **Ans.**

Also, the output resistance with feedback is given by

$$R_{of} = \frac{R_0}{1 + A\beta} = \frac{20 \times 10^3}{(1 + 500 \times 0.01)} = 3.33\ k\Omega \ \textbf{Ans.}$$

3.15 THE EMITTER FOLLOWER

(RGTU, Bhopal, Sem. Exam., 2005-2006)

An emitter follower is also known as a negative feedback circuit and is an example of voltage-series feedback connection. This circuit provides a large input impedance, a small output impedance, and a voltage gain of approximately one. In addition to this, the output voltage is always in phase with the input voltage–hence the name "follower".*

At first glance, one may dismiss such a circuit as useless due to its unity voltage gain. However, its input and output impedances characteristics permit its use as input and output buffer stages in amplifier systems.

The circuit of an emitter follower has been shown in figure 3.15. Note that the main difference between an emitter follower and a conventional amplifier is the absence of collector load and emitter bypass capacitor. Infact, here the emitter resistance R_E itself acts as the load and the a.c. output voltage V_o is obtained across it. The biasing is provided either by base resistor method or potential divider method. The circuit consists of a resistance of the order of 10 kΩ in the emitter itself and the collector is connected directly to the supply. Further, stabilization of operating point is exceptionally good due to high value of emitter resistance R_E.

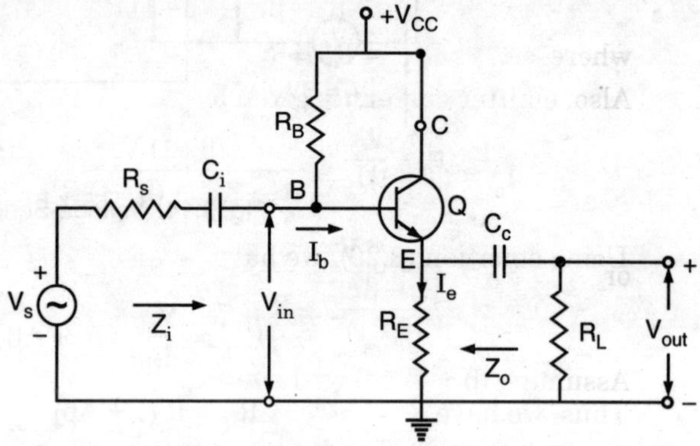

Fig. 3.15. The emitter follower circuit

> **DO YOU KNOW?**
>
> Since the collector is at ac ground, this circuit is also known as common collector (CC) amplifier.

Working Principle

When input signal V_s is applied to the base, the resulting emitter current I_e develops an output voltage V_{out} equal to $I_e R_E$ across the emitter resistance R_E. This voltage opposes the a.c. signal voltage V_s since it is inphase opposition to V_s. Therefore, it provides negative current feedback. In addition to this, voltage (V_{out}) fedback to the input is proportional to the emitter current and hence this circuit is known as a negative current feedback circuit.

It can also be noted that when the input signal voltage V_s goes through its positive half cycle, the output voltage V_{out} also tends to go through its positive half cycle. Therefore, output voltage is in phase with the input signal voltage. This means that the output voltage (emitter voltage) just follows the input voltage and therefore is called emitter follower.

For the emitter follower configuration shown in figure 3.15, let us draw a small-signal a.c. model as illustrated in figure 3.16. For the input section of figure 3.16, the resistance R_B can be neglected because it is usually much larger than source resistance R_s. This means that a Thevenin's equivalent circuit for the configuration of figure 3.16 will result in simply R_s and V_s as shown in figure 3.17. In fact, if current levels are to be determined such as I_i in the original diagram, the effect of R_B has to be considered.

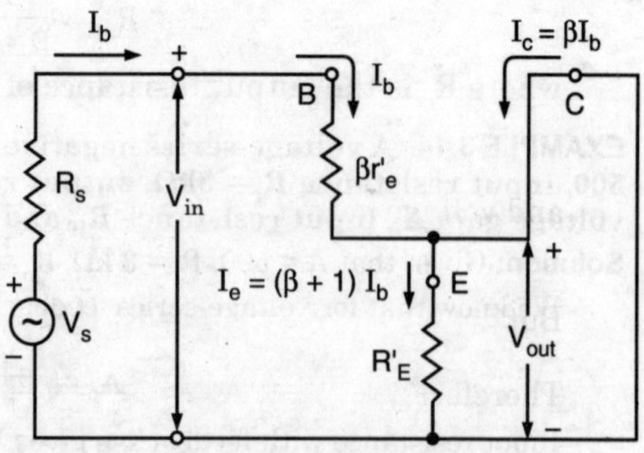

Fig. 3.16. Small signal a.c. Model for emitter follower circuit shown in figure 3.15

Applying KVL to the input side in figure 3.16, we get

* What type of feedback is the emitter follower an example of ? How much negative feedback is used?
(RGTU, Bhopal, Sem. Exam., 2005-2006)

$$V_s - I_b R_s - I_b \beta r'_e - (\beta+1) I_b R'_E = 0$$

or $$V_s - I_b [R_s + \beta r'_e + (\beta+1) R'_E] = 0$$

Simplifying, we get

$$I_b = \frac{V_s}{R_s + \beta r'_e + (\beta+1) R'_E}$$

where $R'_E = R_E || R_L$

Also, emitter current is given by

$$I_e = (\beta+1) I_b = \frac{(\beta+1) V_s}{R_s + \beta r'_e + (\beta+1) R'_E}$$

or $$I_e = \frac{V_s}{\dfrac{R_s + \beta r'_e}{(\beta+1)} + R'_E}$$

Assuming $(\beta + 1) \cong \beta$ we have

$$\text{Emitter current } I_e = \frac{V_s}{\dfrac{R_s}{\beta} + r_e + R_E} \quad ...(3.29)$$

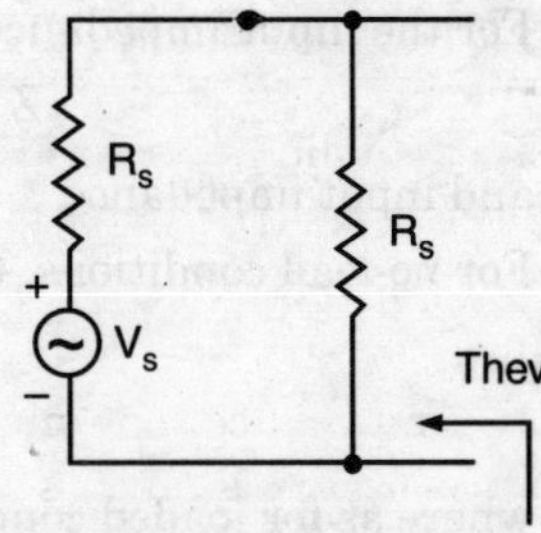

Fig. 3.17. Determination of Thevenin's equivalent circuit for the input circuit shown in fig. 3.15

After this, let us draw a circuit following the above equation. It has been shown in figure 3.18(a). In figure 3.18(b), R_E and load resistance R_L have been subjected to define output impedance Z_o and output current I_o.

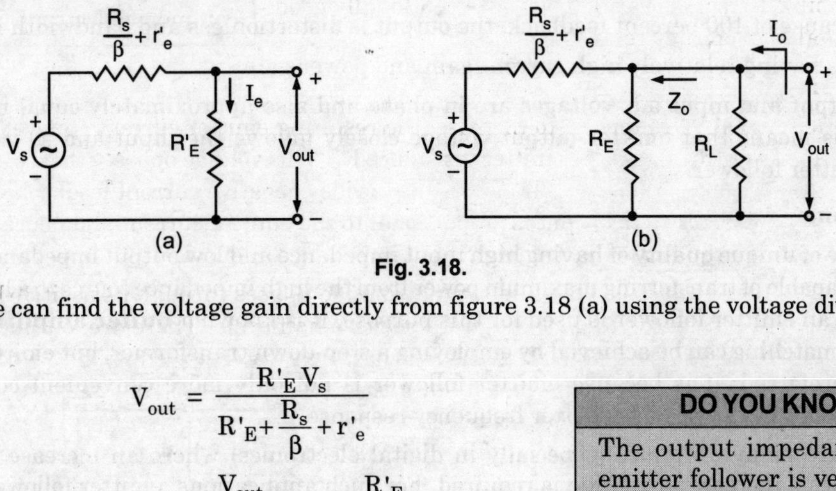

Fig. 3.18.

Now, we can find the voltage gain directly from figure 3.18 (a) using the voltage divider rule as under :

$$V_{out} = \frac{R'_E V_s}{R'_E + \dfrac{R_s}{\beta} + r'_e}$$

and voltage gain, $$A_{vs} = \frac{V_{out}}{V_s} = \frac{R'_E}{R'_E + \dfrac{R_s}{\beta} + r'_e}$$

But $R'_E = R_E || R_L$

Therefore, $$A_{vs} = \frac{R_E || R_L}{R_E || R_L + \dfrac{R_s}{\beta} + r'_e} \quad ...(3.30)$$

$$R'_E = R_E || R_L$$

Now, making $V_s = 0$ and solving for Z_o, we obtain

$$Z_o = R_E \| \left(\frac{R_s}{\beta} + r'_e \right)$$

> **DO YOU KNOW?**
>
> The output impedance of the emitter follower is very low. On the other hand, it has high input impedance. This property makes the emitter follower a perfect circuit for connecting a low impedance load to a high-impedance source.

For the input impedance, we have

$$Z_{i\,(base)} = \beta\,(r'_e + R'_E) = \beta\,(r'_e + R_E\,||\,R_L)$$

and input impedance $\quad Z_i = R_B\,||\,Z_{i(base)} = R_B\,||\,\beta\,(r'_e + R_E\,||\,R_L)$

For no-load conditions, the gain expression becomes

$$A_{vnl} \cong \frac{R_E}{R_E + r'_e} \cong 1^*$$

where as for loaded condition, we have

$$A_v \cong \frac{V_{out}}{V_{in}} = \frac{R_E\,||\,R_L}{R_E\,||\,R_L + r'_e} \cong 1$$

DO YOU KNOW?

In practice, the voltage gain of an emitter follower is between 0.8 and 0.999.

Also, current gain, $\quad A_i = \dfrac{I_e}{I_b} = (1 + \beta) \cong \beta$

From circuit shown in figure 3.15, input voltage to the amplifier, which equals V_{be}, is given by

$$V_{in} = V_{be} = V_s - I_e\,R_E = V_s - V_{out} \quad \text{(neglecting voltage drop in source resistance } R_s)$$

Therefore, the feedback ratio β i.e. feedback is 100%.

Salient Features

The salient features of an emitter follower may be listed as under :

(i) Since it is having high input impedance and low output impedance, therefore it can be used for impedance matching. Sometimes, it is also called impedance transformer.

(ii) Because of 100 percent feedback, the output is distortionless and bandwidth is quite large.

(iii) It is having relatively high current gain and power gain.

(iv) Output and input a.c. voltages are in phase and also approximately equal in magnitude. This means that emitter output voltage closely follows the input and so it is called the emitter follower.

Applications

Because of unique quality of having high input impedance and low output impedance, an emitter follower is capable of transferring maximum power from the high impedance source to a low impedance load. When an emitter follower is used for this purpose, it is known a **buffer amplifier**. Although impedance matching can be achieved by employing a step-down transformer, but emitter follower is generally preferred. It is because emitter follower is not only more convenient compared to a transformer but it also provides better frequency response.

There are many examples (especially in digital electronics) where an increase in current is required but no increase in voltage is required. For such applications, emitter follower, because of its high current gain, can be employed.

EXAMPLE 3.15. Find the input impedance for the emitter follower circuit given in figure 3.19.

Solution : From figure 3.19, we have

Voltage across resistor resistor R_2,

$$V_B = \frac{V_{CC}}{R_1 + R_2} \times R_2$$

Substituting values, we get

$$V_B = \frac{20}{10 + 10} \times 10 = 10 \text{ V}$$

* Since R_E is usually much greater than r'_e.

Voltage across R_E will be

$$V_e = V_B - V_{BE} = 10 - 0.7 = 9.3\text{ V}$$

Therefore, emitter current will be

$$I_e = \frac{V_e}{R_E} = \frac{9.3}{9.3} = 1\text{ mA}$$

Now, A.C. emitter resistance can be found by

$$r_e' = \frac{25\text{ mV}}{I_e} = \frac{25\text{ mV}}{1\text{ mA}} = 25\ \Omega$$

Further, effective emitter resistance becomes

$$R'_E = 9.3\text{ k}\Omega \,||\, 18.6\text{ k}\Omega = 6.2\text{ k}\Omega$$

Input impedance to the base is given by

$$Z_{i\,(base)} = \beta\,(r'_e + R'_E)$$
$$= 100\,(25 + 6{,}200) = 622.5\text{ k}\Omega$$

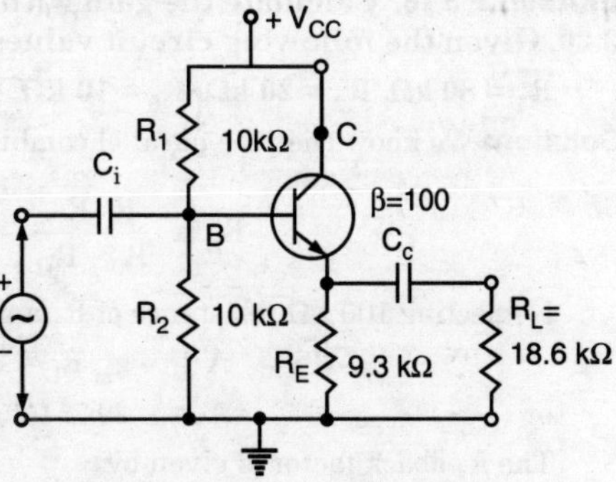

Fig. 3.19. Emitter follower circuit

Now, input impedance of the emitter follower will be

$$Z_i = R_1 \,||\, R_2 \,||\, Z_{i\,(base)} = 10\text{ k}\Omega \,||\, 10\text{k}\Omega \,||\, 622.5\text{ k}\Omega$$

or $Z_i = 4.96\text{ k}\Omega$ **Ans.**

3.16 FEW MORE PRACTICAL VOLTAGE-SERIES FEEDBACK CIRCUITS

In this section, let us discuss few examples of practical voltage-series feedback circuits. In fact, examples of practical feedback circuits will provide a means of demonstrating the effect of voltage-series feedback on usual circuits.

3.16.1. FET Amplifier Circuit with Voltage-Series Feedback

Figure 3.20 shows an FET amplifier stage with voltage series feedback.

Here, in this circuit, a part of the output signal (V_{out}) is obtained using a feedback network of resistors R_1 and R_2. The feedback voltage V_f is connected in series with the source signal V_s, their difference being the output signal V_{in}.

Now, without feedback, the amplifier gain is

$$A = \frac{V_{out}}{V_{in}} = -g_m R_L$$

where R_L is the parallel combination of resistors i.e.,

$$R_L = R_D\ R_0\ (R_1 + R_2)$$

The feedback network provides a feedback factor of

$$\beta = \frac{V_f}{V_{out}} = -\frac{R_2}{R_1 + R_2}$$

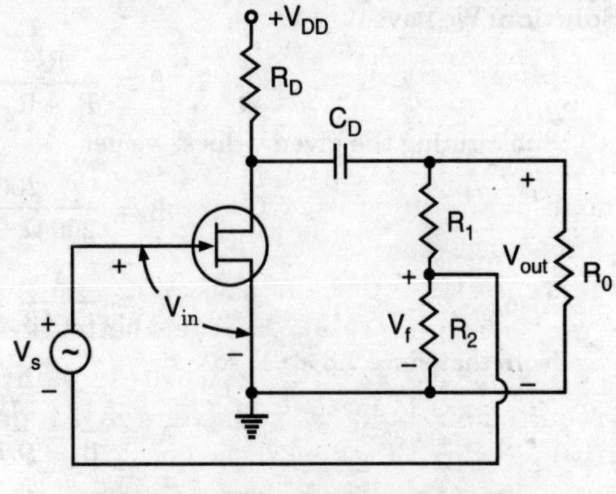

Fig. 3.20. FET amplifier stage with voltage-series feedback

We know that the voltage gain with feedback is expressed as

$$A_f = \frac{V_{out}}{V_s} = \frac{A}{1 + A\beta}$$

Now, substituting the values of A and β in above equation, we find the gain with negative feedback to be

$$A_f = \frac{A}{1 + A\beta} = \frac{-g_m R_L}{1 + [R_2 R_L/(R_1 + R_2)]g_m}$$

If $A\beta >> 1$, then we have

$$A_f \cong \frac{1}{\beta} = -\frac{R_1 + R_2}{R_2}$$

EXAMPLE 3.16. Calculate the gain without and with feedback for the FET amplifier circuit of figure 3.20. Given the following circuit values:

$R_1 = 80\ k\Omega$, $R_2 = 20\ k\Omega$, $R_0 = 10\ k\Omega$, $R_D = 10\ k\Omega$ and $g_m = 4000\ \mu S$.

Solution: We know that the parallel combination of resistors R_L is given by

$$R_L \cong \frac{R_0 R_D}{R_0 + R_D} = \frac{10\ k\Omega\,(10\ k\Omega)}{10\ k\Omega + 10\ k\Omega} = 5\ k\Omega$$

Neglecting 100 kΩ resistance of R_1 and R_2 in series

$$A = -g_m R_L = -(4000 \times 10^{-6}\ \mu s)\,(5\ k\Omega)$$

or $$A = -20$$

The feedback factor is given by

$$\beta = \frac{-R_2}{R_1 + R_2} = \frac{-20\ k\Omega}{80\ k\Omega + 20\ k\Omega} = -0.2$$

Thus, the gain with feedback will be

$$A_f = \frac{A}{1 + A\beta} = \frac{-20}{1 + (-0.2)\,(-20)} = \frac{-20}{5} = -4 \text{ **Ans.**}$$

3.16.2. Voltage-Series Feedback Connection Using an Op-amp

Figure 3.21 shows a voltage-series feedback connection using an op-amp. The gain of the op-amp, A without feedback is reduced by the feedback factor

$$\beta = \frac{R_2}{R_1 + R_2}$$

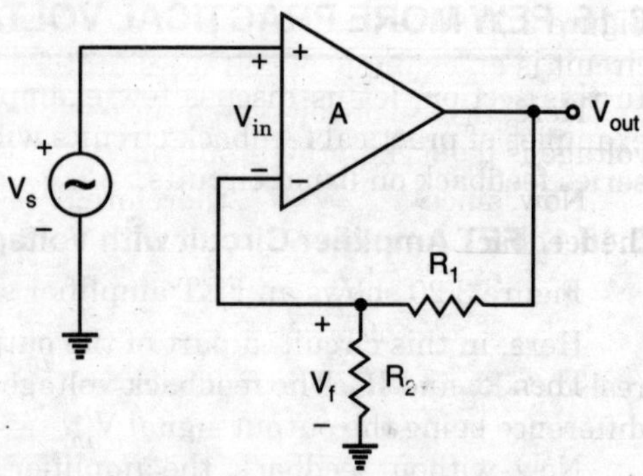

Fig. 3.21. Voltage-series feedback in an op-amp connection

EXAMPLE 3.17. Calculate the amplifier gain of the circuit of figure 3.21 for op-amp gain A = 100,000 and resistance $R_1 = 1.8\ k\Omega$ and $R_2 = 200\ \Omega$.

Solution: We have

$$\beta = \frac{R_2}{R_1 + R_2}$$

Substituting the given values, we get

$$\beta = \frac{200\ \Omega}{200\ \Omega + 1.8\ k\Omega} = 0.1$$

Also, $$A_f = \frac{A}{1 + A\beta} = \frac{100{,}000}{1 + (0.1)\,(100{,}000)} = \frac{100{,}000}{10{,}001} = 9.999$$

Note that since $A\beta >> 1$, so

$$A_f \cong \frac{1}{\beta} = \frac{1}{0.1} = 10 \text{ **Ans.**}$$

3.17 VOLTAGE-SHUNT FEEDBACK **(Very Important)**

Figure 3.22 shows the block diagram of a voltage-shunt feedback amplifier. It is also known as shunt-derived, shunt-fedback connection. In a voltage shunt feedback amplifier, a fraction of the output voltage is supplied in parallel with the input current through the feedback network.

Here, the feedback signal i.e., the current I_f is proportional to the output voltage V_{out}. Hence, the feedback factor is given by $\beta = \dfrac{I_f}{V_{out}}$.

Note that this type of amplifier is also known as a **transresistance amplifier**.

It may be noted that the voltage-shunt feedback provides a stabilized overall gain and decreases both the input as well as the output resistances by a factor $(1 + A\beta)$.

Thus, for a voltage shunt feedback amplifier, we have

Input resistance,

$$R_{if} = \frac{R_i}{(1 + A\beta)}$$

and Output resistance,

$$R_{of} = \frac{R_0}{1 + A\beta}$$

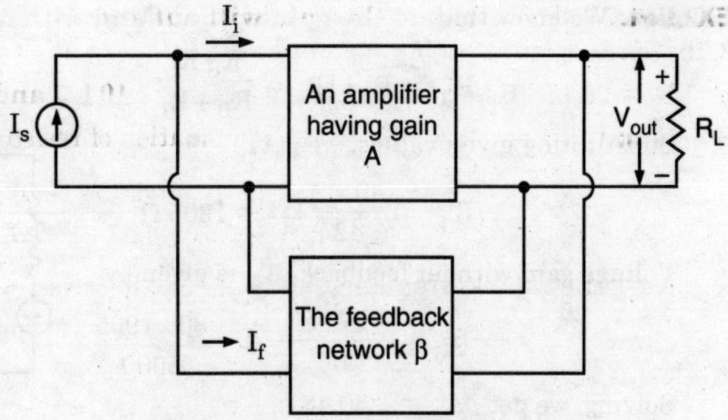

Fig. 3.22. The block diagram of a voltage shunt feedback amplifier

3.18 A COMMON-EMITTER (CE) AMPLIFIER WITH VOLTAGE-SHUNT FEEDBACK

(RGTU, Bhopal, Sem. Exam; 2007-2008)

Figure 3.23 shows the circuit of a collector feedback biased common emitter (CE) amplifier. This circuit is an example of voltage-shunt feedback.

Here, in case of voltage-shunt feedback amplifier, a current which is proportional to the output voltage is fedback to the input side.

Now, since, $V_{out} >> V_{in}$, therefore, the feedback current

$$I_f \cong \frac{V_{out}}{R_B}$$

Then, the feedback ratio, β becomes

$$\beta \cong \frac{1}{R_B}$$

Also, the reduction in input and output resistances occurs because of Miller effect due to R_B.

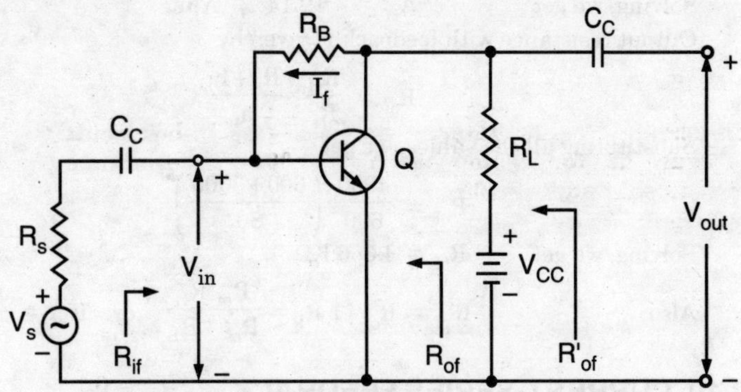

Fig. 3.23. The CE amplifier with voltage-shunt feedback

Thus, we have

$$R_{if} = h_{ie} \| \frac{R_B}{1 - A_v}$$

where, $A_v = \dfrac{V_{out}}{V_{in}} = \dfrac{-h_{fe} R'_L}{h_{ie}}$

and $R'_L = R_B \,||\, R_L$

Also $A_{vf} = \dfrac{V_{out}}{V_s} = A_v \dfrac{R_{if}}{R_s + R_{if}}$

$$R_{of} \cong \frac{R_B}{R_s} \times \frac{R_s\, h_{ie}}{h_{fe}} = R_{of} \,||\, R_L$$

EXAMPLE 3.18. Figure 3.24 shows the circuit of CE amplifier with voltage-shunt feedback. For this circuit, following values of circuit parameters are given :

$R_s = 600\ \Omega$, $h_{ie} = 5k\Omega$, $h_{fe} = 80$, $R_L = 2\ k\Omega$, $R_B = 40\ k\Omega$.

Evaluate the values of A_V, R_{if}, A_{Vf}, R_{of}, R'_{of}.

Solution: We know that

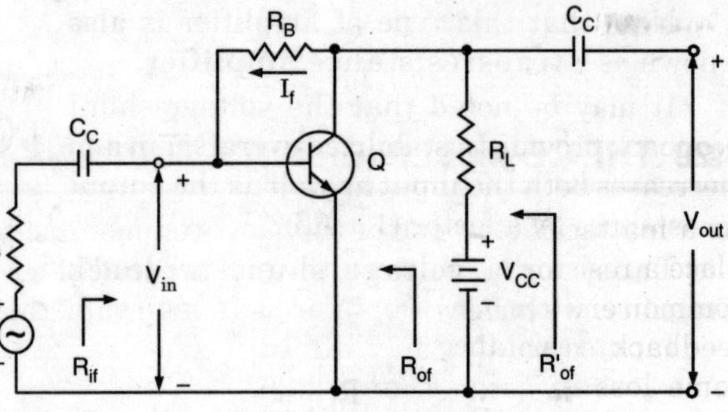

Fig. 3.24. Circuit for example 3.18.

$$R'_L = R_B \,||\, R_L = \frac{R_B . R_L}{R_B + R_L}$$

Substituting given values, we get

$$R'_L = \frac{40 \times 2}{42} \text{ k}\Omega = 1905 \,\Omega$$

Voltage gain without feedback, A_v, is given by

$$A_v = \frac{-h_{fe} . R'_L}{h_{ie}} = -\frac{80 \times 1905}{5000}$$

Solving, we get $A_v = -30.48$

Input resistance with feedback is given by

$$R_{if} = h_{ie} \| \frac{R_B}{1 - A_v} = 5000 \| \frac{40000}{1 + 30.48} = 5000 \,||\, 1270.65$$

Solving, we get $R_{if} = 1013.172 \,\Omega$ **Ans.**

Voltage gain with feedback is given by

$$A_{vf} = \frac{V_{out}}{V_s} = A_v \cdot \frac{R_{if}}{R_s + R_{if}}$$

Substituting all the values, we get

$$A_{vf} = (-30.48) \cdot \frac{(1013.172)}{600 + 1013 \cdot 172}$$

Solving, we get $A_{vf} = -19.14$ **Ans.**

Output resistance with feedback is given by

$$R_{of} = \frac{R_B}{R_s} \times \frac{R_s + h_{ie}}{h_{fe}}$$

Substituting all the values, we get

$$R_{of} = \frac{4000}{600} \cdot \left(\frac{600 + 5000}{80} \right)$$

Solving, we get $R_{of} = 4.666 \text{ K}\Omega$

Also, $R'_{of} = R_{of} \,||\, R_L = \frac{R_{of} . R_L}{R_{of} + R_L}$ or $R'_{of} = \frac{4.666 \times 2}{6.666} \text{ k}\Omega = 1.4 \text{ k}\Omega$ **Ans.**

3.19 CURRENT-SERIES FEEDBACK (RGTU, Bhopal, Sem. Exam., 2005-06)

Figure 3.25 shows the block diagram of a current-series feedback amplifier. In case of a current-series feedback,* a voltage is developed which is proportional to the output current. Note that this is called current feedback even though it is a voltage that subtracts from the input voltage. Due to series connection at the input and output, the input and output resistances are increased. This kind of amplifier is also known as **transconductance** amplifier.

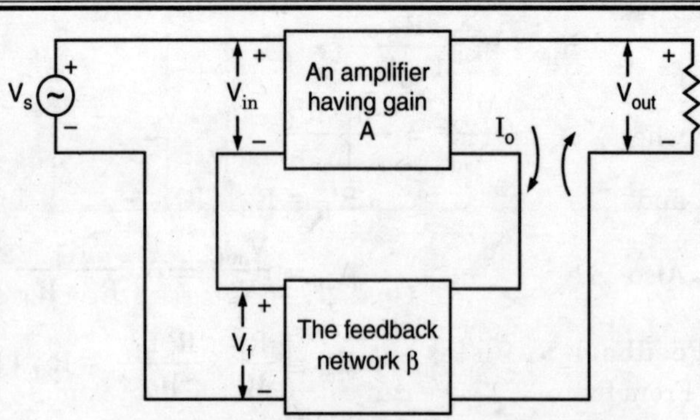

Fig. 3.25. Block diagram of the current-series feedback amplifier

Also, the transconductance feedback factor or feedback ratio is given by

* Derive expression for the effective input and output resistances of current series feedback topology. **(RGTU, Bhopal, Sem. Exam., 2005-06)(10 marks)**

$$\beta = \frac{V_f}{I_0}$$

3.20 THE CE AMPLIFIER WITH CURRENT-SERIES FEEDBACK

As a matter of fact, one of the most common method of applying the current-series feedback is to place a resistor R_E between the emitter lead of a common emitter amplifier and ground. Since, the common emitter (CE) amplifier has a high gain, therefore this is most often used with series negative feedback so that it may afford some loss in gain. Figure 3.26 shows such type of a circuit.

Here, it may be noted that when resistor R_E is properly bypassed with a large capacitor C_E, then the output voltage is V_{out} and the voltage gain without feedback is A. The resistance R_E provides d.c. bias stabilization. However, it provides no a.c. feedback. On the other hand, when the capacitor C_E is removed, an a.c. voltage would be developed across resistance R_E due to the emitter current flowing through R_E and this current is approximately equal to the output collector current.

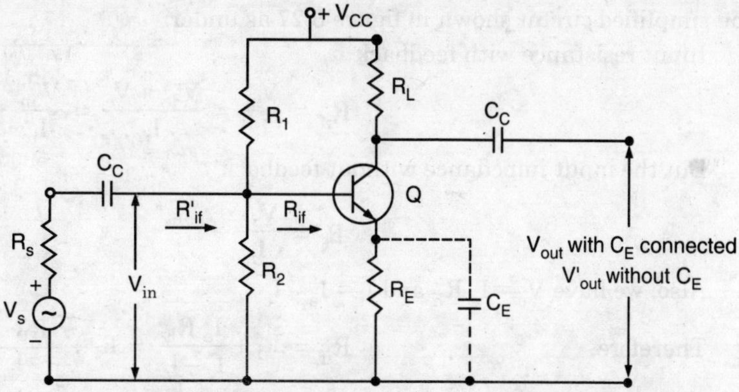

Fig. 3.26. CE amplifier with resistor R_E and when C_E is not present (i.e., removed).

This voltage drop across R_E will serve to decrease the input voltage between base and emitter so that the output voltage would decrease to V_{out}. The gain of the amplifier with negative feedback is now A_f.

Figure 3.27 shows the simplified circuit diagram for a.c. analysis for the circuit shown in figure 3.26.

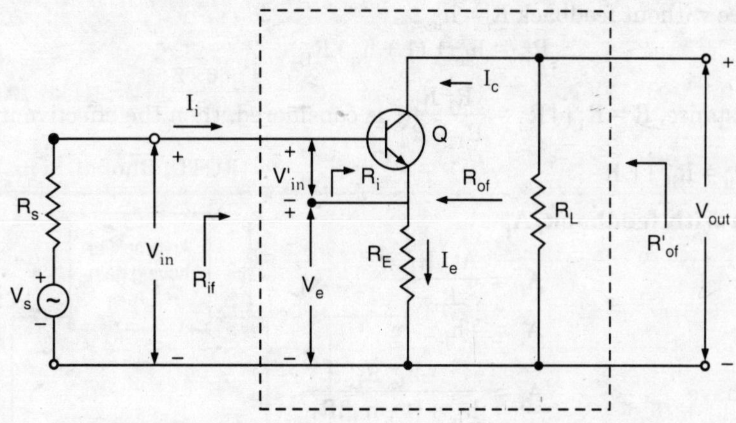

Fig. 3.27. The simplified circuit diagram for a.c. analysis

(i) Feedback Ratio (β)

From figure 3.27, we can calculate the approximate value of feedback ratio β as under:

$$V'_{out} = I_c \times R_L$$

and $\qquad V_e = I_e R_E$

But $\qquad I_e = I_c + I_b$

therefore, $\qquad V_e = (I_c + I_b)R_E$

However, since $I_c >> I_b$

therefore, we have $V_e \cong I_c \cdot R_E$

Thus, β can be written as

$$\beta = \frac{V_e}{V'_{out}} = \frac{I_c R_E}{I_c \cdot R_L} = \frac{R_E}{R_L} \quad ...(3.31)$$

(ii) Input Resistance, R_{if}

For the common-emitter (CE) amplifier shown in figure 3.26, the expression for the input resistance R_{if} and voltage gain A_v with an unbypassed emitter resistance R_E can be derived using the simplified circuit shown in figure 3.27 as under:

Input resistance with feedback

$$R_{if} = \frac{V_{in}}{I_i} = \frac{V'_{in} + V_e}{I_i} = \frac{V'_{in}}{I_i} + \frac{V_e}{I_i}$$

But the input impedance without feedback

$$R_i = \frac{V'_{in}}{I_i}$$

Also, we have $V_e = I_e R_E$ and $I_i = I_e - I_c$

Therefore, $$R_{if} = R_i + \frac{I_e R_E}{I_e - I_c} = R_i + \frac{R_E}{1 - \frac{I_c}{I_e}}$$

Again, we know that $$\frac{I_c}{I_e} = h_{fb} = \frac{h_{fe}}{1 + h_{fe}}$$

where h_{fb} and h_{fe} are common base and common emitter short-circuited current gain, respectively.

Therefore, we write $R_{if} = R_i + (1 + h_{fe})R_E$

Hence, it is clear that there is a large increase in the value of input resistance due to negative feedback.

Input resistance without feedback $R_i = h_{ie}$

hence, $R_{if} = h_{ie} + (1 + h_{fe}) R_E$

If the bias resistance, $R = R_1 || R_2 = \frac{R_1 R_2}{R_1 + R_2}$ is considered, then the effective input resistance with feedback is $R'_{if} = R_{if} || R$.

(iii) Voltage Gain with feedback, A_f

We know that $$A_f = \frac{A_i \cdot R_L}{R_{if}}$$

and $A_i = -h_{fe}$

Therefore, we have $$A_f = \frac{-h_{fe} R_L}{h_{ie} + (1 + h_{fe})R_E}$$

Voltage Gain without feedback, $A = \frac{h_{fe} R_L}{h_{ie}}$

Thus, it is clear that there is a large decrease in voltage gain due to negative feedback.

(iv) Output Resistance, R_{of}

We can find the expression for the output resistance R_{of} looking back into the collector with the help of R_s, R_E and all the h-parameters.

For the values of R_E in the order of R_s and h_{ie}, an approximate expression for R_{of} will be

$$R_{of} = \frac{1+h_{fe}}{h_{oe}} = \frac{1}{h_{ob}}$$

This has usually a large value in the range of MΩ, therefore, the overall output resistance R'_{of} taking the load-resistance R_L into consideration is approximately R_L.

Important Point: As a matter of fact, it can be revealed that the current-series feedback increases the input as well as the output resistance by the same factor (1 + Aβ).

Thus, we have

$$R_{if} = (1 + A\beta)R_i \quad \text{and} \quad R_{of} = (1 + A\beta)R_o$$

EXAMPLE 3.19. For the current-series feedback transistor amplifier shown in figure 3.18, the following parameters are given :

R_1 = 20 kΩ, R_2 = 20 kΩ, h_{ie} = 2 kΩ, R_L = 1 kΩ, R_E = 100Ω and h_{fe} = 80. Find the values of A, β, R_{if}, A_f and the loop gain in dB.

Solution: Voltage gain without feedback is given by

$$A = \frac{h_{fe}\,R_L}{h_{ie}} = \frac{-80\times1\times10^3}{2\times10^3} = -40$$

and $R_i = h_{ie} = 2\text{k}\Omega$

Feedback ratio β is given by $\beta = \frac{R_E}{R_L} = \frac{10}{1000} = 0.1$

Input resistance with feedback is given by

$$R_{if} = h_{ie} + (1 + h_{fe})R_E$$
$$= 2 \times 10^3 + (1 + 80) \times 100 = 10.1\ \text{k}\Omega$$

Voltage gain with feedback is given by

$$A_f = \frac{-h_{fe}\,R_L}{R_{if}} = \frac{-80\times1\times10^3}{10.1\times10^3} = -7.92$$

Also, loop-gain is given by

Loop gain, $A\beta = -40 \times 0.1 = -4$

Also, $A\beta = 12.04$ dB **Ans.**

EXAMPLE 3.20. Calculate the voltage gain of the circuit shown in figure 3.28.

Solution: We have gain without feedback

$$A = \frac{I_0}{V_i} = -\frac{h_{fe}}{h_{ie}+R_E}$$

or, $$A = \frac{-120}{900+510} = -0.085$$

Also, $$\beta = \frac{V_f}{I_0} = -R_E = -510$$

Then, the factor (1 + Aβ) will be given by

$$1 + A\beta = 1 + (-0.085)(-510)$$

or, $$1 + A\beta = 44.35$$

Therefore, the gain with feedback will be

$$A_f = \frac{I_0}{V_s} = \frac{A}{1+A\beta} = -\frac{0.085}{44.35} = -1.92 \times 10^{-3}$$

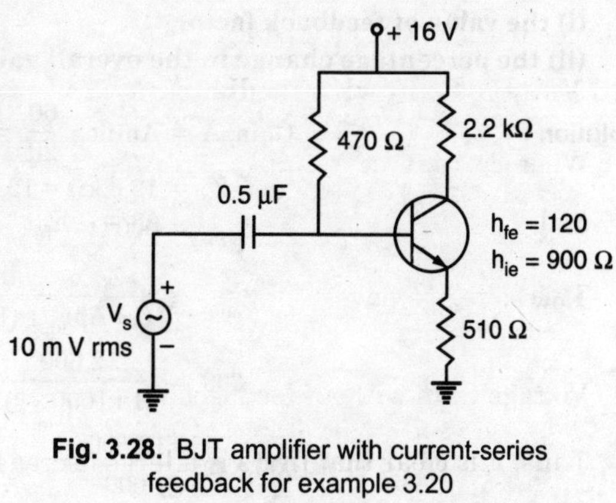

Fig. 3.28. BJT amplifier with current-series feedback for example 3.20

Also, the voltage gain with feedback A_{vf} will be

$$A_{vf} = \frac{V_{out}}{V_s} = A_f R_C = (-1.92 \times 10^{-3})(2.2 \times 10^3) = -4.2 \quad \textbf{Ans.}$$

Further, voltage gain without feedback ($R_E = 0$), the voltage gain will be

$$A_v = -\frac{R_C}{r_e} = \frac{-2.2 \times 10^3}{7.5} = -293.3 \quad \textbf{Ans.}$$

EXAMPLE 3.21. To an amplifier having 60 dB gain, a negative feedback of β = 0.01 is applied. What would be the change in overall gain of the feedback amplifier if the internal amplifier is subjected to a gain reduction 11% ? Comment on the result.

Solution : Given that gain of the basic amplifier,

$$A = 60 \text{ dB} = \text{Antilog}\frac{60}{20} = 1000$$

$$\frac{dA_F}{A_F} = \frac{1}{(1+A\beta)} \times \frac{dA}{A}$$

Substituting all the values, we get

or $$\frac{dA_F}{A_F} = \frac{1}{(1+1000 \times 0.01)} \times 11$$

Solving, we get $$= \frac{11}{11} = 1\% \quad \textbf{Ans.}$$

Important Point: The results clearly show that the % age reduction in overall gain with negative feedback is reduced from 11% to only 1%. That is why we say that the amplifier with negative feedback have stable gain.

EXAMPLE 3.22. The input impedance of an amplifier is 2kΩ and voltage gain 1000. If a negative feedback of β = 0.005 is employed what will be the value of input impedance of overall circuit?

Solution: We know that $$Z_{inf} = (1 + A\beta) Z_{in}$$

$$Z_{inf} = (1 + 1000 \times 0.005)\, 2$$

$$Z_{inf} = (1 + 5) \times 2 = 12 \text{ k}\Omega \quad \textbf{Ans.}$$

EXAMPLE 3.23. An amplifier has gain A = 60 dB and output impedance Z_0 = 12.6 kΩ. It is required to modify its output impedance to 600 Ω by applying negative feedback. Determine

(i) the value of feedback factor

(ii) the percentage change in the overall gain for 10% change in the gain of the basic amplifier.

Solution : $$\text{Gain, } A = \text{Antilog}\frac{60}{20} = 1000$$

$$Z_o = 12.6 \text{ k}\Omega = 12{,}600 \ \Omega$$

$$Z_{o(fb)} = 600 \ \Omega$$

Now $$Z_{o(fb)} = \frac{Z_0}{(1+A\beta)}$$

$$600 = \frac{12{,}600}{(1+1000 \times \beta)}$$

$$(1 + 1000 \times \beta) = \frac{12{,}600}{600}$$

$$\beta = \frac{21-1}{1000} = 0.02 \quad \textbf{Ans.}$$

Now, $$\frac{dA_{fb}}{A_{fb}} = \frac{1}{(1+A\beta)} \times \frac{dA}{A} = \frac{1}{1 \times 1000 \times 0.02} \times 10 = 0.476\% \quad \textbf{Ans.}$$

EXAMPLE 3.24. A negative feedback amplifier is shown in figure 3.29. If the gain of the amplifier without feedback is 5000, find

(i) feedback fraction (ii) overall voltage gain (iii) output voltage if input voltage is 2 mV.

Solution : Here,

$$A = 5000$$
$$R_1 = 1\text{ k}\Omega$$
$$R_2 = 9\text{ k}\Omega$$

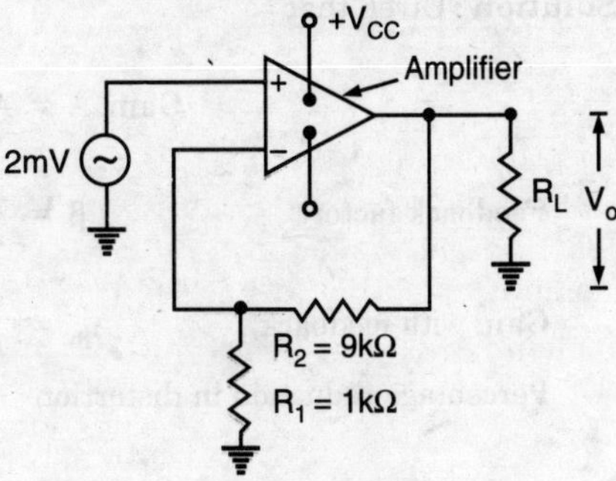

Fig. 3.29.

(i) Feedback fraction, $\beta = \dfrac{R_1}{R_1 + R_2} = \dfrac{1}{1+9} = 0.1$ **Ans.**

(ii) Overall gain, $A_{fb} = \dfrac{A}{1+A\beta}$

or $A_{fb} = \dfrac{500}{1+5000 \times 0.1} = 10$ **Ans.**

(iii) Now, $A_{fb} = \dfrac{V_{out}}{V_s}$

Therefore, output voltage, $V_{out} = A_{fb} \times V_s = 10 \times 2 = 20\text{ mV}$ **Ans.**

EXAMPLE 3.25. A negative feedback amplifier is shown in figure 3.30. If gain without feedback A = 10,000, Z_{in} = 5 kΩ, Z_o = 100 Ω. find (i) feedback fraction, (ii) gain with feedback, (iii) input impedance with feedback, (iv) output impedance with feedback (v) signal voltage if voltage obtained at the output is 10 mV.

Solution : (i) Feedback fraction,

$$\beta = \frac{R_1}{R_1 + R_2} = \frac{2}{2+18} = 0.1 \quad \textbf{Ans.}$$

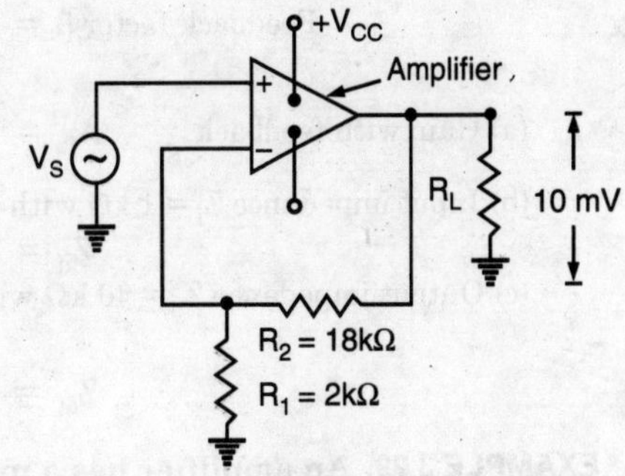

Fig. 3.30.

(ii) Overall gain, $A_{fb} = \dfrac{A}{1+A\beta} = \dfrac{10{,}000}{1+10{,}000 \times 0.1} = 10$ **Ans.**

(iii) Input impedance with feedback,

$$Z_{in\,F} = (1+A\beta)\, Z_{in}$$
$$= (1 + 10{,}000 \times 0.1) \times 5 = 5\text{ M}\Omega \quad \textbf{Ans.}$$

(iv) Output impedance with feedback,

$$Z_{oF} = \frac{Z_o}{1+A\beta} = \frac{100}{1+10{,}000 \times 0.1}$$

$$Z_{oF} = \frac{100}{1001} = 0.1\ \Omega \quad \textbf{Ans.}$$

(v) Now, $A_{fb} = \dfrac{V_{out}}{V_s}$

Therefore, signal voltage, $V_s = \dfrac{V_{out}}{A_{fb}} = \dfrac{10\text{ mV}}{10} = 1\text{ mV}$ **Ans.**

EXAMPLE 3.26. An amplifier has a voltage gain of 60 dB without feedback. What will be the gain if a negative feedback path with a feedback factor 0.009 is connected? If the gain without feedback rises by 20 dB, what will be the new gain with feedback expressed in dB ?

Solution: Given that :

$$\text{Gain, } A = 60\text{ dB} = \text{Antilog}\left(\frac{60}{20}\right) = 1000 \quad \textbf{Ans.}$$

$$\text{Feedback factor, } \beta = 0.009$$

$$\text{Gain with feedback } A_{fb} = \frac{A}{1+\beta A} = \frac{1000}{1+0.009 \times 1000} = 110$$

$$\text{Gain in dB} = 20 \log (110) = 20 \times 2.0414 = 40.83\text{ dB} \quad \textbf{Ans.}$$

EXAMPLE 3.27. An amplifier has a gain of 54.8 dB without feedback. Find the change in gain if 1/50 th of the output voltage is fed back at the input. Determine the percentage reduction in harmonic distortion in the output due to feedback.

Solution : Given that

$$\text{Gain, } A = \text{Antilog}\left(\frac{54.8}{20}\right) = 550$$

Feedback factor $\beta = \frac{1}{50} = 0.02$ **Ans.**

Gain with feedback, $A_{fb} = \frac{A}{1+\beta A} = \frac{550}{1+0.02+550}$ **Ans.**

Percentage reduction in distortion

$$= \frac{1}{1+\beta A} \times 100 = \frac{100}{1+00.2 \times 550} = 8.33\% \quad \textbf{Ans.}$$

EXAMPLE 3.28. An amplifier has a voltage gain of 10,000. Its input impedance is 1 K and output impedance is 40 K. Determine the voltage gain, input and output impedance of the circuit is 5% of the amplifier is fed in the form of series negative voltage feedback.

Solution : Given that

Voltage gain, $A = 10{,}000$

Feedback factor $\beta = 5\% = \frac{A}{100} = 0.05$

(a) Gain with feedback $A_{fb} = \frac{A}{1+A\beta} = \frac{10{,}000}{1+0.05 \times 10{,}000} = 20$ **Ans.**

(b) Input impedance $Z_i = 1\ k\Omega$ with feedback

$$Z_{if} = Z_i\,(1+\beta A) = 10^3\,(1 + 0.05 \times 10{,}000) = 500\ k\Omega \quad \textbf{Ans.}$$

(c) Output impedance $Z_o = 40\ k\Omega$ with feedback

$$Z_{of} = \frac{Z_o}{1+\beta A} = \frac{40k\Omega}{500} = 80\ \Omega \quad \textbf{Ans.}$$

EXAMPLE 3.29. An amplifier has a mid-frequency of 800. Its upper and lower cut-off frequencies f_2 and f_1 are 16 kHz and 40 Hz respectively. Determine the bandwidth of the amplifier. What will be bandwidth after 2% of the signal output is given as negative feedback?

Solution : Given that

Voltage gain $A = 800$

Feedback factor $\beta = 2\% = \frac{2}{100} = 0.02$

Voltage gain with feedback $= A_{fb} = \frac{A}{1+\beta A}$

Therefore, $A_{fb} = \frac{800}{1+0.02 \times 800}$

Bandwidth without feedback $= f_2 - f_2 = 16000 - 40 = 15.96$ kHz

Cut-off frequencies after feedback

$$f_{2(f)} = f_2\,(1+A\beta) = 16000 \times 17 = 272000 = 272 \text{ kHz}$$

$$f_{1(f)} = \frac{f_1}{1+A\beta} = \frac{40}{17} = 2\text{Hz}$$

Bandwidth after feedback $= f_{2(f)} - f_{1(f)} = 272000 - 2 = 272$ kHz **Ans.**

EXAMPLE 3.30. The gain of the amplifier is 100 with its band width of 10 kHz. If 10% of negative feedback is applied in the basic amplifier, determine the feedback gain and new bandwidth of the amplifier.

Solution : Here, gain,

$$A = 100$$

Bandwidth, $BW = 10 \text{ kHz}$

Feedback fraction, $\beta = 10\,\% = \frac{10}{100} = 0.1$ **Ans.**

Gain with negative feedback,

$$A_{fb} = \frac{A}{1+A\beta} = \frac{100}{1+100\times 0.1} = 9.09 \quad \textbf{Ans.}$$

Bandwidth with negative feedback,

$$BW_{fb} = BW\,(1+A\beta) = 10\,(1 + 100 \times 0.1) = 110 \text{ kHz} \quad \textbf{Ans.}$$

3.21 FEEDBACK AMPLIFIER WITH MULTIPLE STAGES

Let us consider a two stage feedback amplifier each with a feedback factor of β and open loop gain of A as shown in figure 3.31.

Then, the overall gain will be given by

$$A_2 = \left(\frac{A}{1+A\beta}\right)\left(\frac{A}{1+A\beta}\right)$$

$$= \frac{A^2}{(1+A\beta)^2}$$

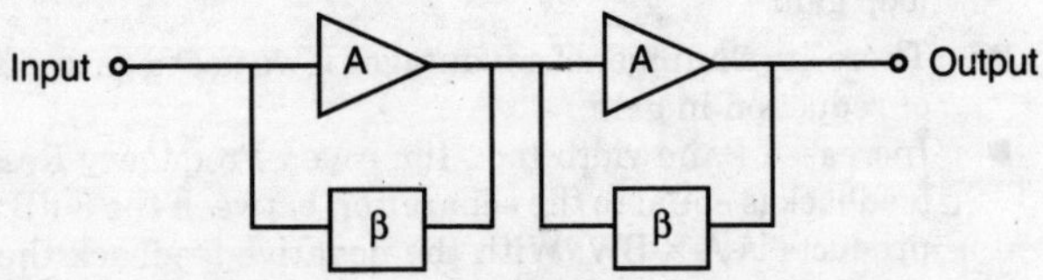

Fig. 3.31. Feedback amplifier with multiple stages

For n stages, the gain will be

$$A_n = \frac{A^n}{(1+A\beta)^n}$$

EXAMPLE 3.31. For a two stage amplifier, the second stage has 10% of the output voltage as negative feedback and has – 200 as forward gain. Find (a) the gain of the first stage (b) the second harmonic distortion, if the second stage introduces 2% second harmonic without feedback. Assume that the first stage does not introduce distortion.

Solution: (a) For second stage, we have

$$D_2' = \frac{D_2}{1+\beta A_2} = \frac{0.02}{1+200\times 0.1} = \frac{0.02}{1+2} = 0.006 = 0.6\%$$

(b) For the second stage, gain with feedback is

$$A_2' = \frac{A_2}{1+\beta A_2} = \frac{200}{1+200\times 0.1} = \frac{200}{20+1} = 9.52$$

Thus $A_1 \times A_2' = 200$

or $A_1 = \frac{200}{9.52} = 21$ **Ans.**

SUMMARY

- The amplifiers, which use the feedback principle, are known as feedback amplifiers.
- A feedback amplifier, basically, consists of two parts namely an amplifier circuit and feedback circuit (or network). The function of the feedback network is to return a fraction of the output energy (i.e., voltage or current) to the input of the amplifier.
- Depending upon whether the feedback signal increases or decreases the input signal. There are two basic types of feedback in amplifiers :
  (i) Positive Feedback, (ii) Negative Feedback.

- If the feedback signal (i.e., voltage or current) is applied in such a way that it is in phase with the input signal and thus increases it, then it is called a positive feedback. Sometimes, it is also known as regenerative feedback or direct feedback.
- Since positive feedback causes excessive distortion and instability, it is seldom used in amplifiers. However, because of its capability of increasing the power of the original signal, it is used in oscillator circuits.
- If the feedback signal (i.e., voltage or current) is applied in such a way that it is out of phase with the input signal and thus decreases it, then it is called a negative feedback. Sometimes, it is also called as degenerative feedback or inverse feedback.
- Negative feedback reduces the amplifier gain but it has a numerous advantages such as gain stability, reduction in non-linear distortion, reduction in noise, increase in bandwidth etc. Therefore, negative feedback is most widely used in amplifier circuits.
- The feedback may also be classified as voltage feedback and current feedback. In voltage feedback, the energy fedback to the input terminals is proportional to the output voltage whereas in current feedback, the energy feedback to the input terminals is proportional to the current through the load.
- Now, both voltage and current can be fedback to the input either in series on in parallel resulting in four basic feedback connections as under :
  (i) Voltage–series feedback (ii) Voltage-shunt feedback
  (iii) Current–series feedback (iv) Current–shunt feedback.
- The term β A is known as the feedback factor whereas β is called the feedback ratio and (1 ± βA) is called loop gain.
- There are a number of advantages if we use negative feedback in amplifier which outweigh its only drawback of reduction in gain.
- Increased Bandwidth (i.e., Improved Frequency Response): The bandwidth (BW) of an amplifier without feedback is equal to the separation between the 3-dB frequencies f_1 and f_2. If A is gain then gain-bandwidth products is A × BW. With the negative feedback the amplifier gain is reduced and since gain bandwidth product has to remain constant in both cases therefore obviously the bandwidth will increase to compensate for the reduction in gain.
- An emitter follower is also known as a negative feedback circuit. This circuit provides a large impedance, a small output impedance, and a voltage gain of approximately unity. More over output voltage is always in phase with the input voltage–hence the term follower.
- Its input and output impedances characteristics permit and output buffer stages in amplifier systems.

SHORT QUESTIONS WITH ANSWERS

Q.1. What is feedback ?

Ans. The feedback may be defined as a process of injecting some energy from the output and then return it back to the input. In other words, when a part or fraction of output is combined to the input, feedback is said to exist.

Q.2. What are the types of feedback?

Ans. Depending upon whether the feedback signal increases or decreases the input signal, there are two basic types of feedback in amplifiers :
(i) Positive Feedback,
(ii) Negative Feedback

Q.3. Write the expressions for voltage gain with negative & positive feedback.

Ans. The voltage gain with negative feedback will be

$$A_f = \frac{V_0}{V_s} = \frac{A}{1+\beta A}$$

Similarly, we can compute the voltage gain with positive feedback i.e.,

$$A_f = \frac{A}{1-\beta A}$$

Q.4. List the merits of negative feedback.

Ans. (i) Stabilization of Gain (Desensitivity of Gain) (ii) Reduction of Non-linear Distortion
(iii) Reduction of Noise

(iv) Increase in Bandwidth (Improved Frequency Response)
(v) Increase in Input Impedance
(vi) Reduction in Output Impedance

Q.5. What are the negative feedback topologies?

Ans. (i) Voltage-series feedback
(ii) Voltage-shunt feedback
(iii) Current-series feedback
(iv) Current-shunt feedback

Q.6. What is an emitter follower?

Ans. An emitter follower is also known as a negative feedback circuit and is an example of voltage-series feedback connection. This circuit provides a large input impedance, a small output impedance, and a voltage gain of approximately one. In addition to this, the output voltage is always in phase with the input voltage–hence the name "follower".

Q.7. What are the salient features of an emitter follower?

Ans. (i) Since it is having high input impedance and low output impedance, therefore it can be used for impedance matching. Sometimes, it is also called impedance transformer.
(ii) Because of 100 percent feedback, the output is distortionless and bandwidth is quite large.
(iii) It is having relatively high current gain and power gain.
(iv) Output and input a.c. voltages are in phase and also approximately equal in magnitude. This means that emitter output voltage closely follows the input and so it is called the emitter follower.

Q.8. What is feedback in amplifiers?

Ans. The process of combining a fraction of output energy back to the input is called the feedback.

Q.9. What is the difference between current and voltage feedback?

Ans. Voltage feedback refers to connecting the output voltage as input to the feedback network while current feedback refers to tapping off some output current through the feedback network.

Q.10. How do series feedback and shunt feedback differ from each other?

Ans. Series feedaback means connecting the feedback signal in series with the input signal voltage while shunt feedback means connecting the feedback signal in parallel with an input current source.

Q.11. Why voltage series feedback is most commonly used in cascaded amplifiers?

Ans. In cascaded amplifiers, high input impedance and low output impedance are usually the main requirements and both of these requirements can be met by voltage series feedback, so, it is most commonly used in cascaded amplifiers.

Q. 12. What is the effect of negat ve feedback on gain in an amplifier?

Ans. The negative feedback in an amplifier decreases the voltage gain.

Q. 13. What is the effect of a negative feedback on the input impedance of a voltage series feedback amplifier?

Ans. Negative feedback increases the input impedance of a voltage series feedback amplifier by a factor (1 + βA) where β A is the feedback factor.

REVIEW QUESTIONS

1. Discuss the effect of negative feedback on voltage gain, stability, distortion, bandwidth, output and input impedance of an amplifier.

   OR

   Discuss the advantages of negative feedback amplifiers.
2. Explain, how is bandwidth increased by using negative feedback.
3. Draw the circuit of a transistor amplifier with negative voltage feedback circuit. Explain how is negative voltage fed back to the input.
4. Draw an emitter follower and explain its working. Why it is called emitter follower ? Where does this amplifier find its use ?
5. Draw the circuit diagram of a common-emitter without by-pass capacitor amplifier circuit. Why it is called a negative current feedback amplifier ? Explain.
6. What is an oscillator ? How does it differ from an amplifier ? What are the essential parts of an oscillator circuit? Explain the function of each part.
7. What do you understand by damped and undamped electrical oscillations ?
8. Explain how L-C tank circuit is used to generate ac oscillations in an electronic oscillator.
9. Distinguish between "voltage" feedback and "current" feedback in amplifier circuits. State the merits of

each and for each, derive the expression for the net output impedance of the amplifier showing the influence of feedback.

10. What is the effect of unbypassed emitter resistance R_E on the gain of the amplifier?
11. Draw the circuit of an emitter follower. List its three most important characteristics.
12. Explain the working of an emitter follower.
13. An emitter follower has a gain less than unity, yet it is used in electronic circuits. Explain why?
14. Enumerate the advantages and disadvantages of negative feedback in a amplifier. Show that negative feedback improves the frequency response of an amplifier. Draw the circuit diagram for an emitter follower.
15. Specify the input and output signals of various feedback amplifiers in terms of voltage and current signals. Obtain expressions for the input and output resistance and voltage gain with feedback for an n p n transistor amplifier with the collector resistance R_L and un-bypassed emitter resistance R_E.
16. Write a short note on: Classification of feedback.
17. What are the different parameters of amplifier that are modified with negative feedback? Define desensitivity D and discuss the effect of large value of D.
18. Derive expression for the effective input and output resistances of current series feedback Topology.
19. What type of feedback is the emitter follower and example of ? How much negative feedback is used?
20. Draw and explain a circuit that uses both current and voltage feedback. What happens to its input resistance and output resistance in case of voltage series and current series feedback.
21. Draw and explain all basic feedback topologies used for negative feedback. What happens to input resistance and output resistance in case of voltage series and current series feedback.
22. Explain how would you determine the loop gain of an amplifier and determine that whether the amplifier is stable or not?

NUMERICAL PROBLEMS

1. The voltage gain an amplifier without feedback is 2000. Calculate the voltage gain of the amplifier if negative feedback is introduced in the circuit. Assume that feedback factor is 0.01. **[Ans.** A_{fb} = 95.24]
2. The gain of an amplifier is 140, when negative feedback is introduced it is reduced to 17.5. Find the value of feedback factor. **[Ans.** β = 0.05]
3. When negative feedback is applied to an amplifier of gain 200, the overall gain falls to 50 :
   (i) Calculate the value of feedback factor.
   (ii) If the fraction i.e. feedback factor remains the same, calculate the value of amplifier gain so that the overall gain becomes 30. **[Ans.** β = 0.015; A = 55.55]
4. To an amplifier of 60 dB gain, a negative feedback of β = 0.005 is applied. What would be the change in overall gain of the feedback amplifier if the internal amplifier is subjected to a gain reduction of 12%. **[Ans.** 2%]
5. For a feedback amplifier, if R_1 = 4kΩ, R_2 = 16 kΩ, whereas, the parameters of amplifier without feedback A = 10000, Z_{in} = 10 kΩ and Z_o = 200 Ω, find ; (i) feedback fraction (ii) gain with feedback (iii) input impedance with feedback (iv) output impedance with feedback and (v) output voltage when the signal voltage is 1 mV. **[Ans.** β = 0.2 ; A_{fb} = 5 ;l $Z_{in(fb)}$ = 20 MΩ ; $Z_{o(fb)}$ = 0.1 Ω ; v_o = 5 mV]
6. An amplifier has a voltage gain of 50. To reduce the distortion present in it 10% negative feedback is employed. Calculate
   (i) Voltage gain with feedback.
   (ii) Voltage gain when feedback ratio is halved of the first case.
7. An amplifier having input impedance 1 KΩ and gain of 60 dB. It is required to modify its output impedance to 600 Ω. How will you attain this characteristics? What will be the percentage change in its overall gain for 10% change in the gain of internal amplifier?

   [Ans. (i) By employing negative feedback, β = 0.019 (ii) $\frac{dA_f}{A_f}$ = 0.005]

□□□

Chapter 4

SINUSOIDAL OSCILLATORS

Syllabus

Sinusoidal Oscillators: L-C (Hartley-Colpitts) Oscillators, RC Phase shift, Wien bridge and Crystal oscillators.

Chapter Outline

4.1 INTRODUCTION

Any circuit which is used to generate a.c. voltage without a.c. input signal is called an oscillator. To generate a.c. voltage, the circuit is supplied energy from a d.c. source. An oscillator is the one of the basic and useful device which is used in electronic and electrical measurements. For example, an oscillator finds various applications in communication systems. As a matter of fact, in amplitude modulation and frequency modulation, superheterodyne receivers local oscillator are used to convert the incoming radio frequency to a lower intermediate frequency. The oscillator circuits are also used in the 'exciter' section of a transmitter to generate the radio frequency (RF) carrier signal. Other applications include their use in the form of clocks in digital systems such as microcomputers and in the sweep circuits used in TV sets and oscilloscopes. Because sinusoidal waveforms are encountered so frequently in electronic measurement, the oscillator represents the largest single category of signal generators. This device covers the frequency range from a few Hz to several GHz.

Generally, it is said that an oscillator generates a sinusoidal signal, however, it does not create energy but simply acts as an energy converter. It simply converts unidirectional current drawn from a dc source of supply into alternating current of desired frequency. The function of an oscillator is just the reverse of a rectifier and, therefore, sometimes called an inverter. However, we generally think of oscillator circuit as providing an ac voltage signal.

4.2 COMPARISON BETWEEN AN AMPLIFIER AND AN OSCILLATOR*

Figure 4.1(a) shows a block diagram of an amplifier whereas figure 4.1(b) shows that of an oscillator. We know that an amplifier is a device which produces an output signal with a same waveform as that of the input. But its power level is generally high. This additional power is supplied by an external d.c. source. Hence an amplifier is essentially an energy conversion device i.e., a device which gets energy from the d.c source and converts it into an a.c energy at the same frequency as that of the input signal. The d.c. to a.c. conversion is controlled by the input signal. This means that if there is no input signal then no energy conversion take place. Thus there is no output signal. On the other hand an oscillator is a device which produces an output signal without any input signal of any desire frequency. It keeps producing an output signal so long as the d.c. power is supplied. It is evident from the above discussion that an oscillator does not require any external signal to start or maintain energy conversion process.

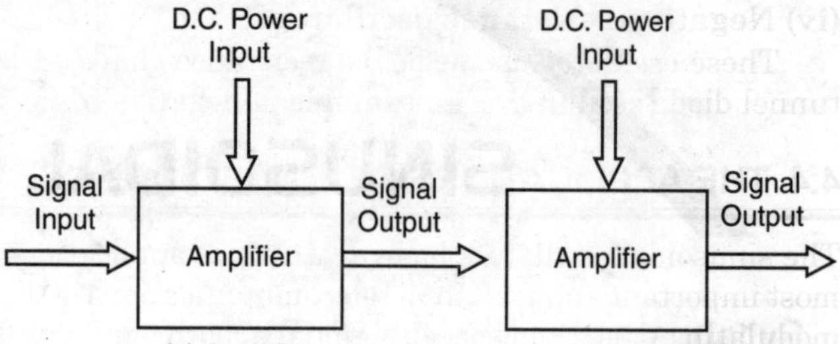

(a) Amplifier (b) Oscillator

Fig. 4.1.

4.3 CLASSIFICATION OF OSCILLATORS (Important)

The electronic oscillators can be broadly classified into the following two categories :

1. Sinusoidal or harmonic-oscillators

The oscillators which provide an output having a sine waveform, known as sinusoi-dal or harmonic oscillators. Such oscillators can provide output at frequency ranging from 20 Hz to 1 GHz (1 GHz = 10^{12} Hz).

2. Non-sinusoidal or relaxation oscillators

The oscillators which provide an output having a square rectangular or saw tooth waveform, are known as non-sinusoidal or relaxation oscillators. Such oscillators can provide output at frequency ranging from zero to 20 MHz (1 MHz = 10^6 Hz).

In this chapter, we shall discuss only the sinusoidal oscillators.

The sinusoidal oscillators can be further sub-divided into the following types :

(i) Tuned circuit oscillators

These oscillators use a tuned-circuit consisting of inductors (L) and capacitors (C) and are used to generate high-frequency signals. Hence, they are also known as radio frequency (R.F) oscillators. Such oscillators are Hartley, Colpitts, Clapp oscillators etc.

(ii) RC oscillators

These oscillators use resistors and capacitors and are used to generate low or audio-frequency

DO YOU KNOW?

To build a sinusoidal oscillator, we need to use an amplifier with positive feedback. For the oscillator to start, the loop gain must be greater than 1 when the phase shift around the loop is 0°.

ALESSANDRO VOLTA (1745-1827)

In 1796 Italian physicist Alessandro Volta (1745-1827) invented the first chemical battery, which provided the first practical source of electricity.

* Every electronic oscillator can be considered to be an amplifier with infinite gain. Clarify it.

(U.P. Tech., Carryover Sem. Exam., 2005-2006) (05 marks)

signal. Thus they are also known as audio-frequency (A.F.) oscillators. Such oscillators are phase-shift and wien bridge oscillators.

(iii) Crystal Oscillators

These oscillators use quartz crystals and are used to generate highly stabilized output signal with frequency up to 10 MHz. The pierce oscillator is an example of crystal oscillator.

(iv) Negative-resistance oscillators

These oscillator use negative resistance characteristic of the devices such as tunnel diodes. A tunnel diode oscillator is an example of negative resistance oscillator.

4.4 THE APPLICATIONS OF SINUSOIDAL OSCILLATORS

The sinusoidal oscillators have a variety of applications in telecommunication and electronics. The most important application in telecommunications is the use of sine wave as a carrier signal (used in modulating music and speech signals) in both wire and wireless transmissions. The sine wave signal are also used in testing frequency response of various electronic systems and equipments such as amplifiers filters analogue communication channels, closed-loop control system etc.

4.5 NATURE OF SINUSOIDAL OSCILLATIONS

The nature of oscillations produced by sinusoidal oscillators can be of two types, namely damped oscillations and undamped oscillations. They can be described as under:

(i) Damped oscillations

The electrical oscillations whose amplitude goes on decreasing with time are known as **damped oscillations**. The waveform of such type of oscillations has been shown in figure 4.2(a). The damped oscillations are produced by those oscillator circuits in which power losses take place continuously during, each oscillation without any means for compensating the same. However, it may be noted that frequency of oscillations remains unchanged because it depends upon the circuit parameters.

DO YOU KNOW?

In most oscillators, the feedback voltage is a fractional part of the output voltage. When this is the case, the voltage gain A_V must be large enough to ensure that $A_V B = 1$. In other words, the amplifier voltage gain must be at least large enough to overcome the losses in the feedback network. However, if an emitter follower is used as the amplifier, the feedback network must provide a slight amount of gain to ensure that $A_V B = 1$.

(ii) Undamped oscillation

The electrical oscillations whose amplitude remains constant with time are known as **undamped oscillations.** The waveform of such oscillations has been shown in figure 4.2 (b). The undamped oscillations are produced by those oscillator circuits in which either there are no power losses or if they have any power loss then they have provision for compensating these losses. It can be noted that an oscillator is required to produce undamped oscillations for utilizing in various telecommunication and electronics applications.

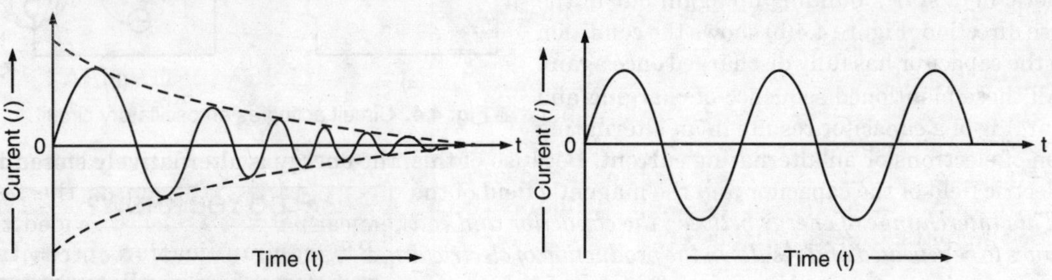

(a) Damped oscillations (b) Undamped oscillations

Fig. 4.2. Illustration of Nature of sinusoidal oscillations

4.6. THE OSCILLATORY CIRCUIT

A circuit which produces electrical oscillations of any desired frequency is known as an oscillatory circuit. This circuit contains two reactive elements namely an inductor (L) and a capacitor (C) which are connected in parallel with each other as shown in figure 4.3. Hence such a circuit is also known as LC circuit or tank circuit. Both the reactive devices are capable of storing electrical energy. It is a well known fact that an inductor stores energy in the form of magnetic field whenever a current flows through it. Similarly, a capacitor stores energy in the form of an electric field whenever a voltage (or potential difference) is applied across its plates. Both the reactive elements are supposed to be an ideal one (i.e., they do not have any power losses). Now let us discuss as to how an LC circuit is able to generate oscillations.

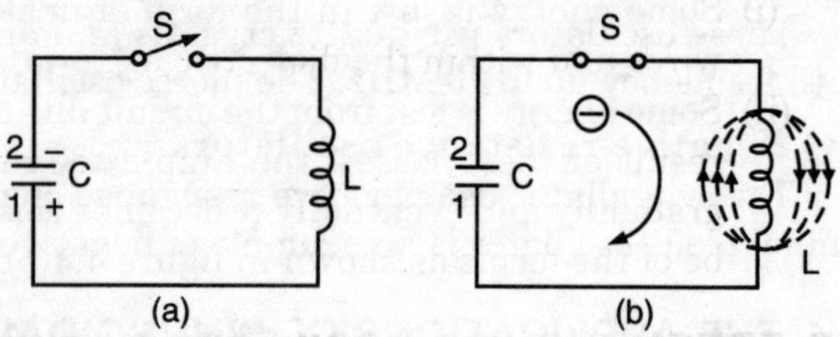

Fig. 4.3. The LC circuit acting as an oscillatory circuit

Suppose that the capacitor has been charged from a d.c. voltage source with a polarity as shown in figure 4.3 (a). As the switch (S) is open it cannot discharge through the coil (L).

When the switch (S) is closed as shown in figure 4.3(b) the capacitor discharges through the coil (L). The discharging occurs due to the flow of electrons from 1 plate to 2 through the coil (L) as indicated by the arrow (here it may be noted, that the conventional current direction is from plate 2 to plate 1). Now, this current flow sets up magnetic field around the coil. This magnetic field stores the energy released by the electric field. Because of the inductive effect, the current grows up slowly towards the maximum value. This situation occurs when the capacitor is fully discharged. At this instant, the electrical energy stored in the capacitor becomes zero and the energy stored in inductor magnetic field is maximum. In other words, we say that electrical energy across the capacitor is completely converted into magnetic field energy around the coil.

Once the capacitor is discharged completely, the magnetic field around the coil begins to collapse and produces a counter e.m.f. According to Lenz's Law the counter e.m.f., keeps the electrons moving in the same direction. This again charges capacitor though in the opposite direction as shown in figure 4.4(a). It can be noted that plate 2 has become positive whereas the plate 1, is still negative. When the capacitor ischarged completely in the opposite direction, the magnetic field around the coils is also collapsed completely. At this instant the energy previously stored in the magnetic field had now converted into the energy stored in the electric field of the capacitor.

After this, capacitor starts discharging in the opposite direction due to which the electrons now move from plate 2 to plate 1 as shown in figure 4.4(b). The electric field starts collapsing whereas magnetic field starts building up again but in the reverse direction. Figure 4.4(b) shows the condition when the capacitor has fully discharged once again.

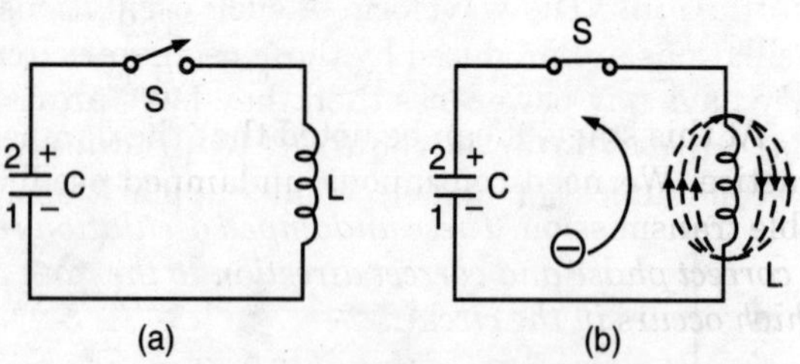

Fig. 4.4. Circuit acting as an oscillatory circuit

All these mentioned sequence of charging and discharging of a capacitor results in an alternating motion of electrons or an alternating current. Because of this, the energy is alternatively stored in the electric field of the capacitor and the magnetic field of the coil. *This interchange of energy between the capacitor and coil continues to be repeated and results in the production of electrical oscillations.*

> **DO YOU KNOW?**
>
> RF communication circuits sometimes use oscillators that contain an emitter follower for the amplifier.

If there are no power losses in the tank circuit the waveform of the electrical oscillations will be undamped (i.e., of constant amplitude). However, in a practical tank circuit the waveform of the oscillations is damped (i.e., of decreasing amplitude). It is because of the energy lose by the L and C element which can take because of the following two reasons :

(i) Some energy is lost in the form of heat produced in the resistance of the coil connecting wires and within the dielectric material of the capacitor.

(ii) Some energy is lost from the circuit due to radiation in the form electromagnetic waves As a result of these losses, the amplitude of oscillating current in the tank circuit decreases gradually and eventually it becomes zero. The waveform of such an oscillating current will be of the forms as shown in figure 4.4(b).

4.7 FREQUENCY OF OSCILLATORY CIRCUIT

The frequency (f) of the oscillatory current produced in the tank circuit can be determined by the following two factors:

(i) Capacitance of the capacitor

The frequency of oscillation (f_o) is inversely proportional to the square root of the capacitance of a capacitor. Hence larger the value of capacitance, greater is the time required for removal of the discharge current and thus lower is its frequency.

Mathematically, the frequency will be

$$f_o \propto 1/\sqrt{C} \qquad \text{...(4.1)}$$

(ii) Self-inductance of the coil

The frequency of oscillation (f_o) is also inversely proportional to the square root of the self-inductance of the coil. Hence, larger the inductance, greater is the inductance effect and this longer the time required by the current to stop flowing during the discharge of the capacitor.

Mathematically, the frequency will be,

$$f_o \propto 1/\sqrt{L} \qquad \text{...(4.2)}$$

Combining the equations (4.1) and (4.2), we get

$$f_o \propto \frac{1}{\sqrt{L.C}} = \frac{1}{2\pi\sqrt{L.C}}$$

or $f_o =$ in Hertz

If L is μH and C in μF, then the frequency will be,

$$f_o = \frac{159}{\sqrt{L.C}} \qquad \text{...(4.3)}$$

At this stage, it can be noted that the damped oscillations, so produced are not useful in actual practice. We need continuous undamped oscillations for practical application such as radio and cable transmission. *These undamped oscillations can be produced if we supply some additional energy in correct phase and correct direction to the tank circuit for making up the power losses continuously which occurs in the circuit.*

4.8 THE FREQUENCY STABILITY OF AN OSCILLATOR **(Very Important)**

The frequency stability of an oscillator can be defined as a measure of its ability to maintain a constant frequency, over a long time interval. However, it has been observed that if an oscillator is set at some particular frequency it does not maintain it for a longer period. In other words, the frequency of an oscillator changes slowly (or drifts away) from the initially set value. Sometimes, the change in frequency is uniform in one direction (i.e., either increasing or decreasing). But at some other times it may be changing quite erratically. The change in oscillation frequency can observed arise due to certain factors

(i) Operating point of the active device

The operating point of the active device (i.e., bipolar transistor or FET) is selected in such a manner so that its operation in non-linear region changes the values of device parameters which in turn, affects the frequency stability of the oscillator.

(ii) Circuit components

The value of circuit components (i.e., resistors, inductors and capacitors) large with the variation in temperature. Because such changes take place quite slowly they also cause a drift in **oscillator frequency**.

(iii) Supply voltage

The changes in d.c supply voltage applied to the active device, shift the oscillator frequency. This problem can be avoided by using a highly regulated power supply.

(iv) Output load

A change in the output load can cause a change in the Q-factor of the tank circuit thereby causing a change in oscillator output frequency.

(v) Inter-element capacitances and stray capacitances

Any change in the interelement capacitances of a transistor (particularly collector-to-emitter capacitance) cause changes in the oscillator output frequency and hence, affect the frequency stability. Similarly, the stray capacitances also affect the frequency stability of an oscillator. The effect of change in interelement capacitances can be neutralized by putting as additional capacitor across the corresponding elements. However, it is difficult to completely avoid the effect of stray capacitances.

4.9 POSITIVE FEEDBACK AMPLIFIER AS AN OSCILLATOR.

(RGTU, Bhopal, Sem. Exam., 2005-06)(5 marks)

As discussed earlier that negative feedback is employed in amplifiers for their stability. On the other hand, the positive feedback is employed in oscillators for their operation. An oscillator is a device which generates a.c. output signal without any input ac signal*. A fraction of output is fedback to the input side. This means that feedback signal is the only input to the internal amplifier.

To understand how an oscillator produces an output signal without an external input signal let us consider a block diagram as shown in figure 4.5. Now we open the feedback loop at mixer as shown in figure 4.5.

then, $X'_f = -X_f = \beta X_{out} = -A\beta X_{in}$

We define Loop gain $= -A\beta$

If we arrange that X'_f is same as X_{in} then X_{in} can be replaced by X'_f. This is possible if

$$-A\beta = 1$$

4.9.1. Barkhausen Criterion

An oscillator is basically an 'amplifier which does not have any ac input but it operates on the principle of positive feedback to generate an ac signal at its output. Thus, it is obvious that an amplifier can work as an oscillator if positive feedback is made to exist. However, positive feedback not always guarantees oscillations. An amplifier will work as an oscillator if and only if it satisfies a set of conditions called the **Barkhausen Criterion**.

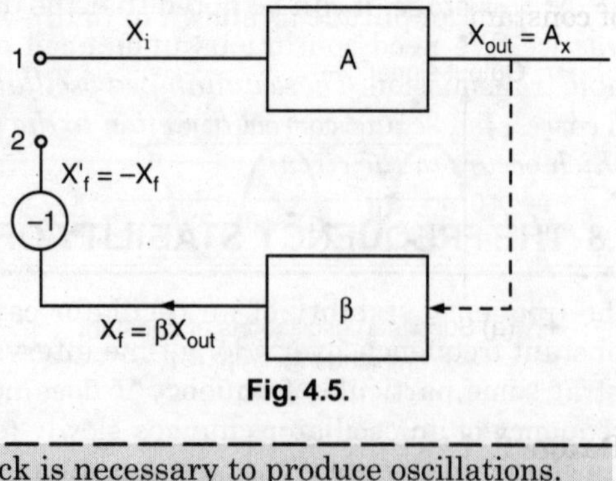

Fig. 4.5.

* Explain why positive feedback and not negative feedback is necessary to produce oscillations.

(RGTU, Bhopal, Sem. Exam., 2005-06) (05 marks)

Barkhausen Criterion states that the frequency of a sinusoidal oscillator determined by the condition that the loop-gain phase shift is zero.

Hence $\qquad -A\beta = 1$

It implies that

(i) $|A\beta| = 1$ i.e., magnitude of feedback factor = unity.

(ii) $\phi = 0$ or multiple of 2π

Another interpretation of Barkhausen Criterion is

$$A_{fb} = \frac{A}{1+A\beta} = \frac{A}{1-1} = \infty$$

Since gain is infinity, we will get faint output with no input.

Two conditions are diagrammatically illustrated in figure 4.6 (a) and (b).

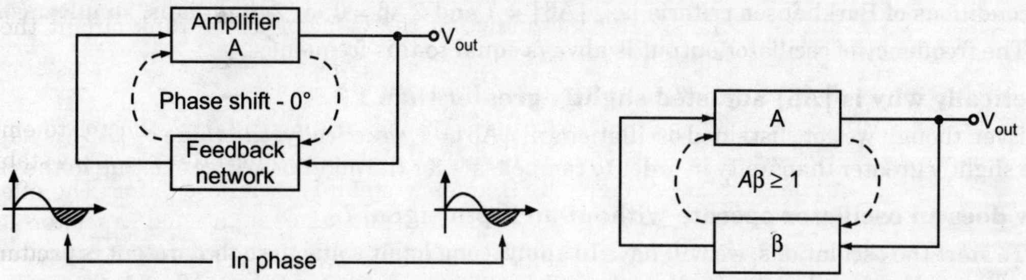

(a) The phase shift around the loop is 0° (b) Loop gain $A\beta \geq 1$

Fig. 4.6. Illustration of Barkhausen criterion

4.9.2. Effect of the Value of |Aβ| on the Nature of Oscillations

The magnitude of the loop gain Aβ can vary arbitrarily. Now, let us observe the effects of magnitude of loop gain on the nature of oscillator output.

There are three possible ranges of the value of $|A\beta|$ as under:

(i) $|A\beta| = 1$

(ii) $|A\beta| < 1$

(iii) $|A\beta| > 1$.

(i) Range 1, $|A\beta| = 1$

Let us consider figure 4.7(a). It appears that if $|A\beta|$ at the oscillator frequency is unity, then, the removal of the external generator will not make any difference, and we get sinusoidal oscillations of constant amplitude as shown in figure 4.7(a). These are known as sustained oscillations.

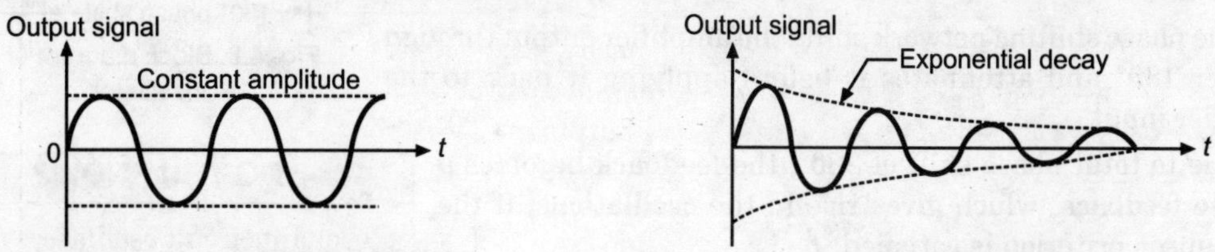

(a) Sustained oscillations of constant amplitude for $|A\beta| = 1$ (b) Exponentially decaying oscillations for $|A\beta| < 1$.

Fig. 4.7.

Range 2, $|A\beta| < 1$

If $|A\beta|$ is less than unity then the feedback signal in absence of the external generator will not be sufficient to sustain the oscillations. Then, the output signal will reduce exponentially as shown in figure 4.7 (b).

Range 3, $|A\beta| > 1$

When $|A\beta| > 1$, and total phase shift is 0°, the oscillations at the output increase exponentially. The amplitude of oscillations will continue to increase continuously without limit as shown in figure 4.7 (c).

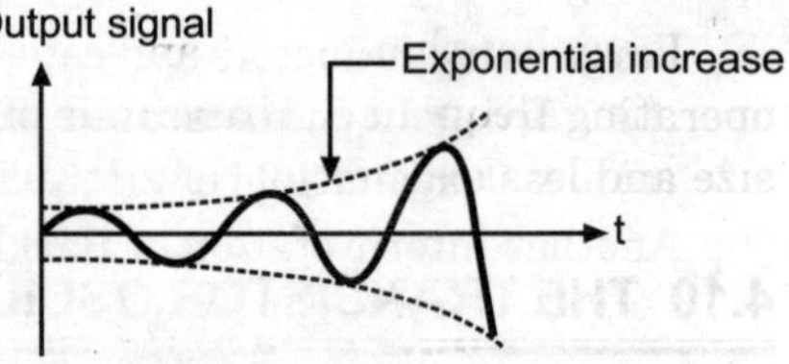

Fig. 4.7. (c) Exponentially increasing oscillations for $|A\beta| > 1$

Conclusion :

From figures 4.7 (a), (b) and (c), it is obvious that the given circuit will oscillate only if $|A\beta| \geq 1$. Sustained oscillations are obtained if $|A\beta| = 1$ and the total phase shift around the loop is zero.

Frequency of Oscillations

Frequency of oscillations is defined as that frequency at which the given circuit satisfies both the conditions of Barkhausen criteria i.e., $|A\beta| = 1$ and $\angle A\beta = 0°$ or $2\pi n$ radians, simultaneously.

The frequency of oscillator output is always equal to this frequency..

Practically why is $|A\beta|$ adjusted slightly greater than 1 ?

Even though we get sustained oscillations at $|A\beta| = 1$, practically, the value of $|A\beta|$ is adjusted to be slightly greater than unity in order to compensate for the non-linearities existing in the circuit.

How does an oscillator operate without an input signal ?

To start the oscillations, we will have to apply some input voltage, so that output is produced. A part of it is fed back and as this feedback is sufficient to satisfy the condition $|A\beta| = 1$, the oscillations are sustained. However, in reality, we do not apply any input voltage to start the oscillations, still the oscillators work. So, we have to find from where does this starting voltage come ?

Due to random movement of electrons inside any electronic device such as resistor, a voltage is generated which is called as noise voltage. This noise voltage is amplified by the amplifier and fed back. If the Barkhausen criterion is satisfied then, sustained oscillations will be obtained at the output. Thus, noise voltage acts as the starting voltage and oscillators can operate without an input signal.

4.9.3. Block Diagram of an Oscillator

The block diagram of an oscillator is shown in figure 4.8.

An oscillator consists basically of an amplifier and a phase shifting network. The amplifier receives the output of the phase shifting network. The amplifier then amplifies it, phase shifts through 180° and applies it to the input of the phase shifting network.

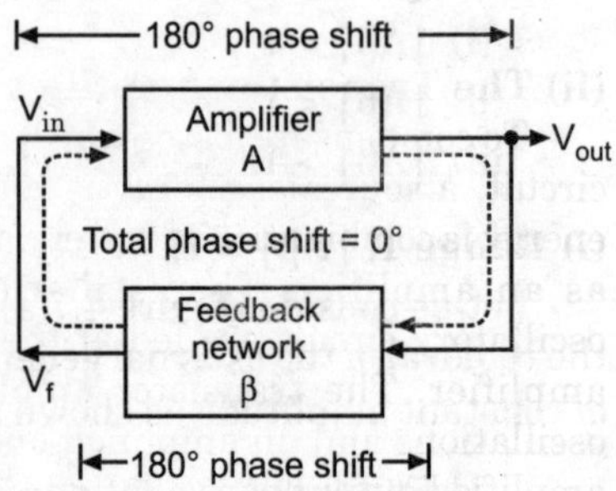

Fig. 4.8. Block diagram of an oscillator

The phase shifting network shifts the amplifier output through another 180° and attenuates it before applying it back to the amplifier input.

Due to total phase shift of 360°, the feedback becomes a positive feedback, which gives rise to the oscillations, if the Barkhausen criterion is satisfied.

4.9.4. Introduction to LC Oscillators

In LC oscillators, the feedback network consists of inductors (L) and capacitors (C), instead of resistors and capacitors in case of RC oscillators. These LC components determine the frequency of oscillations of the LC oscillator.

DO YOU KNOW?

Resonant circuit oscillators (or LC oscillators) are used for generating high-frequency waves as high as a few hundred MHz. This oscillator consists of an amplifier and a tank circuit acting as a feedback network.

The operating principle of LC oscillators is based on the Barkhausen conditions. These oscillators can operate at high frequencies typically from 200 kHz to few GHz. They are not suitable for low operating frequencies because the values of L and C will be large at low frequencies.

Large value inductors and capacitors are bulky (large in size) and expensive as well. But, as the operating frequency is increased we need small value inductors and capacitors which are small in size and less expensive. Now, let us discuss few LC oscillators in the sections to follow:

4.10 THE TRANSISTOR OSCILLATOR

The continuous undamped oscillations can be obtained from an oscillator circuit (L-C circuit) if a correct amount of energy is supplied to it at correct phase. This can be achieved by employing a transistor because it has the ability to amplify a signal. A transistor is a very efficient energy converter i.e., it converts d.c power to a.c power.

When damped oscillations produced by the tank circuit are applied to the base of transistor, it will reproduce the amplified oscillations in the collector circuit. The amplified oscillations will possess more energy in the collector circuit than in the base circuit. If a part of this collector-circuit energy is feedback by some means to the base circuit in proper phase, then the losses will be compensated and the continuous undamped oscillations will be produced.

4.11 THE ESSENTIALS OF A TRANSISTOR OSCILLATOR

Figure 4.9 shows the block diagram of a transistor oscillator. It has the following essential parts:

(i) The Oscillatory Circuit

An oscillatory circuit consists of inductive coil L and capacitor C connected in parallel with each other. The frequency of oscillations depends upon the value of inductance of the coil and capacitance of the capacitor i.e.,

$$f_r = \frac{1}{2\pi\sqrt{LC}} \quad \text{...(4.4)}$$

(ii) The Transistor Amplifier

To compensate losses which occur in the oscillatory circuit, a source of energy is required. This source of energy is combination of battery and transistor working as an amplifier. The oscillations produced by the oscillatory circuit are fed at the input of a transistor amplifier. The transistor amplifier amplifies these oscillations and an amplified output is obtained. This amplified output of oscillations is due to the d.c power supplied by the battery. The output of the transistor thus obtained can now be supplied back to the oscillatory circuit to compensate the losses.

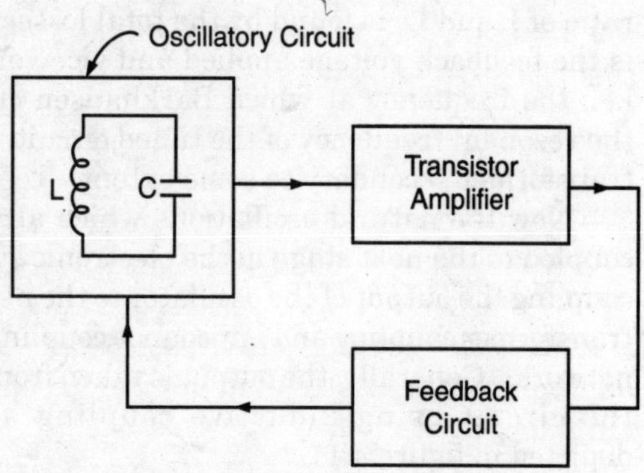

Fig. 4.9. Block diagram of a transistor oscillator

(iii) The Feedback Circuit

This circuit is used to feedback a fraction of amplifier output to the oscillator circuit in correct phase so that it aids the oscillations and compensates the losses. Infact, it provides positive feedback.

4.12 THE TUNED COLLECTOR OSCILLATOR **(Expected)**

Figure 4.10 shows the circuit diagram of a tuned collector oscillator. It is known as a tuned collector oscillator since here the tuned circuit has been connected to the collector. The tuned circuit formed by the capacitor C and the transformer primary coil L_1 forms the load impedance and hence determines the frequency of oscillation. The resistors R_1, R_2 and R_E form the d.c biasing circuit for the transistor. The capacitors C_1 and C_E are called bypass capacitors for R_2 and R_E respectively. Because of these capacitors, the a.c operation of the circuit is not affected. In addition to this, capacitor C_1 provides ac ground for the transformer secondary. The output voltage developed across the tuned circuit is inductively coupled to the base circuit through transformer secondary coil L_1. The feedback voltage appears across the base emitter junction, as the junction point of resistors R_1 and R_2 is at ac ground due to bypass capacitor C_1.

It may be noted that in the absence of C_1 the feeding back voltage induced in the secondary of the transformer will not go directly to the input of the transistor. In fact, some part of this voltage will drop across resistor R_2. The transistor amplifier provides a phase shift of 180° because it is connected in common emitter (CE) configuration. Another phase shift of 180° is provided by the transformer. Hence, a total phase shift of 360° appears between the input and the output voltages. This means that there is a positive feedback between the input and the output voltages. In addition to this, the transistor amplifier provides sufficient gain for oscillator action to take place.

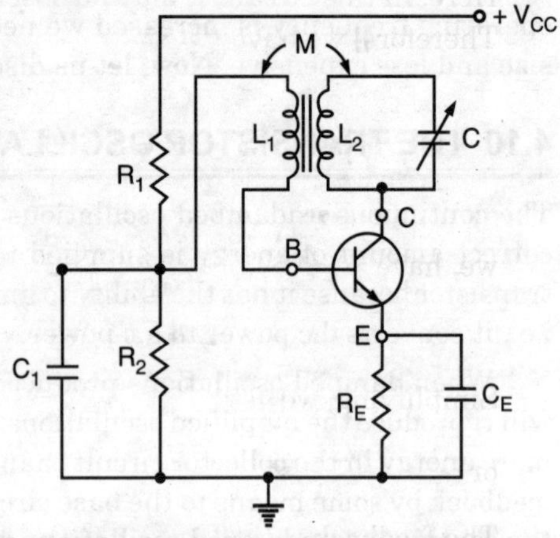

Fig. 4.10. The circuit diagram for tuned collector oscillator

Working operation

In this oscillator, when the power supply V_{CC} is first switched on, a transient current is produced in the tuned L-C circuit. It is because of the increase of collector current to its quiescent value. This transient current starts natural oscillations in the tank circuit. These natural oscillations induce some voltage into coil L_1 by mutual induction which causes corresponding variations in the base current. These variations in base current are amplified β times and appear in the collector circuit. A part of this amplified energy is used to balance the losses which occur in the tank circuit and the rest is radiated out in the form of electromagnetic waves. The turn-ratio of L and L_1 is found by the total losses. Higher is the turn-ratio, lesser is the feedback voltage applied and vice-versa. The frequency of oscillation i.e., the frequency at which Barkhausen criterion is satisfied differs from the resonant frequency of the tuned circuit. This is because of loading of the transformer secondary to some extent.

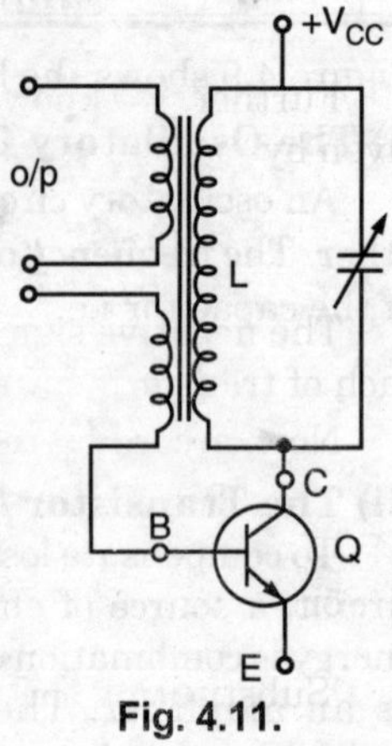

Fig. 4.11.

Now the natural oscillations which are set-up in the tuned circuit are coupled to the next stage in the electronic system. As a matter of fact, there are several methods for coupling the output of the oscillator to the next stage. Few coupling networks are capacitive coupling, transformer coupling and impedance coupling networks. Generally, the output is taken from the circuit using inductive coupling as depicted in figure 4.11.

The output terminals are generally connected to the input terminals of the next stage. This has the loading effect on the circuit causing increase in the circuit losses. For maintaining oscillations, more amount of positive feedback which can be provided by simply increasing the coupling between the primary and the secondary of the transformer is needed.

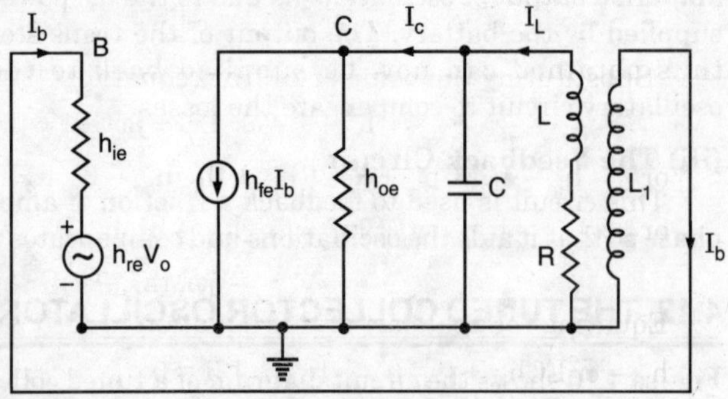

Fig. 4.12. Equivalent Circuit of Tuned collector Oscillator circuit

Analysis of Tuned Collector Oscillator

For the analysis of tuned-collector oscillator, let us first draw the equivalent circuit as shown in figure 4.12.

Here, in this circuit, it may be observed that L-C tuned circuit constitutes the load.

Therefore, we have

$$\frac{1}{Z_L} = \frac{1}{1/j\omega C} + \frac{1}{R + j\omega L}$$

$$\frac{1}{Z_L} = j\omega C + \frac{1}{R + j\omega L}$$

we, have

$$\frac{1}{Z_L} = \frac{1 + j\omega C\,R - \omega^2 LC}{R + j\omega L}$$

Simplifying, we get

or
$$Z_L = \frac{R + j\omega L}{1 - \omega^2 LC + j\omega CR} \quad ...(4.5)$$

The feedback ratio β, which is the ratio of voltage induced in the secondary and voltage across the primary winding, can be given by

$$\beta = \frac{-j\omega M\, I_L}{(R + j\omega L)I_L} = \frac{-j\omega M}{R + j\omega L} \quad ...(4.6)$$

Further, we know that the voltage gain of common emitter (CE) amplifier without feedback is given by

$$A = \frac{-h_{fe}.Z_L}{h_{ie} + (h_{ie}h_{oe} - h_{re}h_{fe})Z_L} \quad ...(4.7)$$

The negative sign in equations (4.6) and (4.7) implies that a phase shift of 180° is introduced by each of transformer and amplifier,

Now, according to Barkhausen Criterion, for sustained oscillation, we have

$$|A\beta| = 1$$

or
$$\frac{1}{|A|} = \beta \quad ...(4.8)$$

Substituting the values of A and β from equations (4.7) and (4.6) in equation (4.8), we get

$$\frac{h_{ie} + (h_{ie}\, h_{oe} - h_{re}\, h_{fe})Z_L}{h_{fe}.Z_L} = \frac{j\omega M}{R + j\omega L}$$

$$\frac{h_{ie}}{Z_L} + h_{ie}\, h_{oe} - h_{re}\, h_{fe} = \frac{j\omega M}{R + j\omega L} h_{f_e} \quad ...(4.9)$$

Substituting the value of Z_L from equation (4.5) into equation (4.9), we get

$$h_{ie}\,[\,1 - \omega^2\, LC + j\omega CR] + [\,h_{ie}\, h_{oe} - h_{re}\, h_{fe}]\,[R + j\omega L] = j\omega Mh_{fe}$$

or
$$h_{ie} - \omega^2 LCh_{ie} + j\omega CRh_{ie} + [h_{ie}\, h_{oe} - h_{re}\, h_{fe}]R + j\,[h_{ie}\, h_{oe} - h_{re}\, h_{fe}]\,\omega L = j\omega Mh_{fe}$$

or
$$h_{ie} - \omega^2 LCh_{ie} + [h_{ie}\, h_{oe} - h_{re}\, h_{fe}]R$$
$$= j\omega Mh_{fe} - j\omega CRh_{ie} - j\omega L\,[h_{ie}\, h_{oe} - h_{re}\, h_{fe}] \quad ...(4.10)$$

Equating the real part of this equation to zero, we get

$$h_{ie} - \omega^2 LCh_{ie} + [h_{ie}\, h_{oe} - h_{re}\, h_{fe}]\, R = 0$$

or
$$\omega^2 = \frac{h_{ie}\, R\Delta h}{LC.h_{ie}}$$

here, $$\Delta h = h_{ie}h_{oe} - h_{re}h_{fe}$$

or $$\omega = \frac{1}{\sqrt{LC}} \cdot \sqrt{1 + \frac{R\Delta h}{h_{ie}}}$$

Hence, the frequency of oscillation is given by

$$f = \frac{\omega}{2\pi} = \frac{1}{2\pi\sqrt{LC}}\sqrt{1 + \frac{R\Delta h}{h_{ie}}} \quad ...(4.11)$$

Because Δh and resistance R of the coil are quite small and h_{ie} is large, hence, the equation (4.11) can be written as

$$f = \frac{1}{2\pi\sqrt{LC}} \quad ...(4.12)$$

This means that the frequency of oscillation can be varied by varying either L or C. However, it is easier to vary C.

Further, equating the imaginary part of equation (4.10) to zero, we shall obtain the condition for sustained oscillations as under :

$$\omega M h_{fe} - \omega C R h_{ie} - \omega L \Delta h = 0$$

or $$M = \frac{RC\,h_{ie} + L.\Delta h}{h_{fe}} \quad ...(4.13)$$

Important Point: Thus, this equation gives the minimum value of the mutual inductance M which is required for sustained oscillations.

EXAMPLE 4.1. A tuned-collector oscillator uses an L-C tuned circuit having L = 29.3 μH and C = 450pF. Obtain the frequency of oscillation.

Solution : Given that inductance of tuned circuit, L = 29.3 μH = 29.3 × 10^{-6} H

Capacitance of tuned circuit, C = 450 pF = 450 × 10^{-12} F

Therefore, frequency of oscillation,

$$f_o = \frac{1}{2\pi\sqrt{LC}} = \frac{1}{2\pi\sqrt{29.3\times10^{-6}\times450\times10^{-12}}} = 1.39 \text{ MHz} \quad \textbf{Ans.}$$

EXAMPLE 4.2. A tuned-collector oscillator has a fixed inductance of 100 μH and has to be tunable over the frequency band of 500 kHz to 1,500 kHz. Find the range of variable capacitor to be used.

Solution : We know that for a tuned collector oscillator, the resonant frequency is given by

$$f_o = \frac{1}{2\pi\sqrt{LC}} \quad ...(i)$$

where L is in henry and C is in farads

From equation (i), we have

$$C = \frac{1}{4\pi^2 f_o^2 L}$$

For frequency of 500 kHz and L = 100 μH, we have

$$C_1 = \frac{1}{4\pi^2(500\times10^3)^2\times100\times10^{-6}} = 1{,}013 \text{ pF}$$

For frequency of 1,500 kHz and L = 100 μH, we have

$$C_2 = \frac{1}{4\pi^2(1{,}500\times10^3)^2\times100\times10^{-6}} = 112.6 \text{ pF}$$

Hence the range of required capacitor is 112.6 pF – 1, 013pF. **Ans.**

EXAMPLE 4.3. In tuned-collector oscillator, the applied voltage is 12 V. The power output of the circuit is 88 m W and total in the tank circuit are 8 m W. Find the transformer winding turn-ratio. Assume g_m to be 10 mA/volt.

Solution : Given that supply voltage, $V_{CC} = 12$ V

Power output of the circuit, $P_o = 88$ mW

Power losses in the circuit, $P_{losses} = 8$ mW

Power input to the circuit, $P_i = P_o + P_{losses} = 88 + 8 = 96$ mW

$$\text{Collector current, } I_C = \frac{P_i}{V_{CC}} = \frac{96\times10^{-3}}{10\times10^{-3}} = 8 \text{ mA}$$

$$\text{Base voltage required, } V_B = \frac{I_C}{g_m} = \frac{8\times10^{-3}}{10\times10^{-3}} = 0.8 \text{ V}$$

Transformer winding turn-ratio,

$$= \frac{\text{voltage across L} - \text{C circuit}}{V_B} = \frac{12}{0.8} = 15 \quad \textbf{Ans.}$$

4.13 THE TUNED-BASE OSCILLATOR

If a parallel tuned L-C circuit is connected in the base-to-ground circuit, then the resulting oscillator is known as the **tuned-base oscillator**. The basic circuit of a tuned-base oscillator has been shown in figure 4.13. The circuit is also known as the tickler oscillator **or Armstrong oscillator** (named after it inventor). In this oscillator, the d.c. bias conditions are set-up by the resistor R_1, R_2 and R_E. The parallel $R_E - C_E$ network in the emitter circuit is a stabilizing circuit and prevents the degeneration of the signal in the circuit. As usual, C_c is the dc blocking capacitor. L_1 and L are the primary and secondary mutually coupled coils of an RF transformer. This RF transformer provides the required feedback between the collector and base circuits. The primary coil L_1 is sometimes known as a tickler coil. The amount of feedback depends on the coefficient of coupling between the two coils. Because, here the transistor is connected in CE configuration, therefore, it itself provides a phase shift of 180° between its input and output terminals. Another phase shift of 180° is provided by the transformer. Hence, a total phase shift of 360° is produced which is a necessary condition for developing oscillations. The parallel tuned L-C circuit connected between the base and emitter determines the frequency of oscillation.

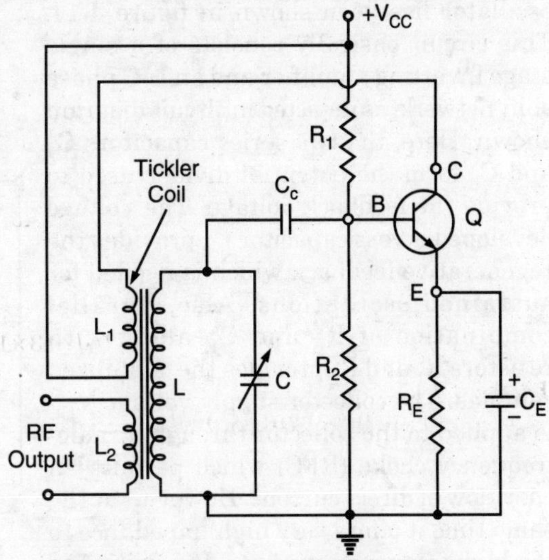

Fig. 4.13. Circuit Tuned-Base Oscillator

The tuned L-C circuit resonates at the frequency $f = \dfrac{1}{2\pi\sqrt{LC}}$.

DO YOU KNOW?

In the colpitts oscillator, the tank inductor is adjusted to vary the frequency of oscillation, whereas in the Hartley oscillator, the tank capacitor is adjusted to vary the frequency of oscillation.

Working Principle

When the supply V_{CC} is switched on the collector current starts increasing. This increasing collector current, which also flows through the tickler coil L_1, produces a varying magnetic field around coil L_1. This varying magnetic field links with the coil L and hence induces a voltage in the tuned circuit. Due to correct phasing to the coils, and sufficient gain of amplifier, the oscillations start building up. Dynamic self bias is provided from the R_2-C_C combination due to flow of base current. The voltage developed across the capacitor C due to its charging, drives the transistor towards cut-off. This decrease the gain of the transistor amplifier, and therefore satisfying the conditions A β = 1. Hence, sustained oscillations are obtained. The output can be obtained by means of a third coil L_2 magnetically coupled to coil L. It has approximately the same waveform as the collector current.

The main drawback of this circuit is that tank circuit gets loaded because of the low base-emitter resistance which comes in shunt with the tuned circuit. This reduces the Q of the tuned circuit and thus the frequency of oscillation differs slightly from the frequency of resonance of the tuned circuit. Because of this reason, the tuned or oscillatory circuit is connected in the collector circuit rather than in base circuit.

However, in case of a vacuum triode or an FET, the impedance is quite large and, therefore, the tank circuit in not loaded by the input impedance if vacuum triode or FET is employed as an employing device. In practice Armstrong oscillator circuit make use of either FETs or a vacuum tubes.

4.14 THE COLPITT'S OSCILLATOR (RGTU, Bhopal, Sem. Exam., 2005-06) (05 marks)

The Colpitt's oscillator is a popular oscillator circuit and is most widely used in commercial signal generators upto frequency range of 100 MHz. The basic circuit of a Colpitt's oscillator has been shown in figure 4.14. This circuit basically consists of a single stage inverting amplifier and an L-C phase shift network, as depicted in circuit diagram shown. Here, the two series capacitors C_1 and C_2 form the potential divider used to provide the feedback voltage. The voltage developed across capacitor C_2 provides the regenerative feedback which is needed for sustained oscillations. Also, parallel combination of R_E and C_E along with resistors R_1 and R_2 provides the stabilized self bias. The collector supply voltage V_{CC} is applied to the collector through a radio-frequency choke (RFC) which permits an easy flow of direct current. However, at the same time it offers very high impedance to the high frequency currents. In addition to this, the presence of coupling capacitor C_C in the output circuit does not allow the dc current to go to the tank circuit (the flow of dc current in a tank circuit reduces its Q). Now, the radio-frequency energy developed across RFC is capacitively coupled to the tank circuit througth the capacitor C_C. The output of the phase-shift L-C network is coupled from the junction of L and C_2 to the amplifier input at base through coupling capacitor C_C, which blocks d.c. but provides path to a.c. As usual, transistor itself produces a phase shift of 180° and another phase shift of 180° is provided by the capacitive feedback. This means that a total

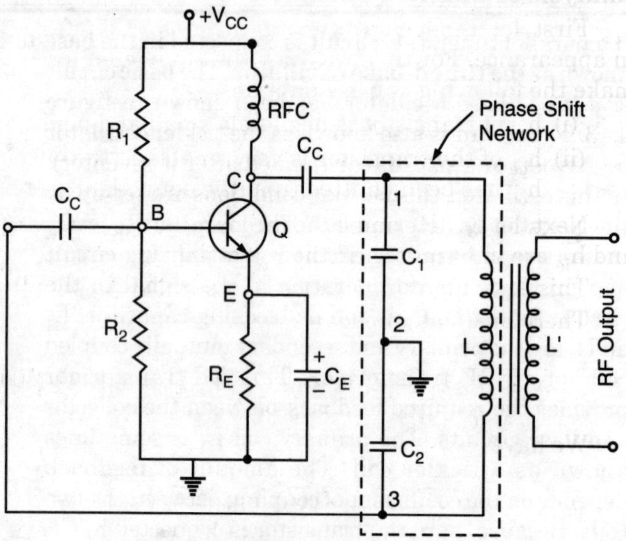

Fig. 4.14. Colpitts oscillator circuit

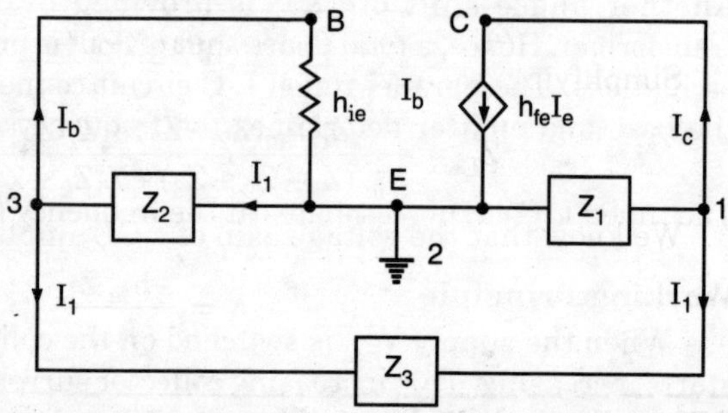

Fig. 4.15. General equivalent circuit for Hartley and Colpitt's oscillators

phase shift of 360° is obtained which is a necessary condition for developing oscillations. The output voltage is derived from a secondary winding L' coupled to the inductance L. The frequency is determined by the tank circuit and is varied by gang-tuning the two capacitors C_1 and C_2. It may be noted that capacitors C_1 and C_2 are ganged. This means that as the tuning is varied, values of both capacitors vary simultaneously, the ratio of the two capacitances remaining the same.

Working Principle

Whenever the collector supply voltage V_{CC} is switched on, the capacitors C_1 and C_2 become charged. These capacitors C_1 and C_2 discharge through the coil L, setting up oscillations frequency $f = \frac{1}{2\pi}\sqrt{\frac{1}{LC_1} + \frac{1}{LC_2}}$. Also, oscillations across capacitor C_2 are made to apply to the base-emitter junction and appear in the amplified form in the collector circuit. Note that, the amplified output in the collector circuit is of the same frequency as that of the oscillatory circuit. This amplified output in the collector circuit is supplied to the tank circuit in order to meet the losses. Hence, the tank circuit is getting continuouly energy from the circuit to make up for the losses occurring in it and, hence, ensures the undamped oscillations. The energy supplied to the tank circuit is of correct phase, as already discussed, and if Aβ exceeds unity, oscillations are sustained in the circuit.

Analysis of Colpitt's Oscillator*

First, let us develop a general theory for Colpitt's and Hartley oscillators because they are same in appearance. For this, we draw the equivalent circuit as shown in figure 4.15. In this circuit, we make the following two assumptions.

(i) h_{re} of transistor is negligibly small and hence, the feedback source $h_{re} V_o$ is negligible.

(ii) h_{oe} of the transistor is very small i.e., the output resistance $1/h_{oe}$ is large and, therefore, $1/h_{oe}$ has been omitted from the equivalent circuit.

Next, let us determine the load impedance between output terminals 1 and 2. Here, note that Z_2 and h_{ie} are in parallel and their resultant impedance is in series with impedance Z_3.

This equivalent impedance is in parallel with impedance Z_1.

Therefore, load impedance between output terminals can be written as

$$Z_L = Z_1 \| \left[(Z_3 + (Z_2 \| h_{ie})\right] = Z_1 \| \left[Z_3 + \frac{Z_2\, h_{ie}}{Z_2 + h_{ie}}\right]$$

We have

$$Z_L = Z_1 \| \left[\frac{Z_3(Z_2 + h_{ie}) + Z_2\, h_{ie}}{Z_2 + h_{ie}}\right] = Z_1 \| \left[\frac{h_{ie}(Z_2 + Z_3) + Z_2 Z_3}{Z_2 + h_{ie}}\right]$$

Again, we write

$$\frac{1}{Z_L} = \frac{1}{Z_1} + \frac{Z_2 + h_{ie}}{h_{ie}(Z_2 + Z_3) + Z_2\, Z_3} = \frac{h_{ie}(Z_1 + Z_2 + Z_3) + Z_1 Z_2 + Z_2\, Z_3}{Z_1[h_{ie}(Z_2 + Z_3) + Z_2\, Z_3]}$$

Simplifying, we have

$$Z_L = \frac{Z_1[h_{ie}(Z_2 + Z_3) + Z_2\, Z_3]}{h_{ie}\,(Z_1 + Z_2 + Z_3) + Z_1 Z_2 + Z_2 Z_3} \quad ...(4.14)$$

We know that the voltage gain of a CE amplifier without feedback is given by

$$A = \frac{-h_{fe}\, Z_L}{h_{ie}} \quad ...(4.15)$$

---

* Draw the circuit diagram of Hartley and Colpitt's Oscillators and obtain the expression for the frequency of oscillation. **(RGTU, Carryover Sem. Exam., 2005-06) (10 marks)**

The output voltage between terminals 1 and 2 is given by

$$V_o = \left[Z_3 + \frac{Z_2 h_{ie}}{Z_2 + h_{ie}}\right] I_1 = \left[\frac{h_{ie}(Z_2 + Z_3) + Z_2 Z_3}{Z_2 + h_{ie}}\right] I_1$$

Now, the voltage feedback to the input terminals 2 and 3 is given by

$$V_f = \frac{Z_2 h_{ie}}{Z_2 + h_{ie}} I_1$$

Therefore, feedback fraction can be found as

$$\beta = \frac{V_f}{V_{out}} = \frac{Z_2 h_{ie}}{h_{ie}(Z_2 + Z_3) + Z_2 Z_3} \quad ...(4.16)$$

Applying Barkhausen criterion of oscillation i.e., $A\beta = 1$, we have

$$\frac{-h_{fe} Z_L}{h_{ie}} \cdot \frac{Z_2 h_{ie}}{h_{ie}(Z_2 + Z_3) + Z_2 Z_3} = 1$$

Manipulating, we get

$$\frac{h_{fe} Z_1 [Z_2 + Z_3) + Z_2 Z_3]}{h_{ie}(Z_1 + Z_2 + Z_3) + Z_1 Z_2 + Z_2 Z_3}\left[\frac{Z_2}{h_{ie}(Z_2 + Z_3) + Z_2 Z_3}\right] = -1$$

or $$\frac{h_{fe} Z_1 Z_2}{h_{ie}(Z_1 + Z_2 + Z_3) + Z_1 Z_2 + Z_2 Z_3} = -1$$

or $$h_{ie}(Z_1 + Z_2 + Z_3) + Z_1 Z_2 + Z_2 Z_3 = -h_{fe} Z_1 Z_2$$

or $$h_{ie}(Z_1 + Z_2 + Z_3) + Z_1 Z_2 (1 + h_{fe}) + Z_2 Z_3 = 0 \quad ...(4.17)$$

This is the general equation for both the oscillators.

Now, in case of the Colpitt's oscillator, we have

$$Z_1 = \frac{1}{j\omega C_1} = -\frac{j}{\omega C_1}$$

$$Z_2 = \frac{1}{j\omega C_2} = \frac{-j}{\omega C_2}$$

$$Z_3 = j\omega L$$

Substituting all these values in equation (4.17), we get

$$h_{ie}\left[\frac{-j}{\omega C_1} - \frac{j}{\omega C_2} + j\omega L\right] + \left[\frac{-j}{\omega C_1} \cdot \frac{-j}{\omega C_2}\right](1 + h_{fe}) + \left[\frac{-j}{\omega C_2}\right] j\omega L = 0$$

or $$-jh_{ie}\left[\frac{1}{\omega C_1} + \frac{1}{\omega C_2} - \omega L\right] - \frac{1 + h_{fe}}{\omega^2 C_1 C_2} + \frac{L}{C_2} = 0 \quad ...(4.18)$$

Equating the imaginary component of the above equation (4.18) to zero we have

$$h_{ie}\left[\frac{1}{\omega C_1} + \frac{1}{\omega C_2} - \omega L\right] = 0$$

or $$\frac{1}{\omega C_1} + \frac{1}{\omega C_2} = \omega L, \quad \text{Since, } h_{ie} \neq 0$$

or $$\frac{C_1 + C_2}{\omega C_1 C_2} = \omega L$$

or $$\omega^2 = \frac{C_1 + C_2}{L C_1 C_2}$$

DO YOU KNOW?

Frequency of oscillation of Hartley and Colpitts oscillators depends on the values of tank circuit parameters. These parameters change with time, temperature, etc. Thus, oscillation frequency is not stable at desired value.

or $$\omega = \sqrt{\frac{C_1 + C_2}{LC_1C_2}} = \sqrt{\frac{1}{LC_1} + \frac{1}{LC_2}}$$

or $$f = \frac{1}{2\pi}\sqrt{\frac{1}{LC_1} + \frac{1}{LC_2}} \qquad \text{...(4.19)}$$

This equation (4.19) gives the frequency of oscillation. Now, equating the real component of equation (4.18) to zero, we have

$$\frac{1 + h_{fe}}{\omega^2 C_1 C_2} = \frac{L}{C_2}$$

or $$1 + h_{fe} = \omega^2 L C_1 = \frac{C_1 + C_2}{LC_1C_2} \times LC_1$$

or $$1 + h_{fe} = \frac{C_1 + C_2}{C_2} = 1 + \frac{C_1}{C_2}$$

Thus, we have $$h_{fe} = \frac{C_1}{C_2} \qquad \text{...(4.20)}$$

As for other oscillator circuits, the loop gain must be greater than unity to ensure that the circuit oscillates.

Therefore, we have $$A\beta > 1 \quad \text{or} \quad A \geq \frac{1}{\beta} \geq \frac{C_2}{C_1} \qquad \text{...(4.21)}$$

Let us discuss one numerical example to illustrate the concept of Colpitt's oscillator.

EXAMPLE 4.4. In a transistor Colpitt's oscillator shown in figure 4.14, $L = L_{RFC} = 0.6\ \mu H$, $C_1 = 0.001\ \mu F$, $C_2 = 0.01\ \mu F$ and $C_C = 10\ \mu F$. Find: (i) operating frequency, (ii) feedback fraction, (iii) minimum gain to sustain oscillations and emitter resistance if $R_C = 2.5\ k\Omega$.

Solution: Given that

Inductance, $L = 100\ \mu H = 100 \times 10^{-6}$ or 1×10^{-4} H

Capacitance, $C_1 = 0.001\ \mu F = 0.001 \times 10^{-6}$ F

Capacitance, $C_2 = 0.01\ \mu F = 0.01 \times 10^{-6}$ F

(i) We know that operating frequency is given by

$$f = \frac{1}{2\pi}\sqrt{\frac{1}{LC_1} + \frac{1}{LC_2}}$$

Substituting all the values, we get

$$f = \frac{1}{2\pi}\sqrt{\frac{1}{10^{-4} \times 0.001 \times 10^{-6}} + \frac{1}{10^{-4} \times 0.01 \times 10^{-6}}} = 528 \text{ kHz} \quad \textbf{Ans.}$$

(ii) Feedback fraction, $$\beta = h_{fe} = \frac{C_1}{C_2} = \frac{0.001 \times 10^{-6}}{0.01 \times 10^{-6}} = 0.1 \quad \textbf{Ans.}$$

(iii) Minimum gain to sustain oscillations is given by

$$A_{min} = \frac{1}{\beta} = \frac{1}{0.1} = 10 \quad \textbf{Ans.}$$

(iv) Again since $$A \cong R_C/R_E$$

therefore, $$R_E \cong \frac{R_C}{A_{min}} = \frac{2.5\ k\Omega}{10} = 250\ \Omega \quad \textbf{Ans.}$$

4.14.1. Colpitt's Oscillator using FET

Figure 4.16 shows the Colpitt's oscillator using FET. Here, R_1, R_2 and R_S are the biasing resistors for the FET. Also, C_{C1} and C_{C2} are the coupling capacitor while C_S is the bypass capacitor. FET is operated in the CS configuration. Therefore, the phase shift introduced is 180°. The feedback network consists of C_1, C_2 and L_3 which introduces an additional phase shift of 180° in order to make the total phase shift to be equal to 0°. Thus, the condition for the positive feedback is satisfied.

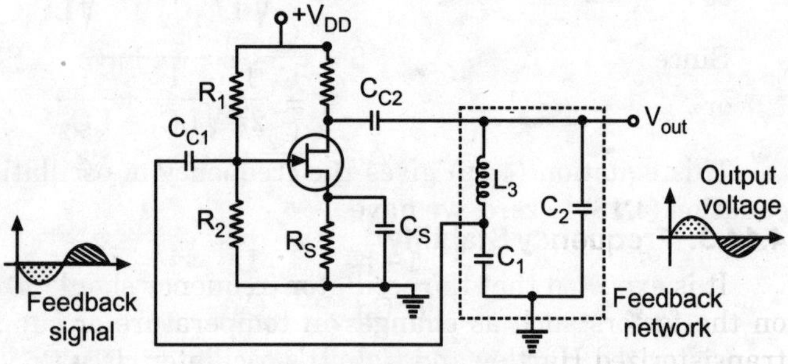

Fig. 4.16. Colpitt's oscillator using FET

The expressions for frequency of oscillations can be written as follows :

$$\omega^2 = \frac{C_1 + C_2}{L_1 C_1 C_2}$$

Let $$C = \frac{C_1 C_2}{C_1 + C_2}$$

Hence, above equation gets converted as,

$$\omega^2 = \frac{1}{CL}$$

or $$\omega = \frac{1}{\sqrt{CL}}$$

and $$f = \frac{1}{2\pi\sqrt{CL}} \qquad \text{... (4.22)}$$

where $$C = \frac{C_1 C_2}{C_1 + C_2}$$

> **DO YOU KNOW?**
>
> RC oscillators usually do not work well above 1 MHz because of the additional phase shift inside the amplifier. This is why LC oscillators are preferred for frequencies between 1 and 500 MHz. This frequency range is beyond the f_{unity} of most op amps, which is why a bipolar junction transistor or FET is commonly used for the amplifying device. The Colpitts oscillator is one of the most widely used LC oscillators.

4.14.2. Colpitt's Oscillator Using OP-AMP

The Colpitt's oscillator using op-amp has been shown in figure 4.17. Here, the Op-Amp is being used as an inverting amplifier. Hence it introduces a phase shift of 180° between its input and output. The feedback network introduces 180° phase shift in order to make the total phase shift equal to 0°. The frequency of oscillations is given by,

$$f = \frac{1}{2\pi\sqrt{LC}}$$

where, $$C = \frac{C_1 C_2}{C_1 + C_2}$$

The resistance R_1 and R_F decide the gain of the inverting amplifier. In order to ensure sustained oscillations, the loop gain should be greater than 1.

Therefore, $|A\beta| > 1$

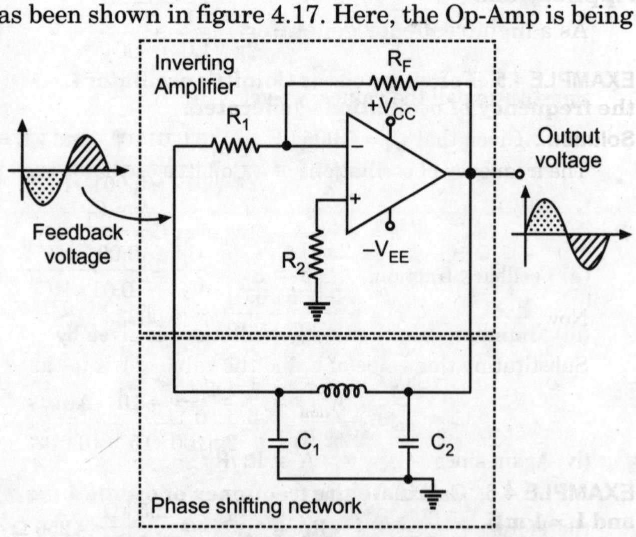

Fig. 4.17. Colpitt's oscillator using op-amp

or $$|A| > \frac{1}{\beta}$$

Since $\beta = \frac{C_2}{C_1}$, so, we have

$$A_V > \frac{C_2}{C_1} \quad \text{... (4.23)}$$

4.14.3. Frequency Stability

It is expected that the oscillator frequency should remain constant. It should not be dependent on the factors such as changes in temperature or supply voltage and so on. Unfortunately, the transistorized Hartley and Colpitt's oscillator circuits do not have a high frequency stability. The reasons for poor frequency stability may be listed as under :

(i) The collector base internal capacitance affects the value of capacitance in the feedback circuit. This will change the frequency of oscillations.

(ii) The transistor parameters depend on the temperature. Due to change in their values, the frequency of oscillations will change.

Remedy

To improve the frequency stability, we can take the following measures :

(i) Use a temperature stabilized chamber to enclose the oscillator circuit. The temperature of chamber is maintained constant at the desired level. This will avoid frequency changes due to drift in temperature.

(ii) Use voltage regulators to keep the supply voltage constant.

(iii) Use a special type of oscillator called Clapp oscillator.

Advantages of Colpitt's Oscillator

(i) Simple construction.

(ii) It is possible to obtain oscillations at very high frequencies.

Drawbacks

(i) It is difficult to adjust the feedback as it demands change in capacitor values.

(ii) Poor frequency stability.

Applications

As a high frequency generator.

EXAMPLE 4.5. For a transistor Colpitt's oscillator L = 100 μH, C_1 = 0.005 μF and C_2 = 0.01 μF. Calculate the frequency of oscillations generated.

Solution : Given that $C_1 = 0.005\ \mu F$, $C_2 = 0.01\ \mu F$ and $L = 100\ \mu H$

The frequency of oscillations of a Colpitt's oscillator is given by

$$f = \frac{1}{2\pi\sqrt{LC}} \quad \text{where , } C = \frac{C_1 C_2}{C_1 + C_2}$$

Now, $$C = \frac{0.005 \times 0.01}{(0.005 + 0.01)} = 0.00333\ \mu F.$$

Substituting this value of C and the value of L into the expression for frequency, we shall have

$$f = \frac{1}{2\pi\sqrt{LC}} = \frac{1}{2\pi[100\times10^{-6}\times0.0033\times10^{-6}]^{1/2}} = 277.05\ \text{kHz.} \quad \textbf{Ans.}$$

EXAMPLE 4.6. Calculate the frequency of oscillations of a Colpitt's oscillator with $C_1 = C_2$ = 500 pF and L = 1 mH.

Solution : Given that

(i) Colpitt's oscillator

(ii) $C_1 = C_2 = 500$ pF

(iii) $L = 1$ mH

The equivalent capacitance for a Colpitt's oscillator is given by,

$$C = \frac{C_1 C_2}{C_1 + C_2} = \frac{500 \times 500}{(500 + 500)} = 250 \text{ pF}$$

The frequency of oscillations is given by

$$f = \frac{1}{2\pi\sqrt{LC}} = \frac{1}{2\pi[1\times10^{-3} \times 250\times10^{-12}]^{1/2}} = 318.3 \text{ kHz.} \quad \textbf{Ans.}$$

EXAMPLE 4.7. In a Colpitt's oscillator, if the desired frequency is 500 kHz, estimate the values of L and C.

Solution : Given that Colpitt's oscillator, frequency of oscillations f = 500 kHz.

(i) The frequency of oscillations of a Colpitt's oscillator is given by,

$$f = \frac{1}{2\pi\sqrt{LC}} \quad \text{or} \quad f^2 = \frac{1}{4\pi^2 LC} \quad \textbf{Ans.}$$

or
$$LC = \frac{1}{4\pi^2 f^2} = \frac{1}{4\pi^2 (500\times10^3)^2} = 1.013 \times 10^{-13} \quad \text{... (i)}$$

(ii) Let us assume $L_3 = 1$ mH, substitute this into equation (i), we get

Therefore,
$$C = \frac{1.013\times10^{-13}}{1\times10^{-3}} = 1.013 \times 10^{-10} = 101.32 \text{ pF} \quad \text{... (ii)}$$

(iii) But C is the equivalent capacitance, which is given by,

$$C = \frac{C_1 C_2}{C_1 + C_2}$$

Assuming $C_1 = C_2 = C$, we get

$$101.32 = \frac{C'^2}{2C'}$$

Solving this, we get $C_1 = C_2 = 202.6$ pF

Thus, $L = 1$ mH, $C_1 = C_2 = 202.6$ pF. **Ans.**

4.15 THE CLAPP OSCILLATOR

(RGTU, Bhopal, Sem. Exam., 2008-09)

Figure 4.18 shows a Clapp oscillator. This oscillator is a refinement of the Colpitt's oscillator in the sense that the single inductor found in the Colpitt's oscillator is replaced by a series L-C combination in a Clapp oscillator.* Addition of capacitor C_3 in series with L improves the frequency stability and eliminates the effect of transistor parameters on the operation of the circuit. The operation of the circuit is the same as that of the Colpitt's oscillator. As the circulating tank current flows through C_1, C_2 and C_3 in series, the equivalent capacitance will be given by

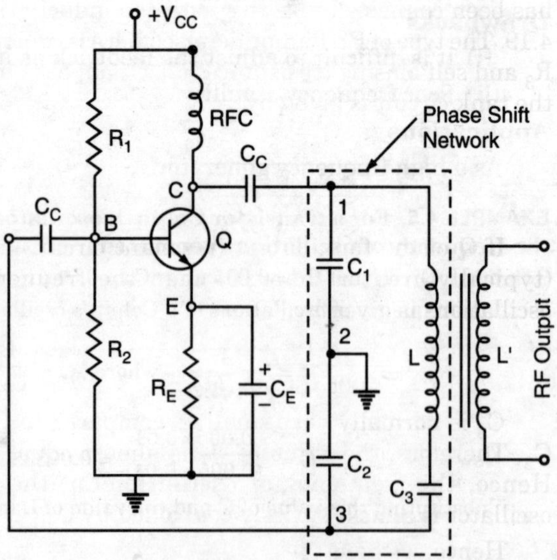

Fig. 4.18. Clapp oscillator circuit

* How does the circuit of a clap oscillator differs from that of a Colpitt's oscillator? Derive frequency of oscillation of any one of them. **(RGTU, Sem. Exam., 2005-06)(10 marks)**

$$C = \frac{1}{\frac{1}{C_1}+\frac{1}{C_2}+\frac{1}{C_3}} = \frac{C_1C_2C_3}{C_1C_2 + C_2C_3 + C_3C_1}$$

The frequency of oscillation can be found by

$$f = \frac{1}{2\pi}\sqrt{\frac{1}{LC_1}+\frac{1}{LC_2}+\frac{1}{LC_3}} \quad ...(4.24)$$

Capacitors C_1 and C_2 are kept fixed where as the capacitor C_3 is employed for tuning purpose.

It may be noted that in a Clapp oscillator C_3 is much smaller than C_1 and C_2. Because of this, the equivalent capacitance C is approximately equal to C_3, and the frequency of oscillation can be written as

$$f = \frac{1}{2\pi\sqrt{LC_3}} \quad ...(4.25)$$

However, we take precaution in selection of C_3. If capacitor C_3 is made too small, the L-C branch would not have a net inductive reactance and under such condition the circuit will refuse to oscillate.

Important Point: In a Colpitt's oscillator, the resonant frequency is affected by the transistor and stray capacitances because the capacitors C_1 and C_2 are shunted by the transistor and stray capacitances and so their values are altered. However, in a Clapp oscillator, the transistor and stray capacitances have no effect on capacitor C_3, therefore the oscillation frequency is more stable and accurate. That is why Clapp oscillator is preferred over a Colpitt's oscillator.
In addition to this, high frequency stability can further be obtained by enclosing the entire circuit in a constant temperature chamber and by maintaining the supply voltage constant with the help of a Zener diode.

4.15.1. Clapp Oscillator Using FET

The circuit diagram of a clapp oscillator using FET has been shown in figure 4.19. The clapp oscillator is a variation of Colpitt's oscillator. The basic difference is that an additional capacitor C_3 has been connected in series with the inductor in the resonant feedback circuit as shown in figure 4.19. The type of FET amplifier used here is common source amplifier with bypassed source resistance R_3 and self biasing (resistance R_1). C_3 appears in series with C_1 and C_2, hence, total capacitance in the tank circuit is given by,

$$C_T = \frac{1}{\frac{1}{C_1}+\frac{1}{C_2}+\frac{1}{C_3}} \quad ... (4.26)$$

If Q of the tank circuit (resonant circuit) is high (typically greater than 10) then the frequency of oscillations is given by,

$$f_r \cong \frac{1}{2\pi\sqrt{LC_T}} \quad ... (4.27)$$

C_3 is normally very small as compared to C_1 and C_2. Therefore, the value of C_T is almost equal to C_3. Hence, the frequency of oscillations of the clapp oscillator is almost entirely controlled by C_3.

Hence, we have

$$f_r \cong \frac{1}{2\pi\sqrt{LC_3}} \text{ (since } C_T = C_3) \quad ... (4.28)$$

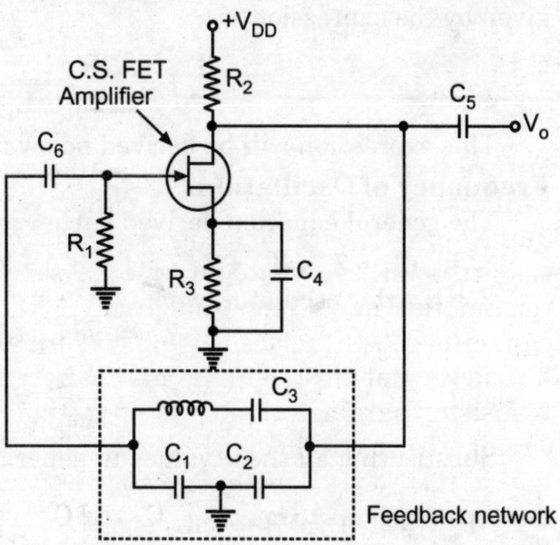

Fig. 4.19. Clapp oscillator using FET

Clapp Oscillator has Better Stability

One end of C_1 and C_2 is connected to ground. Hence, the junction capacitance of transistor and other stray capacitances appear in parallel with C_1 and C_2 to ground. Hence, values of C_1 and C_2 will get altered due to the junction and stray capacitances. But, the value of C_3 is not affected in this manner and frequency of oscillations depends on the capacitor C_3. Hence, clapp oscillators have a better frequency stability.

4.16 THE HARTLEY OSCILLATOR (RGTU, Bhopal, Sem. Exam., 2008-09)(05 marks)

The Hartley oscillator is quite popular like Colpitt's oscillator and is widely used as a local oscillator in radio receivers. The circuit arrangement has been shown in figure 4.20. As a matter of fact, Hartley oscillator circuit is quite similar to Colpitt's oscillator circuit except that here phase-shift network consists of two inductors L_1 and L_2 and a capacitor C instead of two capacitors and one inductor as in a Colpitt's oscillator. The output of the amplifier is applied across inductor L_1 and the voltage across inductor L_2. In fact, this forms the feedback voltage. The coil L_1 is inductively coupled to coil L_2, the combination works as an **auto-transformer**. However, because of direct connection, the junction of L_1 and L_2 cannot be directly grounded. Instead, another capacitor C_L is used. The operation of the circuit is similar to that of the Colpitt's oscillator circuit.

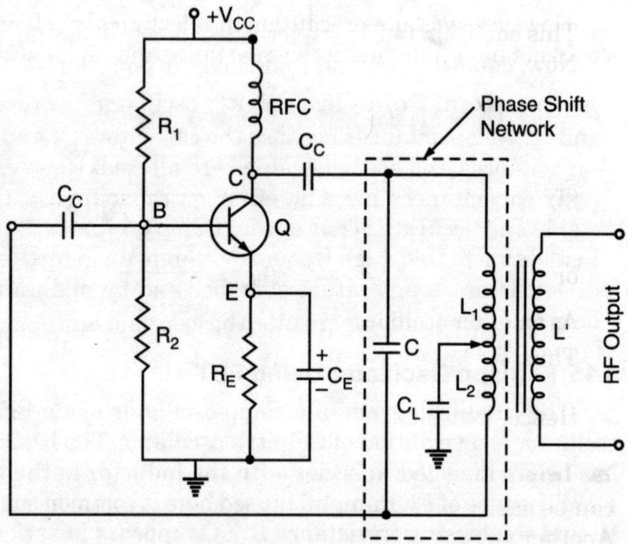

Fig. 4.20. Hartley oscillator circuit

In this circuit, there exists mutual inductance between coils L_1 and L_2 because one coils are wound on the same core, therefore their net effective inductance is increased by mutual inductance M. Hence, in this case, effective inductance is given by the expression

$$L = L_1 + L_2 + 2M$$

and resonant or oscillation frequency is given by the expression

$$f = \frac{1}{2\pi\sqrt{[C(L_1 + L_2 + 2M]}}$$

This expression will be derived below :

Frequency of Oscillation

The general equation derived earlier is reproduced here as under :

$$h_{ie}(Z_1 + Z_2 + Z_3) + Z_1 Z_2 (1 + h_{fe}) + Z_2 Z_3 = 0 \quad \text{...(4.29)}$$

For Hartley oscillator, we have

$$Z_1 = j\omega L_1 + j\omega M \text{ and } Z_2 = j\omega L_2 + j\omega M$$

and
$$Z_3 = \frac{1}{j\omega C} = -\frac{j}{\omega C}$$

Substituting all these values in general equation (4.29), we get

$$h_{ie}\left[(j\omega L_1 + j\omega M) + (j\omega L_2 + j\omega M - \frac{j}{\omega C}\right] + (j\omega L_1 + j\omega M)(j\omega L_2 + j\omega M)(1 + h_{fe})$$
$$+ (j\omega L_2 + j\omega M)\left(\frac{-j}{\omega C}\right) = 0$$

Simplifying, we get

$$j\omega h_{ie}\left[L_1 + L_2 + 2M - \frac{1}{\omega^2 C}\right] - \omega^2 (L_2 + M)\left[(L_1 + M)(1 + h_{fe}) - \frac{1}{w^2 C}\right] = 0 \quad ...(4.30)$$

Equating the imaginary part of above equation (4.30) to zero, we get

$$\omega h_{ie}\left[L_1 + L_2 + 2M - \frac{1}{\omega^2 C}\right] = 0$$

or $$L_1 + L_2 + 2M - \frac{1}{\omega^2 C} = 0 \qquad \text{since, } \omega\, h_{ie} \neq 0$$

or $$\omega^2 C = \frac{1}{L_1 + L_2 + 2M}$$

or $$f = \frac{\omega}{2\pi} = \frac{1}{2\pi\sqrt{C(L_1 + L_2 + 2M)}} \quad ...(4.31)$$

This equation (4.31) gives the frequency of oscillation.

Now, equating the real component of equation (4.30) to zero, we get

$$(L_1 + M)(1 + h_{fe}) - \frac{1}{\omega^2 C} = 0$$

or $$(1 + h_{fe}) = \frac{1}{\omega^2 C(L_1 + M)} = \frac{L_1 + L_2 + 2M}{L_1 + M} = 1 + \frac{L_2 + M}{L_1 + M}$$

or $$h_{fe} = \frac{L_2 + M}{L_1 + M} \quad ...(4.32)$$

As for other oscillator circuits, the loop gain must be greater than 1 to ensure that circuit oscillates.

Thus, we have $$A\,\beta \geq 1$$

Hence, we write $$A \geq \frac{1}{\beta} \geq \frac{L_1 + M}{L_2 + M} \quad ...(4.33)$$

Important Point: Hartley oscillator can also be used to generate RF signals. The frequency can be easily varied by varying the inductances which can be done by making the core movable. Another method of varying frequency is of varying capacitance. Hartley oscillator is not suitable for low frequency work since at low frequency, the value of inductance required becomes large.

4.16.1. Hartley Oscillator using FET

Figure 4.21 shows the Hartley oscillator using FET.

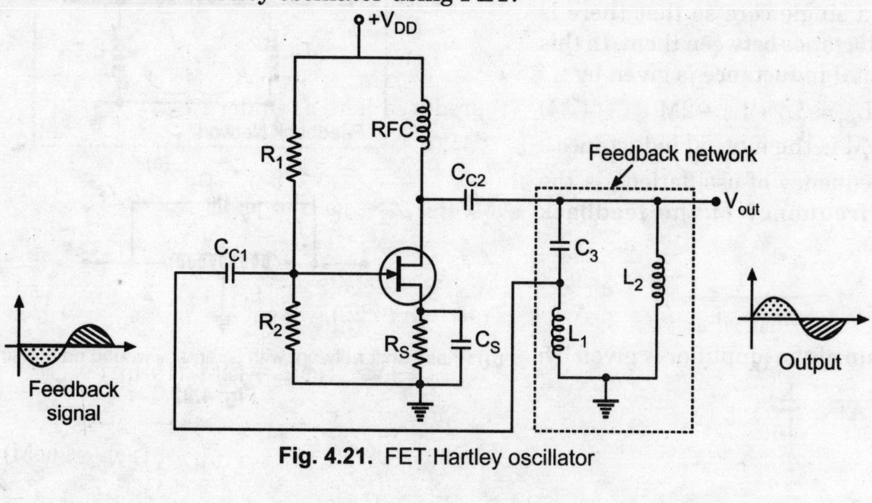

Fig. 4.21. FET Hartley oscillator

Circuit Description

(i) R_1, R_2 and R_3 are used to properly bias the n-channel JFET used here.

(ii) C_1 and C_2 are the coupling capacitors and C_S is the bypass capacitor.

(iii) The amplifier configuration is C_S hence the phase shift between input and output will be 180°. The feedback network provides the additional 180° phase shift required to satisfy the condition for positive feedback..

Expression for the frequency of oscillation

Here, we have,

$$\omega^2 = \frac{1}{(L_1 + L_2) C_3} = \frac{1}{\sqrt{(L_1 + L_2) C_3}}$$

But $\omega = 2\pi f$, so that

$$f = \frac{1}{2\pi\sqrt{C_3(L_1 + L_2)}} \text{ without mutual inductance.}$$

Note that we have not considered the effect of mutual inductance M between the two inductors L_1 and L_2. If the above inductors are not wound on the same core then the mutual inductance between them is zero and equation is valid.

However, if they are wound on the same core than we have to consider the mutual inductance M. The expression for oscillator frequency will then be given by

$$f = \frac{1}{2\pi\sqrt{C_3(L_1 + L_2 + 2M)}} \text{ with mutual inductance.}$$

4.16.2. Hartley Oscillator using Op-Amp

The circuit diagram of a Hartley oscillator using op-amp as an active device has been shown in figure 4.22. (a).

The op-amp is being used as an inverting amplifier. Hence, it introduces a phase shift of 180° between its input and output. The feedback network introduces 180° phase shift between its input and output. Thus, the total phase shift around the circuit will be 0°. Figure 4.22 (b) shows that L_1 and L_2 may be wound on a single core so that there is mutual inductance between them. In this case, the total inductance is given by :

$$L_{eq} = L_1 + L_2 + 2M \quad \dots (4.34)$$

where M is the mutual inductance.

The frequency of oscillations is the resonant frequency of the feedback network i.e.,

$$f = \frac{1}{2\pi\sqrt{C_3 L_{eq}}}$$

The gain of the amplifier is given by:

$$|A| \geq \frac{L_2}{L_1}$$

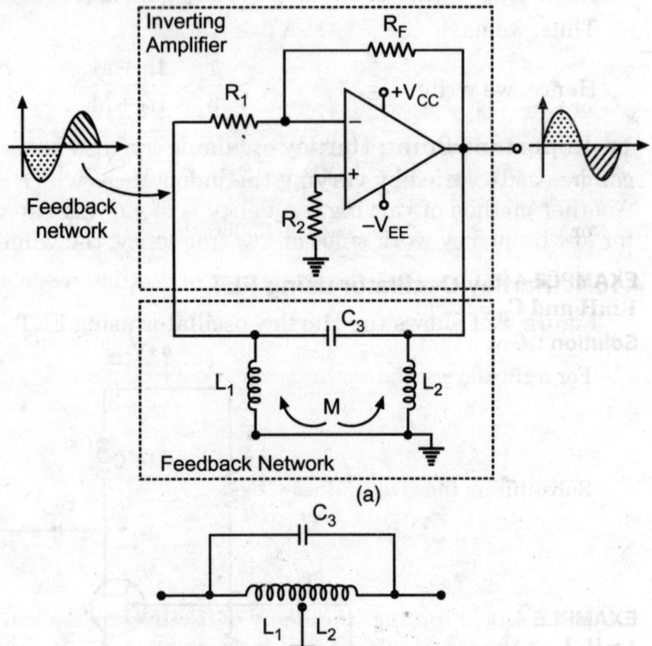

(b) Phase shift network with L_1 and L_2 wound on the single core.

Fig. 4.22.

If we consider the mutual inductance then above equation gets modified to :

$$|A| \geq \frac{L_2 + M}{L_1 + M}$$

Advantages of Hartley Oscillator

(i) It is easy to tune.

(ii) It can operate over a wide frequency range typically from few Hz to several MHz.

(iii) It is easy to change the frequency by means of a variable capacitor.

Applications

(i) It is used as local oscillator in radio and TV receivers.

(ii) In the function generators.

(iii) In RF sources.

Drawback

Poor frequency stability.

EXAMPLE 4.8. A Hartley oscillator circuit has a tank circuit inductance of $L_1 = L_2 = 100\ \mu H$. It is required to design the oscillator to produce oscillations at 50 kHz. Obtain the exact value of the tank circuit capacitance for the above said requirement.

Solution : Given that $L_1 = L_2 = 100\ \mu H$ and $f = 50$ kHz.

The oscillator frequency for a Hartley oscillator is given by :

$$f = \frac{1}{2\pi\sqrt{C_3(L_1 + L_2)}}$$

If $L_1 = L_2 = L$ then, we have

$$f = \frac{1}{2\sqrt{2}\pi\sqrt{LC_3}}$$

or $$f^2 = \frac{1}{8\pi^2 LC_3}$$

or $$C_3 = \frac{1}{8\pi^2 Lf^2} = \frac{1}{8\pi^2 \times 100 \times 10^{-6} \times (50 \times 10^3)^2}$$

or $C_3 = 0.05066\ \mu F$. **Ans.**

EXAMPLE 4.9. Calculate the frequency of oscillations of a Hartley oscillator having $L_1 = 0.5$ mH. $L_2 = 1$ mH and $C_3 = 0.2\ \mu F$.

Solution : Given that $L_1 = 0.5$ mH, $L_2 = 1$ mH and $C_3 = 0.2\ \mu F$.

For a Hartley oscillator, the frequency of oscillations is given by

$$f = \frac{1}{2\pi\sqrt{C_3(L_1 + L_2)}}$$

Substituting the given values we get,

$$f = \frac{1}{2\pi[0.2 \times 10^{-6}(0.5 + 1) \times 10^{-3}]^{1/2}} = 9188.8 \text{ Hz. } \textbf{Ans.}$$

EXAMPLE 4.10. Find the frequency of oscillation of a Hartley oscillator with component values $L_1 = 1$mH, $L_2 = 100\ \mu H$, $M = 50\mu H$ and $C = 100$pF.

Solution : Here, effective inductance, $L = L_1 + L_2 + 2M = 1 \times 10^{-3} + 100 \times 10^{-6} + 2 \times 50 \times 10^{-6} = 1{,}200\mu H$

Capacitance, $C = 100\text{pF} = 10^{-10}\text{F}$

Frequency of oscillation, $$f_0 = \frac{1}{2\pi\sqrt{LC}}$$

or $$f_0 = \frac{1}{2\pi\sqrt{1{,}200\times10^{-6}\times10^{-10}}} = 459 \text{ kHz} \quad \textbf{Ans.}$$

EXAMPLE 4.11. An amplifier with feedback has voltage gain equal to 40. To produce a specified output the input voltage required without feedback is 0.1 V. When feedback has been provided the input must be increased to 2.4 V to produce the same output. Find the value of feedback ratio.

Soution: Given that voltage gain with feedback,

$$A_f = 40$$

Voltage gain without feedback is given by

$$A = A_f \times \frac{\text{Input voltage required}}{\text{Input voltage required without feedback}} = 40 \times \frac{2.4}{0.1} = 960$$

Also, voltage gain with feedback,

$$A_f = \frac{A}{1-\beta A} \quad \text{where } \beta \text{ is the feedback ratio}$$

or feedback ratio, $$\beta = \frac{1-A/A_f}{A} = \frac{1-\frac{960}{40}}{960} = -0.024 \quad \textbf{Ans.}$$

4.17 CRYSTAL OSCILLATORS

(RGTU, Bhopal, Sem. Exam., 2005-06, 2006-07, 2007-08, 2008-09)

1. Principle of Crystal Oscillators

Certain materials such as quartz exhibit a unique property called *piezo electric* property. It states that if mechanical force is applied to a quartz crystal then it generates electric potential. Also if electric field is applied to a crystal it vibrates mechanically. If we apply mechanical vibrations to a quartz crystal then under proper operating conditions we can obtain electrical oscillations from it.

2. Frequency Stability

The biggest advantage of using a crystal oscillator is its high frequency stability. The frequency of a crystal oscillator remains stable inspite of changes in temperature, voltage, humidity or other parameters.

4.17.1. Equivalent Circuit of a Crystal

(RGTU, Bhopal, Sem. Exam., 2008-09, 2009-10)

The AC equivalent circuit of a crystal has been shown in figure 4.23 (b). It shows that the crystal is equivalent to a resonant circuit.

In the AC equivalent circuit of a vibrating crystal, the internal frictional losses are specified by a resistance R. Mass of the crystal and hence its inertia is represented by L and stiffness under the vibrating condition is represented by capacitor C. Due to the mounting arrangement shown in figure 4.23 (a), the crystal is equivalent to a capacitance denoted by C′ in the equivalent circuit. Here, C′ is known as the **mounting capacitance.**

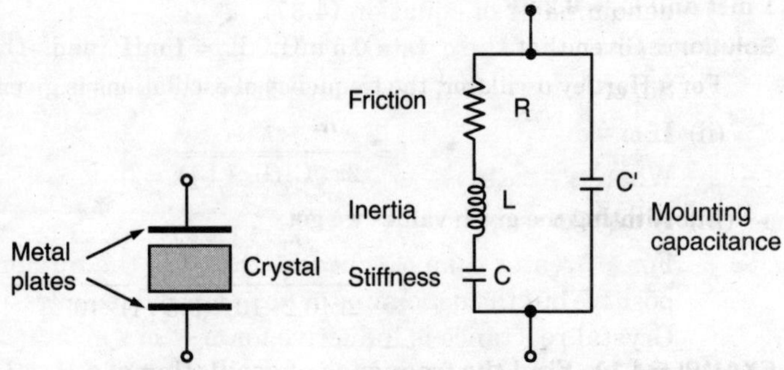

(a) Crystal (b) AC equivalent circuit of a crystal

Fig. 4.23.

4.17.2. Resonant Frequencies

There are two resonant circuits existing in the AC equivalent circuit of the crystal. RLC form a series resonant circuit and RLC in parallel with C′ will form a parallel resonant circuit. The resonant frequency of the series R–L–C series circuit is given by:

$$f_s = \frac{1}{2\pi\sqrt{LC}} \qquad ...(4.35)$$

This is with an assumption that quality factor Q is very large. The resonant frequency of the parallel resonant frequency formed by R–L–C and C′ is given by,

$$f_p = \frac{1}{2\pi\sqrt{LC_{eq}}} \qquad ...(4.36)$$

where, $$C_{eq} = \frac{CC'}{(C + C')}$$

The parallel resonant frequency f_p can be varied by varying the value of C′. The parallel resonant frequency f_p is always higher than the series resonant frequency f_s.

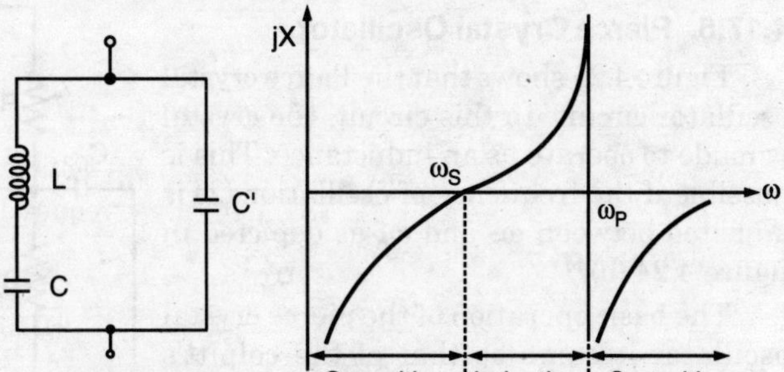

(a) Approximate ac equivalent circuit
(b) Variation in jX with ω

Fig 4.24.

4.17.3. Crystal Impedance

To obtain the expression for crystal impedance, let us use the approximate ac equivalent circuit of figure 4.24(a). The resistance R in the original equivalent circuit has been neglected. Due to the parallel connection, reactance of the crystal is given by

$$jX = \frac{\left[j\omega L + \frac{1}{j\omega C}\right]\frac{1}{j\omega C'}}{j\omega L + \frac{1}{j\omega C} + \frac{1}{j\omega C'}}$$

On simplifying this equation, we get

$$jX = \frac{-j}{\omega C'}\left[\frac{\omega^2 - \omega_s^2}{\omega^2 - \omega_p^2}\right] \qquad ...(4.37)$$

Here, ω is the frequency at which the crystal reactance is to be calculated.

(i) **If $\omega < \omega_S$**

If $\omega < \omega_S$, then, it is less than ω_p as well because $\omega_p > \omega_S$. Hence, both numerator and denominator of equation (4.37), i.e. $(\omega^2 - \omega_s^2)$ and $(\omega^2 - \omega_p^2)$ will be negative. Therefore, the reactance of the crystal i.e. X will be negative i.e. capacitive.

So, crystal reactances is capacitive for $\omega < \omega_S$.

(ii) **If $\omega = \omega_S$**

When $\omega = \omega_S$, $(\omega^2 - \omega^2{}_S) = 0$, hence, jX = 0

(iii) **If $\omega_S < \omega < \omega_P$**

For ω greater than ω_S but less than ω_P, the numerator $(\omega^2 - \omega_s{}^2)$ of equation (4.37) will be positive but the denominator $(\omega^2 - \omega_p^2)$ will be negative. Thus, jX will be positive i.e. inductive. Crystal reactance is inductive for $\omega_s < \omega < \omega_p$.

(iv) **If $\omega > \omega_p$**

Here, both $(\omega^2 - \omega_s^2)$ and $(\omega^2 - \omega_p^2)$ will be positive. Hence, jX will be negative i.e. capacitive.

(v) **If $\omega = \omega_p$**

At $\omega = \omega_p$, the denominator $(\omega^2 - \omega_p^2) = 0$. Hence, $jX = \infty$

(vi) The variation of jX with ω has been shown in figure 4.24 (b).

4.17.4. Types of Crystal Oscillators

The two types of crystal oscillators are:

(i) Pierce crystal oscillator

(ii) Miller crystal oscillator

4.17.5. Pierce Crystal Oscillator

Figure 4.25 shows that the Pierce crystal oscillator circuit. In this circuit, the crystal is made to operate as an inductance. This is possible if the frequency of oscillations ω is adjusted between ω_S and ω_p as depicted in figure 4.24 (b).

The basic operation of the pierce crystal oscillator is same as that of the colpitt's oscillator. RFC is a radio frequency choke which connects the DC supply to the circuit bus isolates the DC supply from the high frequency oscillations generated in the tank circuit (feedback network).

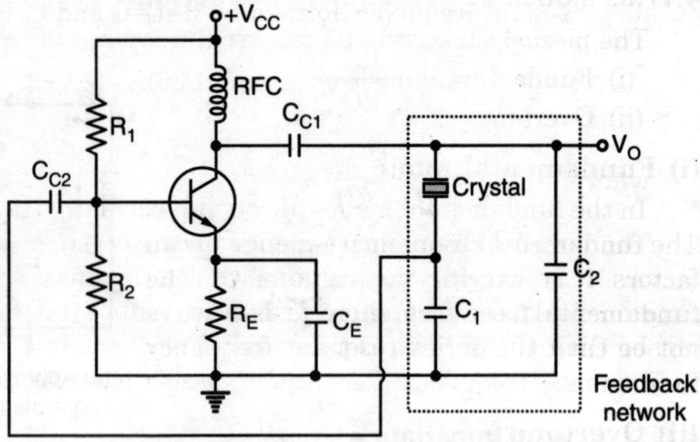

Fig. 4.25. Pierce crystal oscillator

4.17.6 Modified Pierce Crystal Oscillator

Sometimes, the external capacitors C_1 and C_2 are not used. Instead, the internal capacitances of the transistor or FET are used. Such a modified pierce crystal oscillator has been shown in figure 4.26. Both the versions, using a BJT and FET have been shown.

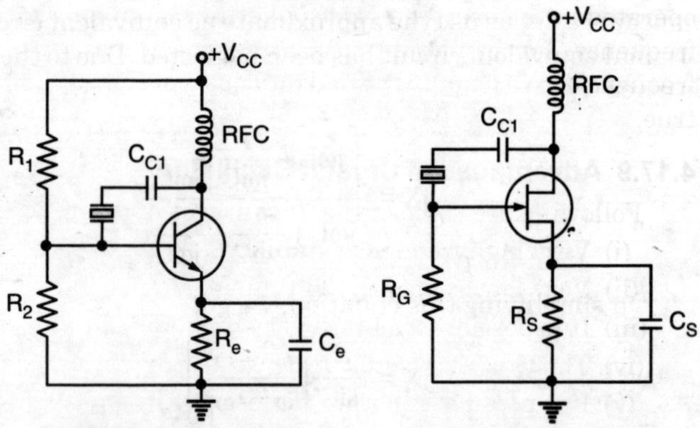

(a) Modified pierce crystal oscillator using transistor

(b) Modified pierce crystal oscillator using FET

Fig. 4.26.

4.17.7. Miller Crystal Oscillator

If the Hartley oscillator circuit is modified by replacing one of the inductors by a crystal, then, we get the Miller crystal oscillator configuration shown in figure 4.27.

The crystal acts like an inductor as the frequency of oscillations ω is adjusted between ω_s and ω_p as depicted in figure 4.24 (b). The Miller crystal oscillator using FET has been shown in figure 4.28.

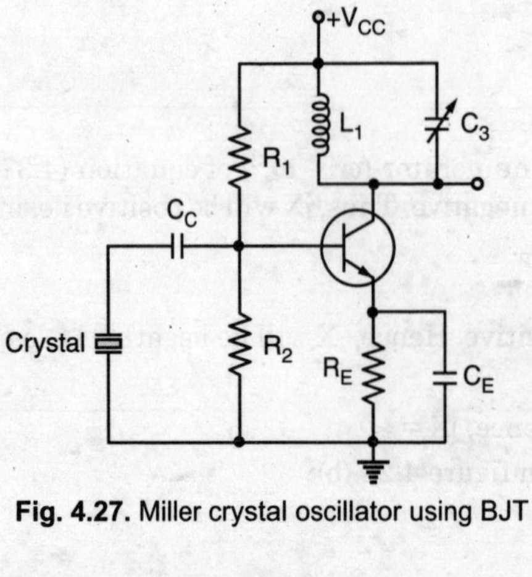

Fig. 4.27. Miller crystal oscillator using BJT

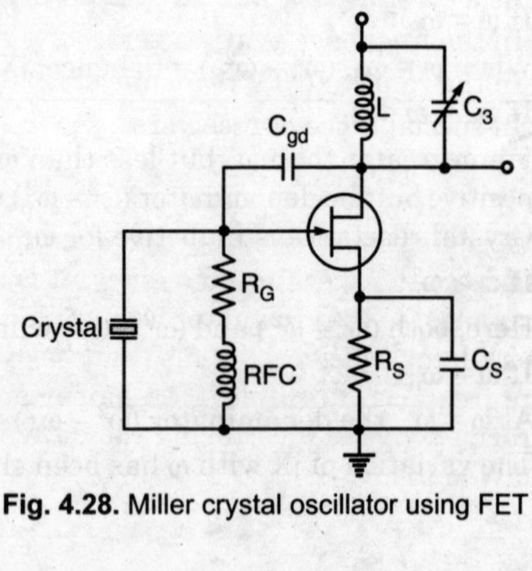

Fig. 4.28. Miller crystal oscillator using FET

C_{gd} is the gate to drain capacitance of FET. The biggest advantages of the pierce and Miller crystal oscillator is the frequency stability. The frequency will remain absolutely stable in spite of variations in the supply voltage or temperature, or the parameters of the active device.

4.17.8. Modes of Operation in the Crystal

The piezoelectric crystals can oscillate in one of the following two modes.

(i) Fundamental mode
(ii) Overtone

(i) Fundamental mode

In the fundamental mode, the crystal oscillator oscillates at the fundamental crystal frequency. The fundamental frequency depends on the crystal's mechanical dimensions, type of cut and other factors. It is inversely proportional to the thickness of the crystal slab. If we want to increase the fundamental frequency, then, we have to reduce the thickness of crystal slab. But, a crystal slab can not be cut too thin due to the possibility of fracturing. Therefore, there is an upper limit on the fundamental frequency. For most crystals the maximum fundamental frequency is about 20 MHz.

(ii) Overtone Mode

For higher frequencies than 20 MHz, we can not use the fundamental mode so the crystal is operated in the overtone mode. Overtones are approximately integer multiples of the fundamental frequency. For example twice, three times, four times the fundamental frequency. The overtone frequencies are usually the odd multiples of the fundamental frequency. However, this is not always true.

4.17.9 Advantages of Crystal Oscillator

Following are some of the advantages of crystal oscillator:

(i) Very high frequency stability.
(ii) Very low frequency drift due to change in temperature and other parameters.
(iii) It is possible to obtain very high, precise and stable frequency of oscillations.
(iv) The Q is very high.
(v) It is possible to obtain frequencies, higher than the fundamental frequency by operating the crystal in the overtone mode.

4.17.10. Drawbacks

(i) These are suitable for high frequency applications.
(ii) Crystals of low fundamental frequencies are not easily available.

4.17.11. Applications of Crystal Oscillators

(i) As a crystal clock in microprocessors.
(ii) In the frequency synthesizers.
(iii) In the radio and TV transmitters.
(iv) In special types of receivers.

Design Tips: Crystal oscillators must be designed to provide a load capacitance on the crystal as per specifications listed by the manufacturer. This requirement is essential for obtaining oscillations at the specified frequency. It is also important from the point of view of limiting the power supplied to the crystal to the specified maximum. Too much crystal power causes distortion in the oscillator waveform. It also causes overheating of the crystal, consequently rendering the resonant frequency unstable. More important is that the thin-plated electrodes may be melted off an overdriven crystal, destroying the device. Typical maximum drive levels for plated crystals varies from 2 mW to 10 mW.

The maximum permissible drive power limits the ac voltages which may be applied across the crystal and consequently affects the design of oscillator circuits. Crystal manufacturers usually specify the resistance of individual crystal, as well as maximum drive power. From these two, the maximum crystal ac voltage may be determined by using the relation $P = V^2/R$.

EXAMPLE 4.12. The parameters of a crystal oscillator equivalent circuit are given as; $L_s = 0.8$ H, $C_s = 0.08$ pF, $R_s = 5$ kΩ and $C_p = 1.0$ pF. Find the resonance frequencies f_s and f_p.

Solution : Given that series resistance,

$$R_s = 5\text{ k}\Omega = 5{,}000\ \Omega$$

Series inductance, $L_s = 0.8$ H

Series capacitance, $C_s = 0.08\text{ pF} = 8 \times 10^{-14}$ F

Parallel capacitance, C_p or $C_M = 1.0\text{ pF} = 1 \times 10^{-12}$ F

We know that

Series resonant frequency,

$$f_s = \frac{1}{2\pi\sqrt{L_s C_s}} = \frac{1}{2\pi\sqrt{0.8 \times 8 \times 10^{-14}}} = 629\text{ kHz} \quad \textbf{Ans.}$$

Also, Parallel resonant frequency,

$$f_p = \frac{1}{2\pi}\sqrt{\frac{1 + C_s/C_p}{L_s C_s}}$$

$$f_p = \frac{1}{2\pi}\sqrt{\frac{1 + 8 \times 10^{-14}/1 \times 10^{-12}}{0.8 \times 8 \times 10^{-14}}} = 654\text{ kHz} \quad \textbf{Ans.}$$

EXAMPLE 4.13. Given that a certain quartz crystal resonates at 450 kHz. It has an equivalent inductance of 4.2 H and an equivalent capacitance of 0.0297 pF. Its equivalent resistance is given to be 60 Ω. Determine its quality factor.

Solution: We know that

Quality factor $$Q = \frac{\omega L}{R} = \frac{2\pi f L}{R}$$

Substituting the given values, we get

$$Q = \frac{2 \times 3.14 \times 450 \times 10^3 \times 4.2}{600}$$

Solving, we get, $Q = 19{,}790$ **Ans.**

EXAMPLE 4.14. Given that $L_S = 800$ mH, $C_S = 0.01$ pF, $R_S = 1000$ Ω and $C_p = 20$ pF are the various values of an ac equivalent circuit of a piezoelectric crystal, calculate the values of f_s and f_p of the crystal.

Solution: We have $f_s = \dfrac{1}{2\pi\sqrt{L_S C_S}}$

Given that

$$L_S = 800\text{ mH} = 0.8\text{ H};$$
$$C_S = 0.01\text{ pF} = 0.01 \times 10^{-12}\text{ F};$$
$$C_p = 20\text{ pF}$$

therefore, $$f_s = \frac{1}{2\pi\sqrt{0.8 \times 0.01 \times 10^{-12}}} = 1779.40\text{ kHz} \quad \textbf{Ans.}$$

We also know that

$$f_p = \frac{1}{2\pi}\sqrt{\frac{1 + C_S/C_P}{L_S C_S}} \quad \textbf{Ans.}$$

Solving, we get $$f_p = \frac{1}{2\pi}\sqrt{\frac{\dfrac{1+0.01\times10^{-12}}{20\times10^{-12}}}{800\times10^{-3}\times0.01\times10^{-12}}} = 1779.85 \text{ kHz} \quad \textbf{Ans.}$$

Thus, the frequency will lie between 1779.40 to 1779.85 kHz if this crystal is used in the crystal oscillator.

EXAMPLE 4.15. A crystal has the following parameters:

$$\mathbf{L_S = 0.33\ H,\ C_S = 0.065\ pF,\ C_p = 1.0\ pF\ and\ R_s = 5.5\ K}$$

(i) Find the series-resonant frequency.

(ii) By what percent does the parallel-resonant frequency exceed the series-resonant frequency?

(iii) Find the Q factor of the crystal.

Solution: Given that $L_S = 0.33$ H

$C_S = 0.065$ pF

$C_p = 1.0$ pF

$R_s = 5.5$ K

(i) We know that series-resonant frequency is given by

$$f_S = \frac{1}{2\pi\sqrt{L_S C_S}}$$

Substituting all the values, we get

$$f_S = \frac{1}{2\times3.14\sqrt{0.33\times0.065\times10^{-12}}} = \frac{1}{2\times3.14\times0.146\times10^{-6}} = 1.087\times10^6 \text{ Hz}$$

Solving, we obtain

$f_S = 1087$ kHz **Ans.**

(ii) Parallel resonant frequency is given by

$$f_p = \frac{1}{2\pi\sqrt{LC_T}}$$

where $$C_T = \frac{C_s C_p}{C_S + C_p} = \frac{0.065\times10^{-12}\times1.0\times10^{-12}}{(0.065+1.0)\times10^{-12}} = 0.06103\times10^{-12} \text{ F}$$

$$f_p = \frac{1}{2\times3.14\sqrt{0.33\times0.06103\times10^{-12}}} = \frac{1}{2\times3.14\times0.142\times10^{-6}} = 1.122\times10^6 \text{ Hz} = 1122 \text{ kHz}$$

Ans.

$$\%\text{ age} = \frac{1122-1087}{1087} = \frac{35}{1087} = 0.032 = 3.2\% \quad \textbf{Ans.}$$

(iii) Q factor of the crystal at f_S will be

$$Q = \frac{\omega_S L}{R} = \frac{2\pi f_S L}{R} = \frac{2\times3.14\times1087\times10^3\times0.33}{5.5\times10^3}$$

Solving, we get $Q = 409.58$ **Ans.**

Now, Q factor of the crystal at f_p will be

$$Q = \frac{\omega_p L}{R} = \frac{2\pi f_p L}{R} = \frac{2\times3.14\times1122\times0.33}{5.5} = 422.76 \quad \textbf{Ans.}$$

EXAMPLE 4.16. Determine the frequency of oscillations of a Clapp oscillator if the component values are as follows :

$$\mathbf{C_1 = 120\ pF,\ C_2 = 1.5\ nF,\ C_3 = 15\ pF\ \ and\ \ L_3 = 10\ \mu H.}$$

Solution : Given that $C_1 = 120$ pF,

$C_2 = 1.5$ nF,

$C_3 = 15$ pF,

$L_3 = 10$ μH.

Equivalent capacitance, $\frac{1}{C} = \frac{1}{C_1} + \frac{1}{C_2} + \frac{1}{C_3} = \frac{1}{120} + \frac{1}{1500} + \frac{1}{15}$

or $C = 13.2$ pF

The frequency of oscillations is given by,

$$f = \frac{1}{2\pi\sqrt{CL_3}} = \frac{1}{2\pi\sqrt{13.2\times10^{-12}\times10\times10^{-6}}} = 13.84 \text{ MHz.} \quad \textbf{Ans.}$$

If we calculate the approximate value of oscillator frequency then let us assume that

$$C = C_3 = 15 \text{ pF}$$

Hence, $$f = \frac{1}{2\pi\sqrt{C_3 L_3}} = \frac{1}{2\pi\sqrt{15\times10^{-12}\times10\times10^{-6}}} = 12.99 \text{ MHz.} \quad \textbf{Ans.}$$

Thus, the oscillator frequency is decided by the additional capacitor C_3.

EXAMPLE 4.17. A quartz crystal has following constants:

$$\mathbf{L = 0.06\ H, \quad C_1 = 0.01\ pF, \quad C_2 = 10\ pF \quad and \quad R = 500\ \Omega.}$$

Calculate the values of series and parallel resonant frequencies.

Solution : Here, C_1 is the capacitor in the equivalent circuit of the crystal and C_2 is the mounting capacitance.

(i) Frequency of series resonance,

$$f_S = \frac{1}{2\pi\sqrt{LC_1}} = \frac{1}{2\pi\sqrt{0.06\times0.01\times10^{-12}}} = 6.497 \text{ MHz.} \quad \textbf{Ans.}$$

(ii) Frequency of parallel resonance, $f_P = \frac{1}{2\pi\sqrt{LC_{eq}}}$

where $$C_{eq} = \frac{C_1C_2}{(C_1+C_2)} = \frac{0.01\times10}{10.01} = 9.99\times10^{-15} \text{ F.}$$

Substituting this, we get,

$$f_P = \frac{1}{2\pi\sqrt{9.99\times10^{-15}\times0.06}} = 6.5007 \text{ MHz.} \quad \textbf{Ans.}$$

EXAMPLE 4.18. What is the frequency of oscillations of a Hartley oscillator if the component values are $L_1 = L_2 = 50\ \mu H$ and $C_3 = 100$ pF.

Solution : The frequency of oscillations is given by :

$$f = \frac{1}{2\pi\sqrt{(L_1+L_2)C_3}} = \frac{1}{2\pi\sqrt{100\times10^{-6}\times100\times10^{-12}}} = 1.59 \text{ MHz.} \quad \textbf{Ans.}$$

EXAMPLE 4.19. Find the frequency of oscillations of a Colpitt's oscillator if $C_1 = 120$ pF, $C_2 = 1.5$ nF and $L_3 = 10\ \mu H$.

Solution : For a Colpitt's oscillator, the equivalent capacitance C is given by

$$C = \frac{C_1C_2}{C_1+C_2} = \frac{120\times10^{-12}\times1.5\times10^{-9}}{(120\times10^{-12}+1.5\times10^{-9})} = 111.11 \text{ pF}$$

The frequency of oscillation is given by,

$$f = \frac{1}{2\pi\sqrt{L_3C}} = \frac{1}{2\pi\sqrt{10\times10^{-6}\times111.11\times10^{-12}}} = 4.77 \text{ MHz.} \quad \textbf{Ans.}$$

EXAMPLE 4.20. For the values of components of previous example calculate the frequency of oscillations for a Clapp oscillator. The additional capacitor $C_3 = 15$ pF.

Solution : Given that $C_1 = 120$ pF, $C_2 = 1.5\ \mu$F, $C_3 = 15$ pF, $L_4 = 10\ \mu$H.

Equivalent capacitance , $\frac{1}{C} = \frac{1}{C_1} + \frac{1}{C_2} + \frac{1}{C_3} = \frac{1}{120} + \frac{1}{1500} + \frac{1}{15}$

or $\quad C = 13.2$ pF.

The frequency of oscillation is given by,

$$f = \frac{1}{2\pi\sqrt{CL_3}} = \frac{1}{2\pi\sqrt{13.2\times10^{-12}\times10\times10^{-6}}} = 13.84 \text{ MHz.} \quad \textbf{Ans.}$$

4.17.12. Performance Comparison of Various LC oscillators

Table 4.1 gives the performance comparison of various LC oscillators.

Table 4.1. Performance Comparison of different LC oscillators

S. No.	Parameter of comparison		Hartley	Colpitt's	Clapp
1.	Circuit diagram		Refer Fig.	Refer Fig.	Refer Fig.
2.	Feedback network Composition	Z_1	Inductor	Capacitor	Capacitor
		Z_2	Inductor	Capacitor	Capacitor
		Z_3	Capacitor	Inductor	Inductor
		C_3	Absent	Absent	Present
3.	Frequency of oscillations		$f = \frac{1}{2\pi\sqrt{(L_1 + L_2) C_3}}$	$f = \frac{1}{2\pi\sqrt{LC}}$ $C = \frac{C_1 C_2}{C_1 + C_2}$	$f = \frac{1}{2\pi\sqrt{LC_{eq}}}$ $\frac{1}{C_{eq}} = \frac{1}{C_1} + \frac{1}{C_2} + \frac{1}{C_3}$
4.	Advantages		Suitable at HF. Small size and low cost.	Suitable at HF. Small size and low cost.	Suitable at HF. Small size and low cost and good frequency stability.
5.	Disadvantages		Poor frequency stability		Additional capacitor needed.

4.17.13. Performance Comparison Between LC Oscillators and Crystal Oscillators

Table 4.2. Performance Comparison between LC oscillators and crystal oscillators

S.No.	Crystal oscillator	LC oscillator
1.	Frequency of oscillations depends on the dimensions of crystal.	Frequency of oscillations is dependent on values of L and C.
2.	Accuracy depends only on fine cut of the crystal.	Accuracy mainly depends on tolerances of L and C.
3.	Q is very high, and it is stable.	Q is less as compared to the crystal.
4.	Miller crystal oscillator, pierce crystal oscillator are the examples of crystal oscillator	Hartley, Colpitt's and Clapp oscillator are the examples of LC oscillator.

4.18 AUDIO OSCILLATORS

Upto this point, we have discussed the oscillators that use L-C tuned circuit which causes a phase shift of 180° due to inductive or capacitive coupling in addition to a 180° phase shift produced by the transistor itself. These oscillators employing L-C elements, called the L-C oscillators, are very popular for generating high frequency oscillations but they cannot be employed for generation of low frequency oscillations because they become too bulky and expensive. R-C oscillators are commonly used for generating audio-frequencies as they provide good frequency stability and waveform. Further, with the advent of IC technology, R-C network is the only feasible solution, as it is very difficult to make a too high value inductance in an integrated circuit. Two commonly used R-C oscillators are as under :

(i) R-C phase shift oscillator, and
(ii) Wien bridge oscillator.

Another commonly used audio-oscillator is the beat frequency oscillator (BFO).

Now, let us discuss, in detail, these audio-oscillators in subsections to follow.

4.18.1. Basic Working Principles of R-C Oscillators

To produce oscillations in an oscillator circuit, we require positive feedback which means that the voltage signal feedback should be in phase with the input signal. To provide a positive feedback at one particular frequency, an inverting amplifier may be used with a feedback network which causes a phase shift of 180° at the desired frequency of oscillation, as shown in figure 4.29(a). The 180° phase shift in the feedback signal can be obtained by a suitable network, which consist of three R-C sections, as shown in figure 4.29(b).

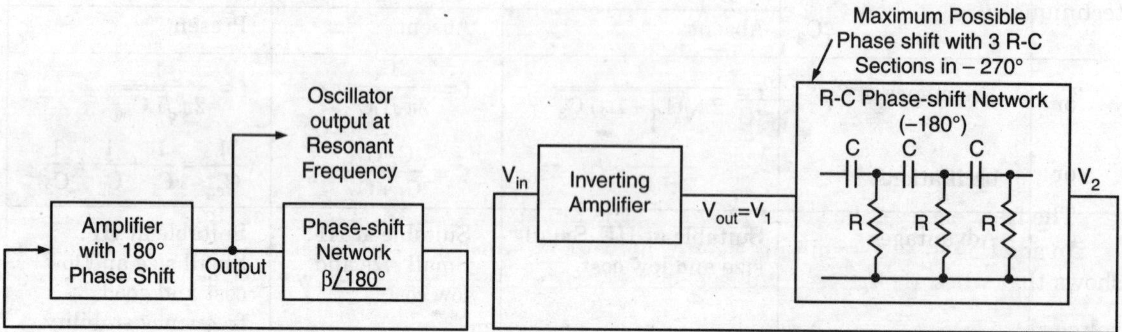

(a) (b)

Fig. 4.29.

When a phase shift network such as that indicated in figure 4.29 is used in phase shift oscillator, the values of R's and C's be selected so as to produce a phase shift of 180° at the desired frequency of oscillation. The output of the voltage amplifier is fed to the input to the phase-shift network. Hence, $V_1 = V_{out}$. The output resistance of the amplifier is designed to be very small in comparison to the input impedance of the phase-shift network. The output voltage of the phase-shift network, V_2 is fed into the input of the amplifier i.e,. $V_2 = V_{in}$. The amplifier's input impedance must be much larger than the output impedance of the phase-shift network.

Alternatively, a positive feedback can be obtained by using two stages of amplifiers each giving a phase shift of 180°. A part of this output is feedback to the input through a feedback network without causing any further phase shift. Wien bridge oscillator operates on this principle.

DO YOU KNOW?

The Armstrong oscillator uses a transformer to produce the feedback signal. The Hartley oscillator uses an inductive voltage divider to produce the feedback signal. The Clapp oscillator has a small series capacitor in the inductive branch of the resonant circuit. This reduces the effect that stray capacitance have on the resonant frequency.

4.19 FET PHASE SHIFT OSCILLATOR

Figure 4.30 shows a *phase shift oscillator*. This circuit consists of a common source FET amplifier which is followed by a three section R-C phase shift network. The amplifier stage is self-biased with a capacitor bypassed source resistor R_S and a drain bias resistance R_D. The output of the last section is supplied back to the gate. If the loading of the phase-shift network on the amplifier is assumed to be negligible, a phase shift of 180° between the amplified output voltage V_o and the input voltage V_i at the gate is produced by the amplifier itself. The three-section R-C phase shift network produces an additional phase shift, which is a function of frequency and equals 180° at some frequency of operation. At this frequency the total phase shift from the gate around the circuit and back to gate will be exactly zero. This particular frequency will be the one at which the circuit will oscillate provided that the magnitude of the amplification is sufficiently large. In a FET phase-shift oscillator voltage series feedback [i.e., feedback voltage proportional to the output voltage V_{out} and supplied in series with the input signal at the gate] is used.

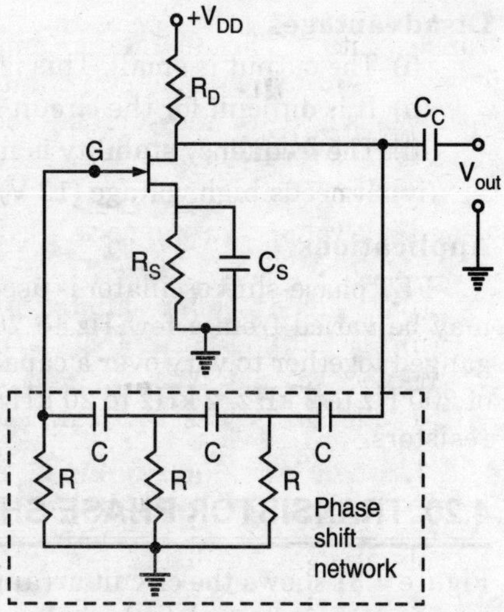

Fig. 4.30. Basic circuit of a FET phase shift oscillator.

The frequency of the oscillator output depends upon the values of capacitors C and resistors R used in the phase shift network. Using basic RC circuit analysis technique, we can show that the network phase shift is 180° when

$$X_C = \sqrt{6}\ R$$

or

$$\frac{1}{2\pi fC} = \sqrt{6}\ R$$

or

$$f = \frac{1}{2\pi RC\sqrt{6}} \qquad ...(4.38)$$

The frequency can be adjusted over a wide range if we use variable capacitors.

In addition to, phase shifting, the R-C network attenuates the amplifier output. Network analysis shows that when the necessary phase shift of 180° is obtained, this network attenuates the output voltage by a factor of $\frac{1}{29}$. This means that the amplifier must have a voltage gain of 29 or more.

When the amplifier voltage gain is 29 and feedback factor of R-C network, $\beta = \frac{1}{29}$ then the loop gain is $A\beta = \frac{1}{29} \times 29 = 1$, the amplifier phase shift of – 180° combined with the network phase shift + 180° gives a loop phase shift of zero. Both of these conditions are essential to satisfy the Barkhausen criteria.

If the amplifier gain is much greater than 29, the oscillator output waveform is likely to be distorted. When the gain is slightly greater than 29, the output is usually a reasonably pure sinusoidal.

Now, let us discuss the advantages and disadvantages of phase shift oscillators as under :

Advantages

(i) It is a cheap and simple circuit because it consists of resistors and capacitors which are not bulky and expensive like high-value inductors.

(ii) It provides good frequency stability.

(iii) The phase shift oscillator circuit is much simpler than the Wien bridge oscillator circuit because it does not need negative feedback and the stabilization arrangements.

(iv) The output is sinusoidal which is quite distortion free.

(v) They have a wide frequency range (from a few Hz to several hundred kHz).

(vi) They are particularly suitable for low frequencies, say of the order of 1 Hz, as these frequencies can be easily obtained by using R and C of large values.

Disadvantages

(i) The output is small. This is due to smaller feedback.

(ii) It is difficult for the circuit to start oscillations because the feedback is usually small.

(iii) The frequency stability is not as good as that of Wien bridge oscillator.

(iv) It needs high voltage (12 V) battery so as develop sufficiently large feedback voltage.

Applications

FET phase-shift oscillator is used to generate signals over a wide frequency range. The frequency may be varied from a few Hz to 200 Hz by employing one set of resistors with three capacitors ganged together to vary over a capacitance range in the 1 : 10 ratio. Similarly the frequency ranges of 200 Hz to 2 kHz, 2 kHz to 20 kHz and 20 kHz to 200 kHz may be obtained by using other sets of resistors.

4.20 TRANSISTOR PHASE SHIFT OSCILLATOR

Figure 4.31 shows the circuit arrangement of a phase-shift oscillator using N-P-N transistor in CE configuration. As usual, the voltage divider R_1-R_2 provides dc emitter base bias R_E and C_E combination provides temperature stability and prevent a.c. signal degeneration and collector resistor R_C controls the collector voltage. The oscillator output voltage is capacitively coupled to the load by coupling capacitor C_C.

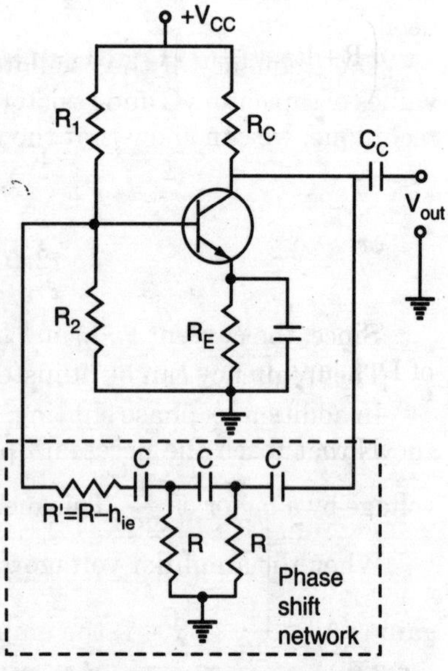

Fig. 4.31. Basic circuit of a transistor phase shift oscillator

In case of a transistor phase shift oscillator, the output of the feedback network is loaded appreciably by the relatively small input resistance (h_{ie}) of the transistor. Hence instead of employing voltage series feedback (as used in case of FET phase shift oscillator), we make use of voltage shunt feedback for a transistor phase shift oscillator, as shown in figure 4.30. In this circuit, the feedback signal is coupled through the feedback resistor R' in series with the amplifier stage input resistance h_{ie}. The value or R' should be such that when added with amplifier stage input resistance h_{ie}, it is equal to R i.e., $R' + h_{ie} = R$.

Working Principle

The circuit is set into oscillations by any random variations caused in base current which may be either due to noise inherent in the transistor or minor variations in voltage of d.c. power supply. This variation in base current is amplified in collector circuit. Now, this small output of the amplifier is supplied to an R-C feedback network. The R-C network produces a phase shift of 180° between output and input voltages. Since CE amplifier produces a phase reversal of the input signal, total phase shift becomes 360° or 0° which is essential for regeneration or for sustained oscillations. Hence, the output of this network is in the same phase as the originally assumed input to the amplifier and is applied to the base terminal of the transistor. Therefore, sustained variation in collector current between saturation and cut-off values are obtained. R-C phase shift network is the frequency determining network, as already explained in case of FET phase-shift oscillator.

Evaluation of Frequency of Oscillation

Figure 4.32(a) shows the equiva-lent circuit for the analysis of a transistor phase shift oscillator. This equivalent circuit, shown in figure 4.32(a), can be simplified if the following assumptions are made:

(i) h_{re} of the transistor is negligibly small and, therefore, $h_{re}V_{out}$ is omitted from the circuit.

(ii) h_{oe} of the transistor is very small i.e., $\frac{1}{h_{oe}}$ is much larger than R_C. Hence, the effect of h_{oe} can be neglected.

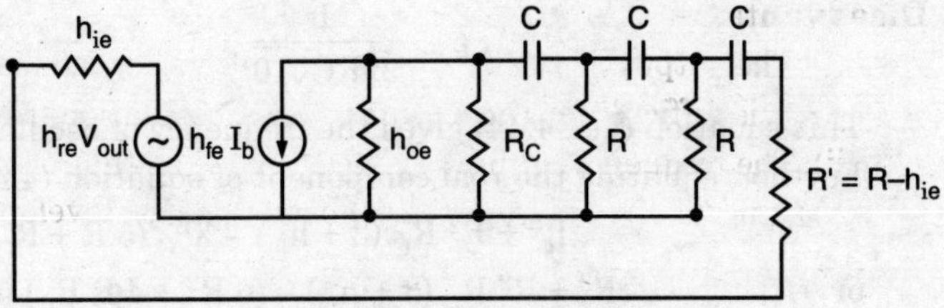

Fig. 4.32 (a). Equivalent circuit of transistor phase-shift oscillator circuit

Making use of these assumptions and replacing current source by equivalent voltage source, the simplified equivalent circuit will be as shown in figure 4.32(b).

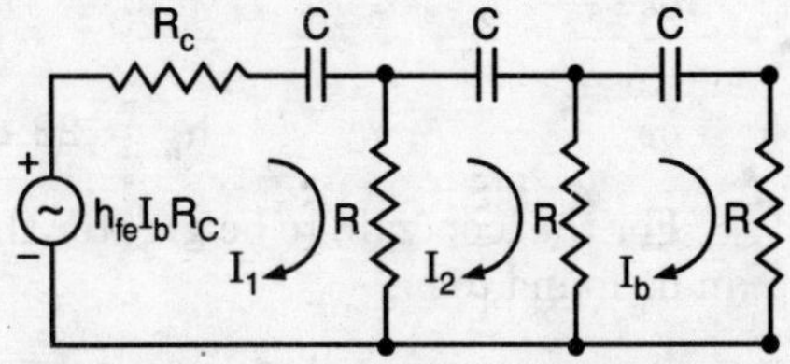

Fig. 4.32 (b). Simplified equivalent circuit for the equivalent circuit shown in figure 4.32(a).

Now, applying Kirchoff's voltage law to the three loops shown in figure 4.32(b), we get

$$\left(R + R_C + \frac{1}{j\omega C}\right)I_1 - RI_2 + h_{fe}\, I_b\, R_C = 0 \quad \text{...(4.39)}$$

$$-RI_1 + \left(2R + \frac{1}{j\omega C}\right)I_2 - RI_b = 0 \quad \text{...(4.39(a))}$$

$$0 - RI_2 + \left(2R + \frac{1}{j\omega C}\right)I_b = 0 \quad \text{...(4.40)}$$

Since, the current I_1, I_2 and I_b and non-vanishing, therefore, the determinant of the coefficients of I_1, I_2 and I_b must be zero.

Now, substituting $\frac{1}{\omega C} = X_C$, we get

$$\begin{vmatrix} (R + R_C - jX_C) & (-R) & (h_{fe}R_C) \\ (-R) & (2R - jX_C) & (-R) \\ 0 & (-R) & (2R - jX_C) \end{vmatrix} = 0 \quad \text{...(4.41)}$$

or $$(R + R_C - j\,X_C)\,[(2R - jX_C)^2 - R^2] + R\,[(-R)\,(2R - jX_C) - h_{fe}\,R_C\,(-R)] = 0$$

or $$(R + R_C - j\,X_C)\,(3\,R^2 - X^2{}_C - j4\,RX_C) - R\,[2R^2 - j\,RX_C - h_{fe}\,R_C\,R] = 0$$

or $$R^3 + R^2\,R_C(3 + h_{fe}) - 5\,R\,X^2{}_C - R_C\,X^2{}_C - 6\,jR^2\,X_C - j\,4\,R\,R_C\,X_C + jX^3{}_C = 0 \quad \text{...(4.42)}$$

Equating the imaginary component of this equation to zero, we get

$$6\,R^2\,X_C + 4\,R\,R_C\,X_C - X^3{}_C = 0$$

or $$X_C = \sqrt{6R^2 + 4R\,R_C}$$

or $$2\pi fC = \frac{1}{\sqrt{6R^2 + 4R\,R_C}} \qquad \left[\text{Since, } X_C = \frac{1}{\omega C} = \frac{1}{2\pi fC}\right]$$

or $$f = \frac{1}{2\pi RC\sqrt{6 + \frac{4R_C}{R}}} \quad \text{...(4.43)}$$

If $R = R_C$, then, we have

$$f = \frac{1}{2\pi RC\sqrt{10}} \qquad ...(4.44)$$

This equation (i.e., 4.44) gives the frequency of oscillation.

Further, equating the real component of equation (4.42) to zero, we get

$$R^3 + R^2 R_C (3 + h_{fe}) - X^2_C (5R + R_C) = 0$$

or $$R^3 + R^2 R_C (3 + h_{fe}) - (6R^2 + 4RR_C)(5R + R_C) = 0$$

or $$-29R^3 - 23R^2 R_C + h_{fe} R^2 R_C - 4RR^2_C = 0$$

or $$\frac{-29R}{R_C} - 23 + h_{fe} - 4\frac{R_C}{R} = 0$$

or $$h_{fe} = 23 + 29\frac{R}{R_C} + 4\frac{R_C}{R} \qquad ...(4.45)$$

For the loop gain to be greater than unity, the requirement of the current gain of the transistor can be found to be

$$h_{fe} > 23 + 29\frac{R}{R_C} + 4\frac{R_C}{R} \qquad ...(4.46)$$

If $R = R_C$, then, we have

$$h_{fe} > (23 + 29 + 4) \text{ i.e., } > 56 \qquad ...(4.47)$$

Main Applications area

The phase shift oscillator is well suited to the range of frequencies from several hertz to several hundred kilohertz (20 Hz to 200 kHz), and hence includes the audio frequency range (upto 20 kHz). To generate different audio-frequencies, variable air capacitors are employed as circuit elements in the phase-shift network. It is possible to vary the frequency in the range of about 1 : 10 because the range of capacitors can be varied in the ratio of 10 : 1 (typically from 40 pF to 450 pF).

For variations of frequency over a large range, the three capacitors are usually ganged so as to vary the capacitance of the three capacitors simultaneously. Such a variation keeps the input impedance to the phase-shift network constant and also keeps constant the magnitudes of β and $A\beta$. Thus, the amplitude of oscillations will remain unaffected as the frequency is adjusted. The phase-shift oscillator is operated in class A so as to keep distortion to the minimum. Frequency range from 20 Hz to 200 Hz, 200 Hz to 2 kHz. 2 kHz to 20 kHz and 20 kHz to 200 kHz may be obtained by using different set of resistors.

Phase-shift oscillators are not suitable for higher frequency operation because at higher frequency, the internal phase shift of the transistor and reduction in h_{fe} cause difficulties in designing the circuit. The frequency of the oscillator cannot be changed easily.

EXAMPLE 4.21. In a phase shift oscillator shown in figure 4.30, $R_1 = R_2 = R_3 = 800$ kΩ and $C_1 = C_2 = C_3 = 100$ pF. Find the frequency of oscillation.

Solution :

$$R_1 = R_2 = R_3 = R = 800 \text{ k}\Omega = 8 \times 10^5 \ \Omega$$

$$C_1 = C_2 = C_3 = C = 100 \text{ pF} = 10^{-10} \text{ F}$$

Frequency of oscillation, $$f_0 = \frac{1}{2\pi RC\sqrt{6}} = \frac{1}{2\pi \times 8 \times 10^5 \times 10^{-10} \times \sqrt{6}} = 812 \text{ Hz} \quad \textbf{Ans.}$$

EXAMPLE 4.22. The gain of the amplifier as a function of frequency is given as $A(j\omega) = \dfrac{-16\times10^6}{j\omega}$. A feedback path is connected around it having a feedback ratio $\beta(j\omega) = \dfrac{10^3}{[2\times10^3 + j\omega]^2}$. Determine:

(i) Will the system oscillate ?

(ii) If so what is the oscillation frequency ?

Solution : For sustained oscillation, we have

$$|A(j\omega)\,\beta(j\omega)| = 1$$

Now,
$$A(j\omega)\,\beta(j\omega) = \left[\frac{-16\times10^6}{j\omega}\right]\left[\frac{10^3}{(2\times10^3+j\omega)^2}\right]$$

or
$$A(j\omega)\,\beta(j\omega) = \frac{-16\times10^9}{j\omega[4\times10^6+j4\omega\times10^3-\omega^2]}$$

or
$$A(j\omega)\,\beta(j\omega) = \frac{-16\times10^9}{j[4\omega\times10^6-\omega^3]-4\omega^2\times10^3}$$

or
$$A(j\omega)\,\beta(j\omega) = \frac{16\times10^9}{4\omega^2\times10^3-j(4\omega\times10^6-\omega^3)}$$

Rationalizing the above term, we have

$$A(j\omega)\,\beta(j\omega) = \frac{16\times10^9[4\omega^2\times10^3+j(4\omega\times10^6-\omega^3)]}{(4\omega^2\times10^3)^2+(4\omega\times10^6-\omega^3)^2} \quad ...(i)$$

Again, for sustained oscillation, we have

$$A(j\omega)\,\beta(j\omega) = -1+j.0 \quad ...(ii)$$

Comparing (i) and (ii) we get that imaginary part in (i) must be zero for sustained oscillation

Thus, $4\omega\times10^6-\omega^3 = 0$ ...(iii)

or $\omega^2 = 4\times10^6$

or $\omega = 2\times10^3$ rad\sec

Therefore, frequency of oscillation is given by

$$f = \frac{2\times10^3}{2\pi} = 318\text{ Hz}$$

Using equations (iii) and (i), we get

$$A(j\omega)\,\beta(j\omega) = \frac{16\times10^9\times4\omega^2\times10^3}{(4\omega^2\times10^3)^2} = 1$$

Therefore, it is obvious that the system will oscillate. **Hence Proved.**

4.21 INTRODUCTION TO WIEN BRIDGE CIRCUIT

Figure 4.33(a) shows the Wien bridge which is used as the feedback network in wien bridge oscillator and figure 4.33 (b) shows the basic Wien Bridge oscillator. The amplifier in figure 4.33 (b) can use a transistor or FET or an OP-AMP as amplifying device. The Wien bridge of figure 4.33 (a) has four arms. The arm AD which consists of the series combination of R_1 and C_1 and the arm CD which consists of the parallel combination of R_2 and C_2 are called as the frequency sensitive arms. This is because the components connected in the arms decide the oscillator frequency. The resistors R_3 and R_4 are used to generate a reference voltage which remains constant independent of frequency. The ac input voltage is applied between points A and C of the bridge. When the Wien bridge is used in the oscillator circuit, the feedback voltage is applied between these points as shown in figure 4.33 (b).

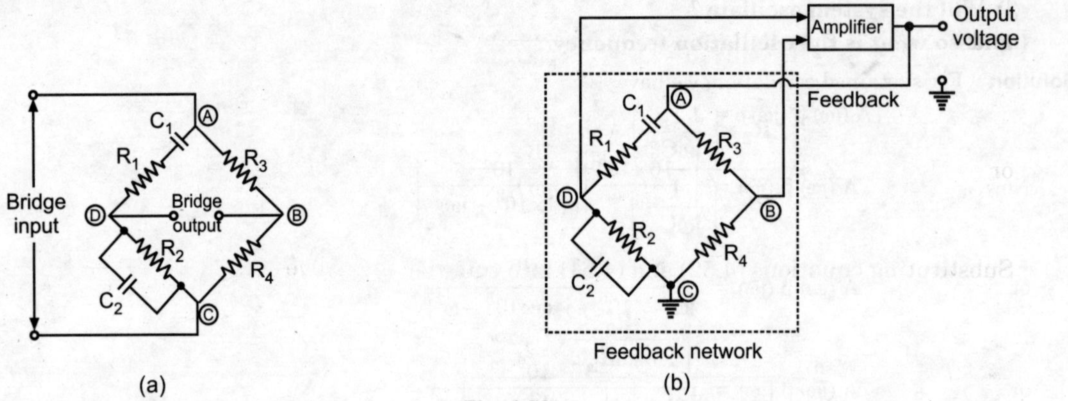

Fig. 4.33.

The a.c. output of the bridge is obtained between points B and D of the bridge. When used in the oscillator circuit, these points are connected to the input of the amplifier as shown in figure 4.33 (b). Wien bridge circuit of figure 4.33 (a) is used as a feedback network in the Wien bridge oscillator circuit as shown in figure 4.33 (b). It is also called the as **Lead-Lag Network** At low frequencies, it acts like a lead network whereas at high frequencies, it acts like a lag network. But, the phase shift introduced at the output frequency is 0°.

> **DO YOU KNOW?**
>
> Wien-bridge oscillator finds extensive use in commercial audio generators in the frequency range from a few Hz to 1 MHz.

4.21.1. Expressions for Feedback Factor (β) and Frequency (f)

(RGTU, Bhopal, Sem. Exam; 2006-07, 2008-09)

In order to obtain the expressions for the feedback factor β and the oscillator frequency f, we shall consider only the frequency sensitive arms AD and CD as shown in figure 4.34. In figure 4.34, it may be noted that V_{in} is actually the output voltage V_{out} which comes from the amplifier output and the voltage across R_2 C_2 acts as a feedback voltage V_F. The feedback factor or gain of the feedback network (β) is defined as :

$$\beta = \frac{V_F}{V_{in}} \quad \text{... (4.48)}$$

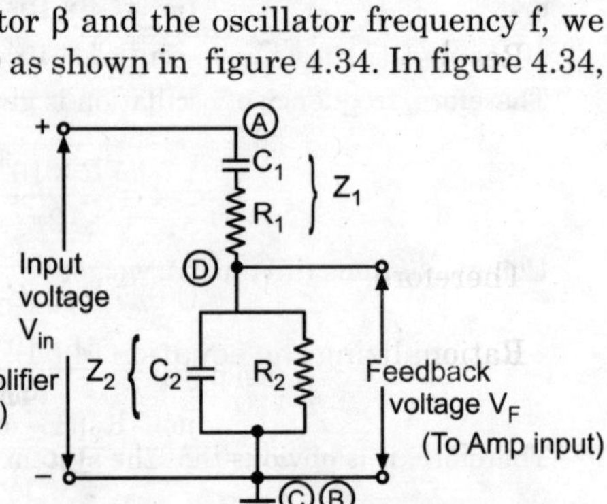

Fig. 4.34. Feedback network of Wien Bridge Oscillator

But, V_F is the voltage across the parallel combination of R_2 and C_2. Let the impedance of the arm AD which has R_1, C_1 in series be Z_1 and that of the arm DC which has R_2, C_2 in parallel be Z_2.

Thus, we have $$V_F = \frac{Z_2}{(Z_1 + Z_2)} \times V_{in} \quad \text{... (4.49)}$$

and $$\beta = \frac{V_F}{V_{in}} = \frac{Z_2}{(Z_1 + Z_2)} \quad \text{... (4.50)}$$

Now, let us obtain the expressions for Z_1 and Z_2

We have $$Z_1 = R_1 + \frac{1}{j\omega C_1} = \frac{1 + j\omega R_1 C_1}{j\omega C_1} \quad \text{... (4.51)}$$

and $$Z_2 = R_2 \,||\, X_{C2} = R_2 \,||\, \frac{1}{j\omega C_2}$$

or $$Z_2 = \frac{R_2 \times \left(\frac{1}{j\omega C_2}\right)}{R_2 + \left(\frac{1}{j\omega C_2}\right)} = \frac{R_2}{1 + j\omega R_2 C_2} \qquad \text{... (4.52)}$$

Substituting equations (4.52) and (4.51) into equation (4.50), we obtain

$$\beta = \frac{\left[\frac{R_2}{1 + j\omega R_2 C_2}\right]}{\left[\frac{1 + j\omega R_1 C_1}{j\omega C_1}\right] + \left[\frac{R_2}{1 + j\omega R_2 C_2}\right]}$$

Substituting $j\omega = s$, in the above expression, we obtain

$$\beta = \frac{\left[\frac{R_2}{1 + sR_2 C_2}\right]}{\left[\frac{1 + sR_1 C_1}{sC_1}\right] + \left[\frac{R_2}{1 + sR_2 C_2}\right]}$$

or $$\beta = \frac{sC_1 R_2}{[(1 + sR_1 C_1)(1 + sR_2 C_2)] + sR_2 C_1} = \frac{sC_1 R_2}{1 + s(R_1 C_1 + R_2 C_2) + s^2 R_1 R_2 C_1 C_2 + sR_2 C_1}$$

or $$\beta = \frac{sC_1 R_2}{1 + s(R_1 C_1 + R_2 C_2 + R_2 C_1) + s^2 R_1 R_2 C_1 C_2} \qquad \text{... (4.53)}$$

Resubstituting $s = j\omega$ and $s^2 = -j^2\omega^2 = -\omega^2$ into equation (4.53), we obtain

$$\beta = \frac{j\omega C_1 R_2}{1 + j\omega(R_1 C_1 + R_2 C_2 + R_2 C_1) - \omega^2 R_1 R_2 C_1 C_2}$$

Therefore, $$\beta = \frac{j\omega C_1 R_2}{(1 - \omega^2 R_1 R_2 C_1 C_2) + j\omega(R_1 C_1 + R_2 C_2 + R_2 C_1)} \qquad \text{... (4.54)}$$

Rationalizing the equation (4.54), we get

$$\beta = \frac{j\omega C_1 R_2[(1 - \omega^2 R_1 R_2 C_1 C_2) - j\omega(R_1 C_1 + R_2 C_2 + R_2 C_1)]}{(1 - \omega^2 R_1 R_2 C_1 C_2)^2 + \omega^2 (R_1 C_1 + R_2 C_2 + R_2 C_1)^2} \qquad \text{... (4.55)}$$

As mentioned earlier, the phase shift introduced by the Wien bridge circuit at the desired output frequency should be 0°. For that the imaginary part of equation (4.55) should have a zero value.

This means that

$$\omega C_1 R_2 (1 - \omega^2 R_1 R_2 C_1 C_2) = 0$$

or $$\omega^2 R_1 R_2 C_1 C_2 = 1$$

or $$\omega^2 = \frac{1}{R_1 R_2 C_1 C_2}$$

or $$\omega = \frac{1}{\sqrt{R_1 R_2 C_1 C_2}}$$

and frequency, $f = \dfrac{1}{2\pi\sqrt{R_1R_2C_1C_2}}$ ... (4.56)

This is the expression for the oscillator frequency.

Now, if we substitute $R_1 = R_2 = R$ and $C_1 = C_2 = C$ in the expression for the oscillator frequency, then equation (4.56) gets modified as follows :

Oscillator frequency, $f = \dfrac{1}{2\pi RC}$ ... (4.57)

Similarly, if we substitute $R_1 = R_2 = R$ and $C_1 = C_2 = C$ into the expression for β then, we have

Feedback factor $$\beta = \frac{\omega^2CR(3RC) + j\omega CR(1-\omega^2R^2C^2)}{(1-\omega^2R^2C^2)^2 + \omega^2(3RC)^2} \quad \text{... (4.58)}$$

This is the required expression for the feedback factor β.

Now, let us obtain the value of feedback factor β at the oscillator output frequency f. For that substituting $\omega = 1/RC$ into equation (4.58), we get

$$\beta = \frac{3}{0 + \frac{1}{C^2R^2}(3RC)^2} = \frac{3}{9} = \frac{1}{3} \quad \text{... (4.59)}$$

Thus, at the oscillator frequency f the value of the feedback β is 1/3.

Gain of the Amplifier for Sustained oscillations

According to Barkhausen criteria, the loop gain should be greater than 1, and phase shift around the loop should be zero. The positive sign of β in equation (4.59) shows that the phase shift introduced at frequency f is zero. To satisfy the other condition, i.e.,

$$|A\beta| \geq 1.$$

Substituting, $\beta = \frac{1}{3}$, we get

$$A \geq 3 \quad \text{... (4.60)}$$

Thus, the amplifier gain should be at least equal to 3 to ensure sustained oscillations.

4.21.2. Wien Bridge Oscillator using a Transistor

(RGTU, Bhopal, Sem. Exam; 2002-03) (05 marks)

Figure 4.35 shows the circuit diagram for a Wien bridge oscillator using a two stage transistor amplifier.

Note that two common emitter amplifiers are being connected in cascade so that the phase shift introduced by the amplifier is zero. The Wien bridge circuit has been shown in a dotted box. The input to this feedback network is the output voltage V_o coupled from the collector of transistor Q_2 through a coupling capacitor.

> **DO YOU KNOW?**
>
> This is the standard oscillator for low to moderate frequencies in the range of 5 Hz to 1 MHz. It produces an almost perfect sine wave. a tungsten lamp or other nonlinear resistance is used to decrease the loop gain to 1.

The voltage developed across the parallel R-C network (arm DC) is used as a feedback voltage. The expression for the oscillation frequency is given by

$$f = \frac{1}{2\pi RC} \quad \text{... (4.61)}$$

Hence, the frequency can be varied by varying either both resistors (R) or both capacitors (C) connected in the frequency sensitive arms of the Wien bridge. In figure 4.35, both the capacitors are ganged so that they can be varied simultaneously to vary the output frequency.

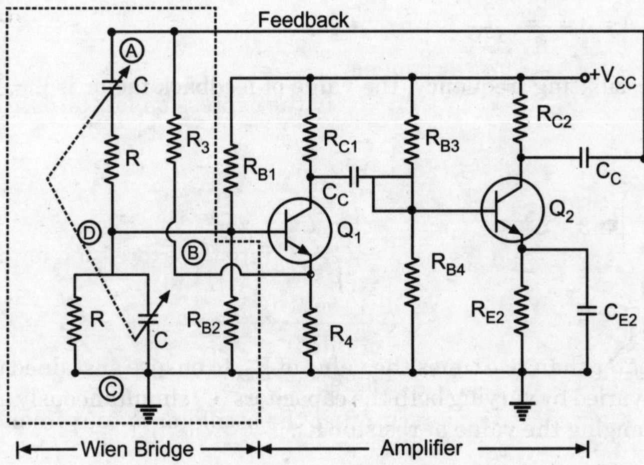

Fig. 4.35. Wien bridge oscillator using transistor

The resistor R_4 is a part of the Wien bridge as well as the amplifier as it serves as the unbypassed emitter resistor of the first stage of the amplifier.

Amplitude stabilization

We know that the required gain of the amplifier for sustained oscillations is $|A| \geq 3$.

Being a two stage amplifier, the amplifier of figure 4.35 can easily provide a much higher gain than 3. The gain of the amplifier should not be too high as it will distort the output voltage waveform of the oscillator. In order to avoid this possibility of distortion, the amplifier gain must be controlled. This is achieved by introducing a negative feedback by keeping resistor R_4 in the emitter of Q_1 unbypassed. This introduces a current series feedback. This process of gain reduction using negative feedback is known as **Amplitude stabilization**.

Another way of amplitude stabilization is to use a nonlinear resistor for R_4. This resistor is called as voltage dependent resistor (VDR). As the amplitude of oscillations increases, the value of R_4 will decrease and the current through it will increase. This will increase the amount of negative feedback and hence will reduce the gain automatically.

This will automatically reduce the amplitude of the output and avoid the waveform distortion.

4.21.3. Wien Bridge Oscillator Using OP-AMP

Figure 4.36 shows the Wien bridge oscillator using op-amp instead of transistor. The Op-Amp output is applied as an input voltage to the Wien bridge between points A and C. The output of the Wien bridge which acts as the feedback network is applied to the Op-Amp input between points D and B.

The R and C components in the frequency sensitive arms of the bridge will decide the oscillator frequency. The expression for oscillator frequency is given by

$$f = \frac{1}{2\pi RC} \qquad \text{... (4.62)}$$

The resistor R_3 gets connected in the feedback path of op-amp where as resistor R_4 gets connected from the inverting (–) terminal to ground. Hence, the amplifier configuration is known as the non-inverting amplifier. The gain of this configuration is given by

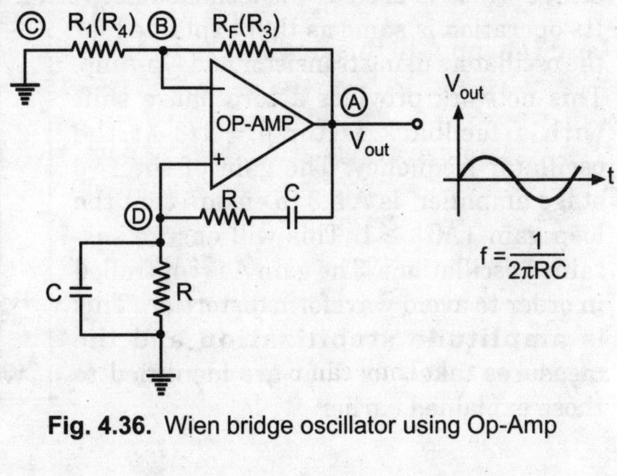

Fig. 4.36. Wien bridge oscillator using Op-Amp

$$A = 1 + \frac{R_3}{R_4}$$

We know that at the oscillating frequency, the value of feedback factor is β = 1/3 and the gain should be A ≥ 3.

Hence, we write

$$\left(1 + \frac{R_3}{R_4}\right) \geq 3$$

or $$\frac{R_3}{R_4} \geq 2 \quad \text{... (4.63)}$$

Thus, R_3 should be greater than two times the value of R_4, to ensure sustained oscillations. The oscillator frequency can be varied by varying both the capacitors (C) simultaneously and the amplifier gain can be changed by changing the value of resistor R_3.

Equivalent Circuit at f = f_0

At f = f_0 (the oscillator frequency), the Barkhausen criterion is satisfied and we obtain sustained sinusoidal oscillations at the output of the oscillator. Figure 4.37 shows the equivalent circuit of the wien bridge oscillator at f = f_0.

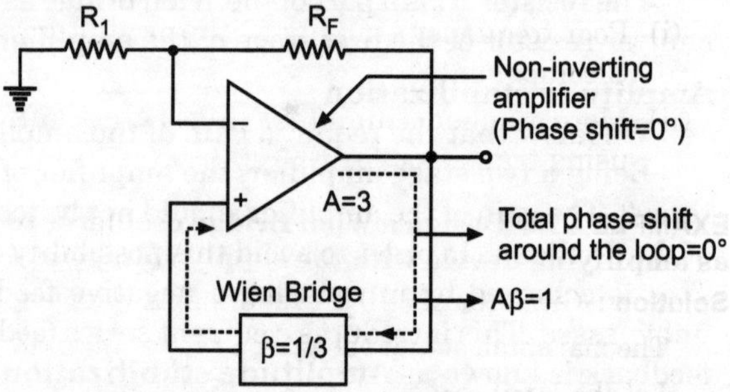

Fig. 4.37. Equivalent circuit of Wien bridge oscillator

Summary of Wien Bridge Oscillator

(i) Gain of the non-inverting amplifier is adjusted at A ≥ 3.

(ii) Wien bridge acts as the lead-lag RC feedback network.

(iii) At a particular frequency called frequency of oscillations, the phase shift introduced by the Wien Bridge is 0° and feedback factor β = 1/3.

(iv) Hence, the Barkhausen criterion is satisfied and we get sustained oscillations at the output..

4.21.4. Wien Bridge Oscillator using FET

Figure 4.38, shows the Wien Bridge oscillator using FET. The amplifier is a two stage RC coupled FET amplifier. The common source type FET amplifiers are cascaded together in order to ensure that the phase shift between the amplifier input and output is 0°. The resistors R_g connected from the gates of FET's to ground are connected for biasing the FETs. This is self biasing. The Wien bridge which is used as the feedback network is connected between the amplifier output and input. Its operation is same as that explained for the oscillators using transistor and Op-Amp. This network provides a zero phase shift with a feedback factor β = 1/3 at the oscillator frequency. The gain of the two stage amplifier is A ≥ 3 to ensure that the loop gain |Aβ| ≥ 1. This will ensure sustained oscillations. The gain A is controlled in order to avoid waveform distortion. This is **amplitude stabilization** and the measures taken for that are identified to those explained earlier.

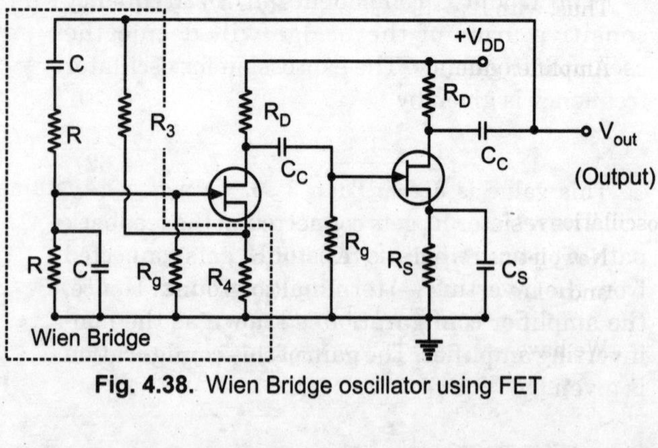

Fig. 4.38. Wien Bridge oscillator using FET

The oscillator frequency is $f = \frac{1}{2\pi RC}$

DO YOU KNOW?

The twin-T oscillator uses an amplifier and *RC* circuits to produce the required loop gain and phase shift at the resonant frequency. It works well at one frequency but is not suitable for an adjustable frequency oscillator. The phase-shift oscillator also uses an amplifier and RC circuits to produce oscillations. An amplifier can act like a phase-shift oscillator because of the stray lead and lag circuits in each stage.

Range of oscillator frequency of a Wien bridge oscillator

Basically, Wien bridge oscillator is an RC oscillator. It is suitable for operation in the low frequency region from 20 Hz to 100 kHz. But they are very popularly used in the audio frequency range. Wien bridge oscillator is used in most of the signal generators used in the laboratories.

Advantages of Wien Bridge Oscillator

(i) Variation in the oscillator frequency is easy to achieve because only two capacitors or resistors are to be simultaneously varied. In phase shift oscillator we have to change three of them at a time.

(ii) A very good quality of sine wave can be obtained at the oscillator output.

Drawbacks

(i) Poor frequency stability as the oscillator frequency is dependent only on the values of R and C components.

(ii) More number of amplifier stages are required to be used for the Wien bridge oscillator circuits using transistors or FETs.

EXAMPLE 4.23. Design a Wien Bridge oscillator to have an output frequency of 10 kHz. Use Op-Amp as amplifying device with the supply voltage of ± 15 V.

Solution : Given that f = 10 kHz and supply voltage ± 15 V.

The maximum output voltage $\cong \pm (V_{CC} - 1) = \pm 14$ V ... (i)

Let the input current drawn by the op-amp be negligible. Therefore current through R_3 is same as that through R_4. Let the current through R_3 and R_4 be I = 1 mA.

Hence, $(R_3 + R) = \frac{V_{out}}{I} = \frac{14}{1 \text{ mA}} = 14 \text{ k}\Omega.$ ...(ii)

From equation (4.63), $R_3 = 2 R_4$. Substituting this into equation (ii), we get

$2R_4 + R_4 = 14 \text{ k}\Omega$

Hence, $R_4 = 4.67$ kΩ (Use 4.7 kΩ) standard value). **Ans.**

And $R_3 = 2 \times 4.7 \text{ k}\Omega = 9.4 \text{ k}\Omega$

(Use 10 kΩ) standard value).**Ans.**

Thus, with $R_3 = 10$ kΩ and $R_4 = 4.7$ kΩ

Amplifier gain, $A = 1 + \frac{R_3}{R_4}$

$A = 1 + \frac{10}{4.7} = 3.12.$

This value is higher than 3 so, we will get sustained oscillations.

Now, let us consider $R_1 = R_2 = R$

and Let $R = R_4 = 4.7$ kΩ

We have $f = \frac{1}{2\pi RC}$

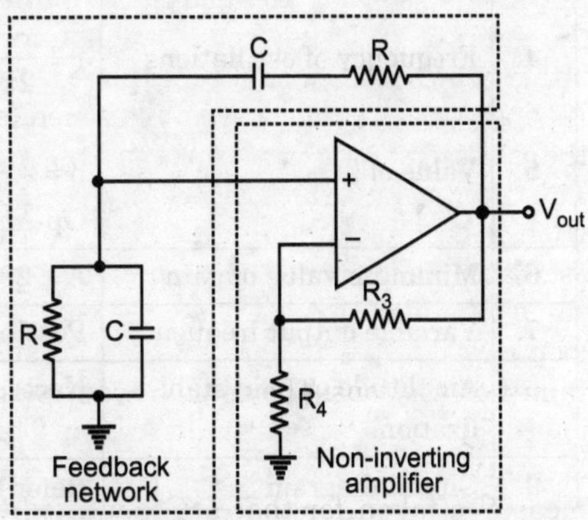

Fig. 4.39. Wien bridge oscillator using Op-Amp

or $$C = \frac{1}{2\pi fR}$$

$$= \frac{1}{2\pi \times 10 \times 10^3 \times 4.7 \times 10^3} = 3.38 \text{ nF.} \quad \textbf{Ans.}$$

Thus, the component values are

$$R_1 = R_2 = R = 4.7 \text{ k}\Omega \quad C_1 = C_2 = C = 3.38 \text{ nF}$$

$$R_3 = 10 \text{ k}\Omega \qquad R_4 = 4.7 \text{ k}\Omega \qquad \textbf{Ans.}$$

EXAMPLE 4.24. For the Op-Amp based Wien bridge shown in figure 4.39 of previous example, if the component values are as follows :

$R = 5.1$ kΩ,

$C = 1$ nF,

$R_3 = 12$ kΩ

and $R_4 = 5.1$ kΩ.

(i) Determine whether the circuit will oscillate or not.

(ii) Obtain the output frequency.

Solution : (i) For the Wien bridge circuit to oscillate, the amplifier gain should be greater than or equal to 3.

Thus, Gain $A = 1 + \frac{R_3}{R_4} = 1 + \frac{12}{5.1} = 3.35.$ **Ans.**

As $A > 3$, the circuit will oscillate.

(ii) Operating frequency, $f = \frac{1}{2\pi RC} = \frac{1}{2\pi \times 5.1 \times 10^3 \times 10^{-9}} = 31.2$ kHz. **Ans.**

4.21.5. Performance Comparison of RC Oscillators

S. No.	Parameter of Comparsion	Phase shift oscillator	Wien bridge oscillator
1.	Feedback network	Consists of three identical RC sections connected in cascade.	Uses Wien bridge circuit as feedback network.
2.	Phase shift introduced by the feedback network.	180° at frequency of oscillations.	0° at frequency of oscillations.
3.	Phase shift introduced by the amplifier.	180° at the frequency of oscillations.	0° at the frequency of oscillations.
4.	Frequency of oscillations	$f = \frac{1}{2\pi\sqrt{6}\,RC}$	$f = \frac{1}{2\pi RC}$
5.	Value of β	$\beta = -\frac{1}{29}$ for oscillator using Op-Amp.	$\beta = +\frac{1}{3}$ for oscillator using Op-Amp.
6.	Minimum value of gain	$A \geq 29$ for sustained oscillations	$A \geq 3$ for sustained oscillations.
7.	Variable output frequency	Possible but difficult	Possible and easy
8.	Amplitude or gain stabilization.	Necessary	Necessary.
9.	Circuit diagram	Refer to figure.	Refer to figure.

EXAMPLE 4.25. Design a Wien bridge oscillator using the op-amp IC 741 to produce sinusoidal output signal of 10 kHz. Use V_{CC} = ± 10 V.

Solution : For 741 IC, $I_{B(max)}$ = 500nA.

Assume current through R_1 = I_1 = 1000 $I_{B(max)}$ = 500 μ A

Maximum output voltage = ± 0.9 V_{CC} = ± 9 V

Thus $$(R_1 + R_F) = \frac{9V}{500\mu A} = 18\ k\Omega$$

Also, we have

$$R_F = 2R_1 \text{ for sustained oscillations.}$$

Hence, $$R_1 = \frac{18k\Omega}{3} = 6k\Omega \text{ (use 5.6 k}\Omega\text{ as standard value)}$$

Thus, $$R_F = 2R_1 = 2 \times 5.6\ k = 11.2\ k\Omega \quad \text{(use } R_F = 12k\Omega \text{ as standard resistor.)}$$

Choose $$R = R_1 = 5.6\ k\Omega$$

Then $$C = \frac{1}{2\pi \times 10 \times 10^3 \times 5.6 \times 10^3} = 2840\ pF.$$

(use C = 2700 pF as a standard capacitor)

Thus designed value of components are

$$R_1 = 5.6k\ \Omega,\ R_F = 12\ k\Omega, \quad \textbf{Ans.}$$

$$R = 5.6k\ \Omega,\ C = 2700\ pF. \quad \textbf{Ans.}$$

EXAMPLE 4.26. Design a BJT R-C phase shift oscillator for a frequency of 100 kHz. Assume for the BJT, h_{fe} = 100, h_{ie} = 1 kW at V_{CE} = 5V and I_C = 1 mA. Available V_{CC} = 20 V. Neglect the loading effect of biasing resistors.

Solution : The circuit is shown in figure 4.40.

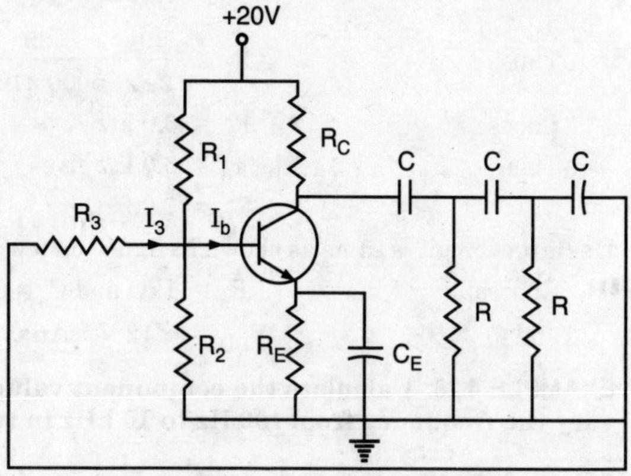

Fig. 4.40. Circuit for example 4.26.

Choose R_E = 1 kΩ, $I_E = I_C$ = 1 mA

Hence $$R_C = \frac{V_{CC} - V_{CE} - I_E R_E}{I_C}$$

or $$R_C = \frac{20 - 5 - 1\times10^{-3} \times 1\times10^3}{1\times10^{-3}}$$

or $$R_C = 14\ k\Omega$$

Choose $$R = 10\ k\Omega$$

Thus $$k = \frac{R_c}{R} = \frac{14}{10} = 1.4$$

and $$h_{fe} \geq \left(23 + \frac{29}{k} + 4k\right) \geq \left(23 + \frac{29}{14} + 4 \times 1.4\right)$$

Thus, $$h_{fe} \geq 49.3$$

Since given BJT has h_{fe} = 100, the gain of the transistor is sufficient for sustained oscillations.

$$\text{Oscillation frequency (f)} = \frac{1}{2\pi RC\sqrt{6+4k}}$$

$$100 \times 10^3 = \frac{1}{2\pi \times 10^4 \times C \times \sqrt{6 + 4 \times 1.4}}$$

Hence, $$C = \frac{1}{2\pi \times 10^4 \times 10^5 \times \sqrt{11.6}} = 66.7\ pF$$

Now $$R = R_3 + R_1 \,||\, R_2 + h_{ie} \cong R_3 + h_{ie}$$

Hence, $$R_3 = R - h_{ie} = (10k - 1k) = 9\ k\Omega$$

V_B = Base voltage = $V_{BE} + I_E R_E = 0.7 + 1.0 = 1.7$ V

Let $R_2 = 10$ kΩ,

Thus current in $R_2 = \dfrac{1.7}{10k} = 0.17$ mA

Drop across $R_1 = (20 - 1.7) = 18.3$ V

Current in $R_1 = I_{R1} = I_{R2} + I_B = \left(0.17 + \dfrac{1}{100}\right)$ mA = 0.18 mA = 180 μA

Thus $R_1 = \dfrac{18.3V}{180 \times 10^{-6}} = 120$ kΩ

Thus Designed values are :

$R_E = 1$ kΩ, $R_C = 14$ kΩ, R = 10 kΩ, C = 46.7 pF **Ans.**

$R_3 = 9$kΩ, $R_2 = 10$ kΩ, $R_1 = 120$ kΩ

EXAMPLE 4.27. Design a phase shift oscillator using FET having a g_m = 5000 k mho, r_d = 40 kΩ and feedback circuit of R = 10 kΩ. Frequency of oscillation required is 1 kHz.

Solution: Oscillation frequency is given by the expression

$$f = \frac{1}{2\pi\sqrt{6}RC} \quad \text{For f = 1 kHz we have}$$

$$C = \frac{1}{10^3 \times 2 \times \pi \times \sqrt{6} \times 10^4} = 6.5 \text{ nF}$$

For sustained oscillation, $|A_V| \geq 29$

We have $R_L = R_D \parallel r_d$ ...(i)

and $|A_V| = |g_m R_L|$

Thus, $R_L \geq \dfrac{29}{g_m} \geq \dfrac{29}{5000 \times 10^{-6}} \geq 5.8$ k

Choose $R_L = 6.0$ kΩ

Then $R_D \parallel r_d = 6.0$ kΩ gives

$R_D = 7.06$ kΩ

Design : Circuit is same as shown in figure 6.49 with R = 10 kΩ C = 6.5 nF, $R_D = 7.06$ kΩ

Choose $R_s = 1$kΩ and $C_s = 0.1$ μF

and $V_{DD} = +12$ V **Ans.**

EXAMPLE 4.28. Calculate the component values of Wien Bridge suitable to be used in oscillator to vary the frequency from 100 Hz to 10 kHz in two stages.

Solution : Oscillation frequency = $f = \dfrac{1}{2\pi RC}$

For $f_{max} = 10 \times 10^3 = \dfrac{1}{2\pi RC}$, Choose R = 100 kΩ

then $C = \dfrac{1}{2\pi \times 10 \times 10^3 \times 100 \times 10^3} = 0.156$ nF.

For $f_{min} = 100$,

we have $C = \dfrac{1}{2\pi \times 100 \times 10^5} = 15.6$ nF

Important Point: Thus, to vary the frequency from 100 Hz to 100 kHz, we must have a bridge with a fixed resistor of 100 kΩ and a variable capacitor. Two capacitors must be varied simultaneously. Alternatively, we can have a fixed capacitor of 1μF and a variable resistor.

4.22 GENERAL APPLICATIONS OF OSCILLATORS

(i) As a local oscillator in radio receivers.
(ii) In T.V. receivers.
(iii) In signal generators.
(iv) As clock generation for logic circuits
(v) AM and FM transmitters
(vi) In phase lock loops.

SUMMARY

- An oscillator is the basic element of all the ac signal sources and generates sinusoidal signal of a known frequency and amplitude.
- The function of an oscillator is just the reverse of a rectifier and, therefore sometimes called an inverter. However, we generally think of oscillator circuit as providing an ac voltage signal.
- The alternator generates large amount of a.c. power but for many applications such as radio-transmitters and receiver, radars etc., an oscillator is always preferred owing to its numerous advantages over alternators for such application.
- An oscillator is a device which produces an output signal without any input signal of any desire frequency.
- The electronic oscillators can be broadly classified into the following two categories :
  (i) Sinusoidal or harmonic oscillators
  (ii) Non-sinusoidal or relaxation oscillator
- The electrical oscillations whose amplitude goes on decreasing with time are known as damped oscillations.
- The electrical oscillations whose amplitude remains constant with time are known as undamped oscillations.
- The frequency stability of an oscillator can be defined as a measure of its ability to maintain a constant frequency, over a long time interval.
- The positive feedback is employed in oscillators for their operation.
- Barkhausen Criterion states that the frequency of a sinusoidal oscillator determined by the condition that the loop-gain phase shift is zero.
- An oscillatory circuit consists of inductive coil L and capacitor C connected in parallel with each other.
- The frequency of oscillations depends upon the value of inductance of the coil and capacitance of the capacitor i.e.,

$$f_r = \frac{1}{2\pi\sqrt{LC}}$$

- This circuit is used to feedback a fraction of amplifier output to the oscillator circuit in correct phase so that it aids the oscillations and compensates the losses. It, provides positive feedback.

SHORT QUESTIONS WITH ANSWERS

Q.1.What is Barkhausen criterion?

Ans. Barkhausen Criterion states that the frequency of a sinusoidal oscillator determined by the condition that the loop-gain phase shift is zero.

Hence $A\beta = 1$

It implies that

(i) $|A\beta| = 1$ i.e., magnitude of feedback factor = unity.

(ii) $\phi = 0$ or multiple of 2π

Another interpretation of Barkhausen Criterion is

$$A_{fb} = \frac{A}{1+A\beta} = \frac{A}{1-1} = \infty$$

Since gain is infinity we will get fainter output with no input.

Q.2. What is a crystal oscillator ?

Ans. In crystal oscillators, the usual electrical resonant circuit is replaced by a mechanically vibrating crystal. As a matter of fact,the crystal (usually quartz) has a high degree of stability in holding constant at whatever frequency the crystal is originally cut to operate. Therefore, the crystal oscillators are, used whenever great stability is required. For example, in communication transmitters, and receivers, digital clocks etc.

Q.3. Write the expressions for two resonant frequencies of a crystal oscillator.

Ans. Two resonant frequencies are given by the expressions

(i) Series resonant frequency, $f_s = \dfrac{1}{2\pi\sqrt{LC}}$

(ii) Parallel resonant frequency, $f_p = \dfrac{1}{2\pi}\sqrt{\dfrac{1+C/C_M}{LC}}$

Q.4. What are the drawbacks of crystal oscillator?

Ans. Following are the drawbacks of crystal oscillators :

(i) The crystal oscillators have a very limited tuning range (or not at all). They are used for frequencies exceeding 100 kHz.

(ii) The crystal oscillators are fragile and, therefore, can only be used in low power circuits.

Q.5. Write the advantages of a Wien bridge oscillator.

Ans. Advantages :

(i) It provides a stable low distortion sinusoidal output over a wide range of frequency.

(ii) The frequency range can be selected simply by using decade resistance boxes.

(iii) The frequency of oscillation can be easily varied by varying capacitances C_1 and C_2 simultaneously.

(iv) The overall gain is high because of two transistors.

Q.6. What is an oscillator?

Ans. Oscillator may be defined as a circuit which generates an ac output signal of very high frequency without requiring any externally applied input signal or it may be defined as an electronic source of alternating current (or voltage) having sinusoidal or non-sinusoidal (square, sawtooth or pulse) waveshape. It can also be defined as a circuit that converts d.c. energy into very high frequency a.c. energy.

Q.7. What is the initial condition for oscillation to start?

Ans. Initial condition for oscillation to start is loop gain $\beta A \geq 1$.

Q.8. What are the Barkhausen conditions of oscillations?

Ans. Barkhausen conditions for sustained oscillations are : (i) the loop gain of the circuit must be ≥ 1 and (ii) the phase shift around the circuit must be zero.

Q 9. Can every electronic oscillator be considered to be an amplifier with an infinite gain? If yes, where from starting signal voltage comes?

Ans. Yes, starting signal voltage comes from more noise inherent in the transistor or variation in voltage of dc power supply.

Q. 10. Why are L-C resonant circuit impractical at audio-frequencies?

Ans. L-C resonant circuits are impractical at audio-frequencies because components required in their construction for low-frequency operation are too bulky and heavy and expensive.

Q. 11. How does Clapp oscillator circuit differ from Colpitt's oscillator circuit in construction?

Ans. Clapp oscillator circuit is a refinement of the Colpitt's oscillator circuit. The only difference is that the single inductor found in the tank circuit of the Colpitt's oscillator is replaced by a series L-C combination (inductor L in series with capacitor C).

Q 12. Why Clapp oscillator is preferred over the Colpitt's oscillator?

Ans. In a Colpitt's oscillator, the resonant frequency is affected by the transistor and stray capacitances but in a Clapp oscillator, the transistor and stray capacitances have no effect on the oscillation frequency. Thus in Clapp oscillator the effect of transistor parameters on the operation of the circuit is eliminated and the frequency stability is improved. That is why, Clapp oscillator is preferred over Colpitt's oscillator.

Q 13. How does Hartley oscillator differ from Colpitt's oscillator in construction?

Ans. Hartley oscillator circuit is similar to Colpitt's oscillator except that phase-shift network consists of two inductors and one capacitor instead of two capacitors and one inductor.

Q. 14. Why is crystal oscillator used in communication transmitters and receivers ?

Ans. Crystal oscillators are used in communication transmitters and receivers because of their greater frequency stability.

Q. 15. Why we use two transistors in a Wien bridge oscillator?

Ans. Wien bridge oscillator employs two transistors so as to obtain 360° or 0° phase shift between output and input voltages.

Q. 16. Sometimes a lamp is used in one of the resistance arms of a Wien bridge oscillator. Why?

Ans. A lamp is used in one of the resistance arms of a Wien bridge oscillator so as to provide a negative feedback, in proportion to the output current, in the circuit and thus ensure constant output over a range of frequencies.

Q. 17. Why beat frequency oscillator (BFO) has become obsolete now?

Ans. Beat frequency oscillator has become obsolete now because its circuit is quite complicated.

Q. 18. How does a signal generator differ from an ordinary oscillator?

Ans. The signal generator, like an ordinary oscillator, is a source of sinusoidal signals but it is also capable of modulating its sinusoidal output signal with other signals.

Q. 19. Why use of regulated power supply is essential for signal generators?

Ans. Frequency of output voltage from a signal generator changes with the change in supply voltage so use of regulated power supply is essential.

Q. 20. Why are buffer amplifiers used in signal generators?

Ans. Buffer amplifiers are used in the signal generators for isolating the oscillator circuit from the output circuit so that any change in the circuit connected to the output does not affect the frequency and amplitude of the oscillator output.

REVIEW QUESTIONS

1. What is the condition for oscillation ?
2. Draw and explain tuned collector oscillator.
3. Draw the circuit and explain the operation of Colpit's oscillator.
4. Discuss in detail Hartley oscillator.
5. Draw and discuss in detail the circuit of an R-C phase shift oscillator.
6. Draw the circuit of a Wienbridge oscillator and explain its working. Why is negative feedback employed in this oscillator circuit in addition to the usual positive feedback ?
7. Explain the properties of quartz crystal which are responsible for its use in an oscillator.
8. Draw the circuit and explain a crystal oscillator.
9. (i) What is Barkhausen criterion for oscillation ?
   (ii) Describe the principle of operation of a Wien Bridge Oscillator and give the condition for sustained oscillation.
10. Explain why :
    (i) Three R-C sections are used in R-C phase shift oscillator.
    (ii) Negative feedback is provided in Wein bridge oscillators.
    (iii) A common emitter circuit without by-pass capacitor is called a negative current feedback circuit.
11. Explain why :
    (i) A negative feedback is always employed in high gain amplifiers.
    (ii) Emitter-follower circuit is also called common-collector amplifier circuit.
    (iii) A common emitter circuit without by-pass capacitor is called a negative current feedback circuit.
12. What are the factors which affect the frequency stability of an oscillator ?
13. Draw the electrical equivalent circuit of quartz crystal and describe its electrical characteristics.
14. What is an oscillator ? How does it differ from an amplifier ? What are the essential parts of an oscillator circuit?

15. Explain how L-C tank Circuit is used to generate ac oscillations in an electronic oscillator.
16. What are the Barkhausen conditions of oscillations in electronic systems ? What are their significance? What are the factors which affect the frequency stability of an oscillator ?
17. What is an oscillator ? Enumerate the different classes of oscillators.
18. How an amplifier can be converted into an oscillator ? Write the oscillation criteria. What is the difference between oscillator and alternator ?
19. Write short notes on Barkhausen criteria. **(RGTU, Sem. Exam; 2003-2004)**
20. Deduce the expression for condition of oscillation and frequency of oscillation of Wien bridge oscillator. **(RGTU, Sem. Exam; 2003-2004, 2007-2007)**
21. What disadvantage does a phase shift oscillator have? How is this removed in a Wien Bridge oscillator. **(RGTU, Sem. Exam; 2007-2008)**
22. Draw the electrical equivalent circuit of the quartz crystal and describe its electrical characteristics. **(RGTU, Sem. Exam; 2001-2002)**
23. What is Piezoelectric effect? Explain the working of Quartz crystal. **(RGTU, Sem. Exam; 2008-2009)**
24. How is the amplitude of Wien bridge oscillator stabilized? Explain. **(RGTU, Sem. Exam; 2008-2009)**
25. How does the circuit of a Clapp oscillator differs from that of a Colpitt's oscillator? Derive frequency of oscillation of any one of them. **(RGTU, Sem. Exam; 2007-2008)**
26. Why do we use of crystal in an oscillator circuit? Obtain the expression of the impedance and show its plot w.r.t. the frequency. **(RGTU, Sem. Exam; 2005-2006)**
27. For the following feedback network, find
    (i) the transfer function
    (ii) the input impedance
    (iii) the frequency of oscillation and minimum amplifier voltage gain, if this network is used in feedback path. **(RGTU, Sem. Exam; 2002-2003, 2004-2005)**

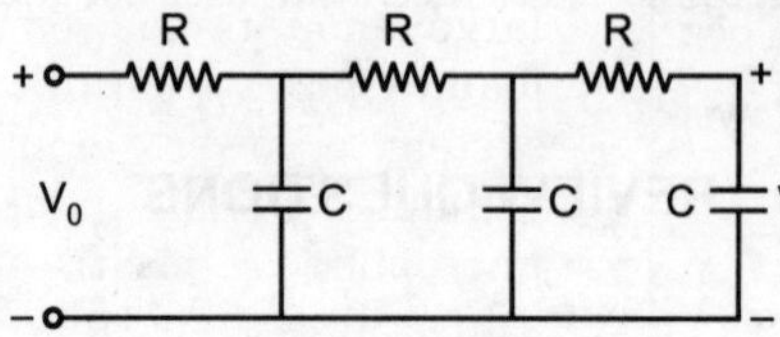

Fig. 4.41.

NUMERICAL QUESTIONS

1. An oscillator circuit has an inductor of 80 mH and a capacitor of 6 pF. Determine the frequency of oscillation. **[Ans.** 230 kHz]
2. The resonant circuit of a tuned-collector transistor oscillator has a resonant frequency of 10 MHz. If the value of capacitance is increased by 40%, calculate the new resonant frequency. **[Ans.** 8.45 MHz]
3. Determine the oscillation frequency of a Colpitt's oscillator if L = 50 μ H, L_{RFC} = 0.8 m H, C_1 = 0.01 μ F, C_2 = 0.02 μ F, C_C = 20 μ F. **[Ans.** 275 kHz]
4. Determine the oscillation frequency of a transistor Hartley oscillator with component values L_1 = 150 μ H, L_2 = 1.5 mH, M = 75 μ H and C = 150 pF. **[Ans.** 306 kHz]
5. In a phase shift oscillator, if $R_1 = R_2 = R_3$ = 400 kΩ and $C_1 = C_2 = C_3$ = 50 pF, determine the frequency of oscillation. **[Ans.** 3.25 kHz]
6. In a Wien bridge oscillator, if $R_1 = R_2$ = 110 kΩ and $C_1 = C_2$ = 1,600 pF, determine the frequency of oscillation. **[Ans.** 904 Hz]

Chapter 5

POWER AMPLIFIERS

Syllabus

Power Amplifiers, class A, class B, class AB, C amplifiers, their efficiency and power dissipation.

Chapter Outline

5.1. INTRODUCTION TO POWER AMPLIFIERS

A power amplifier is an amplifier, which is capable of providing a large amount of power to a load such as loudspeaker. In almost all electronic systems, where a large amount of power is required to be supplied to the load, a power amplifier is required. In an electronic system, a power amplifier is used as a last stage. For example, a public address system (PAS) consists of a microphone, a multistage amplifier, a power amplifier and a loudspeaker as shown in figure 5.1. The microphone converts the sound waves into an electrical signal which is of very low voltage (usually of few millivolts). This low voltage signal is not sufficient to drive the loudspeaker. Therefore, this signal is first raised to a sufficiently high value (i.e., a few volts) by passing it through a multistage small-signal (i.e., voltage) amplifier. This signal is then used to drive the power amplifier, since it is incapable of delivering a large amount of power to the loudspeakers.

A power amplifier is more commonly known as an audio amplifier. The audio amplifiers are used in public address systems, tape recorders, stereo systems, television receivers, radio receivers, broadcast transmitters etc. It would be interesting to know that a power amplifier does not actually amplify the power. In fact, it takes power from the d.c. power supply connected to the output circuit and converts it into the useful a.c. signal power. This power is fed to the load. The type of a.c power developed, at the output of a power amplifier, is controlled by the input signal. Hence, we can say that actually a power amplifier is a d.c. to a.c. power converter, whose action is controlled by the input signal.

Note that the power amplifiers are also called large signal amplifiers*. The term 'large signal' for the power amplifiers arise since these amplifiers make use of a large part of their a.c load line for their operation. It is in contrast to the small signal amplifiers, which use only 10% of their a.c load line for operation. Recall that the small signal amplifiers are more commonly known as voltage amplifiers.

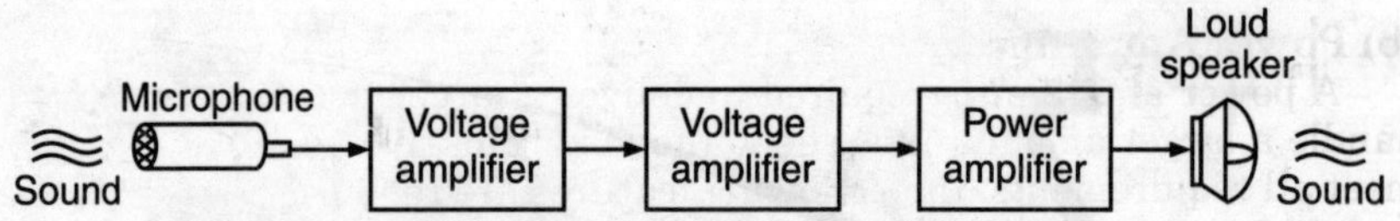

Fig. 5.1. The block diagram of a public address (PA) system

5.2. THE DIFFERENCE BETWEEN THE VOLTAGE AMPLIFIER AND THE POWER AMPLIFIER **

(RGTU, Bhopal, Sem. Exam; Dec.2002, June 2003)

As already discussed in the last article that a small signal amplifier is known as a voltage amplifier and a large signal amplifier is known as a power amplifier. It will be quite interesting to know that a voltage amplifier produces an increased voltage into a high resistance load. On the other hand, the power amplifier produces a large increase in the output power or an appreciable amount of current flow into a relatively low resistance load.

Following are some of the important differences in the construction of a voltage amplifier and a power amplifier :

(a) Voltage amplifier

We have studied that the voltage gain of an amplifier is given by the expression:

$$A_v = \beta \frac{R_C}{R_{in}}$$

Therefore, the following features are required to be incorporated in such amplifiers :

(i) **Higher value of β :** The transistor employed should have higher value of β (>100). This means the transistor should have thin base.

(ii) **Low input resistance (R_{in}) :** The input resistance R_{in} of the transistor should be quite low in comparison to that of collector load resistor R_C.

(iii) **High collector load resistance (R_C) :** The collector load resistance should be relatively high. This condition is obtained by keeping low collector current (≃1 mA). Since collector junction is to carry a small current, therefore no heat sink is required.

(iv) **R-C Coupling :** As this coupling has smaller size, smaller weight, lower cost and occupies less space, therefore it is preferred to couple various stages of voltage amplifiers by this coupling method.

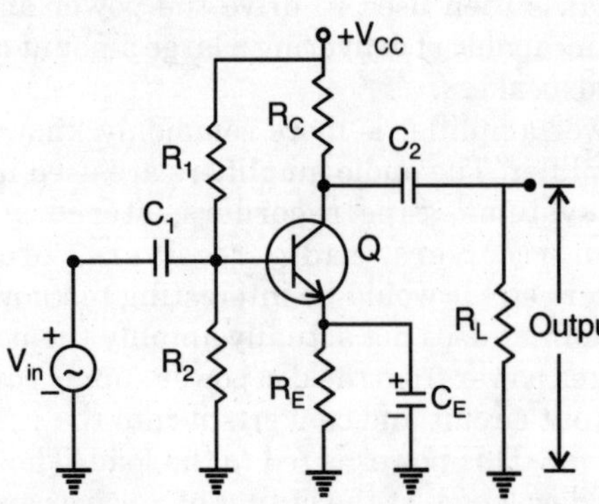

Fig. 5.2. Circuit diagram of a voltage amplifier

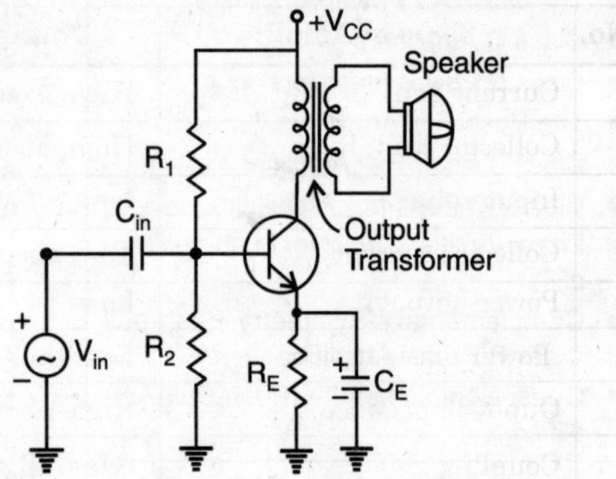

Fig. 5.3. Circuit diagram of a single power amplifier

* The transistors employed in power amplifiers are called power transistors.

** Write short notes on the following : (i) Power Amplifier (ii) Efficiency of power amplifiers.

(RGTU, Bhopal, Sem. Exam; Dec. 2002, June, 2003)

(b) Power Amplifier

A power amplifier is required to deliver a larger amount of power to the load, hence it has to handle a larger amount of power to the load. This means that it has to handle a large current. To meet all requirements, the following features are required to be incorporated in such amplifiers:

(i) Thicker base : Since the transistor employed in power amplifier is to handle larger current, its base should be thick. The transistors having thicker base have smaller value of β.

(ii) Larger size of transistor : In power amplifiers, power transistors are employed which are larger in size since they have to handle more current. When heavy current (> 100 mA) flows through the transistor, more heat is produced at the collector junction. This heat is required to be dissipated hence transistors of larger size are required. In addition to this, heat sinks are employed to improve the heat dissipation.

(iii) Low collector load resistance (R_C) : The collector load resistance R_C should be of smaller value. If its value is high there will be more power loss in the resistor. Moreover, there will be larger voltage drop in it which will reduce the operating voltage. To avoid all these drawbacks, low collector load resistance is employed.

(iv) Transformer coupling : The most important function of a power amplifier is to transfer maximum power to the device i.e., speaker. This is only possible if input impedance is equal to the output impedance. Therefore, we have to match the impedance of loudspeaker, (nearly 8 Ω) to the output impedance of the amplifier or transistor (a few kΩ). This can only be achieved by using a transformer at the output. This turns ratio (N_P/N_s) of the output transformer are designed in such a manner that the load resistance looking at the primary side will be,

$$R'_L = \left(\frac{N_P}{N_s}\right)^2 R_L$$

As a conclusion from the above discussion, let us sum up the following points :

(i) Power amplifier is the last stage of multistage amplifier whereas the first few stages are voltage amplifiers.

(ii) For perfect impedance matching, the transformer coupling is invariably employed with power amplifier whereas in voltage amplifier R-C coupling is generally preferred.

(iii) In power amplifiers, power transistors with suitable heat sink are employed. Since they are capable to carry heavy current and are able to dissipate heat developed at the collector junction.

The comparison between voltage and power amplifiers can be presented in the form of a table 5.1

Table 5.1.

S.No.	Type of Amplifier	Voltage Amplifier	Power Amplifier
1.	Current gain, β	High, exceeding 100	Low, 20 – 50
2.	Collector load, R_C	High, about 10 k Ω	Low, 5 – 20 Ω
3.	Input voltage	Low, a few m V	High, 2 – 4 volts
4.	Collector current I_C	Low, about 1 m A	High, exceeding 100 m A
5.	Power output	Low	High, a few watts to hundred of watts
6.	Power dissipation capacity	Less than 0.5 W	More than 0.5 W
7.	Ouput impedance	High, about 10 k Ω	Low, about 200 Ω
8.	Coupling	Generally R-C coupling	Invariably transformer coupling or tuned circuit.

The important point is that we cannot obtain a power amplifier by simply replacing a transistor of low power dissipation in a voltage amplifier circuit by another transistor with high dissipation.

Now, let us consider a common emitter voltage amplifier circuit using voltage divider baising with replaced transistor as shown in figure 5.2. Let I_{CQ} and V_{CEQ} be the collector current and collector-emitter voltage respectively at the operating point.

Now, collector – emitter voltage will be

$$V_{CEQ} = V_{CC} - I_{CQ} (R_C + R_E) \quad ...(5.1)$$

Total, dc power supplied to the amplifier from collector supply will be

$$= V_{CC} I_{CQ} \quad ...(5.2)$$

DC power input to transistor $= V_{CEQ}I_{CQ}$ ...(5.3)

Out of total dc power of $V_{CC}I_{CQ}$ drawn from the dc supply, $V_{CEQ}I_{CQ}$ is only the power that is converted into useful ac power and rest is wasted.

i.e., power wasted $= V_{CC}I_{CQ} - V_{CEQ} I_{CQ} = I_{CQ} [V_{CC} - V_{CEQ}]$

But using equation (5.1), we have

$$V_{CC} - V_{CEQ} = I_{CQ} (R_C + R_E)$$

therefore, power wasted $= I_{CQ}[I_{CQ} (R_C + R_E) = I^2_{CQ} (R_C + R_E)$ ...(5.4)

This means that $I^2_{CQ} (R_C + R_E)$ is the power which gets wasted unnecessarily in heating of resistors.

Thus, this is the reason why a voltage amplifier cannot be used as a power amplifier.

Now, this Power wastage can be reduced by short-circuiting collector and emitter resistors R_C and R_E,. However, it is not possible. Emitter resistor R_E cannot be eliminated because it is part of biasing network. In the absence of R_E the stabilization of the operating point is adversely affected. Collector resister R_C also cannot be short circuited since in that case load resistor R_L will be short-circuited (from ac point of view) and there will be no transfer of power to the load resistor R_L.

However, collector resistor R_C can be replaced by an inductor whose dc resistance is zero (acts as a short-circuit for dc) and ac impedance is quite high. Doing so, we can achieve two purposes. Firstly the power loss in the amplifier circuit is reduced because dc power loss in the choke (i.e., coil) is almost zero. In addition to this, we can afford to use lower voltage supply V_{CC} for the same amplifier because no dc voltage drop occurs across the choke. Hence, this is the reason that transformer-coupling is usually preferred over R-C coupling in power amplifiers.

Even after this, more improvements can be made in the above circuit so as to make it comparatively more efficient in operation and will be a matter of discussion in this chapter.

5.3. FEW GENERAL TERMS USED FOR A POWER AMPLIFIER

Few important technical terms, generally used for a power amplifier are collector efficiency, power dissipation capacity and distortion. In this section, let us have a quick view of all the three terms.

(i) The Collector Efficiency : As discussed in previous article, a power amplifier converts dc power into ac power at ouput. Because of this, the effectiveness of a power amplifier is measured by its capability of converting dc power from battery into ac power at the ouput side. This is called collector efficiency. Thus, it can be defined as the ratio of ac output signal to the zero signal power (i.e., dc power drawn from the battery of a power amplifier). As a matter of fact, power amplifiers are designed to provide maximum collector efficiency.

(ii) The Power Dissipation Capacity : As discussed earlier, a power transistor handles large amount of currents and becomes heated during its operation. This can cause the junction temperature to rise to a dangerous level. The power dissipation capacity of a transistor can be increased by making a suitable provision for rapid conduction of heat away from the junction. In fact, this can be achieved by attaching a metal case, known as the **heat sink**, to the power transistor case.

Thus, power dissipation capacity of a power transistor is defined as its ability to dissipate the heat developed in it.

(iii) Distortion : It may be defined as the change of output wave-shape from the input wave-shape of the amplifier. Because power amplifier handles large signals and for large input signals, the transistor operation becomes non-linear, therefore, the problem of distortion arises. In fact, lesser the distortion, better is the amplifier.

5.4. CLASSIFICATION OF POWER AMPLIFIERS*

(RGTU, Bhopal, Sem. Exam; June 2004, Nov.Dec. 2007)

As a matter of fact, the power amplifiers can be primarily classified into following two categories:

(i) audio-power amplifiers

(ii) radio-power amplifiers.

Audio-power amplifiers are also called the small signal power amplifiers. These power amplifiers raise the power levels of signals that have audio-frequency range (i.e., 20 Hz – 20 kHz).

On the other hand, radio-power amplifiers, are also called large signal power amplifiers. These power amplifiers raise the power level of signals that have radio frequency range. This means that these amplifiers amplify a specific frequency or a narrow band of frequencies while rejecting all other undesired frequencies.

5.4.1. Classification Based upon Operating Mode

Transistor power amplifiers handle large signals. Most of them are driven so hard by the input large signal that collector current is either cut-off or is in saturation region during a large portion of the input cycle. Hence, we can classify such amplifiers in accordance with their mode of operation. In fact, this classification is based upon the amount of transistor bias and amplitude of the input signal. It takes into account the portion of the cycle for which the transistor conducts. Thus, according to mode of operation, the power amplifiers can be classified as under:

5.4.1.1. Class A Power Amplifiers** (RGTU, Bhopal, Sem. Exam; June 2008)

In a class-A power amplifier, transistor is biased in such a way that the ouput current flows for the entire cycle of the input signal. This means that the operating point is so selected that the transistor operates only over the linear region of its load line. Therefore, such an amplifier can amplify input signal of small amplitude. Since the transistor operates over the linear portion of load line, the output waveform is exactly similar to the input waveform. Hence, class A amplifiers are characterised by a high fidelity of the output. These types of amplifiers are used where freedom from distortion is a prime objective. Thus, operation is restricted only over a small central region of the load line because of which such amplifiers can be used for amplifying signals of small amplitude. Further, ac power output per transistor is small. The maximum possible overall efficiency with resistive load is 25%. The maximum possible collector efficiency with resistive load is 50 %. In case, an output transformer is used, both of these efficiencies are 50%.

5.4.1.2. Class B Power Amplifiers

In class-B power amplifiers, the transistor bias and signal amplitude are such that the output current flows only during positive half cycle of the input signal. At zero signal, the collector current is zero and no biasing system is required in class B amplifiers. The operating point is selected at collector cut-off voltage. Due to total absence of negative half cycle from the output, distortion is high. Zero signal input represents the best condition for class B amplifiers because of zero collector current. The transistor dissipates more power with increase in signal strength. In comparison to class A power amplifiers, average current is less, power dissipation is less. Due to this, overall efficiency is increased. The theoretical efficiency in class B operation is about 78.5% whereas it is only 50% in class A operation.

---

* Give classification of power amplifiers. **(RGTU Bhopal, Sem. Exam; June 2004, Nov./Dec. 2007)**

** What do you mean by class-A, Class-B and Class-C amplifiers ?

(RGTU Bhopal, Sem. Exam; June 2008)

5.4.1.3. The Class AB Power Amplifiers

An amplifier may be biased at a dc level above the zero base current level of class B power amplifiers and above one-half the supply voltage level of class A. This type of bias condition is known as class AB condition. Class AB operation still requires a push-pull connection to achieve a full output cycle, but the dc bias level is usually closer to zero base current level for better power efficiency. For class AB operation, the output signal swing occurs between 180° and 360° and is neither class A nor class B operation.

5.4.1.4. Class C Power Amplifiers

A class C power amplifier is made to bias for operation for less than 180° of the input signal cycle and would operate only with a tuned or resonant circuit which provides a full cycle of operation for the tuned or resonant frequency. Such kind of power amplifiers are, therefore, employed in special areas of tuned circuits, such as radio or communications.

5.4.1.5. Class D Power Amplifiers

Class D power amplifiers are designed to operate with digital or pulse type signals. In fact, digital techniques make it possible to have signal which varies over the entire cycle using sample-and-hold-circuitry to recreate the output from many pieces of input signal. Further, the main advantage of class D power amplifiers is that it is ON (using power) only for short intervals and the overall efficiency can practically be very high (above 90°).

5.4.2. Classification Based upon Deriving Output

Transistor amplifiers, according to the method of deriving output, may be classified as single-ended, double-ended (or push-pull) and complementary symmetry push-pull amplifiers.

Single-ended power amplifier makes use of single transistor and derives output power w.r.t. one end permanently grounded. Double-ended or push-pull amplifier makes use of two transistors in a single stage. It consists of two loops in which the transistor collector currents flow in opposite directions but add in the load. Complementary symmetry push-pull power amplifier makes use of two transistors having complementary symmetry (one N-P-N and another P-N-P). The term complementary arises from the fact that one transistor is the N-P-N type and the other is P-N-P type. They have symmetry as they are made with the same material and technology and are of same maximum ratings.

5.5. PRACTICAL POWER AMPLIFIER STAGES

As discussed earlier, a power amplifier is required to amplify a weak signal until sufficient power is available to operate an output device such as a loudspeaker. Power amplifier, to provide the desired power amplification has usually 3 stages (*i.e.*, voltage amplification stage, driver stage, output stage), as shown by a block diagram in figure 5.4.

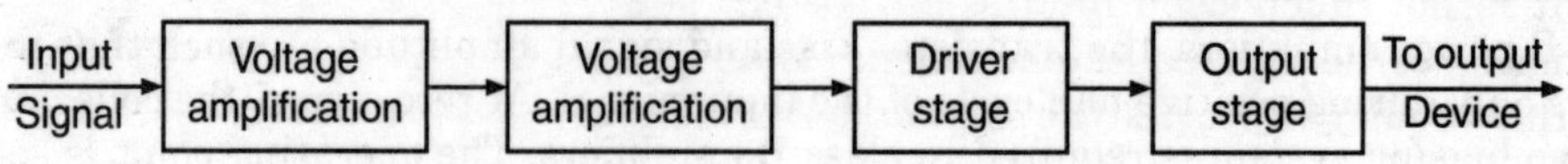

Fig. 5.4. Block diagram of a Practical Power Amplifier.

(i) Voltage Amplification Stage

We know that the signal developed by input transducer is quite small (of the order of few milli-volts) and needs sufficient amplification so as to operate the output device. Therefore for raising the level of the weak input signal, it is amplified in two or more stages, R-C coupling is usually employed.

(ii) Driver Stages

The stage which precedes the output stage is known as the **driver stage**. The output from the last voltage amplification stage is fed to the driver stage and output from the driver stage is fed to the ouput stage. The driver stage renders power amplification as usual. Here, concentrated effort is made to provide maximum power gain and so transformer-coupled class A power amplifier is employed in this stage. The driver transformer is usually a step-down transformer and facilitates impedance matching.

(iii) Output Stage

The output stage essentially consists of a power amplifier and is meant for transferring maximum power to the output device. In order to transfer maximum power at high efficiency, push-pull arrangement is employed in the ouput stage (if a single transistor is used as a class A amplifier in the output stage to provide faithful amplification, the operation efficiency will be quite low). In this arrangement two transistors are used in class B operation and fed from the centre-tapped secondary of the transformer whose primary forms the collector load for the driver stage.

5.6. THE CLASS-A POWER AMPLIFIER* (RGTU Bhopal, Sem. Exam; Dec. 2004)

A class-A power amplifier may be defined as a power amplifier in which the ouput current flows for the full-cycle (i.e., 360°) of the applied input signal as shown in figure 5.5. Thus, we can say that the transistor remains in forward-biased state for the full period of input cycle. Figure 5.6 shows a schematic circuit diagram for a series fed class-A power amplifier or large signal amplifier using resistive load R_C.

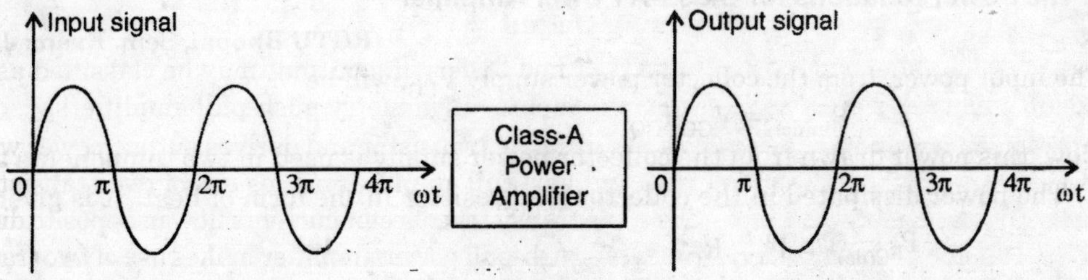

Fig. 5.5.

Note that the term 'series fed' means that the load resistor R_C has been connected in series with the transistor output. The only difference between this circuit and the small signal amplifier circuits discussed previously is that the signals handled by the large-signal circuit are in the range of volts and the transistor used is a power transistor which is capable of operating in the range of few watts. However, this circuit is not much used for the purpose of power amplification since, here the collector efficiency is very poor. However, this circuit can provide the clear understanding of the class A operation to the readers. Figure 5.7, shows the output characteristics with operating point Q. Here, I_{CQ} and V_{CEQ} represent no signal collector current and collector-emitter voltage respectively. It may be observed that when an a.c input signal is applied, the operating point Q shifts up and down causing output current and voltage to vary about It. The output current is increased to $I_{c\,max}$ and falls to $I_{c\,min}$. In the same fashion, the collector emitter voltage increases to $V_{ce\,max}$ and falls to $V_{ce\,min}$.

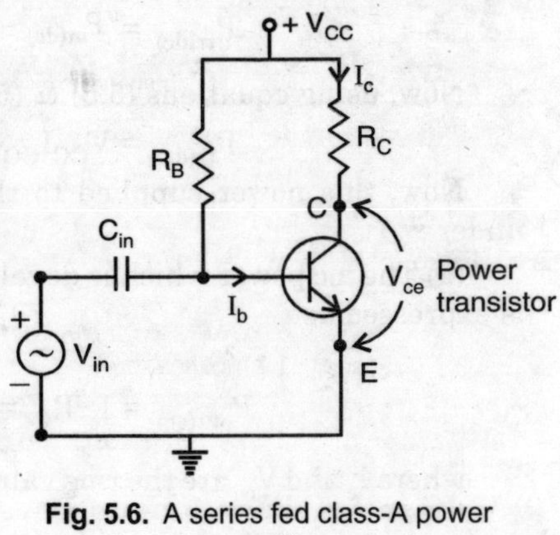

Fig. 5.6. A series fed class-A power amplifier

* Give the characteristics of class-A power amplifeir. Deduce the expressions where necessary. **(RGTU Bhopal, Sem. Exam; Dec. 2004)**

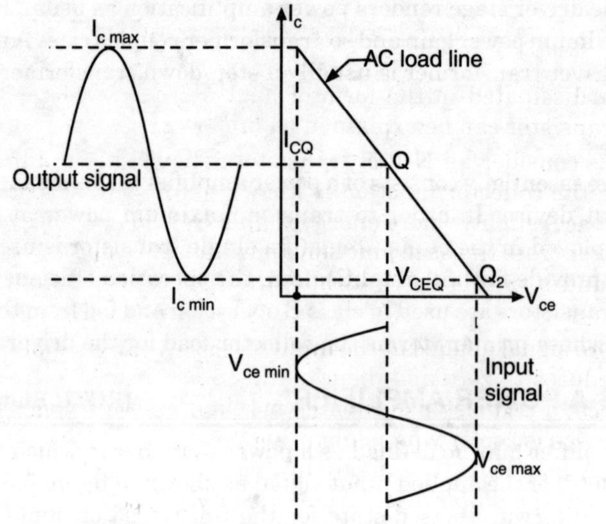

Fig. 5.7. The output characteristics for class-A power Amplifier.

5.6.1. The Power Relations for Class-A Power Amplifier*

(RGTU Bhopal, Sem. Exam; June 2004)

The input power, from the collector power supply V_{CC}, will be

$$P_{in(dc)} = V_{CC}.I_{CQ} \quad ...(5.5)$$

Now, this power drawn from the collector power supply is used in two components as under :

(i) The power dissipated in the collector load resistor in the form of heat. It is given by

$$P_{RC(dc)} = I^2_{CQ}.R_C \quad ...(5.6)$$

(ii) The power supplied to the transistor.

It is given by

$$P_{tr(dc)} = P_{in(dc)} - P_{RC(dc)}$$

Now, using equations (5.5) & (5.6), we get

$$P_{tr(dc)} = V_{CC}I_{CQ} - I_{CQ}{}^2R_C \quad ...(5.7)$$

Now, this power supplied to the transistor $P_{tr(dc)}$ is further subdivided into two parts as under :

(a) The a.c power which is developed across the load resistor constitute a.c. power output and is expressed as

$$P_{out(ac)} = I_c{}^2R_C = \frac{V_{ce}{}^2}{R_C} \quad ...(5.8)$$

where I_c and V_{ce} are the rms values of the collector current and the collector-to-emitter voltage.

$$\text{Thus,} \quad P_{out(ac)} = \left(\frac{I_{c\,max}}{\sqrt{2}}\right)^2 R_C = \frac{V^2_{ce\,max}}{2R_c} \quad ...(5.9)$$

$$\text{or} \quad P_{out(ac)} = \frac{I^2_{c(peak\text{-}to\text{-}peak)}R_C}{8} = \frac{V^2_{ce(peak-to-peak)}R_C}{8R_C} \quad ...(5.10)$$

* Derive an expression for conversion efficiency of class-A power amplifier and find its maximum value. **(RGTU Bhopal, Sem. Exam; June 2004)**

or $$P_{out(ac)} = \frac{I_{c(peak-to-peak)} \times V_{ce(peak-to-peak)}}{8} \quad ...(5.11)$$

(b) The power dissipated in the form of heat by the transistor itself. This cause of power dissipation in a transistor can be explained as under :

For this, let us consider an N-P-N transistor. The potential difference across the depletion layer formed near the collector junction is known as the barrier potential. This potential gives the P-region (i.e., base) slightly more energy than N-region (i.e., collector). Hence, when electrons emitted from emitter cross the base junction and enter the collector region, they give up energy in the form of heat and it is this energy which the transistor has to dissipate to its surrounding atmosphere.

Now, with a zero signal applied at the input of class-A power amplifier, the a.c. power developed across the load reduces to zero and, hence, all the power fed to the transistor is wasted in the form of heat. Hence, a transistor dissipates maximum power under zero signal condition. This means that the device is cooler when delivering power to a load than with zero signal condition. Because, in class-A operation, the maximum power dissipation in the transistor occurs under zero-signal condition, the power dissipation capacity of a power transistor, for class-A operation, must be at least equal to the zero-signal rating. We can show the power distribution in a class-A power amplifier in the form of a power flow diagram as shown in figure 5.8.

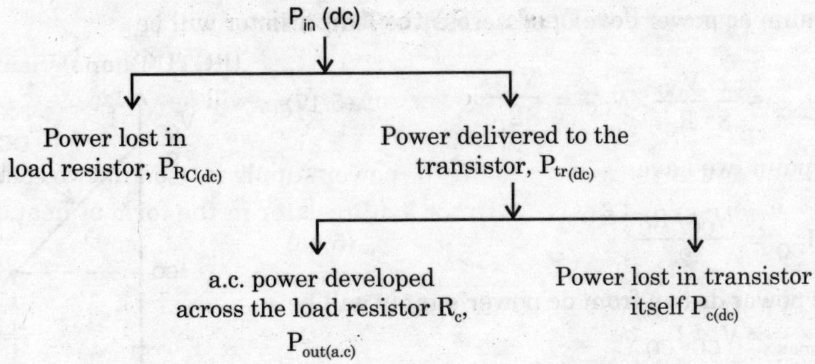

Fig. 5.8. Power flow diagram for a class-A power amplifier.

5.6.2. Expression for Collector Efficiency

Mathematically, the collector efficiency of a power transistor may be defined as

$$\eta_{collector} = \frac{\text{Average a.c. power output, } P_{out(a.c.)}}{\text{Average d.c. power input to the transistor, } P_{tr(d.c.)}} \quad ...(5.12)$$

5.6.3. Expression for Power Efficiency

A measure of the ability of an active device to convert the dc power of supply into the ac (signal) power delivered to the load is known as the power efficiency or conversion efficiency or theoretical efficiency. It is simply denoted by η

Mathematically, we have

$$\eta = \frac{\text{a.c. power delivered to the load, } P_{out(a.c.)}}{\text{total power drawn from d.c. supply } P_{in(dc)}} \quad ...(5.13)$$

Now, as discussed earlier, the ac power delivered to the load is

$$P_{out\ (ac)} = \frac{I_{c(peak-to-peak} \times V_{ce(peak-to-peak)}}{8}.$$

or $$P_{out\,(ac)} = \frac{(I_{c\,max} - I_{c\,min}) \times (V_{ce\,max} - V_{ce\,min})}{8} \quad ..(5.14)$$

Also, the total power drawn from d.c. supply is

$$P_{in}\,(dc) = V_{CC}.I_{CQ} \quad ...(5.15)$$

Using equations (5.13), (5.14) and (5.15) , we get

$$\eta = \frac{P_{out}\,(a.c.)}{P_{in}\,(d.c.)} = \frac{(I_{c\,max} - I_{c\,min}) \times (V_{ce\,max} - V_{ce\,min})}{8V_{CC}.I_{CQ}} \quad ...(5.16)$$

This is the required expression

5.6.4. Evaluation of Maximum Power and Efficiency

If the operating point Q is set at the mid point of the maximum signal swing, then the resulting maximum power condition can be obtained. For the circuit of class-A power amplifier in figure 5.9, this can be determined using

$$\text{Maximum } V_{CE(peak\text{-}to\text{-}peak)} = V_{CC}$$

$$\text{Maximum } I_{CE(peak\text{-}to\text{-}peak)} = \frac{V_{CC}}{R_c}$$

The maximum ac power developed across the load resistor will be

$$P_{out(ac)max} = \frac{1}{8}.\frac{V_{CC}}{R_C} \times V_{CC} = \frac{V_{CC}^2}{8R_C} \quad ...(5.17)$$

For the Q-point, we have

$$I_{CQ} = \frac{V_{CC}/R_C}{2} \quad ...(5.18)$$

and the dc power drawn from dc power supply will be

$$P_{in(dc)max} = V_{CC}.I_{CQ}$$

$$= \frac{V_{CC} \times V_{CC}/R_C}{2}$$

$$= \frac{V_{CC}^{\;2}}{2R_C} \quad ...(5.19)$$

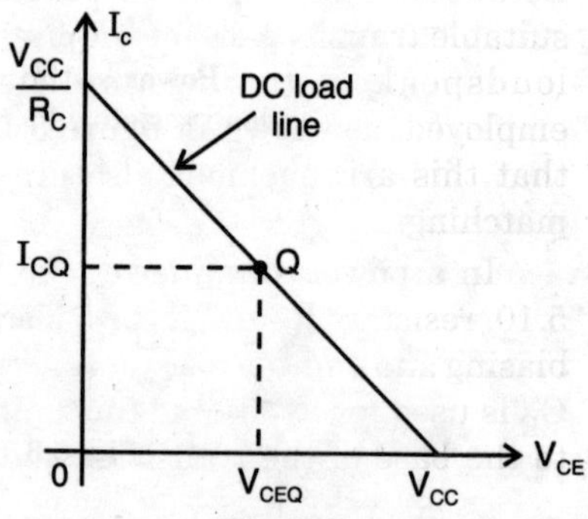

Fig. 5.9

Therefore, the maximum efficiency of a class-A power amplifier is expressed as under:

$$\eta_{max} = \frac{P_{out(ac)\,max}}{P_{in(d.c.)\,max}} = \frac{V_{CC}^2/8R_C}{V_{CC}^2/2R_C} = \frac{2R_C}{8R_C} = 0.25 = 25\%$$

Important Point: *Thus, this is the maximum percent efficiency for a series-fed class A power amplifier. Because, this maximum efficiency would occur only when ideal conditions are fullfilled and for the maximum ac signal swings, therefore, most series-fed class A power amplifiers have power efficiencies much less than 25%*

EXAMPLE 5.1. A power transistor working in class-A operation has zero-signal power dissipation of 8 watts. If the a.c. power output is 2 watts, then find the collector efficiency and the power rating of transistor.

Solution : Given that :

the ac power output $P_{out(ac)}$ = 2 watts

and power delivered to the transistor

$$P_{tr(ac)} = \text{zero-signal power dissipation} = 8 \text{ watts}$$

Therefore, collector efficiency is given by

$$\eta_{\text{collector}} = \frac{P_{out(ac)}}{P_{tr(ac)}} \times 100 = \frac{2}{8} \times 100 = 25\% \ \ \textbf{Ans.}$$

Because zero-signal condition represents the worst condition for the transistor, the power dissipation capacity of the transistor must be at least equal to the zero-signal power dissipation i.e., 8 watts.

5.7. THE TRANSFORMER-COUPLED CLASS-A POWER AMPLIFIER

A transformer coupled class-A power amplifier is also known as single ended power amplifier. Here, term single ended (i.e., one transistor) is used to distinguish it from the push-pull amplifier which makes use of two transistors.

In case of a direct-coupled class-A power amplifier shown in figure 5.10, the quiescent current flows through the collector resistive load and causes great amount of wastage of d.c. power in it. Infact, this d.c. power dissipated in the load resistor does not contribute to the useful a.c. output power. In addition to this, it is generally inadvisable to pass the d.c. through the output device such as in a voice coil of a loudspeaker. Because of these reasons, an arrangement using a suitable transformer for coupling the load (say, a loudspeaker) to the amplifier is generally employed, as shown in figure 5.10. Further, note that this arrangement also provides impedance matching.

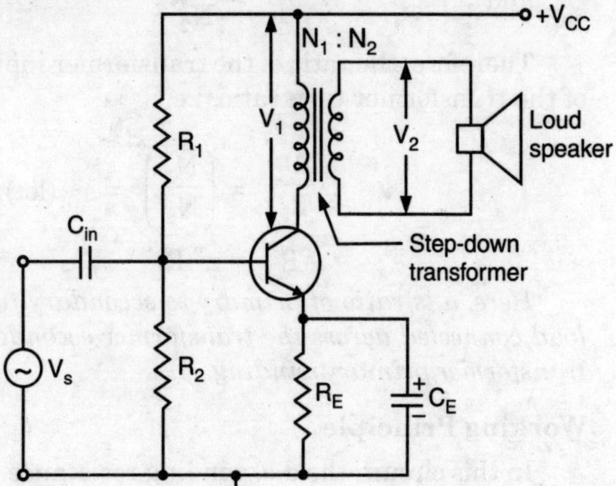

Fig. 5.10. A Transformer-coupled class A power amplifier

In a power amplifier circuit shown in figure 5.10, resistors R_1 and R_2 provide potential divider biasing and emitter resistor R_E is used for bias stablization of the circuit. The emitter bypass capacitor C_E is used for bypassing the emitter resister R_E. The input capacitor C_{in} couples a.c. signal voltage to the base of the transistor and sametimes blocks any dc component from the previous stage. A step-down transformer of suitable turn ratio $\left(a = \frac{N_1}{N_2}\right)$ is used to couple the high impedance collector circuit to low impedance load.

5.7.1. Concept of Impedance Matching

In a transformer-cupled class-A power-amplifier the power transferred from a power amplifier to a load (i.e., a loudspeaker) would be maximum only if the amplifier output impedance R_L equals the load impedance. This statement is in accordance with the maximum power transfer theorem. If we were not able to achieve this condition, lesser power will be transferred to the load R_L, though the amplifier is capable of delivering more power and rest of power developed would be lost in the active device. Hence, for transfer of maximum power from amplifier to the output device, matching of amplifier output impedance with the impedance of output device is essential. This is achieved by using a step-down transformer of suitable turn-ratio. The transformer impedance matching circuit is shown separately in figure 5.11 where R'_L is the resistance looking into the primary of the transformer and is expressed as

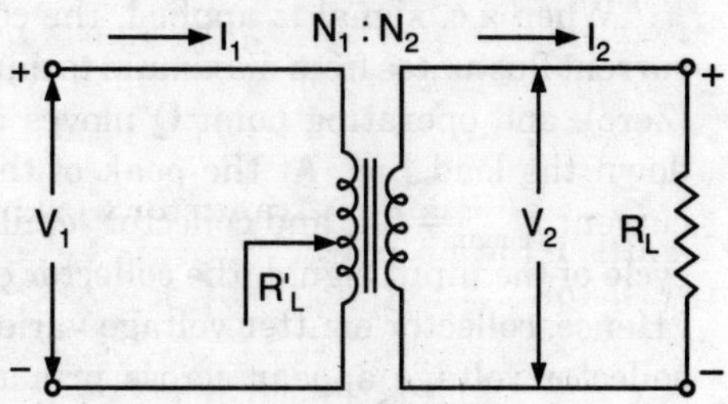

Fig. 5.11.

$$\frac{R'_L}{R_L} = \frac{V_1/I_1}{V_2/I_2} = \frac{V_1}{V_2}\cdot\frac{I_2}{I_1} = \frac{N_1}{N_2}\cdot\frac{N_1}{N_2}$$

or $$\frac{R'_L}{R_L} = \left(\frac{N_1}{N_2}\right)^2$$

where $$\frac{V_1}{V_2} = \frac{N_1}{N_2}$$

and $$\frac{I_2}{I_1} = \frac{N_1}{N_2}$$

Therefore, the ratio of the transformer input and output resistances varies directly as the square of the transformer turn ratio, i.e.

$$\frac{R'_L}{R_L} = \left(\frac{N_1}{N_2}\right)^2 = a^2 \text{ (let)}$$

$$R'_L = a^2 R_L \quad \text{....(5.20)}$$

Here, a is ratio of primary to secondary turns of step-down transformer, R_L is the resistance of load connected across the transformer secondary and R'_L is the effective resistance looking into the transformer primary winding.

Working Principle

In this circuit, the d.c. winding resistance determines the dc load line. This resistance is quite small (assumed to be zero) providing d.c. load line as a vertical line rising from V_{CC} as shown in figure 5.12. Now, when we apply an ac signal to the base of the transistor, the collector current will vary around the operating point Q.

In order to have maximum a.c. power output, the peak value of collector current due to input a.c. signal alone should be equal to the zero-signal collector current. To obtain this, the operating point Q is located at the centre of the a.c. load line. This is accomplished by adjusting the biasing resistors (i.e., R_1, R_2 and R_E).

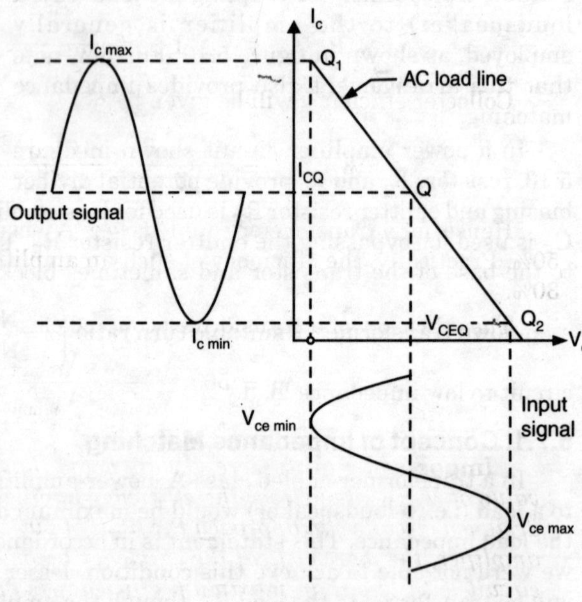

Fig. 5.12.

When a.c. signal is applied, the collector current fluctuates from maximum to minimum (Zero), and operating point Q moves up and down the load line. At the peak of the positive half cycle of the input signal, the total collector current $I_{c\,max} = 2\,I_c$ and collector -emitter voltage $V_{ce\,min} = 0$ while at the peak of the negative half cycle of the input signal, the collector current $I_{c\,min} = 0$ and collector-emitter voltage $V_{ce\,max} = 2\,V_{CC}$. Hence, collector emitter voltage varies in opposite phase to the collector current. The variation of collector voltage appear across primary of the transformer. Now, a.c. voltage is induced in the transformer secondery which in turn develops a.c. power and supplies to the load.

Mathematical Analysis

As a matter of fact, for an ideal transformer, there is no voltage drop in primary. Hence, $V_{CEQ} = V_{CC}$ and power input to the transistor becomes equal to the dc power drawn from the collector supply V_{CC} as power loss in the primary is negligibly small. Therefore, power input to the transistor, P_{tr} = Power drawn from collector supply, $P_{in(dc)}$

$$= V_{CC}\, I_{CQ}$$

and overall efficiency becomes equal to collector efficiency and is

$$= \frac{P_{out(ac)}}{V_{CC} I_{CQ}}$$

Further, under condition of development of maximum a.c. power, voltage swings from $V_{ce\ max}$ to zero and collector current from $I_{c\ max}$ to zero.

Therefore, we have

$$V_{rms} = \frac{R'_L}{R_L}\frac{1}{\sqrt{2}}\left[\frac{V_{ce(max)} - V_{ce(min)}}{2}\right] = \frac{V_{ce\,max}}{2\sqrt{2}} = \frac{2\,V_{CC}}{2\sqrt{2}} = \frac{V_{CC}}{\sqrt{2}}$$

and $$I_{rms} = \frac{1}{\sqrt{2}}\left[\frac{I_{c(max)} - I_{c(min)}}{2}\right] = \frac{I_{c\,max}}{2\sqrt{2}} = \frac{2\,I_{CQ}}{2\sqrt{2}} = \frac{I_{CQ}}{\sqrt{2}}$$

AC power developed across the load will be

$$P_{out(ac)} = V_{rms} I_{rms} = \frac{V_{CC}}{\sqrt{2}} \cdot \frac{I_{CQ}}{\sqrt{2}} = \frac{V_{CC}\, I_{CQ}}{2}$$

Collector efficiency will be given by

$$\eta_{collector} = \frac{P_{out(ac)}}{V_{CC} I_{CQ}} = \frac{V_{CC} I_{CQ}/2}{V_{CC} I_{CQ}} = 0.5 \text{ or } 50\% \qquad ...(5.21)$$

Hence, for a transformer-coupled class-A power amplifier, the maximum theoretical efficiency is 50%. Practically, the efficiency of such an amplifier is somewhat less than 50%. In fact, it is about 30%.

The efficiency of a transformer-coupled class-A power amplifier is expressed as

$$\eta = 50 \times \frac{V_{ce\,max} - V_{ce\,max}}{V_{ce\,max} + V_{ce\,min}}\% \qquad ....(5.22)$$

Important Point: *Here, it may noted that the greater the value of $V_{ce\ max}$ and smaller the value of $V_{ce\ min}$, the closer the efficiency approaches the theoretical limit of 50%. Further, well-designed circuits can approach the limit of 50%. The greater the amount of power handled by the amplifier, the more critical the efficiency becomes. For a few watts of power, a simpler, cheaper circuit with less than maximum efficiency is acceptable (and sometimes desirable). For power levels in the tens to hundreds of watts, efficiency as close as possible to the theoretical maximum will be desired.*

The maximum power dissipated in the transistor is under zero-signal condition i.e., $V_{CC}\, I_{CQ}$ where I_{CQ} is the collector current under zero-signal condition.

Because collector current flows at all times throughout the whole cycle, no part of the signal is cut-off and thus the output waveform obtained in this case is exact replica of the input waveform. Thus, such amplifiers have least distortion. However, they suffer from the following drawbacks:

(i) low output power, and

(ii) low collector efficiency (less than 50%; about 30%).

5.7.2. The Collector Dissipation Curve

In a power amplifier, the main source of power supply is collector dc supply and some of it is converted into ac power and appears in the load and rest of it is dissipated in the form of heat. Hence, in a power amplifier

$$P_{in(dc)} = P_{out(a.c.)} + P_d$$

Neglecting transformer losses, which are very small, power dissipation will be

$$P_d = P_{in\,(dc)} = V_{CC}\, I_{CQ} - P_{out\,(ac)} \quad \text{....(5.23)}$$

From this equation (5.23), it is obvious that power dissipated by the transistor (P_d) is the difference of average dc power input from the collector supply battery, $P_{in\,(dc)}$ (which is constant for a given battery and operating point) and ac output power, P_{out} (a.c.) supplied to the load. For zero output (i.e., when no signal is applied), the transistor is required to dissipate the maximum power i.e., $P_d = V_{CC}\, I_{CQ}$. This is the worst case for a transistor of a power amplifier the important factor which is to be kept in mind is the power dissipation. When the signal is applied, the transistor has to dissipate less power. Let the power dissipation capacity of a transistor be 5 W. For designing a power amplifier, first of all its collector dissipation curve is plotted. For plotting the collector dissipation curve of a power amplifier collector current I_c is determined corresponding to some arbitrary values of collector-emitter voltage V_{ce} from the relation $P_d = V_{ce}\, I_c = 5$ W, the power dissipation capacity of the transistor. Some sample calculations are given below in tabular form. The curve obtained from these values is a hyperbola, as shown in figure 5.13.

S.No.	V_{CE} in volts	I_c in m A	P_d in watts
1	5	1,000	5
2	10	500	5
3	15	333	5
4	20	250	5
5	25	200	5
6	30	167	5

For using this transistor in a power amplifier, its operating point must be below the dissipation curve, otherwise transistor will get damaged. This is the importance of dissipation curve.

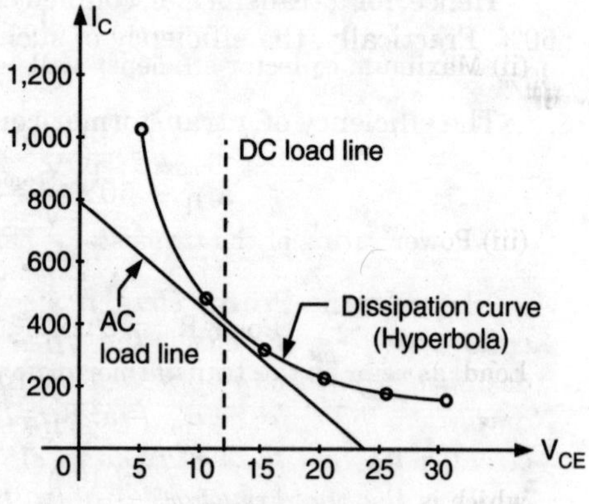

Fig. 5.13.

EXAMPLE 5.2. A transformer-coupled class-A large amplifier has maximum and minimum values of collector-emitter voltage of 25 V and 2.5 V. Determine its collector efficiency.

Solution : Given that

$$V_{ce\,max} = 25 \text{ V}$$

and $$V_{ce\,min} = 2.5 \text{ V}$$

We know that

Collector efficiency, $\eta_{collector}$ is given by

$$\eta_{collector} = 50 \times \frac{V_{ce\,max} - V_{ce\,max}}{V_{ce\,max} + V_{ce\,min}} \% = 50 \times \frac{25 - 2.5}{25 + 2.5} = 40.9\% \quad \textbf{Ans.}$$

EXAMPLE 5.3. A transformer-coupled class A power amplifier supplies the power to an 80 Ω load connected across the secondary of a step-down transformer having turn-ratio 5 : 1. Determine the maximum power output for a zero signal collector current of 120 m A.

Solution : Given that:

Load resistance, $R_L = 80\ \Omega$

and, turn-ratio, $a = 5$

Load as seen by the transformer primary, will be

$$R'_L = a^2R_L = (5)^2 \times 80 = 2{,}000\ \Omega$$

For maximum power output, we have

$$I_{max} = 2\,I_c \quad \text{and} \quad I_{min} = 0$$

So, rms value of current will be

$$I_{rms} = \frac{1}{\sqrt{2}}\left[\frac{I_{max} - I_{min}}{2}\right] = \frac{1}{2\sqrt{2}} \times 2I_c = \frac{I_c}{\sqrt{2}}$$

Maximum power output is given by

$$P_{out\,(ac)} = I^2_{rms}\,R'_L = \frac{I_c^2}{2}R'_L = \frac{(120 \times 10^{-3})}{2} \times 2{,}000 = 12.4\ \text{W} \quad \textbf{Ans.}$$

EXAMPLE 5.4. A transformer-coupled A power amplifier draws a current of 200 mA from a collector supply of 10 V, when no signal is applied to it. Determine: (i) maximum output power, (ii) maximum collector efficiency, and (iii) power rating of the transistor. If the load connected across the transformer secondary is of 2 Ω and transformer turn-ratio is 5 : 1, comment on the impedance matching.

Solution : Given that :

Collector supply voltage,

$$V_{CC} = 10\ \text{V}$$

Zero-signal collector current is

$$I_{CQ} = 200\ \text{m A} = 0.2\ \text{A}$$

(i) Maximum output power is given by

$$P_{out\,(ac)} = \frac{V_{CC}I_{CQ}}{2} = \frac{10 \times 0.2}{2} = 1\ \text{W} \quad \textbf{Ans.}$$

DC power input is

$$P_{in\,(dc)}\,V_{CC}I_{CQ} = 10 \times 0.2 = 2\ \text{W}$$

(ii) Maximum collector efficiency will be given by

$$\eta_{max} = \frac{P_{out\,max}}{P_{in\,(dc)}} \times 100 = \frac{1}{2} \times 100 = 50\% \quad \textbf{Ans.}$$

(iii) Power rating of the transistor = Zero-signal power dissipated

$$= V_{CC}\,I_{CQ} = 10 \times 0.2 = 2\ \text{W} \quad \textbf{Ans.}$$

$$\text{Load, } R_L = 2\ \Omega$$

Load, as seen by the transformer primary,

$$R'_L = a^2\,R_L = (5)^2 \times 2 = 50\Omega,$$

which is also the ratio of $\dfrac{V_{CC}}{I_{CQ}}\left(\dfrac{10}{0.2}\right)$

Hence, the load is properly matched. **Ans.**

EXAMPLE 5.5. A transformer-coupled class A audio-power amplifier driving and 8 W speaker is shown in figure 5.14. If the circuit component values result in a dc base current of 6 mA a and the input signal V_{in} results in a peak base swing of 4 m A.

Determine:

(i) $V_{ce\,max}$,
(ii) $V_{ce\,min}$,
(iii) $I_{c\,max}$,
(iv) $I_{c\,min}$, **(v) rms values of load current and voltage,**
(vi) ac power developed across the load,
(vii) power dissipated by the transistor, and
(viii) the efficiency of the amplifier circuit. Transistor output characteristics are shown in figure 5.15.

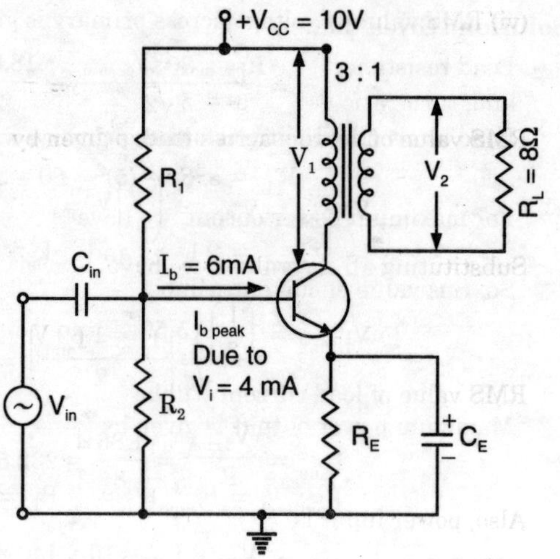

Fig. 5.14.

Solution : $V_{CEQ} = V_{CC} = 10$ V, so dc load line drawn vertically from the voltage point $V_{CEQ} = 10$ V.

For base current $I_b = 6$ mA, and operating point Q in figure 5.14 gives

$$V_{CEQ} = 10 \text{ V} \quad \text{and} \quad I_{CQ} = 140 \text{ m A}$$

Effective a.c. resistance

$$R'_L = a^2 R_L = (3)^2 \times 8 = 72\ \Omega$$

Current swing, $\Delta I_c = \dfrac{\Delta V_{ce}}{R'_L} = \dfrac{10\text{V}}{72\Omega} = 139$ mA

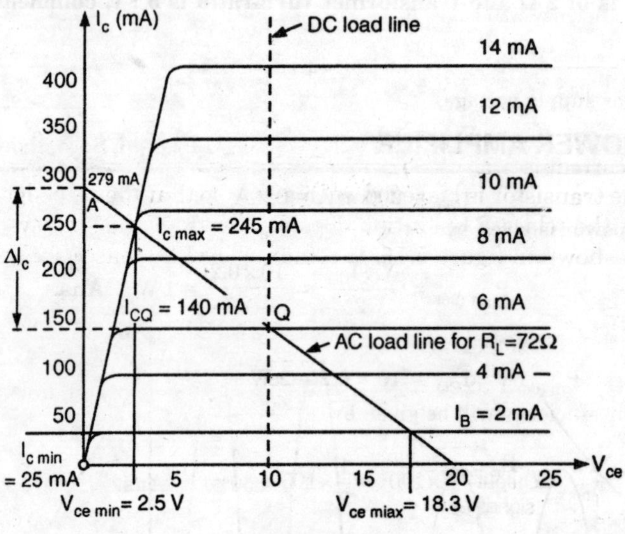

Fig. 5.15.

Now point A is marked corresponding to collector current of 140 + 139 = 279 mA on the current axis. AC load line is now drawn joining points A and Q.

For the given peak base current swing of m A, from figure 5.15,

(i) $V_{ce\,max} = 18.3$ V **Ans.**

(ii) $V_{ce\,min} = 2.5$ V **Ans.**

(iii) $I_{c\,max} = 245$ m A **Ans.**

(iv) $I_{c\,min} = 25$ m A **Ans.**

(v) AC power developed across transformer primary is given by

$$P_{out\,(ac)} = \frac{(V_{ce\,max} - V_{c\,min})(I_{c\,max} - I_{c\,min})}{8}$$

Substituting all the values, we have

$$= \frac{(18.3 - 2.5)(245 - 25) \times 10^{-3}}{8} = 0.434 \text{ W} \quad \textbf{Ans.}$$

(vi) RMS value of voltage across primary is given by

$$V_{1rms} = \frac{V_{ce\,max} - V_{ce\,min}}{2\sqrt{2}} = \frac{18.3 - 2.5}{2\sqrt{2}} = 5.59 \text{ V}$$

RMS value of voltage across load is given by

$$V_{L\,rms} = V_{2rms} = \left(\frac{1}{a}\right) V_{1\,rms}$$

Substituting all the values, we have

$$V_{L\,rms} = \left(\frac{1}{3}\right) \times 5.59 = 1.89 \text{ V} \quad \textbf{Ans.}$$

RMS value of load current will be

$$I_{rms} = \frac{V_{2\,rms}}{R_L} = \frac{1.86}{8} = 232.5 \text{ m A}$$

Also, power input is

$$P_{in\,(d.c.)} = V_{CC} \times I_{CQ} = 10 \times 140 \times 10^{-3} = 1.4 \text{ W}$$

(vii) Power dissipated in the transistor,

$$P_d = P_{in\,(dc)} - P_{out(ac)} = 1.4 - 0.434 \text{ W} = 0.966 \text{ W} \quad \textbf{Ans.}$$

(viii) Amplifier circuit efficiency will be

$$\eta = \frac{P_{out(ac)}}{P_{in(dc)}} \times 100 = \frac{0.434}{1.4} \times 100 = 31\% \quad \textbf{Ans.}$$

5.8. THE CLASS-B POWER AMPLIFIER*

(RGTU Bhopal, Sem. Exam; May/June, 2006)

In a class B operation, the transistor is biased in such a way so that the zero-signal collector current is zero. This means that the class B operation does not need any biasing system. The operating point is set at cut-off, as shown in figure 5.16. It remains forward biased for only half-cycle of the input signal.

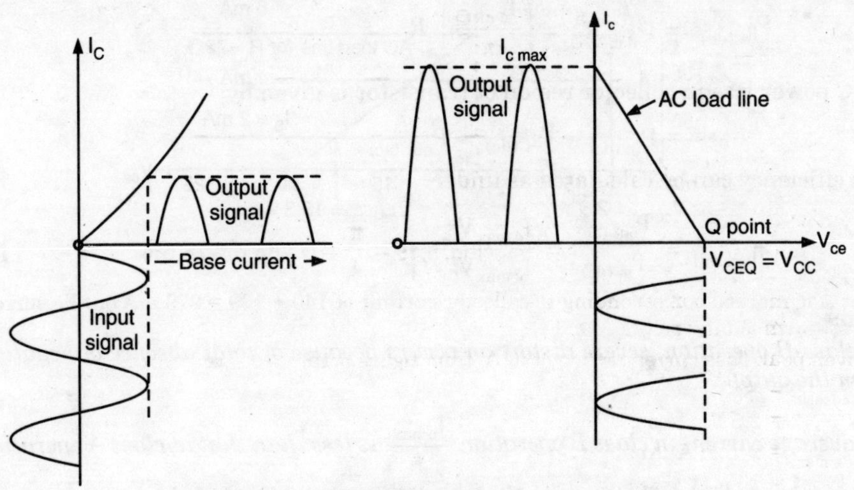

Fig. 5.16. Illustration of class-B operation

As shown in figure 5.16, during the positive half cycle of the input ac signal, the circuit is forward biased and hence, collector current flows. On the other hand, during negative half cycle of the input ac signal, the circuit is reverse biased and no collector current flows.

* What is the difference between power amplifier and voltage amplifier? Draw circuit diagram of a class-B power amplifier and briegfly explain. **(RGTU Bhopal, Sem. Exam; May/June 2006)**

5.8.1. Calculation of Power and Efficiency

Input dc power, $P_{in\ (dc)} = V_{CC}\ I_{dc}$ where I_{dc} is the average of direct current taken from the collector supply.

If $I_{c\ max}$ is the maximum or peak value of collector or output current, then

$$I_{dc} = \frac{1}{2\pi}\int_0^{\pi} I_{c\ max} \sin\theta\ d\theta = \frac{I_{c\ max}}{2\pi}[-\cos\theta]_0^{\pi} = \frac{I_{c\ max}}{\pi}\ *$$

Therefore, $$P_{in\ (dc)} = \frac{V_{CC}\ I_{c\ max}}{\pi} \quad ....(5.24)$$

RMS value of collector current will be

$$I_{c\ rms} = \frac{I_{c\ max}}{\sqrt{2}}$$

RMS value of output voltage will be

$$V_{rms} = \frac{V_{CC}}{\sqrt{2}}$$

Hence, output power during half cycle will be

$$P_{out(ac)} = \frac{1}{2}\frac{I_{c\ max}}{\sqrt{2}} \times \frac{V_{CC}}{\sqrt{2}} = \frac{I_{c\ max}\ V_{CC}}{4} \quad ...(5.25)$$

In above expression, factor $\frac{1}{2}$ is used because power is developed during one half cycle only.

DC power loss in load is given by

$$P_{RC(dc)} = I^2_{d.c.}\ R_C = \left(\frac{I_{c\ max}}{\pi}\right)^2 R_C \quad ...(5.26)$$

Also, DC power loss in collector region of transistor is given by

$$= P_{in\ (dc)} - P_{RC(dc)} - P_{out\ (ac)} \quad ...(5.27)$$

Overall efficiency can be calculated as under

$$\eta_{overall} = \frac{P_{out\ (ac)}}{P_{in\ (dc)}} = \frac{I_{c\ max}V_{CC}/4}{I_{c\ max}V_{CC}/\pi} = \frac{\pi}{4} = 0.785 \text{ or } 78.5\% \quad ...(5.28)$$

Conclusions :

(i) In class-B operation, severe distortion occurs because of total absence of negative half cycle from the output.

(ii) As average current in class B operation $\frac{I_{c\max}}{\pi}$ *is less than that in class A operation, so power dissipation is less and consequently overall efficiency is high**. In general, overall efficiency is given by*

* [Since, $i_c = I_{c\ max} \sin\theta$ for $0 < \theta < \pi$]

** It is 78.5% when peak input signal makes $V_{cemin} = 0$

$$\eta_{overall} = 78.5\left[1 - \frac{V_{ce\,min}}{V_{CC}}\right]\% \qquad ...(5.29)$$

(iii) *Zero signal input represents the best condition for class B operation as collector current is zero. This is quite contrary to class A operation. In class B operation, the transistor dissipation increases with increased input signals.*

(iv) *Half sine-wave type of collector current contains very pronounced even harmonics (particularly the second one). Such impulses can be made useful either by using push-pull arrangement or by employing tuned amplifiers.*

EXAMPLE 5.6. Find the overall efficiency of class B power amplifier when V_{CC} = 20 V and $V_{ce\,min}$ = 2.5 V

Solution : We know that

$$\text{Overall efficiency, } \eta_{overall} = 78.5\left[1 - \frac{2.5}{20}\right] = 68.69\% \quad \textbf{Ans.}$$

5.9. THE CONCEPT OF HARMONIC DISTORTION IN POWER AMPLIFIERS

Output signal variations of less than 360° of the signal cycle are considered to have distortion. This means that the output signal is no longer just an amplifier version of the input signal but in some ways a distorted or changed from that of the input. The poor quality of music coming from a radio or hi-fi system with the music or voice no longer sounding like that which was originally recorded or transmitted is the result of distortion.

A good amplifier should not only give an enlarged version of the input signal at the output, but should also provide a faithful reproduction of the input waveforms. One type of distortion that is common to most of the amplifier is called the amplitude or harmonic distortion, which is caused due to non-linearity of the active device employed for amplification. The active device (semiconductor device or tube) used for amplification may not increase equally all portions of the input signal over its positive excursions. Harmonic distortion increases as we go from class A operation to class C operation. Even in class A operation, some distortion occurs.

When non-linear distortion is present, the output waveform contains components of frequencies which are harmonics (integer multiples) of the signal frequency. There may be present second, third and higher order harmonic components, but the most important in terms of the amount of distortion for the classes of operation, we will consider is the second harmonic. For a signal occurring in class A B or class B, the distortion may be mainly even harmonic of which the second harmonic components is the greatest.

As instrument such as a spectrum analyser permits the measurement of the harmonics present in the signal by providing a display of the fundamental component of signal and a number of harmonics on a CRT screen. Similarly, a wave analyzer instrument allows a more precise measurement of the harmonic components and providing a reading of these components, one at a time. In any case, the technique of considering any distorted signal as containing a fundamental component and harmonic component is practical and useful.

5.9.1. Determination of Second Harmonic Distortion

A collector current waveform is shown in figure 5.17, with the quiescent, minimum and maximum, signal levels, and the time they occur marked on the waveform. An equation which approximately describes the distorted signal waveform is

$$i_C = I_{CQ} + I_0 + I_1 \cos(\omega t) + I_2 \cos(2\omega t) \qquad ...(5.30)$$

The given current waveform contain C the original quiescent current I_{CQ}, that occurs with zero input signal, an additional d.c. current I_0, due to the non-zero average of the distorted signal the fundamental component of the distorted ac signal I_1 and a second harmonic component I_2, at twice the fundamental frequency. Although other harmonics are present, only the second is considered here. Equating the resultant current from equation 5.30 at a few points in the cycle to that shown on the current waveform we get

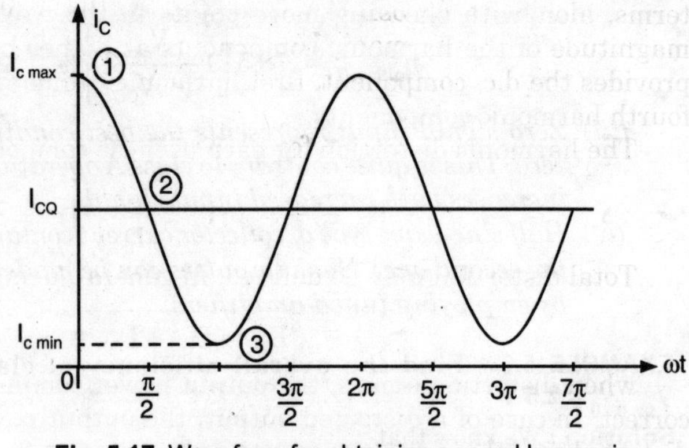

Fig. 5.17. Waveform for obtaining second harmonic distortion

At point ① i.e., when $\omega t = 0$

$$\begin{aligned} i_C &= I_{c\,max} \\ &= I_{CQ} + I_0 + I_1 \cos(0) + I_2 \cos(0) \\ &= I_{CQ} + I_0 + I_1 + I_2 \end{aligned} \qquad ...(5.31)$$

At point ② i.e., when $\omega t = \frac{\pi}{2}$, we have

$$i_C = I_{CQ} = I_{CQ} + I_0 + I_1 \cos\left(\frac{\pi}{2}\right) + I_2 \cos\left(\frac{2\pi}{2}\right) = I_{CQ} + I_0 + I_2 \qquad ...(5.32)$$

At point ③ i.e., when $\omega t = \pi$, we have

$$\begin{aligned} I_C &= I_{cmin} = I_{CQ} + I_0 + I_1 \cos(\pi) + I_2 \cos(2\pi) \\ &= I_{CQ} + I_0 - I_1 + I_2 \end{aligned} \qquad ...(5.33)$$

Solving equations (5.31), (5.32) and (5.33) we get

$$I_0 = I_2 = \frac{I_{c\,max} + I_{c\,min} - 2I_{CQ}}{4} \qquad ...(5.34)$$

and
$$I_1 = \frac{I_{c\,max} - I_{c\,min}}{2} \qquad ...(5.35)$$

By definition, the percent of a second harmonic distortion is given by

$$D_2 = \frac{I_2}{I_1} \times 100\% = \frac{\frac{1}{2}\left(I_{c\,max} + I_{c\,min}\right) - I_{CQ}}{I_{c\,max} - I_{c\,min}} \times 100 \qquad ...(5.36)$$

The second harmonic distortion is the percent of the second harmonic component present in the output current waveform with respect to the amount of the fundamental component. Obviously, 0% distortion is the ideal condition of no distortion.

In a similar way, the percent of second harmonic distortion present in the output voltage waveform w.r.t. the amount of the fundamental component can be determined and is given by

$$D_2 = \frac{V_2}{V_1} \times 100\% = \frac{\frac{1}{2}\left(V_{ce\,max} + V_{ce\,min}\right) - V_{CEQ}}{V_{ce\,max} - V_{ce\,min}} \times 100 \qquad ...(5.37)$$

This is the required expression.

5.9.2. Five Point Method of Determination of Harmonic Distortion in Power Amplifiers

The method of determination of second harmonic distortion is called the 3-point method as it involved equating the assumed waveform of the output current/voltage to the measured voltage at three points in the signal cycle. Using an assumed output signal equation containing more harmonic terms, alongwith choosing more points in the waveform, results in obtaining relations for the magnitude of the harmonic components at higher harmonic frequencies. Using a 5-point method provides the d.c. component, first harmonic (fundamental), second harmonic, third harmonic, and fourth harmonic components.

The harmonic distortion for each of these components is then defined as

$$D_2 = \frac{I_2}{I_1}; \quad D_3 = \frac{I_3}{I_1}; \quad D_4 = \frac{I_4}{I_1} \text{ and so on} \qquad ...(5.38)$$

Total distortion may be defined, in general, using the individual distortion components

$$D = \sqrt{D_2^2 + D_3^2 + D_4^2 + ...} \qquad ....(5.39)$$

when distortion occurs, the output power computed on the basic of non-distortion is no longer correct. In case of a distorted output, the output power due to the fundamental component of the distorted signal is given as

$$P_1 = \frac{I_1^2 R_C}{2}$$

Similarly $$P_2 = \frac{I_2^2 R_C}{2}; \quad P_3 = \frac{I_3^2 R_C}{2} \text{ and so on.}$$

Total power due to all harmonic components of the distorted signal is given by

$$P = (I_1^2 + I_2^2 + I_3^2 + ...)\frac{R_C}{2} = \left[1 + \left(\frac{I_2}{I_1}\right)^2 + \left(\frac{I_3}{I_2}\right)^2 + ...\right]\frac{I_1^2 R_C}{2}$$

or $$P = [1 + D_2^2 + D_3^2 + ...]\, P_1 = (1 + D^2)\, P_1 \qquad ...(5.40)$$

If the total distortion is 10% i.e., 0.1, then the total power is $[1 + (0.1)^2]\, P_1 = 1.01\, P_1$.

Percentage increase in power = $\frac{1.01P_1 - P_1}{P_1} \times 100$ *= 1.0% i.e., a 10% distortion represents a power increase of only 1% of the fundamental. Hence, only a small error in introduced in using only the fundamental term P_1 for determination of the output power.*

EXAMPLE 5.7. An output waveform displayed on an oscilloscope provided the following measured values.

(a) $V_{ce\,min} = 1.2$ V, $V_{ce\,max} = 22$V, $V_{CEQ} = 10$ V

(b) $V_{ce\,min} = 2$ V, $V_{ce\,min} = 18$ V, $V_{CEQ} = 10$ V

Determine the percent second harmonic distortion in each case.

Solution : (a) According to equation (5.37), we have

$$D_2 = \frac{\frac{1}{2}(V_{ce\,max} + V_{ce\,min}) - V_{CEQ}}{V_{ce\,max} - V_{ce\,min}} \times 100$$

or $$D_2 = \frac{\frac{1}{2}(22 + 1.2) - 10}{22 - 1.2} \times 100 = 7.7\% \quad \textbf{Ans.}$$

(b) we have $$D_2 = \frac{\frac{1}{2}(18 + 2) - 10}{18 - 2} \times 100 = 0\% \text{ i.e., no distortion} \quad \textbf{Ans.}$$

EXAMPLE 5.8. A sinusoidal signal $V_s = 1.95 \sin 400t$ is applied to a power amplifier. The resulting current is

$$i_o = 12 \sin 400\,t + 1.2 \sin 800\,t + 0.9 \sin 1{,}200\,t + 0.4 \sin 1{,}600\,t$$

Calculate : (i) the total harmonic distortion

(ii) the percentage increase in powers because of distortion.

Solution : Harmonic distortion of each component is expressed as

$$D_2 = \frac{V_2}{V_1} = \frac{1.2}{12} = 0.1$$

$$D_3 = \frac{V_3}{V_1} = \frac{0.9}{12} = 0.075$$

$$D_4 = \frac{V_4}{V_1} = \frac{0.4}{12} = 0.0333$$

Total distortion will be

$$D = \sqrt{D_2^2 + D_3^2 + D_4^2}$$

$$D = \sqrt{(0.1)^2 + (0.075)^2 + (0.0333)^2}$$

$$D = 0.1294 \quad \text{or} \quad 12.94\% \qquad \textbf{Ans.}$$

Total power, $P = (1 + D^2)\,P_1 = [1 + (0.1294)^2]\,P_1$

Now, percentage increase in power because of distortion will be calculated as

$$= \frac{P - P_1}{P_1} \times 100 = \frac{(1.01674 - 1)}{P_1} \times 100 = 1.674\% \qquad \textbf{Ans.}$$

5.10. THE CLASS AB POWER AMPLIFIERS

In class AB power amplifiers, the biasing circuit is so adjusted that the operating point Q lies near the cut-off voltage. During a small portion of negative half cycle and for complete positive half cycle of the signal, the input circuit remains forward biased and hence collector current flows. But during small portion (less than half cycle) of the negative cycle, the input circuit is reverse biased and, therefore, no collector current flows during this period. Class AB operation needs a push-pull connection to achieve a full output cycle.

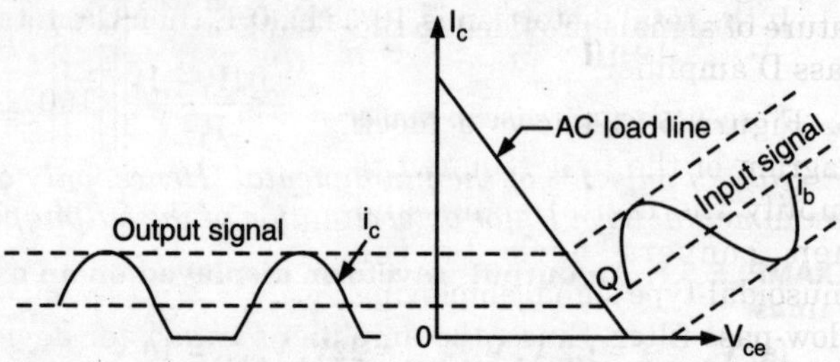

Fig. 5.18. Class AB operation.

5.11. CLASS C AMPLIFIERS

A Class C power amplifier is biased to operate for less than 180° of the input signal cycle as shown in figure 5.19. The tuned circuit in the output, however, will provide a full cycle of output signal for the fundamental or resonant frequency of the tuned circuit (L and C tank circuit) of the output. The use of such amplifiers is, therefore, limited for a fixed frequency, as occurs in communication circuits, for example. Operation of a Class C circuit is not intended primarily for large signal or power amplifiers.

5.12. CLASS D POWER AMPLIFIERS

Historically, audio amplifiers have been configured as class A ,class B or class AB and the art of the design is well known. Also well known is the poor efficiency of these amplifiers compared to that of Class D amplifiers. Whereas the theoretical best efficiency for class B amplifiers is 78.5%, the practical upper limit is more nearly 70% when driving a purely resistive load. But when driving real speaker loads which can have power factor angles of 60° or more, efficiency can comedown to 55% or less.

Class D amplifiers, however, can attain efficiency of 90% and with careful component choices can exceed 95% even. Moreover, the power factor of the load does not affect the on-state power losses in the MOSFET switches normally used in such amplifiers. Using Class D techniques, amplifiers capable of delivering several hundred watts to the load can be designed using small, inexpensive, stamped heat sinks. Efficiency of class D amplifiers open the possibility for powerful, small, light amplifiers with good sound quality to be designed today.

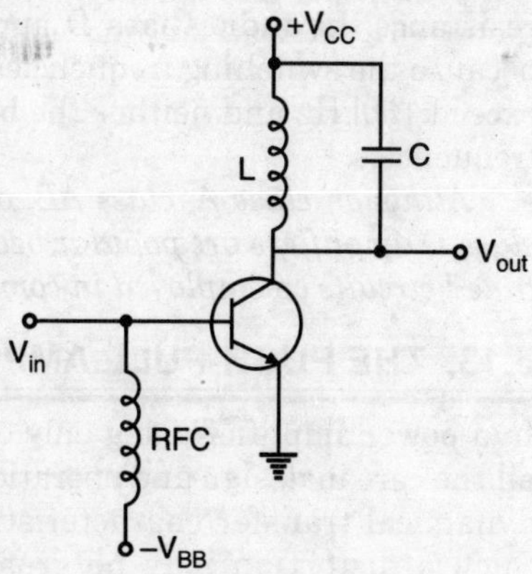

Fig. 5.19. Class C amplifer circuit

A class D amplifier is designed to operate with digital or pulse-type signals. It is necessary, however, to convert any input signal into pulse-type waveform before using it to drive a large power load and to convert the signal back to a sinusoidal-type signal to recover the original signal. Figure 5.20 illustrates how a sinusoidal signal may be converted into a pulse-type signal using some form of sawtooth or chopping waveform to be applied with the input into a comparator type op-amp circuit so that representative pulse type signal is generated. While the letter D is used to the described next type of bias operation after class C, the D could also be considerd to stand for "Digital" since that is the nature of signals provided to the class D amplifier.

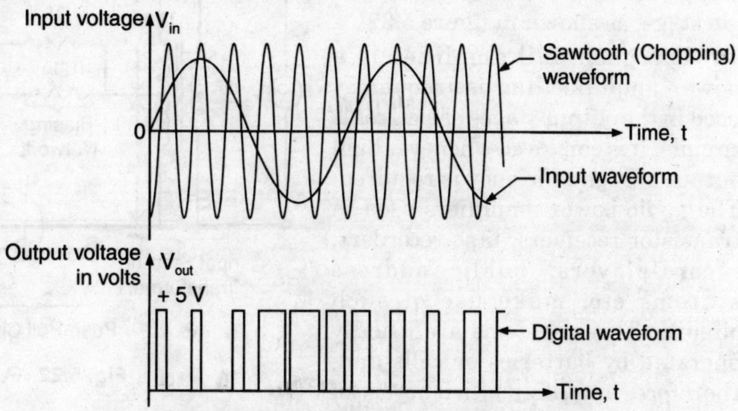

Fig. 5.20. Chopping of sinusoidal waveform for generating digital waveform.

Figure 5.21 shows a block diagram of the unit required to amplify the digital signal and then convert back to the sinusoidal-type signal employing a low-pass filter. Since, the amplifier's transistor devices used to provide the output are basically either OFF or ON, they provide current only when they are turned ON. With little power loss due to their low on-voltage. Since most of the power supplied to the amplifier is transferred to the load, the efficiency of the circuit is typically very high. Power MOSFET devices have been quite popular as the driver devices for the Class D power amplifier.

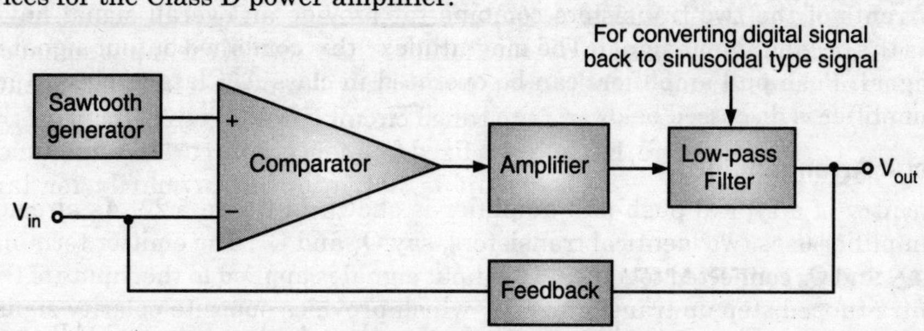

Fig. 5.21. Block diagram of Class D power amplifier.

As a matter of fact, class-D efficiency is largely determined by the ratio of the load resistance to the total dc loop resistance which is the sum of the $r_{DS(ON)}$ of the MOSFETs, wire resistances (including the output filter) and current sense resistor if used , and the load resistance. For highest efficiency, the MOSFET $r_{DS(ON)}$ resistances shunt and filter resistances should be small compared to the load resistance. In audio Class D applications, MOSFETS are employed instead of IGBTs and BJTs because the switching frequencies required, to keep distortion low at 20kHz signal frequencies can exceed 150kHz and neither the bipolar power transistors or IGBTs switch efficiently at such high frequencies.

Although class A, class AB, and Class B, amplifiers are mostly employed as power amplifiers, class D amplifiers are popular because of their very high efficiency. Class C amplifiers do find use in tuned circuits as employed in communications.

5.13. THE PUSH-PULL AMPLIFIERS*

(RGTU Bhopal, Sem. Exam; June 2007)

In a power amplifier using only one transistor, called the signal- ended power amplifier, inspite of all the care in design and operation, appreciable distortion is introduced because of non-linearity of dynamical transfer characteristics. Such a distortion may be greatly reduced by using push-pull operation employing two transistors in a single stage, as shown in figure 5.22.

The push-pull amplifier is a power amplifier and is frequently used in the output stages of electronic circuits. It is employed whenever high output at high efficiency is required. The audio power amplifiers used in transistor receivers, tape-recorders, record-players, public address systems etc. make use of such circuits. These systems are usually operated by batteries or cells and, therefore, amplifier efficiency is of prime importance.

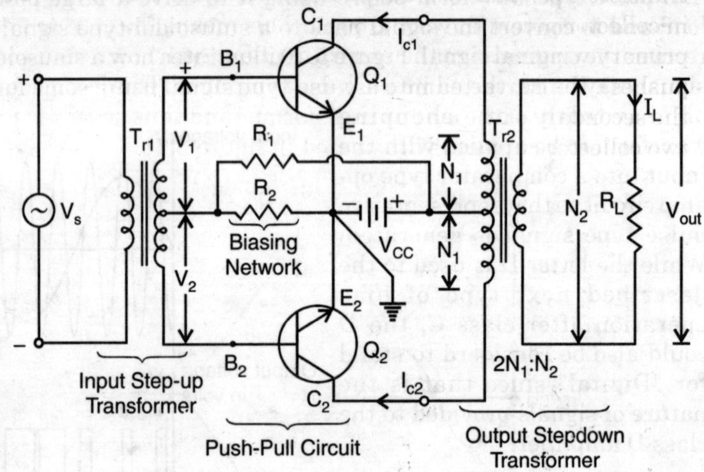

Fig. 5.22. Push-Pull Amplifier Circuit.

5.13.1 Basic Principle of Working of Push-Pull Amplifier

The basic principle on which a push-pull amplifier operates is that the input signal is converted before amplification, into two separate signals, which are identical except for a 180 phase difference. Each of these signals is applied as the input to one of the transistors of the push-pull amplifier. Since the inputs to the two transistors are 180° out of phase, so the output of the transistors. The output transformer is connected in the collector circuits of the transistor in such a way that the collector currents of the two transistors combine to provide an overall signal having the same waveform as the original input signal. The magnitude of this combined output signal is larger than the input signal. Push-pull amplifiers can be operated in class A, Class or class B mode. Class A push-pull amplifier is discussed below.

5.13.2 Circuit Arrangement

The circuitry of a typical push-pull amplifier is shown in figure 5.22. As already mentioned, push-pull amplifier uses two identical transistors, say Q_1 and Q_2. The emitter terminals of the two transistors Q_1 and Q_2 connected together. The input signal is applied to the inputs of two transistors through centre-tapped step up transformer T_{r1} which provides opposite polarity signals to the two transistors. The collector terminals of both the transistors are connected to end terminals of centre-tapped primary of output transformer T_{r2} . The power supply V_{CC} is connected between the emitter terminals and the centre-tap of primary of output transformer. Resistors R_1 and R_2 are used to

* Explain why even harmonics are not present in push-pull transistor? Explain the working of class-A push-pull amplifier. Determine its maximum conversion efficiency. **(RGTU Bhopal, Sem. Exam; June 2007)**

provide the proper bias for the circuit. The load R_L is connected across the secondary of the output transformer Tr_2. The turn-ratio ($2N_1 : N_2$) of the output transformer is chosen so as to match the load with the output impedance of the amplifier and therefore, transfer maximum power. The quiescent currents of the two transistors, which are equal in magnitude, flow in opposite directions through each half of primary of the output transformer T_{r2}, so no saturation of the magnetic core occurs.

5.13.3 Working Operation of Circuit

When the base current of one transistor is being driven positive with respect to the quiescent point Q, the collector current increases, thus causing a decrease in collector potential relative to ground. At the same time, however, a reverse action takes place in the base circuit of the second transistor, i.e., base current decreases causing a drop in the collector current with a consequent rise in collector potential w.r.t ground. This means that the ac current flowing through the transformer primary winding is in the same direction. As I_{c1} increases (i.e. pulls), the current I_{c2} decreases (i.e., pushes). Hence, the name push-pull amplifier. The overall operation results in ac voltage induced in the secondary of the output transformer and thus ac power is delivered to the load. The difference of two collector current is illustrated in figure 5.23.

$$i_{C1} - i_{C2} = i_{C1} + (-i_{C2})$$

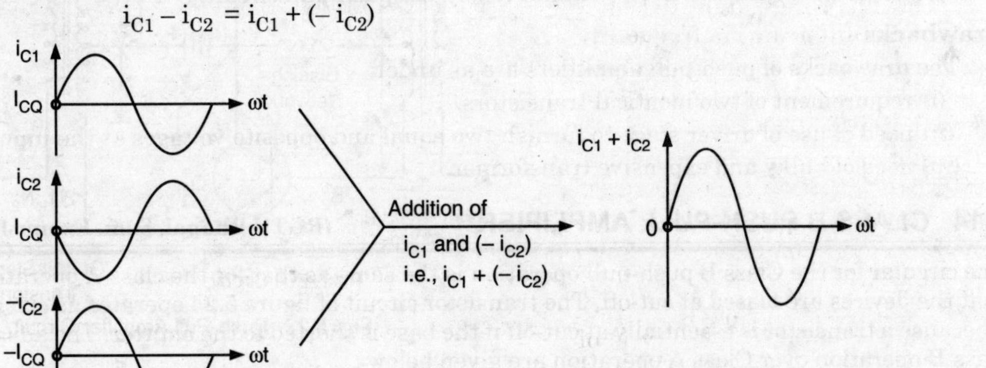

Fig. 5.23. Illustration of the difference of two collector currents.

5.13.4. Distortion

In view of the fact that the load current in the secondary of the output transformer is proportional to the difference of collector currents, it follows that the harmonic content will be less than it is in either of the collector currents because of cancellation effects. Thus the push-pull arrangement yield much less distortion in the output. This is illustrated below.

The base currents i_{b1} and i_{b2} of transistors Q_1 and Q_2 respectively are expressed as

$$i_{b1} = I_b \sin(\omega t)$$

and $$i_{b2} = I_b \sin(\omega t + \pi)$$

Their collector currents are expressed as

$$i_{C1} = I_C + I_1 \sin \omega t + I_2 \sin 2\,\omega t + I_3 \sin \omega t + \ldots$$

$$i_{C2} = I_C + I_1 \sin(\omega t + \pi) + I_2 \sin(2\omega t + 2\pi) + I_3 \sin(3\omega t + 3\pi)$$

$$= I_C - I_1 \sin \omega t + I_2 \sin 2\omega t - I_3 \sin 3\omega t + \ldots$$

and $$i_{C1} - i_{C2} = 2\,I_1 \sin \omega t + 2\,I_3 \sin 3\omega t + \ldots$$

The output voltage induced in the output transformer secondary, being proportional to ($i_{C1} - i_{C2}$),

$$v_{out} = k(i_{C1} - i_{C2}) = 2k\,[I_1 \sin \omega t + I_3 \sin 3\omega t + I_5 \sin 5\omega t + \ldots]$$

From the above equation for output voltage V_{out}, it is obvious that all even harmonics are eliminated. This conclusion was reached on the assumption that the two transistors are identical. If their characteristics differ appreciably, the appearance of even harmonics is not ruled out.

Advantages*

(i) Because of absence of even harmonics in the output of the push-pull amplifier, such a circuit gives more output per active for a given amount of distortion or less distortion for a given power output per transistor.

(ii) As mentioned earlier, the dc components of the collector currents oppose each other magnetically in the transformer core. This eliminates any tendency toward core saturation and consequent non-linear distortion that may arise from the curvature of the transformer magnetization curve.

(iii) Another advantage of this system is that the effects of ripple voltage that may be contained in the power supply because of inadequate filtering will be balanced out. This cancellation results because the currents produced by the ripple voltage are in opposite directions in the transformer winding, and so will not appear in the load. Of course, the power supply hum will also act on the voltage- amplifier stages, and so will be part of the input to the power stage. This hum will not be eliminated by the push-pull circuit.

Drawbacks

The drawbacks of push-pull amplifiers are as under:

(i) requirement of two identical transistors.

(ii) need of use of driver stage to furnish two equal and opposite voltages at the input

(iii) need of bulky and expensive transformer.

5.14. CLASS B PUSH-PULL AMPLIFIER** (RGTU Bhopal, Sem. Exam; June 2004)

The circular for the Class B push-pull operation is the same as that for the class A operation except that the devices are biased at cut-off. The transistor circuit of figure 5.24 operates as class B if $R_2 = 0$ because a transistor is essentially at cut-off if the base is shorted to the emitter. The advantages of class B operation over Class A operation are given below :

(i) It is possible to obtain greater power output. In fact class B push pull amplifier provides practically 4-times the power supplied by a single-ended amplifier provided load resistance remains the same.

(ii) It gives higher operating efficiency (theoretical 78.5%). It is primarily due to the fact that no power is drawn by the circuit under zero-signal condition.

Because of above advantages, a push-pull class B transistor circuit is preferred in systems where the power supply is limited, such as those operating from solar cells or a battery. The automatic cancellation of all even-order harmonics from the output makes class B push-pull amplifier highly desirable for communication sound equipment.

The drawbacks of Class B push-pull amplifiers are that the harmonic distortion is higher, self-bias cannot be used, and the supply voltages must have good regulation.

Class B push-pull amplifiers are widely employed for audio-work in portable record players, as stereo-amplifier and in high-fidelity radio receivers.

5.14.1. Circuit Arrangement

Circuitry for a class B push-pull amplifier is shown in figure 5.24. As illustrated in the figure, the bases of the two CE-connected identical transistors Q_1 and Q_2 have been connected to the opposite ends of the secondary of the input transformer T_{r1}. Load resistance R_L is coupled to the collector terminals of the two transistors through output transformer T_{r2}. The emitter terminals of the

* Discuss the merits of push pull power amplifier over single ended power amplifeir. (RGTU Bhopal, Sem. Exam; June 2008)

** Write a short note on push-pull class-B amplifier. (RGTU Bhopal, Sem. Exam; June 2004)

transistors are connected to the centre tap of the secondary of the input transformer T_{r1} while the supply V_{CC} is connected to the centre tap on the primary of output transformer. This is done in order to have balanced circuits. Since zero-bias is required for cut-off, the two bases of the two transistors have been grounded.

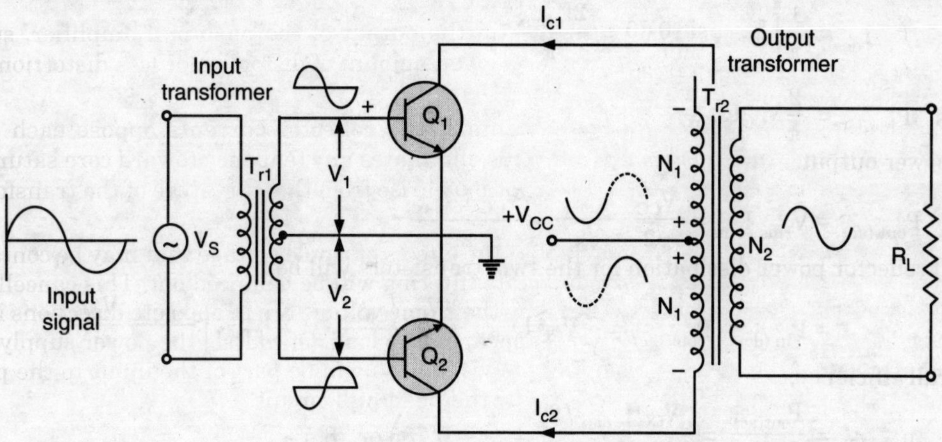

Fig. 5.24. Class B push-pull Amplifier.

5.14.2 Operating Principle* (RGTU Bhopal, Sem. Exam; Dec. 2006)

When the input signal is applied, the centre-tapped secondary of the input transformer develops two signals which are identical but in phase opposition. The transistor, Q_1 and Q_2 are driven by these two signals. Thus, when V_1 is going positive, V_2 is going negative, so that transistor Q_2 is being biased further off when transistor Q_1 is being biased on. As the collector current in Q_1 increases from zero, it produces a half sine wave of voltage across the upper half of the primary of the output transformer, as shown. When the positive half cycle of the input signal to Q_1 begins to go negative, the signal at Q_2 base is commencing to go positive. Thus ,as Q_1 becomes biased off again, Q_2 is biased on and a half cycle of voltage waveform is generated across the lower primary winding of output transformer.

The effect of the two half cycles in separate halves of the primary of output transformer is to produce a magnetic flux in the transformer core, which flows first in one direction and then in the opposite direction. This flux links with the secondary of the output transformer and generates a complete sine-wave output, which is passed on to the load. In the Class B circuit, the two output transistors are said to be operating in push-pull.

As explained above, a full sine-wave is developed at the output of the transformer. When no signal is present, both transistors remain off, and thus no current is drawn from dc supply source V_{CC}. Hence, there is no dissipation of power with zero-signal. Some power is dissipated in each transistor only while the transistor is conducting.

The load resistance (usually a loudspeaker) is connected across the secondary of the output transformer. The turn-ratio $2N_1 : N_2$ of the transistor is chosen so that load resistance R_L is matched with the output impedance of the transistor . Under matched conditions, maximum power is delivered to the load by the amplifier.

5.14.3. Power and Efficiency Calculations

The power provided to the load R_L coupled to the amplifier is drawn from the dc power supply V_{CC} and is considered an input dc power. The input dc power if given as

$$P_{in\ (dc)} = V_{CC}\, I_{dc}$$

* Explain the working of class-B push pull amplifier with the help of circuit diagram. Show that its maximum conversion efficiency is 78.5%. (RGTU Bhopal, Sem. Exam; Dec. 2006)

where I_{dc} is the average or direct current taken from the dc supply V_{CC}.

In class B operation the current drawn from a single power supply is a full-wave rectified signal, thus, we have

$$I_{dc} = \frac{1}{\pi}\int_0^{\pi} I_{c\,max} \sin\theta \, d\theta = \frac{2I_{c\,max}}{\pi}$$

Thus $$P_{in\,(dc)} = \frac{2}{\pi} V_{CC} I_{c\,max} \qquad ...(5.41)$$

AC power output,

$$P_{out\,(ac)} = V_{rms}\, I_{rms} = \frac{V_{CC}}{\sqrt{2}} \cdot \frac{I_{c\,max}}{\sqrt{2}} = \frac{V_{CC}\, I_{c\,max}}{2} \qquad ...(5.42)$$

Total collector power dissipation for the two transistors will be

$$= P_{in\,(dc)} - P_{out\,(ac)} = \frac{2}{\pi} V_{CC}\, I_{c\,max} - \frac{V_{CC}\, I_{c\,max}}{2} = 2\, I_{c\,max}\left(\frac{V_{CC}}{\pi} - \frac{V_{CC}}{4}\right)$$

Overall efficiency,

$$\eta_{overall} = \frac{P_{out(ac)}}{P_{in(dc)}} = \frac{V_{CC}\, I_{c\,max}/2}{2\,V_{CC}\, I_{c\,max}/\pi} = \frac{\pi}{4} = 0.785 \text{ or } 78.5\% \qquad ...(5.43)$$

5.14.4. Cross-Over Distortion

In addition to the distortion introduced due to the non-linearity of the collector characteristics and due to non-matching of the two transistors, there is one source of distortion, that caused by non-linearity of the input characteristic.

Recall that silicon transistors must have at least 0.5 V to 0.6 V of forward base emitter bias before they will go into conduction. Since in class B push-pull amplifiers, the forward bias is produced by the input signal, both of transistors will be non-conducting (or off), when the input signal is approximately ± 0.5 V. This forms a "dead band" in the input and produces cross-over distortion in the output. In sample words, cross-over conducting, as illustrated in figure 5.25 (a) The distortion introduced is called the cross-over distortion because it occurs during the time operation cross-over from one transistor to the other in the push-pull amplifier. The same is shown in figure 5.25(b) by using transfer characteristics of the two transistors.

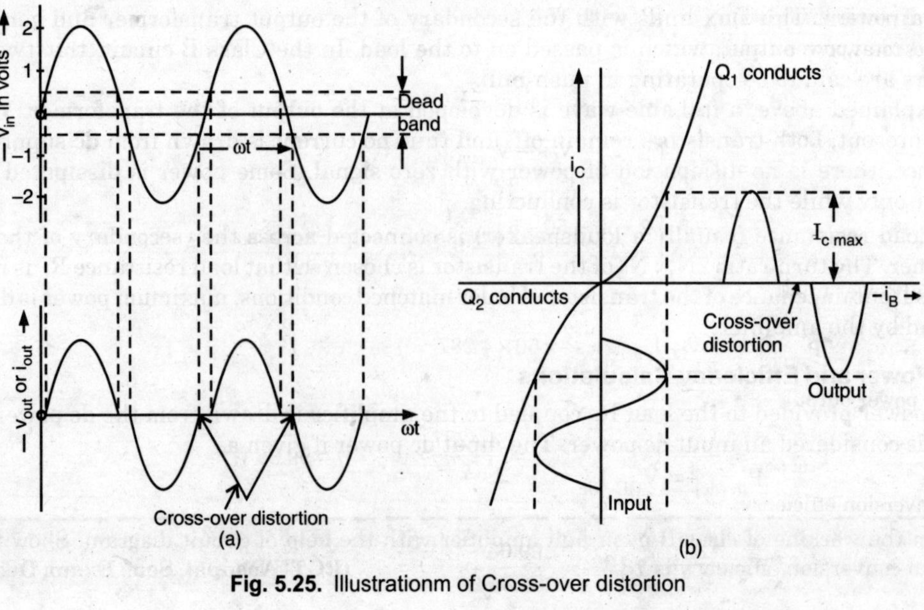

Fig. 5.25. Illustrationm of Cross-over distortion

To eliminate cross-over distortion, it is necessary to add a small amount of forward bias to take the transistors to the average of conduction or slightly beyond. This does slightly lowers the efficiency of the circuit and there is a waste of stand by power, but it alleviates the cross-over distortion problem. Technically, the operation of transistors lies between class B and class A mode. Therefore, the circuit operation is often referred to as being class AB operation.

EXAMPLE 5.9. A class B push-pull amplifier must deliver 10 W of audio-power to the output load. (i) If the output transformer is 80% efficient, what is the minimum power drain on the power supply under optimum conditions and (ii) what is the minimum average dissipation rating required for each transistor ?

Solution : Power delivered to load,

$$P_L = 10 \text{ W}$$

AC power output of amplifier,

$$P_{out\,(ac)} = \frac{P_L}{\text{Output transformer efficiency}} = \frac{10}{0.8} = 12.5 \text{ W}$$

Efficiency of a class B push-pull amplifier under optimum conditions,

$$\eta = 78.5\% = 0.785$$

(i) Minimum power drain on the power supply

$$= P_{in\,(dc)} = \frac{P_{out\,(ac)}}{\eta} = \frac{12.5}{0.785} = 15.9 \text{ W} \quad \textbf{Ans.}$$

(ii) Minimum average power dissipation,

$$P_d = P_{in\,(dc)} - P_{out\,(ac)} = 15.9 - 12.5 = 3.4 \text{ W}$$

Minimum average dissipation rating required for each transistor

$$= \frac{P_d}{2} = \frac{3.4}{2} = 1.7 \text{ W} \quad \textbf{Ans.}$$

EXAMPLE 5.10. A class B push-pull power amplifier is supplied with V_{CC} = 50 V. The signal swings the collector voltage down V_{min} = 5 V. The total dissipation in both transistors is 40 W. Find the total power and conversion efficiency.

Solution : DC supply voltage,

$$V_{CC} = 50 \text{ V}; \quad V_{min} = 5 \text{ V}$$

Total power dissipation, $P_d = 40$ W

Also, total power dissipation,

$$P_d = P_{in\,(dc)} - P_{out\,(ac)} = 40 = \frac{2}{\pi} V_{CC} I_{c\,max} - \frac{I_{c\,max}}{2}(V_{CC} - V_{c\,min})$$

$$P_d = I_{c\,max}\left[\frac{2V_{CC}}{\pi} - \frac{V_{CC} - V_{c\,min}}{2}\right]$$

$$P_d = I_{c\,max}\left[\frac{2 \times 50}{\pi} - \frac{50 - 5}{2}\right] = 9.33\, I_{c\,max} = \frac{40}{9.33} = 2.287 \text{ A}$$

Total power input,

$$P_{in\,(dc)} = \frac{2}{\pi} V_{CC} I_{c\,max} = \frac{2}{\pi} \times 50 \times 4.287 = 136.45 \text{ W} \quad \textbf{Ans.}$$

AC power output,

$$P_{out\,(ac)} = \frac{I_{c\,max}}{2}(V_{CC} - V_{min}) = \frac{4.4287}{2} \times (50 - 5) = 99.65 \text{ W}$$

Conversion efficiency,

$$\eta = \frac{P_{out\,(ac)}}{P_{in\,(dc)}} \times 100 = \frac{99.65}{136.345} \times 100 = 73.03\% \quad \textbf{Ans.}$$

5.15. CLASS AB PUSH-PULL AMPLIFIERS

The basic circuit of class A B push-pull amplifier is the same as that of class A Push-pull amplifier shown in figure 5.26 except that the voltage drop across resistor R_2 is so adjusted that it is approximately equal to cut-in voltage (0.5 V for Si and 0.1 for Ge). Now operation becomes A B type i.e., the collector current flows for more than half the cycle of the input signal but less than the complete cycle. The distortion introduced in class A B push-pull amplifier is less than that in class B push-pull amplifier but more than in Class B push-pull amplifier are lower conversion efficiency and wastage of stand by power.

5.16. PHASE-SPLITTER CIRCUIT FOR PUSH-PULL AMPLIFIER

In a push-pull amplifier circuit an input transformer is used to provide the polarity inversion between the two input signal to the push-pull circuit, but the transformer is quite expensive and off bulky. The same result may be achieved by using a circuit called the phase-splitter or phase inverter. Such a circuit is shown in figure 5.26.

The signal appearing at the collector is of polarity opposite to that of the input signal. During positive half cycle of input voltage, base current, and therefore, collector current increases causing increases in $I_c R_C$ drop and consequently decrease in collector-emitter voltage V_{ce}. Thus we get negative half cycle of V_{ce} i.e., phase of signal at collector of a CE transistor is reversed. The output from the emitter is of the same polarity as the input. Thus the two output signal obtained from this circuit are of opposite polarity. Values of resistors R_C, R_E and β are chosen so as to make the voltage gain for the output signal at collector to be unity. The gain for signal at emitter is 1. Thus the circuit gives two signals of equal magnitude but of opposite particles. The use of input transformer can thus be eliminated.

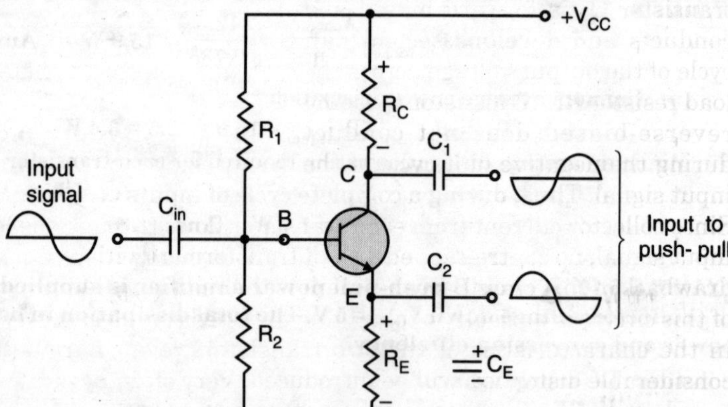

Fig. 5.26. Phase-splitter circuit for push-pull amplifier.

5.17. COMPLEMENTARY SYMMETRY PUSH-PULL CLASS B POWER AMPLIFIER

The use of transformer, at input as well as at output ends in the push-pull amplifier shown in figure 5.27, makes it bulky and expensive especially in this age of the integrated circuits. Another drawbacks of the circuit given in figure 5.27, is that it needs two out-of-phase signals which necessitates an input tapped transformer or phase inverter, and thus makes the circuitry quite complicated. The above two drawbacks of an ordinary push-pull amplifier have been overcome in the complementary symmetry push-pull amplifier. This arrangement uses two transistors having complementary symmetry (one N-PN and another P-NP). The term complementary arises from the fact that one transistor is the N-P-N type and the other is P-N-P

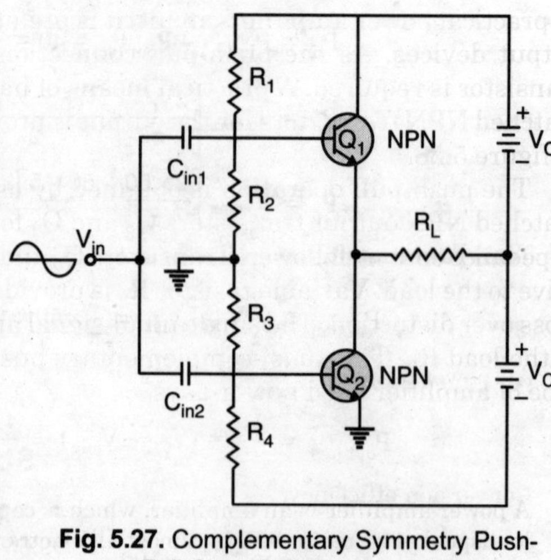

Fig. 5.27. Complementary Symmetry Push-Pull Class B Power Amplifier

type. They have symmetry as they are made with the same meterial and technology and are of same maximum rating.

The circuit of a complementary push-pull Class A B power amplifier is shown in figure 5.28. The resistors R_1 and R_2 provide the voltage divider bias to forward bias the emitter-base junction of transistor Q_1 and similarly resistors R_3 and R_4 provide the that under zero signal condition, the operating point is cut-off and so no collector current flows. The signal applied at the input goes to the base of both the transistor. The two transistors conduct in opposite half cycles of the input. For example during the positive half cycle of the input signal, the NPN transistor Q_1 is forward biased and so does not conduct. This results in half cycle of output voltages across the load, resistor R_L. Similarly during the negative half cycle only the PNP transistor Q_2 is forward biased and conducts and develops second half cycle of the output voltage across the load resistor RL. Transistor Q_1 being reverse-biased does not conduct during the negative half cycle of the input signal. Thus, during a complete cycle of inputs complete cycle of the output signal is devolved. Since collector current from each transistor flows through the load during alternate half cycle of the input signal, no centre-tapped output transformer (without centre tapping) can be used. One obvious drawback of this circuit arrangement is the requirement for two supply voltages. Another drawback of this circuit is difficulty of obtaining matched complementary transistors. If there is an unbalance in the characteristics of the two transistors, even harmonics will no longer be cancelled and considerable distortion will be introduced. Very often negative feedback is used in power amplifiers to reduce non-linear distortion.

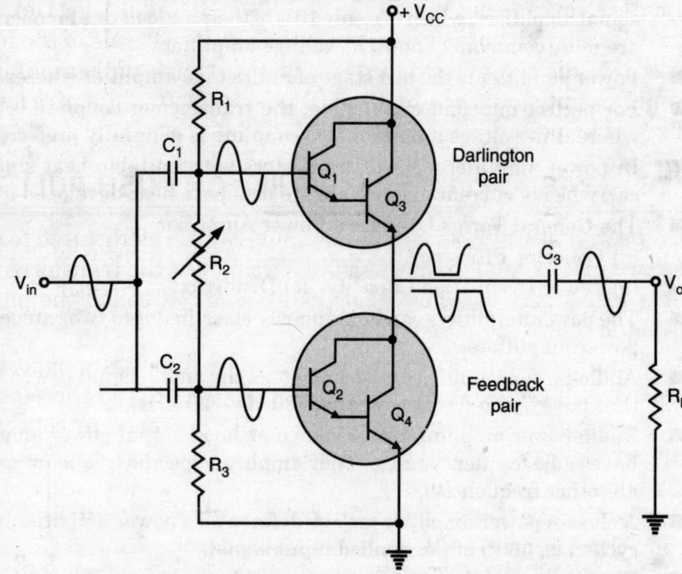

Fig. 5.28. Quasi-Complementary Push-Pull Power Amplifier.

5.18. QUASI-COMPLEMENTARY PUSH-PULL AMPLIFIER

In practical power amplifier-circuit, it is preferable to employ NPN transistors for both high current output devices. As the push-pull connections need complementary devices, a PNP high power transistor is required. A practical means of having complementary operation while using the same, matched NPN transistors for the output is provided by a quasi-complementary circuit, as illustrated in figure 5.28.

The push-pull operation is obtained by using complementary transistors Q_1 and Q_2 before the matched NPN output transistors Q_3 and Q_4 form a darlington pair that provides output from a low-impedance emitter follower. Transistors Q_2 and Q_4 form a feedback pair that provides a low-impedance drive to the load. Variable resistor R_2 is provide for adjusting the d.c. bias condition so as to minimize cross over distortion. The single input signal applied to the push-pull stage provides full-cycle output to the load R_L. The quasi-complementary push-pull amplifier is considered to be the most popular type of amplifier used now-a-days.

SUMMARY

- A power amplifier is an amplifier, which is capable of providing a large amount of power to a load such as loudspeaker or motor etc. In almost all electronic systems where a large amount of power is required to be supplied to the load, a power amplifier is essential.

- A power amplifier is more commonly known as an audio amplifier. The audio amplifiers are used in public address systems, tape recorders, stereo systems, television receivers, radio receivers, broadcast transmitters etc.
- The power amplifiers are also called large signal amplifiers. The term 'large signal' for the power amplifiers arises since these amplifiers use a large part of their a.c load line for operation. It is in contrast to the small signal amplifiers, which use only 10% of their a.c load line for operation. Recall that the small signal amplifiers are more commonly known as voltage amplifiers.
- Power amplifier is the last stage of multistage amplifier whereas the first few stages are voltage amplifiers.
- For perfect impedance matching, the transformer coupling is invariably employed with power amplifier whereas in voltage amplifier R-C coupling is generally preferred.
- In power amplifiers, power transistors with suitable heat sink are employed. Since they are capable to carry heavy current and are able to dissipate heat developed at the collector junction.
- The General Terms Used for a Power Amplifier
  (a) Collector Efficiency,
  (b) Power Dissipation Capacity, (c) Distortion
- The power amplifiers can be primarily classified into two categories viz., audio-power amplifiers and radio-power amplifiers.
- Audio-power amplifiers, also known as the small signal power amplifiers, raise the power levels of signals that have audio-frequency range (20 Hz – 20 kHz).
- Radio-power amplifiers, also known as large signal power amplifiers raise the power level of signals that have radio frequency range. They amplify a specific frequency or narrow band of frequencies while rejecting all other frequencies.
- A class-A power amplifier may be defined as a power amplifier in which the ouput current flows for the full-cycle (i.e., 360°) of the applied input signal.
- That the transistor remains in forward-biased state throughout the period of input cycle.
- The collector efficiency of a power transistor may be defined as

$$\eta_{\text{collector}} = \frac{\text{Average a.c. power output, } P_{out}\text{(a.c.)}}{\text{Average d.c. power input to the transistor, } P_{tr(d.c.)}}$$

- The power efficiency of a power transistor may be defined as

$$\eta = \frac{\text{a.c. power delivered to the load, } P_{out}\text{(a.c.)}}{\text{total power drawn from d.c. supply } P_{in(dc)}}$$

SHORT QUESTIONS WITH ANSWERS

Q.1. What are the general terms used for a power amplifier? Explain.

Ans. (i) **The Collector Efficiency :** A power amplifier converts dc power into ac power at ouput. Because of this, the effectiveness of a power amplifier is measured by its capability of converting dc power from battery into ac power at the ouput side. This is called collector efficiency. Thus, it can be defined as the ratio of ac output signal to the zero signal power (i.e., dc power drawn from the battery of a power amplifier). As a matter of fact, power amplifiers are designed to provide maximum collector efficiency.

(ii) **The Power Dissipation Capacity :** A power transistor handles large amount of currents and becomes heated during its operation. This can cause the junction temperature to rise to a dangerous level. The power dissipation capacity of a transistor can be increased by making a suitable provision for rapid conduction of heat away from the junction. In fact, this can be achieved by attaching a metal case, known as the **heat sink**, to the power transistor case.

Thus, power dissipation capacity of a power transistor is defined as its ability to dissipate the heat developed in it.

(iii) **Distortion :** It may be defined as the change of output wave-shape from the input wave-shape of the amplifier. Because power amplifier handles large signals and for large input signals, the transistor operation becomes non-linear therefore the problem of distortion arises. In fact, lesser the distortion, better is the amplifier.

Q.2. What is the formula for collector efficiency ?

Ans. The collector efficiency of a power transistor may be defined as

$$\eta_{\text{collector}} = \frac{\text{Average a.c. power output, } P_{out(a.c.)}}{\text{Average d.c. power input to the transistor, } P_{tr(d.c.)}}$$

REVIEW QUESTIONS

1. Distinguish between voltage amplifier and power amplifier.
2. Discuss the classification of power amplifiers.
3. Why does a power amplifier always form the last part of an amplifier circuit.
4. Explain class A, B and C amplifiers, giving their relative merits and applications.
5. Derive expression for efficiency of class-A amplifier. Show that its maximum value is 50%.
6. Draw the circuit for a single-ended class-A amplifier. Explain its working. What will be its collector efficiency?
7. What is a power amplifier ? In what respects does it differ from a voltage amplifier? Why heat sinks are needed?
8. Explain collector efficiency, distortion and power dissipation. Discuss the function of a heat sink.
9. Differentiate between voltage and power amplifier. Give the classiciation of power amplifiers and the use of each class.
10. Write the classification of amplifiers on the basis of period of conduction and define each type.
11. Distinguish amongst class A, class B and class C operation of amplifier.
12. Draw the circuit of a simple class A amplifier which can provide power amplification and explain its working.
13. Explain how ac power is developed across a loudspeaker in a single ended power amplifier. Explain why a power amplifier is also known as the large signal amplifier?
14. What are the typical characteristics of a power amplifier ? Draw a basic circuit of a power amplifier.
15. Draw the circuit of a transformer-coupled single stage transistor power amplifier and explain the need for impedance matching.

NUMERICAL PROBLEMS

1. A power transistor operating in class A operation has zero-signal powr dissipation of 12 W and ac power output is 4W. Determine collector efficiency and power rating of the transistor. **[Ans.** 33.3%; 12 W]
2. For a class A power amplifier, the maximum and minimum values of collector-emitter voltage are 20V and 4V respectively. V_{CC} = 25 V. Determine overall efficiency of the amplifier. **[Ans.** 16%]
3. For class A, CE large signal amplifier, V_{CC} = 25 V, V_{DEQ} = 12.5 V, I_{CQ} = 0.5A, R_C = 25Ω. If the output current varies by ±0.25 A when an input signal is applied to the base, determine (i) input power, (ii) output power, (iii) power delivered to the transistor (iv) power lost in transistor, (v) collector efficiency and (vi) overall efficiency.
   [Ans. (i) 12.5 W, (ii) 0.78125 W, (iii) 6.25W, (iv) 5.47 W, (v) 12.5%, (vi) 6.25%]
4. Maximum and minimum values of collector-emitter voltage for a transformer-coupled class A power amplifier are 27 and 2 volts respectively. Determine its overall efficiency. **[Ans.** 43.1%]
5. A transformer-coupled class A power amplifier supplies the power to a 100Ω load. Determine the maximum power output for a zero signal collector current of 80 mA if the transformer turn-ratio is 10 : 1.
   [Ans. 32 W]

□□□

UNIT III

Switching characteristics of diode and transistor, turn ON, OFF time, reverse recovery time transistor as switch, Multivibrators, Bistable, Monostable, Astable multivibrators, Clippers and clampers. Differential amplifier, Calculation of differential, Common mode gain and CMRR using h-parameters, Darlington pair, Boot strapping technique, Cascade and cascode amplifier.

Above syllabus has been covered in following chapters :

Chapter 6

WAVESHAPING CIRCUITS : MULTIVIBRATORS, CLIPPERS AND CLAMPERS

Syllabus

Switching characteristics of diode and transistor, turn ON, turn OFF time, reverse recovery time, Transistor as a Switch, Multivibrators : Bistable, Monostable, Astable Multivibrators, clippers and clampers.

Chapter Outline

- Introduction
- Diode Switching Times
- Transistor as a Switch
- Transistor Switching Times
- Standard Tests for Saturation
- Application of Transistor as a Switch
- Multivibrators Using Transistors
- Monostable Multivibrator
- Astable Multivibrator
- Bistable Multivibrator Using a Transistor
- Schmitt Trigger Circuit
- Diode Circuits
- Clipper Circuits
- Clampers

6.1. INTRODUCTION

Multivibrators are the electronic circuits having one, two or no stable states of operation. Depending on the number of stable states, the multivibrators are classified into astable, monostable and bistable multivibrators. The transistors used for these circuits operate as switches i.e. they are biased to operate either in cutoff or saturation. In this chapter all types of multivibrators have been discussed.

6.2. DIODE SWITCHING TIMES

When a diode is driven from reverse condition to forward condition or in the opposite direction, it does not change its state instantaneously. Instead it takes a finite time to switch from off to on state or from on to off state. This time consists of a transient time and the time interval elapsed before the diode recovers to its steady state.

The diode characteristics corresponding to the interval in which the change of state is taking place is known as the **switching characteristics**. Diodes are used in applications such as logic gates. In such applications, diodes are used as switches and not for the purpose of rectification. The switching characteristics and the parameters derived from it are useful for such applications.

In the switching applications we define two time durations namely the turn ON time and the turn OFF time of the diode. The time required to turn OFF is longer than the time required to turn ON a diode. Hence, the turn OFF time and turn OFF characteristics is more important.

6.2.1. Forward and Reverse Recovery Times

Now, let us define two important delays related to the switching of a diode. They are as under :

(i) Forward recovery time t_{rf}.

(ii) Reverse recovery time t_{rr}.

6.2.2. Forward Recovery Time T_{rf}

The forward recovery time (t_{rf}) is defined as the time required for the diode current to change from 10 to 90 percent of its final value when the diode is switched from OFF to ON state. In practice, the forward recovery time is very short as compared to the reverse recovery time. Hence, we can neglect it.

The turn ON time of a diode is always shorter than its turn OFF time. Manufacturers normally specify the reverse recovery time t_{rr} of a diode.

6.2.3. Diode Reverse Recovery Time (t_{rr})

In order to understand the reverse recovery process, let us consider the minority carrier density distribution on the two sides of the P-N junction as a function of distance x shown in figure 6.1 Figure 6.1 shows the minority carrier density distribution when the external voltage forward biases the diode.

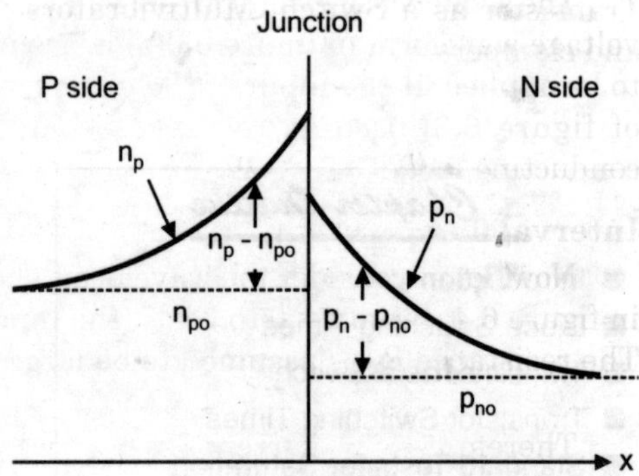

Fig. 6.1. Minority carrier density distribution in a forward biased diode

In the forward biased state, a large number of electrons diffuse from N-side to P-side. When they enter the P side, they become minority carriers. Similarly the holes diffusing from P to N side become minority carrier on the N side.

Therefore, in the forward biased diode, the number of minority carriers is very large. Now, let us define various terms shown in figure 6.1. These are :

n_p = Concentration of electrons on the P-side

p_n = Concentration of holes on the N-side

n_{po} = Concentration of electrons on P-side at thermal equilibrium

p_{no} = Concentration of holes on N-side at thermal equilibrium

Thus, n_p and p_n are the minority carrier concentrations and figure 6.1 shows their variation as a function of distance x from the junction. It shows that the minority carrier distribution is maximum at the junction (x = 0) and it reduces at we move away from the junction (as x increases).

Now, if the external voltage reverse biases the junction, then the steady state density of minority carriers has been shown in figure 6.2.

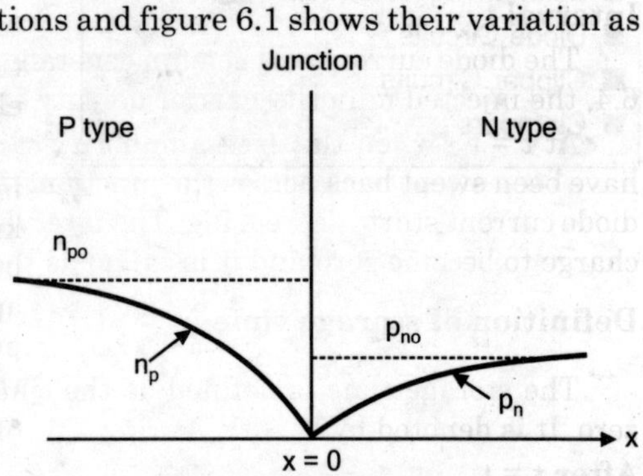

Fig. 6.2. Minority carrier density distribution in the reverse biased diode

It shows that far from the junction, the concentration of minority carriers is equal to the thermal equilibrium value. This means that $n_p = n_{po}$ and $p_n = p_{no}$. This is same as that shown in figure 6.1. But, the minority carrier concentration reduces as we approach the junction (x = 0). This happens because, as the minority carriers approach the junction, they are rapidly swept across the junction and the density minority carriers diminishes rapidly to zero at the junction as shown in figure 6.2.

If the external voltage is suddenly reversed to reverse bias the forward biased diode, then the forward diode current will not fall instantaneously to its steady state value in the reverse direction. The diode current cannot reach its steady state value in the reverse direction until the minority carrier distribution in the forward biased state (figure 6.1) does not completely change to that in the reverse biased state as shown in figure 6.2.

Till such time as the injected or excess, minority carrier density ($p_n - p_{no}$) or ($n_p - n_{po}$) has dropped nominally to zero, the diode will continue to conduct easily and the current will be determined by the external resistance in the diode circuit.

6.2.4. Storage and Transition Times

If the conducting diode is suddenly reverse biased by applying the voltage waveform shown in figure 6.3(a), then the sequence of events will take place.

The sequence of events are as follows. The voltage waveform in figure 6.3(a) is assumed to be applied at the input of the diode circuit of figure 6.3(b) so as to reverse bias the conducting diode.

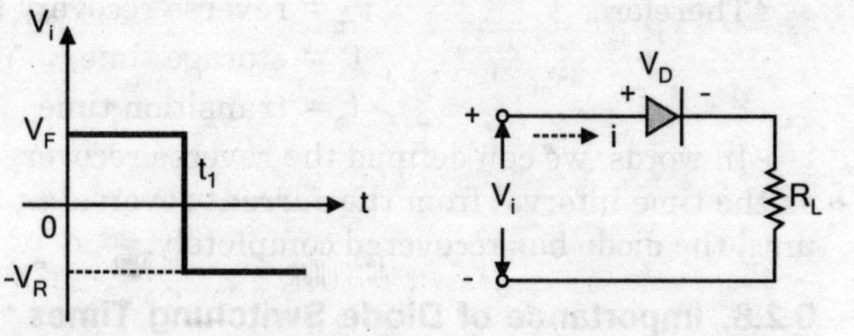

(a) Applied voltage at the input of diode circuit to reverse bias it (a) Simple diode circuit

Fig. 6.3.

Interval 0 – t_1

Now, let us consider the waveforms shown in figure 6.4. From t = 0 to $t = t_1$, the input voltage $V_i = V_F$ and the diode remains forward biased. The resistance R_L is assumed to be large as compared to the resistance of diode.

Therefore, $$\text{current } i = +\frac{V_F}{R_L} = +I_F \quad ...(6.1)$$

At $t = t_1$

At $t = t_1$, input voltage reverses abruptly from $+V_F$ to $-V_R$. Due to the presence of large number of minority carriers on either sides, the current will simply reverse. The new value of reverse current is,

$$i = \frac{-V_R}{R_L} = -I_R$$

Interval $t_1 - t_2$

The diode current will remain constant at $-I_R$ from t_1 to t_2. At instant, $t = t_2$ as shown in figure 6.4, the injected minority carrier density at x = 0 has reached its equilibrium state.

At $t = t_2$, when the excess minority carriers in the immediate neighbourhood of the junction have been swept back across the junction, the diode voltage begins to reverse and the magnitude of diode current starts decreasing. The interval t_1 to t_2 is the duration required for the stored minority charge to become zero and it is called as the storage time t_s.

Definition of storage time

The storage time is defined as the time required for the stored minority charge to become zero. It is denoted by t_s.

After $t = t_2$

As shown in figure 6.4, the diode voltage starts becoming negative and increases towards the maximum value $-V_R$. The magnitude of reverse current decreases towards I_o. This is an indication that the diode has fully recovered.

Definition of transition time t_t

The time elapsed between $t = t_2$ and the time when diode is completely recovered is called as the transition time. It is denoted by t_t.

6.2.5. Reverse Recovery Time (t_{rr})

Manufacturers usually specify the reverse recovery time t_{rr}. The reverse recovery time is the sum of these two time intervals.

Where, $t_{rr} = t_s + t_t$ ...(6.2)

Therefore, t_{rr} = reverse recovery time,

t_s = storage time,

t_t = transition time

In words, we can defined the reverse recovery time as the time interval from the current reversal at $t = t_1$ until the diode has recovered completely.

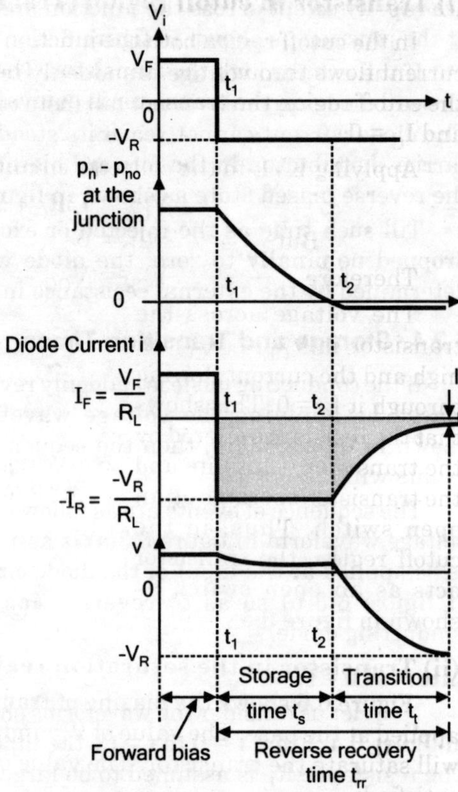

Fig. 6.4. Diode switching waveforms

6.2.6. Importance of Diode Switching Times

The reverse recovery time decides the time required to switch the diode off. So it decides, how fast or at what maximum rate can we switch the diode on and off. Thus reverse recovery time decides the highest switching frequency of a diode. The reverse recovery time depends on the transition capacitance of the diode. So the transition capacitance of switching diodes is small. The maximum switching frequency is given by,

$$f_{max} = \frac{1}{10t_{rr}} \quad ...(6.3)$$

Thus, as t_{rr} decreases, f_{max} will increase.

Switching Diodes

Commercial switching type diodes are available with times t_{rr} ranging from 1 nanosecond (ns) upto as high as 1 microsecond (μs). Higher values of t_{rr} correspond to the diodes intended to handle larger currents. IN 4153 is a switching diode. It has a reverse recovery time of a few nanoseconds under the test conditions. IN 4153 is high speed switching diode for computer and general purpose applications. Some of the important features of this switching diode are as under:

Important Characteristics of switching diode IN 4153

(i) Reverse breakdown voltage : V_{br} = 75 Volts.

(ii) Static forward voltage : 0.5 to 0.8 volts depending on the forward current.

(iii) Total capacitance : 2 pF

(iv) Reverse recovery time : 4 ns

6.3. TRANSISTOR AS A SWITCH

(RGTU, Bhopal, Sem. Exam., Dec. 2006, June 2007, Dec. 2007)

The transistor can be used for two types of applications viz. amplification and switching. For the amplification applications, a transistor is biased in its active region. Whereas, for switching applications it is biased to operate in the saturation (fully on) or cutoff (fully off) regions.

(i) Transistor in cutoff region (Transistor as an open switch)

In the cutoff region both the junction of a transistor are reverse biased and a very small reverse current flows through the transistor. The voltage drop across the transistor (V_{CE}) is high. Thus, in the cutoff region the transistor is equivalent to an open switch. The input voltage is zero, so, $I_B = 0$ and $I_C = 0$.

Applying KVL to the collector circuit of figure 6.5, we get.

$$V_{CC} = I_C R_C + V_{CE}$$

But $\quad I_C = 0$

Therefore, $\quad V_{CE} = V_{CC}$

The voltage across the transistor is $V_{CE} = V_{CC}$ i.e. high and the current flowing through it $I_C = 0$. This shows that the resistance offered by the transistor is infinite and the transistor operates as an open switch. Thus, in the cutoff region, the transistor acts as an open switch as shown in figure 6.5.

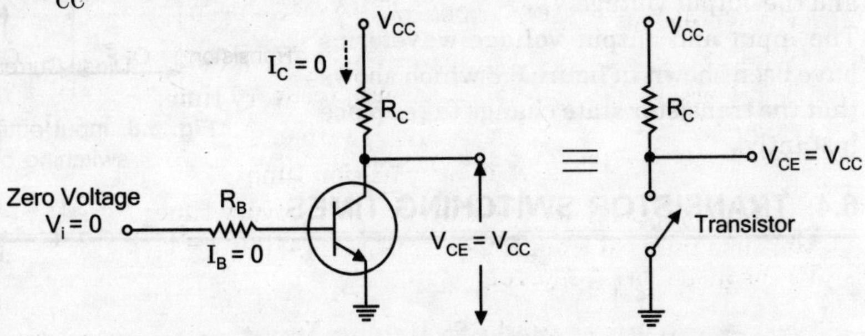

Fig. 6.5. Transistor in cutoff is equivalent to an open switch

(ii) Transistor in the saturation region (transistor as closed switch)

Figure 6.6 shows the biasing of transistor in the saturation region. A high input voltage V_{in} is applied at the base. The value of V_{in} and resistance R_B are adjusted such that a large I_B flows. This will saturate the transistor. The value of I_B should be such that the following condition should be satisfied:

$$I_B > \frac{I_{C(sat)}}{\beta} \quad \text{...(6.4)}$$

In the saturation region both the junction of a transistor are forward biased. The voltage drop across the transistor (V_{CE}) is very small of the order of 0.2 V to 1 V depending on the type of transistor and collector current is very large.

A low (almost 0) voltage drop across the transistor and a large current through it indicates that the transistor is equivalent to a closed switch.

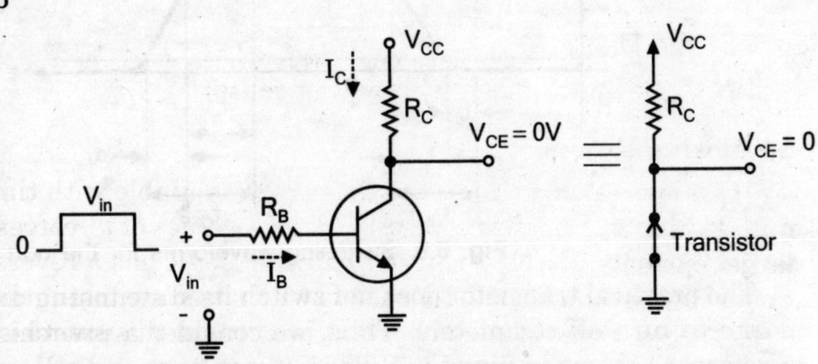

Fig. 6.6. Transistor in saturation is equivalent to a closed switch

6.3.1. Switching Waveforms of an Ideal Transistor Switch

Let us, consider the circuit diagram of figure 6.7 which shows the biasing arrangement for a transistor to operate as a switch. A rectangular waveform is applied at the base and the

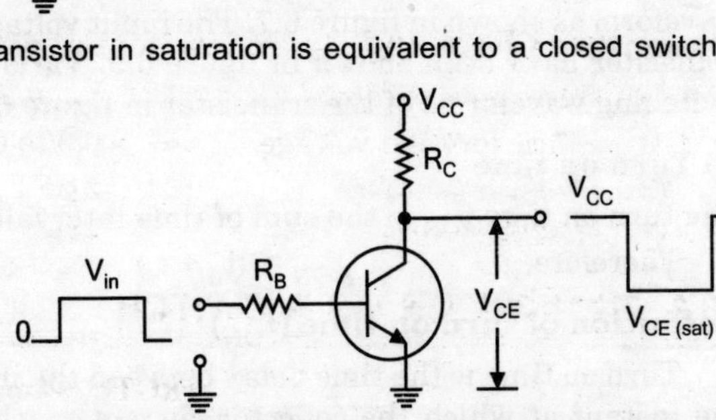

Fig. 6.7. Circuit diagram showing transistor biased to operate as a switch

transistor is assumed to be an ideal one. Hence, the transistor will change its state from ON to OFF or vice versa instantaneously.

When the input is low (0), the transistor is off, it acts as an open switch and the output voltage is $V_{CE} = V_{CC}$. When the input voltage goes high, the transistor is driven into saturation. It acts as a closed switch and the output voltage $V_{CE} = V_{CE(sat)} = 0$ V. The input and output voltage waveforms have been shown in figure 6.8 which shows that the transistor state change takes place instantly.

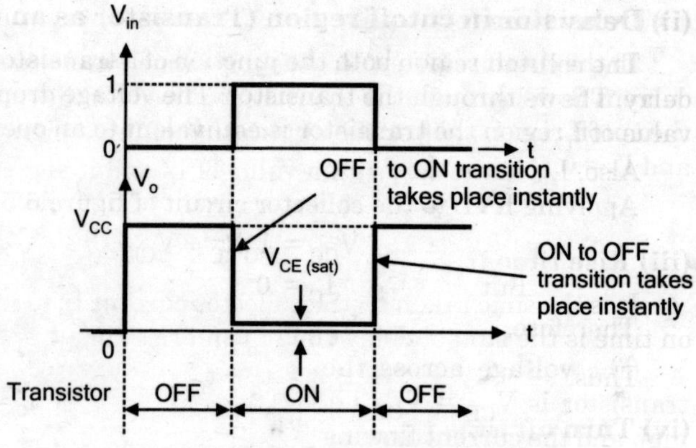

Fig. 6.8. Input output waveforms for transistor switching of an ideal transistor

6.4. TRANSISTOR SWITCHING TIMES

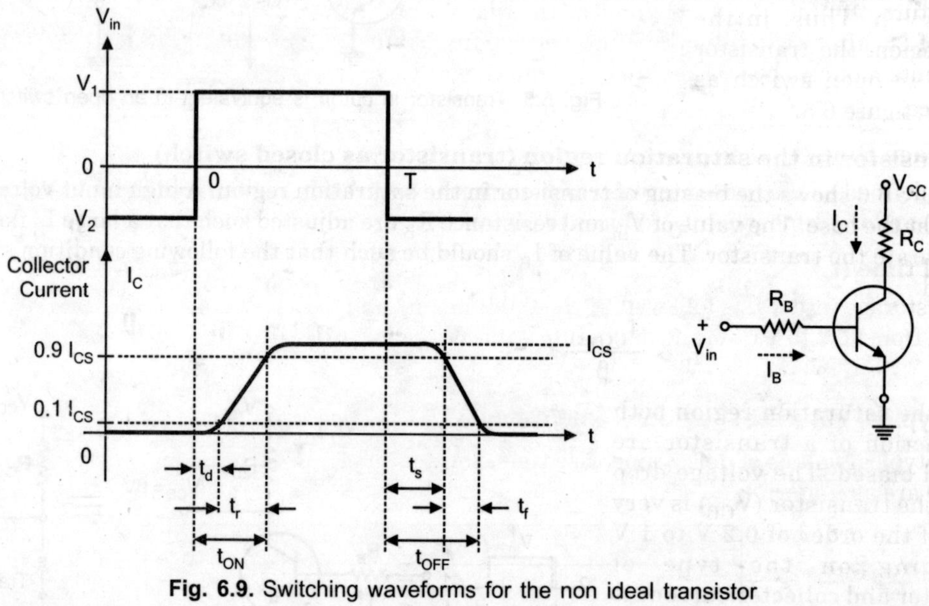

Fig. 6.9. Switching waveforms for the non ideal transistor

The practical transistor does not switch its state instantaneously. It needs a finite time to turn on and to turn off completely. Thus, we consider a practical transistor being driven by a pulse waveform as shown in figure 6.7. The input voltage and collector current waveforms for a non ideal transistor have been shown in figure 6.9. Various delays and time intervals are involved in the switching waveforms of the transistor in figure 6.9.

(i) Turn on time **(RGTU, Bhopal, Sem. Exam; Dec. 2006)**

The turn on time t_{ON} is the sum of time intervals namely, the delay time t_d and the rise time t_r.

Therefore, $$t_{ON} = (t_d + t_r) \qquad ...(6.5)$$

Definition of turn on time (t_{ON})

Turn on time is the time delay between the instant at which a high base voltage is applied and the instant at which the collector current reaches 90% of its maximum value. Turn on time is typically a few microseconds.

(ii) Delay time (t_d)

The collector current I_C does not immediately respond to the input signal. Instead, there is a delay. The delay time t_d is the time required for the collector current to attain 10% of the maximum value of I_C i.e. I_{CS}, after the application of input signal.

Also, I_{CS} is the saturation value of I_C and it is given by,

$$I_{CS} = V_{CC}/R_C \quad \text{...(6.6)s}$$

(iii) Rise time (t_r)

It is the time taken by the collector current I_C to increase from 0.1 I_{CS} to 0.9 I_{CS}. The total turn on time is the sum of delay time t_d and rise time t_r.

Thus, $$t_{ON} = t_d + t_r$$

(iv) Turn off time t_{OFF} **(RGTU, Bhopal, Sem. Exam; Dec. 2006)**

The non-ideal transistor does not turn off instantaneously instead it takes a finite time to turn off completely after the removal of the input signal.

Definition

The turn off time (t_{OFF}) is defined as the time taken by the collector current to reduce from I_{CE} to 10% of I_{CS} after switching the input voltage from V_1 to V_2. The turn off time is the sum of two components namely the storage time (t_s) and the fall time (t_f).

Hence, $$t_{OFF} = t_s + t_f \quad \text{...(6.7)}$$

(v) Storage time (t_s)

It is defined as the time interval which elapses between the transition of input waveform (from V_1 to V_2) and the time when I_C reduces to 90% of I_{CS}.

(vi) Fall time (t_f)

The storage interval is followed by the fall time interval t_f. It is defined as the time required for I_C to fall from 90% to 10% of the maximum value of I_C i.e. I_{CS}. The turn off time is the sum of t_s and t_f..

6.4.1. Typical values of Various Time Delays

The typical values of delay time, rise time, storage time and fall time for the N-P-N transistor 2N 3830 are as under:

Sr.No.	Time	Value
1.	Delay time t_d	10 n sec
2.	Rise time t_r	50 n sec
3.	Storage time t_s	40 n sec
4.	Fall time t_f	30 n sec

6.4.2. Importance of Turn ON and Turn OFF Time

The turn on and turn off times are important in deciding the maximum switching frequency of the transistor. In order to operate the transistor at a high frequency the turn on and turn off times should be as short as possible.

6.5. STANDARD TESTS FOR SATURATION

It is often important to know whether or not a transistor is in saturation. The two tests to be carried out to decide about the saturation of a transistor are:

Test No. 1: Determine I_C and I_B separately from the circuit configuration under consideration. The transistor is in saturation if.

$$|I_B| \geq \frac{|I_C|}{\beta}$$

Test No. 2: Both the junctions of a transistor should be forward biased for operation in the saturation region. Measure V_{CB}. If it is positive for P-N-P transistor and negative for N-P-N transistor, then the transistor is in saturation.

6.6. APPLICATIONS OF TRANSISTOR AS A SWITCH

Some of the applications of the transistor as a switch are:

(i) In the logic circuits.
(ii) Temperature control.
(iii) Chopper circuits used for dc motor speed control.
(iv) Inverters for ac motor control.
(v) SMPS and UPS.

EXAMPLE 6.1. For the circuit shown in figure 6.10, calculate $I_{C(sat)}$, and the value of R_1 to maintain the transistor in saturation. Assume $V_{CE(sat)} = 0.2$ V, $\beta = 50$ and $V_{BE(sat)} = 0.8$ V.

Solution: (i) We have $I_{C\,(sat)} = \dfrac{V_{CC} - V_{CE(sat)}}{R_C} = \dfrac{12 - 0.2}{1.2\text{ k}} = 9.83$ mA **Ans.**

(ii) The minimum value of the base current to ensure transistor saturation is given by,

$$I_{B(min)} = \frac{I_{C(sat)}}{\beta} = \frac{9.83\text{ mA}}{50} = 0.196\text{ mA}$$

The maximum value of R_1 which will ensure the transistor operation in saturation is given by,

$$R_{1max} = \frac{V_i - V_{BE(sat)}}{I_{B(min)}} = \frac{3 - 0.8}{0.196 \times 10^{-3}}$$

$= 11.18$ kΩ. **Ans.**

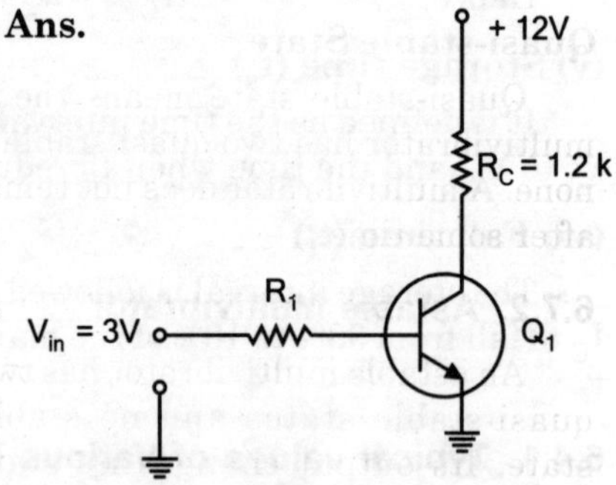

Fig. 6.10.

EXAMPLE 6.2. For the circuit shown in figure 6.11. the transistor is to be operated as a switch. Calculate the minimum value of h_{FE} or β_{dc} to operate the transistor as a switch.

Solution: Assume that the transistor is in saturation.

(i) Applying KVL to the collector circuit to get,

$$V_{CC} = I_C R_C + V_{CE(sat)}$$

Therefore, $I_C = \dfrac{V_{CC} - V_{CE(sat)}}{R_C} = \dfrac{10\text{V} - 0.2}{1\text{k}} = 9.8\text{mA}$ **Ans.**

(ii) Apply KVL to the base circuit to write,

$$V_{in} = I_B R_1 + V_{be(sat)}$$

or, $I_B = \dfrac{V_{in} - V_{be(sat)}}{R_1} = \dfrac{5 - 0.8}{100\text{ k}} = 42\,\mu\text{A}$

(iii) $h_{FE(min)} = \dfrac{I_C}{I_B} = \dfrac{9.8\text{ mA}}{42\,\mu\text{A}} = 233.33$ **Ans.**

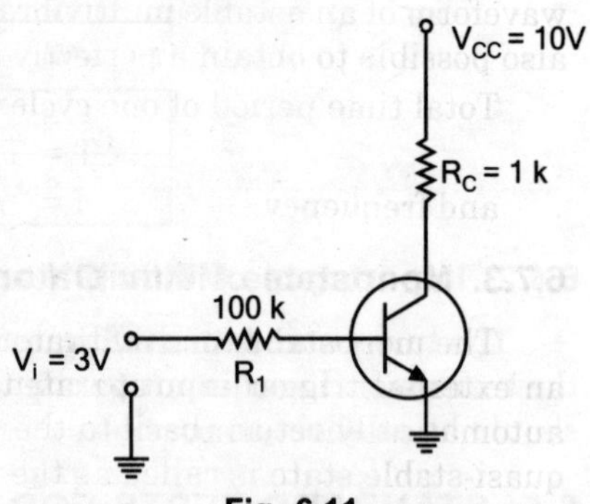

Fig. 6.11.

6.7. MULTIVIBRATORS USING TRANSISTORS

Mulivibrator is an electronic circuit which can have no, one or two stable states of operation. Depending on the number of stable states we can classify the multivibrators as shown in figure 6.12.

6.7.1. Classification of Multivibrators

Figure 6.12 shows the classification of multivibrators.

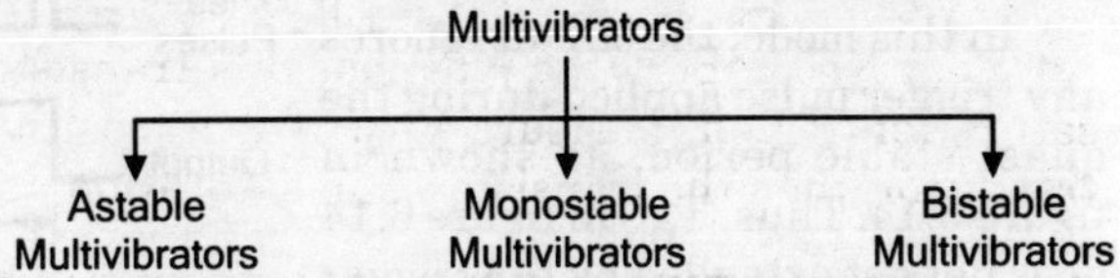

Fig. 6.12. Classification of multivibrators

Astable multivibrator

In this multivibrator, output does not have any stable state. Its output changes its state from high to low and low to high repeatedly.

Monostable multivibrator

In the multivibrator output has only one stable state.

Bistable multivibrator

It is a multivibrator circuit, the output of which has two stable states. We have to apply external inputs in order to change the existing state of output. Bistable multivibrator is also called as a flipflop.

Quasi-stable State

Quasi-stable state means the unstable or temporarily stable state of a flipflop. An astable multivibrator has two quasi-stable states, monostable has one quasi stable state and bistable has none. A multivibrator does not remain in quasi-stable state permanently. It comes out of this state after some time.

6.7.2. Astable Multivibrator

An astable multivibrator has two quasi-stable states and no stable state. Its output changes its state from one quasi state to the other. It does not require any external trigger input to change its state. Hence, it is also called as a **free running multivibrator**. The output voltage waveform of an astable multivibrator is as shown in figure 6.13. It is a rectangular waveform. It is also possible to obtain a perfectly square waveform at the output of an astable multivibrator.

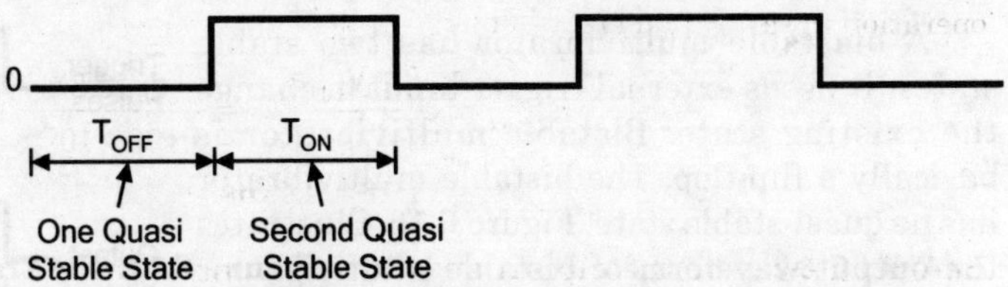

Fig. 6.13. Output waveform of an astable multivibrator

Total time period of one cycle of the output is given by,

$$T = T_{ON} + T_{OFF}$$

and frequency $f = 1/T$.

6.7.3. Monostable Multivibrator

The monostable multivibrator has one stable state and another quasi stable state. It requires an external trigger input to change its state from stable to quasi stable. After some time it will automatically return back to the stable state. The duration for which the output remains in the quasi-stable state is called as the pulse width.

Types of monostable

There are two types of monostable multivibrators:

1. Non-retriggerable monostable multivibrator.
2. Retriggerable monostable multivibrator.

1. Non-retriggerable Monostable Multivibrator

This is the normal monostable multivibrator.

In this mode, the circuit ignores any trigger pulse applied during the quasi stable period, as shown in figure 6.14. Thus, T_{ON} in figure 6.14 does not get extended by the trigger pulses appearing during T_{ON}.

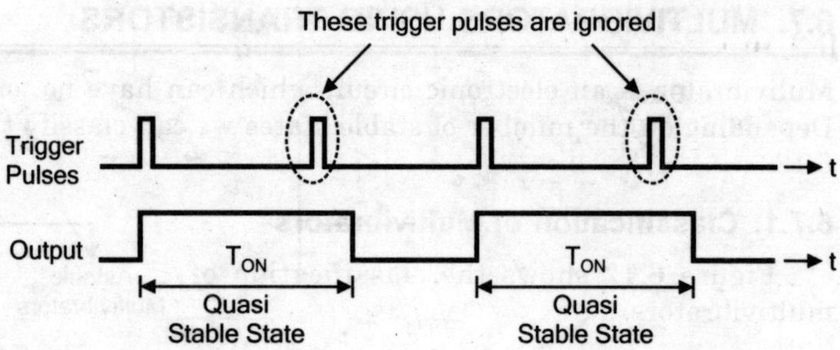

Fig. 6.14. Non retriggerable monostable multivibrator

2. Retriggerable Monostable Multivibrator

This type of monostable circuit responds to every trigger pulse applied at its inputs even when the trigger pulse appears during the quasi stable period as shown in figure 6.15. The quasi stable duration goes on extending due to the trigger pulses arriving in between the quasi stable period. The monostable multivibrator is said to be retriggered by every trigger input pulse.

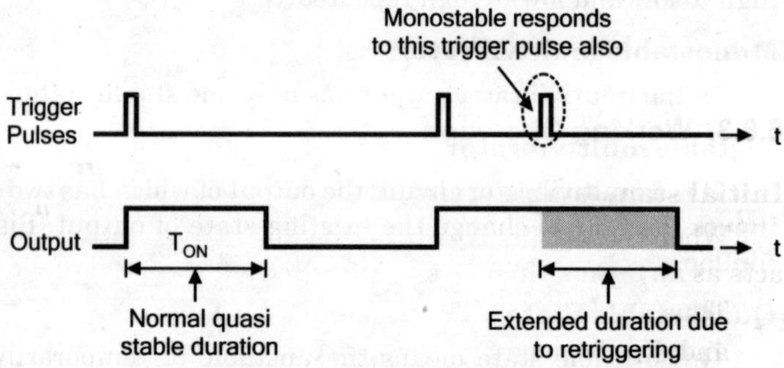

Fig. 6.15. Waveforms of retriggerable monostable multivibrators

The normal monostable multivibrators are also called as *one shot*. They are also called as **pulse stretchers.**

6.7.4. Bistable Multivibrator

(RGTU, Bhopal, Sem. Exam., June 2009)

A biastable multivibrator has two stable states. It needs external trigger input to change the existing state. Bistable multivibrator is basically a flip-flop. The bistable multivibrator has no quasi-stable state. Figure 6.16, illustrates the output waveform of bistable multivibrator circuit.

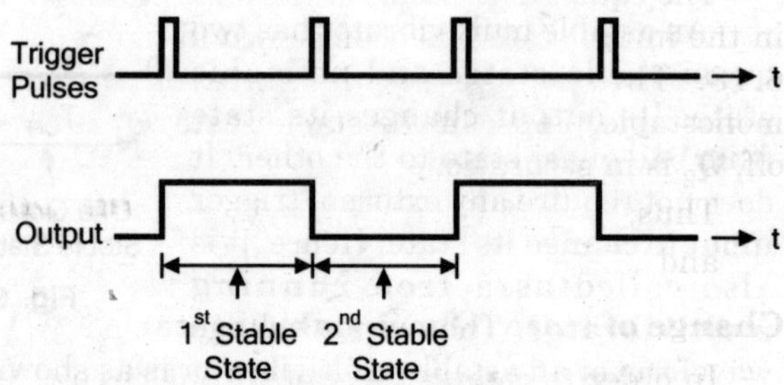

Fig. 6.16. Waveforms of a bistable multivibrator

6.8. MONOSTABLE MULTIVIBRATOR

6.8.1. Basic Concept

The monostable circuit has only one permanently stable state and one quasi-stable state. For a monostable circuit, we require and external triggering signal to induce a transition of its output from the stable state to the quasi-stable state. The circuit remains in the quasi-stable state for some time and eventually returns to its stable state on its own.

6.8.2. Circuit Diagram

The circuit diagram of the collector coupled transistorised monostable multivibrator is shown in figure 6.17. The collector of transistor Q_1 is connected to base of Q_2 through capacitor C. Similarly, collector of Q_2 is coupled to the base of Q_1 through C_1. That is why the circuit is called as **collector coupled circuit**. A negative voltage $-V_{BB}$ (typically -1.5 V) is applied to the base of Q_1. Hence, by default the transistor Q_1 is in the off state.

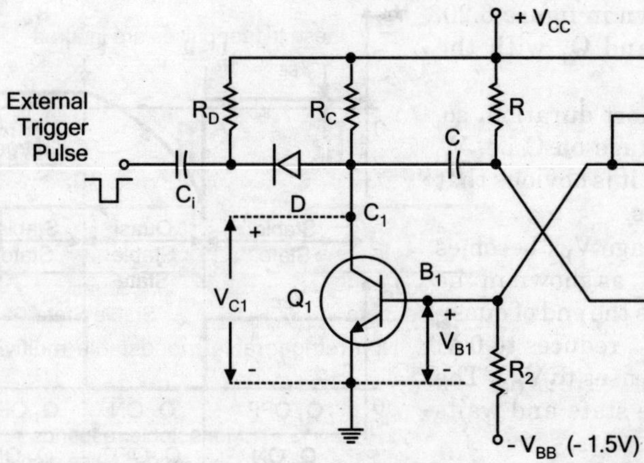

Fig. 6.17. A monostable multivibrator

6.8.3. Working Operation

Initial state (stable state)

When the circuits is switched on due to $-V_{BB}$ applied at B_1, transistor Q_1 is in the off state and acts as an open switch.

Thus, $I_{C1} = 0$

and, V_{C1} is given by, $V_{C1} = V_{CC} - I_C R_C = V_{CC}$

Hence, capacitor C is inactive and Q_2 receives its base current through R. The value of R is such that Q_2 goes into saturation.

The equivalent circuit of the monostable in the initial state has been shown in figure 6.18. This is the stable state of the monostable. Thus, in the stable state, Q_1 is off, Q_2 is in saturation,

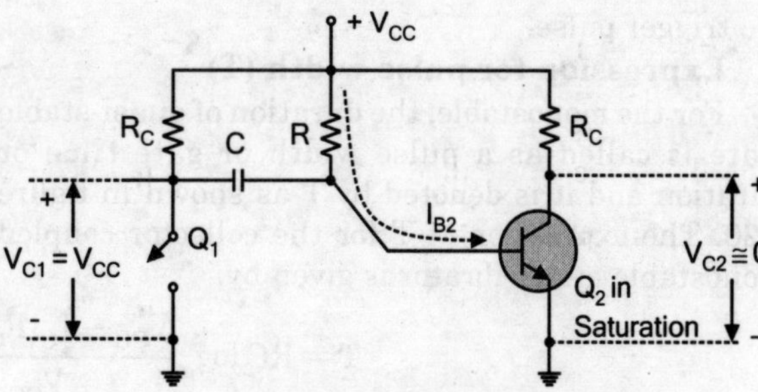

Fig. 6.18. Equivalent circuit in the initial state (stable state)

Thus, $V_{C1} = V_{CC}$

and $V_{C2} = V_{CE(sat)} = 0V$

Change of state (Quasi-stable State)

In order to change the state of the circuit i.e. to force the circuit from its stable state to quasi-stable state, we have to apply a low going rectangular trigger pulse at the trigger input. Let us consider the waveforms of figure 6.20 for better understanding. The trigger pulse is applied at t = 0. It gets coupled through capacitor C_i to the base of Q_2 is B_2. This will force the conducting transistor Q_2 into cutoff, and the collector voltage V_{C2} will increase from 0 to V_{CC} suddenly.

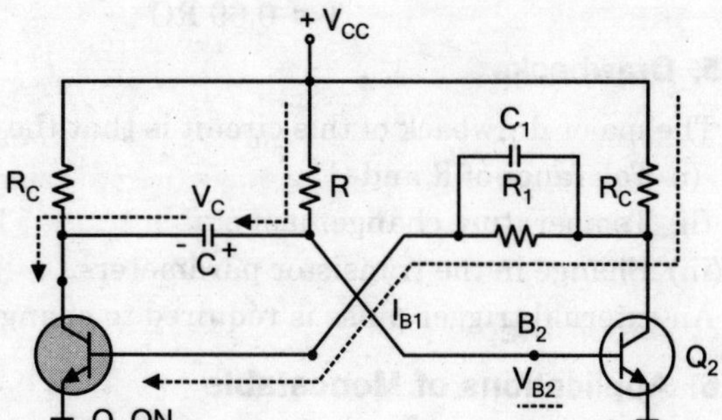

Fig. 6.19. Equivalent circuit in the quasi stable state

This sudden positive change is coupled to B_1 via capacitor C_1 and Q_1 is pulled out of cutoff and driven into saturation. Hence, V_{C1} reduces to almost zero volts. The base current for Q_1 starts flowing through R_C and R_1 as shown in figure 6.20. Capacitor C charges through R and Q_1 with the polarities shown in figure 6.19.

The trigger pulse is of very short duration, so, very soon it extinguishes. The voltage on C i.e. V_C keeps increasing. From figure 6.19, it is obvious that $V_C = V_{B2}$. Hence, V_{B2} also increases.

At instant t = T, the base voltage V_{B2} becomes sufficiently positive and turns on Q_2 as shown in the waveforms of figure 6.20. This marks the end of quasi-stable state. Q_2 is turned on, so V_{C2} reduces to 0 V. Thus, Q_1 is turned off and V_{C1} increases to V_{CC}. The circuit again enters into the stable state and waits for the next trigger pulse.

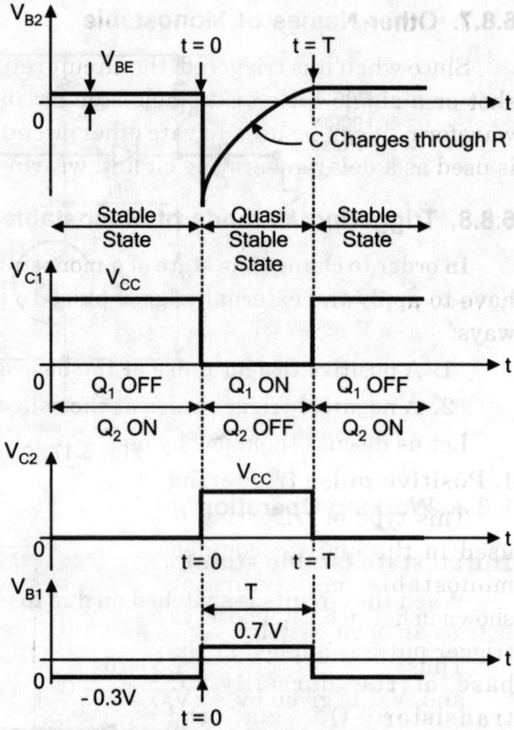

Fig. 6.20. Waveforms of the monostable multivibrator

6.8.4. On Time (T)

The interval 0 to T in figure 6.20 for which the monostable remains in the quasi-stable state is also called as the on time of the monostable circuit. This duration is dependent on the time taken by C to charge i.e. T depends on the RC time constant value. The on time T is much longer than the duration of the trigger pulse.

Expression for pulse width (T)

For the monostable, the duration of quasi-stable state is called as a pulse width or gate time or duration and it is denoted by T as shown in figure 6.20. The expression of T for the collector coupled monostable multivibrator is given by,

$$T = RC \log_e \frac{V_{CC} - I_C R_C - V_{BE(sat)}}{V_{CC} - V_\gamma} \quad ...(6.8)$$

Where, $V_\gamma = 0.1$ V for Germanium and 0.5 V for silicon. The expression for T shows that we can vary the pulse width by varyng the RC time constant or V_{CC}.

Simplified expression for T

For the collector coupled monostable multivibrator, the expression for T can be simplified as under:

$$T = 0.69\ RC \quad ...(6.9)$$

6.8.5. Drawbacks

The major drawback of this circuit is that the on time T is dependent on the following factors:

(i) Tolerance of R and C

(ii) Temperature change.

(iii) Change in the transistor parameters.

An external trigger pulse is required to change the stable state of circuit.

6.8.6. Applications of Monostable

(i) The main application of monostable multivibrator is as a timer.

(ii) It can be used to gate another circuit.

(iii) As a delay generator.

6.8.7. Other Names of Monostable

Since when it is triggered, the circuit returns to its steady state after time T, it is known as one shot or a single cycle or a single step circuit or a univibrator. Since it generates a rectangular waveform, it can be used to gate other circuit. Hence, it is also called as a **gating circuit**. Since it is used as a delay generating circuit, we can call it as **delay circuit**.

6.8.8. Triggering Methods of Monostable Multivibrator

In order to change the state of a monostable multivibrator from stable to quasi-stable state, we have to apply the external trigger pulse to it. This trigger pulse can be applied in two different ways:

1. A positive trigger pulse at the base of OFF transistor.
2. A negative trigger pulse at the collector of OFF transistor.

Let us discuss them one by one.

1. Positive pulse triggering

This type of triggering is used in the emitter coupled monostable multivibrator shown in figure 6.21. A positive trigger pulse is applied at the base of the normally off transistor Q_1 via the differentiator C_T R_T. Diode D_1 will allow only the positive spikes to pass through to the base of Q_1. As soon as the pulse is applied, Q_1 is turned on, its collector voltage will reduce to a low value, so Q_2 is turned off and output goes high. This is the quasi-stable state. The duration of this state depends on the R_BC time constant.

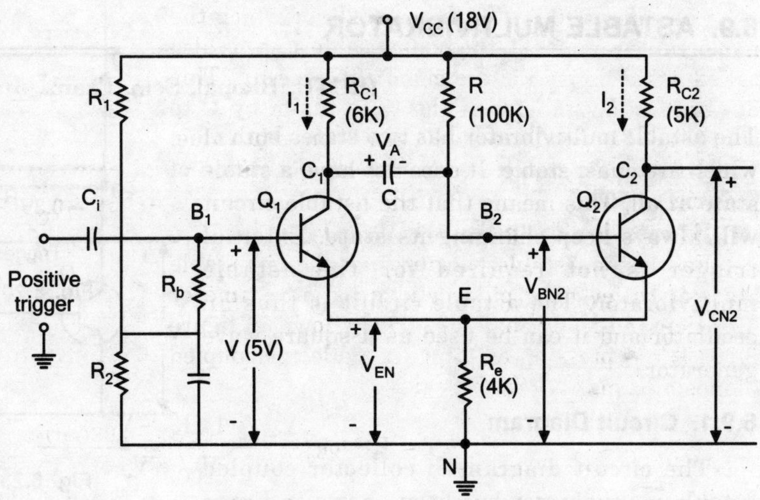

Fig. 6.21. Positive pulse triggering

Disadvantages of Positive Pulse Triggering

(*i*) The source of triggering pulses is loaded heavily due to low input impedance of Q_1. Hence, it is suitable only for the emitter coupled monostable.

(*ii*) This type of triggering has low sensitivity.

2. Negative Pulse Triggering

As shown in figure 6.22, a negative trigger pulse is applied to the collector of normally off transistor Q_1. The pulse is applied through a differentiator and the diode D will allow only the negative pulses (spikes) to pass through to the collector of Q_1. The negative trigger pulse will force Q_1 into ON state and Q_2 into OFF state which is the quasi-stable state for the monostable circuit.

Advantages of negative pulse triggering

(*i*) It does not load the trigger pulse source.

(*ii*) It is more sensitive than positive pulse triggering.

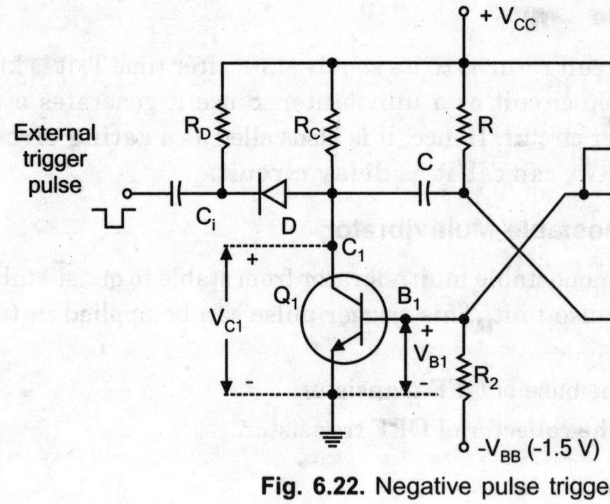

Fig. 6.22. Negative pulse triggering

6.9. ASTABLE MULTIVIBRATOR

(RGTU, Bhopal, Sem. Exam., June 2005, Dec. 2006, Dec. 2007)

The astable multivibrator has two states both of which are quasi stable. It does not have a stable state at all. This means that the astable circuit will always keep changing its state. External trigger is not required for the astable multivibrator. The astable circuit is thus an oscillator and it can be used as a square wave generator.

6.9.1. Circuit Diagram

The circuit diagram of collector coupled astable multivibrator has been shown in figure 6.23. The collector of Q_1 is coupled to the base of Q_2 via capacitor C_2 whereas the collector of Q_2 is coupled to the base of Q_1 via capacitor C_1. Due to the capacitive coupling used between the two transistors, none of the two transistors can remain permanently off. Instead the circuit will have two quasi stable states and it will make periodic transition between these states.

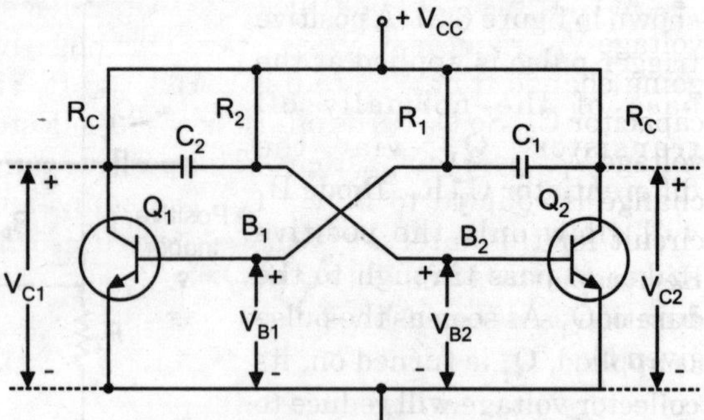

Fig. 6.23. Astable multivibrator

6.9.2. Waveforms

The waveform at the bases and collectors of the astable multivibrator have been shown in figure 6.24. Let us make use of these waveforms to understand the circuit operation.

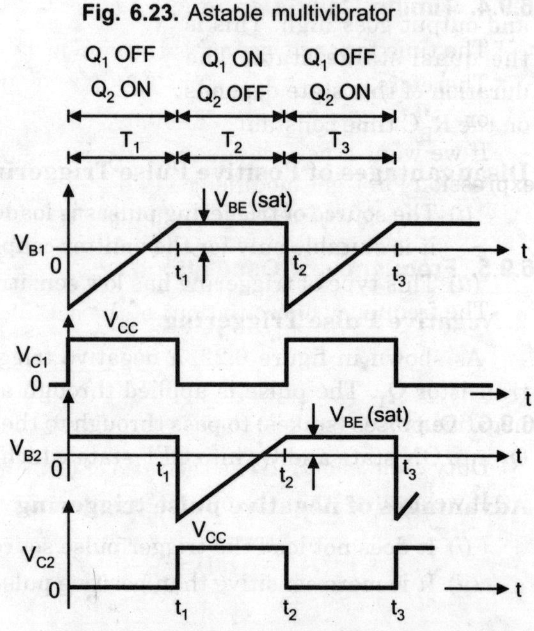

Fig. 6.24. Waveforms of astable multivibrator

6.9.3. Working Operation

Working Operation in the Interval $(0 - t_1)$

At instant $t = 0$, let Q_1 be in the off state and Q_2 be in the on state. Hence, $V_{C1} = V_{CC}$ and $V_{C2} = V_{CE} = V_{CE(sat)} = 0V$. The transition 0 to V_{CC} of V_{C1} gets coupled through C_2 to base of Q_2 and Q_2 is forced to saturation. The transition V_{CC} to 0 of V_{C2} gets coupled through C_1 to base of Q_1 and Q_1 is forced to off state. The capacitor C_1 starts charging through R_1 and Q_2 as shown in the equivalent circuit of figure 6.25. As C_1 is connected to base of Q_1, the voltage at B_1 will increase. At $t = t_1$, V_{B1} reaches $V_{BE(sat)}$ and Q_1 is turned on. The duration 0 to t_1 is denoted by T_1. Hence, T_1 is dependent on the value of R_1 and C_1. The expression for T_1 is given by,

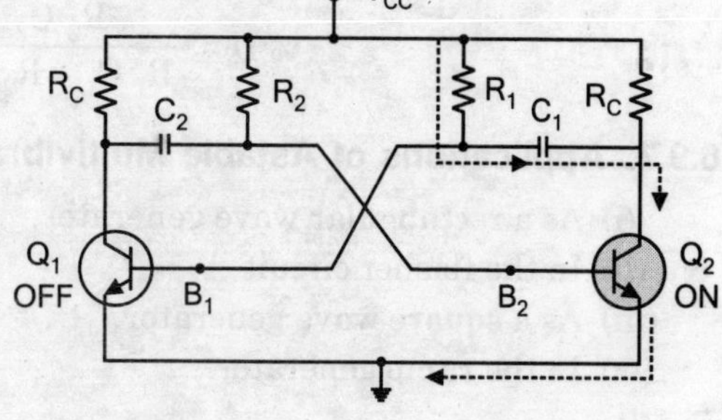

Fig. 6.25. Equivalent circuit for the interval $(0 - t_1)$

$$T_1 = 0.693\ R_1\ C_1 \quad ...(6.10)$$

Working Operation during interval $(t_1 - t_2)$

At t_1, transistor Q_1 is turned on so its collector voltage V_{C1} reduces to $V_{CE(sat)} = 0$. This negative going change from V_{CC} to 0 is coupled to B_2 via capacitor C_2. So Q_2 turns off. Hence, the collector voltage V_{C2} rises to V_{CC} from 0. This positive going change is coupled to B_1 via C_1. The equivalent circuit for this interval is shown in figure 6.26. Here, capacitor C_2 charges through R_2 and Q_1. This duration is denoted by T_2 and its expression is given by,

$$T_2 = 0.693\ R_2\ C_2 \quad ...(6.11)$$

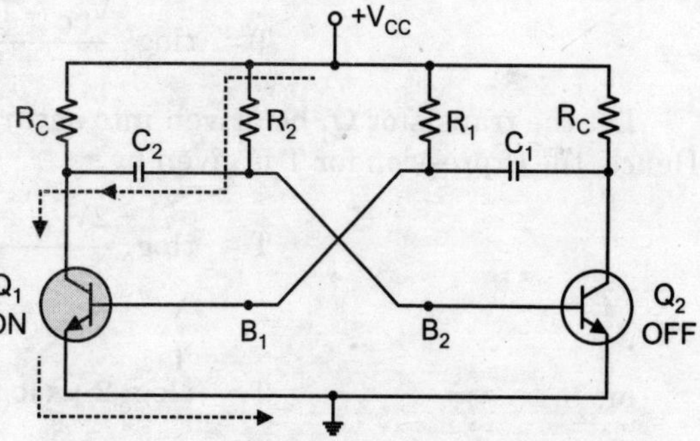

Fig. 6.26. Equivalent circuit for interval $(t_1 - t_2)$

6.9.4. Timing Considerations

The time for each cycle T is the sum of T_1 and T_2.

Therefore, $T = T_1 + T_2 = 0.693\ R_1\ C_1 + 0.693\ R_2\ C_2$

or $T = 0.693\ (R_1\ C_1 + R_2\ C_2)$

If we want a perfect square wave at the output, the $R_1 = R_2 = R$ and $C_1 = C_2 = C$. Then, the expression for T get modified as under:

$$T = 0.693\ (RC + RC) = 1.38\ RC \quad ...(6.12)$$

6.9.5. Frequency of Oscillations

The frequency of oscillations of the astable multivibrator is given by

$$f = \frac{1}{T} = \frac{1}{1.38\ RC} \quad ...(6.13)$$

6.9.6. Duty Cycle

Duty cycle is defined as the ratio of ON time to the total time.

If we assume that the output is obtained at the collector of Q_2, then the duty cycle is given by,

$$D = \frac{T_2}{T_1 + T_2} = \frac{0.69\ R_2\ C_2}{0.69\ (R_1\ C_1 + R_2\ C_2)} \quad ...(6.14)$$

$$D = \frac{R_2 C_2}{R_1 C_1 + R_2 C_2} \quad ...(6.15)$$

Duty cycle varies between 0 and 1. It can also be expressed as a percentage as under:

$$\% D = \frac{R_2 C_2}{R_1 C_1 + R_2 C_2} \times 100 \quad ...(6.16)$$

6.9.7. Applications of Astable Multivibrator (RGTU, Bhopal, Sem. Exam, Dec. 2009)

(i) As a rectangular wave generator.
(ii) In the flasher circuit.
(iii) As a square wave generator.
(iv) In the ramp generator.

6.9.8. Derivation for T_1, T_2 and T

For a monostable multivibrator, we have written the exact expression for the pulse width T as follows:

$$T = \tau \log_e \frac{V_{CC} + I_C R_C - V_{BE(sat)}}{V_{CC} - V\gamma}$$

Let the transistor Q_1 be driven into saturation. Then $I_C R_C = V_{CC} - V_{CE(sat)}$ and $V_\gamma = V_{BE(sat)}$. Hence, the expression for T is given by,

$$T = \tau \log_e \frac{2V_{CC} + V_{CE(sat)} - V_{BE(sat)}}{V_{CC} - V\gamma}$$

or

$$T = \tau \log_e 2 + \tau \log_e \frac{V_{CC} - \frac{V_{CE(sat)} + V_{BE(sat)}}{2}}{V_{CC} - V\gamma}$$

At room temperature, $V_{CE(sat)} + V_{BE(sat)} = 2V_\gamma$ for either silicon or germanium. Hence, the second term in the above expression is zero.

Therefore, $T = \tau \log_e 2 = 0.69\,\tau$ or, $T = 0.69\,RC$ ...(6.17)

We can apply the same logic to obtain T_1 and T_2 for the astable multivibrator.

Thus, $T_1 = 0.69\,R_1C_1$ and $T_2 = 0.69\,R_2C_2$

Hence, time period $T = T_1 + T_2 = 0.69\,(R_1 C_1 + R_2 C_2)$ ...(6.18)

6.10. BISTABLE MULTIVIBRATOR USING A TRANSISTOR

(RGTU, Bhopal, Sem. Exam, Dec. 2004, June 2006-07)

The name **bistable** suggests that this circuit has two possible stable states of operation. Once the multivibrator enters in one stable state, it continues to be in that state, until it is forced to change the state by an external excitation. Figure 6.27 shows the circuit diagram for the bistable multivibrator using two N-P-N transistors.

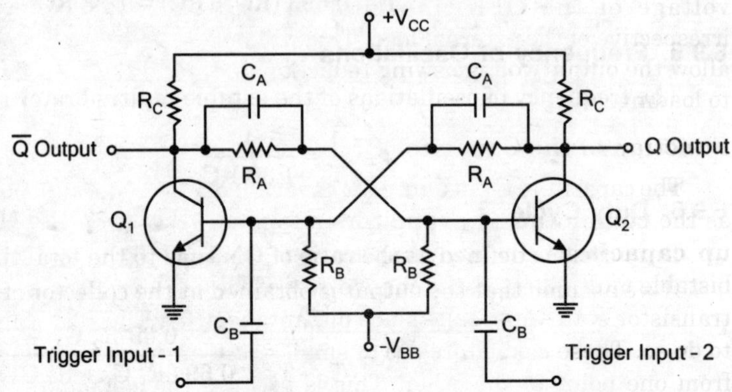

Fig. 6.27. A bistable multivibrator using N-P-N transistors

Working Operation of the Circuit

We assume that both the transistors are identical and one of the transistors say Q_1 is conducting. The biasing resistors are so adjusted that Q_1 will be saturated. This is one stable state. Here, $\overline{Q}$ will be low and Q will be high. To make Q low, we apply a negative trigger pulse of an adequate amplitude to trigger input-1. Due to the application of a negative trigger pulse, transistor Q_1 comes out of conduction. Its collector voltage increases to + V_{CC}. Due to cut off, the collector voltage of Q_1 will increase to +V_{CC}. This high going change in level of collector voltage of Q_1 is coupled to the base of Q_2 through the capacitor C_A, due to which Q_2 will be forced into on state. Infact, Q_2 will saturate and Q output will become low.

As Q_1 is turned off, its collector voltage rises to + V_{CC}. This voltage will then ensure that Q_2 will remain in saturation. This is the second stable state of operation. If we want to turn on Q_1 again then we have to apply a negative trigger pulse at trigger pulse input – 2. The operation has been summarized in table 6.1.

Table 6.1.

Sr. No.	Q_1	Q_2	Q output	$\overline{Q}$ output
1.	ON	OFF	HIGH	LOW
2.	OFF	ON	LOW	HIGH

The circuit shown in figure 6.27 is also called as a **fixed bias transistor bistable circuit.**

Loading and its Effects

The Q and $\overline{Q}$ outputs of the multivibrator are connected to some external load. Due to this external load, the collector voltage of the OFF transistor decreases (In absence of external load, it is + V_{CC}). This is called as loading and it is undesired feature of the bistable circuit. The major effect of loading is reduced output voltage swing. Even, the resistor R_A will load the OFF transistor.

Remedy

To overcome the effects of loading, we can take the following action :

We increase the value of R_A so that they do not load the OFF transistor. We use additional diodes as shown in figure 6.28. Alongwith an auxiliary power supply of voltage V_{DC}. V_{DC} should be less than the supply voltage V_{CC}. The diodes will clamp the collector voltage of the OFF transistor to V_{DC}, irrespective of the external load. This will not allow the output voltage swing reduction due to loading.

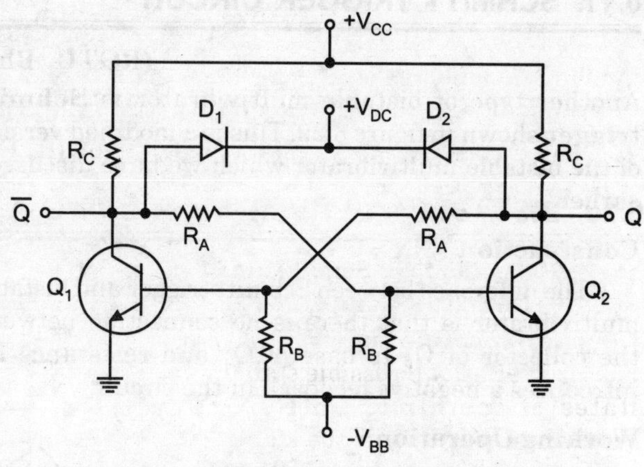

Fig. 6.28. Use of diodes and an auxiliary source to reduce loading effect

Function of the Capacitor C_A

The capacitors C_A in figure 6.27 are called as the **commutating capacitors** or **speed up capacitors**. In many applications of a bistable multivibrator, the change over of the transistor state must take place instantaneously. The speed up capacitors will help the transistor to do so. These capacitors are of small value and they couple the sudden change in voltage level, from one point to the other. This is because the capacitors will act as short circuit for sudden changes in voltage.

The action of these capacitors is as under:

Let us, Q_1 is ON and Q_2 is OFF. Therefore, Q is at + V_{CC} and $\overline{Q}$ is at $V_{CE(sat)}$. If we apply trigger pulse to turn Q_2 ON. Then, its collector voltage drops down suddenly from + V_{CC} to $V_{CE(sat)}$. This sudden change in voltage is coupled instantaneously by the capacitor C_A, to the base of Q_1 and Q_1 is turned OFF instantaneously. A similar action can take place when Q_1 is turned on, to turn OFF Q_2.

6.10.1. Performance Comparison of Astable, Monostable and Bistable Multivibrator

Sr.No.	Astable multivibrator	Monostable multivibrator	Bistable multivibrator
1.	There are no stable states of output.	There is only one stable state of the output.	There are two stable states of the output
2.	Trigger input is not necessary for changing the state of the output	Trigger pulse is required for changing the state of output.	Trigger input is required for changing the state of output.
3.	Used as rectangular, square wave or ramp generator.	Used as timer.	Used as flip-flop.
4.	Number of quasi-stable states is 2.	Number of quasi-stable states is 1.	No quasi-stable states
5.	Time for the two quasi-stable states depends on RC time constant. The two quasi-stable states can have different intervals.	Time for the quasi-stable state depends on RC time constant	No quasi-stable states.

6.11. SCHMITT TRIGGER CIRCUIT

(RGTU, Bhopal, Sem. Exam, Dec. 2007, June 2010)

Another type of bistable multivibrator is **Schmitt trigger** shown in figure 6.29. This is a modified version of the bistable multivibrator which we have discussed earlier.

Construction

The difference between Schmitt trigger and bistable multivibrator is that there is no connection between the collector of Q_2 to base of Q_1 and resistance R_E introduces a negative feedback in the circuit.

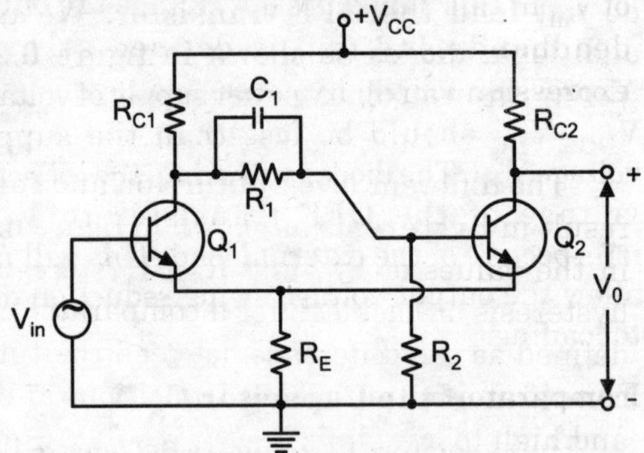

Fig. 6.29. Schmitt trigger circuit

Working Operation

A sine wave is applied at the base of transistor Q_1 as shown in figure 6.29. When V_{in} = 0, transistor Q_1 will be OFF. Its collector voltage will be equal to V_{CC} which will turn on and saturate transistor Q_2. The collector voltage of Q_2 i.e. output voltage is given by,

$$V_o = V_{CE(sat)} + V_{RE} \quad ...(6.19)$$

The voltage across R_E reverse biases Q_1. Therefore even when V_{in} becomes positive Q_1 will remain off till V_{in} is less than ($V_{RE} + V_{BE}$).

As soon as ($V_{in} = V_E + V_{\gamma 1}$), transistor Q_1 will be turned on. This input voltage is called as **Upper threshold Point or UTP**. As soon as transistor Q_1 turns on, its collector voltage will decrease to $[V_{CE(sat)} + V_E]$. This will force the transistor Q_2 into off state. Its collector voltage will now increase to V_{CC}. Thus, at UTP, V_0 increases to $+ V_{CC}$.

Now, when Q_1 is conducting, the voltage across R_E is given by,

$$V_{RE} = I_{C1} R_E \quad \text{....(6.20)}$$

Here, Q_1 will remain on as long the input voltage is higher than $(I_{C1} R_E + V_{BE})$ as shown in the waveforms of figure 6.30.

As soon as V_{in} decreases below $(I_{C1} R_E + V_{BE})$, transistor Q_1 will be turned OFF. This level is called as the **Lower threshold Point (LTP)**. As Q_1 turns OFF, its collector potential increases, which will turn the transistor Q_2 ON. As Q_2 saturates, the output voltage will reduce to a voltage stated by equation (6.19). The waveforms for Schmitt trigger have been shown in figure 6.30.

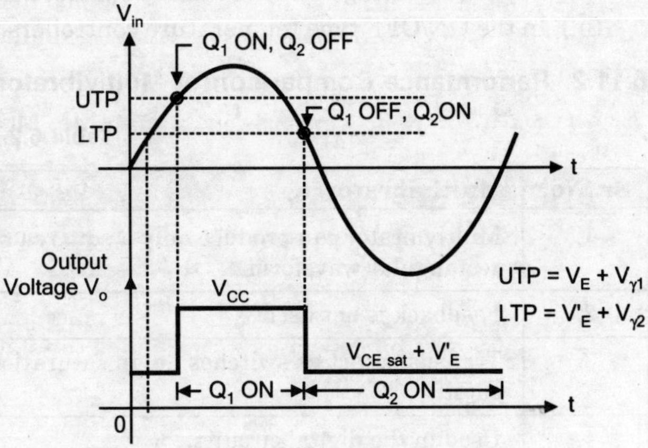

Fig. 6.30. Waveforms for schmitt trigger

Transfer Characteristics and Hysteresis

Transfer characteristics is the graph of input voltage V_{in} versus the output voltage V_0 as shown in figure 6.31. The transfer characteristics is as shown in figure 6.31. Due to the positive feedback the comparator is said to exhibit **hysteresis**. The hysteresis voltage is equal to the difference between V_{UT} and V_{LT}. It is also called as **hysteresis width**.

The interval $V_{LT} < V_{in} < V_{UT}$ in figure 6.31 is called as dead zone or dead band. This is because the variation of V_{in} in this band does not change the output at all. This deadband condition is defined as hysteresis. The expression for the hysteresis voltage is given by,

$$H_V = V_{UTP} - V_{LTP} \quad \text{...(6.21)}$$

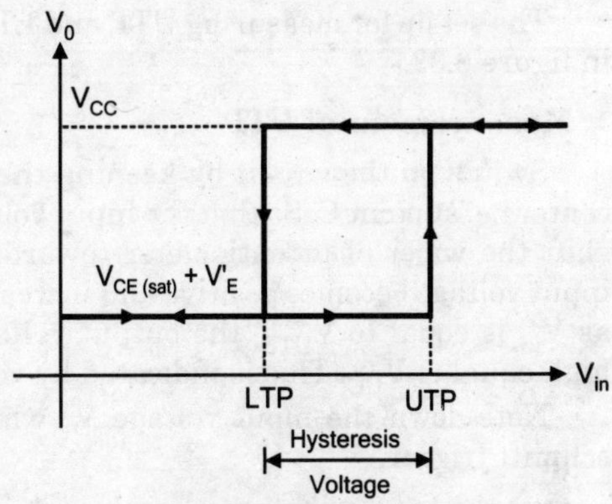

Fig. 6.31. Hysteresis loop or transfer characteristics

The different levels of turn on and turn off threshold voltages (LTP and UTP) at the input will result in **hysteresis** as shown in figure 6.31. The value of hysteresis will change with the changes in the values of R_{C1} and R_{C2}. Hysteresis loop is also called as **transfer characteristics**. The hysteresis or backlash for a comparator comes into existence when positive feedback is used. It is defined as the difference between the turn on and turn off input voltages. Due to hysteresis, the comparator output makes transitions at different points when it changes its state from low to high and high to low state.

Effects of Hysteresis

(i) Hysteresis improves the noise immunity.

(ii) It reduces the response time.

(iii) Increased hysteresis will make the sensitivity poor.

(iv) Hysteresis reduces the possibility of false triggering produced by noise.

6.11.1. Applications of Schmitt Trigger

Some of the important applications of Schmitt trigger are:

(i) To convert any waveform into a square wave.

(ii) In the overvoltage and overcurrent protection circuit.

(iii) In the ON/OFF type temperature controllers.

6.11.2. Performance Comparison of Multivibrator and Oscillator

Table 6.2.

Sr. No.	Multivibrator	Oscillator
1.	Multivibrator can produce only a square or rectangular waveform.	Oscillator can produce a square or sinusoidal output.
2.	Feedback is not used	Positive feedback is used.
3.	Transistors act as switches i.e. in saturation or cutoff.	Transistors operate in the active region
4.	Used in the digital circuits.	Used in the analog circuits.

6.11.3. Set up to Measure UTP and LTP

The set up for measuring UTP and LTP practically is shown in figure 6.32.

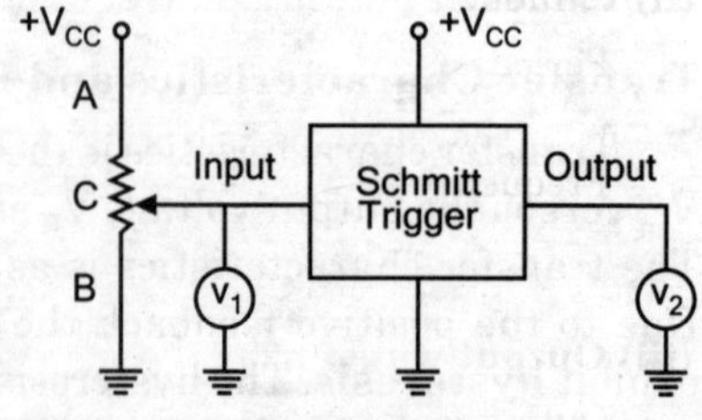

Fig. 6.32. Set up to measure UTP and LTP

1. Measurement of UTP

Switch on the circuit by keeping the potentiometer at the center i.e. at point C. So that the input voltage is zero. Gradually shift the wiper of potentiometer towards point A, so that the input voltage becomes positive and increases gradually. As soon as V_{in} is equal to V_{UTP}, the output voltage suddenly becomes high equal to V_{CC}. This is indicated by voltmeter V_2.

Note down the input voltage V_1 which is the UTP of the schmitt trigger.

2. Measurement of LTP

Now reduce the input voltage by moving the wiper of the potentiometer towards B. As soon as the input voltage is equal to LTP, the output voltage makes a sudden transition from V_{CC} to $[V_{CE(sat)} + V_E]$. Note down the reading V_1 as LTP.

IMPORTANT SOLVED EXAMPLES

EXAMPLE 6.3. Find the value of resistor, if C = 100 μF for monostable multivibrator to give output pulse width of 400 ns. What will be the resistor, if C is changed to 40 pF keeping pulse width constant.

Solution: Given C = 100 μF, T = 400 ns. R = ?

(i) Value of R

We know that $T = 0.69\ RC$

Therefore, $$R = \frac{T}{0.69C} = \frac{400\times10^{-9}}{0.69\times100\times10^{-6}} = 5.797 \times 10^{-3}\ \Omega \quad \textbf{Ans.}$$

(ii) Calculate new value of R

$$T = 400 \text{ ns}$$

and $$C = 40 \text{ pF}$$

Therefore, $$R = \frac{400 \times 10^{-9}}{0.69 \times 40 \times 10^{-12}} = 14.492 \text{ k}\Omega \quad \textbf{Ans.}$$

EXAMPLE 6.4. Calculate the time period of quasistable state t_1, t_2, T and frequency of oscillation if $R_1 = 1$ kΩ, $R_2 = 20$ kΩ and $C_1 = 0.01$ μF, $C_2 = 0.06$ μF. Draw the waveform at the output.

Solution: Given
$$R_1 = 1 \text{ k}\Omega$$
$$R_2 = 20 \text{ k}\Omega$$
$$C_1 = 0.01 \, \mu\text{F}$$
$$C_2 = 0.06 \, \mu\text{F}$$

(i) Values of T_1 and T_2.

$$T_1 = 0.69 \, R_1 \, C_1 = 0.69 \times 1 \times 10^3 \times 0.01 \times 10^{-6}$$

or, $$T_1 = 6.9 \times 10^{-6} \text{ sec or } 6.9 \, \mu\text{sec.} \quad \textbf{Ans.}$$

Also, $$T_2 = 0.69 \, R_2 \, C_2 = 0.69 \times 20 \times 10^3 \times 0.06 \times 10^{-6}$$

or, $$T_2 = 8.28 \times 10^{-4} \text{ sec or } 828 \, \mu\text{sec.} \quad \textbf{Ans.}$$

(ii) Values of period and frequency

Period $$T = T_1 + T_2 = 6.9 \, \mu\text{s} + 828 \, \mu\text{s} = 834.9 \, \mu\text{s} \quad \textbf{Ans.}$$

Frequency $$f = \frac{1}{T} = \frac{1}{834.9 \times 10^{-6}}$$
$$= 1197.74 \text{ Hz} \quad \textbf{Ans.}$$

(iii) Output waveform

Output waveform has been shown in figure 6.33.

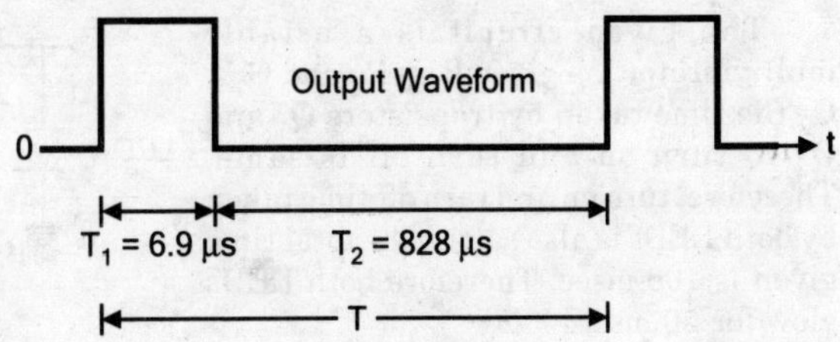

Fig. 6.33.

EXAMPLE 6.5. Derive the expression for the time period T and frequency of oscillation for astable multivibrator.

If $R_1 = R_2 = 7.5$ k ohm

and $C_1 = C_2 = 0.15$ μF.

Calculate the time period for astable multivibrator.

Solution:

(i) Time period $$T = 0.69 \, (R_1 \, C_1 + R_2 \, C_2) = 0.69 \times 2 \, RC = 1.38 \times 7.5 \times 10^3 \times 0.15 \times 10^{-6}$$

Hence, $$T = 1.5525 \times 10^{-3} \text{ sec or } 1.5525 \text{ ms} \quad \textbf{Ans.}$$

(ii) Frequency $$f = \frac{1}{T} = \frac{1}{1.5525 \times 10^{-3}} = 644.12 \text{ Hz} \quad \textbf{Ans.}$$

EXAMPLE 6.6. A collector coupled monostable multivibrator using NPN transistor has following parameters : $V_{CC} = 12$ V, $V_{BB} = 3$ V, R = 20 k, C = 100 pF, $R_{C1} = R_{C2} = 2$k Ω. Find the width of the output and draw the collector and base waveform for Q_1 OFF and Q_2 ON.

Solution: Given $V_{CC} = 12$ V, $V_{BB} = 3$ V, R = 20 k
$$C = 100 \text{ pF},$$
$$R_{C1} = 2 \text{ k}\Omega,$$
$$R_{C2} = 2 \text{ k}\Omega$$

The expression for pulse width is given by
$$T = 0.69 \, RC = 0.69 \times 20 \times 10^3 \times 100 \times 10^{-12} = 1380 \times 10^{-9} = 1380 \text{ ns} \quad \textbf{Ans.}$$

For waveform, please go through the text.

EXAMPLE 6.7. Draw the collector. waveforms for NPN transistor is Bistable multivibrator with Q_1 OFF and Q_2 ON.

Solution:

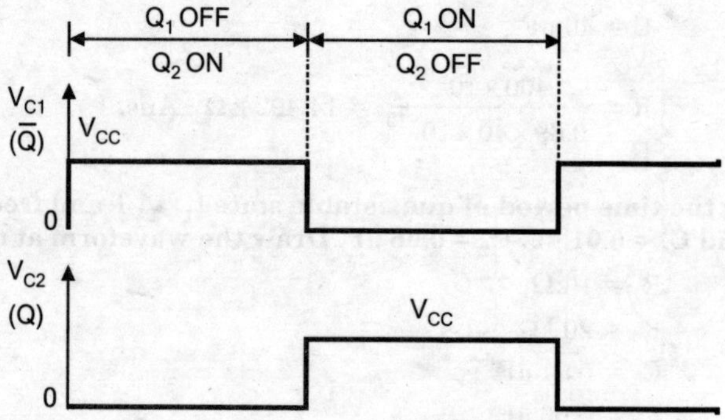

Fig. 6.34. Collector waveforms of bistable multivibrator

EXAMPLE 6.8. The circuit diagram shown in figure 6.35 to answer the following: When V_{CC} = 10 V, $C_1 = C_2$ = 0.1 mF, $R_1 = R_2$ = 1 MΩ and T = 100 msec. Suppose LED_2 glows at t = 0 sec, find out time period taken by LED_1 to glow and justify the answer. Name circuit diagram.

Solution: Given that

$$V_{CC} = 10 \text{ V}$$
$$C_1 = C_2 = 0.1 \text{ mF}$$
$$R_1 = R_2 = 1 \text{ M}\Omega$$
$$T = 100 \text{ msec}$$

The given circuit is a astable multivibrator. As given $R_1 = R_2$ and $C_1 = C_2$ the time taken by transistors Q_1 and Q_2 to turn on and turn off is same. Therefore, turn on and turn off time taken by both LEDs is also same. The total time given is 100 msec. Therefore both LEDs glow for 50 msec.

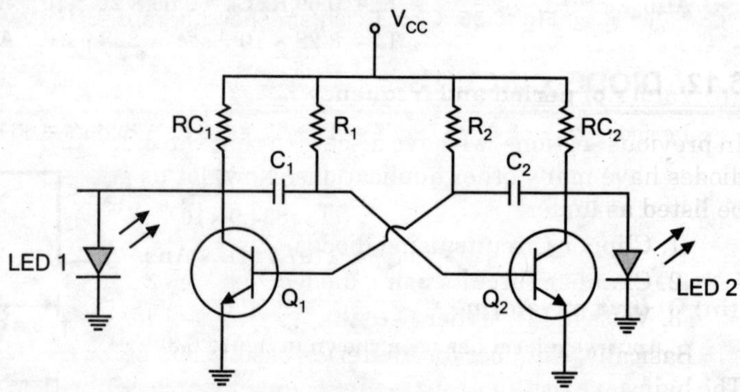

Fig. 6.35.

EXAMPLE 6.9. Calculate the time period and frequency of oscillations for astable multivibrator with component value of $R_1 = R_2$ = 4.7 kΩ and $C_1 = C_2$ = 270 pF. if R is changed to 10 kΩ what is the frequency of oscillations?

Solution: Given : Astable multivibrator

$$R_1 = R_2 = 4.7 \text{ k}\Omega \quad \text{and} \quad C_1 = C_2 = 270 \text{ pF}$$

Time period
$$T = 0.69\,(R_1C_1 + R_2C_2)$$
$$T = 0.69 \times 2RC = 0.69 \times 2 \times 4.7 \times 10^3 \times 270 \times 10^{-12}$$
$$T = 1.75 \times 10^{-6} \text{ or } 1.75\ \mu\text{sec} \quad \textbf{Ans.}$$

Frequency
$$f = \frac{1}{T} = \frac{1}{1.75 \times 10^{-6}} = 571 \text{ kHz}$$

If R is changed to 10 kΩ, then,

$$T = 0.69 \times 2\,RC = 0.69 \times 2 \times 10 \times 10^3 \times 270 \times 10^{-12}$$
$$T = 3.72 \times 10^{-6} = 3.72\ \mu\text{sec}$$

Also, frequency
$$f = \frac{1}{T} = \frac{1}{3.72 \times 10^{-6}} = 268.38 \text{ kHz} \quad \textbf{Ans.}$$

EXAMPLE 6.10. Draw and explain transistor as switch with capacitive load.

Solution: Figure 6.36 (a) shows the transistor switch with a capacitive load. When the transistor is ON, it short circuits the capacitor and V_C = 0. But, when the transistor turns off, capacitor charges exponentially through resistor R_C.

The waveforms have been shown in figure 6.36. (b)

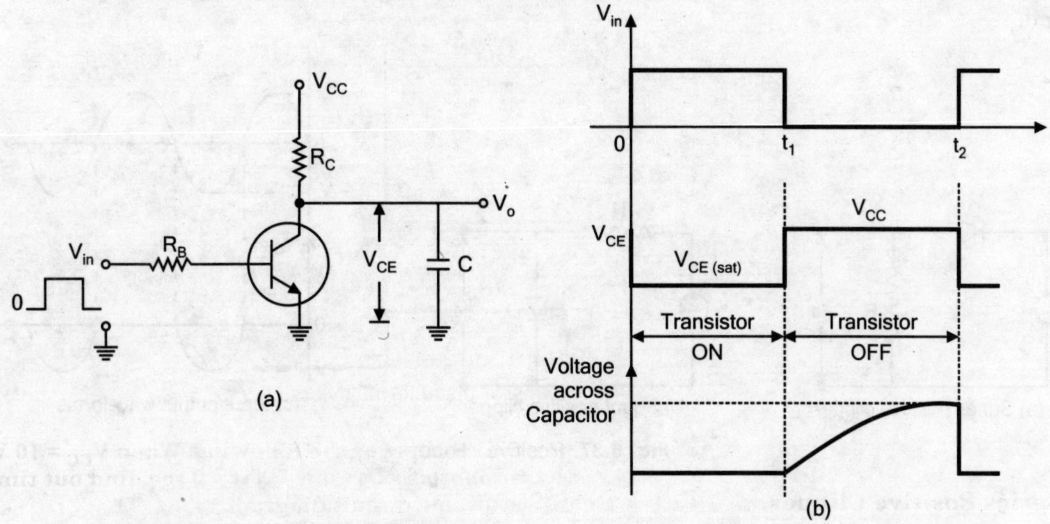

Fig. 6.36. Circuit diagram showing transistor biased to operate as a switch

6.12. DIODE CIRCUITS

In previous sections, we have discussed that the diodes can be used as rectifiers. In addition to this, diodes have many other applications. Now, let us study few such applications of diode which may be listed as under:

1. Clipper Circuits using diodes
2. Clamper Circuits using diodes
3. Voltage Multiplier Circuits

Basically, a clipper (or limiter) is used to clip off or remove a particular portion of an a.c. signal. The half wave rectifier is basically a clipper which eliminates one of the alternations of an a.c. signal. On the other hand, a clamper (or dc restorer) is used to restore or change the dc reference of an ac signal. As an example, we may have a 10 V_{pp} a.c. signal that varies equally above and below 2 V_{dc}.

6.13. CLIPPER CIRCUITS

A clipper is a circuit with which the waveform is shaped by removing or clipping a portion of the applied input signal waveform without distorting the remaining part. Clipper can remove signal voltages above or below a specified level. Clippers find extensive use in radar, digital and other electronic systems. Although several clipping circuits have been developed to change the waveshape, we shall confine our attention to diode clippers. As a matter of fact, the clipper circuits may be classified as under:

(i) Positive clipper (ii) Negative clipper
(iii) Biased clipper (iv) Combination clipper

6.13.1. Positive Clipper

A positive clipper removes the positive half-cycles of the input voltage waveform. Positive clipper may also be classified into following two ways:

(i) Series positive clipper and
(ii) Shunt positive clipper

Both series positive clipper and shunt positive clipper have been shown in figure 6.37. In both the circuits, the diodes used have been assumed as ideal. This means that a diode works as short circuit or closed switch when forward biased and as an open circuit or open switch when reverse biased.

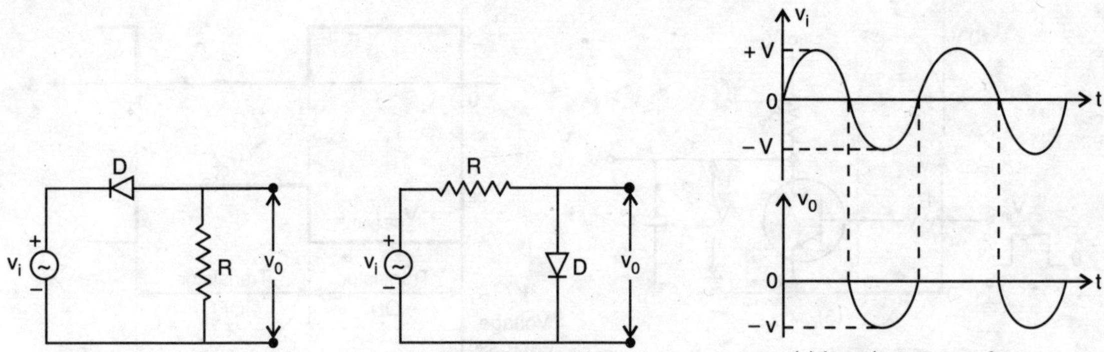

Fig. 6.37. Positive clipper

(i) Series Positive Clipper

Figure 6.37(a) shows the circuit of series positive clipper. If the input voltage is positive, the diode will be reverse biased and it will not conduct. In this condition, the diode will work as an open circuit and hence the positive half-cycle does not appear at the output (across the resistor). All the positive half-cycle voltage appears across the open circuit diode. Hence, the positive half-cycle is clipped off. When the input signal is negative, the diode is forward biased and it conducts. In this condition, the diode acts as short-circuit. This means that the voltage drop across the diode is zero and all the negative cycle voltage appears at the output (across the resistor).

(ii) Shunt Positive Clipper

Figure 6.37(b) shows the shunt positive clipper. If the input voltage is positive, the diode conducts and acts as a short-circuit. So, the voltage drop across the diode is zero and hence there is zero signal at the output. Therefore, positive cycle is clipped off. All the positive half cycle voltage appears across the resistor. When the input signal is negative, the diode is reverse biased and does not conduct and acts as an open circuit. Hence, all the negative half-cycle voltage appears across the diode or at the output.

Figure 6.37(*c*) shows the input and output waveforms.

EXAMPLE 6.11. Figure 6.38 shows the circuit of a shunt positive clipper and the waveform of input voltage applied. Determine the value of output voltage for positive and negative cycles. Assume the diode to be ideal.

Solution: Case (i) For positive half-cycle

During the positive half-cycle of the input signal, the diode is forward biased. Because diode is ideal, it will work as a short-circuit in forward biased. Short-circuit means zero voltage. Since the load voltage is connected in parallel with the diode, the output voltage will also be zero. Hence positive cycle is clipped off.

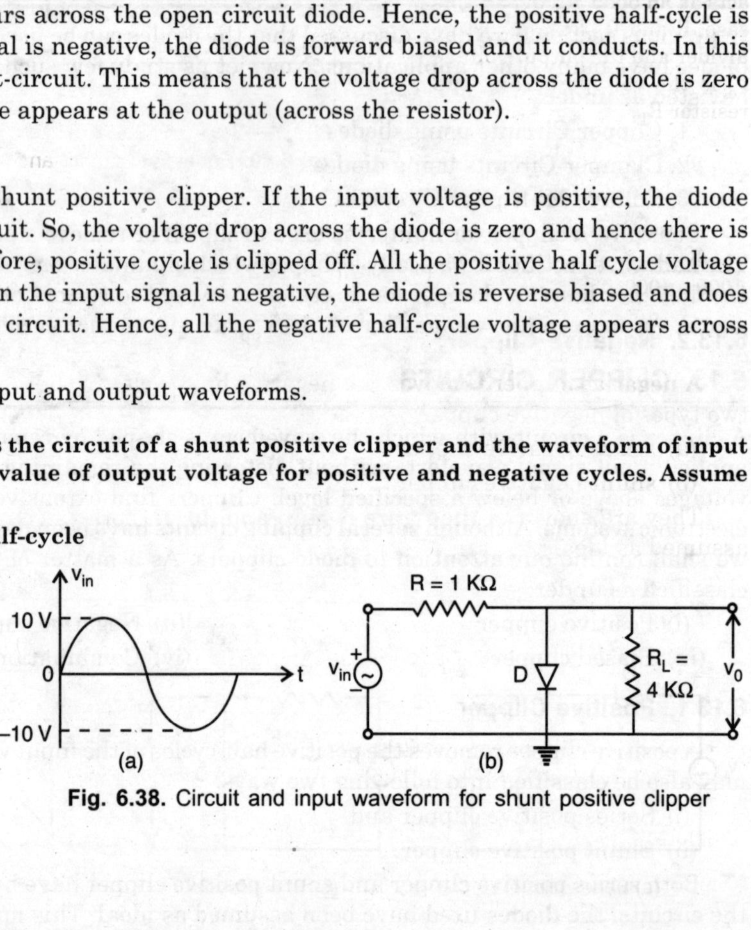

Fig. 6.38. Circuit and input waveform for shunt positive clipper

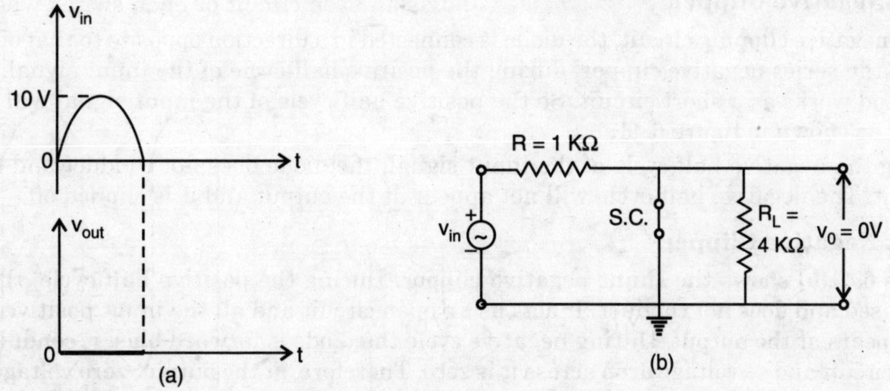

Fig. 6.39. Equivalent circuit and input and output waveform for positive half-cycle

Case (ii) For negative half-cycle

During the negative half-cycle of the input a.c. signal, the diode is reverse biased and hence it conducts no current. It acts as an open circuit. In this case, R and R_L form a voltage divider and the output voltage v_o is the voltage across load resistor R_L.

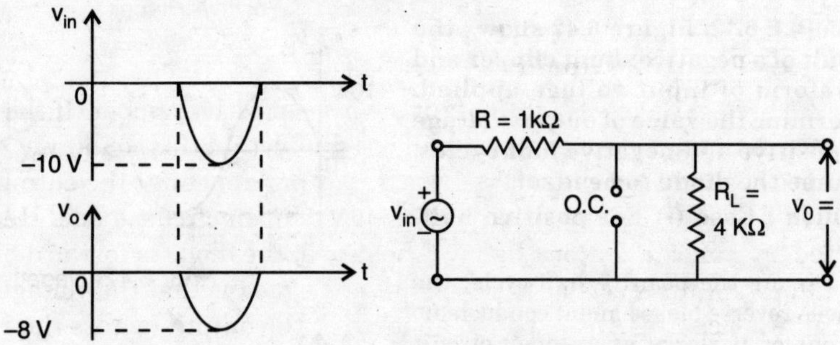

Fig. 6.40. Equivalent circuit and input and output waveforms for negative half-cycle

$$v_o = \frac{R_L}{R + R_L} . v_{in}$$

$$v_{o(peak)} = \frac{4000}{1000 + 4000} . v_{in(peak)} = \frac{4000}{5000}(-10) = 0.8 \times (-10) = -8 \text{ Volt. } \textbf{Ans.}$$

6.13.2. Negative Clipper **(Important)**

A negative clipper removes the negative half-cycles of the input voltage waveform. There are two types of negative clipper

(a) series negative clipper

(b) shunt negative clipper

They are shown in the figure 6.41(a) and (b). In both the circuits, the diodes used have been assumed as ideals.

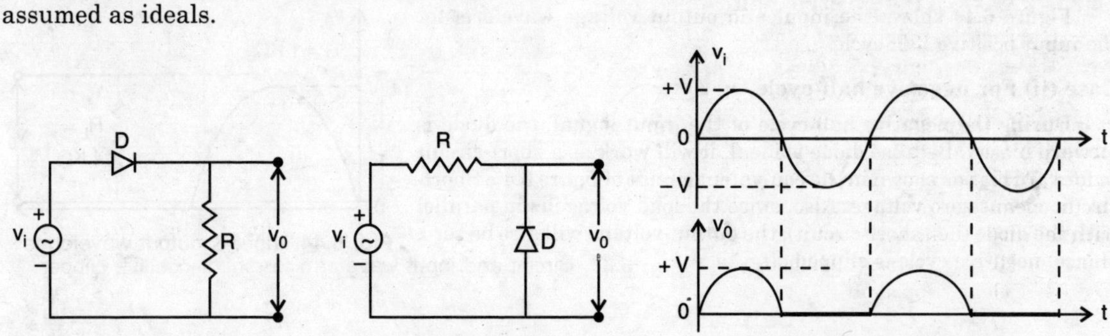

(a) Series clipper (b) Shunt clipper (c) Input/Output Waveforms

Fig. 6.41. Negative clipper

(i) Series Negative Clipper

In the negative clipping circuit, the diode is connected in a direction opposite to that of a positive clipper. In the series negative clipper, during the positive half-cycle of the input signal, the diode conducts and works as a short-circuit. So the positive half-cycle of the input signal will appear at the output as shown in figure 6.41.

During the negative half-cycle of the input signal, the diode does not conduct and acts as an open circuit. The negative half-cycle will not appear at the output and it is clipped off.

(ii) Shunt Negative Clipper

Figure 6.41(b) shows the shunt negative clipper. During the positive half-cycle, the diode is reverse biased and does not conduct. It acts as an open circuit and all the input positive half-cycle voltage appears at the output. During negative cycle the diode is forward bias, it conducts. It acts as a short circuit and so voltage drop across it is zero. Therefore, at the output, zero voltage appears. All the negative half-cycle voltage appears across the resistor R. Hence, negative half-cycle is clipped-off.

EXAMPLE 6.12. Figure 6.42 shows the circuit of a negative shunt clipper and waveform of input voltage applied. Determine the value of output voltage for positive and negative half-cycles. Assume the diode to be ideal.

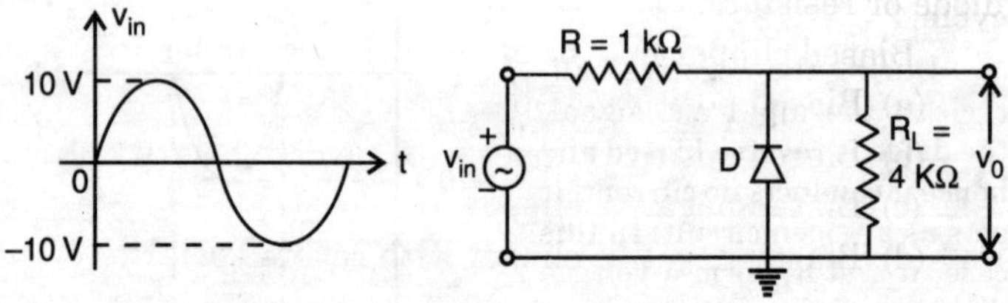

Fig. 6.42. Negative shunt clipper

Solution : Case (i) For positive half-cycle

During the positive half-cycle, the diode is reverse biased and it conducts no current or it works as an open circuit. Hence, from equivalent circuit in figure 6.43, it may be observed that when diode is replaced with an open circuit, the resistors R and R_L form a voltage divider. Thus, the voltage across the load resistor R_L will be the output voltage.

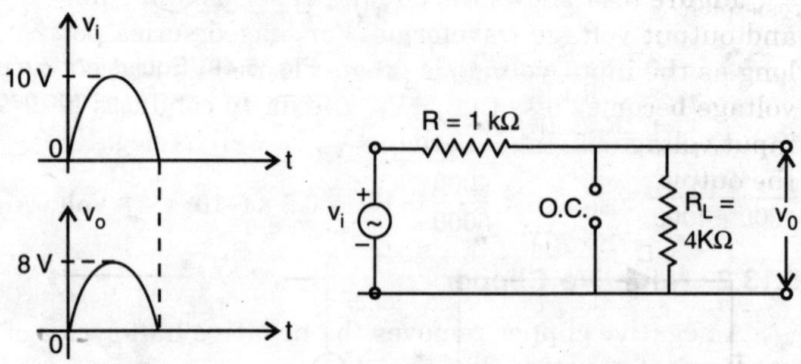

Fig. 6.43. Equivalent circuit for positive half-cycle

$$v_{0(peak)} = \frac{R_L}{R + R_L} v_{in} \text{ (peak)}$$

$$v_{0(peak)} = \frac{4000}{1000 + 4000} \times (+10 \text{ V})$$

$$v_{0(peak)} = +\frac{40{,}000}{5000} = +8 \text{ V} \quad \textbf{Ans.}$$

Figure 6.44 shows the input and output voltage waveform for the input positive half-cycle.

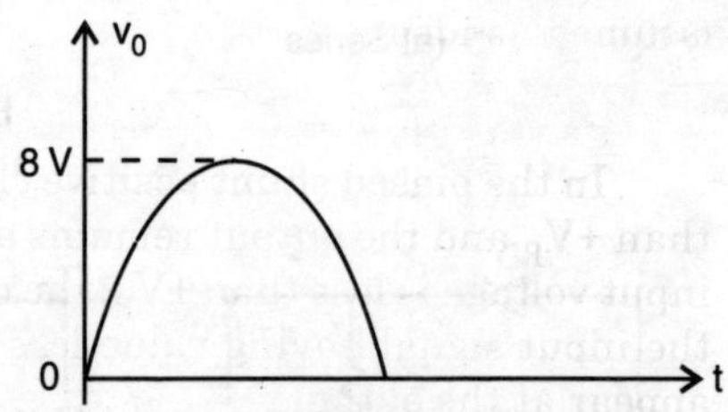

Fig. 6.44. Input & output waveforms

Case (ii) For negative half-cycle

During the negative half-cycle of the input signal, the diode is forward biased. Because diode is ideal, it will work as a short-circuit in forward bias as shown in the equivalent circuit of figure 6.45. Short-circuit means zero voltage. Also, since the load voltage is in parallel with the diode (i.e., short-circuit), the output voltage will also be zero. Hence, negative cycle is clipped off.

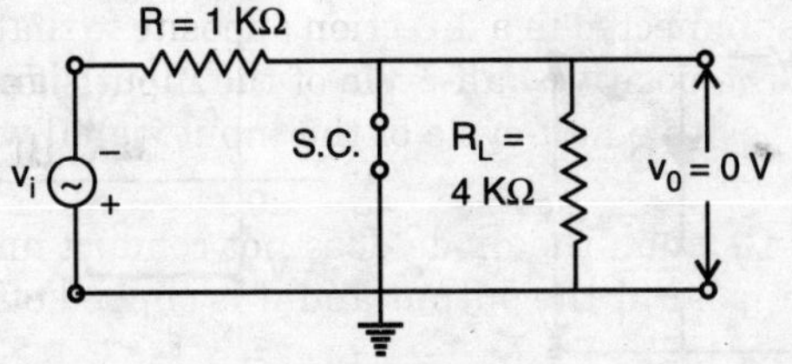

Fig. 6.45. Equivalent circuit for negative half cycle

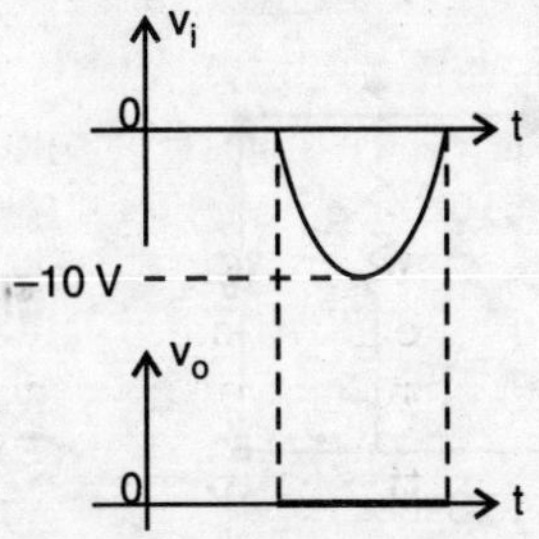

Fig. 6.46. Input and output waveforms

6.13.3. Biased Clipper

In some applications, there may be a requirement to remove a small portion of positive or negative half-cycle of the signal voltage and hence the biased clipper is used. The name bias means that the adjustment of the clipping level is achieved by adding a biasing voltage in series with the diode or resistor.

Biased clipper may be of four types as under:

(a) Biased positive clipper
(b) Biased positive clipper with reverse polarity of the battery
(c) Biased negative clipper
(d) Biased negative clipper with reverse polarity of the battery.

(a) Biased positive clipper

Figure 6.47 shows the circuits of series and shunt type biased positive clipper with the input and output voltage waveforms. For biased series positive clipper, the diode does not conduct as long as the input voltage is greater than $+V_R$ and so the output remains at $+V_R$. When the input voltage becomes less than $+V_R$, the diode conducts and acts as a short-circuit. Therefore, all the input voltage having less than $+V_R$ as well as negative half-cycle of the input wave will appear at the output.

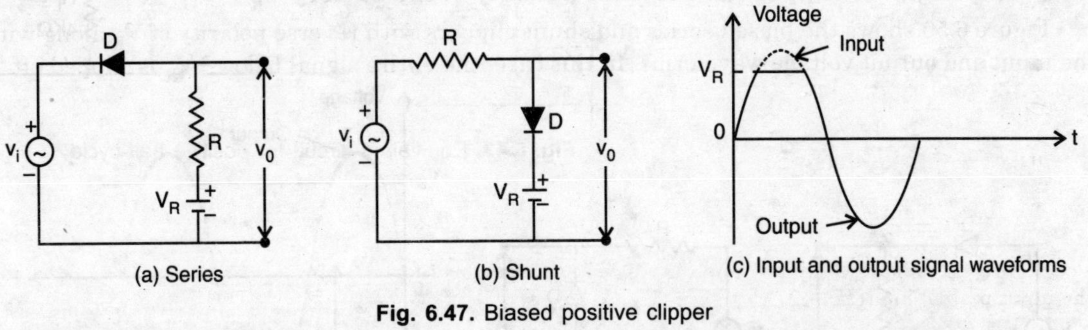

Fig. 6.47. Biased positive clipper

In the biased shunt positive clipper, the diode conducts as long as the input voltage is greater than $+V_R$ and the output remains at $+V_R$ until the input voltage becomes less than $+V_R$. When the input voltage is less than $+V_R$, the diode does not conduct and acts as an open switch. Therefore, all the input signal having value less than $+V_R$ as well as negative half-cycle of the input wave will appear at the output.

(b) Biased positive clipper with reverse polarity of the battery

Figure 6.48 shows the biased series and shunt clippers with reverse polarity of V_R along with the input and output voltage waveforms. In this case the entire signal above voltage $-V_R$ is clipped off.

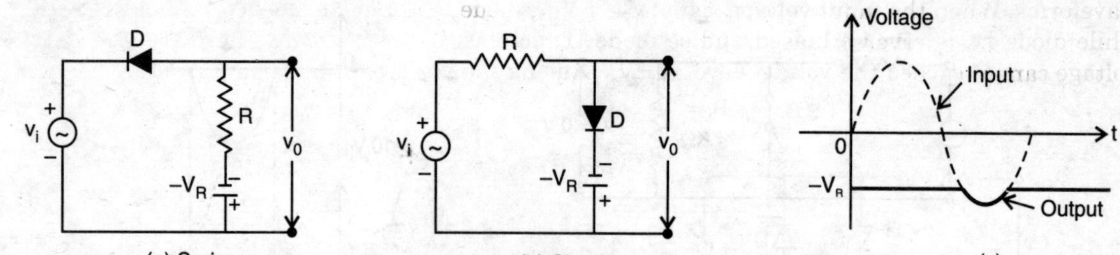

(a) Series (b) Shunt (c)

Fig. 6.48. Biased positive clipper with reverse polarity of V_R

(c) Biased negative clipper

For biased series negative clipper shown in figure 6.49 (a), when the input voltage $v_i \leq -V_R$, the diode does not conduct and clipping takes place.

For the biased shunt clipper shown in figure 6.49 (b), when the input voltage $v_i \leq -V_R$, the diode conducts and clipping takes place. The clipping can be shifted up and down, by varying the bias voltage ($-V_R$).

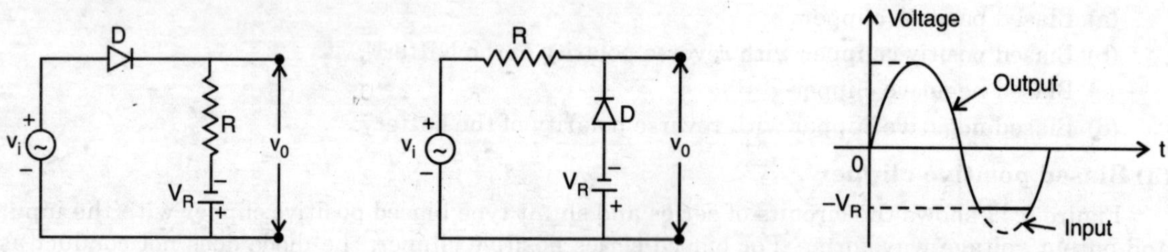

(a) Biased series negative clipper (b) Biased shunt negative clipper (c) Input and output signal waveform

Fig. 6.49.

(d) Biased negative clipper with reverse polarity of the battery V_R

Figure 6.50 shows the biased series and shunt clippers with reverse polarity of V_R along with the input and output voltage waveforms. In this case, the entire signal below $+V_R$ is clipped off.

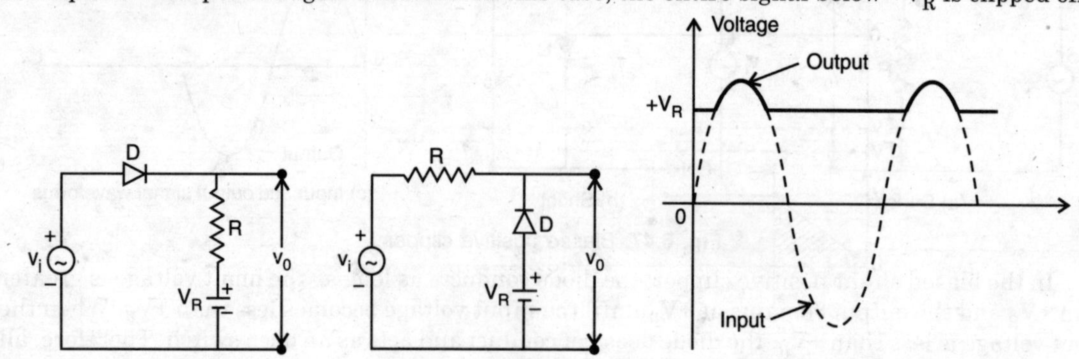

(a) Series (b) Shunt (c) Input and output signal waveforms

Fig. 6.50.

6.13.4. Combination Clipper

Combination clipper is the combination of a biased positive clipper and a biased negative clipper. Figure 6.51 shows the circuit diagram of combination clipper with input and output voltage waveforms. When the input voltage signal $v_i \geq +V_{R1}$, diode D_1 conducts and acts as a closed switch while diode D_2 is reverse biased and so diode D_2 acts as an open circuit. Therefore, the output voltage cannot exceed the voltage level of $+V_{R1}$ during the positive half cycle.

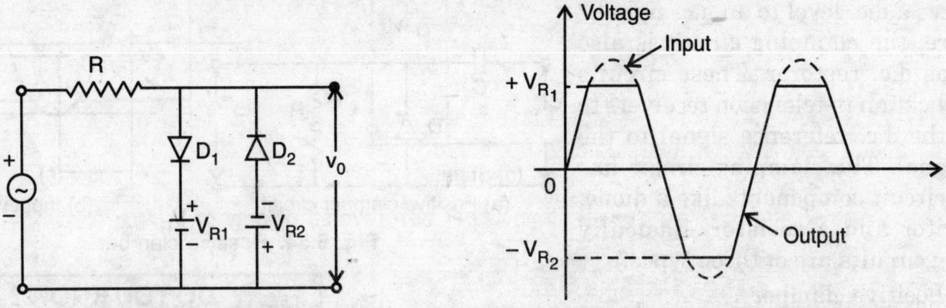

(a) Combination clipper

(b) Input and output waveform

Fig. 6.51. Combination clipper

EXAMPLE 6.13. Sketch I_R and v_0 w.r.t. time for the network shown in figure 6.52 for the input v_i shown in the same figure.

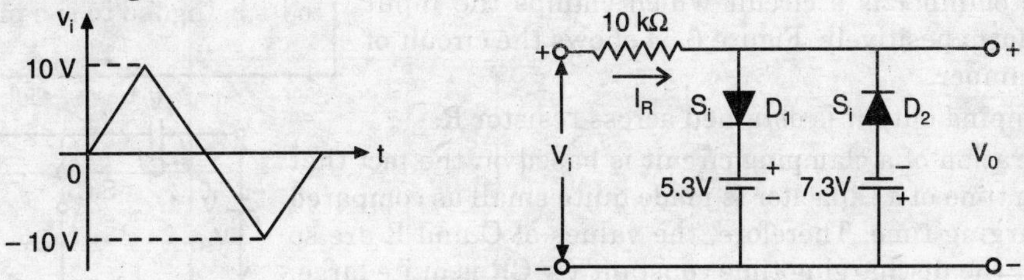

Fig. 6.52.

Assume that both diodes are silicon type with $R_f = 0\ \Omega$ and $R_r = \infty$ with $V_f = 0.7$ V.

Solution: The waveforms have been shown in figure 6.53.

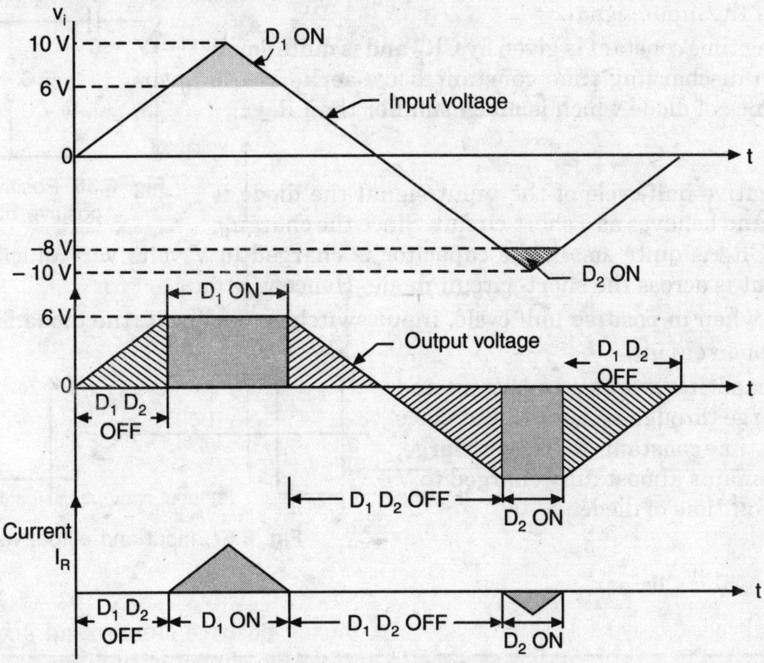

Fig. 6.53. Waveforms

6.14. CLAMPERS (RGTU, Bhopal, Sem. Exam., 2006-07)

Clamping circuit shifts or clamps a signal to a different d.c. level. In other words, clamping circuit introduces a d.c. level to an a.c. signal.* Therefore, the clamping circuit is also known as d.c. restorer. These circuits find application in television receivers to restore the d.c. reference signal to the video signal. The clamping circuit has various circuit components like a diode, a capacitor and a resistor. Basically clamping circuits are of three types:

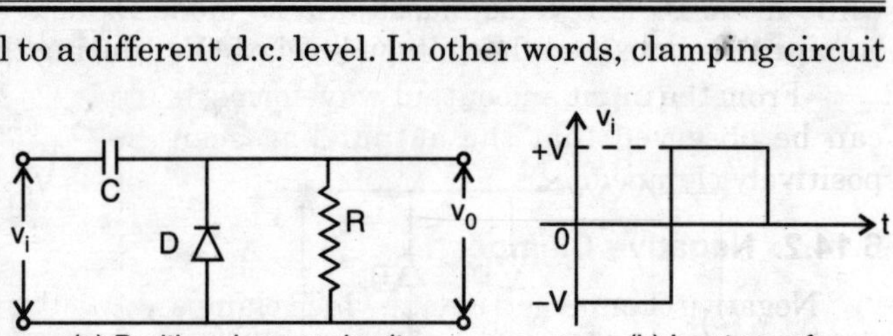

(a) Positive clamper circuit (b) Input waveform

Fig. 6.54. Positive clamper

(i) Positive clamper
(ii) Negative clamper
(iii) Biased clamper.

DO YOU KNOW?

Clampers are commonly used in analog television receivers for the purpose of restoring the dc component of the video signal prior to its being fed to the picture tube.

6.14.1. Positive Clamper

Positive clamper is a circuit which clamps the input signal waveform positively. Figure 6.54 shows the circuit of a positive clamper.

The clamping output is obtained across resistor R.

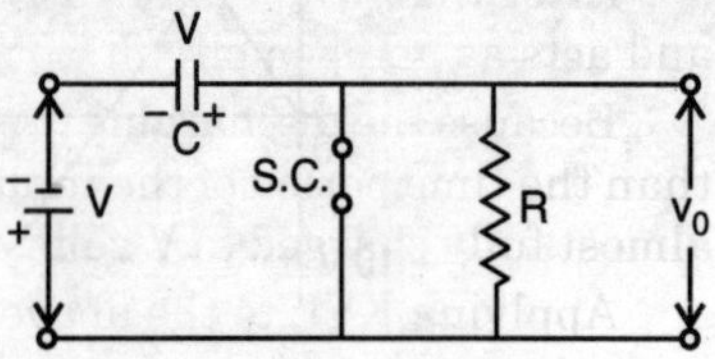

Fig. 6.55. Positive clamper for negative half-cycle

The operation of a clamping circuit is based on the fact that the charging time of a capacitor is made quite small as compared to its discharging time. Therefore, the values of C and R are so selected that the discharging time constant $\tau = CR$ is quite large. This implies that the voltage across the capacitor will not discharge significantly during the interval the diode is non-conducting. Discharging time constant is also made much larger than the time period T of the input signal.

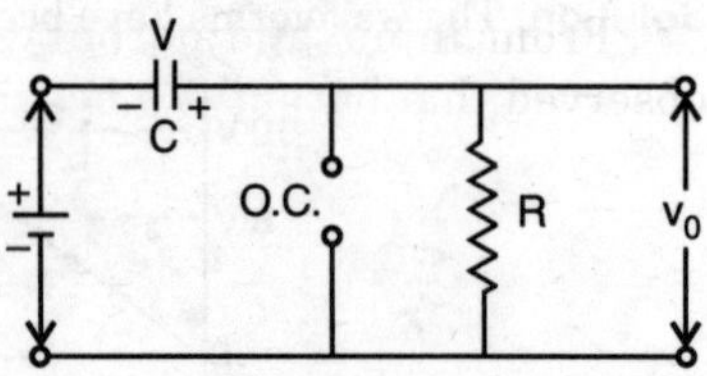

Fig. 6.56. Positive clamper for positive half-cycle

The charging time constant is given by CR_F and is quite small as compared to discharging time constant because R_F is the forward resistance of diode which is much smaller than R.

Working

During the negative half-cycle of the input signal the diode is forward biased and behaves as a short-circuit. Since the charging time constant CR_F is quite small, the capacitor is charged to V volts very quickly. During this period, the ouput is across the short- circuit diode. Hence $v_0 = 0$.

After this when in positive half cycle, input switches to +V volt, the diode is reverse biased and acts as an open circuit.

Now, the capacitor works as a battery and it should discharge through resistor R. But since the discharging time constant (RC) is very large, the capacitor remains almost fully charged to V volt during the off time of diode.

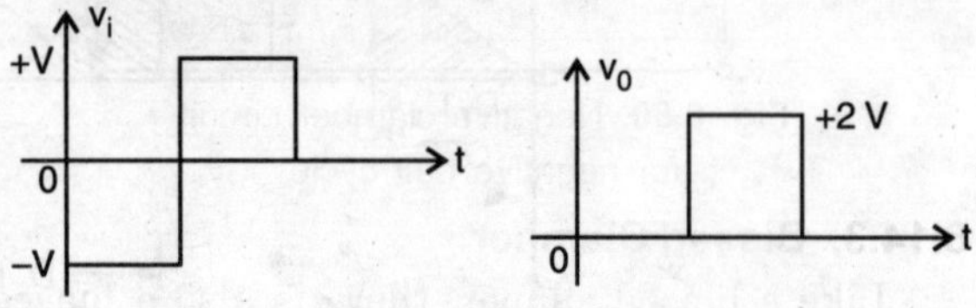

Fig. 6.57. Input and output waveforms

* Distinguish between clipping and clamping circuits. (RGTU, Bhopal, Sem. Exam., 2006-07)

Applying KVL to the above circuit, we have

$$V + V - v_o = 0$$

or $$v_o = 2\text{ V}$$

From the input and output waveforms, it can be observed that the output has been positively clamped.

6.14.2. Negative Clamper

Negative clamper is a circuit which clamps the input signal waveform negatively. Figure 6.58 shows the circuit of a negative clamper.

In negative clamper, the only change from positive clamper is that the diode is reversed.

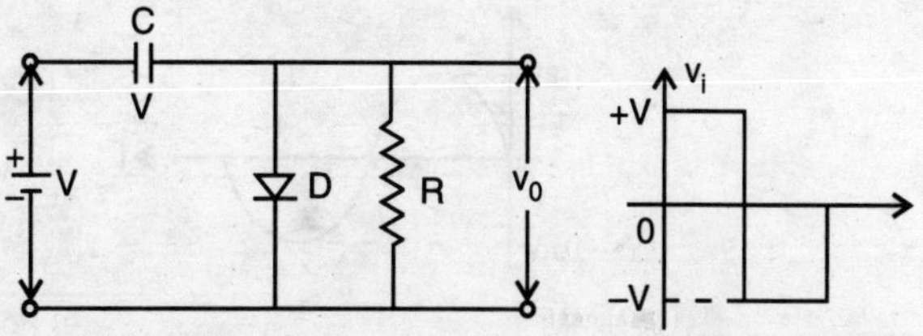

(a) Negative clamper circuit (b) Input waveform

Fig. 6.58. Negative clamper circuit

Working

During the positive half-cycle of the input signal, the diode is forward biased and so acts as a short-circuit. The capacitor charges to V volts very soon because the charging time constant (CR_F) is very small. The output is obtained across the short-circuit diode.

Hence $v_o = 0$

After this, when in negative half-cycle, input switches to – V Volt, the diode is reverse biased and acts as an open circuit.

Because the discharging time constant (CR) is much greater than the time period of the input signal, the capacitor remains almost fully charged to V volts during the off time of the diode.

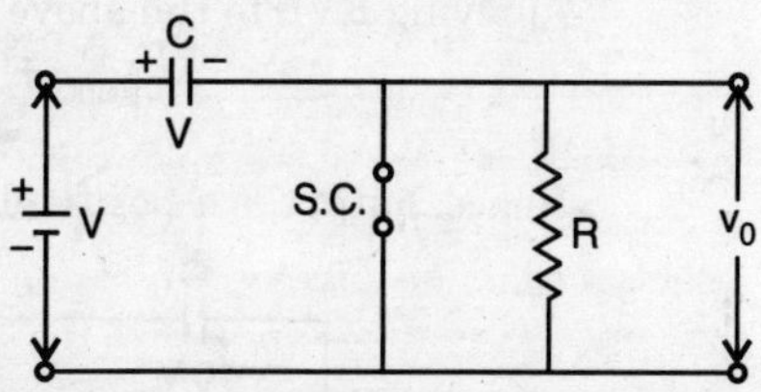

Fig. 6.59. Negative clamper circuit for positive half-cycle

Applying KVL to the above circuit

$$-V - V - v_0 = 0$$

$$v_0 = -2\text{ V}$$

From input and output voltage waveforms, it may be observed that output is a negatively clamped signal.

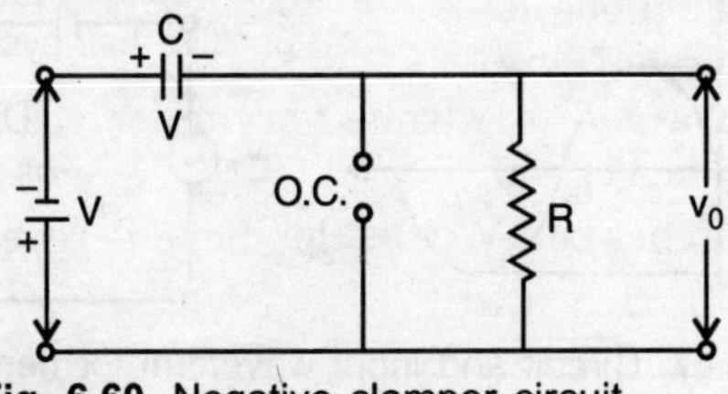

Fig. 6.60. Negative clamper circuit for negative half cycle

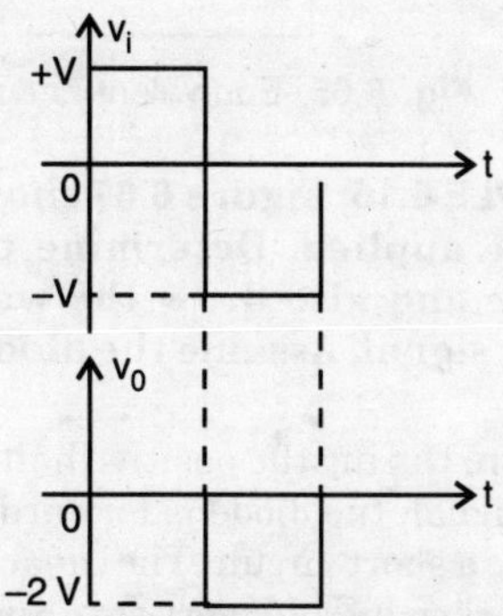

Fig. 6.61. Input/output waveforms for negative clamper circuit

6.14.3. Biased Clamper

Like a biased clipper, there is also a biased clamper. In case of a biased clamper, the clamping may be done at any voltage level other than zero shown in figure 6.62.

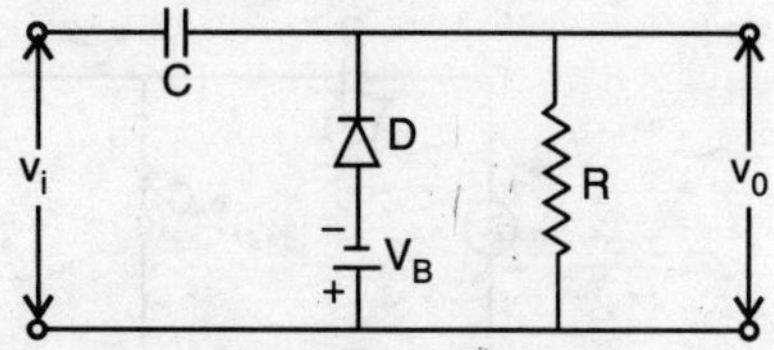

Fig. 6.62. Biased clamper

EXAMPLE 6.14. Figure 6.63 shows the circuit of a positive clamper and the waveform of input voltage applied. Determine the output voltage and also draw the waveform of output signal. Assume the diode to be the ideal.

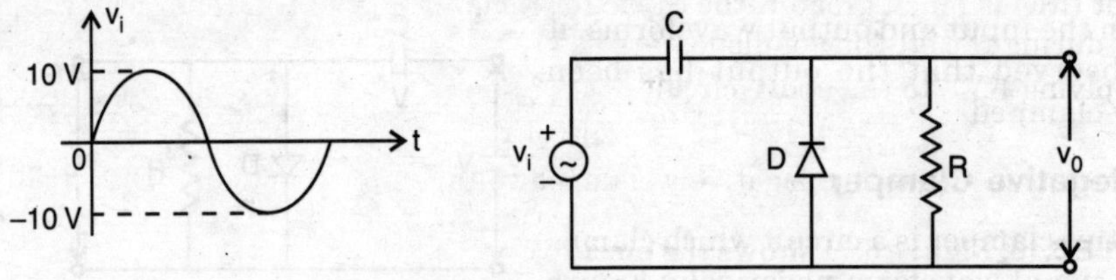

Fig. 6.63. Circuit and input waveform for positive clamper

Solution: During the negative half-cycle of the input signal, the diode acts as a short-circuit and it conducts. The capacitor charges upto 10 V (peak value).

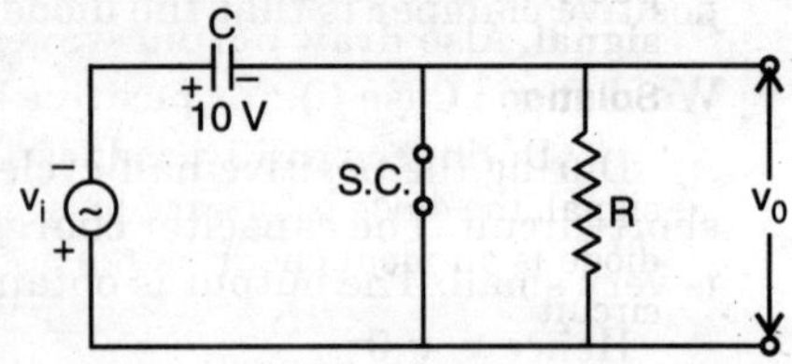

Fig. 6.64. Equivalent circuit for negative half-cycle

The output appears across the short-circuit diode $v_0 = 0$ V

During the positive half-cycle of the input, the diode does not conduct and acts as an open circuit. Now the capacitor acts as a battery and it should discharge through resistor R. But since the discharging time constant (RC) is kept very large, the capacitor remains almost fully charged to 10 volt during the off time of diode.

Applying KVL to the above circuit, we have

$$v_{i(peak)} + 10 = v_0$$

or

$$v_0 = 10 + 10 = 20 \text{ V}$$

Hence, output is a positively clamped voltage.

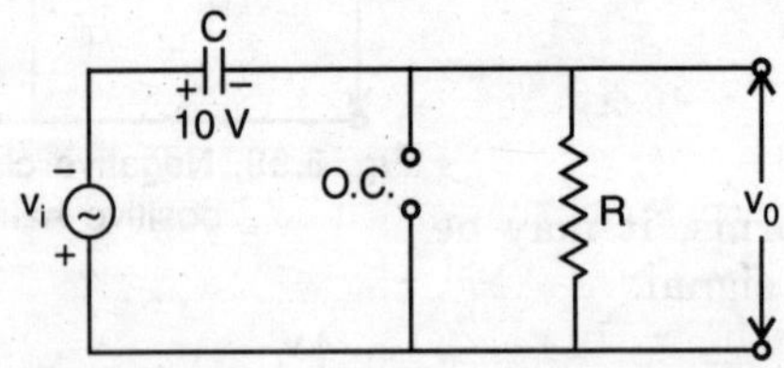

Fig. 6.65. Equivalent circuit for positive half-cycle

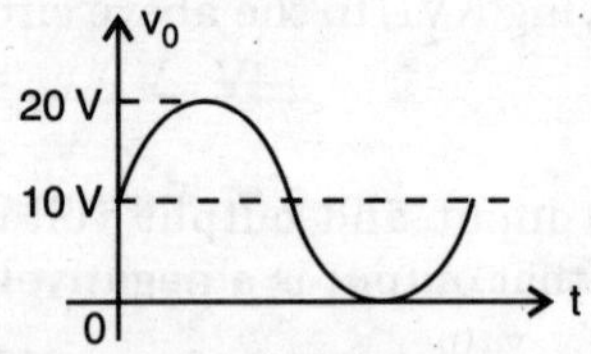

Fig. 6.66. Output voltage waveform

EXAMPLE 6.15. Figure 6.67 shows the circuit of a negative clamper and the waveform of input voltage applied. Determine the output voltage and also draw the waveform of output signal. Assume the diode to be the ideal.

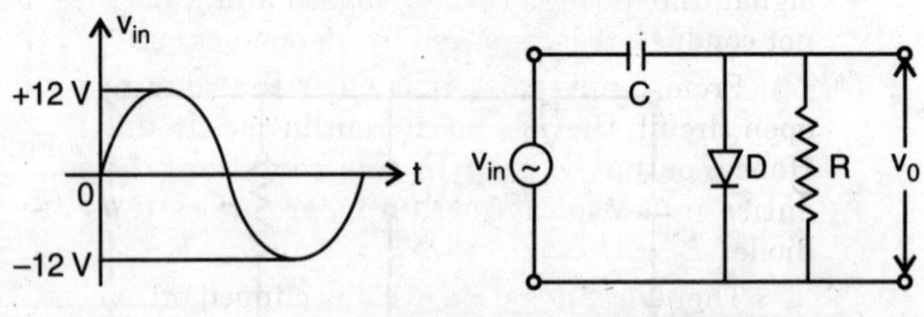

Fig. 6.67. Circuit and input waveform for negative clamper

Solution: During the positive half-cycle of the input signal, the diode is forward biased and acts like a short-circuit. The capacitor charges up to peak value 12 V very soon and the output is obtained across the short-circuit diode.

$$v_0 = 0 \text{ V}$$

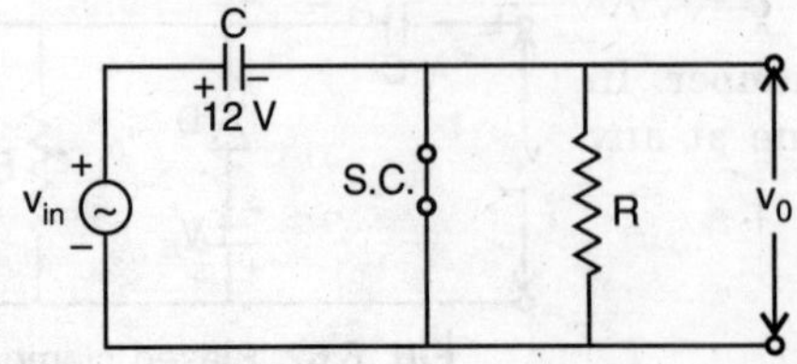

Fig. 6.68. Equivalent circuit for positive half-cycle

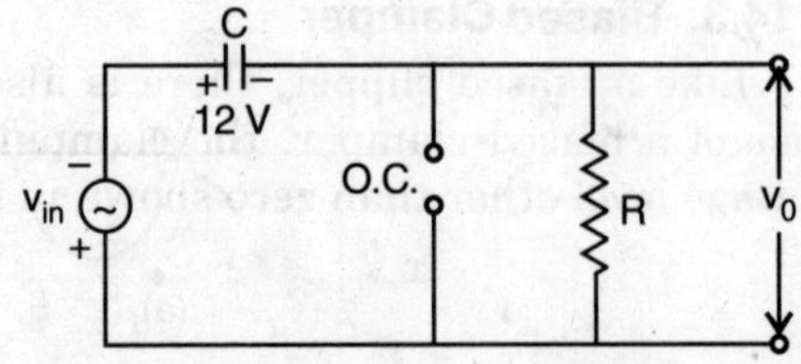

Fig. 6.69. Equivalent circuit for negative half-cycle

During the negative half-cycle of the input signal, the diode does not conduct and acts like an open circuit. Now the capacitor acts as a battery and it should discharge through resistor R. But since the discharging time constant (RC) is much greater, the capacitor remains almost fully charged to 12 V during the off time of diode.

Applying KVL to the above circuit

$$V_{0(peak)} = -12 - 12 = -24 \text{ volt}$$

Hence, the output is a negatively clamped voltage shown in figure 6.70

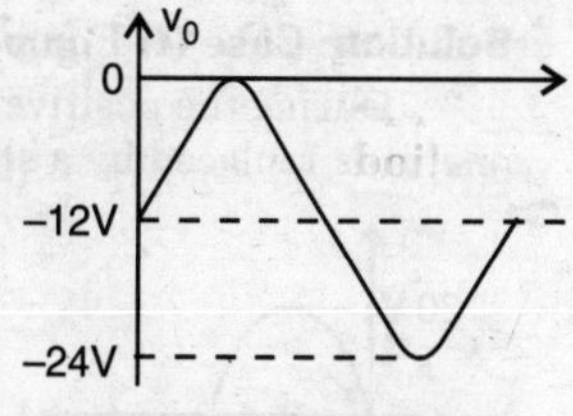

Fig. 6.70. Output voltage waveform

EXAMPLE 6.16. Figure 6.71 shows the circuit of a negative series clipper and input waveform. Assuming the diode to be ideal one, obtain the output peak value for positive and negative half cycles of input signal. Also draw output waveform.

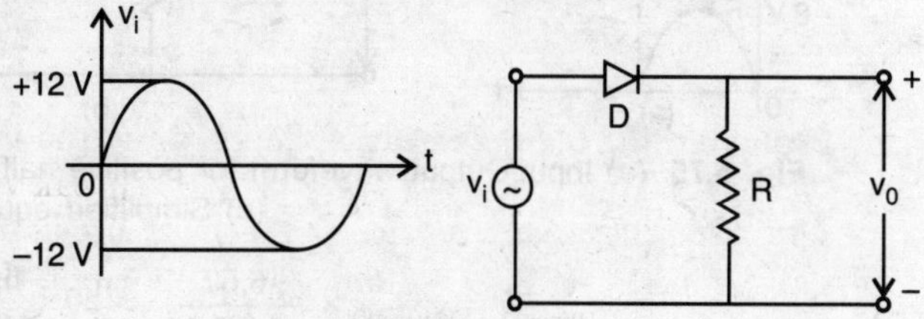

Fig. 6.71.

Solution : Case (i) For positive half cycle

During the positive half-cycle of the input signal, the diode is forward-biased. Because the diode is an ideal one, it is replaced by a short-circuit

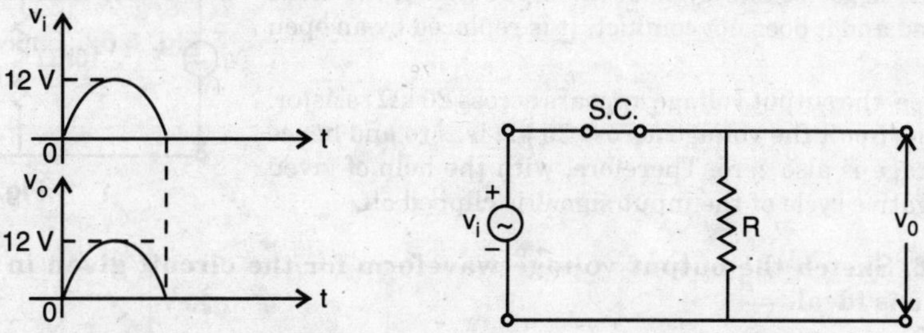

Fig. 6.72. Equivalent circuit and input/output waveform for positive half-cycle

From figure 6.72, the output voltage is

$$v_0 \text{ (peak)} = v_i \text{ (peak)} = 12 \text{ Volt}$$

Case (ii) For negative half-cycle

During the negative half-cycle of input signal, the diode is reverse biased and it does not conduct. It is replaced by an open circuit.

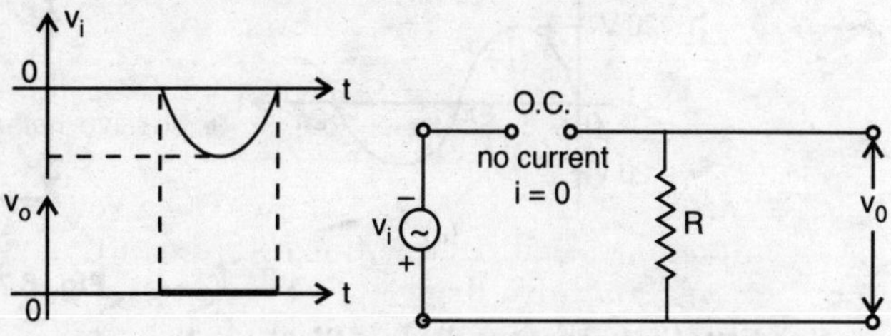

Fig. 6.73. Equivalent circuit and input/output waveform for negative half-cycle

From figure 6.73, it is clear that due to open circuit, there is no current in the circuit. Hence, output $v_0 = 0$. In this condition, the entire input voltage will appear across the diode.

Therefore, negative cycle is clipped off.

EXAMPLE 6.17. Sketch the output voltage waveform for the following circuit (Figure 6.74). Assume that the diode is ideal.

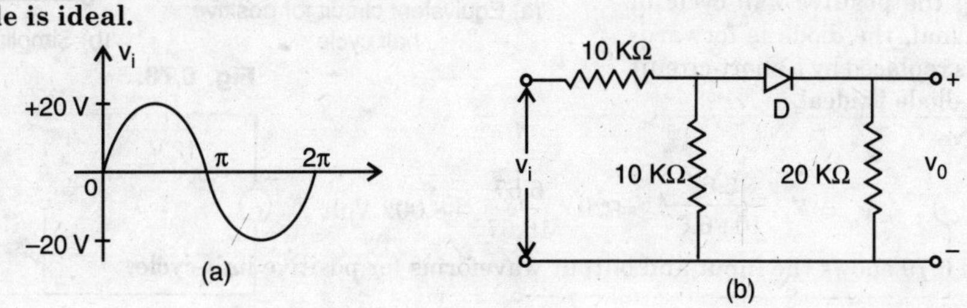

Fig. 6.74.

Solution: Case (i) For positive half-cycle.

During the positive half-cycle of the input signal, the diode is forward-biased. Because the diode is ideal one, it is replaced by a short-circuit.

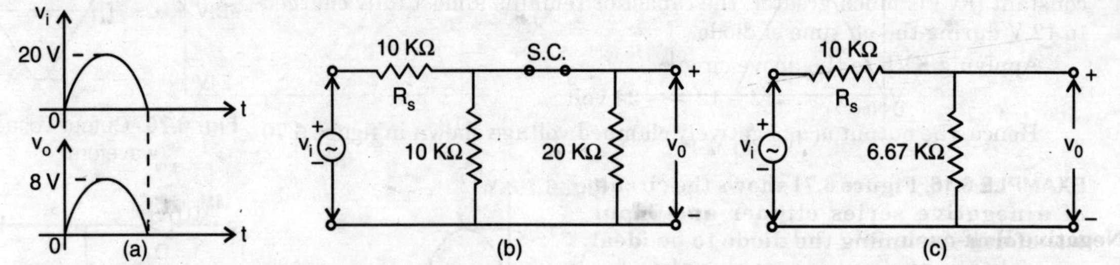

Fig. 6.75. (a) Input/Output waveform for positive half-cycle (b) Equivalent circuit for positive half-cycle (c) Simplified equivalent circuit

$$v_{0(peak)} = v_{i(peak)} \times \frac{6.67}{10+6.67} = 20 \times \frac{6.67}{16.67} = \frac{133.4}{16.67} = 7.998 \text{ Volt} \quad \textbf{Ans.}$$

Case (ii) For negative half-cycle

During the negative half-cycle of the input signal, the diode is reverse-biased and it does not conduct. It is replaced by an open circuit.

In this case, the output voltage appears across 20 kΩ resistor. But due to open circuit the voltage across 20 kΩ is zero and hence the output voltage is also zero. Therefore, with the help of given circuit, the negative cycle of the input signal is clipped off.

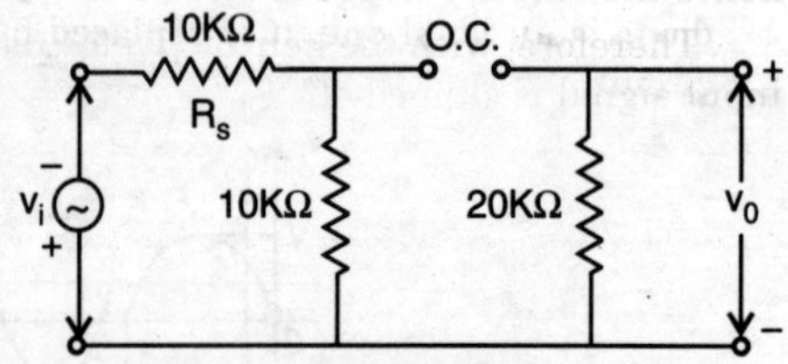

Fig. 6.76.

EXAMPLE 6.18. Sketch the output voltage waveform for the circuit given in figure 6.77. Assume that the diode is ideal.

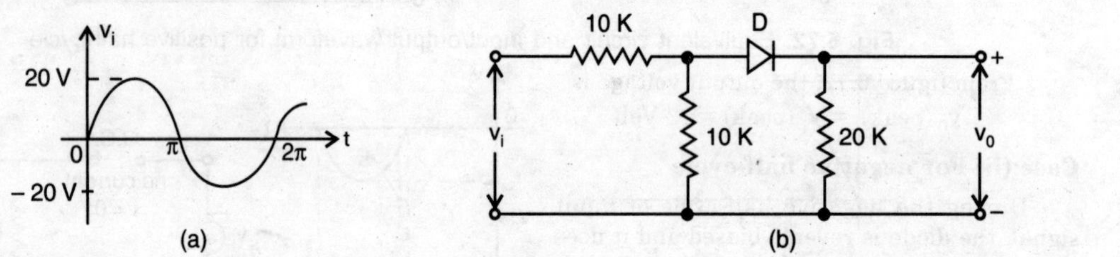

Fig. 6.77.

Solution: Given that the diode is ideal.

Let us solve this question separately for positive and negative half-cycles.

Positive Half-Cycle

During the positive half-cycle of the input signal, the diode is forward-biased and is replaced by a short-circuit because the diode is ideal.

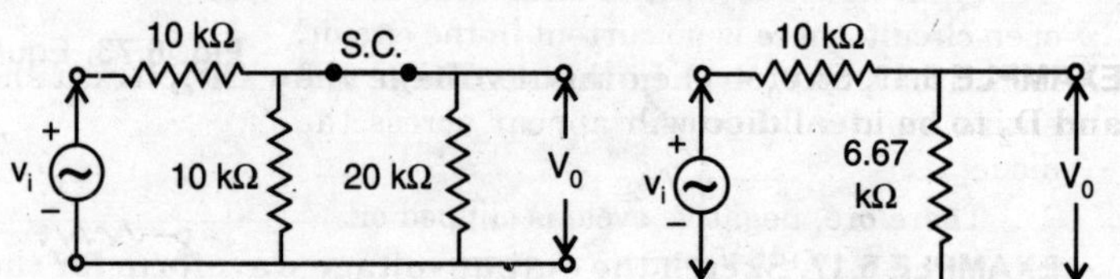

(a) Equivalent circuit for positive half cycle (b) Simplified equivalent circuit

Fig. 6.78.

We have

$$v_o = v_i \cdot \frac{6.67}{10+6.67} = 20 \times \frac{6.67}{16.67} = 8.002 \text{ Volt}$$

Figure 6.79 shows the input and output waveforms for positive half-cycle.

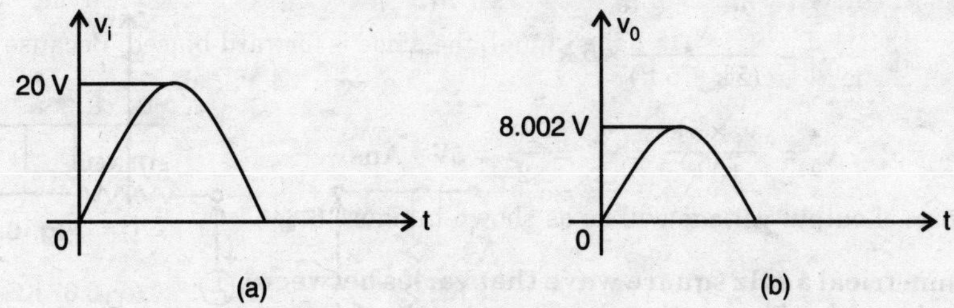

Fig. 6.79.

Negative Half-cycle

For negative half-cycle, the diode is reverse biased and it does not conduct and hence can be replaced by an open circuit as shown in figure 6.80.

In this case, the output voltage appears across 20 kΩ resistor. But, due to open circuit, the voltage across 20 kΩ resistor is zero and hence the output voltage is also zero.

Therefore, with the help of given circuit, the negative cycle of the input signal is clipped off.

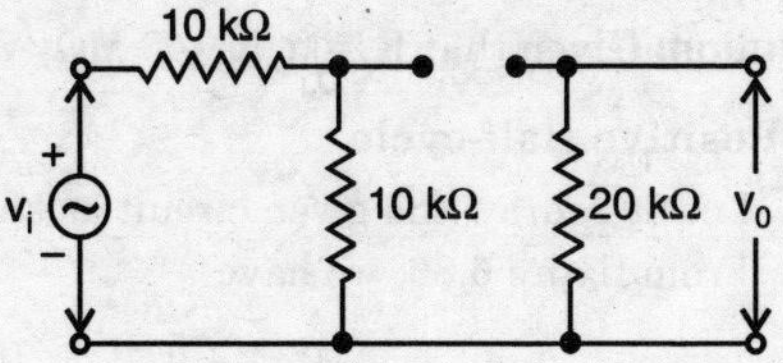

Fig. 6.80.

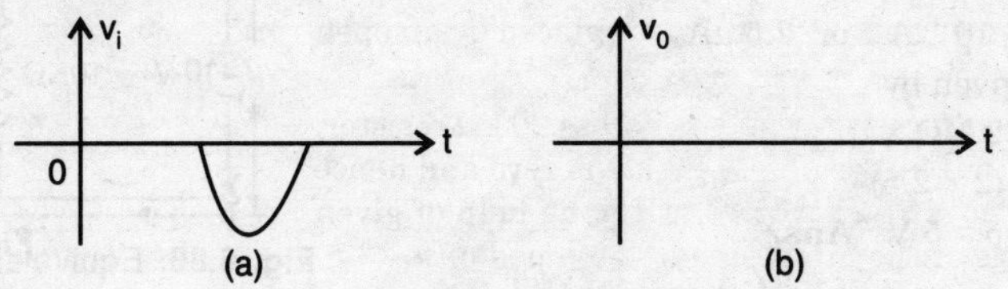

Fig. 6.81.

Finally, the output waveform will be as shown in figure 6.82.

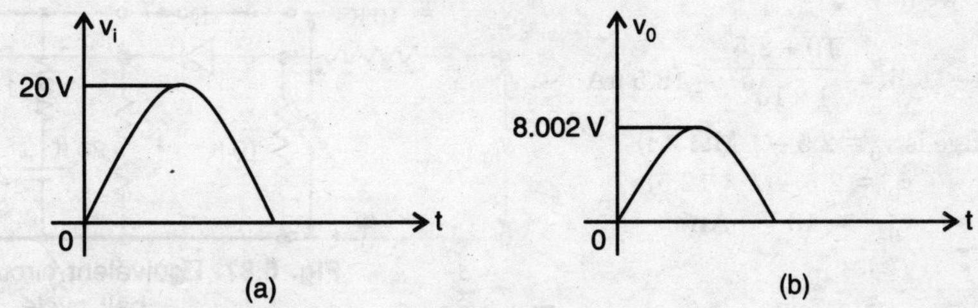

Fig. 6.82.

EXAMPLE 6.19. Sketch the output voltage v_o for the circuit shown in figure 6.83. Assume diodes D_1 and D_2 to be ideal diodes.

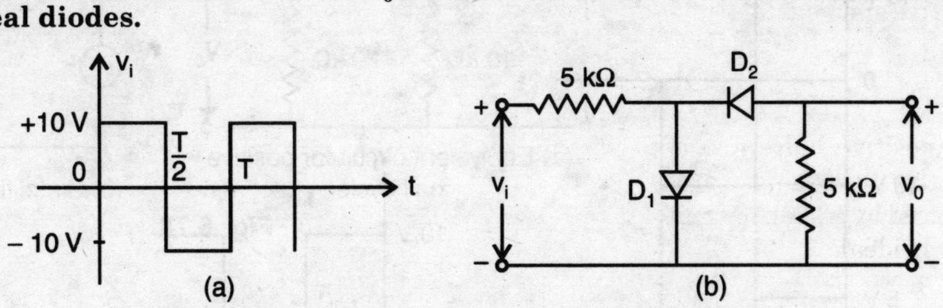

Fig. 6.83. Waveform and circuit for Example 6.19

Solution: We have

Output voltage $$v_0 = \frac{v_i}{(5\,k + 5\,k)} \times 5\,k$$

or $$v_0 = \frac{v_i \times 5k}{10\,k} = \frac{v_i}{2} = \frac{10}{2} = 5V \quad \textbf{Ans.}$$

Therefore, waveform of output voltage will be as shown in figure 6.84.

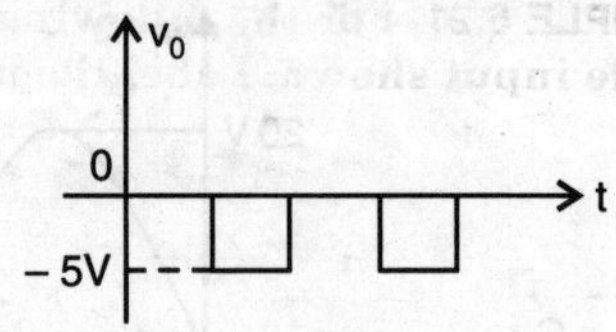

Fig. 6.84.

EXAMPLE 6.20. A symmetrical 5 kHz square wave that varies between + 10 V and – 10 V is impressed upon the clipping circuit of figure 6.85. Assume $R_f = 0$, $R_r = 2\ M\Omega$, $V_j = 0$. Sketch the steady state output waveform indicating the values of minimum, maximum and constant portions.

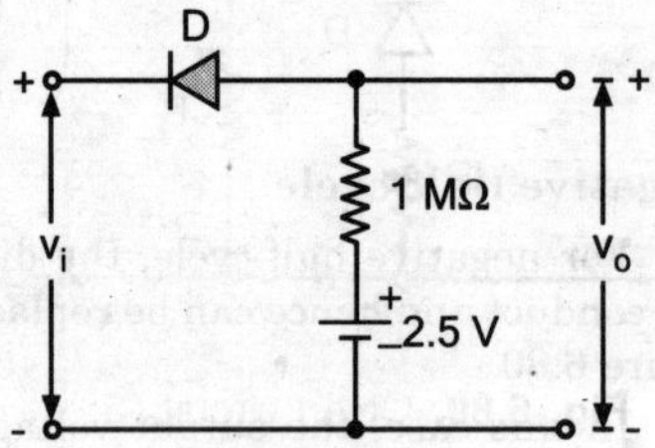

Fig. 6.85.

Solution: Given that $R_f = 0$, $R_r = 2\ M\Omega$, $V_j = 0$

(i) Positive Half-cycle

Let us redraw the given circuit as shown in figure 6.86.

From figure 6.86, we have

$$i = \frac{v_i - 2.5}{2\ M\Omega + 1\ M\Omega} = \frac{10 - 2.5}{3 \times 10^6}$$

or $$i = 2.5 \times 10^{-6}\ A \quad \text{or}\ 2.5\ \mu A$$

The output voltage is given by

$$v_0 = v_i - (2\ M\Omega \times i)$$

or $$v_0 = 10 - (2 \times 2.5)$$

or $$v_0 = 10 - 5 = 5\ V \quad \textbf{Ans.}$$

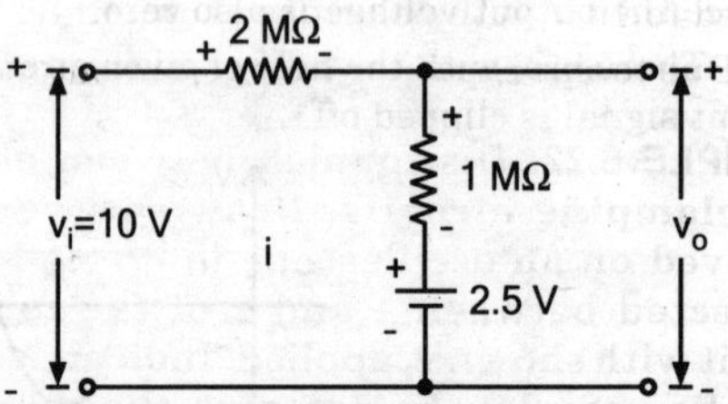

Fig. 6.86. Equivalent circuit in positive half-cycle

(ii) Negative Half-Cycle

Again, let us redraw the given circuit in negative half-cycle as shown in figure 6.87.

From figure 6.87, we have

$$i = \frac{10 + 2.5}{1 \times 10^6} = 12.5\ \mu A$$

The output voltage is $v_0 = 2.5 - (1\ M\Omega \times i)$

$$v_0 = 2.5 - (1 \times 12.5)$$

$$v_0 = -10\ V \quad \textbf{Ans.}$$

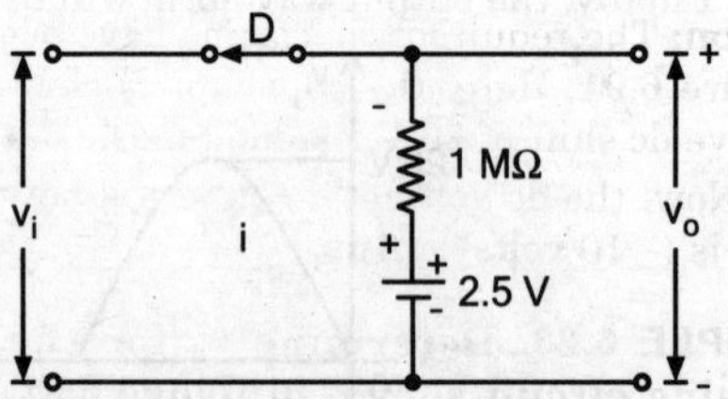

Fig. 6.87. Equivalent circuit in negative half cycle

(iii) Waveforms

Figure 6.88 shows the input and output voltage waveforms.

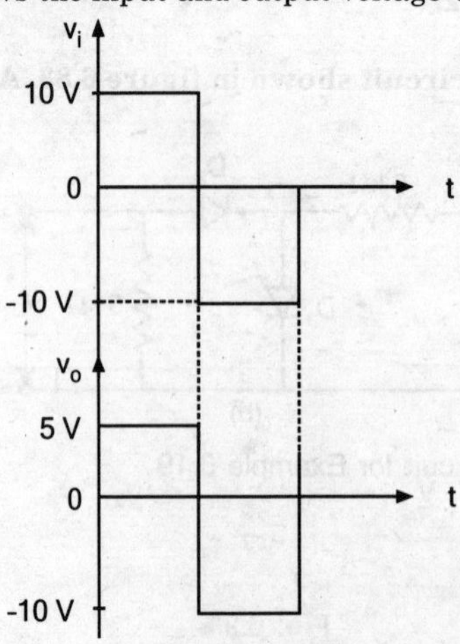

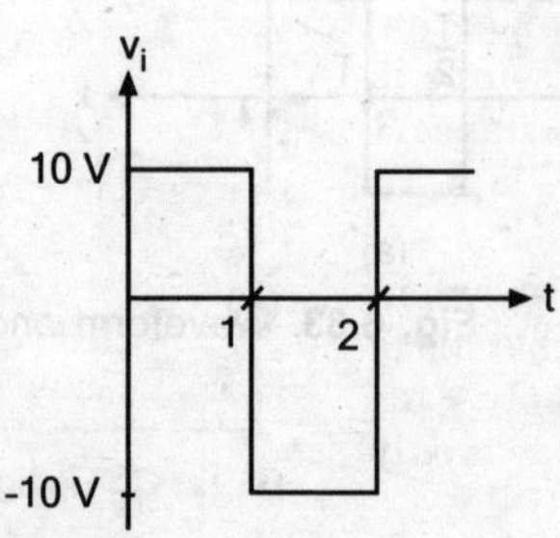

Fig. 6.88. Input and output voltage waveforms

EXAMPLE 6.21. For the following circuit using ideal diode shown in figure 6.89, sketch the output for the input shown. Label the most positive and most negative output levels. Assume RC >> T.

(RGTU, Bhopal, Sem. Exam. 2005-06)

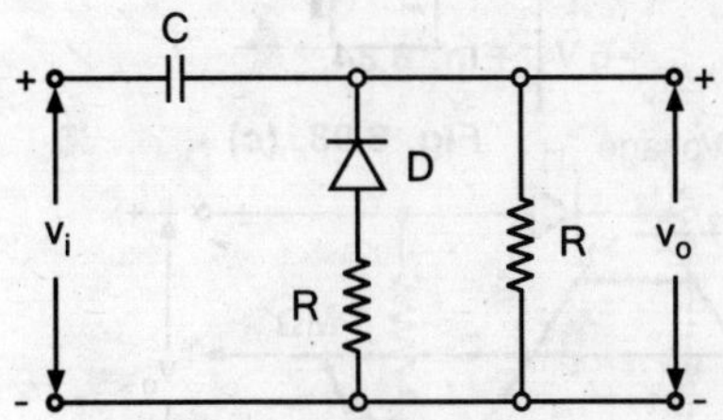

Fig. 6.89. Given circuit

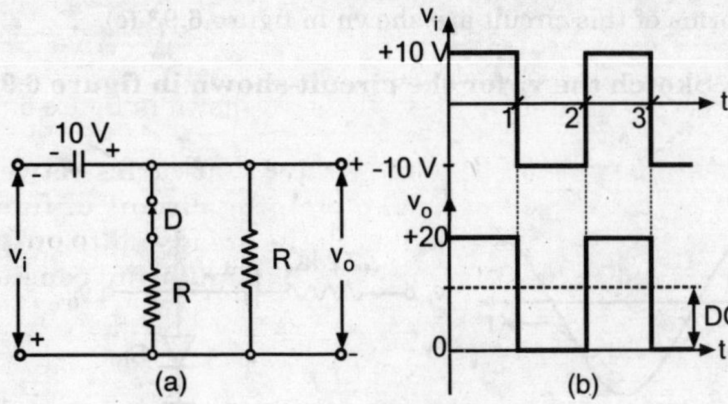

Fig. 6.90.

Solution: Assume that the diode is ideal. In the negative half-cycle, the diode is ON and the capacitor is charged to 10 V with polarities shown in figure 6.90 (a).

Due to this capacitor voltage, the output voltage will have a dc shift of + 10 V as shown in figure 6.90 (b).

EXAMPLE 6.22. Distinguish between clipping and clamping circuits. Draw the waveform observed on an oscilloscope in dc mode when connected between 1 and 2 of the following circuit with shown v_i applied. Indicate voltages and the zero level. Also give the numerical values of the reading of a d.c. voltmeter connected across 1 and 2. Find the PIV of the diode.

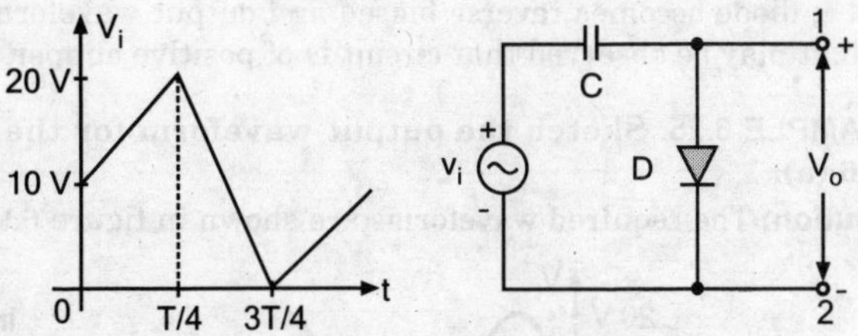

Fig. 6.91. Given waveform and circuit diagram

Solution: The required waveforms have been shown in figure 6.91. Here, the capacitor is charged upto + 20 V with a polarity shown in figure 6.91. Hence, a negative dc shift of 20 V is added in the output waveform.

Now, the dc voltmeter connected between the output terminals will show the average value of output which is (– 10 volts). **Ans.**

EXAMPLE 6.23. Determine v_0 for the following circuit shown in figure 6.92. Also name the configuration.

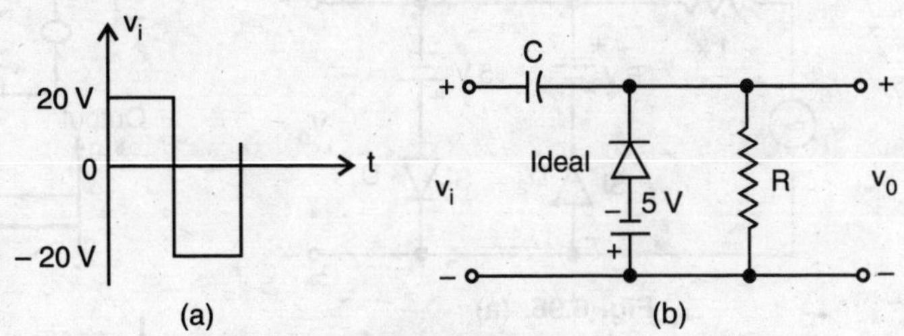

Fig. 6.92.

Solution: It may be observed that the given circuit is a Clamper circuit with an additional dc source. In negative half-cycle of the input, the diode will be forward biased and the capacitor gets charged to a voltage (V_m – V) volts with the polarities shown in figure 6.93 (a).

The capacitor acts as a battery. However, an additional dc source is already connected across resistance R so that the output v_0 is – 5 V in negative half-cycle.

The equivlent circuit is shown in figure 6.93 (b) which shows that diode will remain permanently off thereafter. The job of the diode is only to charge the capacitor.

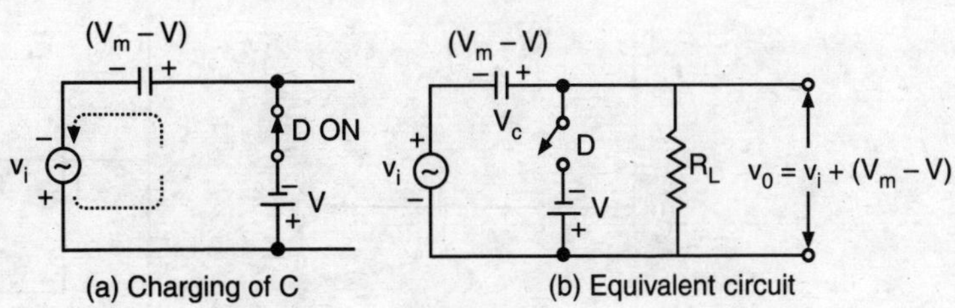

Fig. 6.93.

From figure 6.93 (b), the expression for output voltage is given by

$$v_0 = (V_m - V) + v_i \quad ...(i)$$

or

$$v_0 = (20 - 5) + v_i = 35 \text{ V}$$

The waveforms of this circuit are shown in figure 6.93 (c).

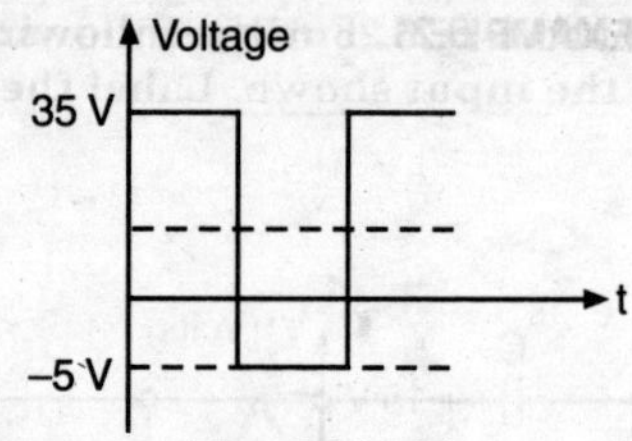

Fig. 6.93. (c)

EXAMPLE 6.24. Sketch the v_0 for the circuit shown in figure 6.94.

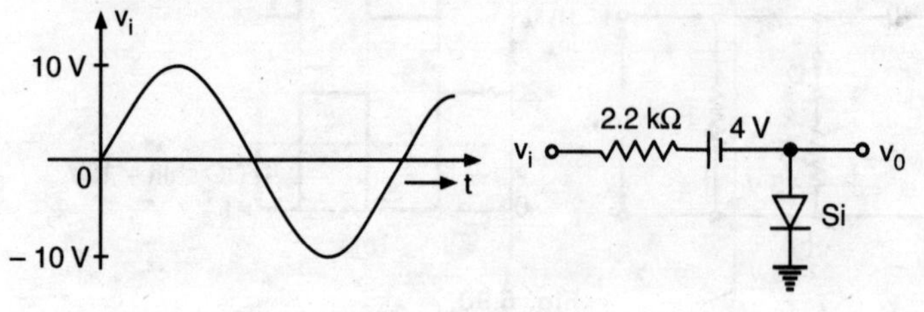

Fig. 6.94.

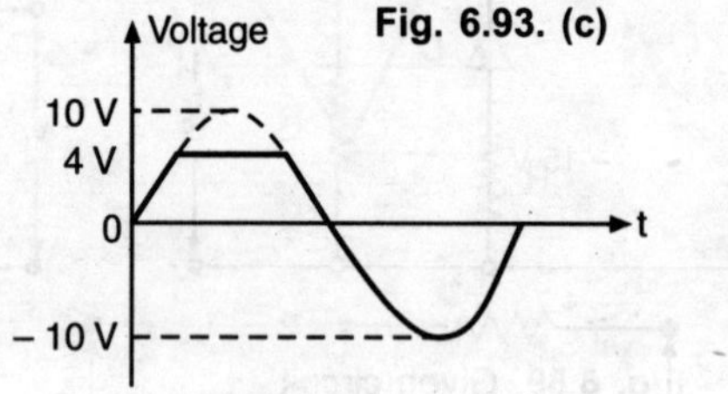

Fig. 6.95.

Solution: For input voltage $v_i \geq 4$, the diode is forward biased and hence acts as short-circuit. As input voltage $v_i < 4$, diode becomes reverse biased and output waveform is obtained as shown in figure 6.95. From figure 6.95, it may be observed that circuit is of positive clipper.

EXAMPLE 6.25. Sketch the output waveform for the circuit and input waveform shown in figure 6.96 (a).

Solution: The required waveforms are shown in figure 6.96 (b).

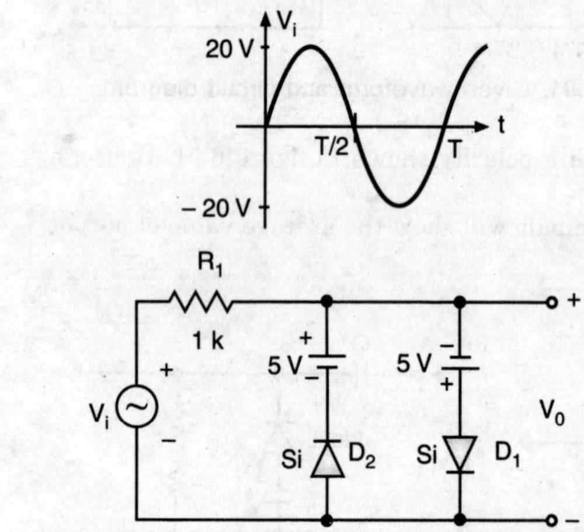

Fig. 6.96. (a)

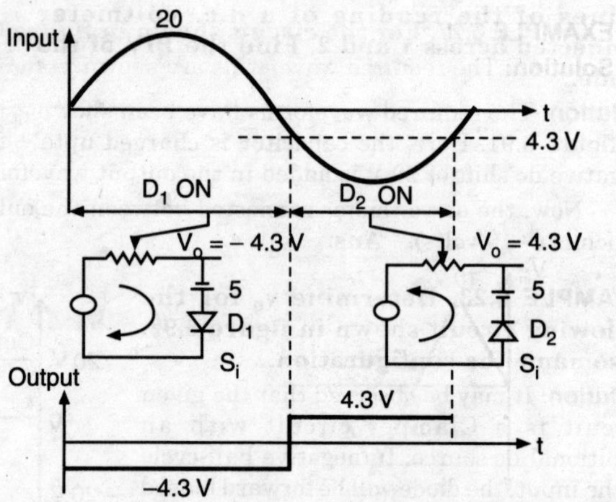

Fig. 6.96. (b) Waveforms

EXAMPLE 6.26. Sketch the output waveform for the circuit and input waveform shown in figure 6.97(a).

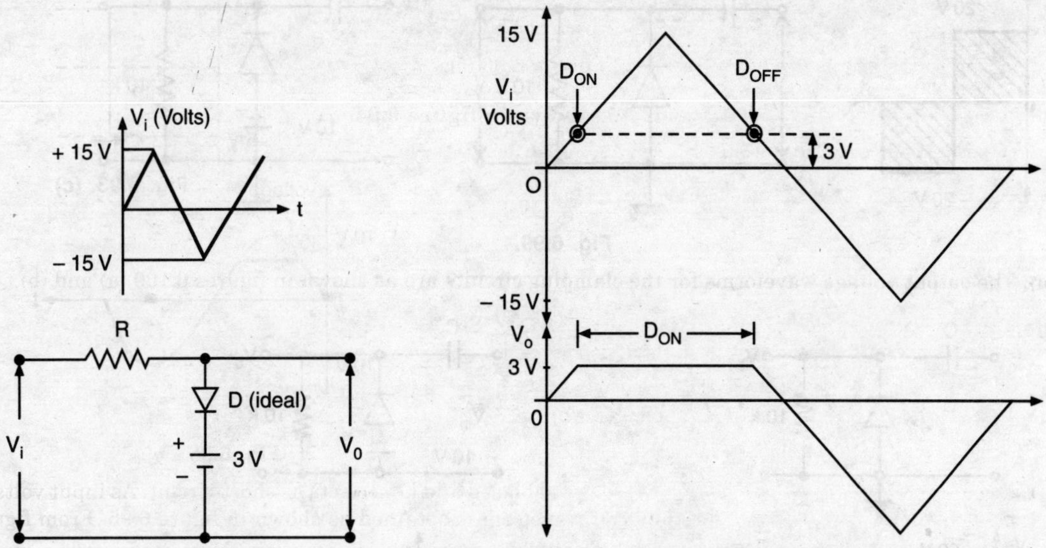

(a) (b) Input/Output waveforms

Fig. 6.97.

Solution: The waveforms for the given circuit are as shown in figure 6.97 (b).

EXAMPLE 6.27. For the circuit shown in figure 6.98, sketch I_R and V_0.
Solution: The required waveforms are shown in figure 6.98 (b).

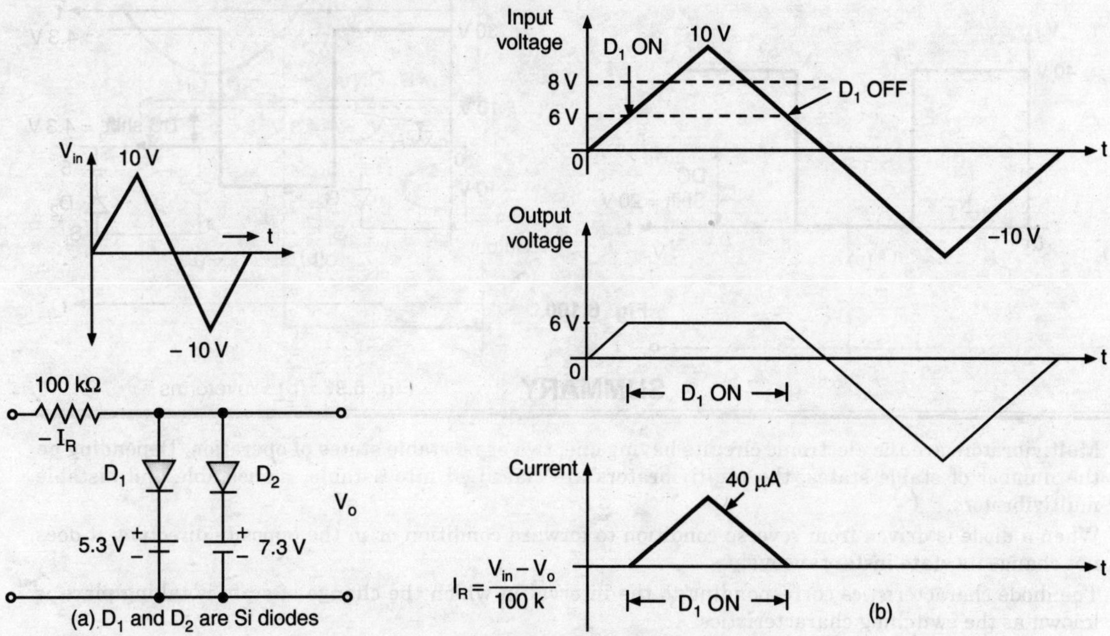

Fig. 6.98.

EXAMPLE 6.28. For the clamping circuit shown in figure 6.99, sketch for V_0.

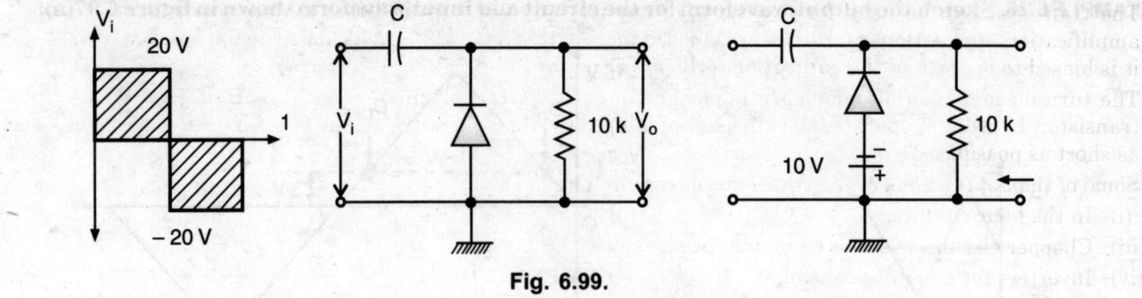

Fig. 6.99.

Solution: The output voltage waveforms for the clamping circuits are as shown in figures 6.100 (a) and (b).

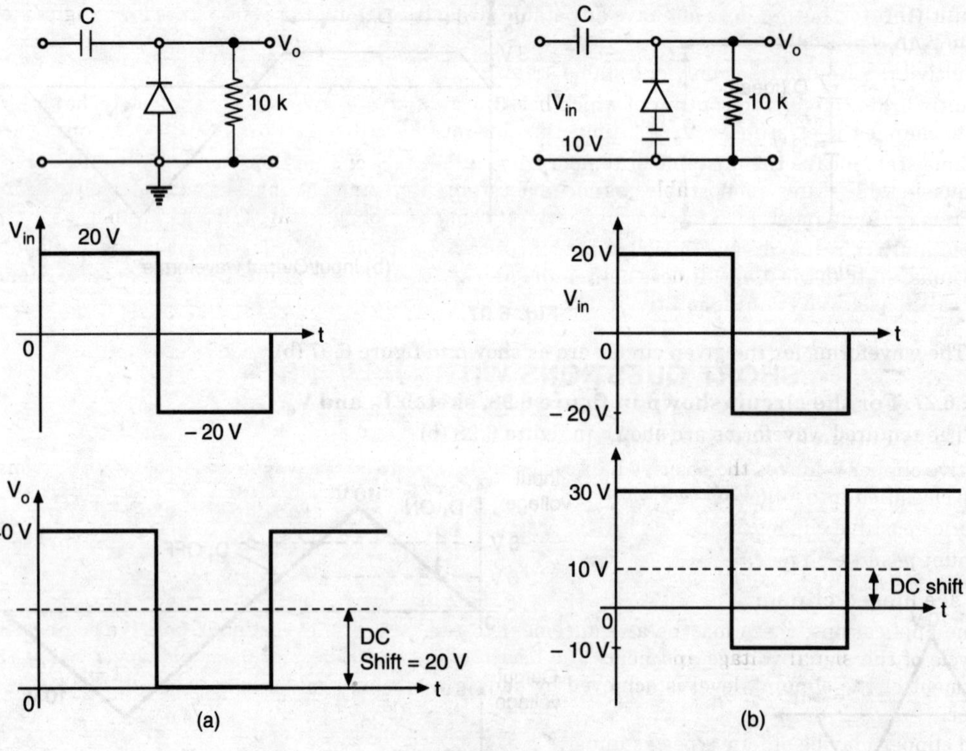

Fig. 6.100.

SUMMARY

- Multivibrators are the electronic circuits having one, two or no stable states of operation. Depending on the number of stable states, the multivibrators are classified into astable, monostable and bistable multivibrators.
- When a diode is driven from reverse condition to forward condition or in the opposite direction, it does not change its state instantaneously.
- The diode characteristics corresponding to the interval in which the change of state is taking place is known as the switching characteristics.
- The forward recovery time (t_{rf}) is defined as the time required for the diode current to change from 10 to 90 percent of its final value when the diode is switched from OFF to ON state. In practice, the forward recovery time is very short as compared to the reverse recovery time. Hence we can neglect it.
- The storage time is defined as the time required for the stored minority charge to become zero. It is denoted by t_s.

- The transistor can be used for two types of applications viz. amplification and switching. For the amplification applications, a transistor is biased in its active region. Whereas, for switching applications it is biased to operate in the saturation (fully on) or cutoff (fully off) regions.
- The turn on and turn off times are important in deciding the maximum switching frequency of the transistor. In order to operate the transistor at a high frequency the turn on and turn off times should be as short as possible.
- Some of the applications of the transistor as a switch are:
  - (i) In the logic circuits.
  - (ii) Temperature control.
  - (iii) Chopper circuits used for dc motor speed control.
  - (iv) Inverters for ac motor control.
  - (v) SMPS and UPS.
- Mulivibrator is an electronic circuit which can have no, one or two stable states of operation. Depending on the number of stable states we can classify the multivibrators.
- In this multivibrator, output does not have any stable state. Its output changes its state from high to low and low to high repeatedly.
- In the multivibrator output has only one stable state.
- It is a multivibrator circuit, the output of which has two stable states. We have to apply external inputs in order to change the existing state of output. Bistable multivibrator is also called as a flipflop.
- Quasi-stable state means the unstable or temporarily stable state of a flipflop. An astable multivibrator has two quasi-stable states, monostable has one quasi stable state and bistable has none. A multivibrator does not remain in quasi-stable state permanently. It comes out of this state after some time.
- An astable multivibrator has two quasi-stable states and no stable state. Its output changes its state from one quasi state to the other. It does not require any external trigger input to change its state. Hence it is also called as a free running multivibrator.

SHORT QUESTIONS WITH ANSWERS

Q.1. What do you mean by positive clipper.

Ans. A positive clipper removes the positive half-cycles of the input voltage waveform. Positive clipper may also be classified into following two ways:

(i) series positive clipper, and

(ii) shunt positive clipper

Q.2. What are biased clipper.

Ans. In some applications, there may be a requirement to remove a small portion of positive or negative half-cycle of the signal voltage and hence the biased clipper is used. The name bias means that the adjustment of the clipping level is achieved by adding a biasing voltage in series with the diode or resistor.

Biased clipper may be of four types as under:

(a) Biased positive clipper

(b) Biased positive clipper with reverse polarity of the battery

(c) Biased negative clipper

(d) Biased negative clipper with reverse polarity of battery.

Q.3. What is meant by negative clamper.

Ans. Negative clamper is a circuit which clamps the input signal waveform negatively.

Q.4. Write applications of transistor as a switch.

Ans. Some of the applications of the transistor as a switch are:

(i) In the logic circuits. (ii) Temperature control.

(iii) Chopper circuits used for dc motor speed control.

(iv) Inverters for ac motor control. (v) SMPS and UPS.

Q.5. Define astable multivibrator.

Ans. An astable multivibrator has two quasi-stable states and no stable state. Its output changes its state from one quasi state to the other. It does not require any external trigger input to change its state. Hence it is also called as a free running multivibrator.

Q.6. Define monostable multivibrator.

Ans. The monostable multivibrator has one stable state and another quasi stable state. It requires an external trigger input to change its state from stable to quasi stable. After some time it will automatically return back to the stable state. The duration for which the output remains in the quasi-stable state is called as the pulse width.

REVIEW QUESTIONS

1. Explain the operation of a transistor as switch.
2. Give classification of multivibrators.
3. What is a quasi stable state?
4. Define astable, monostable and bistable multivibrators.
5. What is a retriggerable monostable?
6. With the help of circuit diagram and waveforms, explain the operation of collector coupled monostable.
7. Drive the expression for on time (T) of the output of collector coupled monostable multivibrator.
8. State merits demerits and applications of monostable multivibrator.
9. Explain the operation of emitter coupled monostable multivibrator and draw its waveforms.
10. Derive the expression for output pulse width of an emitter coupled monostable multivibrator.
11. State and explain various triggering methods of monostable multivibrator.
12. Explain the operation of collector coupled astable and draw waveforms.

NUMERICAL PROBLEMS

1. What is the difference between Clipping circuits and Clamping circuits? Determine V_0 for the network of figure 6.101.

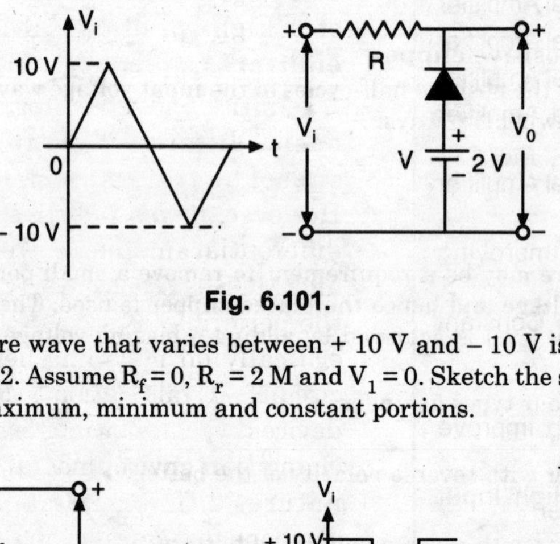

Fig. 6.101.

2. A symmetrical 5 kHz square wave that varies between + 10 V and – 10 V is impressed upon the clipping circuit shown in figure 6.102. Assume $R_f = 0$, $R_r = 2$ M and $V_1 = 0$. Sketch the steady state output waveform, indicating values of the maximum, minimum and constant portions.

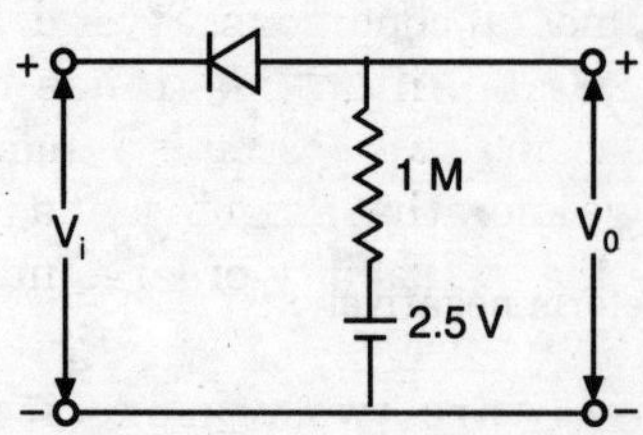

Fig. 6.102.

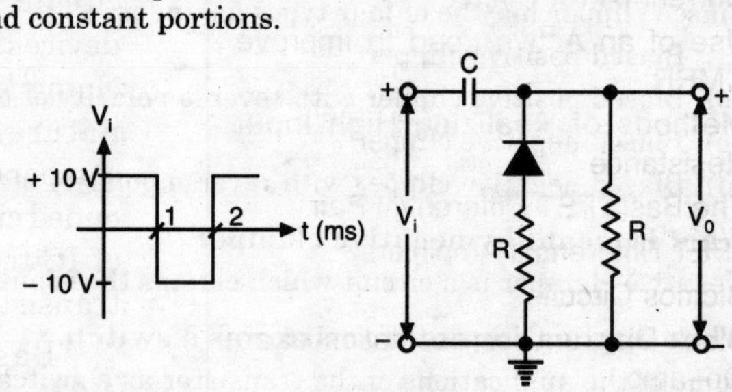

Fig. 6.103.

3. For the following circuit using ideal diode, sketch the output for the input shown. Label the most positive and most negative output levels. Assume RC >> T.

Chapter 7

DIFFERENTIAL AMPLIFIERS

Syllabus

Differential Amplifier: Calculation of differential, common mode gain and CMRR using h-parameters.

Chapter Outline

7.1. INTRODUCTION

The differential amplifier is a fundamental building block of analog circuits. It is the input stage of virtually every operational amplifier (op-amp), and is the basis of a high-speed digital logic circuit family, called **emitter-coupled logic (ECL)**.

Initially invented for use with vacuum tubes, the basic differential-amplifier configuration was subsequently implemented with discrete BJTs. However, it was the advent of ICs that has made the differential amplifier extremely popular in both bipolar and MOS technologies. First, as we shall shortly see, the performance of the differential amplifier depends critically on matching between the two sides of the circuits. IC fabrication is capable of providing matched devices whose parameters track over wide ranges of changes in environmental conditions. Second, by their nature, differential amplifiers make use of more components (approaching twice as many) than single-ended circuits and we know that a significant advantage of IC technology is the availability of large number of transistors at relatively low cost.

Basically, there are two reasons for using differential amplifiers in preference to single ended amplifiers. First, the differential amplifiers are much less sensitive to noise and interference than single ended circuits. The second reason for preferring differential amplifiers is that the differential configuration enables us to bias the amplifier and to couple amplifier stages together without the necessity of bypass and coupling capacitors such as those used in the designs of discrete-circuit amplifiers.

Differential amplifier at the input stage of an op-amp provides very high input impedance, noise rejection and differential amplification capabilities. Since the differential amplifier is the basic building block of an op-amp, the analysis of differential amplifier will not only clarify the operation of an op-amp but also set a ground work to follow the performance characteristics of the op-amp.

MICHAEL FARADAY (1791-1867)

The unit of measure for capacitance, the farad (F), was named for Michael Faraday (1791-1867), an English chemist and physicist who discovered the principle of induction (1 F is the unit of capacitance that stores 1 coulomb (C) of charge when 1 volt (V) is applied).

7.2 BASICS OF DIFFERENTIAL AMPLIFIER

Definition

A differential amplifier amplifies the difference between two input voltage signals. Therefore, it is also called difference amplifier. Let us consider an ideal differential amplifier shown in figure 7.1. Here, V_1 and V_2 are the two input signals while V_{out} is the single ended output. Each signal is measured with respect to the ground. In an ideal differential amplifier, the output voltage V_{out} is proportional to the difference between the two input signals. This means that we can write,

$$V_{out} \propto (V_1 - V_2) \quad ...(7.1)$$

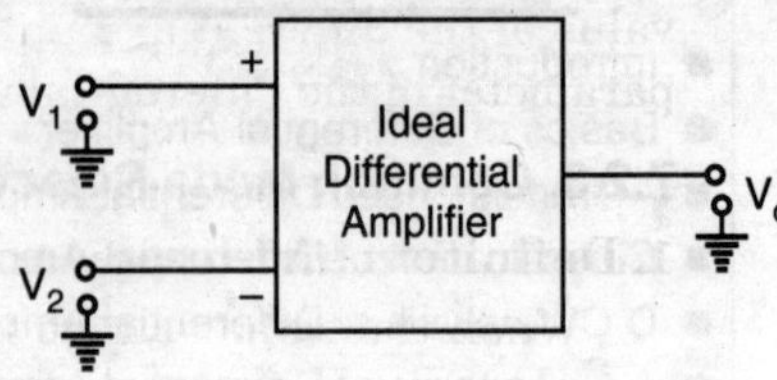

Fig. 7.1. Ideal differential amplifier

7.2.1. Differential Gain A_{id}

Using equation (7.1), we can write,

$$V_{out} = A_{id}(V_1 - V_2) \quad ...(7.2)$$

where A_{id} is the constant of proportionality. The A_{id} is the gain with which differential amplifier amplifies the difference between two input signals. Hence, it is called **differential gain** of the differential amplifier.

Thus, A_{id} = differential gain

The difference between the two inputs $(V_1 - V_2)$ is generally called difference voltage and denoted as V_{id}.

$$V_{out} = A_{id} V_{id} \quad ...(7.3)$$

Hence, the differential gain can be expressed as,

$$A_{id} = \frac{V_{out}}{V_{id}} \quad ...(7.4)$$

Generally, the differential gain is expressed in its decibel (dB) value as under :

$$A_{id} = 20 \text{ Log}_{10} (A_{id}) \text{ in dB} \quad ...(7.5)$$

7.2.2. Common Mode Gain A_{cm}

If we apply two input voltages which are equal in all the respects to the differential amplifier i.e., $V_1 = V_2$ then ideally the output voltage $V_{out} = (V_1 - V_2)\, A_{id}$, must be zero. However, the output voltage of the practical differential amplifier not only depends on the difference voltage but also depends on the average common level of the two inputs. Such an average level of the two input signals is called **common mode signal** denoted as V_{cm}.

$$V_{cm} = \frac{V_1 + V_2}{2} \quad ...(7.6)$$

Practically, the differential amplifier produces the output voltage proportional to such common mode signal.

> **Important Point :** The gain with which it amplifies the common mode signal to produce the output is called common mode gain of the differential amplifier denoted as A_{cm}.

$$V_{out} = A_{cm}\, V_{cm} \quad ...(7.7)$$

Thus, there exists some finite output for $V_1 = V_2$ due to such common mode gain A_{cm}, in case of practical differential amplifiers. So, the total output of any differential amplifier can be expressed as,

$$V_{out} = A_{id}\, V_{id} + A_{cm}\, V_{cm} \quad ...(7.8)$$

This shows that if one input is + 25 μV and other is – 25 μV then the output of the amplifier will not be same, with the inputs as 600 μV and 650 μV, though the difference between the two sets of the inputs is 50 μV.

> **Important Point :** For an ideal differential amplifier, the differential gain A_{id} must be infinite while the common mode gain must be zero. This ensures zero output for $V_1 = V_2$.

But due to mismatch in the internal circuitry, there is some output available for $V_1 = V_2$ and gain A_{cm} is not practically zero. The value of such common mode gain A_{cm} is very small while the value of the differential gain A_{id} is always very large. At this stage, we can define one important parameter of the differential amplifier known as common mode rejection ratio (CMRR).

7.2.3. Common Mode Rejection Ratio (CMRR)

1. Definition and Importance

When the same voltage is applied to both the inputs, the differential amplifier is said to be operated in a common mode configuration. Many disturbance signals, noise signals appear as a common input signal to both the input terminals of the differential amplifier. Such a common signal should be rejected by the differential amplifier. The ability of a differential amplifier to reject a common mode signal is expressed by a ratio called **common mode rejection ratio** denoted as CMRR.

2. Mathematical Expression

It is defined as the ratio of the differential voltage gain A_{id} to common mode voltage gain A_{cm}.

$$\text{CMRR} = \rho = \left|\frac{A_{id}}{A_{cm}}\right| \quad ...(7.9)$$

> **Important Point :** Ideally the common mode voltage gain is zero, hence the ideal value of CMRR is infinite.

For a practical differential amplifier, A_{id} is large and A_{cm} is small, hence, the value of CMRR is also very large.

$$\text{CMRR in dB} = 20\log\left|\frac{A_{id}}{A_{cm}}\right| \text{ dB} \quad ...(7.10)$$

The output voltage can be expressed in terms of CMRR as below :

$$V_{out} = A_{id}\, V_{id} + A_{cm}\, V_{cm} = A_{id} V_{id}\left[1 + \frac{A_{cm}\, V_{cm}}{A_{id}\, V_{id}}\right]$$

$$V_{out} = A_{id} V_{id}\left[1 + \frac{1}{\left(\frac{A_{id}}{A_{cm}}\right)}\frac{V_{cm}}{V_{id}}\right] = A_{id} V_{id}\left[1 + \frac{1}{CMRR}\frac{V_{cm}}{V_{id}}\right] \quad ...(7.11)$$

The above equation explains that as CMRR is practically very large, though both V_{cm} and V_{id} components are present, the output is mostly proportional to the difference signal only. The common mode component is greatly rejected.

7.2.4. Salient Features of Differential Amplifier (Important)

The various features of a differential amplifier may be listed as under:

(i) High differential voltage gain
(ii) Low common mode gain
(iii) High CMRR
(iv) Two input terminals
(v) High input impedance
(vi) Large bandwidth
(vii) Low offset voltage and currents
(viii) Low output impedance.

EXAMPLE 7.1. Determine the output voltage of a differential amplifier for the input voltages of 300 μV and 240 μV. The differential gain of the amplifier is 5000 and the value of the CMMR is (i) 100 and (ii) 10^5.

Solution: The differential amplifier is represented as shown in figure 7.2.

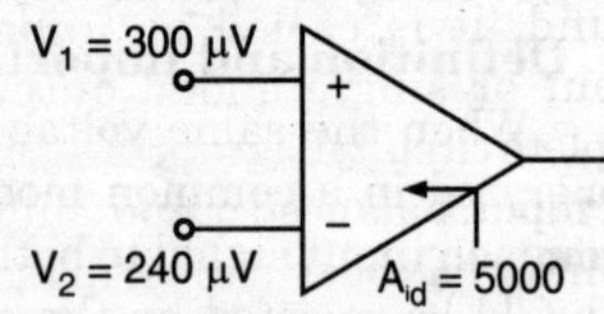

Fig. 7.2.

(i) Given that CMRR = 100

$$V_{id} = V_1 - V_2 = 300 - 240 = 60\ \mu V$$

$$V_{cm} = \frac{V_1 + V_2}{2} = \frac{300 + 240}{2} = 270\ \mu V$$

$$CMRR = \frac{A_{id}}{A_{cm}}$$

or $$100 = \frac{5000}{A_{cm}}$$

$$A_{cm} = 50$$

$$V_{out} = A_{id} V_{id} + A_{cm} V_{cm} = 5000 \times 60 + 50 \times 270 = 313500\ \mu V = 313.5\ mV \quad \textbf{Ans.}$$

(ii) Given that $CMRR = 10^5$

We have $$A_{cm} = \frac{A_{id}}{CMRR} = \frac{5000}{10^5} = 0.05$$

$$V_{id} = A_{id} V_{id} + A_{cm} V_{cm} = 5000 \times 60 + 0.05 \times 270$$

$$V_{id} = 300013.5\ \mu V = 300.0135\ mV \quad \textbf{Ans.}$$

Ideally, A_{cm} must be zero and output should be only $A_{id}\ V_{id}$ which is $5000 \times 60 \times 10^{-6}$ i.e., 300 mV.

Important Point : It can be seen that higher the value of CMRR, the output is almost proportional to the difference voltage V_{id}, rejecting the common mode signal.

7.3 A TRANSISTORISED DIFFERENTIAL AMPLIFIER

The transistorised differential amplifier basically makes use of the emitter biased circuits which are identical in characteristics. Such two identical emitter biased circuits are shown in figure 7.3. The two transistors Q_1 and Q_2 have exactly matched characteristics. The two collector resistances R_{C1} and R_{C2} are equal while the two emitter resistance R_{E1} and R_{E2} are also equal.

Thus, $R_{C1} = R_{C2}$ and $R_{E1} = R_{E2}$

The magnitudes of $+V_{CC}$ and $-V_{EE}$ are also same. The differential amplifier can be obtained by using such two emitter biased circuits. This is achieved by connecting emitter E_1 of Q_1 to the emitter E_2 of Q_2. Due to this, R_{E1} appears in parallel with R_{E2} and the combination can be replaced by a single resistance denoted as R_E. The base B_1 of Q_1 is connected to the input 1 which is V_{in_1} while the base B_2 of Q_2 is connected to the input 2 which is V_{in_2}. The supply voltages are measured with respect to ground. The balanced output is taken between the collector C_1 of Q_1 and the collector C_2 of Q_2. Such an amplifier is called **emitter coupled differential amplifier**. The two collector resistances are same and hence can be denoted as R_C.

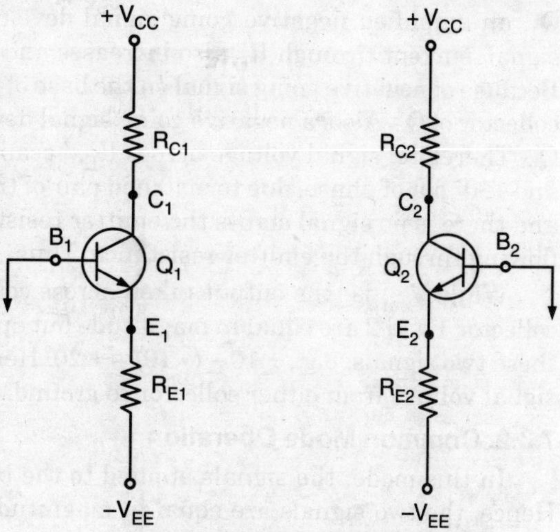

Fig. 7.3. Emitter biased circuits

The output can be taken between two collectors or in between one of the two collectors and the ground. When the output is taken between two collectors, none of them is grounded. Then, it is called as balanced output double ended output or floating output. When the output is taken between any of the collectors and ground, it is called as unbalanced output or single ended output. The complete circuit diagram of such a basic dual input, balanced output differential amplifier has been shown in figure 7.4. As the output is taken between two output terminals, none of them is grounded, it is called balanced output differential amplifier.

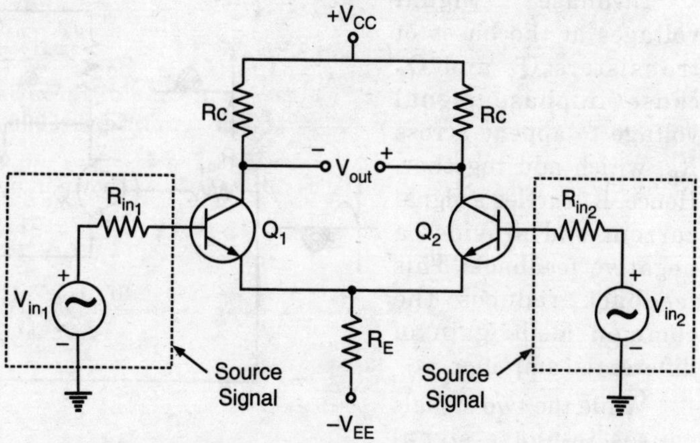

Fig. 7.4. Dual input, balanced output differential amplifier

Let us discuss the circuit operation in the following two modes :

(i) Differential mode operation
(ii) Common mode operation.

7.3.1. Differential Mode Operation

In differential mode, the two input signals are different from each other. Let us consider the two input signals which are same in magnitude but 180° out of phase. These signals with opposite phase can be obtained from the center tap transformer. The circuit used in differential mode operation has been shown in figure 7.5. Let us assume that the sine wave on the base of transistor Q_1 is positive going while on the base of transistor Q_2 is negative going. With a positive going signal on the base of

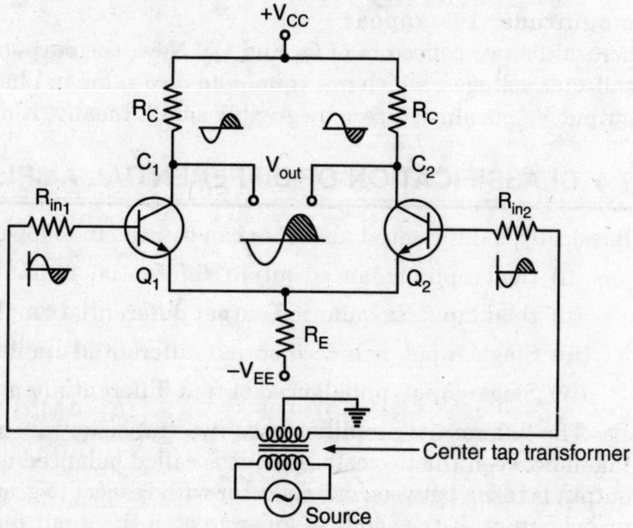

Fig. 7.5. Illustration of Differential mode operation

Q_1, an amplified negative going signal develops on the collector of Q_1. Because of positive going signal, current through R_E also increases and hence a positive going wave is developed across R_E. Because of negative going signal on the base of Q_2, an amplified positive going signal develops on the collector of Q_2. Also, a negative going signal develops across R_E, because of emitter follower action of Q_2. Therefore, signal voltage across R_E, because of the effect of Q_1 and Q_2 are equal in magnitude and 180° out of phase, due to matched pair of transistors. Hence, these two signals cancel each other and there is no signal across the emitter resistance. This means that there is no a.c. signal current flowing through the emitter resistance. Thus, R_E in this case does not introduce negative feedback.

While V_{out} is the output taken across collector of Q_1 and collector of Q_2. The two outputs on collector 1 and 2 are equal in magnitude but opposite in polarity. And V_{out} is the difference between these two signals, e.g., + 10 – (– 10) = + 20. Hence, the difference output V_{out} is twice as large as the signal voltage from either collector to ground.

7.3.2. Common Mode Operation

In this mode, the signals applied to the base of Q_1 and Q_2 are derived from the same source. Hence, the two signals are equal in magnitude as well as in phase. The circuit diagram has been shown in the figure 7.6.

In-phase signal voltages at the bases of transistors Q_1 and Q_2 cause in-phase signal voltage to appear across R_E, which add together. Hence, R_E carries a signal current and provides a negative feedback. This feedback reduces the common mode gain of differential amplifier.

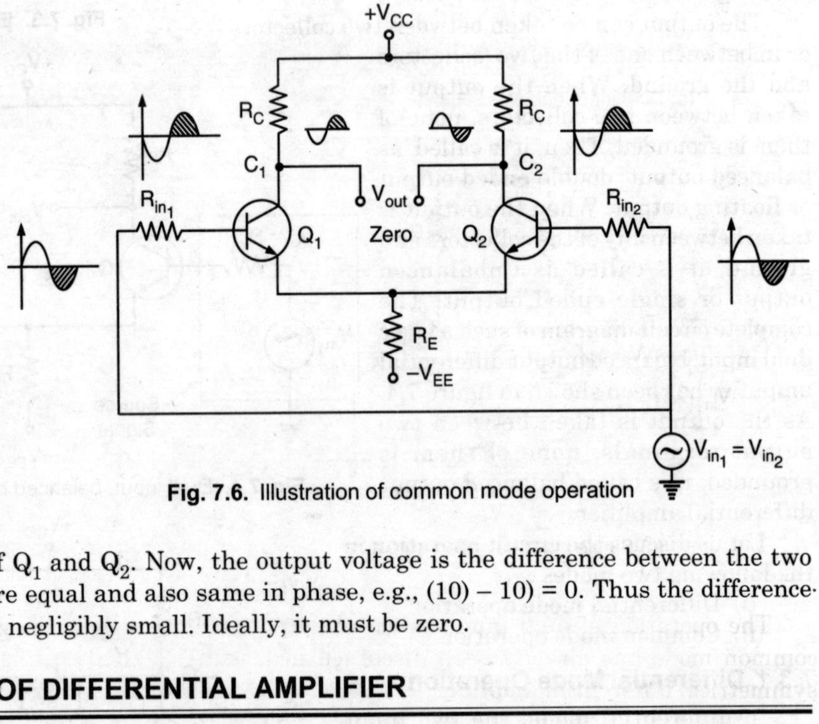

Fig. 7.6. Illustration of common mode operation

While the two signals causes in-phase signal voltages of equal magnitude to appear across the two collectors of Q_1 and Q_2. Now, the output voltage is the difference between the two collector voltages, which are equal and also same in phase, e.g., (10) – 10) = 0. Thus the difference output V_{out} is almost, zero, negligibly small. Ideally, it must be zero.

7.4 CLASSIFICATION OF DIFFERENTIAL AMPLIFIER

Bassically, a differential amplifier can be used in following four configurations:

(i) Dual input, balanced output differential amplifier

(ii) Dual input, unbalanced output differential amplifier

(iii) Single input, balanced output differential amplifier.

(iv) Single input, unbalanced output differential amplifier.

The differential amplifier uses two transistors in common emitter configuration. If output is taken between the two collectors, it is called balanced output or double ended output. While if the output is taken between one collector with respect to ground it is called unbalanced output or single ended output. If the signal is given to both the input terminals it is called dual input, while if the signal is given to only one input terminal and other terminal is grounded it is called single input, or single ended input. Out of these four configurations, the dual input, balanced output is the basic differential amplifier configuration. This has been shown in figure 7.7 (a). The dual input, unbalanced output differential amplifier has been shown in figure 7.7 (b). The single input, balanced output differential amplifier has been shown in figure 7.7 (c), and the single input, unbalanced output differential amplifier has been shown in figure 7.7 (d).

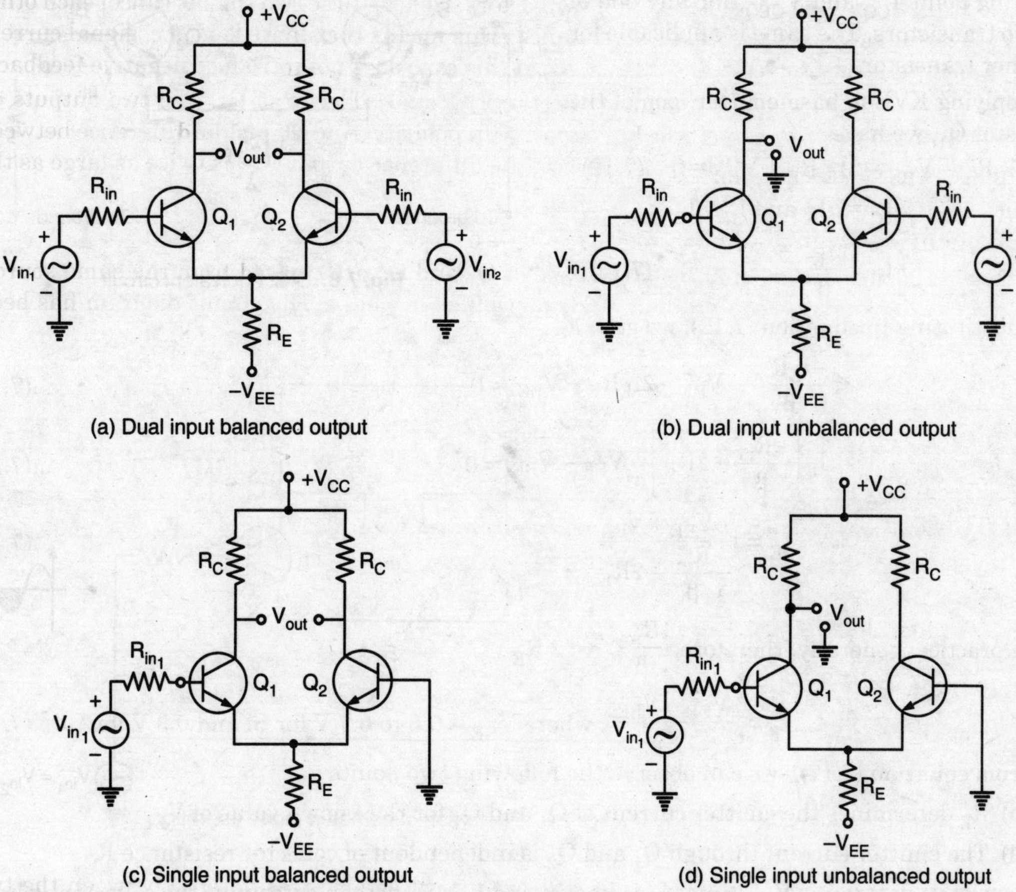

(a) Dual input balanced output (b) Dual input unbalanced output (c) Single input balanced output (d) Single input unbalanced output

Fig. 7.7.

The operation of dual input balanced output differential amplifier in differential mode and common mode has already been discussed in last article. This configuration is also known as symmetrical differential amplifier.

7.5 D.C. ANALYSIS OF DIFFERENTIAL AMPLIFIER

The d.c. analysis means to obtain the operating point values i.e., I_{CQ} and V_{CEQ} for the transistors used. The supply voltages are d.c. while the input signals are a.c., so d.c. equivalent circuit can be obtained simply by reducing the input a.c. signals to zero. The d.c. equivalent circuit thus obtained has been shown in figure 7.8. Assuming $R_{in1} = R_{in2}$, the source resistance is simply denoted by R_{in}.

The transistors Q_1 and Q_2 are matched transistors and hence for such a matched pair we can assume :

(i) Both the transistors have the same characteristic

(ii) $R_{E1} = R_{E2}$ hence $R_E = R_{E1} \| R_{E2}$

(iii) $R_{C1} = R_{C2}$ hence denoted as R_C

(iv) $|V_{CC}| = |V_{EE}|$ and both are measured with respect to ground

Since the two transistors are matched and circuit is symmetrical, it is enough to find out operating point I_{CQ} and V_{CEQ}, for any one of the two transistors. The same is applicable for the other transistor.

Applying KVL to base-emitter loop of the transistor Q_1, we have

$$-I_B R_{in} - V_{BE} - 2I_E R_E + V_{EE} = 0 \quad ...(7.12)$$

But, $I_C = \alpha I_E$ and $I_C \cong I_E$

$$I_B = \frac{I_E}{\beta} \quad ...(7.13)$$

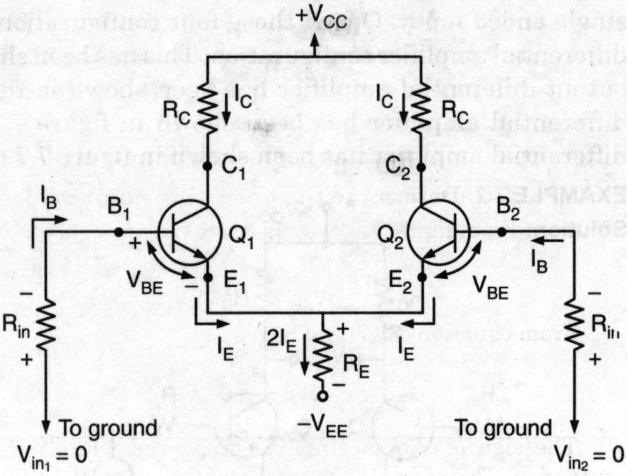

Fig. 7.8. D.C. equivalent circuit

Substituting in equation (7.12), we get

$$\frac{-I_E R_{in}}{\beta} - V_{BE} - 2I_E R_E + V_{EE} = 0 \quad ...(7.14)$$

$$I_E\left[\frac{-R_{in}}{\beta} - 2R_E\right] + V_{EE} - V_{BE} = 0 \quad ...(7.15)$$

$$I_E = \frac{-V_{EE} + V_{BE}}{-\frac{R_{in}}{\beta} - 2R_E} \quad ...(7.16)$$

In practice, generally, therefore, $\frac{R_{in}}{\beta} << 2R_E$,

$$I_E = \frac{V_{EE} - V_{BE}}{2R_E}, \text{ where } V_{BE} = 0.6 \text{ to } 0.7 \text{ V for Si and } 0.3 \text{ V for Ge} \quad ...(7.17)$$

From equation (7.17), we can observe the following two points :

(i) R_E determines the emitter current of Q_1 and Q_2 for the known value of V_{EE}

(ii) The emitter current through Q_1 and Q_2 is independent of collector resistance R_C

Now let us determine V_{CE}. Since I_E is known and $I_E \cong I_C$, we can determine the collector voltage of Q_1 as under:

$$V_C = V_{CC} - I_C R_C \quad ...(7.18)$$

Neglecting the drop across R_E, we can say that the voltage at the emitter of Q_1 is approximately equal to $-V_{BE}$. Therefore, the collector to emitter voltage will be

$$V_{CE} = V_C - V_E = (V_{CC} - I_C R_C) - (-V_{BE})$$

$$V_{CE} = V_{CC} + V_{BE} - I_C R_C \quad ...(7.19)$$

Hence, $I_E = I_C = I_{CQ}$ while $V_{CE} = V_{CEQ}$ for given values of V_{CC} and V_{EE}.

✍ **Important Point** : In the equation (7.17), the sign of V_{EE} is already considered to be negative, while deriving it. Hence while using this equation to solve the problem, only the magnitude of V_{EE} should be used and negative sign of V_{EE} should not be used again.

This means that for both the transistors, we can determine operating point values, using equations (7.17) and (7.19). With the same biasing arrangement, the d.c. analysis remains same for all the four possible configurations of differential amplifier.

$$I_E = \frac{V_{EE} - V_{BE}}{\frac{R_s}{\beta} + 2R_E} = \frac{V_{EE} - V_{BE}}{2R_E}$$

$$V_{CEQ} = V_{CC} + V_{BE} - I_{CQ}R_C$$

EXAMPLE 7.2. Determine the operating point values for the circuit shown in figure 7.9.

Solution: From figure 7.9, we can write

$$R_C = 4.7\ k\Omega,\ R_E = 3.3\ k\Omega,$$
$$V_{CC} = 12\ V,\ V_{EE} = -12\ V$$

From equation (7.17), we have

$$I_E = \frac{V_{EE} - V_{BE}}{2\ R_E}$$

The sign of V_{EE} is already considered while deriving the expression. Therefore, V_{EE} must be replaced by 12 V. Further, let us assume $V_{BE} = 0.7$ V.

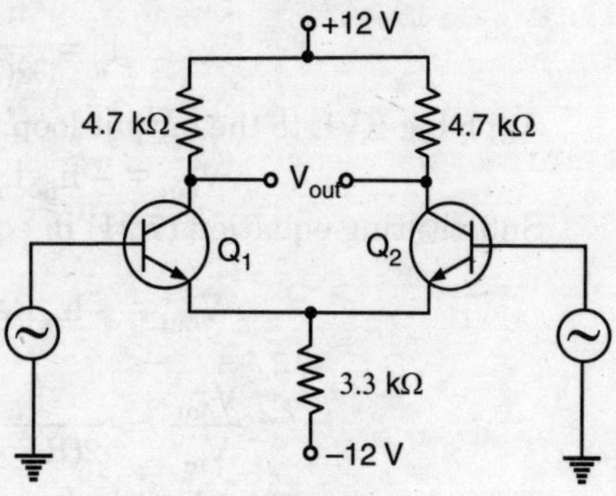

Fig. 7.9.

Thus, we have $I_E = \frac{12 - 0.7}{2 \times 3.3 \times 10^3} = 1.712$ mA

Now $I_C = I_E = 1.712$ mA

Therefore, $I_{CQ} = 1.712$ mA

Also, we write $V_{CEQ} = V_{CC} + V_{BE} - I_{CQ}R_C$

$= 12 + 07 - 1.712 \times 10^{-3} = 4.653$ V **Ans.**

Hence, Q point is (1.712 mA, 4.653 V) **Ans.**

7.6 A.C. ANALYSIS OF DIFFERENTIAL AMPLIFIER USING h-PARAMETERS

In a.c. analysis, let us evaluate the differential gain A_{id}, common mode gain A_{cm}, input resistance R_i and the output resistance R_0 of the differential amplifier circuit using the h-parameters.

7.6.1. Differential Gain (A_{id})

For the determination of differential gain, the two input signals must be different from each other. Let the two a.c. input signals be equal in magnitude but having 180° phase difference in between them. The magnitude of each a.c input voltage V_{in_1} and V_{in_2} be $\frac{V_{in}}{2}$.

The two a.c. emitter currents I_{e1} and I_{e2} are equal in magnitude and 180° out of phase. Hence, they cancel each other to get resultant a.c. current through the emitter as zero. Therefore, for a.c. purpose emitter terminal can be grounded. The a.c small signal differential amplifier circuit with grounded emitter terminal has been shown in figure 7.10. Since the two transistors are matched, the a.c. equivalent circuit for other transistor is identical to the one shown in figure 7.10.

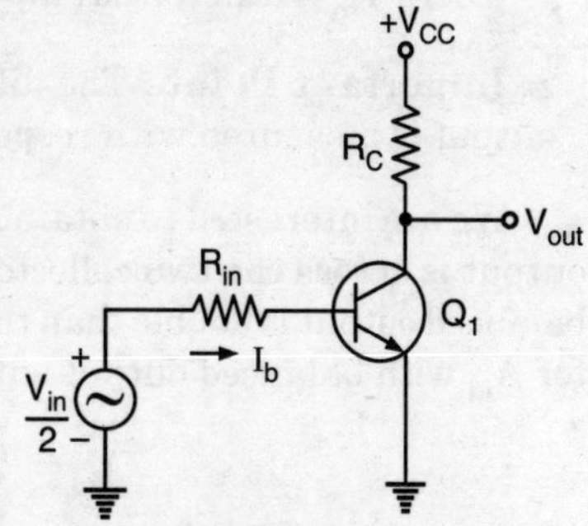

Fig. 7.10. A.C. equivalent for differential operation

Thus, the circuit can be analysed by considering only one transistor. This is called as half circuit concept of analysis.

The approximate hybrid model for the above circuit can be shown as in the figure 7.11, neglecting h_{oe},

Figure 7.11 illustrates an approximate hybrid model

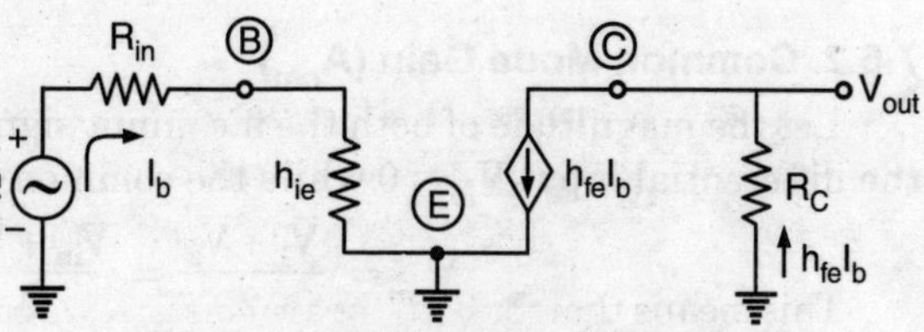

Fig. 7.11. Approximate hybrid model

Applying KVL to the input loop, we obtain

$$-I_b R_{in} - I_b h_{ie} + \frac{V_{in}}{2} = 0 \quad ...(7.20)$$

$$-I_b(R_{in} + h_{ie}) = -\frac{V_{in}}{2}$$

$$I_b = \frac{V_{in}}{2(R_{in} + h_{ie})} \quad ...(7.21)$$

Applying KVL to the output loop, we get

$$V_{out} = -h_{fe} I_b R_C \quad ...(7.22)$$

Substituting equation (7.21) in equation (7.22), we get

$$V_{out} = -h_{fe} R_C \frac{V_{in}}{2(R_{in} + h_{ie})}$$

$$\frac{V_{out}}{V_{in}} = \frac{-h_{fe} R_C}{2(R_{in} + h_{ie})} \quad ...(7.23)$$

The negative sign indicates the phase difference between input and output.

Now, two input signal magnitudes are $\frac{V_{in}}{2}$, but, they are opposite in polarity, as 180° out of phase.

Hence, we have $\quad V_{id} = V_1 - V_2 = \frac{V_{in}}{2} - \left(-\frac{V_{in}}{2}\right) = V_{in}$

The magnitude of the differential gain A_{id} is given by

$$A_{id} = \frac{V_{out}}{V_{in}} = \frac{h_{fe} R_C}{2(R_{in} + h_{ie})} \quad ...(7.24)$$

where V_{in} = differential input.

Important Point : The differential gain obtained in equation (7.24) is for the circuit where output is measured with respect to ground. Therefore, it is unbalanced output.

We are interested to obtain A_{id} for the differential amplifier with the balanced output. Balanced output is across the two collectors of the transistors Q_1 and Q_2, which are perfectly matched. Such balanced output is double than that obtained above, with unbalanced output. Therefore, the expression for A_{id} with balanced output will take the following form:

$$A_{id} = 2 \times \frac{h_{fe} R_C}{2(R_{in} + h_{ie})}$$

$$A_{id} = \frac{h_{fe} R_C}{(R_{in} + h_{ie})} \text{ (magnitude)} \quad ...(7.25)$$

This is the differential gain for balanced output dual input differential amplifier circuit.

7.6.2. Common Mode Gain (A_{cm})

Let the magnitude of both the a.c. input signals be V_{in} and are in phase with each other. Hence, the differential input $V_{id} = 0$ while the common mode input V_{cm} is the average value of the two.

$$V_{cm} = \frac{V_1 + V_2}{2} = \frac{V_{in} + V_{in}}{2} = V_{in} \quad ...(7.26)$$

While the output can be expressed as

$$V_{out} = A_{cm} V_{in} \quad ...(7.27)$$

$$A_{cm} = \frac{V_{out}}{V_{in}} \quad \text{...(7.28)}$$

But, now, both the emitter currents $I_{e1} = I_{e2} = I_e$, flows through R_E in the same direction. Therefore, the total current flowing through R_E is 2 I_e. Since, the two transistors are matched, a.c. equivalent of common mode operation can be shown, considering only one transistor, as in the figure 7.12. The approximate hybrid model for the above circuit has been shown in figure 7.13.

Since the current through R_E is 2 I_e, for simplicity of derivation, the current can be assumed to be I_e and effective emitter resistance as 2 R_E. Therefore, the emitter resistance is shown 2 R_E in figure 7.13.

Hence, current through R_C = load current I_L
Effective emitter resistance = 2 R_E
Current through emitter resistance = $I_L + I_b$
Current through h_{oe} = $(I_L - h_{fe} I_b)$

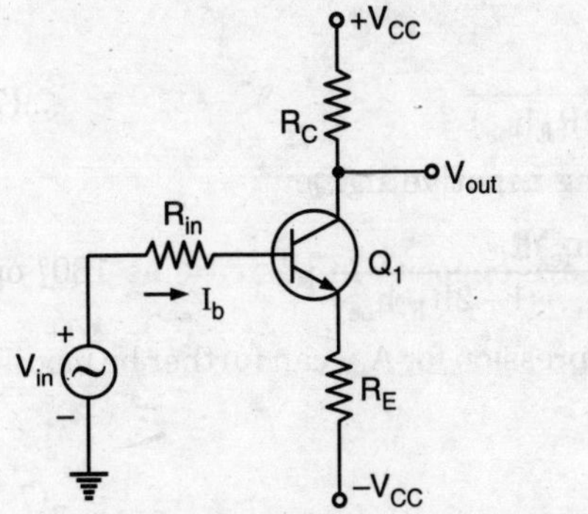

Fig. 7.12. A.C. equivalent for common mode operation

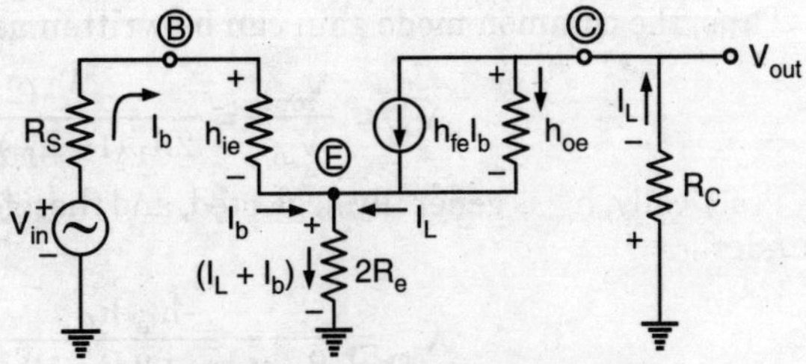

Fig. 7.13. Approximate hybrid model

Applying KVL to the input side, we get

$$-I_b R_{in} - I_b h_{ie} - 2 R_E(I_L + I_b) + V_{in} = 0$$

or $$V_{in} = I_b (R_{in} + h_{ie} + 2 R_E) + I_L (2 R_E) \quad \text{...(7.29)}$$

While $$V_{out} = -I_L R_C \quad \text{...(7.30)}$$

In above expression, there is a negative sign due to the assumed direction of current. Applying KVL to the output loop, we get

$$\frac{-(I_L - h_{fe} I_b)}{h_{oe}} - 2R_E(I_L + I_b) - I_L R_C = 0$$

or, $$-\frac{I_L}{h_{oe}} + \frac{h_{fe}}{h_{oe}} I_b - 2 R_E I_L - 2 R_E I_b - I_L R_C = 0$$

or, $$I_b\left[\frac{h_{fe}}{h_{oe}} - 2R_E\right] = I_L\left[\frac{1}{h_{oe}} + 2R_E + R_C\right]$$

or, $$I_b [h_{fe} - 2 R_E h_{oe}] = I_L [1 + h_{oe} (2 R_E + R_C)]$$

$$\frac{I_L}{I_b} = \frac{[h_{fe} - 2R_E h_{oe}]}{[1 + h_{oe}(2R_E + R_C)]} \quad \text{...(7.31)}$$

Substituting value of I_b into equation (7.29), we obtain

$$V_{in} = \frac{I_L[1 + h_{oe}(2R_E + R_C)](R_{in} + h_{ie} + 2R_E)}{[h_{fe} - 2R_E h_{oe}]} + I_L(2R_E)$$

$$\frac{V_{in}}{I_L} = \frac{[1+h_{oe}(2R_E+R_C)](R_{in}+h_{ie}+2R_E)}{[h_{fe}-2R_E\, h_{oe}]} + (2R_E)$$

Finding LCM and adjusting the terms, we obtain

$$\frac{V_{in}}{I_L} = \frac{2R_E(1+h_{fe})+R_{in}(1+2R_E\, h_{oe})+h_{ie}(1+2R_E h_{oe})+h_{oe}R_C\,[2R_E+R_{in}+h_{ie}]}{[h_{fe}-2R_E\, h_{oe}]}$$

or
$$\frac{V_{in}}{I_L} = \frac{2R_E(1+h_{fe})+(R_{in}+h_{ie})\,(1+2R_E\, h_{oe})+h_{oe}R_C[2\,R_E+R_{in}+h_{ie}]}{[h_{fe}-2R_E\, h_{oe}]} \qquad ...(7.32)$$

Neglecting the terms h_{oe} R_C, since, in practice, h_{oe} $R_C << 1$.

$$\frac{V_{in}}{I_L} = \frac{2R_E(1+h_{fe})+(R_{in}+h_{ie})\,(1+2R_E\, h_{oe})}{[h_{fe}-2\,R_E\, h_{oe}]} \qquad ...(7.33)$$

Substituting the value of I_L, in equation (7.30), we get

$$V_{out} = \frac{-V_{in}(h_{fe}-2R_E\, h_{oe})R_C}{2R_E(1+h_{fe})+(R_{in}+h_{ie})\,(1+2R_E h_{oe})}$$

Thus, the common mode gain can be written as (absorbing negative sign)

$$A_{cm} = \frac{V_{out}}{V_{in}} = \frac{(2\,R_E\, h_{oe}-h_{fe})R_C}{2R_E(1+h_{fe})+(R_{in}+h_{ie})\,(1+2R_E\, h_{oe})} \qquad ...(7.34)$$

Practically, h_{oe} is generally neglected, and therefore, the expression for A_{cm} can further be modified as under :

$$A_{cm} = \frac{-h_{fe}\, R_C}{R_{in}+h_{ie}+2\,R_E(1+h_{fe})} \qquad ...(7.35)$$

> **Important Point :** The expression for A_{cm} remains same whether the output is balanced or unbalanced.

7.6.3. Common Mode Rejection Ratio (CMRR)

Once the differential and common mode gains are obtained, the expression for the CMRR can be obtained as,

$$\text{CMRR} = \left|\frac{A_{id}}{A_{cm}}\right|$$

$$\text{CMRR} = \frac{R_{in}+h_{ie}+2\,R_E(1+h_{fe})}{(R_{in}+h_{ie})} \qquad ...(7.36)$$

This is CMRR for dual input balanced output differential amplifier circuit.

7.6.4. Differential Input Impedance (R_i) **(RGTU, Bhopal, Sem. Exam., 2007-08)**

It is the equivalent resistance between one of the input and the ground when the other input terminal is grounded. Hence, we have

$$R_i = \frac{V_{in}}{I_b} \qquad ...(7.37)$$

But referring to figure 7.11 and equation (7.21), we can write

$$R_i = 2\,(R_{in}+h_{ie}) \qquad ...(7.38)$$

For one transistor and input pair, the input resistance is $R_{in} + h_{ie}$, hence, for dual input circuit the total input resistance is 2 ($R_{in} + h_{ie}$), as the two circuits are perfectly matched.

Important Point : This input resistance is not dependent on whether output is balanced or unbalanced.

7.6.5. Output Impedance (R_0)

It is defined as the equivalent resistance between one of the output terminals with respect to ground. As seen from the figure 7.11, the resistance between output terminal with respect to ground is R_C.

$$R_0 = R_C \qquad ...(7.39)$$

In the a.c. analysis of dual input, balanced output differential amplifier, it can be observed that the results are derived by considering only one transistor and one output at the time of analysis. And then due to matched pair of transistors, the results are derived for dual input, balanced output configuration. Hence, the results for the remaining configurations can be directly obtained from the results derived earlier. And hence the summary of the differential amplifier circuit configurations is directly provided in the table 7.1.

Table 7.1.

S. No.	Name of Configuration	Circuit diagram	Voltage gain, A_{id}	Input resistance, R_i	Output resistance, R_o
1.	Dual Input Balanced Output	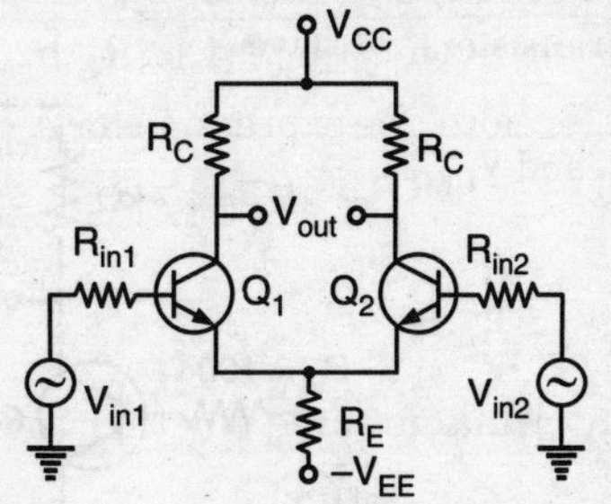	$\frac{h_{fe} R_C}{R_{in} + h_{ie}}$	$2(R_E + h_{ie})$	R_C
2.	Dual Input Unbalanced Output	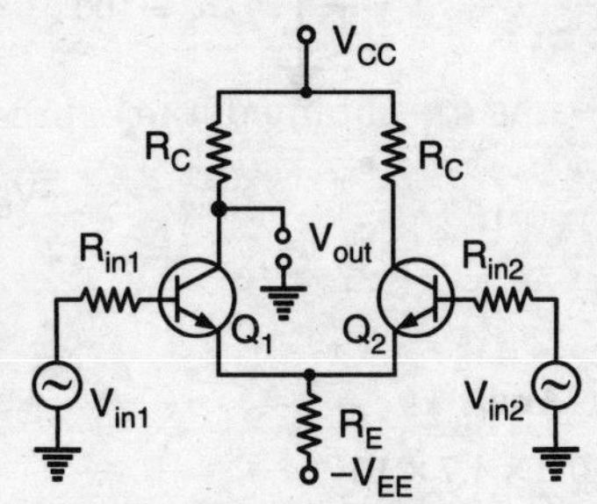	$\frac{h_{fe} R_C}{2(R_{in} + h_{ie})}$	$2(R_E + h_{ie})$	R_C
3.	Single Input Balanced Output	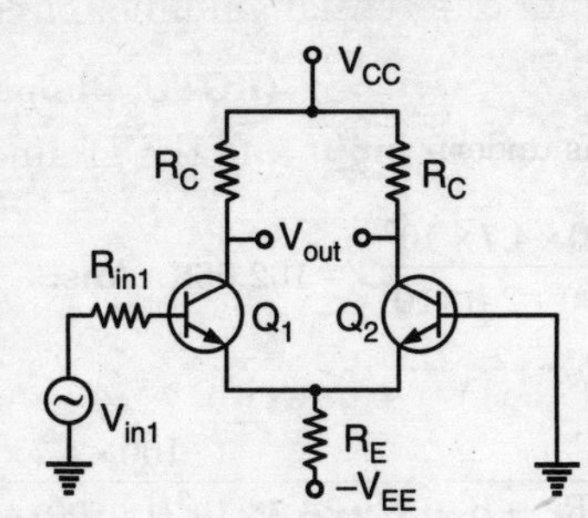	$\frac{h_{fe} R_C}{R_{in} + h_{ie}}$	$2(R_{in} + h_{ie})$	R_C

4.	Single Input Unbalanced output	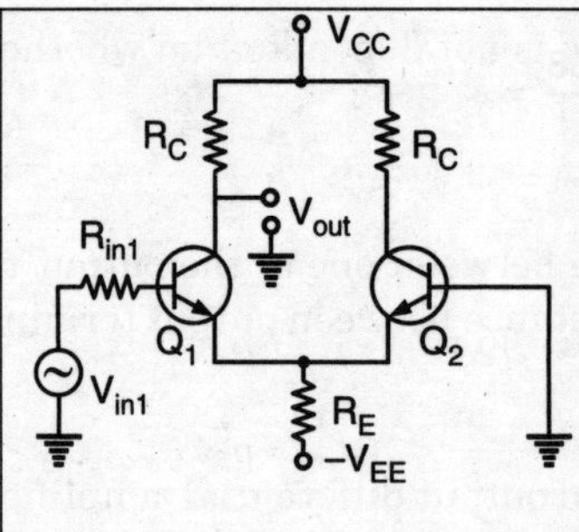	$\dfrac{h_{fe}\ R_C}{2(R_{in}+h_{ie})}$	$2\ (R_{in} + h_{ie})$	R_C

EXAMPLE 7.3. Figure 7.14 illustrates dual input balanced output differential amplifier configuration. Assuming silicon transistors with h_{ie} = 2.8 kΩ, determine

(i) Operating point values

(ii) Differential gain

(iii) Common mode gain

(iv) CMRR

(v) Output if V_{in1} = 70 mV peak to peak at 1 kHz and V_{in2} = 40 mV peak to peak at 1 kHz.

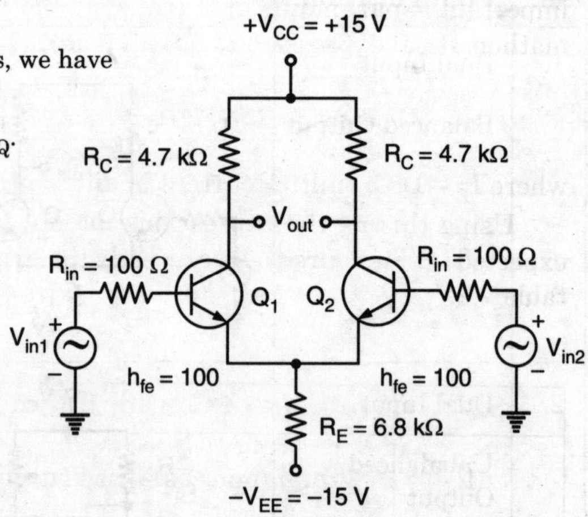

Fig. 7.14.

Solution: Since the given transistors are silicon transistors, we have

$$V_{BE} = 0.7 \text{ V}$$

(i) Operating point values are denoted by I_{CQ} and V_{CEQ}.

We have

$$I_E = \frac{V_{EE} - V_{BE}}{2R_E + \dfrac{R_{in}}{\beta}}$$

$$\beta = h_{fe} - 100$$

$$I_E = \frac{15 - 0.7}{2 \times 6.8 \times 10^3 + \dfrac{100}{100}}$$

$$I_E = 1.051 \text{ mA}$$

Now

$$I_C \cong I_E = 1.051 \text{ mA}$$

$$I_{CQ} = 1.051 \text{ mA}$$

Now

$$V_{CE} = V_{CC} + V_{BE} - I_C R_C \quad \textbf{Ans.}$$

or

$$V_{CE} = 15 + 0.7 - 1.051 \times 10^{-3} \times 4.7 \times 10^3$$

$$V_{CE} = 10.758 \text{ V}$$

$$V_{CEQ} = 10.758 \text{ V} \quad \textbf{Ans.}$$

(ii) Differential gain A_{id} can be determined as under:

$$A_{id} = \frac{h_{fe}\ R_C}{R_{in} + h_{ie}} = \frac{100 \times 4.7 \times 10^3}{100 + 2.8 \times 10^3} = 162.068 \quad \textbf{Ans.}$$

(iii) Common mode gain can be found as under:

$$A_{cm} = \frac{h_{fe}\ R_C}{2R_E(1 + h_{fe}) + R_{in} + h_{ie}} = \frac{100 \times 4.7 \times 10^3}{2 \times 6.8 \times 10^3 (1 + 100) + 100 + 2.8 \times 10^3}$$

or

$$A_{cm} = 0.3414 \quad \textbf{Ans.}$$

(iv) $$\text{CMRR} = \frac{A_{id}}{A_{cm}} = \frac{162.068}{0.3414} = 474.652$$

$$\text{CMRR in dB} = 20 \log (474.652) = 53.527 \text{ dB} \quad \textbf{Ans.}$$

(v) Output voltage is given by

$$V_{out} = A_{id} V_{id} + A_{cm} V_{cm}$$

Now, we have $$V_{id} = V_{in1} - V_{in2} = 70 - 40 = 30 \text{ mV peak to peak}$$

$$V_{cm} = \frac{V_{in1} + V_{in2}}{2} = \frac{70+40}{2} = 55 \text{ mV peak to peak}$$

$$V_{out} = 162.068 \times 30 \times 10^{-3} + 55 \times 10^{-3} \times 0.3414 = 4.86204 + 0.0187$$

$$V_{out} = 4.88 \text{ V peak to peak} \quad \textbf{Ans.}$$

Important Point : Practically, $A_{cm} V_{cm}$ can be neglected as compared to $A_{id} V_{id}$ since common mode gain is always very small compared to differential gain

7.7 A.C. ANALYSIS OF DIFFERENTIAL AMPLIFIER USING r-PARAMETERS

The a.c. analysis can be obtained using r parameters of transistors instead of h parameters. An important r parameter of a transistor is called **a.c. emitter resistance denoted as r_e.** The mathematical expression to obtain r_e is,

$$r_c = \frac{26 \text{ mV}}{I_E} = \frac{26 \times 10^{-3}}{I_E} \Omega$$

where I_E = D.C. emitter current of differential amplifier.

Using this r_e, the expressions for A_{id}, R_i an R_o can be obtained. The detailed analysis is not expected in this course hence directly the expressions for A_{id}, R_i and R_o in terms of r_e are given in the table 7.2.

Table 7.2.

S.No.	Name of Configuration	A_{id}	R_i	R_o
1.	Dual input balanced output	$\frac{R_c}{r_e}$	$2r_e\beta_{ac}$	R_C
2.	Dual input unbalanced output	$\frac{R_c}{2 r_e}$	$2r_e\beta_{ac}$	R_C
3.	Single input balanced output	$\frac{R_C}{r_e}$	$2r_e\beta_{ac}$	R_C
4.	Single input unbalanced output	$\frac{R_C}{2 r_e}$	$2r_e\beta_{ac}$	R_C

Alternatively, we can find out h_{ie} from r_e and use the expressions in terms of h-parameters. The approximate relation between r_e and h_{ie} is given by

$h_{ie} = h_{fe} r_e$ where $h_{fe} = \beta$ for transistor

Here, it may be noted that the above relation is approximate, hence, there is little difference in the value of A_{id} obtained using r-parameters and h-parameters. However, this difference is small and practically acceptable.

If h_{ie} of the transistor is not given in the problem, find r_e and use the expressions in terms of r_e to obtain A_{id}, r_o and r_i.

EXAMPLE 7.4. For the differential amplifier shown in figure 7.15, determine;

(i) Operating point i.e., I_{CQ} and V_{CEQ}

(ii) Voltage gain

(iii) Input and output impedance

Assume β = 100.

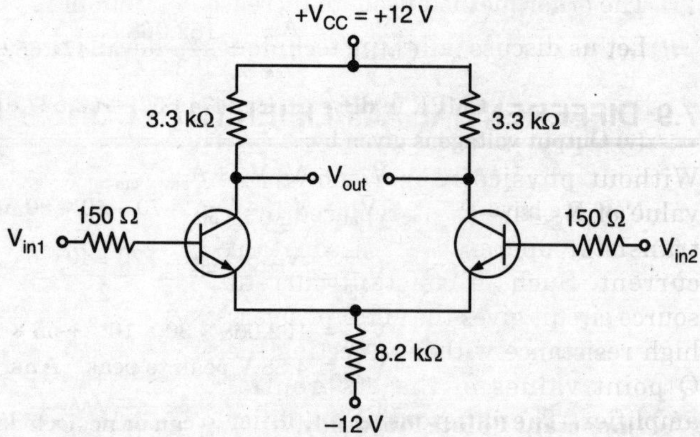

Fig. 7.15.

Solution: Since, h_{ie} is not given, let us use r parameters to obtain voltage gain.

We have, $R_C = 3.3$ kΩ, $R_{in} = 150$ Ω, $R_E = 8.2$ kΩ, $V_{CC} = 12$ V $= V_{EE}$, $\beta = 100$

(i) We know that I_E is given by,

$$I_E = \frac{V_{EE} - V_{BE}}{\frac{R_{in}}{\beta} + 2R_E}$$

$$= \frac{12 - 0.7}{\frac{150}{100} + 2 \times 8.2 \times 10^3} = 0.688 \text{ mA}$$

$I_{CQ} = I_E = 0.688$ mA **Ans.**

$V_{CEQ} = V_{CC} + V_{BE} - I_{CQ} R_C = 12 + 0.7 - 0.688 \times 10^{-3} \times 3.3 \times 10^3$

$V_{CEQ} = 10.4296$ V **Ans.**

(ii) For voltage gain, we write

$$r_e = \frac{26 \text{ mA}}{I_E} = \frac{26 \times 10^{-3}}{0.688 \times 10^{-3}} = 37.7906 \ \Omega$$

$$A_{id} = \frac{R_C}{r_e} = \frac{3.3 \times 10^3}{37.7906} = 87.32 \quad \textbf{Ans.}$$

Since the configuration is dual input balanced output, therefore, input and output impedance can be obtained as follows :

(iii) $R_i = 2\, r_e\, \beta = 2 \times 37.7906 \times 100 = 7.558$ kΩ **Ans.**

(iv) $R_0 = R_C = 3.3$ kΩ **Ans.**

7.8 VARIOUS TECHNIQUES FOR IMPROVING CMRR

As discussed earlier, higher the value of CMRR better is the performance of differential amplifier. Hence in practice the efforts are always to improve the CMRR of the differential amplifier. Alongwith the basic circuit, various other circuits are used to improve the performance of differential amplifier.

7.8.1. Effect of Emitter Resistance (R_E)

To improve the CMRR, the common mode gain A_{cm} must be reduced. The common mode gain A_{cm} approaches zero as R_E tends to infinity. This is because R_E introduces a negative feedback in the common mode operation which reduces the common mode gain A_{cm}. Thus, **higher the value of R_E, lesser is the value of A_{cm} and higher is the value of CMRR.** The differential gain A_{id} is not dependent on R_E. However, in practice, R_E can not be selected very high due to certain limitations such as

(i) Large R_E needs higher biasing voltage to set the operating Q point of the transistors.

(ii) This increases the overall chip area

Hence, practically instead of increasing R_E various other methods are used which provide effect of increased R_E without any limitations. Such two methods are:

(i) Constant current bias method

(ii) Use of current mirror circuit.

The other method used to increase A_{id} to improve CMRR is known as use of an active load.

Let us discuss all these techniques in details in the sections to follow.

7.9 DIFFERENTIAL AMPLIFIER WITH CONSTANT CURRENT SOURCE

Without physically increasing the value of R_E, the R_E is replaced by a transistor operated at a constant current. Such a constant current source circuit gives the effect of a very high resistance without affecting the Q point values of the differential amplifier. The differential amplifier using constant current bias circuit instead of R_E is shown in the figure 7.16. The transistor used is Q_3 and the values of R_1, R_2 and R_3 are selected so as to give the same operating point values for the two transistors Q_1 and Q_2.

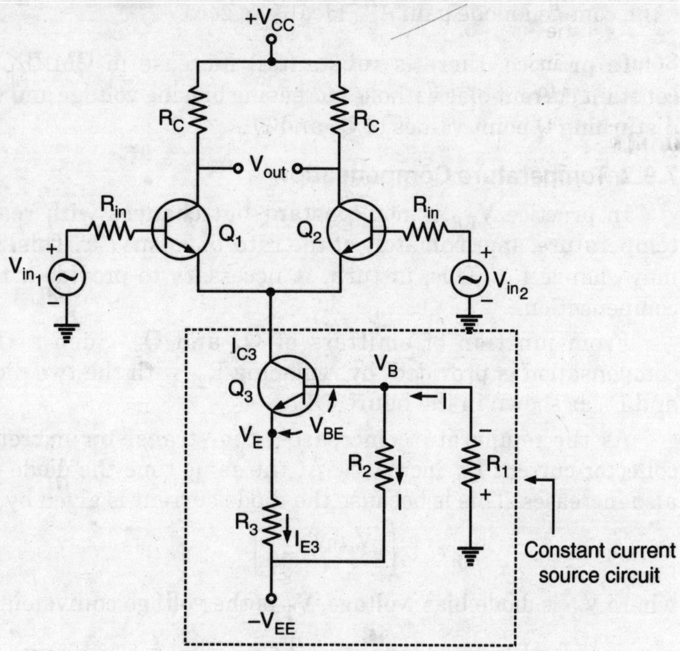

Fig. 7.16. Use of constant current source

7.9.1. Circuit Analysis

Let current through R_3 be I_{E3} while current through R_1 is I. Neglecting the base current of Q_3 which is very small due to large β_{ac}, we can assume that current through R_2 is also I.

Applying Kirchoff's law, we obtain

$$-IR_1 - IR_2 + V_{EE} = 0$$

$$I = \frac{V_{EE}}{R_1 + R_2} \quad \text{...(7.40)}$$

Now $\quad V_B = -IR_1$

Negative sign according to the direction of current, we have

$$V_B = -\frac{V_{EE}\,R_1}{R_1 + R_2} \quad \text{...(7.41)}$$

Now $\quad V_E = V_B - V_{BE}$

and $$I_{E3} = \frac{V_E - (-V_{EE})}{R_3} \quad \text{...(7.42)}$$

Substituting expressions for V_B and V_E, we get

$$I_{E3} = \frac{\dfrac{-V_{EE}\,R_1}{R_1 + R_2} - V_{BE} - V_{EE}}{R_3} = \frac{V_{EE}\left[\dfrac{R_2}{R_1 + R_2}\right] - V_{BE}}{R_3} \quad \text{...(7.43)}$$

Neglecting I_{B3}, we can write

$$I_{C3} = I_{E3}$$

Thus, as V_{EE}, R_1, R_2, R_3 and V_{BE} are constants, current I_{C3} is almost equal to I_{E3} and also constant. Thus circuit with transistor Q_3, acts as a constant current source.

Important Point : The internal resistance of a constant current source is very high, ideally infinite. Hence this circuit makes the value of emitter resistance ideally infinite which reduces the common mode gain A_{cm} ideally to zero.

In practice, there is substantial increase in CMRR due to constant current bias without increasing biasing voltage and without disturbing Q point values of Q_1 and Q_2.

7.9.2. Temperature Compensation

In practice V_{BE} is not constant but changes with respect to temperature, approximately at the rate of 2.5 mV/°C. This, in turn, may change I_{E3}. This, in turn, is necessary to provide a thermal compensation.

From junction of emitters of Q_1 and Q_2. Such a thermal compensation is provided by replacing R_2 with the two diodes D_1 and D_2 as shown in the figure 7.17.

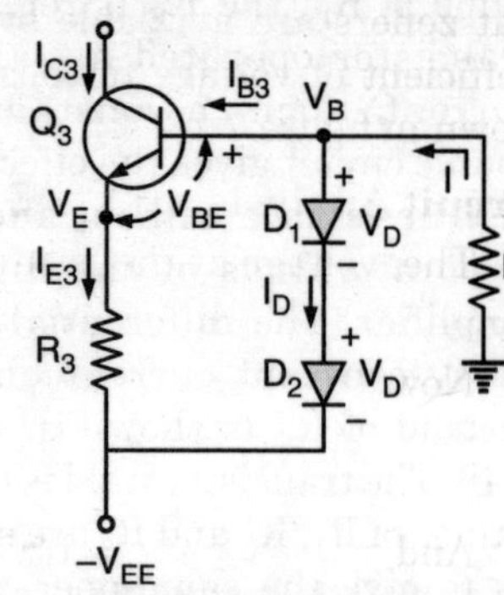

Fig. 7.17. Temperature Compensation

As the temperature increases, the transistor current i.e., collector current I_{C3} increases. At the same time the diode current also increases. This is because the diode current is given by

$$I_D = I_0\left[e^{V_0/\eta V_T} - 1\right] \quad ...(7.44)$$

where V_0 is diode bias voltage, V_T is the voltage equivalent of temperature and η is constant.

$$V_T = \frac{T}{11600}$$

$$\eta = 1 \text{ for germanium and } \eta = 2 \text{ for silicon}$$

Because of increased diode current I_D, the base current of Q_3 decreases. This compensates for the increase in I_{C3}.

Circuit Analysis

Let us assume that the voltage drop across the diodes is same as V_D.

We have $\quad V_B = -V_{EE} + 2\,V_D$

Now $\quad V_E = V_B - V_{BE}$ ...(7.45)

$$V_E = -V_{EE} + 2\,V_D - V_{BE}$$

And
$$I_{E3} = \frac{V_E - (-V_{EE})}{R_3} = \frac{-V_{EE} + 2\,V_D - V_{BE} + V_{EE}}{R_3} \quad ...(7.46)$$

$$I_{E3} = \frac{2\,V_D - V_{BE}}{R_3} \quad ...(7.47)$$

If the diode and transistor characteristics are same, then V_D is same as V_{BE}.

Therefore,
$$I_{E3} = \frac{V_D}{R_3} = I_{C3} \quad ...(7.48)$$

Thus, for a given value of R_E, the emitter current I_{E3} depends on the voltage drop across diodes. But drop across the diodes is a function of the current I_D, which is a part of current I, which is determined by the value of R_1. Thus, the value of I_{E3} can be varied by changing the value of R_1 or R_E, as per the requirement.

The value of R_1 can be obtained as under:

$$R_1 = \frac{V_{EE} - 2\,V_D}{I_{E3}} = \frac{V_{EE} - 1.4(V)}{I_{E3}} \quad ...(7.49)$$

7.9.3. Constant Current Bias using Zener

The zener diode is often used in place of the two diodes. Then problem of temperature dependent characteristics of the transistor can be solved by using zener diode. This happens because of the fact that zeners are available over a wide range of voltages and can have a matching temperature coefficient of voltage to those of transistors. Constant current bias circuit using zener has been shown in figure 7.18.

Circuit Analysis

The voltage at the base of transistor Q_3 is given by

$$V_B = -V_{EE} + V_Z \quad ...(7.50)$$

Now
$$V_E = V_B - V_{BE}$$
$$V_E = -V_{EE} + V_Z - V_{BE} \quad ...(7.51)$$

And
$$I_{E3} = \frac{V_E - (-V_{EE})}{R_3}$$
$$I_{E3} = \frac{-V_{EE} + V_Z - V_{BE} + V_{EE}}{R_3}$$
$$I_{E3} = \frac{V_Z - V_{BE}}{R_3} \quad ...(7.52)$$

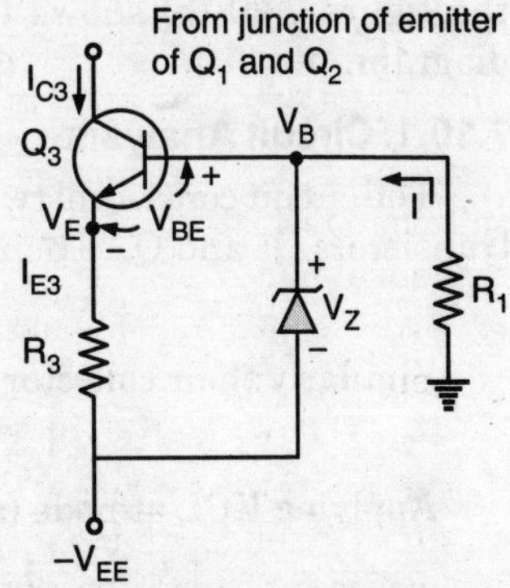

Fig. 7.18. Constant current bias using zener.

Here, the value of resistor R_1 is selected in such a way that zener diode always conducts in reverse region. It can be calculated as under:

$$R_1 = \frac{V_{EE} - V_Z}{I} \quad ...(7.53)$$

> **Important Point :** The common mode gain is almost zero, providing very large value of CMRR, with a constant current bias. To offer extremely large resistance under a.c. conditions, a simple circuit of constant current source, with less number of components is now a days used. This circuit is called current mirror or current repeater circuit.

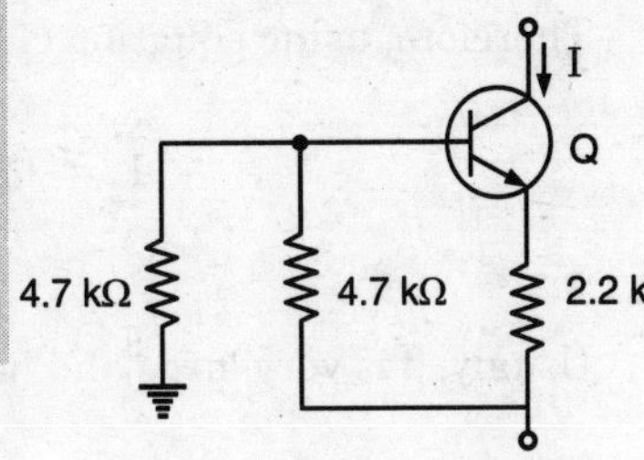

Fig. 7.19 (a).

EXMAPLE 7.5. Evaluate the constant current I in the circuit shown in figure 7.19(a).

Solution: The circuit has been shown again in figure 7.19 (b) with various voltages.

$$V_B = \frac{-V_{EE}\,R_1}{R_1 + R_2} = \frac{-20 \times 4.7 \times 10^3}{(4.7 \times 10^3 + 4.7 \times 10^3)}$$

or $V_B = -10\text{ V}$ ... (neglecting transistor base current)

$$V_E = V_B - V_{BE} = -\ 10 - 27$$

or $V_E = -10.7\text{ V}$

$$I_E = \frac{V_E - (-V_{EE})}{R_3} = \frac{-10.7 + 20}{2.2 \times 10^3} = 4.22\text{ mA}$$

$$I = I_E = 4.22\text{ mA}$$

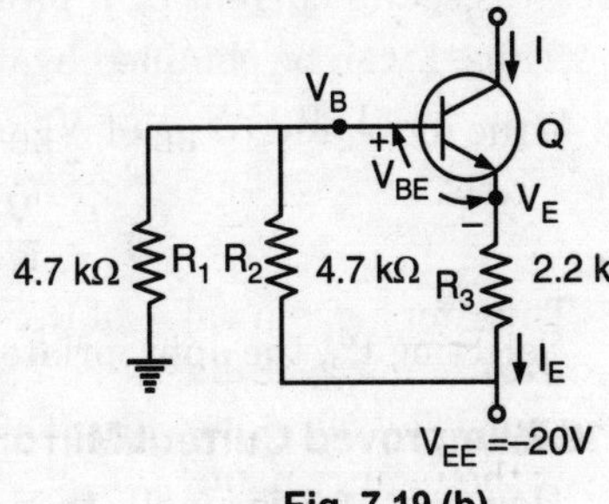

Fig. 7.19 (b).

7.10 CURRENT MIRROR CIRCUIT

The circuit in which the output current is forced to equal the input current is called as current mirror circuit. In a current mirror circuit, the output current is the mirror image of input current. The basic block diagram has been shown in the figure 7.20 (a) while the figure 7.20 (b) shows the circuit diagram.

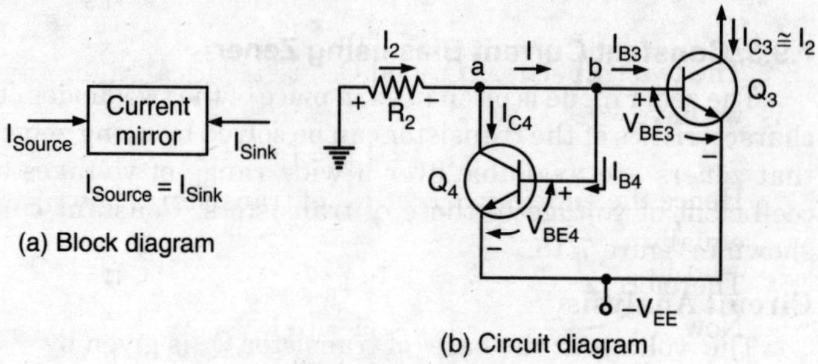

Fig. 7.20. Illustration of current mirror

7.10.1. Circuit Analysis

The circuit consists of two matched transistors Q_3 and Q_4. Their base-emitter voltage and base current are same. This means that

$$V_{BE3} = V_{BE4} \quad \text{and} \quad I_{B3} = I_{B4}$$

Similarly their collector currents are also same

$$I_{C3} = I_{C4}$$

Applying KCL at node (a), we obtain

$$I_2 = I_{C4} + I \qquad \text{...(7.54)}$$

Applying KCL at node (b), we obtain

$$I = I_{B3} + I_{B4} = 2\,I_{B4} = 2\,I_{B3} \qquad \text{...(7.55)}$$

$$I_2 = I_{C4} + 2\,I_{B4} = I_{C3} + 2\,I_{B3} \qquad \text{...(7.56)}$$

Now, we know that $I_{B3} = \frac{I_{C3}}{\beta}$

Therefore, using equation (7.56), we have

$$I_2 = I_{C3} + I_{C3}\left(\frac{2}{\beta}\right) \qquad \text{...(7.57)}$$

Usualy, β is very large and hence $\left(\frac{2}{\beta}\right)$ is negligibly small.

Therefore, we have

$$I_2 \cong I_{C3} \qquad \text{...(7.58)}$$

Thus, the collector current of Q_3 is nearly equal to the current I_2. Hence, once current mirror circuit is set for current I_2, it provides constant current bias to the differential amplifier.

Thus, I_2 can be obtained by writing KVL for the base-emitter loop of transistor Q_3.

$$-I_2 R_2 - V_{BE3} + V_{EE} = 0$$

$$I_2 = \frac{V_{EE} - V_{BE3}}{R_2} \qquad \text{...(7.59)}$$

Selecting R_2, the appropriate I_2 can be set for the current mirror circuit.

7.10.2. Improved Current Mirror Circuit

However, if β is small, then I_2 cannot be equal to I_{C3}. In such case modified current mirror circuit is used. The improved current mirror circuit is shown in figure 7.21.

Circuit Analysis

Applying KCL at node 'b' we obtain

$$I_1 = I_{C1} + I_{B3} \quad ...(7.60)$$

The two transistors Q_1 and Q_2 are identical, therefore, we have

$$I_{B1} = I_{B2} = I_B \quad ...(7.61)$$

Hence the emitter current I_{E3} of transistor Q_3 gets divided equally

Therefore, $I_{E3} = 2\, I_B$ ...(7.62)

Now $I_{E3} = (1 + \beta)\, I_{B3}$

or $2\, I_B = (1 + \beta)\, I_{B3}$ ...(7.63)

Now $I_{C1} = I_{E1} = \beta I_B$ ...(7.64)

and $I_{C2} = I_{E2} = \beta I_B$ ...(7.65)

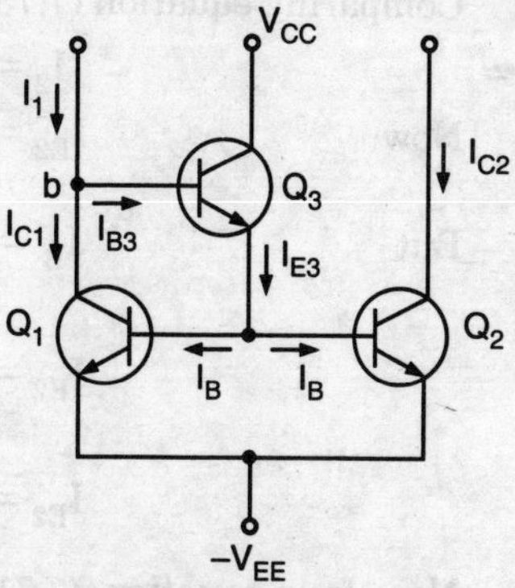

Fig. 7.21. Improved current mirror circuit

Substituting equations (7.63) and (7.64) in equation (7.60), we obtain

$$I_1 = \beta I_B + \frac{2I_B}{(1+\beta)} = I_B\left[\frac{\beta(1+\beta)+2}{(1+\beta)}\right] \quad ...(7.66)$$

Substituting in equation (7.65), we obtain

$$I_{C2} = \beta\left[\frac{I_1(1+\beta)}{\beta(1+\beta)+2}\right] \quad ...(7.67)$$

Thus, we have $I_{C2} = I_{C1} = I_2 = \dfrac{I_1\, \beta(1+\beta)}{\beta(1+\beta)+2}$ ...(7.68)

The ratio $\dfrac{I_2}{I_1}$ is much less dependent upon β for this modified circuit.

7.10.3. Wilson Current Source Circuit

Another widely used current source using current mirror technique is **Wilson current source.** High output resistance is the feature of Wilson current source. The Wilson current source circuit is shown in the figure 7.22. The high output resistance is achieved due to the negative feedback through Q_3. It also provides base currents cancellation making I_2 nearly equal to I_{ref}.

Circuit Analysis

Applying KCL at node (b), we get

$$I_{ref} = I_{C1} + I_{B2} \quad ...(7.69)$$

and $I_{E2} = I_{C3} + I_{B1} + I_{B3}$ ...(7.70)

But as all the transistors Q_1, Q_2 and Q_3 are identical hence their base-emitter voltage and base currents are also equal.

Therefore, we have

$$I_{B1} = I_{B2} = I_{B3} = I_B$$

or $I_{E2} = I_{C3} + 2\, I_B$ ...(7.71)

Also $I_{E2} = I_{C2} + I_{B2}$

i.e., $I_{E2} = I_{C2} + I_B$

Now $I_{C2} = I_{E2} - I_B$ ...(7.72)

$$= I_{C3} + 2\, I_B - I_B = I_{C3} + I_B$$

But, $I_{C3} = I_{C1}$ as transistor are identical, therefore,

$$I_{C2} = I_{C1} + I_B \quad ...(7.73)$$

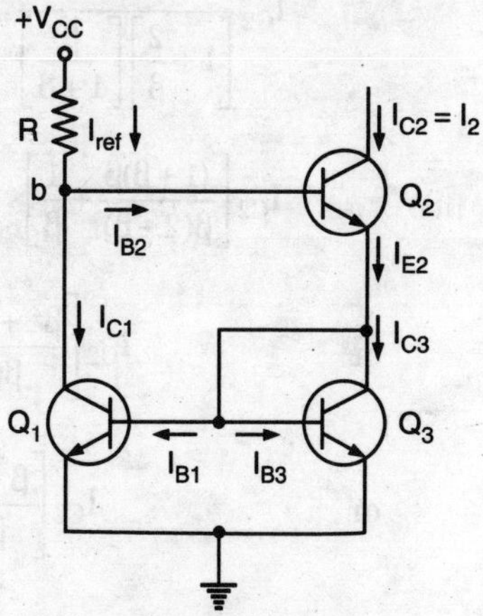

Fig. 7.22. Wilson current source

Comparing equation (7.73) and (7.69), we can say that,

$$I_2 = I_{C2} = I_{ref}$$

Now $$I_{E2} = I_{C3} + 2\,I_B$$

But $$I_B = \frac{I_{C3}}{\beta}$$

$$I_{E2} = I_{C3} + \frac{2I_{C3}}{\beta}$$

$$I_{E2} = I_{C3}\left[1+\frac{2}{\beta}\right] \qquad ...(7.74)$$

Now, from equation (7.71), we have

$$I_{C2} = I_{E2}\left[\frac{\beta}{1+\beta}\right] \qquad ...(7.75)$$

Substituting equation (7.74) in equation (7.75), we obtain

$$I_{C2} = I_{C_3}\left[1+\frac{2}{\beta}\right]\left[\frac{\beta}{1+\beta}\right] \qquad ...(7.76)$$

$$I_{C3} = I_{C_2} \times \frac{1}{\left(1+\frac{2}{\beta}\right)\left(\frac{\beta}{1+\beta}\right)} \qquad ...(7.77)$$

From equation (7.69), we can write,

$$I_{ref} = I_{C1} + I_{B2}$$

i.e., $$I_{C1} = I_{ref} - I_B$$

$$I_{C1} = I_{ref} - \frac{I_{C2}}{\beta} \qquad ...(7.78)$$

But as two transistors are identical, therefore, we have

$$I_{C3} = I_{C1}$$

Hence, from the equation (7.77) and (7.78), we have

$$I_{C2}\frac{1}{\left[1+\frac{2}{\beta}\right]\left[\frac{\beta}{1+\beta}\right]} = I_{ref} - \frac{I_{C2}}{\beta}$$

$$I_{C2}\left[\frac{(1+\beta)\beta}{\beta(2+\beta)} + \frac{1}{\beta}\right] = I_{ref}$$

or $$I_{C2}\left[\frac{\beta^2+\beta+2+\beta}{\beta(2+\beta)}\right] = I_{ref}$$

or $$I_{C2}\left[\frac{\beta^2+2\beta+2}{\beta(2+\beta)}\right] = I_{ref}$$

or $$I_{C2} = I_{ref}\left[\frac{\beta(2+\beta)}{\beta^2+2\beta+2}\right] = I_{ref}\left[1-\frac{2}{\beta^2+2\beta+2}\right] \qquad ...(7.79)$$

Important Point : The equation (7.79) shows that the output current ($I_{C2} = I_2$) and the reference current I_{ref} differ by only a factor which is of the order of $2/\beta^2$. Thus, low sensitivity of the circuit towards the transistor base currents is verified.

7.10.4. Widlar Current Source Circuit

In operational amplifier low input current is required. Hence input stage is biased at very low current, typically at a collector current of the order of 5 µA. Currents of such low magnitude can be obtained with a modified circuit called Widlar current source. The circuit is shown in the figure 7.23.

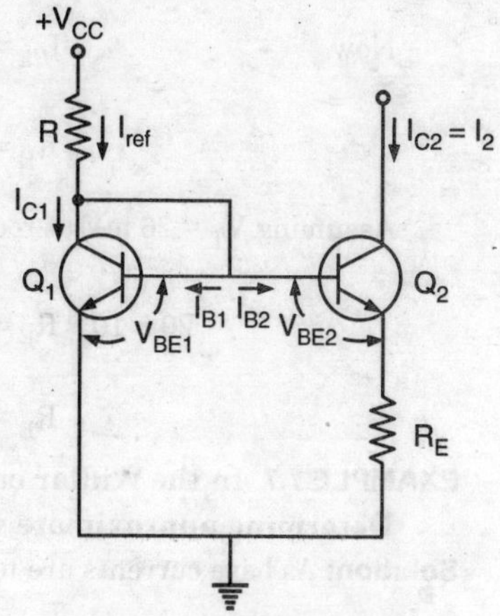

Fig. 7.23. Widlar current source

Circuit Analysis

The two transistor are identical to each other, but due to the emitter resistance R_E, V_{BE1} and V_{BE2} are different. In fact $V_{BE2} < V_{BE1}$ and hence $I_{C2} < I_{C1}$. Due to asymmetric nature of the base emitter loop the circuit is called **'lens'** rather than a **'mirror'**.

Applying KVL to the base emitter loop, we have

$$V_{BE1} = V_{BE2} + (I_{B2} + I_{C2})R_E \quad \text{...(7.80)}$$

i.e.,

$$V_{BE1} - V_{BE2} = (I_{B2} + I_{C2})\, R_E \quad \text{...(7.81)}$$

For a transistor, we can write,

$$I_{C1} = I_s e^{V_{BE1}/V_T}$$

And

$$I_{C2} = I_s e^{V_{BE2}/V_T}$$

Where I_s is the reverse saturation current, same for the two transistors and V_T is voltage equivalent of temperature. These equations are called **Ebers-Moll** equations for the transistor.

$$\frac{I_{C1}}{I_{C2}} = e^{(V_{BE1} - V_{BE2})/V_T} \quad \text{...(7.82)}$$

Taking natural logarithm of both sides, we get

$$\ln\left(\frac{I_{C1}}{I_{C2}}\right) = \frac{V_{BE1} - V_{BE2}}{V_T}$$

Hence, we get

$$V_{BE1} - V_{BE2} = V_T \ln\left(\frac{I_{C1}}{I_{C2}}\right) \quad \text{...(7.83)}$$

Equation (7.81) and (7.83), we obtain

$$(I_{B2} + I_{C2})\, R_E = V_T \ln\left(\frac{I_{C1}}{I_{C2}}\right) \quad \text{...(7.84)}$$

Neglecting I_{B2} for large value of β, we have

$$I_{C2}\, R_E = V_T \ln\left(\frac{I_{C1}}{I_{C2}}\right) \quad \text{...(7.85)}$$

This is **transcendental equation** and must be solved by trial and error method. For design purposes, I_{C1} and I_{C2} are usually known and equation (7.85) provides the value of R_E required to achieve the desired value of I_{C2}.

EXAMPLE 7.6. For the Widlar current source shown in figure 7.24, design the value of R_E to get I_{C2} as 20 µA. Assume, V_{BE} = 0.7 V. Neglect base current.

Solution: Since, base current is neglected, therefore

$$I_{C1} = \frac{V_{CC} - V_{BE}}{R}$$

$$I_{C1} = \frac{30 - 0.7}{33.3 \times 10^3} = 0.8798 \text{ mA}$$

Now $I_{C2} = 20\ \mu A$

$$I_{C2} R_E = V_T \ln\left(\frac{I_{C1}}{I_{C2}}\right)$$

Assuming $V_T = 26$ mV at room temperature of 27°C, we have

$$20 \times 10^{-6} R_E = 26 \times 10^{-3} \ln\left(\frac{0.8798 \times 10^{-3}}{20 \times 10^{-6}}\right)$$

$$R_E = 4.92 \text{ k}\Omega \quad \textbf{Ans.}$$

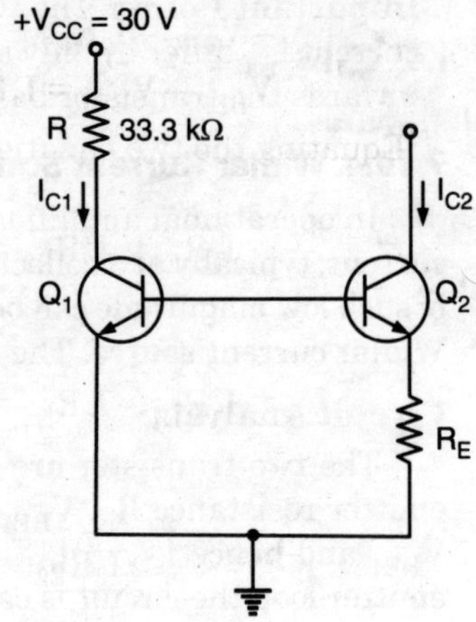

Fig. 7.24.

EXAMPLE 7.7. In the Widlar current source, reference current is 1 mA and R_E is selected as 5 kΩ. Determine approximate value of I_{C2} neglecting the base currents.

Solution: As base currents are to be neglected,

$$I_{ref} = I_{C1} = 1 \text{ mA}$$

$$R_E = 5 \text{ k}\Omega$$

Assuming $V_1 = 26$ mV at 300° K,

$$I_{C2} R_E = V_T \ln\left(\frac{I_{C1}}{I_{C2}}\right)$$

$$26 \times 10^{-3} \ln\left(\frac{1 \times 10^{-3}}{I_{C2}}\right) - I_{C2} \times 5 \times 10^3 = 0$$

This is to be solved by trial and error method.

Let $I_{C2} = 12\ \mu A$, L.H.S. = 0.05499

At $I_{C2} = 22\ \mu A$, L.H.S. = – 0.01076

Hence, actual value is less than but closer to 22 μA.

At $I_{C2} = 20\ \mu A$, L.H.S = 0.0017

Hence, LHS is almost zero, so, the value of I_{C2} is 20 μA. **Ans.**

7.10.5. Current Mirror with Magnification

Sometimes it is essential to have a circuit in which ratio of biasing currents in two transistors is required to be fixed. Such a circuit is called as **current mirror circuit with magnification.** The circuit is improved version of Widlar circuit. Such a circuit is shown in the figure 7.25. In this circuit, emitter resistances R_{E1} and R_{E2} are added to the transistors Q_1 and Q_2.

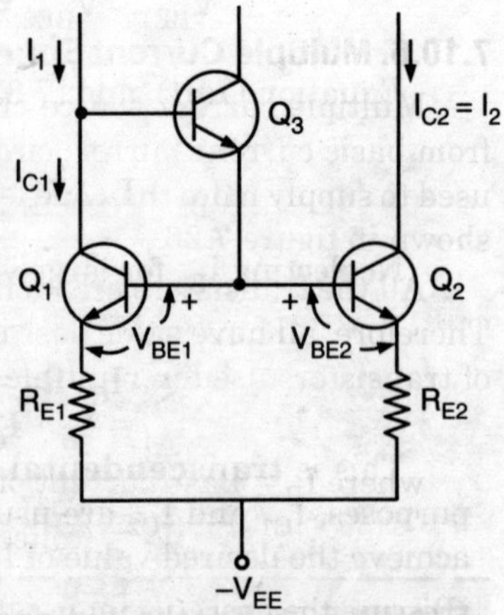

Fig. 7.25. Current mirror with magnification

Circuit Analysis

Neglecting the base currents, similar to equation (7.82), we have

$$\frac{I_{C1}}{I_{C2}} = e^{(V_{BE1} - V_{BE2})/\eta V_T}$$

Assuming η = 1 and rearranging the terms, we can write

$$V_{BE1} - V_{BE2} = V_T \ln\left(\frac{I_{C1}}{I_{C2}}\right) \quad \text{...(7.86)}$$

Applying KVL to the loop containing R_{E1} and R_{E2}, we get

$$-V_{BE2} - I_2R_{E2} + I_1 R_{E1} + V_{BE1} = 0$$

$$V_{BE2} - V_{BE1} = I_1R_{E1} - I_2R_{E2} \quad ...(7.87)$$

Equating the two equations (7.86) and (7.87), we obtain

$$I_1R_{E1} - I_2R_{E2} = -\left[V_T \ln\left(\frac{I_{C1}}{I_{C2}}\right)\right]$$

$$I_1R_{E1} - I_2R_{E2} = V_T \ln\left(\frac{I_{C2}}{I_{C1}}\right)$$

$$1 - \frac{I_2R_{E2}}{I_1R_{E1}} = \frac{V_T \ln\left(\frac{I_{C2}}{I_{C1}}\right)}{I_1R_{E1}}$$

or,
$$\frac{I_2\,R_{E2}}{I_1R_{E1}} = 1 - \frac{V_T \ln\left(\frac{I_{C2}}{I_{C1}}\right)}{I_1R_{E1}} \le$$

or,
$$\frac{I_2}{I_1} = \frac{R_{E1}}{R_{E2}}\left[1 - \frac{V_T \ln\left(\frac{I_{C2}}{I_{C1}}\right)}{I_1R_{E1}}\right] \quad ...(7.88)$$

For the range of $0.1 < \frac{I_{C2}}{I_{C1}} < 10$, $\ln\left(\frac{I_{C2}}{I_{C1}}\right)$ can be neglected and hence, we obtain

$$\frac{I_2}{I_1} = \frac{R_{E1}}{R_{E2}} \quad ...(7.89)$$

Hence, selecting proper values of R_{E1} and R_{E2}, large ratio of I_2 and I_1 and be obtained by this modified circuit.

7.10.6. Multiple Current Source*

Multiple current source circuits can be obtained from basic current mirror circuit. This circuit can be used to supply more than one load. The circuit has been shown in figure 7.26.

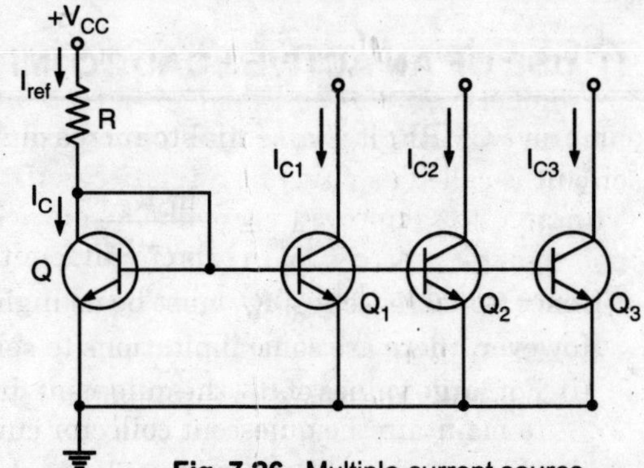

Fig. 7.26. Multiple current source

All the transistors are identical in characteristics. Therefore, all have same base currents. If n is number of transistors used in multiple current source then

$$I_{ref} = I_B + I_C + n\,I_B \quad ...(7.90)$$

where I_B = Base current of Q

I_C = Collector current of Q

*Discuss the operation and significance of a multiple output transistor current source.

(U.P. Tech, Sem. Exam, 2007-08) (05 marks)

Therefore, $I_{ref} = (n + 1) I_B + \beta I_B$

or $I_{ref} = (n + 1 + \beta) I_B$ ...(7.91)

Now, we know that $I_C = \beta I_B$

so that $$I_C = I_{ref}\left(\frac{\beta}{n+1+\beta}\right)$$ ...(7.92)

For large value of β, we obtain

$$I_C = I_{ref} = \frac{V_{CC} - V_{BE}}{R}$$ ...(7.93)

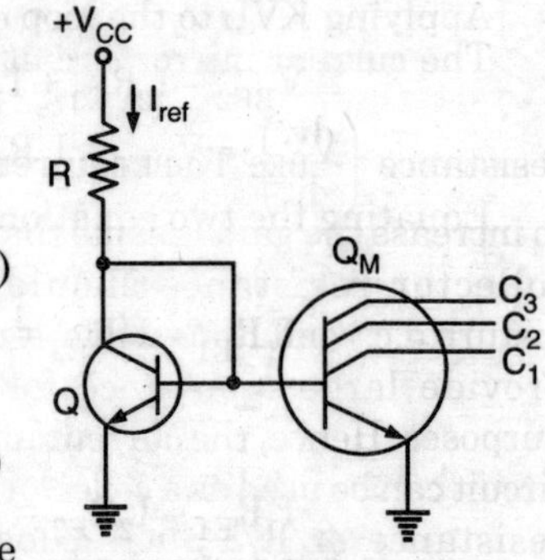

Fig. 7.27. Multiple collector merged multiple current source

Practically, the transistors like Q_1, Q_2, Q_3 are fabricated as a single multiple-collector merged device. Such a circuit has been shown in figure 7.27 where, Q_M represents the transistors Q_1, Q_2 and Q_3. In such multiple current source, more general relationship involves the area. The current in each transistor is proportional to the area of the emitter. Therefore, we write

$$I_C = I_{ref} = \frac{A_Q}{A}$$ ...(7.94)

where A_Q = Area of emitter of transistor from which current is to be calculated.

A = Area of emitter of transistor Q as depicted in figure 7.27.

Hence, in figure 7.27, if the transistor Q_1, Q_2, Q_3 are scaled 2, 4 and 8, then the currents I_{C1}, I_{C2}, I_{C3} are also scaled 2, 4 and 8. Such current sources are quite useful in digital to analog converters.

7.10.7. Salient Features

The various salient features of current mirror circuit are,

(i) Provides very high emitter resistance R_E,

(ii) Requires less components that constant current bias

(iii) Simple to design

(iv) Easy to fabricate

(v) With properly matched transistors, collector current thermal stability is achieved.

Thus, constant current bias can be easily replaced by current mirror circuit to improve CMRR. Because of its advantages, current mirror circuit is most commonly used in the integrated circuit op-amps.

7.11 USE OF AN ACTIVE LOAD TO IMPROVE CMRR*

To improve CMRR, it is essential to increase A_{id}.

While, $$A_{id} = \frac{h_{fe} R_C}{R_{in} + h_{ie}}$$

Hence, to increase A_{id}, R_C must be as high as possible.

However, there are some limitations to select maximum value of R_C such as:

(i) For large values of R_C, the quiescent drop is more, hence, higher biasing voltage is necessary to maintain the quiescent collector current.

(ii) Higher value of R_C requires a large chip area. Hence it is not possible to increase the value of R_C beyond a particular limit.

*Discuss the operation of an active load in differential amplifier. What is primary advantage of using an active load. **(U.P. Tech, Sem. Exam, 2008-09) (05 marks)**

The current mirror circuit discussed earlier has very low d.c. resistance $\left(\frac{dV}{dI}\right)$ and higher a.c. resistance $\left(\frac{dv}{di}\right)$. The requirement to increase the gain is same that the collector resistance should not disturb d.c. conditions while it must provide large resistance for a.c. purposes. Hence, the current mirror circuit can be used as a collector load resistance or R_C. Such a load is called an **active load**. The quiescent voltage across current mirror is the fraction of supply voltage. This eliminates the need of high biasing supply voltage. It basically acts as a current source and provides large a.c. resistance. The differential amplifier using a current mirror as an active load is shown in the figure 7.28.

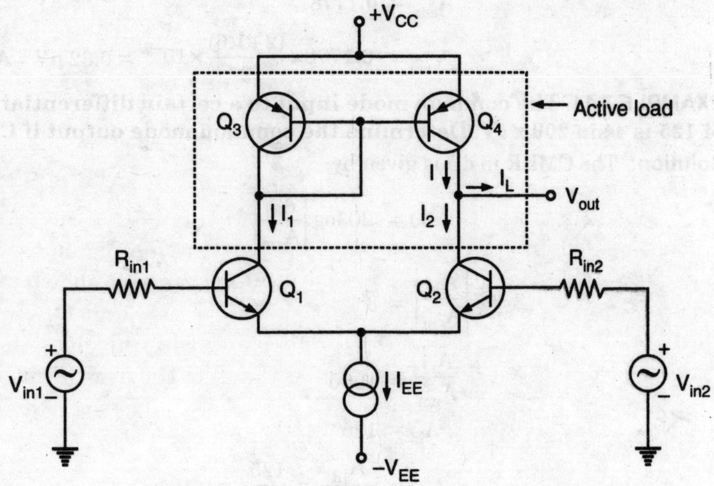

Fig. 7.28. Differential amplifier with an active load

Under the d.c. conditions, $V_{in1} = V_{in2} = 0$. As Q_1 and Q_2 are matched transistor hence $I_1 = I_2 = I_{EE}/2$, where base currents of Q_1 and Q_2 are neglected. The transistors Q_3 and Q_4 form a current repeater hence $I = I_1 = I_2$.

The load current I_1 entering the next stage is,

$$I_L = I - I_2 = 0 \quad \text{...(7.95)}$$

But when V_{in1} increases over V_{in2}, the current I_1 increases whereas I_2 decreases as $I_1 + I_2 = I_{EE}$ constant. Also the current I always remains equal to I_1 due to the current mirror action.

✍ **Important Point :** Thus the active load provides very high a.c. resistance and hence high differential mode voltage gain. Thus, A_{id} becomes high, CMRR gets improved.

EXAMPLE 7.10. An op-amp has a differential gain of 80 dB and CMRR of 95 dB. If $V_1 = 2\ \mu V$ and $V_2 = 1.6\ \mu V$, then calculate the differential and common mode output values.

Solution: We have, A_{id} = 80 dB and CMRR = 95 dB

$$A_{id} \text{ in dB} = 20 \log A_{id}$$

$$80 = 20 \log A_{id}$$

$$A_{id} = 10000 \text{ (absolute)}$$

$$\text{CMRR in dB} = 20 \log \text{CMRR}$$

$$95 = 20 \log \text{CMRR}$$

$$\text{CMRR} = 5.6234 \times 10^4$$

Differential output can be calculated as under :

$$V_{id} = A_{id}(V_1 - V_2) = 1 \times 10^4\ (2 - 1.6) \times 10^{-6} = 4 \text{ mV} \quad \textbf{Ans.}$$

Also, common mode output can be calculated as under:

$$V_{cm} = A_{cm} \frac{(V_1 + V_2)}{2}$$

Now $$\text{CMRR} = \frac{A_{id}}{A_{cm}}$$

$$5.6234 \times 10^4 = \frac{1\times10^4}{A_{cm}}$$

$$A_{cm} = 0.1778$$

$$V_{cm} = 0.1778 \times \frac{(2+1.6)}{2} \times 10^{-6} = 0.32\ \mu V \quad \textbf{Ans.}$$

EXAMPLE 7.11. The common mode input to a certain differential amplifier, having differential gain of 125 is 4 sin 200 π tV. Determine the common mode output if CMRR is 60 dB.

Solution: The CMRR in dB is given by,

$$60 = 20 \log \left|\frac{A_{id}}{A_{cm}}\right|$$

$$\log \left|\frac{A_{id}}{A_{cm}}\right| = 3$$

$$\frac{A_{id}}{A_{cm}} = 1000$$

Now $A_{id} = 125$

$$A_{cm} = \frac{A_{id}}{1000} = \frac{125}{1000} = 0.125$$

Hence, the common mode output is given by

$$A_{cm} V_{cm} = 0.125\ (4 \sin 200\ \pi t)$$

or $\mathbf{A_{cm} V_{cm} = 0.5 \sin (200\ \pi\ t)\ V}$ **Ans.**

EXAMPLE 7.12. For the differential amplifier shown in figure 7.29, determine

(i) The voltage at the collector of each transistor

(ii) The differential voltage gain,

Assume $V_{BE} = 0.7$ V, $h_{fe} = 100$, $h_{ie} = 3.9$ kΩ and the source resistance as 1 kΩ.

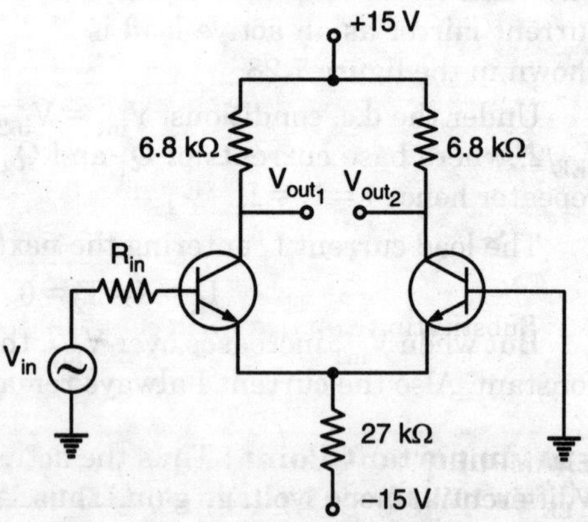

Fig. 7.29.

Solution: We have redrawn the given circuit for various voltage and current as shown in figure 7.30.

(i) Applying KVL to the base emitter loop of Q_1, we get

$$-I_B R_{in} - V_{BE} - 2 I_E R_E + V_{EE} = 0$$

We have $I_B = \frac{I_E}{\beta}$ where $\beta = h_{fe}$

$$\frac{-I_E R_{in}}{\beta} - V_{BE} - 2 I_E R_E + V_{EE} = 0$$

$$I_E \left[\frac{R_{in}}{\beta} + 2R_E\right] + V_{BE} - V_{EE} = 0$$

or
$$I_E = \frac{V_{EE} - V_{BE}}{\frac{R_{in}}{\beta} + 2R_E}$$

or
$$I_E = \frac{15 - 0.7}{\frac{1\times10^3}{100} + 2\times27\times10^3} = 0.264 \text{ mA}$$

$$I_C = I_E = 0.264 \text{ mA}$$

$$V_{out_1} = V_{out_2} = V_{CC} - I_C R_C$$

$$= 15 - 0.264 \times 10^{-3} \times 6.8 \times 10^3$$

$$= 13.2 \text{ V} \quad \textbf{Ans.}$$

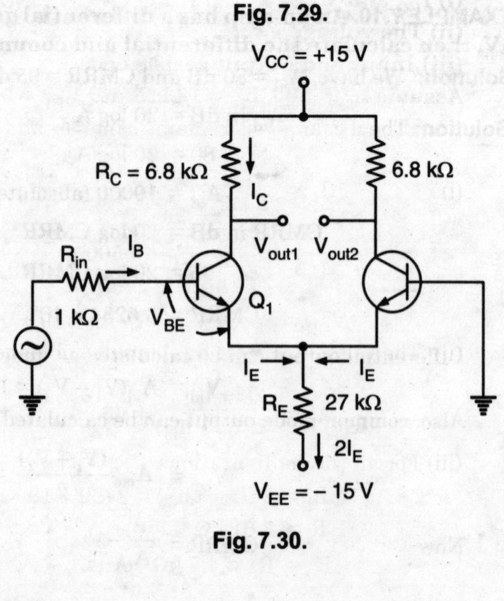

Fig. 7.30.

(ii) Differential gain A_{id}

$$A_{id} = \left| \frac{h_{fe}\, R_C}{(R_{in} + h_{ie})} \right| = \left| \frac{100 \times 6.8 \times 10^3}{[1 \times 10^3 + 3.9 \times 10^3]} \right|$$

$= 34.69$ **Ans.**

EXAMPLE 7.13. For a differential amplifier circuit operated from ± 10 V, assume matched pair of transistors with V_{BE} of 0.7 V. The collector resistance for each is 47 kΩ. Calculate the value of R_E so that the quiescent collector emitter voltage for each transistor is 8.6 V. Assume h_{fe} = 100.

Solution: From d.c. analysis of differential amplifier, we have

$$I_E = \frac{V_{EE} - V_{BE}}{2\, R_E}$$

And $$I_E = \frac{(1 + h_{fe})}{h_{fe}} I_C$$

And $$V_{CEQ} = V_{CC} + V_{BE} - I_C\, R_C$$

$$V_{CC} = 10 \text{ V},\ V_{EE} = -10 \text{ V},\ V_{BE} = 0.7 \text{ V}$$

$$R_C = 47 \text{ k}\Omega,\ V_{CEQ} = 8.6 \text{ V},\ h_{fe} = 100$$

$$I_E = \frac{10 - 0.7}{2\, R_E} \quad \text{But} \quad I_E = \frac{(1 + 100)}{100} I_C$$

$$\frac{101}{100} I_C = \frac{9.3}{2\, R_E} \quad \text{or} \quad R_E = \frac{4.6039}{I_C} \qquad \text{...(i)}$$

Now, $$V_{CEQ} = V_{CC} + V_{BE} - I_C\, R_C$$

$$86 = 10 + 0.7 - I_C \times 47 \times 10^3$$

$$I_C = \frac{10.7 - 8.6}{47 \times 10^3} = 4.468 \times 10^{-5} \text{ A} \qquad \text{...(ii)}$$

Substituting equation (i), in equation (ii), we get

$$R_E = \frac{4.6039}{4.468 \times 10^{-5}} = 103.03 \text{ k}\Omega \quad \textbf{Ans.}$$

EXAMPLE 7.14. For a dual input unbalanced output differential amplifier, V_{CC} = 10 V, V_{EE} = – 10V, R_C = 4.7 kΩ, R_E = 6.8 kΩ and R_{in} = 50 Ω. Determine

(i) I_{CQ} and V_{CEQ}

(ii) The voltage gain

(iii) Input and output resistance

Assume h_{fe} = 500, h_{ie} = 18 kΩ and V_{BE} = 0.712 V.

Solution: The d.c. analysis remains same for all the configurations.

(i) $$I_E = \frac{V_{EE} - V_{BE}}{2R_E} = \frac{10 - 0.712}{2 \times 6.8 \times 10^3} = 6.829 \times 10^{-4} \text{ A}$$

$$I_C = \left[\frac{h_{fe}}{1 + h_{fe}} \right] I_E = \frac{500}{501} \times 6.829 \times 10^{-4} = 0.6815 \text{ mA}$$

$$V_{CE} = V_{CC} + V_{BE} - I_C\, R_C = 10 + 0.712 - 0.62815 \times 10^{-3} \times 4.7 \times 10^3 = \mathbf{7.508\ V}$$

(ii) The voltage gain for dual input, unbalanced output is

$$A_d = \frac{h_{fe} R_C}{2(R_{in} + h_{ie})} = \mathbf{65.09}$$

(iii) For all the configurations

$$R_i = 2\,(R_{in} + h_{ie}) = 2\,(50 + 18 \times 10^3)$$

$$R_i = 3.61 \text{ k}\Omega$$

$$R_0 = R_C = 4.7 \text{ k}\Omega \quad \textbf{Ans.}$$

SUMMARY

- The differential amplifier is a fundamental building block of analog circuits. It is the input stage of virtually operational amplifier (op-amp), and is the basis of a high-speed digital logic circuit family, called **emitter-coupled logic (ECL)**.
- Initially invented for use with vacuum tubes, the basic differential-amplifier configuration was subsequently implemented with discrete BJTs. However, it was the advent of ICs that has made the differential amplifier extremely popular in both bipolar and MOS technologies.
- Basically, there are two reasons for using differential amplifiers in preference to single ended amplifiers. First, the differential amplifiers are much less sensitive to noise and interference than single ended circuits. The second reason for preferring differential amplifiers is that the differential configuration enables us to bias the amplifier and to couple amplifier stages together without the necessity of bypass and coupling capacitors such as those used in the designs of discrete-circuit amplifiers.
- Differential amplifier at the input stage of an op-amp provides very high input impedance, noise rejection and differential amplification capabilities. Since the differential amplifier is the basic building block of an op-amp, the analysis of differential amplifier will not only clarify the operation of an op-amp but also set a ground work to follow the performance characteristics of the op-amp.
- A differential amplifier amplifies the difference between two input voltage signals. Therefore, it is also called difference amplifier.
- If we apply two input voltages which are equal in all the respects to the differential amplifier i.e., $V_1 = V_2$ then ideally the output voltage $V_{out} = (V_1 - V_2) A_{id}$, must be zero.

  However, the output voltage of the practical differential amplifier not only depends on the difference voltage but also depends on the average common level of the two inputs. Such an average level of the two input signals is called **common mode signal** denoted as V_{cm}.

  $$V_{cm} = \frac{V_1 + V_2}{2}$$

  Practically, the differential amplifier produces the output voltage proportional to such common mode signal, also.
- The gain with which it amplifies the common mode signal to produce the output is called common mode gain of the differential amplifier denoted as A_{cm}.

  $$V_{out} = A_{cm} V_{cm}$$

  Thus there exists some finite output for $V_1 = V_2$ due to such common mode gain A_{cm}, in case of practical differential amplifiers.
- When the same voltage is applied to both the inputs, the differential amplifier is said to be operated in a common mode configuration. Many disturbance signals, noise signals appear as a common input signal to both the input terminals of the differential amplifier. Such a common signal should be rejected by the differential amplifier.

  The ability of a differential amplifier to reject a common mode signal is expressed by a ratio called **common mode rejection ratio** denoted as CMRR.
- The various features of a differential amplifier may be listed as under :

  (i) High differential voltage gain, (ii) Low common mode gain, (iii) High CMRR, (iv) Two input terminals, (v) High input impedance, (vi) Large bandwidth, (vii) Low offset voltage and currents, (viii) Low output impedance
- Basically, a differential amplifier can be used in following four configurations:

  (i) Dual input, balanced output differential amplifier, (ii) Dual input, unbalanced output differential amplifier, (iii) Single input, balanced output differential amplifier, (iv) Single input, unbalanced output differential amplifier.
- To improve the CMRR, the common mode gain A_{cm} must be reduced. The common mode gain A_c approaches zero as R_E tends to infinity. This is because R_E introduces a negative feedback in the common mode operation which reduces the common mode gain A_{cm}. Thus, **higher the value of R_E, lesser is the value of A_{cm} and higher is the value of CMRR**. The differential gain A_{id} is not dependent on R_E. However, in practice, R_E can not be selected very high due to certain limitations such as

  (i) Large R_E needs higher biasing voltage to set the operating Q point of the transistors, (ii) This increases the overall chip area

Hence, practically instead of increasing R_E various other methods are used which provide effect of increased R_E without any limitations. Such two methods are:

(i) Constant current bias method, (ii) Use of current mirror circuit.

SHORT QUESTIONS WITH ANSWERS

Q.1. Define Common-mode Gain.

Ans. It may be defined as the gain parameter for an op-amp providing the ratio of common-mode output voltage to common mode input voltage.

Q.2. What is Common-mode Voltage?

Ans. It is a voltage that appears the same for both magnitude and phase at two separate terminals.

Q.3. What do you mean by a Common - Mode Rejection Ratio?

Ans. It is the ratio of differential gain to common-mode gain.

Q.4. What is a Differential Amplifier?

Ans. It is an amplifier in which the output voltage is proportional to the difference between two external input voltages.

Q.5. Define Differential Input Resistance.

Ans. It is the Input resistance of op-amp between the two input terminals.

Q.6. What do you mean by Differential Gain?

Ans. It is a gain parameter for an op-amp giving the ratio of differential output voltage to differential input voltage.

Q.7. What is the use of difference amplifier stage in an op-amp?

Ans. The main purpose of the difference amplifier stage is to provide high gain to the difference mode signal and cancel the common mode signal, i.e., it must have high CMRR.

Q.8. How do you classify ICs?

Ans. The classification of ICs are: (i) Digital IC's (ii) Linear IC's.

Q.9. Advent of ICs made the differential amplifier extremely popular in both BJT and MOS technologies. Why?

Ans. The main requirements of differential amplifiers are matched devices and use of more components and IC fabrication is capable of providing matched devices and is also capable of providing large number of transistors at relatively low cost.

Q.10. Why are differential amplifiers preferred over single ended amplifiers?

Ans. Basically there are two reasons for using differential amplifiers in preference to single ended amplifiers. First, the differential amplifiers are much less sensitive to noise and interference than single ended circuits. The second reason for preferring differential amplifiers is that the differential configuration enables us to bias the amplifier and to couple amplifier stages together without the necessity of bypass and coupling capacitors.

Q.11. What is the basis of classification of differential amplifier configurations?

Ans. The differential amplifier configurations are classified on the basis of the number of inputs and the way an output voltage is measured.

Q.12. What are the four differential amplifier configurations?

Ans. The four differential amplifier configurations are:

1. Dual-input, balanced-output.
2. Dual-input, unbalanced-output.
3. Single-input, balanced-output.
4. Single-input, balanced-output.

Q.13. Which one of the four configurations is not commonly used and why?

Ans. Single-input, unbalanced-output differential amplifier configuration is rarely used because of the following wo reasons:

1. This configuration is identical to the CE amplifier but it requires comparatively more components and provides less voltage gain than that of CE amplifier.
2. There exists a dc output voltage without any input signal applied.

Q.14. What is difference between balanced and unbalanced output?

Ans. When the output voltage is measured between two collectors, the configuration is referred to as a balanced output, and if it is measured across any one collector (either C_1 or C_2) with respect to ground, the configuration is then said to be unbalanced output.

Q.15. Why dc analysis of a differential amplifier circuit is required?

Ans. DC analysis of a differential amplifier circuit is required for determination of operating point (quiescent collector current I_{CQ} and quiescent collector-emitter voltage V_{CEQ}).

Q.16. Why dc analysis made for dual-input, balanced-output differential amplifier configuration is also applicable for the remaining configurations too?

Ans. DC analysis for all the four configurations of the differential amplifiers is the same as long as same biasing arrangement is used for each of them.

Q.17. What is CMRR?

Ans. CMRR is defined as the ratio of differential voltage gain to common-mode voltage gain and it is given by the expression,

$$\text{CMRR} = \frac{A_{id}}{A_{cm}}$$

REVIEW QUESTIONS

1. What is meant by differential amplifier ?
2. Explain the operation of a basic differential amplifier.
3. Give the four differential amplifier configuration. Which is most commonly used and why ?
4. What is the meaning of balanced and unbalanced output ?
5. Draw the circuit diagram of emitter coupled differential amplifier and obtain its d.c. analysis.
6. Using the h-parameters, carry out the a.c. analysis of the differential amplifier. Find A_{id}, A_{cm}, R_i and R_o for dual input balanced output.
7. Define CMRR.
8. Obtain expression for CMRR.
9. Expalin the method of improving CMRR.
10. Draw the circuit diagram of a symmetrical emitter coupled differene amplifier and show that a very high CMRR will result if the differential amplifier is supplied by a constant current bias.
11. What is the active load ? Where it is used and why ?
12. Explain the operation of differential amplifier with swamping resistors.

Chapter 8

MULTISTAGE AMPLIFIERS

Syllabus

Multistage amplifiers: Darlington pair, Boot strapping technique, Cascade and cascode amplifier.

Chapter Outline

8.1. INTRODUCTION

It has been observed that the voltage or power gain, obtained from a single stage small signal amplifier, is limited. Also, it is insufficient for all practical applications. Therefore, in order to achieve greater voltage and power gain, we must use more than one stage of amplification. Such an amplifier is known as a **multistage amplifier**. In multistage amplifier, we feed the output of one stage to the input of the next stage as shown in figure 8.1. It means that much higher gains (i.e., thousands and millions times compared to a single stage amplifier) can be achieved with the help of a multistage amplifier.

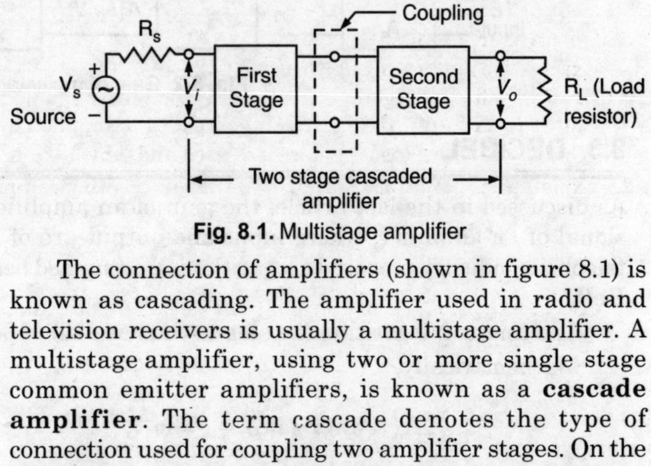

Fig. 8.1. Multistage amplifier

The connection of amplifiers (shown in figure 8.1) is known as cascading. The amplifier used in radio and television receivers is usually a multistage amplifier. A multistage amplifier, using two or more single stage common emitter amplifiers, is known as a **cascade amplifier**. The term cascade denotes the type of connection used for coupling two amplifier stages. On the other hand, a multistage amplifier with common emitter amplifier as the first stage and common base (or common collector) amplifier as the second stage is known as a

cascode amplifier. It will be interesting to know that the cascode and cascade connections are not only used in transistor amplifiers but also in field-effect transistors.

Before discussing the multistage amplifiers in detail, let us first discuss few importance terms related to multistage amplifiers. These are gain, decibel, frequency response and bandwidth.

8.2. GAIN

The ratio of output to the input of an amplifier is known as gain of the amplifier. The output quantity may be current, voltage or power. Generally, it is represented by letter 'G' or 'A'.

The gain may be current gain (A_i), voltage gain (A_v) or power gain (A_p). As a matter of fact, the gain is taken as voltage gain (A_v) if not mentioned otherwise.

8.2.1. Gain of a Multi-stage Amplifier

Let V_{in} be the input voltage to the first stage and A_1, A_2, A_3 ..... A_n be the gain of first, second, third ... and nth stage of the amplifiers respectively as shown in figure 8.2.

Then

The output of the first stage $= A_1 V_{in}$

The output of second stage $= A_2 \times (A_1 V_{in}) = A_1 \times A_2 \times V_{in}$

The output of the third stage $= A_3 (A_1 A_2 V_{in}) = A_1 \times A_2 \times A_3 \times V_{in}$

Similarly, the output of the nth stage,

$$V_{out} = A_n \times (A_1 \times A_2 \times A_3 \times .... \times V_{in})$$
$$= (A_1 \times A_2 \times A_3 \times ..... \times A_n) V_{in}$$

Hence, according to the definition,

The overall gain of a multistage amplifier is given by

$$A = \frac{V_{out}}{V_{in}} = \frac{(A_1 \times A_2 \times A_3 \times .... \times A_n)V_{in}}{V_{in}}$$

Simplifying, we get $A = A_1 \times A_2 \times A_3 \times ... \times A_n$ ...(8.1)

Therefore, the overall gain of a multistage amplifier is equal to the product of gains of individual stages. However, in practice the overall gain A is always less than $A_1 \times A_2 \times ... \times A_n$ due to the loading effect of the next stages.

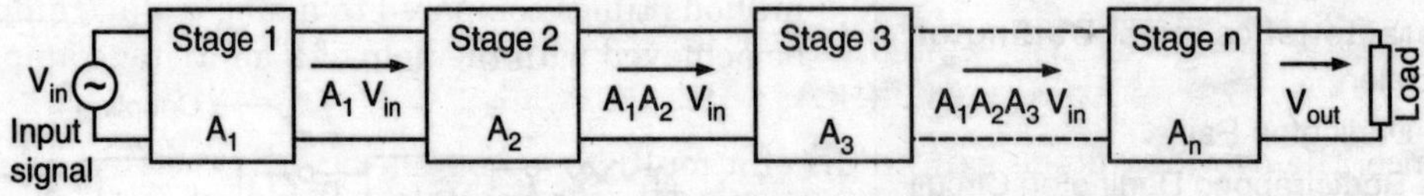

Fig. 8.2. Gain of multistage amplifier

8.3. DECIBEL

As discussed in the last article, the gain of an amplifier is the ratio of output signal to the input signal of an amplifier, where input and output are of similar quantities having same units. The telephone industry proposed a logarithmic unit named bel **(or decibel dB)** after **Alexander Graham Bell.**

The common logarithm (i.e., log to the base 10) of power gain is called as bel power gain Mathematically,

$$A_p \text{ (Power gain)} = \log_{10} \frac{P_{out}}{P_{in}} \text{ bel}$$

Optimism, is the faith that leads to achievement. Nothing can be done without hope and confidence.

— **Helen Keller**

In other words, bel is the unit of gain and is defined as the common logarithm of power gain. Thus, we can write

$$\text{Number of bels} = \log_{10} \frac{P_{out}}{P_{in}}$$

However, for practical purposes, it has been found that the unit bel is quite a large unit. Therefore, one-tenth of this unit is always considered to be more convenient. This smaller unit is known as **decibel (dB).**

Thus, 1 bel = 10 dB

$$\text{Therefore,} \quad A_P \text{ (Power gain)} = 10 \log_{10} \frac{V_{out}}{V_{in}} \text{ dB} \qquad ...(8.2)$$

As P α V^2* or I^2, therefore, the voltage or current gains are expressed in dB as under

$$A_v^{**} \text{ (voltage gain)} = 20 \log_{10} \frac{V_{out}}{V_{in}} \text{ dB} \qquad ...(8.3)$$

$$\text{and, } A_i \text{ (Current gain)} = 20 \log_{10} \frac{V_{out}}{V_{in}} \text{ dB} \qquad ...(8.4)$$

These expressions are quite useful for determining power, voltage and current gains in dB. For example, if the voltage gain of an amplifier is 10, it can be denoted on the dB scale as under:

$$\text{Voltage gain in dB} = 20 \log_{10} \frac{V_{out}}{V_{in}}$$

$$= 20 \log_{10} 10 = 20 \times 1 = 20 \text{ dB}$$

8.3.1. Gain of Multistage Amplifier in dB

If voltage gains of different stages are given as A_1, A_2, A_3, ..., A_n, then the overall gain of the multistage amplifier is given by

$$A = A_1 \times A_2 \times A_3 \times .... , A_n.$$

The overall gain of the multi-stage amplifier can be computed in dB as under:

$$20 \log_{10} A = 20 \log_{10} A_1 + 20 \log_{10} A_2 + 20 \log_{10} A_3 + ... + 20 \log_{10} A_n$$

$$\text{or} \quad A \text{ dB}^{***} = A_1 \text{ dB} + A_2 \text{ dB} + A_3 \text{ dB} + ... + A_n \text{ dB.} \qquad ....(8.5)$$

Hence, the overall voltage gain in dB of a multistage amplifier**** is the sum of the decibel voltage gain of the individual stages.

8.3.2. Reasons for Expressing the Gain in dB

In this subsection, let us discuss various reasons for expressing the gain in dB. These are :

(i) When the gains are expressed in dB, the overall gains of a multistage amplifier can be determined by simply adding them according to equation (8.5). On the other hand, the absolute gain is obtained by multiplying the gains of individual stages.

* Use of logarithms changes multiplication into an addition.

** It is also true for power and current gains.

*** As $P = \frac{V^2}{R}$ or $I^2 R$

**** Voltage gain = $10 \log_{10} \left(\frac{V_{out}}{V_{in}} \right)^2$ dB = $20 \log_{10} \frac{V_{out}}{V_{in}}$ dB

(ii) Use of dB permits us to denote very large gains on linear scale by conveniently small figures on dB (logarithmic) scale. For instance, a power gain of 12000 is just 40.79 dB.

(iii) The output of most of the amplifiers is ultimately converted into sound. Experiments reveal the fact that our ears do not respond according to the intensity of sound but they respond according to the log of intensity of sound. Hence, if the intensity of sound given by speaker increases by 100 times then the hearing level will increase only by 2 times ($\log_{10} 100 = 2$). Since dB unit tallies with the natural response of human ears, therefore, it has become more popular compared to linear scales.

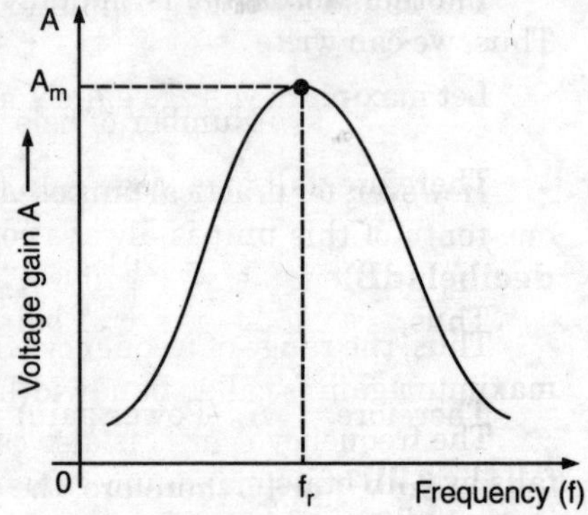

Fig. 8.3. Frequency response curve.

8.4. FREQUENCY RESPONSE

A curve drawn between voltage gain and signal frequency of an amplifier is known as the frequency response of the amplifier.

It is generally observed that the voltage gain of an amplifier is not constant, rather it varies with the signal frequency. It is because of the fact that the reactances of the capacitors connected in the circuit change with signal frequency and hence affect the output voltage or gain. The frequency response of a typical amplifier has been shown in figure 8.3. It can be noted that voltage gain increases with the increase in frequency and becomes maximum at resonant frequency f_r. However, if the frequency of signal increases beyond f_r, the gain decreases rapidly.

However, from practical point of view, this type of frequency response is not desirable since it will introduce distortion in the output signal of the amplifier. Therefore, suitable steps must be taken while designing the amplifier circuit so that a constant voltage gain may be obtained over some specified frequency range. For example in case of audio amplifiers, it is necessary that all the frequencies of sound spectrum (i.e., 20 Hz to 20 kHz) must be amplified uniformly so that distortion free output may be obtained to drive the loudspeakers.

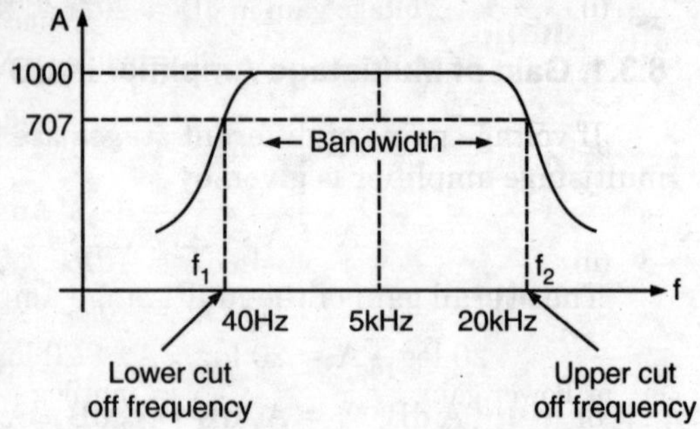

Fig. 8.4. Gain versus frequency plot showing band-width

8.5. THE BANDWIDTH

The range of frequency over which the gain of an amplifier is equal to or greater than 70.7%* of its maximum gain is known as bandwidth of amplifier.

To understand the concept of bandwidth more clearly, let us observe figure 8.4. Here gain is 70.7% or more within the frequency range f_1 to f_2. This range of frequencies from f_1 to f_2 is called bandwidth. The lower value of frequency f_1 where gain is 0.707 A_m is called lower cut-off frequency and the higher value of frequency f_2 where gain is 0.707 A_m is called higher cut-off frequency. For distortionless amplification, the signal frequency must lie within the bandwidth of the amplifier.

* This range of frequency over which the gain is greater than or equal to 70.7% of the maximum gain is considered since this change of gain cannot be detected by human ears. This means that the distortion cannot be detected over this range.

The bandwidth of an amplifier can also be defined in terms of dB as under :

Let maximum voltage gain of an amplifier be A_m. Then, 70.7% of it is $0.707\ A_m \left(\text{i.e., } \frac{70.7}{100} \times A_m\right)$

Therefore, fall in voltage gain from maximum gain will be

$$= 20 \log_{10} \frac{A_m}{0.707 A_m} \text{ dB} = 20 \log_{10} 1.4142 \text{ dB} = 3 \text{ dB}$$

Thus, the range of frequency within which the voltage gain of an amplifier falls by 3 dB from its maximum gain is called bandwidth of that amplifier.

The frequency f_1 or f_2 is also called 3 dB frequency because at this frequency, the voltage gain falls by 3 dB of its maximum value. Sometimes, it is also called **half-power frequency**. This term is used because when voltage gain falls to 0.707 of its maximum value, the power falls to one-half (therefore, $P \alpha V^2 \alpha (0.707)^2$ i.e., 0.5) of its maximum value.

EXAMPLE 8.1. Determine the gain in dB for the following cases :

(i) Voltage gain = 40, (ii) Power gain = 317.

Solution : (i) Voltage gain in dB = $20 \log_{10} 40 = 20 \times 1.602 = 32.04$. **Ans.**

(ii) Voltage gain = 40 dB, = $10 \log_{10} 317 = 10 \times 2.5 = 2.5$. **Ans.**

EXAMPLE 8.2. Determine the gain in number for the following cases :

(i) Voltage gain = 40 dB, (ii) Power gain = 44.772 dB.

Solution: Let the voltage gain be X and power gain be Y.

(i)
$$\text{Voltage gain in dB} = 20 \log_{10} X$$
$$40 = 20 \log_{10} X$$

or
$$\text{Voltage gain, } X = \text{antilog} \frac{40}{20}$$
$$= 100.$$ **Ans.**

(ii)
$$\text{Power gain in dB} = 10 \log_{10} Y$$
$$44.772 = 10 \log_{10} Y$$

or Power gain,
$$Y = \text{antilog} \frac{44.772}{10}$$
$$= 30000.$$ **Ans.**

EXAMPLE 8.3. A multistage amplifier consists of three stages. The voltage gains of three stages are 40, 60 and 80 respectively.

(i) Find the overall gain in dB,

(ii) If the power gains of the three individual stages are 100, 400 and 800 respectively, what would be the overall power gains in dB.

Solution :

(i) Voltage gain of first stage = $20 \log_{10} 40 = 32.04$ dB

Voltage gain of second stage = $20 \log_{10} 60 = 35.56$ dB

Voltage gain of third stage = $20 \log_{10} 80 = 38.06$ dB

Therefore, Overall voltage gain = 32.04 + 35.56 + 38.06 dB

= 105.66 dB **Ans.**

(ii) Power gain of first stage = $10 \log_{10} 100 = 20.00$ dB

Power gain of second stage = $10 \log_{10} 400 = 26.04$ dB

Power gain of third stage = $10 \log_{10} 800 = 29.03$ dB

Therefore, Overall power gain = 20.00 + 26.02 + 29.03 = 75.5 dB

EXAMPLE 8.4. A multi-stage amplifier employs four stages each of which has a voltage gain of 40.
(i) What is the overall gain of the amplifier?
(ii) It one stage of the four stages gives negative feedback, what will be the resultant gain?
(iii) If fifth stage gives a negative feedback of 10 dB is employed in (i), what will be the overall gains?

Solution: The voltage gain of one stage = 20 $\log_{10}$ 40 = 20 × 1.602 = 32.04 dB

(i) Overall voltage gain of 4 stages

$$= 4 \times 32.04 = 128.16 \text{ dB}$$

(ii) Now, when one stage is supplying negative feedback, the overall voltage gain

$$= 32.04 + 32.04 + 32.04 - 32.04 = \mathbf{64.08\ dB}$$

(iii) Also, when one additional stage of voltage gain 10 dB is supplying negtive feedback, then

Overall voltage gain = 32.04 × 4 – 10 = **118.16 dB** **Ans.**

EXAMPLE 8.5. A certain amplifier has power gain of 20 dB if the input power is 10 mW, what will be the output power ?

Solution: Let P_{out} be the output power

Therefore, Power gain $= 10 \log_{10} \frac{P_{out}}{P_{in}}$ dB

Substituting values, we get

$$20 = 10 \log \frac{P_{out}}{10 \text{ mW}}$$

or $\frac{P_{out}}{10 \text{ mW}} = \text{antilog} \frac{20}{10} = 100$

or $P_{out} = 10 \text{ mW} \times 100 = 1 \text{ W}$ **Ans.**

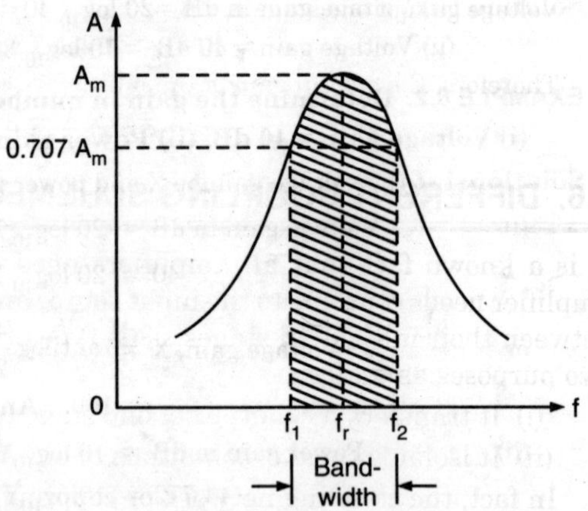

Fig. 8.5.

EXAMPLE 8.6. In a certain amplifier, the maximum voltage gain is 1000 and it occurs at 5 kHz. It falls to 707 at 20 kHz and 40 Hz. Find :

(i) lower cut-off frequency, (ii) upper cut-off frequency, (iii) bandwidth, (iv) resonant frequency.

Solution : Figure 8.5 shows the frequency response curve

Given that :

Maximum gain = 1000

$$70.7\% \text{ of max. gain} = \frac{70.7}{100} \times 1000 = 707$$

Therefore,

(i) Lower cut-off frequency = 20 Hz **Ans.**

(ii) Upper cut-off frequency = 40 kHz **Ans.**

(iii) Band width = 40 Hz to 20 kHz **Ans.**

(iv) Resonant frequency i.e., the frequency where the gain is maximum = 5 kHz. **Ans.**

EXAMPLE 8.7. The voltage gain of cascaded amplifier is 100 dB. Find the value of input signal if 20 V output is desired from this amplifier.

Solution: Here, overall voltage gain of cascaded amplifier is

$$A = 100 \text{ dB}$$

Output signal, $V_{out} = 20$ V

Overall voltage gain, $A_v = 20 \log_{10} \frac{V_{out}}{V_{in}}$

or $$100 = 20\log_{10}\frac{20}{v_{in}}$$

or $$\frac{20}{V_{in}} = \text{antilog}\,\frac{100}{20}$$

or $$\frac{20}{V_{in}} = 10^5$$

or $$V_{in} = \frac{20}{10^5} = 0.2 \text{ mV} \quad \textbf{Ans.}$$

EXAMPLE 8.8. A multistage amplifier consists of 3 similar stages. Determine the overall voltage gain of the amplifier if each stage has

$$\mathbf{R_C = 2.4\ k\Omega;\quad R_{in} = 1\ k\Omega \text{ and } \beta = 60.}$$

Solution: We know that the voltage gain of single stage is given by the relation

$$A_1 = \beta\frac{R_C}{R_{in}} = 60\times\frac{2.4}{1} = 144$$

Voltage gain of one stage in dB

$$= 20\log_{10} A_1 = 20\log_{10} 144 = 20 \times 2.158 = 43.17 \text{ dB}$$

Therefore, the overall gain of the multistage amplifier is

$$A = 3 \times A_1 = 3 \times 43.17 = 129.5 \text{ dB} \quad \textbf{Ans.}$$

8.6. DIFFERENT COUPLING SCHEMES USED IN AMPLIFIERS

It is a known fact that all amplifiers need some kind of coupling network. Even a single stage amplifier needs coupling to the input source and output load. The multistage amplifiers need coupling between their individual stages. This type of coupling is known as **interstage coupling**. It serves two purposes as under:

(i) It transfers a.c. output of one stage to the input of the next stage.

(ii) It isolates the d.c. conditions of one stage to the next stages.

In fact, the coupling network or coupling device must ensure that both the above purposes are fullfilled, when an a.c. signal is to be amplified. Following are the four coupling schemes, often used in amplifiers :

(i) Resistance-capacitance (RC) coupling,

(ii) Impedance coupling,

(iii) Transformer coupling,

(iv) Direct-coupling.

Now, we shall discuss all these coupling schemes, in detail, in the sub-sections to follow.

8.6.1. Resistance-Capacitance (RC) Coupling

This is the most important method of coupling the signal from one stage to the next stage and has been shown in figure 8.6. In this method, the signal developed across the collector resistor of each stage is coupled through the capacitor into the base of the next stage.

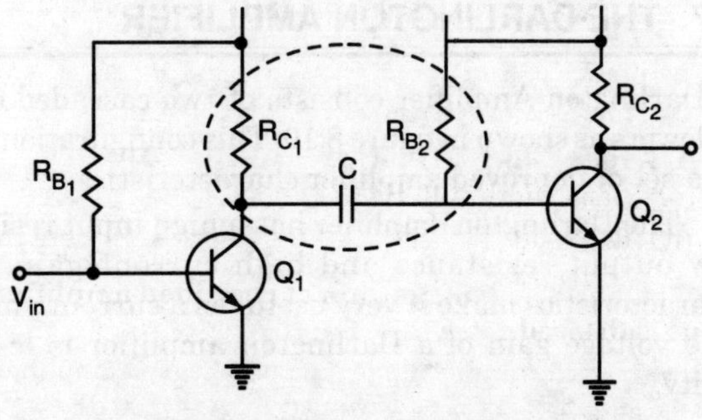

Fig. 8.6. Resistance-capacitance coupling

The cascaded stages amplify the signal and the overall gain is equal to the product of individual stage gains. The amplifiers, using this coupling scheme are known as **RC-coupled amplifiers.**

8.6.2. Impedance Coupling

Figure 8.7 shows the impedance coupling. It can be noted that the collector resistor (R_C) is replaced by an inductor (L). As the frequency increases, the inductive reactance X_L (equal to ωL) approaches infinity and each inductor appears open. In other words, the inductors pass direct current but block alternating current. The amplifier, using this coupling scheme, are known as **impedance-coupling amplifiers.**

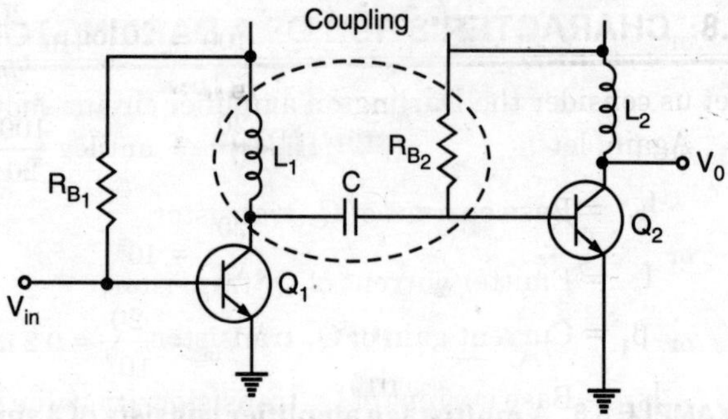

Fig. 8.7. Impedance coupling

8.6.3. Transformer Coupling

Figure 8.8 shows the transformer coupling. In this method, the primary winding of the transistor acts as a collector load and the secondary winding conveys the a.c. output signal directly to the base of the next stage. It can be noted that there is no need of coupling capacitor in the transformer coupling. The amplifiers, using this coupling scheme, are known as **transformer-coupled amplifiers.**

Figure 8.9 shows the direct-coupling. In this method, the a.c. output signal is fed directly to the next stage. This type of coupling is used where low frequency signals are to be amplified. The coupling devices such as capacitors, inductors and transformers can not be used at low frequencies because their size becomes very large. The amplifiers, using this coupling scheme, are known as direct **coupled amplifiers or d.c. amplifiers.**

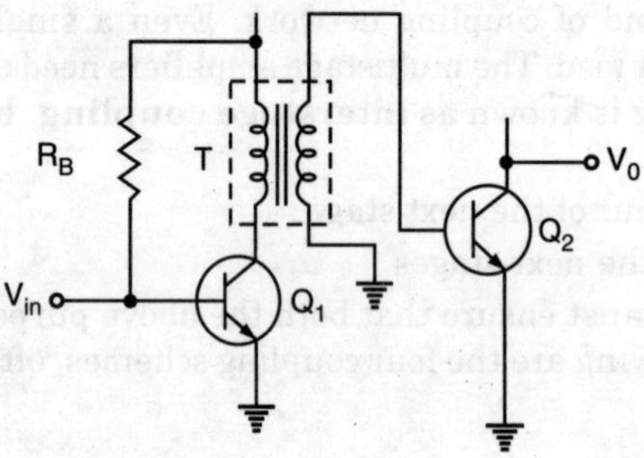

Fig. 8.8. Transformer coupling

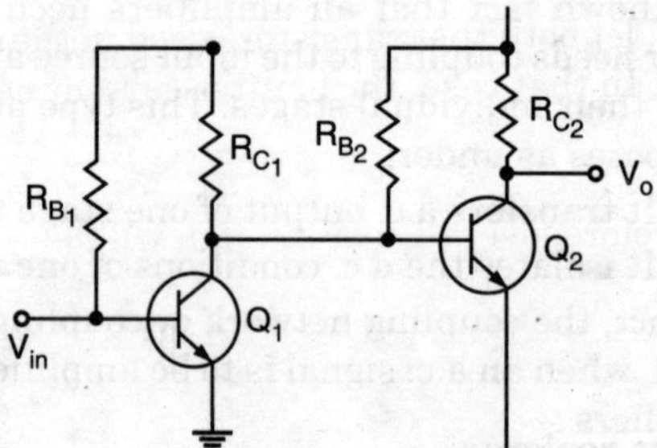

Fig. 8.9. Direct coupling

8.7. THE DARLINGTON AMPLIFIER

A Darlington Amplifier consists of two cascaded emitter followers as shown in figure 8.10. This configuration results in a set of improved amplifier characteristics.

The Darlington amplifier has a high input resistance, low output resistance and high current gain. These characteristics make it very useful as a current amplifier. The voltage gain of a Darlington amplifier is less than unity.

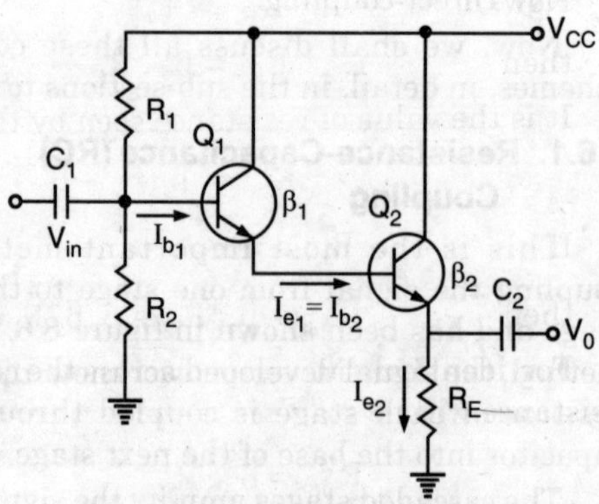

Fig. 8.10. Darlington amplifier

8.8. CHARACTERISTICS OF A DARLINGTON AMPLIFIER

Let us consider the Darlington amplifier circuit shown in figure 8.10

Again, let

I_{b_1} = Base current of Q_1 transistor,

I_{e_1} = Emitter current of Q_1 transistor,

β_1 = Current gain of Q_1 transistor,

I_{b_2} = Base current of Q_2 transistor. Its value is equal to the emitter current of Q_1 transistor

I_{e_2} = Emitter current of Q_2 transistor, and

β_2 = Current gain of Q_2 transistor.

(i) Current gain

We know that the emitter current of Q_2 transistor is given as

$$I_{e_1} = \beta_1 . I_{b_1}$$

and emitter current of Q_2 transistor is given as

$$I_{e_2} = \beta_2 . I_{b_2} = \beta_2 . I_{e_1} \qquad [\because I_{b_2} = I_{e_1}]$$

or $$I_{e_2} = \beta_2 (\beta_1 . I_{b_1}) \qquad [\because I_{e_1} = \beta_1 . I_{b_1}]$$

or $$I_{e_2} = \beta_1 . \beta_2 . I_{b_1} = \beta^2 . I_{e_1}$$

Now, let both the transistors used in the amplifier be identical. Then current gain β_1 and β_2 will be equal. In that case, the emitter current of Q_2 transistor will be

$$I_{e_2} = \beta^2 . I_{b_1} \qquad [\because \beta_1 . \beta_2 = \beta^2]$$

Therefore, overall current gain will be

$$A_i = \frac{I_{e_2}}{I_{b_1}} = \beta^2$$

(ii) Input resistance

We know that input resistance of the second stage is given by the expression

$$R_{i_2} = \beta_2 (r'_{e_2} + R_E)$$

Now, if $Q_E >> r'_{e2}$

then $$R_{i_2} = \beta_2 . R_E$$

It is the value of resistance seen by the emitter of Q_1 transistor and is given by the expression

$$R'_{i_1} = \beta_1 (r'_{e_1} + R_{i_2})$$

If $R_{i_2} >> r'_{e_1}$

then $$R'_{i_1} = \beta_1 . R_{i2} = \beta_1 . \beta_2 . R_E \qquad [\because R_{i_2} = \beta_2 . R_E]$$

For identical transistor, we know that current gains β_1 and β_2 are also equal. Hence, input resistance will be

$$R'_{i_1} = \beta^2 . R_E$$

and input resistance of the amplifier stage will be

$$R_{i_1} = (R_1 \,||\, R_2) \,||\, R'_{i_1} = (R_1 \,||\, R_2 \,||\, \beta^2.R_E$$

or $$R_{i_1} = R_1 \,||\, R_2 \qquad [\because (R_1 \,||\, R_2) << \beta^2.R_E]$$

(iii) Output resistance

The output resistance of the first stage is given by the expression

$$R_{o_1} = r'_{e_1} + \frac{R_1 \,||\, R_2}{\beta_1}$$

and output resistance of the second stage will be

$$R_{o_2} = r'_{e_2} + \frac{R_{o_1}}{\beta_2} = r'_{e_2} + \frac{r'_{e_1} + \dfrac{R_1 \,||\, R_2}{\beta_1}}{\beta_2}$$

$$= r'_{e_2} + \frac{r'_{e_1}}{\beta_2} + \frac{R_1 \,||\, R_2}{\beta_1.\beta_2}$$

or $$R_{o_2} = r'_{e_2} \qquad \left[\because \frac{r'_{e_1}}{\beta_2} \text{ and } \frac{(R_1 \,||\, R_2)}{\beta_1.\beta_2} \text{ are negligible}\right]$$

(iv) Voltage gain

The voltage gain of Darlington amplifier is given by the expression

$$A_V = \frac{V_o}{V_{in}}$$

The value of output voltage of a Darlington amplifier will be

$$V_o = I_{e_2}.R_E$$

and the input voltage will be

$$V_{in} = I_{e_1}.r'_{e_1} + I_{e_2}(r'_{e_2} + R_E)$$

The above expression is obtained by applying Kirchhoff's voltage Law to the input circuit of the Darlington amplifier. Substituting the value of I_{e_1} [equal to I_{e_2}/β_2] in last equation, we get

$$v_{in} = \frac{I_{e_2}}{\beta_2} r'_{e_1} + I_{e_2}(r'_{e_2} + R_E)$$

Therefore, voltage gain will be

$$A_V = \frac{I_{e_2}.R_E}{\dfrac{I_{e_2}}{\beta_2}.r'_{e_1} + I_{e_2}(r'_{e_2} + R_E)}$$

$$= \frac{R_E}{\dfrac{r'_{e_1}}{\beta_2} + (r'_{e_2} + R_E)}$$

or $$A_V \cong 1 \qquad [\because R_E >> r'_{e2} \text{ or } r'_{e_1}/\beta_2]$$

8.8.1. Analysis of Darlington Amplifier Using h-parameters

Let us again consider the circuit of Darlington amplifier with voltage divider bias as shown in figure 8.11.

Figure 8.12 shows the Darlington amplifier circuit in its small signal a.c. form.

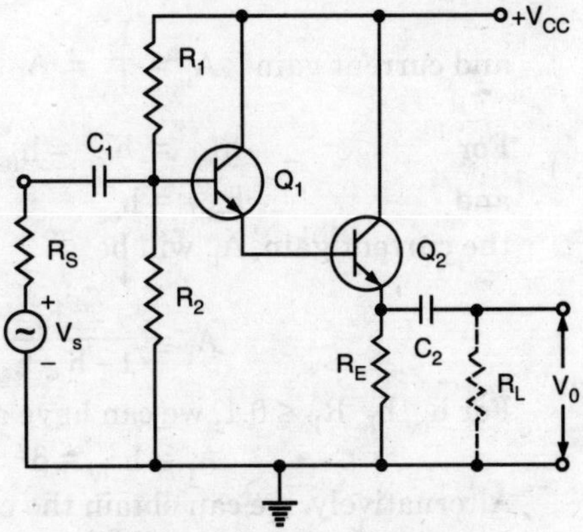

Fig. 8.11. Darlington Amplifier

8.8.1.1. Characteristics of Darlington Amplifier

1. Current Gain

For the second stage of darlington amplifier, we have

Input impedance

$$Z_{in_2} \cong h_{fe_2} R_E$$

and current gain,

$$A_{i_2} = \frac{I_0}{I_2} = \frac{I_{e_2}}{I_{b_2}} \cong h_{fe_2}$$

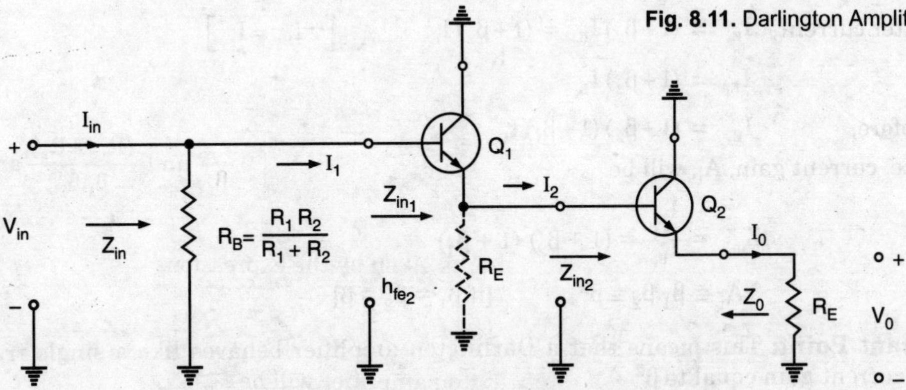

Fig. 8.12. Small Signal a.c. circuit for Darlington Amplifier

Based upon a good approximation, we cannot apply these equations to the first stage. The obvious reason is the closeners with which Z_{in2} compares with $\frac{1}{h_{oe_1}}$.

We have observed that $1/h_{oe}$, could be eliminated in the majority of the cases since the load impedance $Z_L << 1/h_{oe}$.

For the Darlington circuit under consideation, the input impedance Z_{in2} is close enough in magnitude to $\frac{1}{h_{oe_1}}$ to necessitate considering the effects of h_{oe_1}.

We know that for a single stage grounded emitter transistor amplifier, with $\frac{1}{h_{oe}}$ taken into consideration,

$$A_i = \frac{h_{fe}}{1 + h_{oe} Z_L}$$

Applying the above equation to this situation, we have

$$Z_L = Z_{in2} \cong h_{fe2} R_E$$

and

$$A_{i1} = \frac{I_2}{I_1} = \frac{I_{e_1}}{I_{b_1}} \cong \frac{h_{fe_1}}{1 + h_{oe_1} (h_{fe_2} R_E)}$$

DO YOU KNOW?

The input resistance of the Darlington pair is in the mega ohm range, which should minimize severe loading effects.

and current gain $A_i = \frac{I_0}{I_1} = A_{i_1}.A_{i_2} = \frac{h_{fe_1}.h_{fe_2}}{1 + h_{oe_1}(h_{fe_2} R_E)}$

For $h_{fe_1} = h_{fe_2} = h_{fe}$

and $h_{oe1} = h_{oe}$

the current gain, A_i, will be

$$A_i \cong \frac{h_{fe}^2}{1 + h_{oe} h_{fe} R_E}$$

For $h_{oe} h_{fe} R_E \leq 0.1$, we can have a very good approximation (within 10%) as under :

$$A_i \cong h^2{}_{fe} = \beta^2$$

Alternatively, we can obtain the current gain as under :

Emitter current, $I_{E_2} = (1 + \beta_2) I_{B_2} = (1 + \beta_2) I_{E_1}$ $\quad [\because I_{B_2} = I_{E_1}]$

But $I_{E_1} = (1 + \beta_2) I_{B_2}$

therefore, $I_{E_2} = (1 + \beta_2)(1 + \beta_1) I_{B_1}$

Hence, current gain, A_i, will be

$$Z_{in_1} = \frac{I_{E_2}}{I_{B_1}} = (1 + \beta_2)(1 + \beta_1)$$

or $A_i \cong \beta_1 \beta_2 \cong \beta^2$ $\quad$ [if $\beta_1 = \beta_2 = \beta$]

Important Point: This means that a Darlington amplifier behaves like a single transistor having current gain equal to β^2.

2. Input Impedance

Because $Z_{in_2} \cong h_{fe_2} R_E$

is the emitter resistance of the first stage, therefore, the input impedance to the first stage will be

$$Z_{in_1} \cong h_{fe_1}\left[Z_{in_2} \left\| \frac{1}{h_{oe_1}}\right.\right]$$

Since $Z_{in_2} = h_{fe_2} R_E$, and $\frac{1}{h_{oe_1}}$ would appear in parallel in the small signal equivalent circuit.

Therefore, $Z_{in_1} = h_{fe_1}\left[h_{fe_2} R_E \left\| \frac{1}{h_{oe_1}}\right.\right] = \frac{h_{fe_1} h_{fe_2} \frac{R_E}{h_{oe_1}}}{h_{fe_2} R_E + \frac{1}{h_{oe_1}}}$

or $Z_{in_1} = \frac{h_{fe_1} h_{fe_2} R_E}{h_{oe_1} h_{fe_2} R_E + 1}$

Now, if $h_{fe_1} = h_{fe_2}$

and $h_{oe_1} = h_{oe}$ then, we have

Input Impedance,

$$Z_{in_1} = \frac{h_{fe}^2 R_E}{1 + h_{oe} h_{fe} R_E}$$

For $h_{oe}\, h_{fe}\, R_E \le 0.1$, we have

$$Z_{in_1} \cong h_{fe}^2 R_E = \beta^2 R_E$$

Alternatively, we can obtain the input impedance as under :

Input impedance of second stage is given by

$$Z_{in_2} = \beta_2 (R_E + r'_{e_2})$$

where r'_{e_2} is the a.c. emitter resistance of the second transistor.

If $R_E >> r'_{e_2}$

then, we have $Z_{in_2} = \beta_2 \cdot R_E$

Z_{in_2} is also the output impedance while looking into the emitter of the first transistor. Hence, the input impedance looking into the base of the first transistor, will be

$$Z_{in_1} = \beta_1 (r'_{e_1} + Z_{in_2}) = \beta_1 (r'_{e_1} + \beta_2 R_E)$$

or $Z_{in_1} \cong \beta_1 \beta_2 R_E \cong \beta^2 \cdot R_E \quad [\text{If } \beta_1 = \beta_2 = \beta]$

This input impedance is very high because of the product of two betas, therefore input impedance

$$Z_{in} = R_1 || R_2 || Z_{in1} \cong R_1 || R_2 \qquad [\because Z_{in1} >> R_1 || R_2]$$

Also, if there is a load resistance R_L coupled to the emitter of second transistor, then, we have

$$Z_{in_1} = \beta^2 (R_E || R_L) \cong \beta^2 R_E$$

3. Output Impedance

The output impedance Z_0 can be determined directly from the emitter-equivalent circuits as under :

$$Z_{0_1} = \frac{R_S + h_{ie_1}}{h_{fe_1}} \quad \text{and} \quad Z_{0_2} = \frac{Z_{0_1} \left\| \frac{1}{h_{oe_1}} \right. + h_{ie_2}}{h_{fe_2}}$$

Alternatively, we can determine the output impedance Z_0 as under :

The a.c. Thevenin's impedance at the input will be

$$r_{th} = R_S || R_1 || R_2$$

Since by shorting all voltage source R_1 and R_S come in parallel with R_2.

Thus, output impedance of first stage is

$$Z_{0_1} = r'_{e_1} + \frac{r_{th}}{\beta_1}$$

Output impedance of second stage is

$$Z_{0_2} = r'_{e_2} + \frac{Z_{0_1}}{\beta} = r'_{e_2} + \frac{r'_{e_1} + \frac{r_{th}}{\beta_1}}{\beta_2}$$

Z_0 is quite smaller than Z_{0_1} i.e., output impedance is lowered.

Thus, due to this, the Darlington amplifier can be used to isolate high impedance source from low impedance load. If high impedance source is directly connected to a low impedance load, most of the signal voltage will be dropped across the high impedance of the source and the remaining source signal available may not be able to drive the load.

4. Voltage Gain

Applying KVL to the circuit shown in figure 8.12, we have

$$V_0 = V_{in} - V_{be_1} - V_{be_2}$$

This means that the output potential is the input potential less the base-to-emitter potential of each transistor. This clearly indicates that $V_0 < V_{in}$. Voltage gain is closer in magnitude to one than to zero.

Further, on an approximate basis, it is given as

$$A_v \cong \frac{1}{1 + \dfrac{h_{ie_2}}{h_{fe_2} R_E}}$$

Also, which is on an approximate basis, the voltage gain can be determined from the following relation

$$A_v \cong \frac{R_E}{r'_e + R_E} = \frac{1}{1 + \dfrac{r'_e}{R_E}} \cong 1$$

which is slightly less than unity as in case of emitter follower.

8.9. THE DARLINGTON PAIR

These days, the semiconductor device manufactures are supplying a pair of identical transistors, connected suitably, in a single package. It is called a Darlington pair. The two transistors are connected in such a way that the emitter of one transistor is connected to the base of the other as shown in figures 8.13(a) and (b). The Darlington pairs shown in figures 8.13 (a) and (b) and are known as PNP Darlington pair and NPN Darlington pair respectively.

A Darlington pair is a three terminal device namely base (B) emitter (E) and collector (C). It acts like a transistor, with an extremely high current gain (β). The Darlington pairs are available for both NPN and PNP transistors. Even photo-transistor Darlington pairs are also available. Such a device is β times more sensitive than that of a photo transistor.

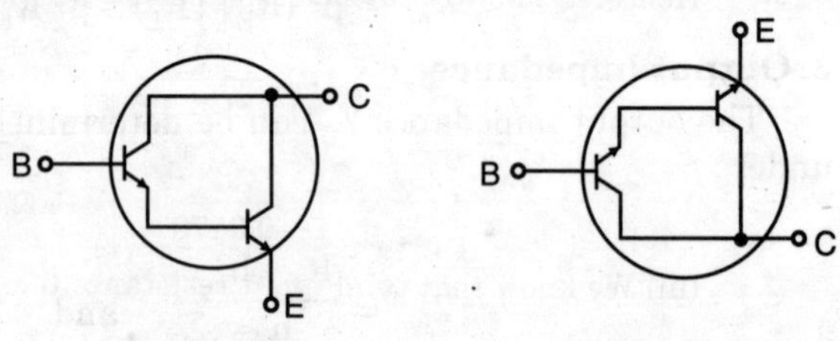

(a) NPN transistor Darlington pair

(b) PNP transistor Darlington pair

Fig. 8.13.

EXAMPLE 8.14. Figure 8.14 shows the circuit diagram of a Darlington amplifier.

Find (i) the overall gain; (ii) the a.c. emitter diode resistance for each transistor; (iii) total input resistance (iv) the overall voltage gain.

Solution: Given that

$$V_{CC} = 10 \text{ volt}$$
$$R_1 = 30 \text{ k}\Omega$$
$$R_2 = 20 \text{ k}\Omega$$
$$R_E = 1.5 \text{ k}\Omega = 1.5 \times 10^3 \ \Omega$$
$$\beta_1 = 150, \text{ and}$$
$$\beta_2 = 100$$

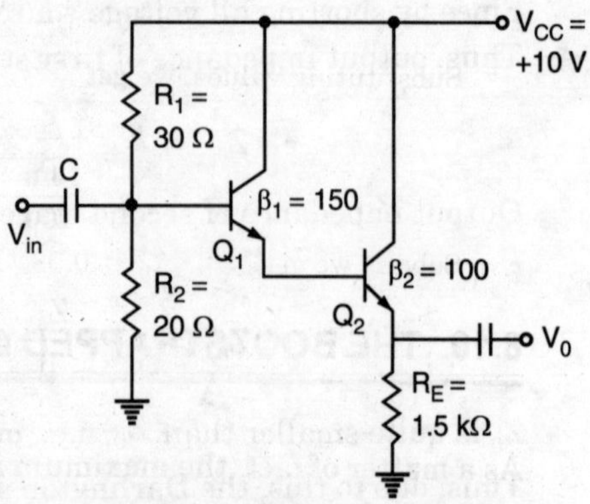

Fig. 8.14. Circuit for example 8.14

(i) We know that the overall current gain is given as

$$A_i = \beta_1 \cdot \beta_2$$

or $A_i = 150 \times 100 = 15000$ **Ans.**

(ii) We know that the voltage drop across resistor R_2 will be

$$V_{R_2} = V_{CC} \times \frac{R_2}{R_1 + R_2} = 10\left(\frac{20}{30+20}\right) = 4 \text{ V}$$

and the voltage at the base of Q_2 (or emitter of Q_1), will be

$$V_{B_2} = V_{R_2} - V_{BE_1} = 4 - 0.7 = 3.3 \text{ V}$$

Similarly, voltage at the emitter of Q_2 will be

$$V_{E_2} = V_{B_2} - V_{BE_2} = 3.3 - 0.7 = 2.6 \text{ V}$$

and the value of emitter current of Q_2 will be

$$I_{E_2} = \frac{V_{E_2}}{R_E} = \frac{2.6}{1.5\times10^3} = 1.73 \times 10^{-3} \text{ A} = 1.73 \text{ mA}$$

Hence, a.c. emitter diode resistance of Q_2 transistor will be

$$r'_{e_2} = \frac{25}{I_{E_2}(\text{in mA})} = \frac{25}{1.73} = 14.5\ \Omega \quad \textbf{Ans.}$$

Now, the base current of Q_2, will be

$$I_{b_2} = \frac{I_{E_2}}{\beta_2} = \frac{1.73}{100} = 0.0173 \text{ mA}$$

and the emitter of Q_1 will be

$$I_{E_1} = I_{b_2} = 0.0173 \text{ mA}$$

Hence, a.c. emitter diode resistance of Q_1 will be

$$r'_{e_1} = \frac{25}{I_{E_1}(\text{in mA})}$$

$$= \frac{25}{0.0173} = 1445\ \Omega \quad \textbf{Ans.}$$

(iii) We know that total input resistance (i.e., input resistance of the amplifier stage will be given by

$$R_{i_1} = R_1 || R_2 = 30 || 20 = 12 \text{ k}\Omega \quad \textbf{Ans.}$$

(iv) We also know that overall voltage gain is expressed as

$$A_V = \frac{R_E}{\frac{r'_{e_1}}{\beta_2} + (r'_{e_2} + R_E)}$$

Substituting values, we get

$$A_V = \frac{1.5\times10^3}{\frac{1445}{100} + [14.4 + (1.5\times10^3)]}$$

Solving we get $A_V = 0.98$ **Ans.**

8.10. THE BOOTSTRAPPED DARLINGTON CIRCUIT.

As a matter of fact, the maximum input resistance of a Darlington circuit is limited to $\frac{1}{h_{ob}} \cong 2\ \text{M}\Omega$

as $1/h_{oe}$ is the resistance between the base and the collector. However, the input resistance can be largely increased by bootstrapping the Darlington circuit through the addition of capacitor C' bet-ween the first collector terminal C_1 and the second emitter ter-minal E_2 as shown in figure 8.15. It may be noted that the resistor R_c is essential because in absence of R_c, resistor R_E will be shorted to ground. If the input signal changes by V_{in} then E_2 change by $A_v \cdot V_{in}$ (assuming that reactance of C' is negligible) and the collector changes by the same amount. This means that $\frac{1}{h_{ob}}$ is now effectively increased to $\frac{1}{h_{ob}(1 - A_v)}$ $\cong$ 400 MΩ for a voltage gain of 0.995.

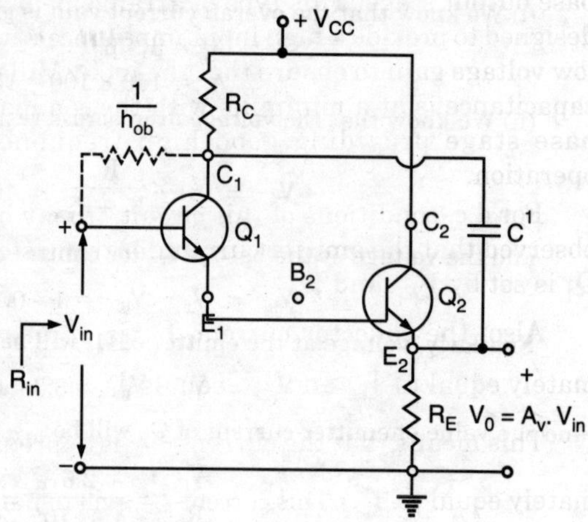

Fig. 8.15. A Bootstrapped Darlington circuit

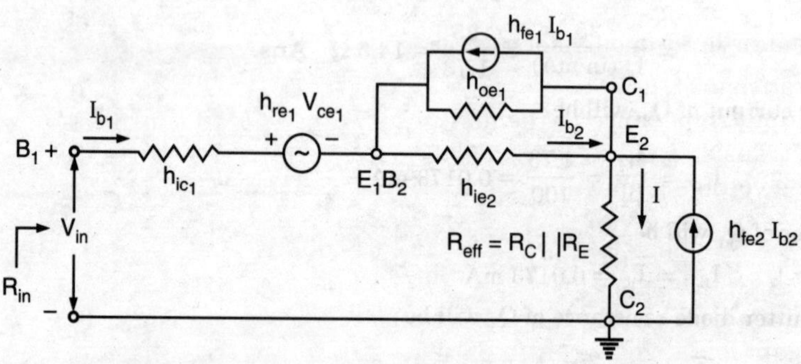

Fig. 8.16. Equivalent circuit for the circuit shown in figure 8.15

We can obtain an expression for the input resistance R_{in} of the bootstrapped Darlington circuit by using the equivalent circuit shown in figure 8.16. The effetive resistance R_{eff} between terminal E_2 and ground will be

$$R_C || R_E.$$

If h_{oe} $R_{eff} \leq 0.1$, then the transistor Q_2 can be represented by approximate hybrid-model. But, we must use exact hybrid model for transistor Q_2 as shown in figure 8.16. Because $\frac{1}{h_{oe_1}} >> h_{ie_2}$ therefore, h_{oe_1} can be omitted from figure 8.16.

Thus, solving for $\frac{V_{in}}{I_{b_1}}$, we get

$$R_{in} = h_{fe_1} \cdot h_{fe_2} \cdot R_{eff}$$

This equation shows that the input resistance of the bootstrapped Darlington circuit is essentially equal to the product of the short-circuit current gains and the effective emitter resistance. If the transistor with the current gains of the order of magnitude of 100 are used and effective resistance as 5 kΩ, then the input resistance obtained will be of the order of 50 MΩ.

8.11. THE CASCODE AMPLIFIER

As a matter of fact, a cascade amplifier has one transistor on the top of (in series with) another. This has been illustrated in figure 8.17. In this figure, transistor Q_1 and its associated components operate as a common emitter input stage whereas the circuit of transistor Q_2 operates as a common base output stage. This type of arrangement is designed to provide a high input impedance with low voltage gain to ensure that the input Miller capacitance is at a minimum with the common base stage providing good high frequency operation.

For d.c. conditions of this circuit, it may be observed that the emitter current for transistor Q_1 is set by V_{E_1} and R_4.

Also, the collector current I_{C_1} is approximately equal to I_{E_1} and I_{E_2} equals I_{C_2}.

This means that the current I_{C_2} is approximately equal to I_{E_1}. This current I_{C_2} will remain constant irrespective of the value of voltage V_{B_2}, as long as V_{CE_1} remains large enough for correct operation of the transistor Q_1.

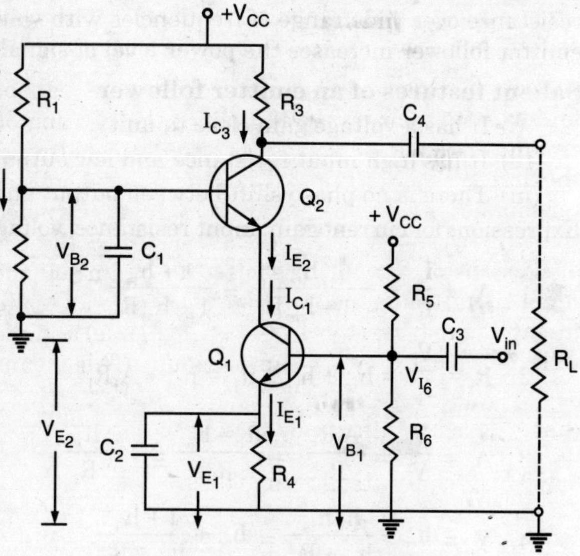

Fig. 8.17. A cascode Amplifier

Further, the input impedance to the emitter of transistor Q_2 constitutes the a.c. load in the collector circuit of transistor Q_1.

For stage 1, the voltage gain will be

$$A_{v_1} = -\frac{h_{fe}}{h_{ie}} \times (Z_{in} \text{ to } Q_2) = -\frac{h_{fe}}{h_{ie}}\left[\frac{h_{ie}}{1+h_{fe}}\right] \cong -1$$

Thus, with a stage gain of only 1, no Miller effect occurs at transistor Q_1.

Now, voltage gain for stage 2 is given by

$$A_{v_2} = \frac{h_{fb} \times (R_3 \| R_L)}{h_{ib}}$$

Converting to common emitter parameters, the overall voltage gain for the cascode amplifier will be same as that for a CE amplifier.

Thus, overall voltage gain,

$$A_v = \frac{-h_{fe} \times (R_3 \| R_L)}{h_{ie}}$$

This means that a cascade amplifier has high input impedance, the same voltage gain as a CE amplifier but very low input capacitance.

8.12. THE EMITTER FOLLOWER

As a matter of fact, common collector transistor amplifier is also known as since emitter follower its voltage gain is close to unity and due to which a change in base voltage appears as an almost equal change across the load voltage at the emitter (i.e., output) terminals. this means that in emitter following, the emitter follows the input signal. Further, input resistance (R_i) of the emitter follower is very high and output resistance (R_0) is very low. Therefore, common collector or emitter follower configuration is most commonly used as a buffer stage which performs the function of resistance transformation from the high to low

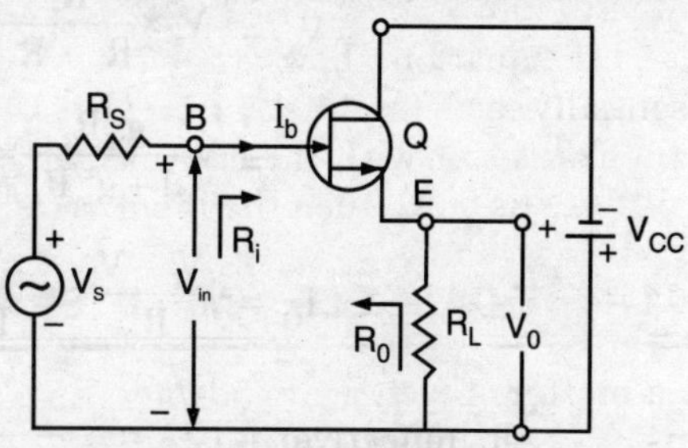

Fig. 8.18. A common collector or emitter follower configuration

resistance over wide range of frequencies with voltage gain close to unity. In addition to this, the emitter follower increases the power level of signal.

Salient features of an emitter follower

(i) It has a voltage gain close to unity.

(ii) It has high input resistance and low output resistance.

(iii) There is no phase shift between output and input in either voltage or current.

Expressions for current gain, input resistance, voltage gain and output resistance of emitter follower:

1. $A_I = \frac{I_e}{I_b} = -\frac{h_{fe}}{1+h_{oe}R_L} = \frac{1+h_{fe}}{1+h_{oe}R_L}$

2. $R_i = \frac{V_i}{I_b} = h_{ie} + h_{rc}A_I R_L = h_{ie} + A_I R_L$

3. $A_V = \frac{V_o}{V_i} = \frac{A_i R_L}{R_i} = \frac{R_i - h_{ie}}{R_i} = 1 - \frac{h_{ie}}{R_i}$

4. $Y_o = h_{oc} - \frac{h_{fc}h_{rc}}{h_{ie}+R_s} = h_{oc} + \frac{1+h_{fe}}{h_{ie}+R_s}$

5. $A_{vs} = \frac{A_v R_i}{R_i + R_s}$

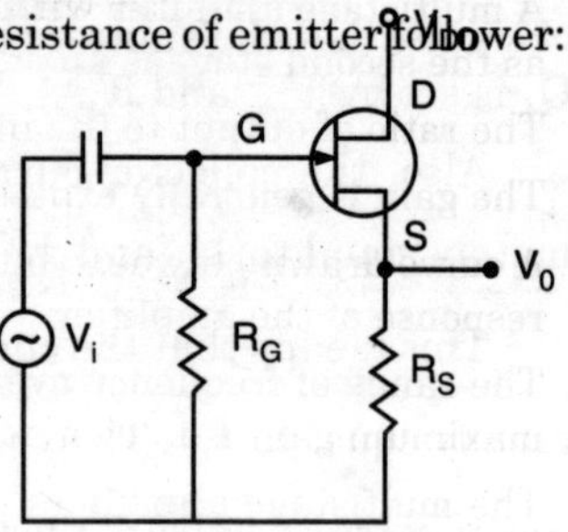

Fig. 8.19. Circuit diagram of a source follower

8.12.1. The Source Follower

This is common drain amplifier emitter to common collector amplifier. The output voltage of a common drain amplifier is approximately equal to and in phase with the input voltage. Because of this fact, the circuit is known as source follower, (when drain is common and output is taken from source, the amplifier is called source follower).

The circuit of source follower is shown in figure 8.19.

From circuit diagram of source follower, we have

The voltage gain $\quad A_v = \frac{V_o}{V_i} = \frac{g_m R_s}{1+g_m R_s}$

If $g_m R_s >> 1$ then $\quad A_v = 1$

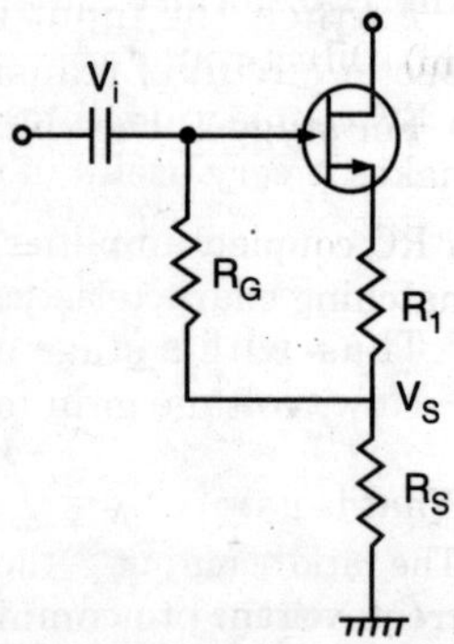

Fig. 8.20. Boot-strapping of R_G

A voltage gain of unity means that the output (at the source) follows the input (at the gate). Hence, the common drain configuration is called a source follower.

Bootstrapping of R_G

In order to reduce the loading effect of R_G, its lower end is connected to point V_x as shown in figure 8.20.

$$V_x = V_s \times \frac{R_s}{R_s + R_1}$$

$$= V_i \frac{g_m R_s}{1+g_m R_s} \times \frac{R_s}{R_s + R_1}$$

$$I_G = \frac{V_i - V_x}{R_G} = \frac{V_i}{R_G}\left(1 - \frac{g_m R_s}{1+g_m R_s} \times \frac{R_s}{R_s + R_1}\right)$$

$$R_G\text{ (effective)} = \frac{V_i}{I_G} = \frac{R_G}{1 - \frac{g_m R_s}{1+g_m R_s} \times \frac{R_s}{R_s + R_1}}$$

SUMMARY

- In order to achieve greater voltage and power gain, we must use more than one stage of amplification. Such an amplifier is known as a multistage implifier.
- A multistage amplifier using two or more single stage common emitter amplifier, is known as a cascade amplifier.
- A multistage amplifier with common emitter amplifier as the first stage and common base amplifier as the second stage is known as a cascode amplifier.
- The ratio of output to the input of an amplifier is known as gain of the amplifier.
- The gain is generally expressed in dB.
- A curve drawn between voltage gain and signal frequency of an amplifier is known as the frequency response of the amplifier.
- The range of frequency over which the gain of an amplifier is equal to or greater than 70.7% of its maximum gain is known as bandwidth of amplifier.
- The multistage amplifiers need coupling between their individual stages. This types of coupling is known as interstage coupling.
- Following are the four coupling, schemes which are often used in amplifiers :
  (i) Resistance-capacitance (RC) coupling
  (ii) Transformer coupling
  (iii) Direct-coupling
- A RC coupled amplifier has an excellent frequency response from 50 Hz to 20 kHz. This property makes it very useful in the initial stages of all the public address (PA) system.
- A RC coupled amplifier cannot be used as a final stage of the amplifier due to its poor resistance matching characteristics.

SHORT QUESTIONS WITH ANSWERS

Q.1. What is gain?

Ans. The ratio of output to the input of an amplifier is known as gain of the amplifier. The output quantity may be current, voltage or power. Generally, it is represented by letter 'G' or 'A'.

Q.2. What do you mean by Decibel?

Ans. the gain of an amplifier is the ratio of output signal to the input signal of an amplifier, where input and output are of similar quantities having same units. The telephone industry proposed a logarithmic unit named bel **(or decibel dB)** after Alexander Graham Bell.

Q.3. Explain the reasons for expressing the gain in dB.

Ans.

(i) When the gains are expressed in dB, the overall gains of a multi-stage amplifier can be determined by simply adding them. On the other hand, the absolute gain is obtained by multiplying the gains of individual stages.

(ii) Use of dB permits us to denote very large gains on linear scale by conveniently small figures on dB (logarithmic) scale. For instance, a power gain of 12000 is just 40.79 dB.

(iii) The output of most of the amplifiers is ultimately converted into sound. Experiments reveal the fact that our ears do not respond according to the intensity of sound but they respond according to the log of intensity of sound. Hence, if the intensity of sound given by speaker increases by 100 times then the hearing level will increase only by 2 times ($\log_{10} 100 = 2$). Since dB unit tallies with the natural response of human ears, therefore, it has become more popular compared to linear scales.

Q.4. What do you mean by frequency response?

Ans. A curve drawn between voltage gain and signal frequency of an amplifier is known as the frequency response of the amplifier.

Q.5. What do you mean by bandwidth of an amplifier?

Ans. The range of frequency over which the gain of an amplifier is equal to or greater than 70.7% of its maximum gain is known as bandwidth of amplifier.

Q.6. What are the requirements of an ideal coupling network?

Ans. An ideal coupling network should satisfy the following requirements.

(i) It must not disturb the d.c. bias conditions of the amplifiers being coupled. This means that the direct currents must not pass through the coupling network.

(ii) The coupling network must transfer ac signal waveform from one amplifier to the next amplifier without any distortion.

(iii) Although a small amount of voltage loss of signal cannot be avoided in the coupling network but this loss must be minimum.

(iv) The coupling network must offer equal impedance to the various frequencies of signal wave. Following are the four coupling schemes, often used in amplifiers.

Q.7. What is main advantage of an impedance coupled amplifier?

Ans. The main advantage of an impedance coupled amplifier is that it can be operated at low collector supply voltage, since there is hardly any drop across inductor (L). The drawback of an impedance coupled amplifier is that it is relatively expensive because of the use of inductors. In addition to this, the impedance of the inductor drops off at lower frequencies. Hence, as a rule, an impedance coupled amplifier is suitable only at frequency above 20 kHz (also called radio frequencies).

Q.8. What are the advantages of a transformer coupled amplifier?

Ans. Advantages :

(i) No signal power is lost in the collector or base resistors because of the low winding resistance of the transformer.

(ii) It provides a higher voltage gain than the RC coupled amplifier.

(iii) It provides an excellent resistance (or impedance) matching between the stage. The resistance matching is desirable for maximum power transfer.

Q.9. What are the drawbacks of transformer coupled amplifier?

Ans. Disadvantages :

(i) The coupling transformer is expensive and bulky, particularly when operated at audio frequencies.

(ii) At radio frequencies, the winding inductance and distributed capacitance produces reverse frequency distortion.

(iii) It tends to produce 'hum' in the circuit which is highly undesirable.

Q.10. Write the advantages of a direct coupled amplifier?

Ans. Advantages :

(i) The circuit arrangement is very simple because it uses a minimum number of resistors.

(ii) The circuit cost is low because of the absence of expensive coupling devices.

(iii) It can amplify very low frequency signals down to zero frequency.

Q.11. Write the drawbacks of a direct coupled amplifier?

Ans. Disadvantages :

(i) It cannot amplify high frequency signals.

(ii) It has a poor temperature stability. Because of this, its, Q-point shifts. In a multistage direct coupled amplifier, the Q-point shifts are amplified in succeeding stages. Thus a small d.c. shift, in the first stage can cause the final stage to be either saturated or cut-off. All integrated circuit amplifiers are direct coupled because of the difficulty of fabricating large integrated capacitors. It leads to special problems in their design.

Q.12. What are the applications of a direct coupled amplifier?

Ans. The direct coupled amplifier are used in several electronic systems that handle signals which change very slowly with time. Some of the important applications can be listed as under:

(i) Analog computation.

(ii) Power supply regulators.

(iii) Bioelectric measurements.

(iv) Linear integrated circuits (LICs).

REVIEW QUESTIONS

1. Why do you need more than one stage of amplifiers in practical circuits ?
2. Define and explain the following terms:
   (i) Decibel gain, (ii) Frequency response, (iii) Bandwidth.
3. We prefer to express the gain of an amplifier on a logarithmic scale rather than on a linear scale. State the reasons.
4. Define the unit decibel for expressing voltage, current and power gain.
5. What are the various methods of cascading a two stage transistor amplifier ? Discuss their relative advantages and disadvantages ?
6. Draw the circuit of an R-C coupled amplifier. Draw its gain v/s frequency characteristics and indicate cut-off frequencies and bandwidth.
7. Compare the various methods of coupling on the basis of frequency response and applications.
8. Why does the gain of an R-C coupled amplifier falls in
   (i) low frequency range, (ii) high frequency range
9. Draw the circuit diagram of a direct coupled amplifier. Explain its working. Discuss its applications.
10. What is the necessity of impedance matching in amplifiers ?
11. Draw a circuit diagram for a transformer coupled amplifier and explain its working.
12. Write a short note on the importance of half-power points in the frequency response curve of an amplifier.
13. How will you obtain impedance matching with transformer coupling ?
14. In a multistage amplifier, the input impedance of a amplifier stage should be very high and output impedance must be very low. Justify this statement.

NUMERICAL PROBLEMS

1. The three amplifier stages are cascaded to provide an overall gain of 10,000. The first two stages have a gain of 40 dB and 26 dB. Determine the gain of last stage in numerical values as well as in terms of dB.
   [Ans. 5 or 14 dB]
2. The three identical stages are cascaded and has a voltage gain of 120 dB. Determine the voltage gain in dB and number of each stage. **[Ans.** G_V = 40 dB ; A_V = 100]
3. A multi-stage amplifier consists of three stages. The voltage gain of the stages is 30, 50 and 80. Calculate the overall gain in dB. **[Ans.** 101.58 dB]
4. Determine the following gains in number
   (i) Power gain of 40 dB,
   (ii) Power gain of 43 dB **[Ans.** A_p = 10,000 ; A_p = 20,000]
5. A multistage amplifier contains five stages each stages each of which have a voltage gain of 100.
   (i) What is the total gain of the amplifier in dB?
   (ii) If a negative feedback of 20 dB is employed with it, find the resultant gain. **[Ans.** 200 dB ; 180 dB]
6. A multistage amplifier contains 4 similar stages. If each stage has R_C = 2.5 kΩ, R_{in} = 1.5 kΩ and β = 50, find the overall voltage and power gain. **[Ans.** G_V = 174 dB ; G_p = 155 dB]
7. An amplifier has voltage gain of 15 dB. If the input signal voltage is 0.8 V, what is the output voltage ?
   [Ans. v_{out} = 2.5 V]
8. In an amplifier, the maximum voltage gain is 2400 and occurs at 5 kHz. It falls to 1697 at 10 kHz and 20 kHz. Find (i) lower cut-off frequency, (ii) upper cut-off frequency, (iii) bandwidth.
   [Ans. A_2 = 200 ; A_1 = 66 ; A = 13,2000;
   A = 82.41 dB]

❑❑❑

UNIT IV

Operational amplifier characteristics, slew rate, full power bandwidth, offset voltage, bias current, application, inverting, non-inverting amplifier, summer, averager, differentiator, integrator, differential amplifier, instrumentation amplifier, log and antilog amplifier, voltage to current and current to voltage converters, comparators, Schmitt trigger, active filters, 555 timer and its application.

Above syllabus has been covered in following chapters :

Chapter 9

OPERATIONAL AMPLIFIER CHARACTERISTICS

Syllabus

Operational amplifier characteristics : slew rate, full power bandwidth, offset voltage, bias current.

Chapter Outline

- Introduction
- Basics of an Amplifier
- Review of Block Diagram of an Op-Amp
- Packaging and Pin Outs
- Circuit Symbol of Operational Amplifier
- OP-AMP Characteristics
- Equivalent Circuit of an Op Amp
- Equivalent circuit of an ideal op-amp
- Ideal Voltage Transfer Curve
- Open-Loop Op-Amp Configurations
- Limitations of Open-Loop Op-Amp Configurations
- Practical simplifying assumptions about op-amp
- Effect of input offset current on output voltage
- Power Supply Rejection Ratio (PSRR)
- Some Basic Aspects about open-loop op-amp
- Effect of Finite Open Loop Gain and Bandwidth On Circuit Performance
- Large Signal Operation of OP AMPs

9.1 INTRODUCTION

As a matter of fact, the operational amplifier was introduced in 1940s. The first operational amplifier was designed in 1948 using vacuum tubes. In early days, the op-amps found applications in analog computers to perform various mathematical operations such as addition, subtraction, multiplication etc. Because of the use of vacuum tubes, the early op-amps were quite bulky, power consuming and also expensive.

Fobrt J. Widlar at Fairchild brought out the popular 741 integrated circuit (IC) op-amp between 1964 to 1968. The IC version of op-amp made use of BJTs and FETs which can be fabricated along with the other supporting components, on a single semiconductor chip or wafer of very small size. With the use of IC op-amp, the circuit design has become very simple. A wide range of useful circuits can be designed without the necessity of knowing about the complex internal circuit. In addition, IC op-amps are inexpensive, take up less space and consume less power. Now-a-days, the IC op-amp has become an integral part of almost every electronic circuit which uses linear integrated circuit (LIC).

Also, the modern linear IC op-amp works at lower voltages. Because of their low cost, small size, versatility, flexibility, and dependability, op-amps are finding applications, now-a-days, in the fields of process control, communications, computers, power and signal sources, displays and measuring systems. Basically, the op-amp is an excellent high gain d.c. amplifier.

In this chapter, we shall discuss the basic characteristics of the operational amplifier along with the detailed discussion of various other aspects. We shall discuss the basic op-amp symbol and types of op-amp ICs. Before we start actual study of an op-amp, let us review the fundamental concepts of the amplification and the loading effect.

CHARLES AUGUSTIN COULOMB (1736-1806)

French natural philosopher *Charles Augustin Coulomb (1736-1806)* invented a method for measuring the force of attraction and repulsion between two electrically charged spheres. Coulomb established a law of inverse squares and defined the basic unit of charge quantity, the coulomb.

9.2 BASICS OF AN AMPLIFIER

An amplifier may be defined as a two port device, which accepts an input signal and produces an output signal proportional to the input. The proportionality constant between input and output is termed as **gain** of an amplifier. There are variety of amplifiers available depending upon the nature of the input and output signals. These types are the voltage amplifier (voltage input, voltage output), the current amplifier (current input, current output), the transconductance amplifier (voltage input, current output) and the transresistance amplifier (current input, voltage output). Let us discuss voltage amplifier since an op-amp is basically a voltage amplifier.

Let us consider a voltage amplifier having a constant gain A as shown in figure 9.1(a), supplying power to the load resistance R_L.

Let R_s = Source resistance

R_i = Input resistance of an amplifier

Figure 9.1 (b) shows an equivalent circuit of the input side.

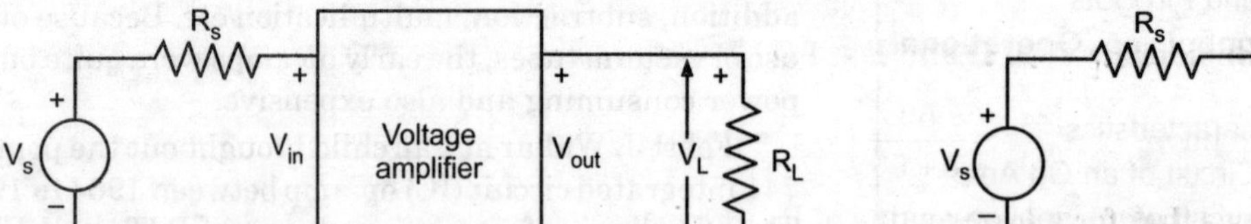

(a) (b)

Fig. 9.1.

According to figure 9.1 (b), the expression for V_{in} is given as*

$$V_{in} = \left(\frac{R_i}{R_s + R_i}\right) V_s \qquad \text{...(9.1)}$$

Now, if we substitute R_s = 50 Ω and R_i = 500 Ω, with V_s = 10 mV, then, we get

$$V_{in} = \left(\frac{500}{50 + 500}\right) 10 = 9.09 \text{ mV}$$

It is obvious from above that when source voltage is 10 mV, we get less voltage as an input to the amplifier.

Now, if R_i = 5 kΩ with same values of R_s, V_s as before, we obtain

$$V_{in} = \left(\frac{5 \times 10^3}{50 + 5 \times 10^3}\right) 10 = 9.9 \text{ mV}$$

* Here, we are making use of popular 'voltage-divider' rule.

In this case, input signal to the amplifier is almost same as that of source signal.

Important Point: Thus, from above, it is obvious that to get an input voltage same as that of source voltage, the input impedance of the amplifier must be as large as possible i.e., ideally infinite. This is one of the basic requirement of the op-amp that its input impedance must be as large as possible, as the op-amp is basically a voltage amplifier.

Now, let us consider the output side where the output voltage of the amplifier V_{out} is AV_{in}. Figure 9.1 (c) shows the equivalent circuit of the output side.

In this circuit, R_o = Output resistance of an amplifier.

Applying the voltage divider rule to the output circuit, we obtain

$$V_L = \left(\frac{R_L}{R_0 + R_L}\right)V_{out} \quad ...(9.2)$$

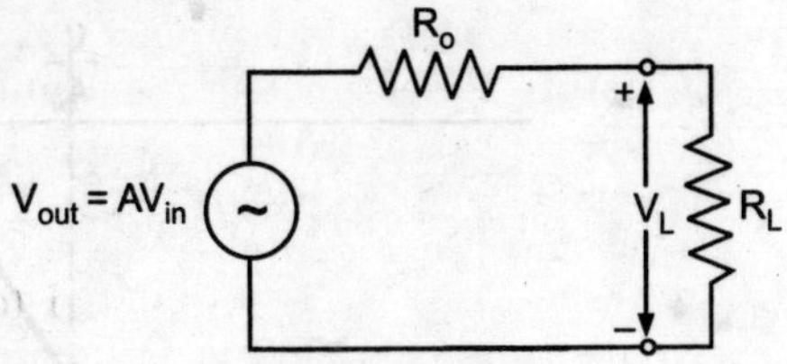

Fig. 9.1 (c)

Now, if we substitute $R_L = 50\ \Omega$ and $R_0 = 100\ \Omega$, then, we obtain

$$V_L = \left(\frac{50}{100 + 50}\right)V_{out} = 0.33\ V_{out}$$

Hence, we conclude that the actual load voltage is just 33.33% of the output voltage of the amplifier. This has happened due to the drop across the output resistance R_0.

Such an effect is known as **loading effect.** In other words, we can say that the amplifier output is loaded. In electronics circuits, loading always creates the problems since everytime as we change the source or the load, there has to be a change in the loading effect that alters the calculation of the overall gain of amplifier. Hence, it is essential to reduce the loading effect and if possible to eliminate it.

Let us consider $R_L = 50\ \Omega$ and $R = 1\ \Omega$

Then, we have $\quad V_L = \left(\frac{50}{1 + 50}\right)V_{out} = 0.9803\ V_{out}$

This shows that for a lower value of R_0, the loading effect reduces considerably and we get the load voltage same as the output voltage.

Important Point: *Hence, the output impedance of the amplifier must be as small as possible, ideally zero to eliminate the loading effect. This is another basic requirement of the op-amp. Thus, in short, we can say that an ideal voltage amplifier has $R_i = \infty$ i.e., no current flowing into or out of the input port, and $R_0 = 0$.*

Voltage Gain and Bandwidth (BW) of Amplifier

Further, the main objective of an amplifier is to provide a maximum voltage gain. Hence, ideal voltage amplifier has voltage gain of infinity. In addition, an amplifier must be able to amplify the signal without any distortion over the entire range of frequency. Such a range of frequency over which the amplifier performance is satisfactory is known as its **bandwidth.** Thus, an ideal voltage amplifier has infinite bandwidth.

The op-amp is a voltage amplifier having extremely high gain.

9.2.1. Amplifier Saturation

In practice, the transfer characteristic of an amplifier remains linear over only a limited range of input and output voltages. For an amplifier having two power supplies, the output voltage cannot exceed a specified positive value and cannot decrease below a specified negative value. The resulting transfer characteristic has been shown in figure 9.2, with the positive and the operation to be no longer restricted to an almost linear segment of the transfer curve. This in turn results in a distorted output signal waveform. Such non-linear distortion is undesirable. The output signal consists of additional spurious information which is not part of the input.

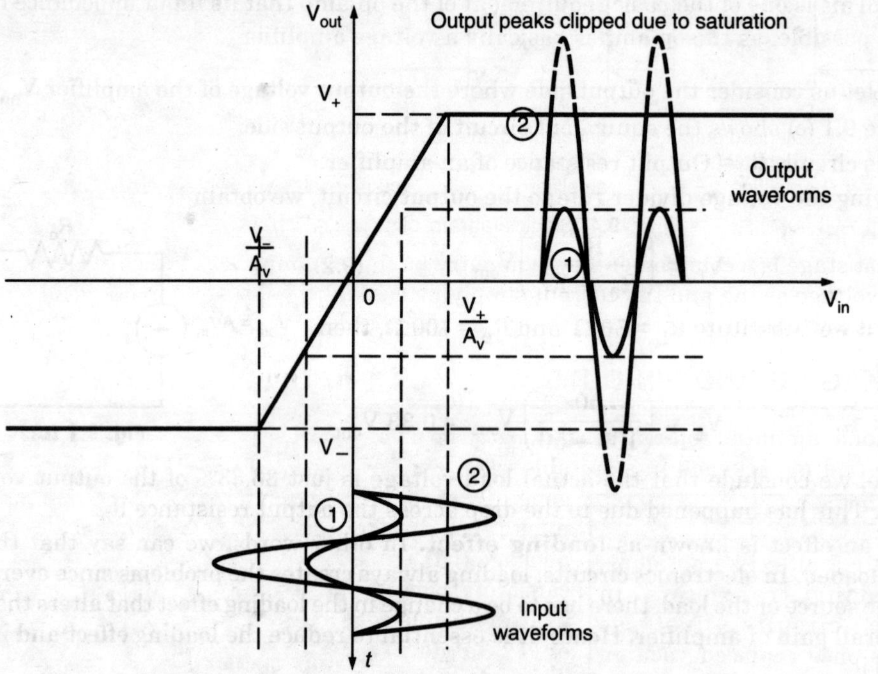

Fig. 9.2. An amplifier transfer characteristic that is linear except for output saturation

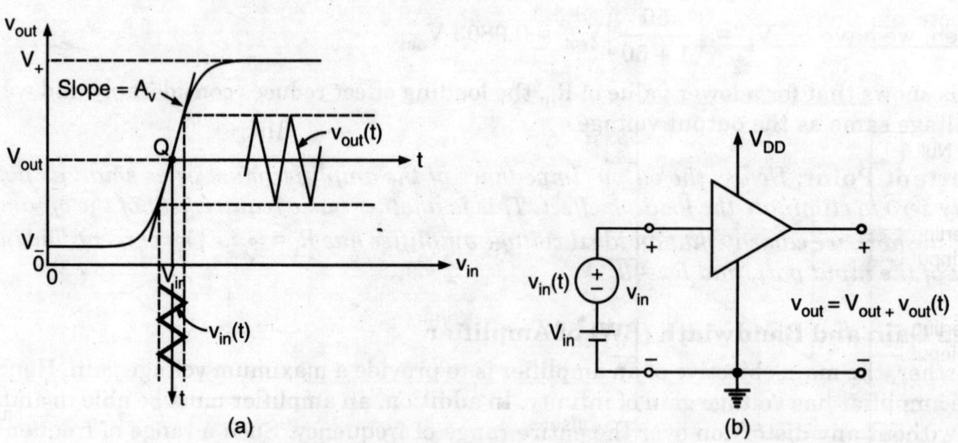

Fig. 9.3. (a) An amplifier transfer characteristic illustrating non-linearity, (b) Biasing of amplifier to obtain linear operation

9.3. REVIEW OF BLOCK DIAGRAM OF AN OP-AMP

The input stage is a dual-input, balanced output differential amplifier. This stage provides most of the voltage gain of the amplifier and also establishes the input resistance of the op-amp. In most of the amplifiers, an intermediate stage is a dual input, unbalanced output differential amplifier.

This stage increases the overall gain of the op-amp. The level shifting stage is used after the intermediate stage to shift the d.c. level at the output of the intermediate stage downward to zero volts with respect to ground. Usually this stage is an emitter follower using constant current source.

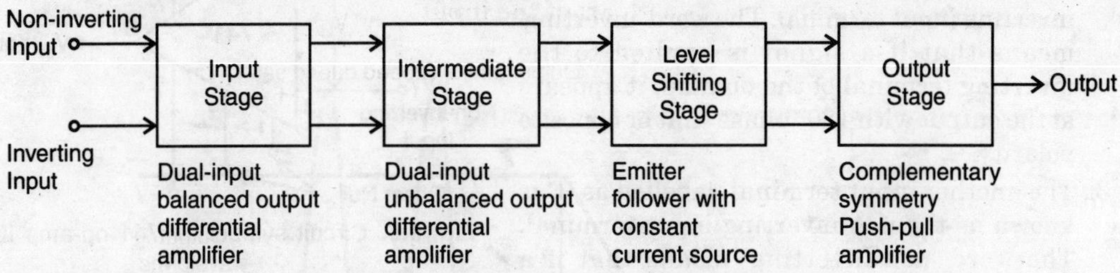

Fig. 9.4. Block diagram of a typical op-amp

The last stage is a complementary symmetry push-pull amplifier. The output stage increases the output voltage swing and the current supplying capability of the op-amp. A good output stage also provides low output resistance.

9.4. PACKAGING AND PIN OUTS

The operational amplifier (op-amp) is fabricated on a very small silicon chip and is packaged in a suitable case. The op-amp is generally available in two packages, dual-in-line package (DIP) and metal can. Metal cans (type TO-5) are available with 8, 10 or 12 leads. The DIP packages of op amps have 8 or 14 pins. The DIP package is most widely used.

> **DO YOU KNOW?**
>
> Integrated circuits, like op-amps, are replacing transistors in electronic circuits, just as transistors once replaced vacuum tubes. Op amps and linear ICs, however, are actually microelectronic circuits.

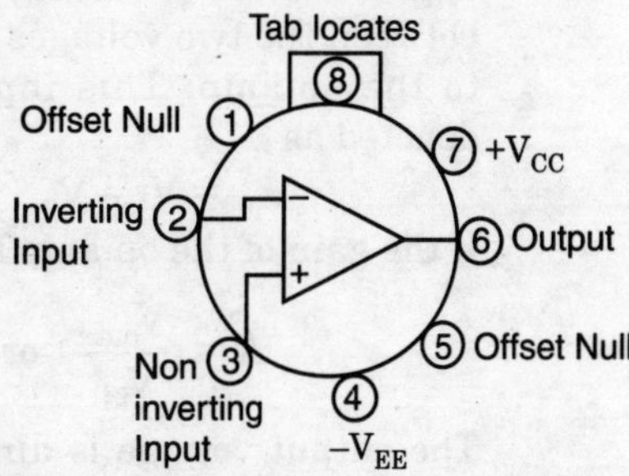

Fig. 9.5. Metal can (TO-5) package

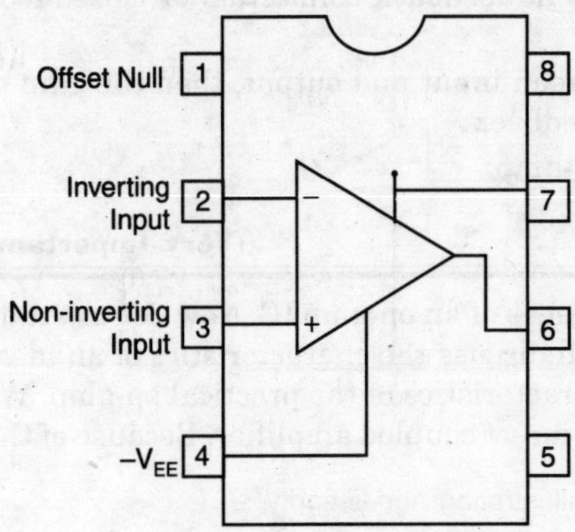

Fig. 9.6. 8-pin DIP package

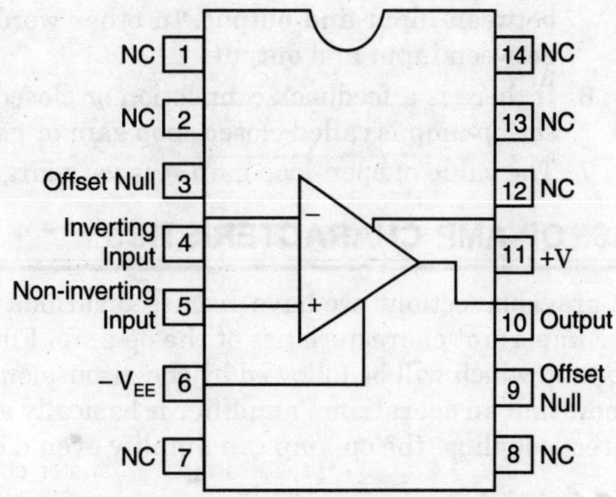

Fig. 9.7. 14-Pin DIP Package

9.5. CIRCUIT SYMBOL OF OPERATIONAL AMPLIFIER (U.P. Tech., Tutorial Question Bank)

Figure 9.8 shows the circuit symbol for 741 op-amp IC.

From this symbol, we can draw few important points about an op-amp as listed ahead:

1. The op-amp is provided with two input terminals.
2. One of the terminals, labelled '–' is known as inverting input terminal. The word 'inverting' means that if a signal is applied to the inverting terminal of the op-amp, it appears at the output with 180° phase shift or opposite polarity.
3. The another input terminal, labelled as '+' is known as the non-inverting input terminal. The word 'non-inverting' means that if a signal is applied at this terminal of the op-amp it appears at the output without any phase shift.

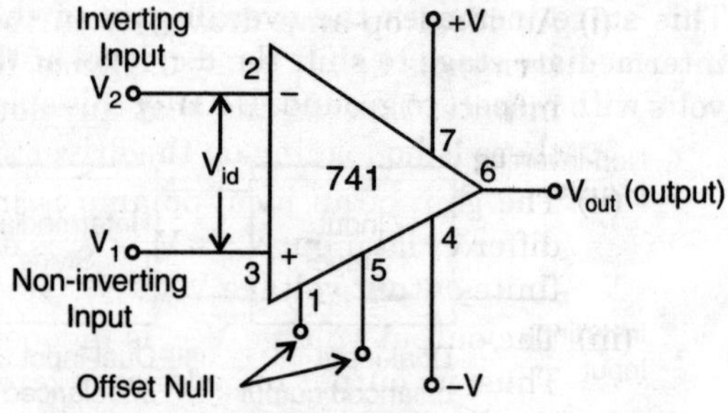

Fig. 9.8. Circuit symbol for 741 op-amp IC

4. The op-amp amplifies the difference between the voltages applied at the non-inverting input and inverting input. In figure 9.8, V_1 is voltage applied at the non-inverting terminal and V_2 is the voltage applied at the inverting terminal. The difference between the two voltages $(V_1 - V_2)$ acts as an input to the op-amp. This input difference voltage is denoted as

$$V_{id} = V_1 - V_2$$

If the gain of the op-amp is A, then

$$A = \frac{V_{out}}{V_{id}} \quad \text{or} \quad V_{out} = A.\ V_{id}$$

The output voltage is directly proportional to the algebraic difference between the two input voltages. Hence, the op-amp amplifies the difference between the two input voltages. It does not amplify the input voltages themselves.

DO YOU KNOW?

Although the 741 is usually connected to both a positive and a negative supply voltage, the op amp can still be operated using a single supply voltage. For example, the $-V_{EE}$ input could be grounded, and the $+V_{CC}$ input could be connected to a positive dc supply voltage.

5. Gain A is called open-loop gain of the op-amp because in figure 9.8, there is an open loop between input and output. In other words, there is no feedback connection or closed-loop between input and output.
6. If there is a feedback connection or closed-loop between input and output, then the gain of the op-amp is called closed-loop gain or gain with feedback.
7. The value of open-loop gain A is very large for an op-amp.

9.6. OP-AMP CHARACTERISTICS **(Very Important)**

In previous sections, we have discussed various internal stages of an op-amp IC. Now, let us study the important characteristics of the op-amp. First, we shall discuss the characteristics of an ideal op-amp which will be followed by the discussion of the characteristics of the practical op-amp. We know that an operational amplifier is basically a high gain direct coupled amplifier. Because of the direct coupling, the op-amp can amplify even d.c. voltages.

9.6.1. Ideal Op-amp

Figure 9.9 shows an ideal op-amp. An ideal op-amp has two input signals V_1 and V_2 applied to non-inverting and inverting terminals, respectively. The two input currents namely I_1 and I_2 are shown zero.

Few Important Points

The following points may be noted for the ideal op-amp shown in figure 9.9 :

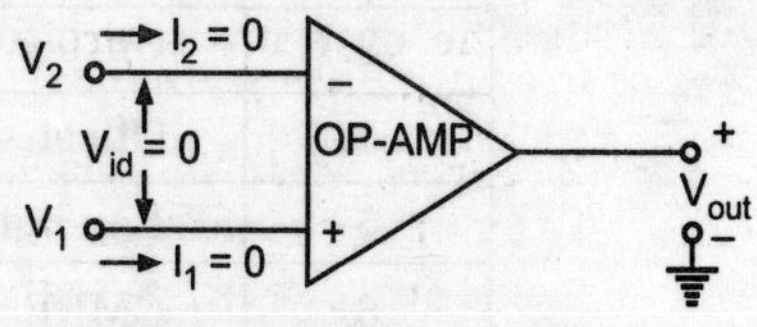

Fig. 9.9. Ideal op-amp

(i) An ideal op-amp draws no current at both the input terminals i.e., $I_1 = I_2 = 0$. Hence, its input impedance is infinite. This means that any source can drive it and there is no loading on the driver stage.

(ii) The gain of an ideal op-amp is infinite (∞), hence the differential input $V_{id} = V_1 - V_2$ is essentially zero for the finite output voltage V_{out}.

(iii) The output voltage V_{out} is independent of the current drawn from the output terminals. Thus, its output impedance is zero and hence output can drive an infinite number of other circuits.

These properties are expressed generally as the characteristics of an ideal op-amp.

9.6.1.1. Characteristics or Properties of an Ideal op-amp

The various characteristics or properties of an ideal op-amp may be listed as under :

1. Infinite voltage gain

It is denoted as A. It is the differential open loop gain and is infinite for an ideal op-amp.

2. Infinite input impedance

The input impedance is denoted as R_i and is infinite for an ideal op-amp. This ensures that no current can flow into an ideal op-amp.

3. Zero output impedance

The output impedance is denoted as R_0 and is zero for an ideal op-amp. This ensures that the output voltage of the op-amp remains same, irrespective of the value of the load resistance connected.

4. Zero offset voltage

The presence of the small output voltage though $V_1 = V_2 = 0$ is called as an offset voltage. It is zero for an ideal op-amp. This ensures zero output for zero input signal voltage.

5. Infinite bandwidth

The range of frequency over which the amplifier performance is satisfactory is called as its bandwidth. The bandwidth of an ideal op-amp is infinite. This means that the operating frequency range is from 0 to ∞. This ensures that the gain of the op-amp remains constant over the frequency range from d.c. (zero frequency) to infinite frequency. Therefore, an op-amp can amplify d.c. as well as a.c. signals.

6. Infinite CMRR

The ratio of differential gain and common mode gain is defined as CMRR. Thus, infinite CMRR of an ideal op-amp ensures zero common mode gain. Because of this, common mode noise output voltage is zero for an ideal op-amp.

7. Infinite slew rate

This ensures that the changes in the output voltage occur simultaneously with the changes in the input voltage.

Table 9.1 summarizes the characteristics of an ideal op-amp

Table 9.1. Ideal op-amp characteristics.

S.No.	Characteristics	Representation	Values
1.	Open loop voltage gain	A	∞
2.	Input impedance	R_i	∞

3.	Output impedance	R_o	0
4.	Offset voltage	V_{oos}	0
5.	Bandwidth	B.W.	∞
6.	CMRR	ρ	∞
7.	Slew rate	S	∞

9.6.2. Characteristics of Practical Op-Amp

The characteristics of an ideal op-amp can be approximated closely enough, for many practical op-amps. However, basically, the practical op-amp characteristics are little bit different than the ideal op-amp characteristics. Let us study the equivalent circuit of a practical op-amp which has finite output and input resistances and a finite open loop voltage gain. Figure 9.10 shows the equivalent circuit of a practical op-amp.

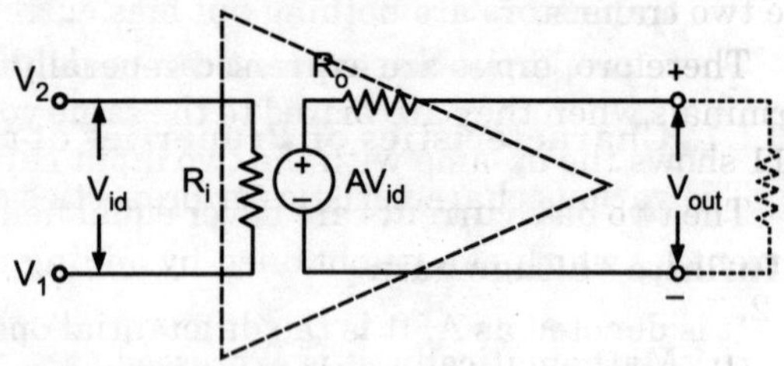

Fig. 9.10. Equivalent circuit of practical op-amp

In the practical op-amp, input impedance is not infinite and output impedance is not zero.

The output voltage is given by

$$V_{out} = AV_{id} = A(V_1 - V_2)$$

where A = open loop gain with open output terminals

The op-amp always amplifies the difference between the two input voltages.

The various characteristics or properties of a practical op-amp can be described as under :

1. Open loop gain

It is the voltage gain of the op-amp when no feedback is applied. Practically, it is several thousands.

2. Input impedance

It is finite and typically greater than 1 MΩ. However, if we use FETs for the input stage, it can be increased upto several hundred MΩ.

3. Output impedance

It is typically few hundred ohms. With the application of negative feedback, it can be reduced to a very small value like 1 or 2 ohms.

4. Bandwidth

The bandwidth of a practical op-amp in open loop configuration is very small. But, if we apply negative feedback, it can be increased to a desired value.

5. Input offset voltage

Whenever both the input terminals of the op-amp are grounded, ideally, the output voltage must be zero. However, in this condition, the practical op-amp always shows a small non zero output voltage. To bring this output voltage to zero, a small voltage in milli volts is needed to be applied to one of the input terminals. Such a voltage makes the output exactly zero. This d.c. voltage which makes the output voltage zero, when the other terminal is grounded is known as **input offset voltage** and it is V_{ios}. How much voltage, to which terminal and with what polarity, to be applied, is specified by the manufacturer in the data sheet. The input offset voltage depends upon the temperature.

6. Input bias current

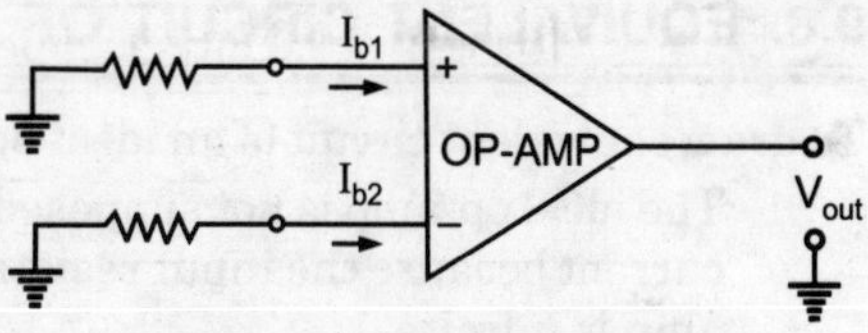

Fig. 9.11. Input bias currents

For ideal op-amp, no current flows into the input terminals. The practical op-amps do have some input currents which are very small, of the order of 10^{-6} A to 10^{-14} A. Most of the op-amps use differential amplifier as the input stage. The two transistors of the differential amplifier must be biased correctly. However, in practice, it is not possible to get exact matching of the two transistors. Hence, the inputs terminals which are the base terminals of the two transistors, conduct the small d.c. current. These small base currents of the two transistors are nothing but bias currents represented by I_{b1} and I_{b2}.

Therefore, input bias current can be defined as the current flowing into each of the two input terminals when they are biased to the same voltage level i.e., when the op-amp is balanced. Figure 9.11 shows the op-amp with the two input currents, when op-amp is balanced.

The two bias currents are never equal hence the manufacturers specify the average input bias current I_b, which can be obtained by adding the magnitudes of I_{b1} and I_{b2} and dividing the sum by 2.

Mathematically, it is expressed as,

$$\text{Input bias current } I_b = \frac{|I_{b1}| + |I_{b2}|}{2} \qquad ...(9.3)$$

7. Input offset current

The difference in magnitudes of I_{b1} and I_{b2} is known as a input offset current and is denoted as I_{ios}. Hence, we have

Input offset current $I_{ios} = | I_{b1} - I_{b2} |$

The magnitude of this current is very small, of the order of 20 to 60 nA. It is measured under the condition that input voltage to op-amp is zero. If we supply equal d.c. currents to the two inputs, output voltage of op-amp must be zero. However, in practice, there exists some voltage at the output terminal to make it zero, the two input currents are made to differ by small amount. This difference is nothing but the input offset current. Both input bias as well as input offset currents depend upon the temperature.

> **DO YOU KNOW**
>
> The utility of an op-amp can be greatly increased by providing negative feedback. The output in this case is not driven into saturation and the circuit behaves in a linear manner.

9.7. EQUIVALENT CIRCUIT OF AN OP AMP

Figure 9.12 shows equivalent circuit of an op-amp. The equivalent circuit of an op-amp is helpful in analyzing the operating principle of an op-amp and in observing the effects of feedback. The voltage source $A.V_{id}$ is an equivalent Thevenin voltage source and R_o is the Thevenin equivalent resistance looking back into the output terminal of the op-amp.

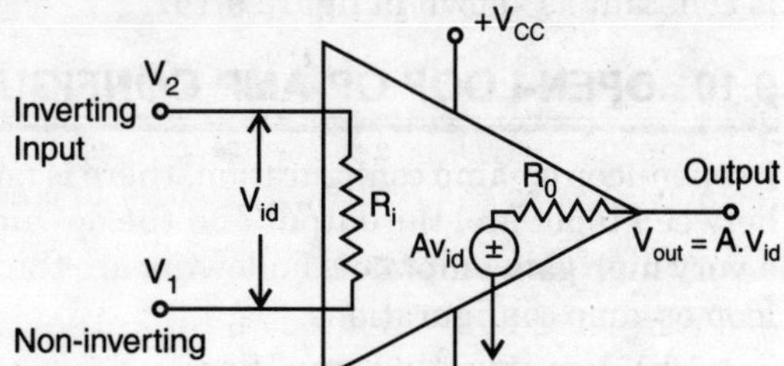

Fig. 9.12. Equivalent circuit of an op-amp

The output voltage is

$$V_{out} = A.V_{id}$$

But $\quad V_{id} = V_1 - V_2$

Therefore, $\quad V_{out} = A(V_1 - V_2)$

where A = large-signal voltage gain

V_1 = voltage at the non-inverting terminal

V_2 = voltage at the inverting terminal

V_{id} = difference input voltage.

9.8. EQUIVALENT CIRCUIT OF AN IDEAL OP-AMP

To draw equivalent circuit of an ideal op-amp, we first notes few points about ideal op-amp as follows:

1. The ideal op-amp is not supposed to draw any input current because the input resistance of an ideal op-amp is infinite.

   Since, $V = IR$

   Therefore, $I = \frac{V}{R}$

   or $I = \frac{V}{\infty} = 0$

   Hence, infinite resistance or zero current means there is an open-circuit between the two terminals.

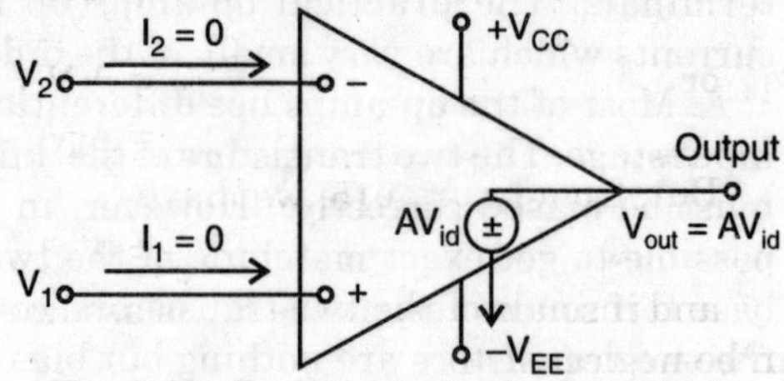

Fig. 9.13. Equivalent circuit of an ideal op-amp

2. For an ideal op-amp, output resistance $R_0 = 0$ means that the output voltage V_{out} is not dependent on the load resistance connected across the output. Since output resistance R_0 is zero, it is replaced by a short-circuit in the output portion of the equivalent circuit. With the help of above two points, the equivalent circuit of an ideal op-amp may be drawn as shown in figure 9.13.

9.9. IDEAL VOLTAGE TRANSFER CURVE

(RGTU, Tutorial Question Bank)

A voltage transfer curve may be defined as a curve which is drawn between input and output voltages. For op-amp, the input voltage is V_{id} and output is V_{out}. For an op-amp we know that

$$V_{out} = AV_{id} = A(V_1 - V_2)$$

Graphical representation of this equation is called the transfer curve of an op-amp because the output voltage is plotted against input difference voltage V_{id}, keeping gain A constant. This transfer curve for ideal op-amp is plotted in figure 9.14. From curve, it is clear that output voltage cannot exceed the positive and negative saturation voltage. Positive and negative saturation voltages are specified for given value of supply voltages. From curve, it is also observed that output voltage is proportional to input difference voltage only before saturation and after saturation output is constant as shown in figure 9.14.

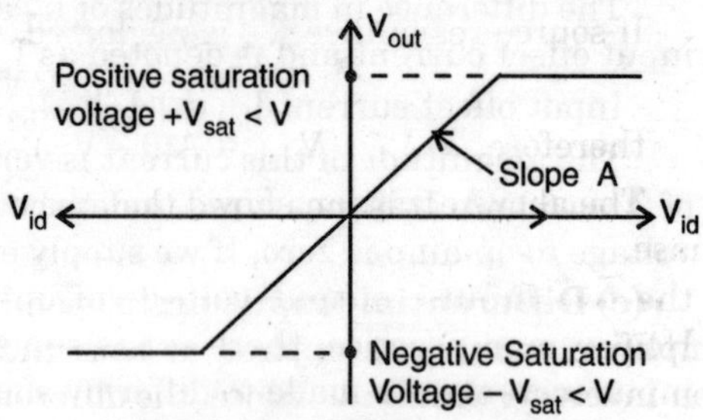

Fig. 9.14. Transfer curve

9.10. OPEN-LOOP OP-AMP CONFIGURATIONS

In open-loop op-amp configuration, there is no connection between input and the output and the op- amp works as a very high gain amplifier. Following are the three open-loop op-amp configurations:

(a) Non-inverting amplifier

(b) Inverting amplifier

(c) Differential amplifier.

(a) Non-inverting amplifier : Figure 9.15 shows the open-loop non-inverting amplifier. In non-inverting configuration, input is applied at the non-inverting terminal of the op-amp and inverting terminal is grounded. Also the output of non-inverting amplifier is in phase with the input.

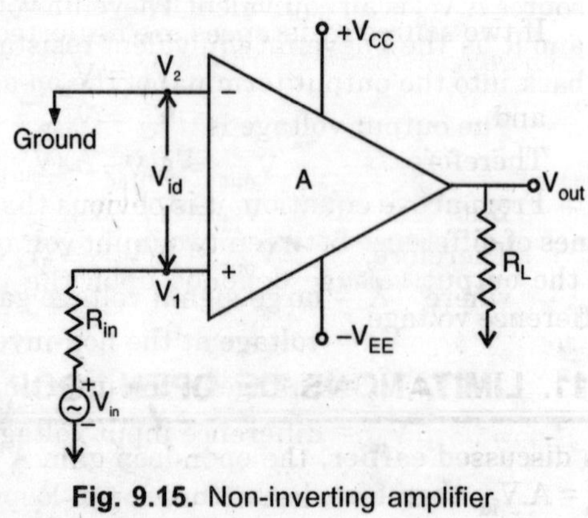

Fig. 9.15. Non-inverting amplifier

We know that for an op-amp

$$\text{open-loop gain } A = \frac{V_{out}}{V_{id}}$$

or
$$V_{out} = A.V_{id}$$
$$V_{out} = A(V_1 - V_2)$$

But, from figure 9.15, we have

$$V_2 = 0$$

and if source resistance R_{in} is assumed very small then it can be neglected.

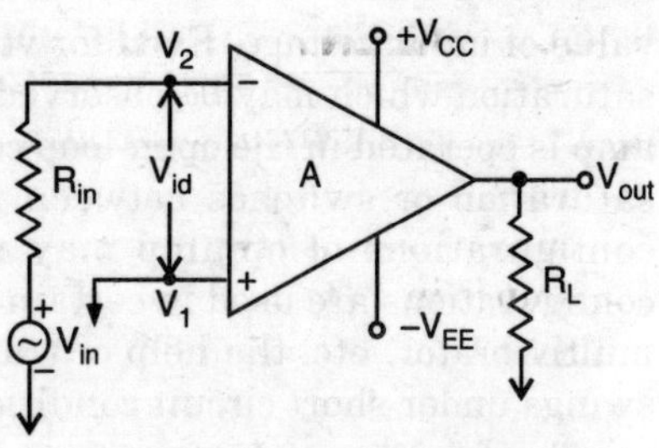

Fig. 9.16. Inverting amplifier

Therefore, $V_1 = V_{in}$

So that, $V_{out} = A(V_{in} - 0) = A.V_{in}$

The above equation shows that the output is A times larger than input and is in phase with the input

(b) Inverting amplifier: In inverting configuration, the input is applied at the inverting terminal and the non-inverting terminal is grounded. Also, the output of inverting amplifier is out of phase with the input.

$$\text{We know that } A = \frac{V_{out}}{V_{id}}$$

$$V_{out} = A.V_{id} = A(V_1 - V_2)$$

If source resistance R_{in} is neglected, then

$$V_2 = V_{in} \text{ and } V_1 = 0$$

therefore, $V_{out} = A(0 - V_{in}) = -AV_{in}$

The above equation shows that the output is A times larger than the input and is opposite in phase.

(c) Differential amplifier: In open-loop differential amplifier, inputs are applied at both the inverting and non-inverting terminals. Since, the difference between the two input signals is amplified, the configuration is called the differential amplifier.

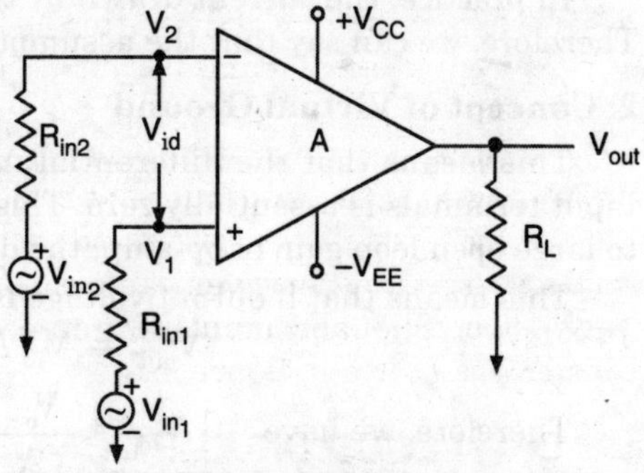

Fig. 9.17. Differential amplifier

We know that
$$A = \frac{V_{out}}{V_{id}}$$
$$V_{out} = A.V_{id}$$
$$V_{out} = A.(V_1 - V_2)$$

If two source resistances are neglected, then

$$V_1 = V_{in1}$$

and $V_2 = V_{in2}$

Therefore, $V_{out} = A.(V_{in1} - V_{in2})$

From above equation, it is obvious that the output is A times of difference between two input voltages. The polarity of the output voltage depends upon the polarity of input difference voltage.

DO YOU KNOW?

Basically, any op-amp circuit that does not use negative feedback is considered too unstable to be useful.

9.11. LIMITATIONS OF OPEN-LOOP OP-AMP CONFIGURATIONS

As discussed earlier, the open-loop gain A is very large and since the output voltage is given as $V_o = A.V_{id}$, therefore, in the three open-loop op-amp configurations, we can apply only a very small value of input voltage. Even for voltage levels slightly greater than zero, the output is driven into saturation which may be observed from ideal transfer characteristics of op-amp. Hence, when op-amp is operated in the open-loop configuration, the output is either negative saturation or positive saturation or switches between positive and negative saturation levels. Therefore, open-loop configurations of op-amp may not be used in linear applications.* But open-loop op-amp configurations are used in certain non-linear** applications like square-wave generation, a stable multivibrator, etc. the help of transistors is preferred which has no effect on the output voltage swings under short circuit condition.

EXAMPLE 9.1. The bias currents for the emitter coupled transistors of a differential amplifier are 18 μA and 22 μA. Determine:

(i) Input bias current, (ii) Input offset current for an op-amp.

Solution: The two input bias currents are as under :

$$I_{b1} = 18\ \mu A \quad \text{and} \quad I_{b2} = 22\ \mu A$$

(i) We know that the input bias current is given by

$$I_b = \frac{I_{b1} + I_{b2}}{2} = \frac{18+22}{2} = 20\ \mu A \quad \textbf{Ans.}$$

(ii) Also, the input offset current is given as

$$I_{ios} = |\ I_{b1} - I_{b2}\ | = |\ 18 - 22\ | = 4\ \mu A \quad \textbf{Ans.}$$

9.12. PRACTICAL SIMPLIFYING ASSUMPTIONS ABOUT OP-AMP

We shall make two assumptions which are realistic and simplify the analysis of op-amp circuits upto a great extent.

1. Zero input current

The current drawn by either of the input terminals (inverting and non-inverting) is zero.

In practice, the current drawn by the input terminals is very small, of the order of μA or nA. Therefore, we can say that the assumption of zero input current is realistic.

2. Concept of Virtual Ground

This means that the differential input voltage V_{id} between the non-inverting and inverting input terminals is essentially zero. This is obvious because even if output voltage is few volts, due to large open loop gain of op-amp, the difference voltage V_{id} at the input terminals is almost zero.

This means that if output voltage is 10 V and the A i.e. open loop gain is 10^4 then, we have

$$V_{out} = V_{id} A$$

Therefore, we have

$$V_{id} = \frac{V_{out}}{A} = \frac{10}{10^4} = 1\ \text{mV}$$

Hence, V_{id} is very small. As $A \to \infty$, the difference voltage $V_{id} \to 0$ and realistically assumed to be zero for analysing the circuits.

Thus, we write

$$V_{id} = \frac{V_{out}}{A}$$

or

$$(V_{in1} - V_{in2}) = \frac{V_{out}}{\infty} = 0$$

* Linear applications means that those applications in which there is a linear relationship between input and output.

** Since in saturation, a device acts as a non-linear device.

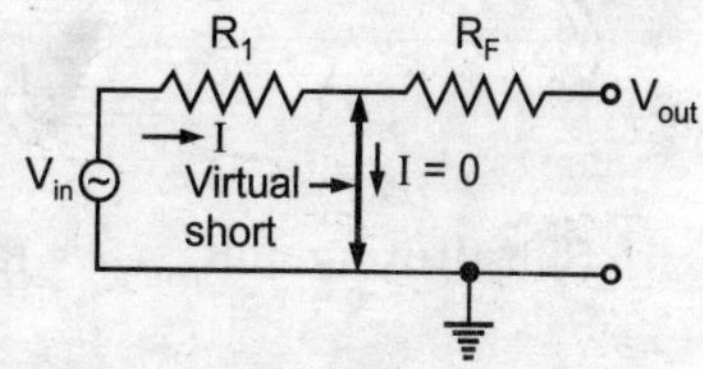

Fig. 9.18. Concept of virtual ground in an op-amp

or $$V_{in1} = V_{in2} \quad ...(9.4)$$

Thus, we can say that under linear range of operation, there is virtually short circuit between the two input terminals, in the sense that their voltage are same. No current flows from the input terminals to the ground. Figure 9.18 shows the concept of the virtual ground. The thick line indicates the virtual short circuit between the two input terminals.

Now, if the non-inverting terminals is grounded, by the concept of virtual short, the inverting terminal is also at ground potential, though there is no physical connection between the inverting terminal and the ground. This is the **principle of virtual ground.**

Thus, from the equation (9.4), **the voltage at the non inverting input terminal of an op-amp can be realistically assumed to be equal to the voltage at the inverting input terminal.**

The concept of virtual ground, helps in determining the closed loop gain, in various practical circuits.

9.13. EFFECT OF INPUT OFFSET CURRENT ON OUTPUT VOLTAGE

Although the effect of input bias current is compensated with R_{comp}, the input offset current I_{ios} still contributes the output voltage. Let us discuss the effect of input offset current on the output voltage.

Let us consider the op-amp used in the closed loop configuration with R_{comp} as shown in figure 9.19.

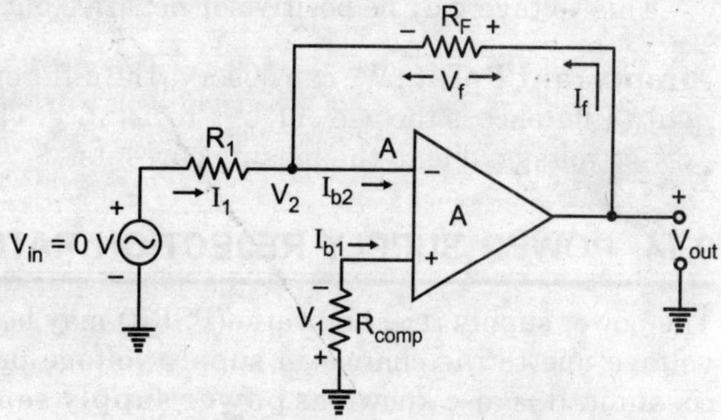

Fig. 9.19. Effect of input offset current.

Neglecting V_{ios}, we can write,

$$V_1 = I_{b1}R_{comp} \quad ...(9.5)$$

$$I_1 = \frac{V_2}{R_1} \quad ...(9.6)$$

But we know that $V_2 = V_1$ *

Therefore, $$I_1 = \frac{V_1}{R_1} = \frac{I_{b1}R_{comp}}{R_1} \quad ...(9.7)$$

Now, $$I_f = I_{b2} - I_1 \quad ...(9.8)$$

$\therefore$ $$I_f = I_{b2} - \frac{I_{b1}R_{comp}}{R_1} \quad ...(9.9)$$

DO YOU KNOW?

Inputs offset current results due to non-identical transistors and is minimized by using a T-network in the feedback path.

Applying KVL to input feedback-output loop, we obtain

$$-V_2 + V_f - V_{out} = 0 \quad ...(9.10)$$

or $$V_{out} = V_f - V_2 = V_f - V_1 \text{ as } V_1 = V_2$$

or $$V_{out} = I_f R_F - I_{b1} R_{comp} \quad ...(9.11)$$

Substituting I_f, we obtain

$$V_{out} = \left[I_{b2} - \frac{I_{b1}R_{comp}}{R_1} \right] R_F - I_{b1}R_{comp}$$

or $$V_{out} = \frac{I_{b2}R_1R_F - I_{b1}R_{comp}R_F - I_{b1}R_{comp}R_1}{R_1} \quad ...(9.12)$$

* Because of virtual short-circuit.

or $$V_{out} = I_{b2}R_F - I_{b1}\frac{R_{comp}}{R_1}(R_F + R_1)$$

Substituting $R_{comp} = R_1 \parallel R_F = \dfrac{R_1R_F}{R_1 + R_F}$, we have

$$V_{out} = I_{b2}R_F - I_{b1}.\frac{R_1R_F}{(R_1+R_F)}.\frac{1}{R_1}.(R_F+R_1) \quad ...(9.13)$$

or $$V_{out} = I_{b2}\,R_F - I_{b1}R_F \quad ...(9.14)$$

or $$V_{out} = R_F\,(I_{b2} - I_{b1})$$

or $$V_{out} = R_F\,I_{ios} \quad ...(9.15)$$

Thus, there exists an output voltage due to input offset current I_{ios}.

This means that to reduce the effect of input offset current, the feedback resistance R_F must be selected small. The **total output offset voltage,** for an opa-amp which is compensated for input bias current using R_{comp} is given by

$$V_{oos} = V_{ios}\left(1+\frac{R_F}{R_1}\right) + R_FI_{ios} \quad ...(9.16)$$

DO YOU KNOW?

The input offset voltage and current drift with the change in temperature, thereby producing a d.c. error voltage in the output.

This voltage may be positive or negative but its value must be as small as possible.

Important Point : We can observe the difference between the equations (9.15) and (9.16). The only difference is the use of R_{comp} resistance. Thus, $I_{ios} < I_b$, using R_{comp}, we get reduced output offset voltage, due to the biasing currents.

9.14. POWER SUPPLY REJECTION RATIO (PSRR)

The power supply rejection ratio (PSRR) may be defined as the ratio of the change in input offset voltage due to the change in supply voltage producing it, keeping other power supply voltage constant. It is also known as **power supply sensitivity (PSV).**

Now, if V_{EE} is constant and due to certain change in V_{CC}, there is change in input offset voltage then PSRR is given by

$$PSRR = \frac{\Delta V_{ios}}{\Delta V_{CC}}$$

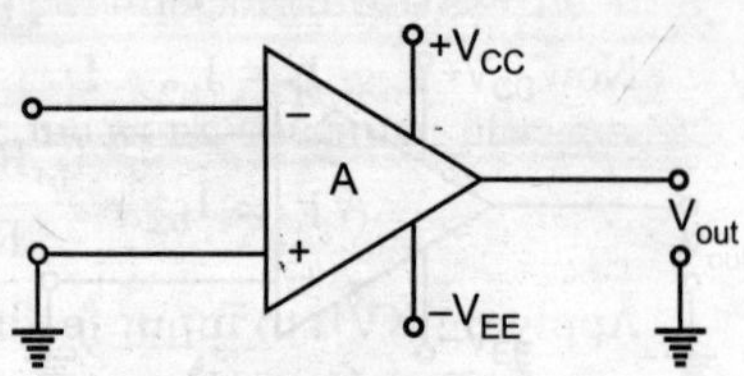

Fig. 9.20.

For a fixed V_{CC}, if there is a change in V_{EE}, then we write

$$PSRR = \frac{\Delta V_{ios}}{\Delta V_{EE}}$$

Since input offset voltage is very small, PSRR is expressed in mV/V or μV/V.

The typical value of PSRR for IC 741 op-amp is 30 μV/V.

9.15. SOME BASIC ASPECTS ABOUT OPEN-LOOP OP-AMP

1. Open-loop op-amp

An open-loop op-amp offers very high voltage gain [ranging from 10^5 to 10^6 at dc (or 0 Hz)] and narrower bandwidth (band of frequencies for which the gain remains constant-almost zero. Besides being large, the voltage gain of the open-loop op-amp is not a constant. The voltage gain varies with variations in temperature and power supply voltage as well as mass production techniques.

2. Open-Loop Op-Amp are not Suitable for Linear Applications

Because of these features, the open-loop op-amps are not suitable for many linear applications as explained below:

(i) Because of the very high voltage gain of open-loop op-amp, only the smaller signals (of the order of microvolts or less) having very low frequency may be amplified accurately without distortion. For instance, if the input signal is of 1 mV, then output voltage will be equal to $1 \times 10^5 \times 1$ mV i.e., 100 V (assuming open-loop gain = 10^5). Such a large output cannot be obtained because the output from the op-amp can never exceed the saturation levels $\pm V_{sat}$. However, the signals of small amplitudes are very susceptible to noise and almost impossible to obtain in the laboratory.

(ii) The large variations in voltage gain of an open-loop op-amp makes it unsuitable for many linear applications.

(iii) Because of narrower bandwidth, the open-loop op-amp is impractical in a.c. application. For instance, the open-loop bandwidth of the 741 IC op-amp is roughly 5 Hz but in almost all a.c. applications, a bandwidth exceeding 5 Hz is required.

3. Remedy

However, the gain of the op-amp can be controlled and stabilized and the bandwidth can be increased if necessary modification is introduced in the op-amp basic circuit. This modification involves the use of feedback. The process of combining a fraction of output energy (voltage or current) back to the input is called feedback. When feedback voltage (or current) is so applied as to increase the input signal (i.e., when the feedback is in phase with input signal), it is called the *positive* or *direct* or *regenerative feedback.* Since positive feedback causes excessive distortion and instability, it is seldom used in amplifiers. However, because of its capability of increasing the power of the original signal, it is used in oscillator circuits.

4. Negative feedback and its features

When feedback voltage (or current) is so applied as to weaken the input signal (i.e., when feedback is in phase opposition to the input signal), it is called negative, inverse or degenerative feedback. Negative feedback reduces the amplifier gain but it has numerous advantages (such as gain stability, reduction in non-linear distortion, reduction in noise, increase in bandwidth or improvement in frequency response, increase in input impedance and decrease in output impedance) and is frequently used in amplifier circuits. Negative feedback also reduces the effects of input offset voltage and variations in temperature and supply voltages on the output of the op-amp.

9.16. EFFECT OF FINITE OPEN LOOP GAIN AND BANDWIDTH ON CIRCUIT PERFORMANCE

In this topic, we have assumed that the op-amp is ideal. In many applications, such an assumption is all right but as a designer we must know the characteristics of a practical op-amp and the effects of such characteristics on the performance of op-amp circuits. In this section we will consider some of the important non-ideal characteristics of the op-amp.

9.16.1. Frequency Dependence of the Open Loop Gain

Some of the characteristics of an op-amp are frequency dependent. One of them is the open loop gain A_v. The open loop gain is ideally supposed to be constant but in practice it decreases with increase in frequency.

9.16.2. Frequency Response

The op-amp open loop gain is a complex number. So at a given frequency, it will have a specific magnitude and phase angle. The magnitude and phase angle, both will change with change in frequency. The variation in gain plotted against frequency is called as frequency response The gain may be expressed in dB or numerical value or in dB but the frequency is always plotted on the log scale.

9.16.3. Compensating Networks

For an amplifier, the increase in frequency will have two effects:

(a) Reduction in gain

(b) Increase in phase shift between input and output.

In case of an op-amp these changes are due to the internal and stray capacitances. Such capacitances correspond to the physical characteristics of semiconductor devices and the constructional features of an op-amp.

9.16.3.1. Magnitude Plot

It is a graph of op-amp gain plotted on the y-axis versus frequency plotted on the x-axis.

9.16.3.2. Phase Angle Plot

It is a graph of phase shift on the y-axis versus frequency plotted on the x-axis.

9.16.3.3. Compensating Network

The magnitude and phase plots of an op-amp may be altered as per requirements by using external resistors and capacitors alongwith it. The network formed by such components is known as **compensating network.**

The most commonly used compensating networks are:

(i) Phase lag network

(ii) Phase lead network

The main objective of compensating network is to modify the op-amp performance over the desired frequency range by controlling its gain phase shift.

9.16.3.4. Types of Compensating Networks

The compensating network may be of following types:

(i) External compensating networks

(ii) Internal compensating networks

The earlier op-amps such as 709 needed external compensating network. Because of this, they were called as externally compensated op-amps.

The later generation op-amps like 741 have an internal compensating network. Because of this, they are called as internally compensated op-amps.

9.16.4. Frequency Response of an OP-AMP

Basically, the frequency response of an op-amp is the graph of its open loop gain versus frequency. The open loop gain is generally expressed in dB and frequency is plotted on the logarithmic scale. Ideally, the open loop gain of the op-amp is expected to remain constant irrespective of frequency. This means that ideally, the bandwidth of op-amp should be infinite but practically it is not so. The open loop gain does change with frequency. To plot the frequency response of op-amp, we require to consider the high frequency model of op-amp which has been shown in figure 9.21.

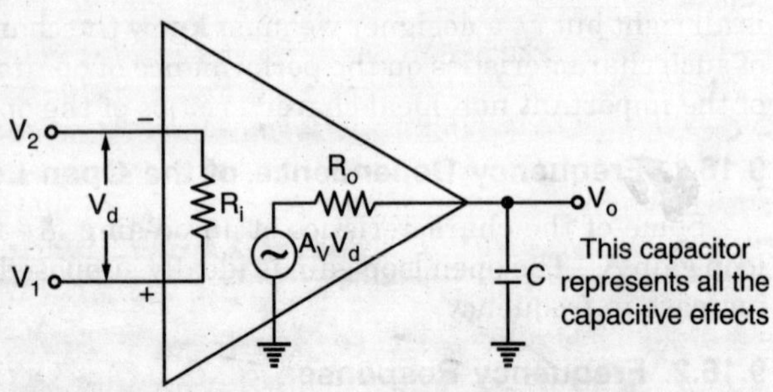

Fig. 9.21. HIgh frequency model of an op-amp with a single corner frequency

It has been observed that op-amp gain rolls off (decreases) after a certain frequency is reached. This means that there must be a capacitor involved in the equivalent circuit. This is the capacitor C shown in figure 9.21. This capacitor is not an externally connected one but it is due to the physical characteristics of the device (BJT or FET) used and the internal construction of an op-amp. For an op-amp with a single break frequency, all the capacitor effects can be represented by a single capacitor C as shown in figure 9.21.

Here, A_v represents the open loop gain and V_d is the differential input voltage. Considering figure 9.21, we can write

$$V_o = \frac{-jX_C}{R_o - jX_C} A_v V_d \qquad ...(9.17)$$

But, $X_C = 1/2\ \pi fC$

therefore,
$$V_o = \frac{1/j2\pi fC}{R_o + \frac{1}{j2\pi fC}} A_v V_d \text{ or } V_o = \frac{A_v V_d}{1 + j2\pi f R_o C} \qquad ...(9.18)$$

Let the open loop gain which is the function of frequency be denoted by A.

or
$$A = \frac{V_o}{V_d} \qquad ...(9.19)$$

Therefore, from equation (9.19), we get,

or
$$A = \frac{V_o}{V_d} = \frac{A_v}{(1 + j2\pi f C R_o)} \qquad ...(9.20)$$

$$\text{Let } f_1 = \frac{1}{2\pi R_o C}$$

or
$$2\pi R_o C = \frac{1}{f_1}$$

Substituting this value into equation (9.20), we obtain

$$A = \frac{A_v}{1 + j\left(\frac{f}{f_1}\right)} \qquad ...(9.21)$$

In this expression, we have

A = Open loop voltage gain as a function of frequency.

A_v = Open loop voltage gain at f = 0 Hz or dc.

f = Operating frequency.

f_1 = Break frequency or cut off frequency.

The open loop gain *A* of equation (9.21) is a complex number which is the function of frequency. We can plot two graphs called *magnitude plot* and *phase plot* from the expression for A. Magnitude plot represents the variation in the magnitude of output voltage and phase plot represents the variation in phase shift with variation in frequency. Magnitude plot is obtained by taking the magnitude of A.

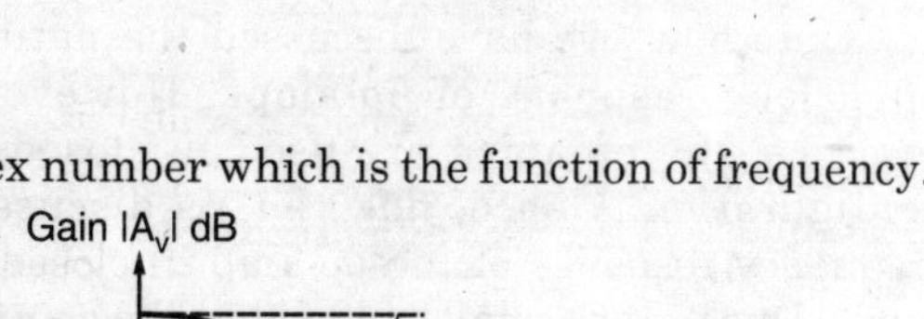

Fig. 9.22. Frequency response of an op-amp

Therefore,
$$|A| = \frac{A_v}{\sqrt{1 + \left(\frac{f}{f_1}\right)^2}} \qquad ...(9.22)$$

and the phase plot may be obtained as under :

$$\phi = \tan^{-1}\left(\frac{f}{f_1}\right) \qquad ...(9.23)$$

To draw the magnitude plot, we consider equation (9.22). There are three operating conditions possible. They are as under :

(i) For frequency $f << f_1$, the magnitude of gain $|A|$ is $(20 \log A_v)$ dB.

(ii) At frequency $f = f_1$ i.e. at the corner frequency, the gain $|A|$ is 3 dB below the dc value of gain A_v is dB.

(iii) For $f >> f_1$, the gain $|A|$ rolls off at a rate of –20 dB/decade or – 6 dB/octave

The frequency response has been shown in figure 9.22.

The corner frequency $f_1 = 1/2 \pi R_o C$ is also known as the **3 dB frequency** as the gain drops by 3 dB as depicted in figure 9.22. The frequency at which the gain reduces to 0 dB [or $A_v = 1$ (numerical value)] is called as the **Gain cross over frequency** or **Unity Gain Bandwidth** (UGB). For op-amp IC 741, this frequency is about 1 MHz. In a practical op-amp, there are many number of stages and each stage produces its own capacitive component. Usually, there will be a number of different break frequencies, f_1, f_2, f_3...... etc. However, for the internally compensated op-amps such as IC 741, the frequency response has been shown in figure 9.23 which has only one break frequency.

9.16.5. Frequency Response of IC-741

The frequency response of IC 741 has been depicted in figure 9.23. IC 741 is an internally compensated operational amplifier, therefore, it does not need external compensating components. IC 741 consists of a capacitor C_1 of 30 pF which reduces the output signal at higher frequencies. Due to this internal capacitance, which is an internal compensating component, the open loop gain will roll off at a rate of – 20 dB/ decade. This will assure stability of the circuit. The 741 op-amp has a 1 MHz gain-bandwidth product. This means that the product of gain and frequency of any point on the frequency response curve is about 1 MHz. Thus, if the op-amp is wired for closed loop gain of 10^4 or 80 dB then bandwidth will be 100 Hz. Now, if the gain is reduced to 10^2 then the bandwidth will increase to 10 kHz and for a gain of 1, the bandwidth will be 1 MHz.

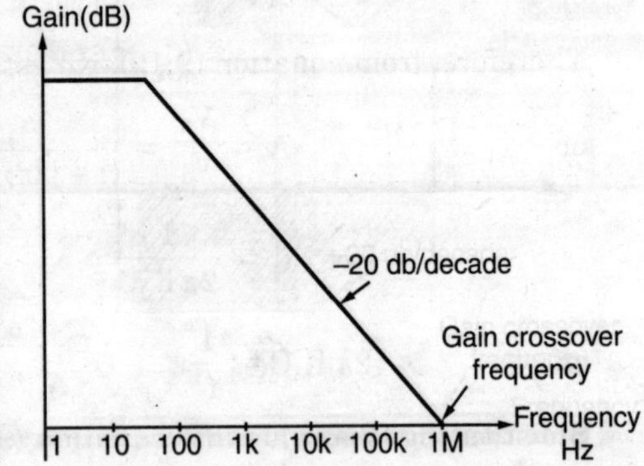

Fig 9.23. Frequency response of op-amp IC 741

9.16.6. Frequency Response of Closed Loop Amplifiers

Upto now, we have discussed the nature of open loop response of op-amps. However, in practice, the op-amps are used in closed loop configuration. Hence, now, let us discuss the frequency response of the op-amp in closed-loop mode. Due to very small bandwidth, the op-amp is rarely used in the open loop configuration. To increase bandwidth, we have to use the negative feedback. The maximum bandwidth of a single break frequency op-amp (such as 741 C) is equal to its unity gain bandwidth (UGB) which is 1 MHz. That means the product of gain and frequency at any point beyond f_c (break frequency) for 741 C is about 1 MHz. Hence, if the voltage gain is given,

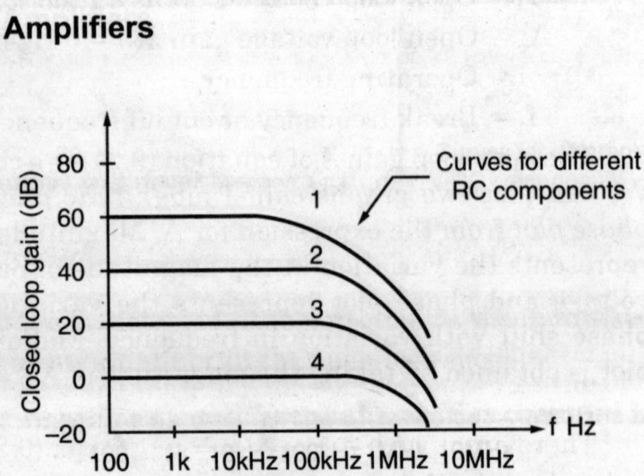

Fig. 9.24. Frequency response of various closed loop gains

we can determine the closed loop bandwidth by using the frequency response curve. The manufacturers data sheets provide the frequency response of various closed loop gains as depicted in figure 9.24.

9.17. LARGE SIGNAL OPERATION OF OP AMPs

In this article, let us discuss the limitations on the performance of op-amp circuits when large output signals are present.

9.17.1. Output Voltage Saturation

The maximum possible value of output voltage is dependent on the supply voltages V_{CC} and V_{EE}. This has been depicted in figure 9.25 which indicates that the op-amp can operate in three different regions namely,

(i) Lower saturation region

(ii) Linear region

(iii) Upper saturation region.

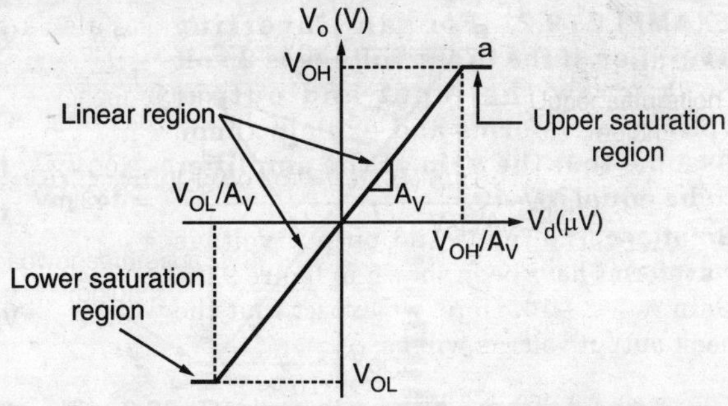

Fig. 9.25. Regions of op-amp operation and corresponding approximate models

(i) Linear region

In the linear region, the transfer characteristics (V_o versus V_d) is approximately linear and the slope of the linear region is equal to the open loop gain A_v. As A_v is very large (typically 2×10^4) the curve is so steep that it coincides with the y-axis. If V_o is measured in volts and V_d in microvolts, then, the slope becomes 0.2 V/ μV for $A_v = 2 \times 10^4$. Hence, in the linear range, the op-amp behaves like a dependent source of value $A_v V_d$.

(ii) Positive Saturation

In linear region, the output voltage is proportional to the differential input voltages (V_d). But as V_d is increased beyond point *a* in figure 9.25, the internal transistor saturates and the voltage transfer characteristics is flattened out. For V_d more positive than at *a*, the output voltage remains constant at V_{OH} or $+ V_{sat}$. This is known as the **positive saturation region.** The typical value of V_{OH} or $+ V_{sat}$ is 2 volts less than V_{CC}. Thus, if $V_{CC} = 15$ V, then, $V_{OH} = V_{sat} = 13$V. In the saturation region, the op-amp acts as an independent voltage source.

(iii) Negative saturation

As shown in figure 9.25, the linear region on the negative side extends upto point *b*. However, as V_d becomes more negative, the internal transistor saturates and the output voltage remains constant at V_{OL} or $- V_{sat}$. The voltage transfer characteristics will be flattened out and the op-amp is said to be operating in the **negative saturation region.** The typical value of V_{OL} or $- V_{sat}$ is 2 volts above V_{EE}. Therefore, if V_{EE} is – 15, then, V_{OL} or $- V_{sat}$ is – 13 V.

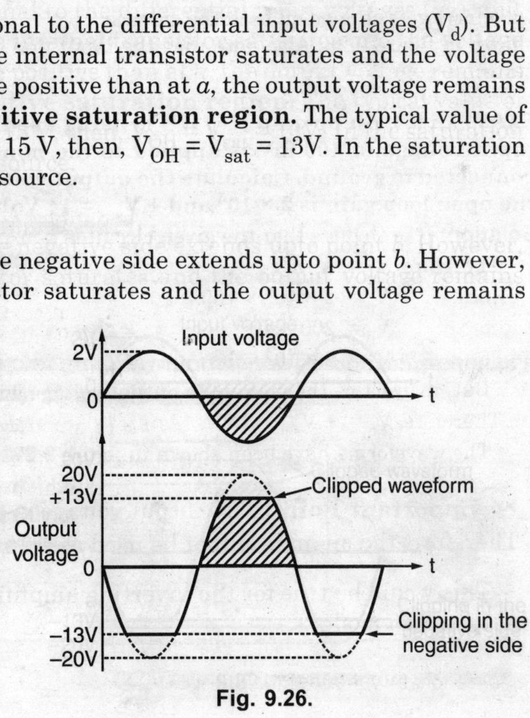

Fig. 9.26.

9.17.2. Differential Voltage Corresponding to Saturation

We know that the open loop voltage gain of a practical op-amp such as IC 741 is 2×10^5. Therefore, even for very small magnitudes of V_d, the op-amp output reach the positive or negative saturation voltage $+V_{sat}$ or $-V_{sat}$. The polarity of output is dependent on the polarity of V_d. If $+V_{sat} = +13$ V, then, let us calculate the required differential input voltage V_d to obtain $+V_{sat}$ at the output.

Therefore, $$V_o = A_v \times V_d \quad ...(9.24)$$

$$+V_{sat} = A_v \times V_d$$

or, $$V_d = \frac{\pm V_{sat}}{A_v} = \frac{\pm 13V}{2\times10^5}$$

or, $$V_d = +65\ \mu\ V \quad ...(9.25)$$

Equation (9.25) shows that if V_d is greater than or equal to 65 μ V, then, the output voltage will be $+V_{sat}$ shown in the voltage transfer characteristics in figure 9.26.

EXAMPLE 9.2. For an inverting amplifier, if the input voltage is 2 volt peak, draw the input and output voltage waveforms and explain them. Assume that the gain of the amplifier to be equal to – 10.

Solution: The input and output voltage waveforms have been shown in figure 9.26. Gain $A_{VF} = -10$. Here, we expect that the peak output voltage will be:

$$V_{o\ (peak)} = -10 \times 2 = -20\ V$$

However, it does not happen so. The maximum output voltage that we can get is restricted to $+V_{(sat)}$ i.e. the saturation voltage. If we assume that $V_{sat} = V_{CC} - 2$ V, then, the maximum output voltage is restricted to + 13 V. Hence, the output waveform gets clipped on both the sides as shown in figure 9.26. Also note that the input and output waveforms are out of phase with respect to each other.

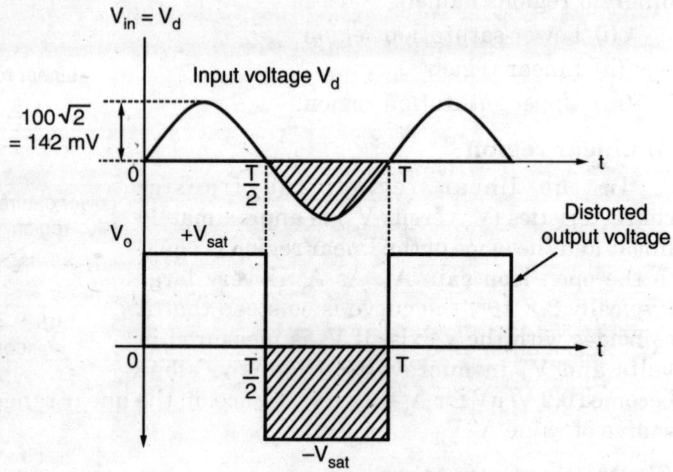

Fig. 9.27.

EXAMPLE 9.3. An op-amp IC 741C is being used as a voltage amplifier in the open loop mode. The input voltage of 100 mV is applied to the non-inverting terminal and the inverting terminal is connected to ground. Calculate the output voltage and plot the input output voltage waveforms if the open loop gain is 2×10^5 and $+V_{sat} = 14$ Volts.

Solution: It is obvious that the given circuit is non-inverting amplifier operating in the open loop mode.

Therefore, $V_o = A_v \times V_d$

or, $V_o = 2 \times 10^5 \times 100 \times 10^{-3}$

or, $V_o = 200 \times 10^2$

or, $V_o = 2 \times 10^4$ volts

But, in practice the maximum output voltage will never be greater than $+V_{sat}$.

Therefore, $V_o = +V_{sat} = 14$ V **Ans.**

The waveforms have been shown in figure 9.27.

Important Point: The output voltage is distorted. It is not same as the input (sine wave). Therefore, the op-amp cannot be used as an amplifier in the open-loop mode.

This would be true for the inverting amplifier as well as the differential amplifier operated in the open loop mode.

9.17.3. Output Current Limits

Another important limitation on the operation of op-amps is that their output current is limited to a specific maximum. For example, the op amp IC 741 has a maximum output current of + 20 mA. Hence, a designer has to ensure that the op-amp should never be required to supply more than 20 mA current in either direction. The output current includes the current supplied to the load as well as the current supplied to the feedback loop. If the output current exceeds the maximum current value, the op-amp output voltage will saturate at the level corresponding to the maximum allowed output current.

9.17.4. Transient Response (Slew Rate)

After performing the frequency domain analysis, now, let us come back to the time-domain analysis. The time-domain analysis can be performed with the help of transient response.

9.17.4.1. Rise time t_R

For the transient response, we consider the voltage follower circuit as depicted in figure 9.28 (a).

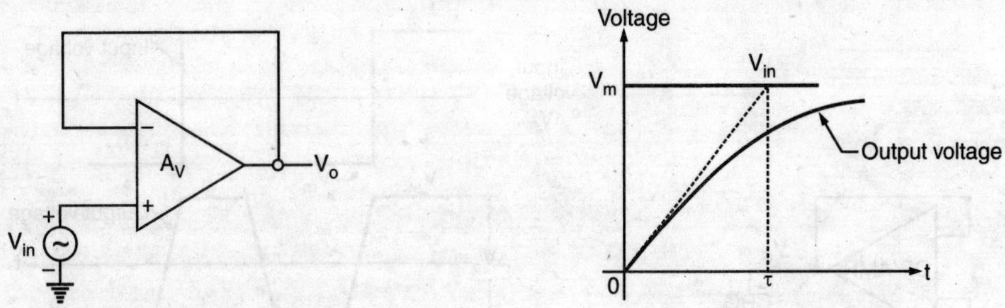

(a) Voltage follower circuit

(b) Small signal step response

Fig. 9.28.

We know that the small signal bandwidth of a voltage follower is f_t. Hence the frequency response of the voltage follower circuit is given by,

$$A_{VF}(jf) = \frac{1}{1+(jf/f_t)} \quad ...(9.26)$$

At the input of the voltage follower, if we apply a step voltage of amplitude V_m, then, the output voltage will increase exponentially. The output voltages is given by,

$$V_o(t) = V_m(1 - e^{-t/\pi}) \quad ..(9.27)$$

where, τ is the time constant and it is given by,

$$\tau = \frac{1}{2\pi f_t} \quad ...(9.28)$$

The rise time t_R may be defined as the time taken by the output voltage to increase from 10% to 90% of the maximum possible value V_m. Rise time should be as small as possible. It indicates how rapid the change in output voltage takes place.

Thus, $$t_R = \frac{0.35}{f_t} \quad ...(9.29)$$

The above expression shows the relation between time-domain and frequency domain. It shows that with the increase in the value of f_t, the rise time goes on decreasing.

The value of t_R for 741 op-amp is 350 ns which is quite high as compared to some other fast op-amps.

9.17.4.2. Slew rate

Slew rate may be defined as the maximum rate of change of output voltage per unit time and it is expressed in volts/microseconds.

$$\text{Therefore,} \quad S = \left.\frac{dV_o}{dt}\right|_{\text{maximum}} \quad \text{Volts}/\mu\text{sec} \qquad \text{...(9.30)}$$

Importance of Slew rate

Slew rate decides the capability of an op-amp to change its output rapidly, therefore, it decides the highest frequency of operation of a given op-amp. Slew rate changes with the change in voltage gain. Therefore, it is generally specified at unity (+ 1) gain. Slew rate should be ideally ∞ and, practically, as high as possible. Slew rate of 741 op-amp is only 0.5 V/μS, which is its biggest drawback. Hence, it cannot be used for high frequency applications.

The circuit used for slew rate measurement has been shown in figure 9.29 (a) and the input output voltage waveforms have been depicted in figure 9.29 (b).

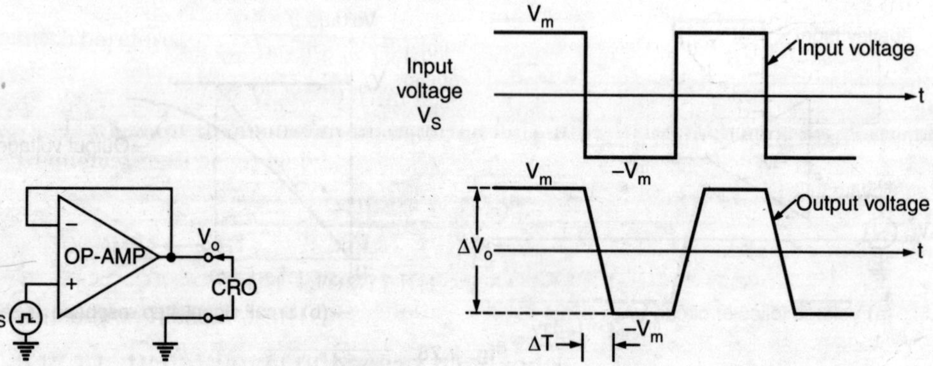

Fig. 9.29.

The slew rate can be measured from the output voltage waveform as

$$\text{Slew rate } S = \frac{\Delta V_o}{\Delta t} \qquad \text{...(9.31)}$$

Ideally, the slew rate should be infinite and practically, it should be as large as possible.

9.17.4.3. Calculation of f_m by Slew Rate

Let us assume that the input signal V_S is a sine wave, we can obtain the value of maximum frequency for which the amplifier produces an undistorted output.

$$\text{Let} \quad V_S = V_m \sin \omega t \qquad \text{...(9.32)}$$

As we have used a unity gain non-inverting amplifier, the output is exactly equal to the input.

$$\text{Therefore,} \quad V_o = V_m \text{ Sin } \omega t$$

$$\text{or} \quad \frac{dV_o}{dt} = \omega V_m \cos \omega t$$

Now, $\frac{dV_o}{dt}$ will be maximum when cos ω t = 1 and maximum value of $\frac{dV_o}{dt}$ is nothing but slew rate S.

$$\text{Therefore} \quad S = \frac{dV_o}{dt}(\text{max}) = \omega V_m = 2\pi f_m V_m \text{ V/sec.}$$

$$\text{or} \quad f_m = \frac{S}{2\pi V_m} \qquad \text{...(9.33)}$$

This is the maximum frequency f_m for which the amplifier produces an undistorted output. It is also known as full power bandwidth.

EXAMPLE 9.4. The output of an op-amp voltage follower is a triangular wave as shown in figure 9.62 for a square wave input of frequency 1 MHz and 4V peak to peak amplitude. What is the slew rate of the op-amp?

Solution: The slew rate is defined as the maximum rate of change of the output.

Hence, from figure 9.30, we have

$$S = \frac{4V}{0.5\mu s} = 8\ V/\mu s$$

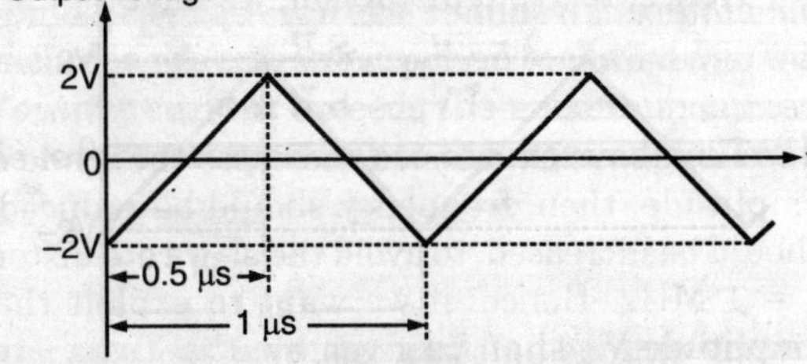

Fig. 9.30.

9.17.4.4. Slew rate of op-amp in inverting mode

When an op-amp is operated in the inverting mode, the slew-rate corresponding to the positive going swing is same as that corresponding to the negative going swing. This means that we have a symmetrical slew rate in the inverting mode of operation.

9.17.4.5. Slew rate of op–amp in the non-inverting mode

When the op-amp operates in the non-inverting mode, the additional parasitic capacitances due to the common mode input swing results in an asymmetrical slew rate.

Important Point: Rise time t_R is a linear small signal parameter, whereas, slew rate is a non-linear large signal parameter.

9.17.4.6. Effect of slew rate limiting

The effect of slew rate limitation is distortion in the output signal. This means that if the slew rate capabilities of an op-amp are exceeded, the output is no more an exact replica of the input.

9.17.5. Full Power Bandwidth

If the rate of change of input voltage is higher than the rate prescribed by slew rate, then, the output waveform will get distorted as illustrated in figure 9.31 (b).

Thus, $\left(\frac{dV_o}{dt}\right) <$ slew rate ...(9.34)

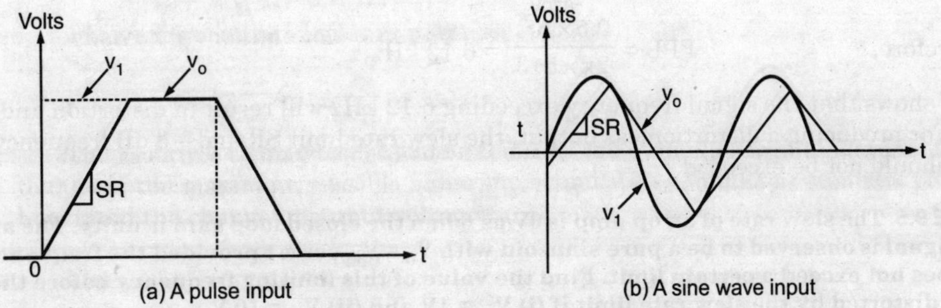

Fig. 9.31. Effect of slew rate limitation

If the input is a sine wave, then, the instantaneous output voltage is given by

$$V_o(t) = V_m \sin(2\pi ft)$$

Therefore, $$\frac{dV_o(t)}{dt} = 2\pi f V_m \cos(2\pi ft)$$

or, $\left[\frac{dV_0(t)}{dt}\right]_{max} = 2\pi fV_m$ for $\cos(2\pi f t) = 1$

Hence, to avoid distortion, we have

$$2\pi f V_m < S.R.$$

$$f V_m < S.R./2\pi \qquad ...(9.35)$$

The above expression indicates the trade off between frequency and amplitude. If we increase amplitude, then, frequency should be reduced and if amplitude is decreased, then, the frequency should be increased, to avoid the slew rate distortion. For op-amp 741, the full small signal bandwidth f_t = 1 MHz. Hence, if we want to exploit the full small-signal bandwidth, then, the maximum amplitude V_m shall be given by

$$V_m \leq \frac{S.R.}{2\pi f_t}$$

But, S.R. = 0.5 V/μ s and f_t = 1 MHz

Therefore, $V_m < \frac{0.5\times10^6}{2\pi\times1\times10^6}$

or V_m < 79.57 mV ...(9.36)

If we want to ensure an undistorted output at V_m = 1V, then, the corresponding maximum frequency shall be given by,

$$f < \frac{S.R}{2\pi V_m} < \frac{0.5\times10^6}{2\pi\times1}$$

or, f < 79.577 KHz ...(9.37)

which is well below 1 MHz.

9.17.5.1. Definition of full power bandwidth

It may be defined as the maximum frequency at which the op-amp will yield an undistorted ac output with the largest possible signal amplitude. This amplitude is dependent on the type of op-amp and the power supplies. The full-power bandwidth (FPB) will be given by

$$FPB = \frac{S.R.}{2\pi V_{sat}} \qquad ...(9.38)$$

For op-amp 741, we have, V_{sat} = 13 V

Therefore, $FPB = \frac{0.5\times10^6}{2\pi\times 513} = 6.12$ kHz ...(9.39)

This shows that the signal frequency exceeding 6.12 kHz will result in distortion and reduced output. For producing a distortionless output, the slew rate limit SR and – 3 dB frequency f_B of an op-amp should not be exceeded.

EXAMPLE 9.5. The slew rate of an op-amp is 6V/μs when the closed loop gain is unity. The amplified output signal is observed to be a pure sinusoid with $V_o = V_m \cos \omega t$ provided the frequency of this signal does not exceed a certain limit. Find the value of this limiting frequency before the output signal is distorted by the slew rate limit if (i) V_m = 1V and (ii) V_m = 10 V.

Solution: Given that slew rate S = 6V/μs

We know that the maximum frequency f_m for which the amplifier produces an undistorted output is given by

$$f_m = \frac{S}{2\pi V_m} \qquad ...(i)$$

(i) For V_m = 1V, we have

$$f_m = \frac{6V/\mu s}{2\pi\times 1V} = \frac{6\times 10^6 V/sec}{2\pi\times 1V} = 954.9 \text{ kHz} \quad \textbf{Ans.}$$

This is the value of limiting frequency.

(ii) For V_m = 10 V, we have

$$f_m = \frac{6\times 10^6 V/sec}{2\pi\times 10V} = 95.49 \text{ kHz} \quad \textbf{Ans.}$$

Important Point: As the peak value of the input signal increases, the maximum frequency for which the amplifier produces an undistorted output, decreases.

EXAMPLE 9.6. Find the slew rate of a frequency-compensated op-amp at room temperature which has a unity cross over frequency of 5 MHz.

Solution: Given that :

Crossover frequency f = 5 MHz

Slew rate $S = 2\pi V_m \times f_m$ ...(i)

Substituting f_m = 5 MHz and V_m = 1V, in equation (i), we get

$$S = 2\pi \times 1 \times 5 \times 10^6$$

$$S = 31.4 \text{ V/}\mu\text{s}$$

At unity cross-over frequency, gain is equal to 1, hence, output is equal to input. If $V_{in} = V_m \sin \omega t$, then

$$V_o = V_m \sin \omega t$$

Therefore, $$\frac{dV_o}{dt} = \omega V_m \cos \omega t$$

Now, $\frac{dV_o}{dt}$ will be maximum, when cos ωt = 1

Therefore, $$\left.\frac{dV_o}{dt}\right|_{max} = \omega V_m = 2\pi f_m V_m$$

This is nothing but the Slew rate.

Thus, $$S = 2\pi f_m V_m \quad \textbf{Ans.}$$

Hence, we have used this equation to find the slew rate.

EXAMPLE 9.7. The input and output waveforms of a voltage follower have been shown in figure 9.32. Calculate the slew rate.

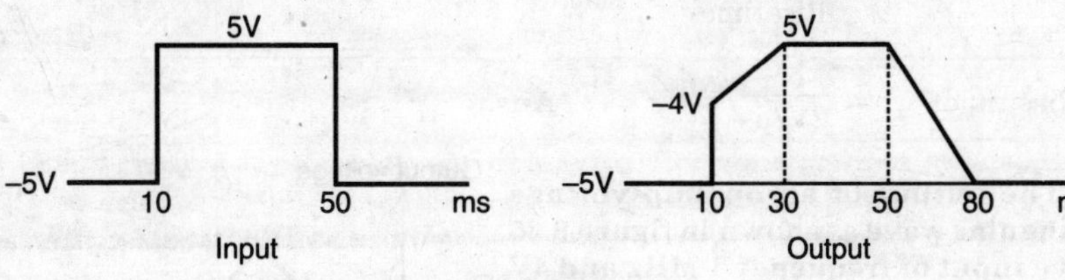

Fig. 9.32.

Solution: The Slew rate $S = \frac{dV_o}{dt} = \frac{\Delta V_o}{\Delta t}$ (V/ μs)

Looking at the output voltage waveform shown in figure 9.32, we can calculate two values of S as under:

For the rising edge, we have $$S_1 = \frac{5-(-4)\text{Volt}}{(30-10)\mu s}$$

Therefore, $$S_1 = 0.45 \text{ V/}\mu\text{s}$$ ...(i)

For the falling edge, we have $$S_2 = \frac{5-(-5)\text{Volt}}{(80-50)\mu s}$$

Therefore, $S_2 = 0.33\ V/\mu s$ ...(ii)

Out of the two, we have to specify the slower rate i.e. S_2.

Hence, slew rate $S_2 = 0.33\ V/\mu\ sec$ **Ans.**

EXAMPLE 9.8. A square wave with negligible rise time and peak to peak amplitude of 500 mV must be amplified to peak to peak amplitude of 3V with rise of 4 μs or less. Can the 741 op-amp which has a slew rate of 0.5 V/μs be used?

Solution: The input and output voltage waveforms have been shown in figure 9.33. Since the output has a peak amplitude greater than 1V, the slew rate is the limiting factor.

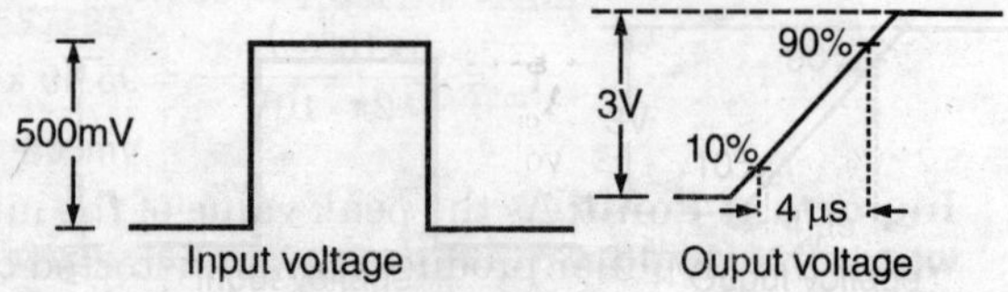

Fig. 9.33.

The required slew rate = $\dfrac{\Delta V}{\Delta t}$

From the definition of rise time, the change in output voltage Δ V in 4 μs is equal to

$$\Delta V = (0.9 - 0.1) \times 3\ V = 2.4\ V$$

Therefore, slew rate $S = \dfrac{2.4V}{4\mu s} = 0.6\ V/\mu s$ **Ans.**

Since, the slew rate of 741 is 0.5 V/μs, it is too slow and cannot be used.

EXAMPLE 9.9. An op-amp has a slew rate of 2V/μs. Find the rise time for an output voltage of 10V amplitude resulting from a rectangular pulse input if the op-amp is slew rate limited. Derive the relation used.

Solution: The input and output voltage waveforms have been shown in figures 9.34 (a) and (b) respectively.

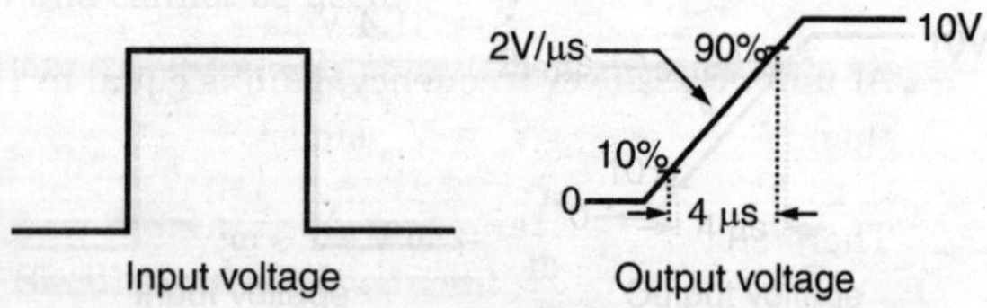

Fig. 9.34.

(i) Rise time = Time difference between the instant at which the output voltage reaches 90% of its final value and the instant at which it reaches 10% of its final value.

(ii) Since the op-amp is slew rate limited, the output voltage will change at the slew rate i.e. at 2V/μs rate.

Slew rate $S = \dfrac{\Delta V}{\Delta t}$ ...(i)

(iii) Now corresponding to the rise time Δt, the change in output voltage i.e. ΔV = (0.9 – 0.1) × 10V = 8

Therefore, $2V/\mu\ s = \dfrac{8V}{\text{Rise time}}$

Hence, Rise time $= \dfrac{8V}{2V/\mu s} = 4\ \mu\ sec.$ **Ans.**

EXAMPLE 9.10. The output of an op-amp voltage follower is a triangular wave as shown in figure 9.35 for a square wave input of frequency 1 MHz and 4V peak to peak amplitude. What is the slew rate of the op-amp?

Solution: The slew rate is defined as the maximum rate of change of the output.

Hence, from figure 9.35, we have

$$S = \frac{4V}{0.5\mu s} = 8\ V/\mu s \quad \textbf{Ans.}$$

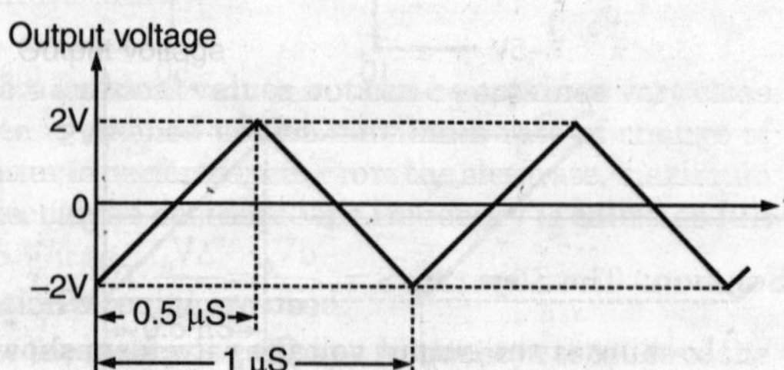

Fig. 9.35.

EXAMPLE 9.11. A differential d.c. amplifier has a differential mode gain of 100 and a common mode gain 0.01. What is its CMMR in dB?

Solution : Differential mode gain, $A_{id} = 100$

Common-mode gain, $A_{cm} = 0.01$

Therefore, $$\text{CMRR} = \frac{A_{id}}{A_{cm}} = \frac{100}{0.01} = 10^4$$

CMRR in dB, $\text{CMRR}' = 20 \log_{10} 10^4 = 20 \times 4 = 80$ dB **Ans.**

EXAMPLE 9.12. For a given op-amp, CMRR = 10^5 and differential gain $A_{id} = 10^5$. Determine the common-mode gain, A_{cm} of the op-amp.

Solution : Differential voltage gain, $A_{id} = 10^5$

Therefore, $$\text{CMRR} = \frac{A_{id}}{A_{cm}}$$

Therefore, common-mode gain $$A_{cm} = \frac{A_{id}}{\text{CMRR}} = \frac{10^5}{10^5} = 1 \quad \textbf{Ans.}$$

EXAMPLE 9.13. A differential amplifier has inputs $V_1 = 10$ mV and $V_2 = 9$ mV. It has a differential mode gain of 60 dB and a CMRR of 80 dB. Find the percentage error in the output voltage and the error voltage.

Solution : Inputs $V_1 = 10$ mV and $V_2 = 9$ mV

Difference signal, $V_{id} = V_1 - V_2 = 10 - 9 = 1$ mV

Common mode signal, $$V_{cm} = \frac{V_1 + V_2}{2} = \frac{10+9}{2} = 9.5 \text{ mV}$$

Differential voltage gain, $$A_{id} = 60 \text{ dB or antilog } \frac{60}{20} = 1{,}000$$

$$\text{CMMR} = 80 \text{ dB or antilog } \frac{80}{20} = 10{,}000$$

Output voltage, $$V_{out} = A_{id} V_{id} \left(1 + \frac{1}{\text{CMMR}} \cdot \frac{V_{cm}}{V_{id}}\right)$$

or $$V_{out} = 1{,}000 \times 1 \text{ mV} \left[1 + \frac{1}{10{,}000} \times \frac{9.5}{1}\right] = 1.00095 \text{ V}$$

We have $A_{id}\, V_{id} = 1{,}000 \times 1 \text{ mV} = 1\text{V}$

Hence, error voltage = $V_{out} - A_{id}\, V_{id} = 1.00095 - 1.0 = 0.00095$ V or 0.95 mV **Ans.**

Also, % error $$= \frac{\text{Error voltage}}{V_{out}} \times 100 = \frac{0.00095}{1.00095} \times 100 = 0.095\% \quad \textbf{Ans.}$$

EXAMPLE 9.14. The two input terminals of an op-amp are connected to voltage signals of strength 745 μV and 740 μV respectively. The gain of the op-amp in differential mode is 5×10^5 and its CMRR is 80 dB. Calculate the output voltage and percentage error due to common mode.

Solution: Input voltage $V_1 = 745\ \mu\text{V}$ and $V_2 = 740\ \mu\text{V}$

Differential voltage, $V_{id} = V_1 - V_2 = 745 - 740 = 5\ \mu\text{V}$

Common mode signal, $$V_{cm} = \frac{V_1 + V_2}{2} = \frac{745+740}{2} = 742.5\ \mu\text{V}$$

Differential voltage gain, $A_{id} = 5 \times 10^5$

Also, $$\text{CMRR} = 80 \text{ dB} = \text{Antilog } \frac{80}{20} = 10{,}000$$

We know that output voltage,

$$V_{out} = A_{id} V_{id} \left[1 + \frac{1}{\text{CMRR}} \frac{V_{cm}}{V_{id}}\right]$$

or $$V_{out} = 5\times10^5\times5\times10^{-6}\left[1+\frac{1}{10,000}\times\frac{742.5\times10^{-6}}{5\times10^{-6}}\right]V = 2.537125\text{ V} \quad \textbf{Ans.}$$

Hence, error voltage $= V_{out} - A_{id} V_{id} = 2.537125 - 5 \times 10^5 \times 5 \times 10^{-6} = 0.037125$ V

Also, percentage error $= \frac{\text{Error voltage}}{V_{out}}\times100 = \frac{0.037125}{2.537125}\times100 = 1.4633\%$ **Ans.**

SUMMARY

- A two port device, which accepts an input signal and produces an output signal proportional to the input, is called as an amplifier.
- An ideal voltage amplifier has $R_i = \infty$, i.e, no current flow into or out of the input port while $R_0 = 0$
- The op-amp is a voltage amplifier having extremely high gain.
- The op-amp works on dual supply. A dual supply consists of two supply voltages both d.c. whose middle point is generally grounded.
- The output taken between output terminal and the ground called as single ended or unbalanced output. The output taken between the two output terminals available, is called as double ended or balanced output. It is also called as floating output.
- The functional block diagram of op-amp consists of four stages namely input stage, intermediate stage, level shifting stage and the output stage.
- The input stage requirements are
  (i) High input resistance
  (ii) Low input bias current
  (iii) Small input offset voltage
  (iv) Small input offset current
  (v) High CMRR
  (vi) High differential input range
  (vii) High common mode range
  (viii) High open loop voltage gain
  (ix) Low output impedance.
- The output stage requirements are
  (i) Large output voltage swing capability
  (ii) Large output current swing capability
  (iii) Low output impedance
  (iv) Low quiescent power dissipation
  (v) Short circuit protection.
- A push-pull amplifier in class AB or class B operation satisfies all these requirements. The diodes are used to eliminate cross over distortion. The circuit which uses both NPN transistors can be used to increase current carrying capacity.
- The characteristics of ideal op-amp are :
  (i) Infinite voltage gain
  (ii) Infinite input impedance
  (iii) Zero output impedance
  (iv) Zero offset voltage
  (v) Infinite bandwidth
  (vi) Infinite CMRR
  (vii) Infinite slew rate
- The characteristics of practical op-amp differ slightly from its ideal values but can be obtained very close to ideal one using feedback arrangement. The slew rate is defined as the maximum rate of change of output voltage with time. Higher the value of slew rate, better is performance. From the slew rate, maximum signal frequency for which undistorted output is possible, can be decided. This frequency is called as full power bandwidth. The slew rate of IC 741 op-amp is 0.5 V/μ sec.
- To minimise offset voltage and current, some compensation circuits are used.
- The graph showing the change in gain with respect to the frequency is called as frequency response of the op-amp.
- The frequency at which gain of the op-amp becomes unity i.e., 0 dB is called as unity gain bandwidth (UGB). For IC 741 op-amp it is 1 MHz.
- The stability of the circuit can be decided by calculating gain margin and the phase margin from the frequency response.

- The IC 741 op-amp is internally compensated which uses builts in lag compensation. Miller effect compensation is commonly used for the internally compensated op-amps.
- The various op-amp packages are
  (i) Metal can (ii) Dual in line package, and
  (iii) Flat pack.
- Op-amp is used in many linear and non-linear applications. In linear applications output varies linearly with input. Such applications are adder, subtractor, voltage follower etc. In non-linear circuits, non-linear elements like diodes, transistors are used. Such applications are precision rectifiers, log and antilog amplifiers, comparators etc.

SHORT QUESTIONS WITH ANSWERS

Q.1. Define Input Bias.

Ans. It is defined as the average of the two DC currents at the inputs of an op-amp.

Q.2. Define Input Impedance.

Ans. It is defined as the impedance seen by a signal source at the input terminals of an amplifier.

Q.3. Define input offset current.

Ans. It is defined as the difference between the two op-amp input currents.

Q.4. Define Input Offset Voltage.

Ans. It is defined as the magnitude of DC voltage which, if applied to input terminals, would reduce op-amp output DC voltage to zero.

Q.5. What is an operational amplifier? Explain.

Ans. It is a high-gain integrated-circuit, direct coupled amplifier with differential input terminals.

Q.6. Define Output Impedance.

Ans. It is the impedance seen looking back at the output terminals of an amplifier, i.e., the thevenin equivalent impedance seen at the output.

Q.7. What are Saturation Voltages?

Ans. The maximum and minimum voltage levels at which the output of an op-amp can assume based upon given power supply voltage.

Q.8. What is Slew Rate?

Ans. It is the maximum rate at which the output of an op-amp can change. It is usually expressed in volts/microsecond.

Q.9. What do you mean by Slew Rate Limiting Frequency?

Ans. It is the highest frequency that can be processed in a circuit without distortion as a result of the slew rate.

Q.10. What is a DIP?

Ans. It is an acronym for dual-in-line package, a package configuration having equal number of pins spaced 0.1 inch apart in two parallel rows.

Q.11. What is meant by open-loop mode?

Ans. It represents the situation in which an operational amplifier is operated without external feedback.

Q.12. Define open-loop gain.

Ans. It is the voltage gain of the operational amplifier without any external feedback.

Q.13. What is meant by Gain Bandwidth Product?

Ans. It is a specification for an op-amp giving product of the DC gain and the 3-dB bandwidth. It is equal to the unity-gain frequency.

Q.14. What is current gain?

Ans. It is the ratio of output current to input current.

Q.15. What is voltage gain?

Ans. It is the ratio of output voltage to input voltage.

Q.16. What do you mean by operating temperature?

Ans. It is the ambient temperature for which the operational amplifier operates within the manufacturer's specifications.

Q.17. Define output voltage swing.

Ans. It is defined as the maximum peak output voltage which the operational amplifier can produce without saturation or clipping. It is typically about two diode voltage drops less than the corresponding supply voltages.

Q.18. Define power-supply rejection ratio (PSRR).

Ans. It is defined as the ratio of the change in input offset voltage of an operational amplifier to a corresponding change in the supply voltage. It is also known as the power-supply sensitivity.

Q.19. What is meant by unity-gain frequency?

Ans. It is the frequency at which the operational amplifier's open-loop gain is unity. It is also known as either the unity-gain crossover frequency or the small-signal unity-gain bandwidth.

Q.20. Mention the various stages in an op-amp.

Ans. The various stages in an op-amp are as under :

(i) Input stage

(ii) Intermediate stage

(iii) Level translator (or) Buffer stage

(iv) Output stage

Q.21. What is the gain of an ideal op-amp?

Ans. The gain of an ideal op-amp is infinity.

Q.22. If the output voltage simultaneously changes with the input then what is the value of slew rate?

Ans. The value tends to infinity.

Q.23. Define common mode rejection ratio (CMRR) and give an expression.

Ans. The relative sensitivity of an op-amp to a difference signal as compounded to a common mode signal is called as common mode rejection ratio (CMRR). It is the ratio of differential gain (A_{id}) to the common mode gain (A_{cm}).

$$\text{CMRR} = \frac{A_{id}}{A_{cm}}$$

It is usually expressed in decibels (dB). Higher is the value of CMRR, better is the op-amp

REVIEW QUESTIONS

1. Draw the block diagram of internal construction of op-amp, and explain the function of each block in detail.
2. What is CMRR ? Find an expression.
3. How can we improve CMRR ? Explain.
4. Explain the concept of virtual ground in op-amp.
5. Write a note on frequency response and explain UGB.
6. Write a note on packages of op-amp.
7. Why open loop op-amp configuration is not used in linear applications.
8. Why differential amplifier is used as an input stage of op-amp ?
9. Name the important terminals of an integrated op-amp.
10. Define and explain the significance of the gain margin and phase margin.

□□□

Chapter 10

BASIC APPLICATIONS OF OP-AMPS

Syllabus

Operational amplifier applications : inverting, non-inverting amplifier, summer, averager, differentiator, integrator, differential amplifier, instrumentation amplifier, log and antilog amplifier, voltage to current and current to voltage converters, comparators, schmitt trigger, active filter, 555 timer and its application.

Chapter Outline

10.1. INTRODUCTION

1. Open-loop op-amp

As discussed in last chapter, an open-loop op-amp offers very high voltage gain [ranging from 10^5 to 10^6 at dc (or 0 Hz)] and narrower bandwidth (band of frequencies) for which the gain remains constant–almost zero. Besides being large, the voltage gain of the open-loop op-amp is not a constant. The voltage gain varies with variations in temperature and power supply voltage as well as mass production techniques.

2. Open-loop op-amp not suitable for linear applications

Because of these features, the open-loop op-amps are not suitable for many linear applications as explained below:

1. Because of the very high voltage gain of open-loop op-amp, only the smaller signals (of the order of microvolts or less) having very low frequency may be amplified accurately without distortion. For instance, if the input signal is of 1 mV, then output voltage will be equal to $1 \times 10^5 \times 1$ mV i.e., 100 V (assuming open-loop gain = 10^5). Such a large output cannot be obtained because the output from the op-amp can never exceed the saturation levels $\pm V_{sat}$. However, the signals of small amplitudes are very susceptible to noise and almost impossible to obtain in the laboratory.

2. The large variations in voltage gain of an open-loop op-amp makes it unsuitable for many linear applications.
3. Because of narrower bandwidth, the open-loop op-amp is impractical in a.c. application. For instance, the open-loop bandwidth of the 741 IC op-amp is roughly 5 Hz but in almost all ac applications, a bandwidth exceeding 5 Hz is required.

3. Remedy

However, the gain of the op-amp can be controlled and stabilized and the bandwidth can be increased if necessary modification is introduced in the op-amp basic circuit. This modification involves the use of feedback. The process of combining a fraction of output energy (voltage or current) back to the input is called feedback. When feedback voltage (or current) is so applied as to increase the input signal (i.e., when the feedback is in phase with input signal), it is called the positive or direct or regenerative feedback. Since positive feedback causes excessive distortion and instability, it is seldom used in amplifiers. However, because of its capability of increasing the power of the original signal, it is used in oscillator circuits.

4. Negative feedback and its features

When feedback voltage (or current) is so applied as to weaken the input signal (i.e., when feedback is in phase opposition to the input signal), it is called negative or inverse or degenerative feedback. Negative feedback reduces the amplifier gain but it has numerous advantages (such as gain stability, reduction in non-linear distortion, reduction in noise, increase in bandwidth or improvement in frequency response, increase in input impedance and decrease in output impedance) and is frequently used in amplifier circuits. Negative feedback also reduces the effects of input offset voltage and variations in temperature and supply voltages on the output of the op-amp.

10.2. BLOCK DIAGRAM REPRESENTATION OF FEEDBACK CONFIGURATIONS

An op-amp using the feedback is known as a **feedback amplifier** and sometimes referred to as a **closed-loop amplifier** because the feedback forms a closed loop between the input and the output. A feedback amplifier essentially consists of two parts viz. an amplifier (an op-amp) and a feedback network as a illustrated in figure 10.1. The function of the feedback network is to return a fraction of the output energy (voltage or current) to the input. The feedback network may be made of either passive components (resistors, inductors and capacitors), active components or combination of both. The feedback network is usually in the form of a passive two port network and is usually formed of resistors, inductor and capacitors.

The feedback may also be classified as voltage feedback and current feedback. In voltage feedback, the energy fedback to the input terminals is proportional to the output voltage while in current feedback, the energy fedback to the input terminals is proportional to the current through the load.

Both voltage and current can be fed back to the input either in series or in parallel resulting in four basic feedback connections, given below

1. Voltage-series feedback, as shown in figure 10.1 (a).
2. Voltage-shunt feedback, as shown in figure 10.1 (b)
3. Current-series feedback, as shown in figure 10.1 (c).
4. Current-shunt feedback, as shown in figure 10.1 (d).

In the list above, voltage refers to connecting the output voltage as input to the feedback network and current refers to tapping off some output current through the feedback network. Series refers to connecting the feedback signal in series with the input signal voltage; and shunt refers to connecting the feedback signal in parallel (shunt) with an input current source

Series feedback connections tend to increase the input impedance of the amplifier. This is because in this case, feedback signal is returned to the input in series to oppose the applied voltage causing input current I_f and therefore, input impedance falls. Voltage feedback tends to reduce the output impedance while current feedback tends to increase the output impedance.

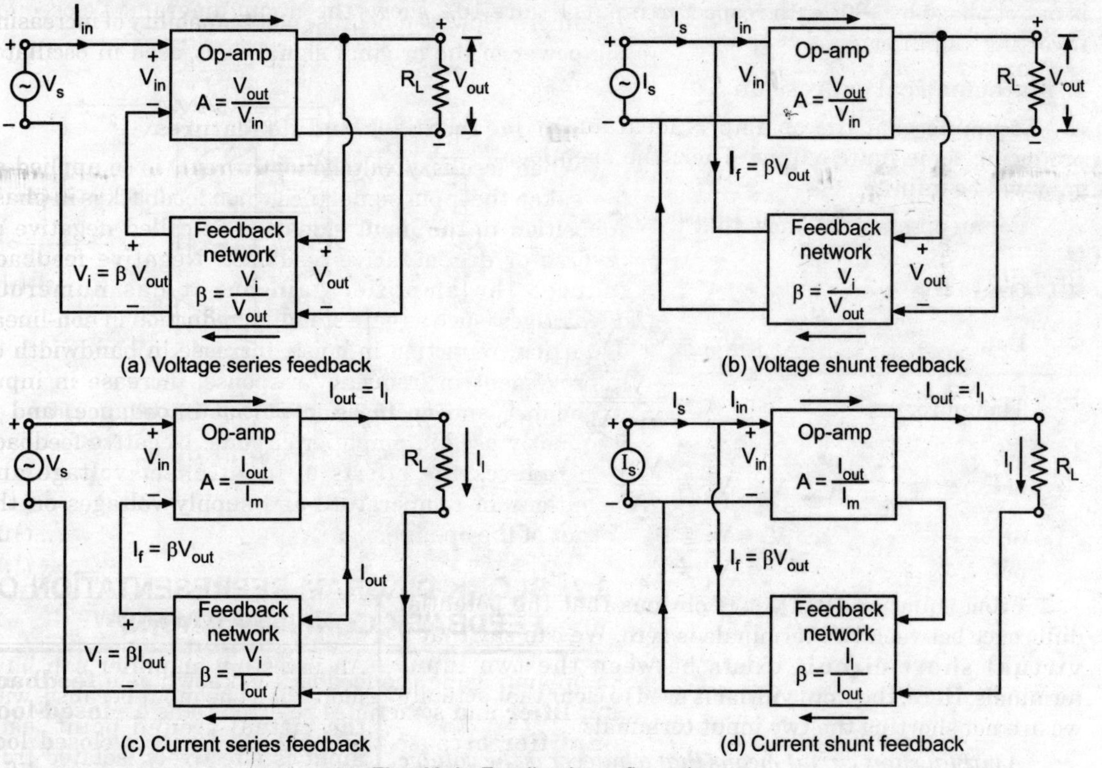

Fig. 10.1. Feedback configuration

In all four of the configurations depicted in figure 10.1, the direction of signal through the op-amp is from the input to output while in an ideal case, the direction of signal through the feedback circuit is exactly opposite i.e., from the output to input.

The most widely used closed configuration are :

1. Voltage-series feedback configuration
2. Voltage-shunt feedback configuration

Hence, only these two configurations will be discussed here in details.

10.3. CLOSED-LOOP OP-AMP CONFIGURATIONS

We know that an open-loop op-amp cannot be used in linear applications. However, an op-amp can be effectively used in linear applications by introducing feedback* from the output to the input. If the signal fed back is out of phase by 180° with respect to the input, then the feedback is called negative feedback. On the other hand, if the signal fed back is in phase with that at the input, then the feedback is called positive feedback. An op-amp that uses feedback is called a closed-loop amplifier. The most widely used closed-loop configurations are as under:

(i) Inverting Amplifier
(ii) Non-inverting Amplifier.

10.4. CLOSED-LOOP INVERTING AMPLIFIER OR INVERTING VCVS

1. Definition

An inverting amplifier or inverting voltage controlled voltage source (VCVS) is a closed-loop amplifier in which the input is applied at the inverting terminal. The output of inverting amplifier is out of phase by 180° with respect to input. Figure 10.2 shows the circuit diagram of closed loop inverting amplifier.

2. Mathematical expression

Assuming that the op-amp is ideal and it is producing some finite output. Then, the open loop gain will be infinite.

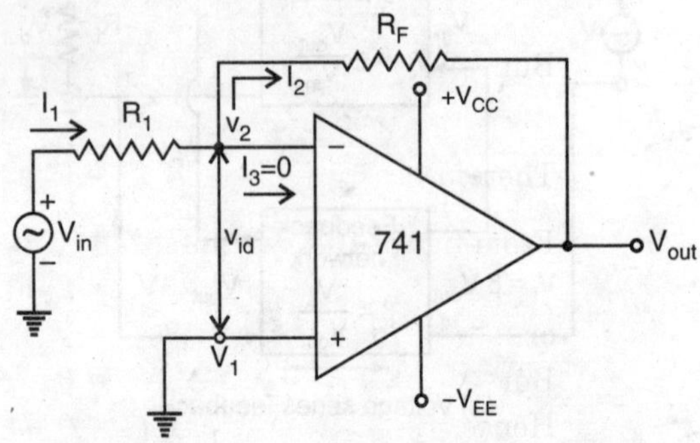

Fig. 10.2. Closed loop inverting amplifier

For an op-amp, we know that

Gain $$A = \frac{V_{out}}{V_{id}}$$

But $$V_{id} = V_1 - V_2$$

Therefore, $$A = \frac{V_{out}}{V_1 - V_2}$$

or $$V_1 - V_2 = \frac{V_{out}}{A} = \frac{V_{out}}{\infty} = 0$$

or $$V_1 - V_2 = 0 \quad ...(10.1)$$

or $$V_1 = V_2 \quad ...(10.2)$$

From equation (10.1), it is obvious that the potential difference between two terminals is zero. We can say that a **virtual short-circuit** exists between the two input terminals. Here, the word virtual is used to clear that actually we are not shorting the two input terminals.

A virtual short circuit means that whatever is the voltage at non-inverting terminal, it will automatically appear at the inverting terminal because of the infinite voltage gain A.

DO YOU KNOW?

An inverting amplifier can have more than one input because, with the virtual ground point, each input is effectively isolated from the other. Each input sees its own input resistance and nothing else.

* Feedback means that a portion of the output is applied at the input.

Here, in this case

Since, $V_1 = 0$

So, from equation (10.2), we have

$$V_2 = 0.$$

Therefore, potential at inverting terminal is zero and it is called **virtual ground**. Virtual ground means that the terminal is not connected to the ground actually even then the voltage at the terminal is zero.

Applying ohm's law to find the current I_1 through R_1, we have

$$I_1 = \frac{V_{in} - V_2}{R_1}$$

But, $V_2 = 0$

Therefore, $$I_1 = \frac{V_{in} - 0}{R_1} \quad \text{or} \quad I_1 = \frac{V_{in}}{R_1} \qquad ...(10.3)$$

Let current inside the op-amp be $I_3 0$.

Now, applying KCL at the node V_2, we get

$$I_1 = I_3 + I_2 \qquad ...(10.4)$$

But here $I_3 \simeq 0$ because ideal op-amp has an infinite input impedence and hence draws zero current.

Therefore, using equation (10.4), we get

$$I_1 = 0 + I_2$$

or $$I_1 = I_2$$

But $$I_1 = \frac{V_{in}}{R_1}$$

Therefore, $$I_1 = I_2 = \frac{V_{in}}{R_1}$$

From figure 10.2, we note that

$$V_2 = I_2 R_F + V_{out}$$

or $$V_{out} = V_2 - I_2 R_F$$

But $$V_2 = 0$$

Hence, $$V_{out} = 0 - I_2 R_F$$

or $$V_{out} = -I_2 R_F$$

But $$I_2 = \frac{V_{in}}{R_1}$$

Thus, $$V_{out} = -\frac{V_{in} R_F}{R_1}$$

$$\frac{V_{out}}{V_{in}} = \frac{-R_F}{R_1}$$

The factor $\frac{V_{out}}{V_{in}}$ is called the closed loop gain A_F.

Thus, $$A_F = \frac{-R_F}{R_1}$$

Important Point: Hence, the closed-loop gain is simply the ratio of the two resistances R_2 and R_1. This means that the gain of the closed-loop inverting op-amp can be changed by simply changing the values of two resistors. In other words we can design a closed-loop inverting op-amp having a particular gain depending upon the application. The minus sign means that the closed-loop amplifier provides signal inversion.

3. Observations from Ideal Inverting Amplifier

(i) The negative sign indicates that the output is inverted with respect to input.

(ii) The voltage gain is independent of open loop gain of the op-amp, which is assumed to be large.

(iii) The closed-loop voltage gain depends on the ratio of the two resistances. Hence, selecting R_F and R_1, the required value of gain can be easily obtained.

(iv) If $R_F > R_1$, the gain is greater than 1.
If $R_F < R_1$, the gain is less than 1.
If $R_F = R_1$, the gain is unity.
Thus, the output voltage can be greater than, less than or equal to the input voltage, in magnitude.

(v) If the ratio of R_F and R_1 is K which is other than one, the circuit is called as **scale changer** while for $R_F/R_1 = 1$, it is called as **phase inverter.**

4. Bias Compensation

Most of the times, the resistance $R_{comp} = R_1 \,||\, R_F$ is connected between op-amp's non-inverting input terminal and the ground. This resistance provides the necessary bias current compensation. As op-amp input current is assumed zero, no current passes through R_{comp} ideally and hence due to R_{comp}, there is no change in the expression for the closed loop voltage gain of the inverting amplifier. Figure 10.3 shows an inverting amplifier with input bias current compensation.

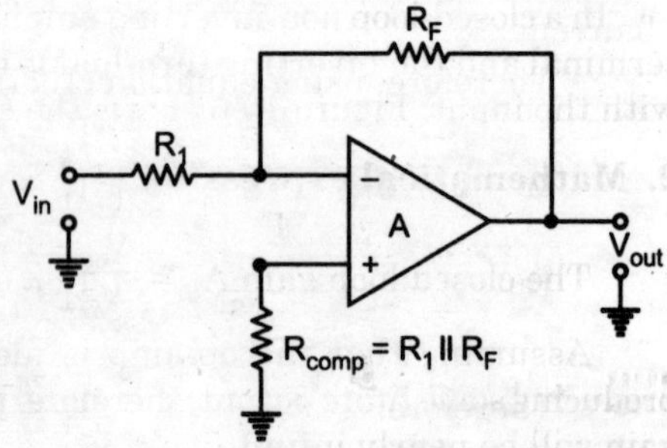

Fig. 10.3. Inverting amplifier with bias compensation

5. Closed-loop Inverting amplifier as VCVS

Because the magnitude of the output voltage can be controlled by the selection of R_F and R_1 $\left(\text{Since, } V_{out} = \frac{-R_F}{R_1} V_{in}\right)$, hence, this is also called as a voltage-controlled voltages source (VCVS). Thus the output of an op-amp in the inverting mode can also be made to serve as dependent voltage source. The corresponding model for this mode is drawn in figure 10.4. A comparison of the model for the non-inverting mode indicates that the difference lies mainly in the input resistance encountered. For the non-inverting mode, it is so high that the model on the input side indicates an open-circuit. For the inverting mode, the value of the input resistance is finite and is accordingly included in the model.

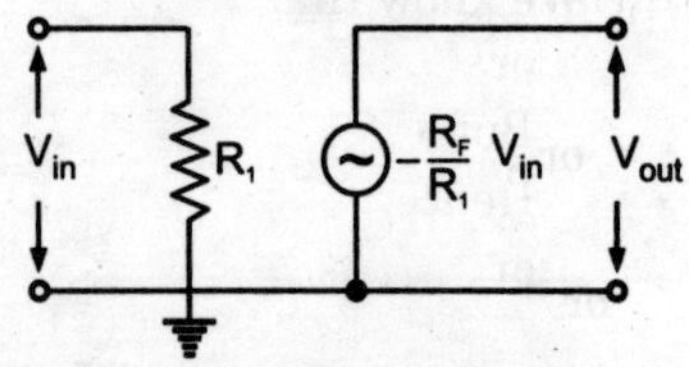

Fig. 10.4. Circuit Model For VCVS (Op-amp in Inverting Mode)

EXAMPLE 10.1. When will the output get saturated in figure 10.5.

Solution: It may be observed that the given circuit is an ideal integrator. The problems associated with the integrator are:

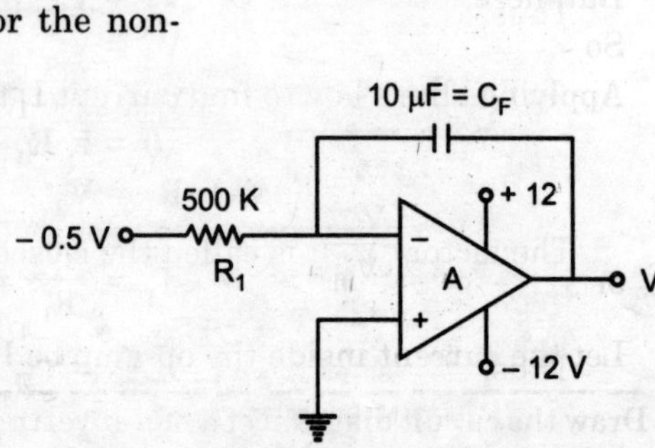

Fig. 10.5.

In an ideal integrator, the capacitor C_F is connected in the feedback path. Hence, its reactance decides the gain of the inverting amplifier as,

$$A_F = \frac{-R_F}{R_1} = \frac{-X_F}{R_1}$$

where X_F is the reactance of the capacitor C_F.

In the absence of any ac input, the capacitive reactance will be ∞. Hence, the gain of the integrator will be ∞. Hence, it will amplify any input offset voltage (V_{ios}) present at the Op-Amp input. Thus, in absence of any input voltage also, a large output voltage is present which is obviously the error voltage. This is one of the problems.

The other problem is that the bias current can charge the integrator capacitor to charge or discharge forcing the integrator output to ramp up or ramp down.

Because of these two problems, the output voltage is likely to ramp up or down and due to very high gain, it may reach the $+V_{sat}$ or $-V_{sat}$ level. This is the major drawback of an ideal integrator.

10.5. CLOSED-LOOP NON-INVERTING AMPLIFIER OR NON-INVERTING VCVS*

(RGTU, Bhopal, Sem. Exam. 2005-06)(05 marks)

1. Definition

In a closed-loop non-inverting amplifier, the external input signal is applied at the non-inverting terminal and the inverting terminal is grounded. The output of non-inverting amplifier is in phase with the input. Figure 10.6 shows the circuit diagram of closed-loop non-inverting amplifier.

2. Mathematical expression

The closed loop gain $A_F = \frac{V_{out}}{V_{in}}$ ...(10.5)

Assuming that the op-amp is ideal and it is producing some finite output, therefore, the open loop gain will be nearly infinite.

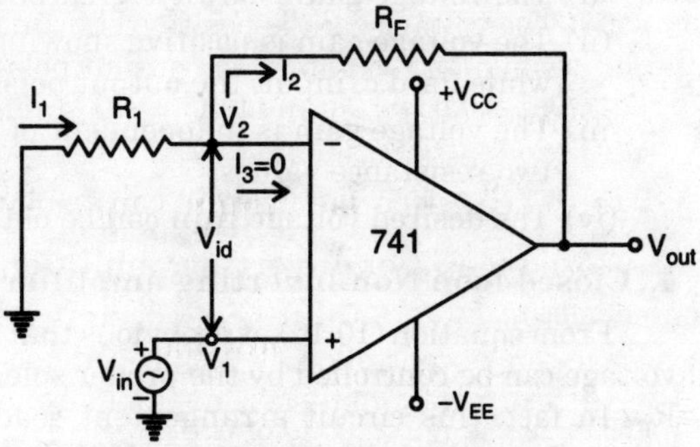

Fig. 10.6. Non-inverting amplifier

We know that $A = \frac{V_{out}}{V_{Id}}$

or $A = \frac{V_{out}}{V_1 - V_2}$

or $V_1 - V_2 = \frac{V_{out}}{A} \cong \frac{V_{out}}{\infty} = 0$

or $V_1 - V_2 = 0$

or $V_1 = V_2$ ...(10.6)

But, here, $V_1 = V_{in}$

So $V_2 = V_{in}$ ...(10.7)

Applying ohm's law to find current I_1 through resistor R_1, we have

$$0 = I_1 R_1 + V_2$$

$$-I_1 R_1 = V_2$$

or $I_1 = \frac{-V_2}{R_1} = \frac{-V_{in}}{R_1}$ ...(10.8)

Let the current inside the op-amp be I_3

* Draw the circuit diagram of a non-inverting amplifier. Explain its working and derive the expression for voltage gain. **(RGTU, Bhopal, Sem. Exam. 2005-06)**

Therefore, $I_1 = I_2 + I_3$

But $I_3 = 0$ since input impedance is infinite

Therefore, $I_1 = I_2 + 0 = I_2$

Using equation (10.8), we get

$$I_2 = I_1 = \frac{-V_{in}}{R_1} \quad ...(10.9)$$

Again, from figure, we have

$$V_2 = I_2 R_F + V_{out}$$

Using equations (10.7) and (10.9), we get

$$V_{in} = -\frac{V_{in} R_F}{R_1} + V_{out}$$

$$V_{out} = V_{in} + \frac{V_{in} R_F}{R_1} = V_{in}\left(1 + \frac{R_F}{R_1}\right)$$

or $$\frac{V_{out}}{V_{in}} = 1 + \frac{R_F}{R_1}$$

or $$A_F = 1 + \frac{R_F}{R_1} \quad ...(10.10)$$

3. Observations

(i) The voltage gain is always greater than one.

(ii) The voltage gain is positive showing that for a.c. input, the output and input are in phase while for d.c. input, the output polarity is same as that of input.

(iii) The voltage gain is independent of open loop gain of an op-amp, but depends only on the two resistance values.

(iv) The desired voltage gain can be obtained by selecting proper values of R_F and R_1.

4. Closed-loop Non-inverting amplifier as VCVS

From equation (10.10), it is obvious that the magnitude of output voltage can be controlled by the proper selection of resistors R_1 and R_F. In fact, this circuit arrangement readily provides us with a voltage-controlled voltage source (VCVS) as shown in figure 10.7. Accordingly, if R_F is set equal to 24 kΩ and R_1 is 1 kΩ, the output voltage of the op-amp will be 25 times the value of input voltage.

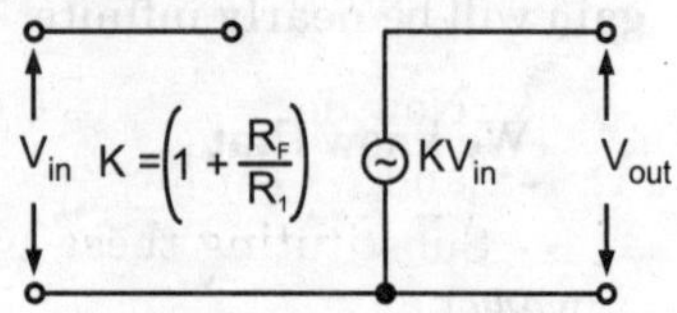

Fig. 10.7. Circuit model for non-inverting VCVS

10.5.1. Performance Comparison of Closed Loop Inverting and Non-inverting Amplifier

The table 10.1 provides the performance comparison of the inverting and non-inverting amplifier op-amp circuits.

Table 10.1.

S.No.	Inverting Amplifier	Non-inverting Amplifier
1.	Voltage gain = $-\frac{R_F}{R_1}$	Voltage gain = $1 + \left(\frac{R_F}{R_1}\right)$
2.	The output is inverted with respect to input	No phase shift between input and output.
3.	The voltage gain can be adjusted as greater than, equal to or less than one.	The voltage gain is always greater than one.
4.	The input impedance is R_1.	The input impedance is extremely large.

10.6. FEW IMPORTANT APPLICATIONS OF OP-AMP

As discussed earlier, the closed-loop op-amp is widely used in a number of linear applications. Here, linear means that the output and input of op-amp are linearly related. Op-amp may be used as a unity gain amplifier, as an integrator, as a differentiator etc.

Now, let us discuss some of the important applications.

10.6.1. Buffer Amplifier

The gain for buffer amplifier is

$$A_F = 1 \text{ so that } \frac{V_{out}}{V_{in}} = 1$$

or $$V_{out} = V_{in}$$

Because of high input resistance, unity transmission gain and extremely low output resistance, the above circuit is an ideal circuit device to serve as a buffer amplifier (or isolation unit).

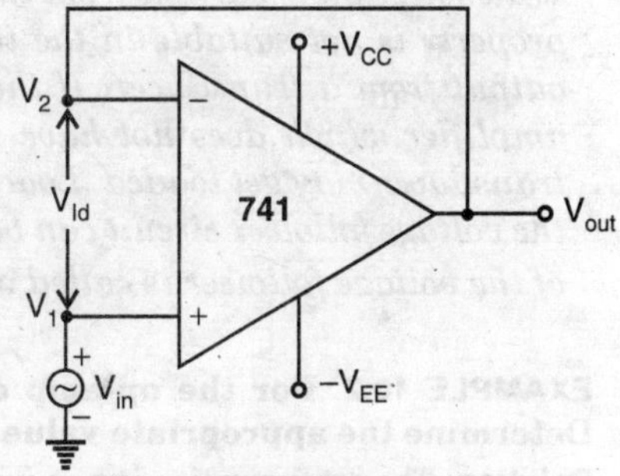

Fig. 10.8. Buffer amplifier

The buffer amplifier provides a means of isolating an input signal from a load.

Thus, it can be made to prevent the disturbance of one part of a circuit on another as might be encountered.

10.6.2. Voltage Follower*

To discuss voltage follower, let us first draw the circuit of a closed-loop non-inverting amplifier. Figure 10.9 shows the closed-loop non-inverting amplifier.

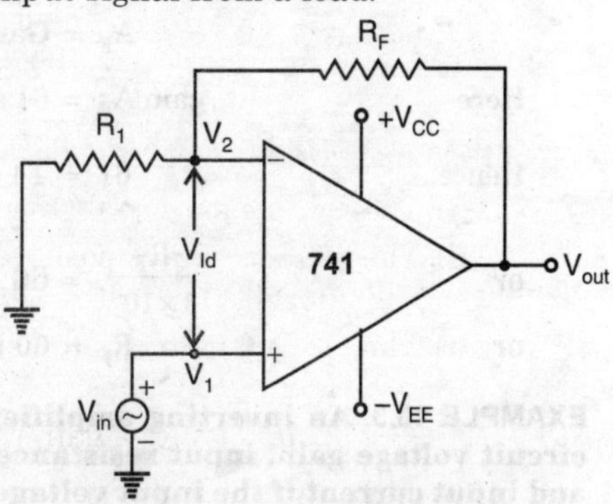

Fig. 10.9. Non-inverting amplifier

We know that the gain for a closed-loop non-inverting amplifier is expresed as

$$A_F = 1 + \frac{R_F}{R_1} \quad \text{...(10.11)}$$

Now, let $R_F = 0$ (shorting)

and $R_1 = \infty$ (opening)

Substituting these two conditions in equation (10.11), we get

$$A_F = 1 + 0 = 1$$

$$\frac{V_{out}}{V_{in}} = 1$$

or $$V_{out} = V_{in}$$

DO YOU KNOW?

Using an op amp as a voltage follower (buffer amplifier) is very common in electronic circuits.

Thus, if in a closed-loop non-inverting amplifier, resistor R_F is shorted (i.e., $R_F = 0$) and resistor R_1 is open (i.e. $R_1 = \infty$), then according to above equation, the output will be equal to the input all the times. If these two conditions (i.e., $R_F = 0$ and $R_1 = \infty$) are applied to figure 10.9, then a new circuit is formed as shown in figure 10.10.

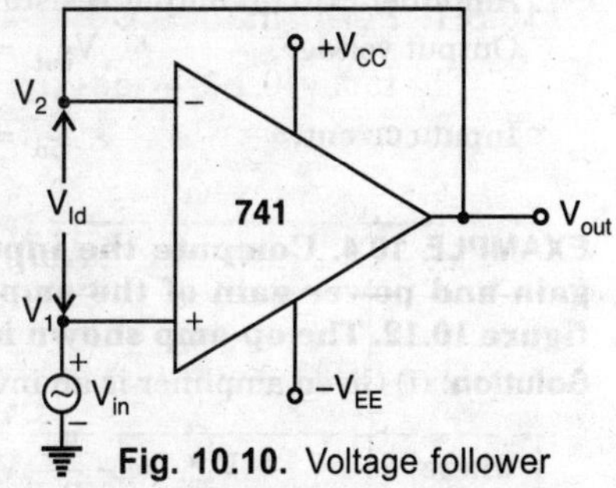

Fig. 10.10. Voltage follower

This circuit (figure 10.10) is called a **voltage follower** because the output voltage is equal to and in phase with the

* With neat sketch, explain the working of :
(*i*) non-inverting amplifier using op-amp, (*ii*) integrator using op-amp, (*iii*) unity gain amplifier.
(RGTU, Bhopal, Sem. Exam. 2004-05)

input. In other words, in voltage follower **the output follows the input.**

Hence voltage follower is a special case of the non-inverting amplifier. Voltage follower is also used as a buffer amplifier.

> **Important Point:** *Because the input resistance of the voltage follower is extremely large, it draws negligible current from the signal source. Therefore, the signal source is not loaded. This property is not suitable in the instrumentation amplifier, which is used to amplify the signal output from a transducer. If the output of a transducer is directly applied to the input of an amplifier which does not have high input impedance, then there is every possibility that the transducer may get loaded. Loading alters the characteristics of the transducer. To prevent this, the voltage follower circuit can be used in between the transducer and the amplifier. Such an use of the voltage follower is called as* ***voltage follower as an isolator or a buffer.***

EXAMPLE 10.2. For the op-amp configuration shown in figure 10.11, the gain required is 61. Determine the appropriate value of feedback resistance R_F.

Solution: The configuration is non-inverting amplifier.

Therefore, we have

$$A_F = \text{Gain} = \frac{V_{out}}{V_{in}} = 1 + \frac{R_F}{R_1}$$

Here gain $A_F = 61$ and $R_1 = 1\ k\Omega$

Hence, $61 = 1 + \dfrac{R_F}{1+10^3}$

or $\dfrac{R_F}{1\times10^3} = 60$

or $R_F = 60 \times 1 \times 10^3\ \Omega = 60\ \Omega$ **Ans.**

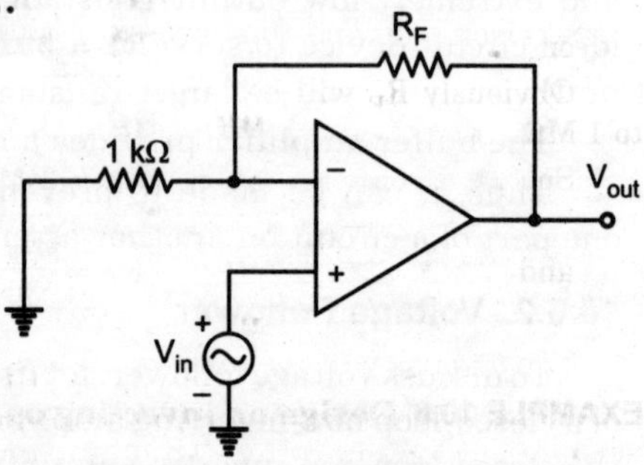

Fig. 10.11.

EXAMPLE 10.3. An inverting amplifier has R_F = 500 kΩ and R_1 = 5 kΩ. Determine the amplifier circuit voltage gain, input resistance and output resistance. Determine also the output voltage and input current if the input voltage is 0.1 V. Assume op-amp to be ideal one.

Solution: Resistance $R_1 = 5\ k\Omega$

Feedback resistance, $R_F = 500\ k\Omega$

Input voltage, $V_{in} = 0.1\ V$

Amplifier circuit voltage gain, $A_F = -\dfrac{R_F}{R_1} = -\dfrac{500k\Omega}{5k\Omega} = -100$ **Ans.**

Amplifier circuit input resistance, $R_i = R_1 = 5\ k\Omega$ **Ans.**

Amplifier cxircuit output resistance, R_o = Output resistance of op-amp = 0 Ω **Ans.**

Output voltage, $V_{out} = A_F V_{in} = -100 \times 0.1 = -10\ V$ **Ans.**

Input current, $I_{in} = \dfrac{V_{in}}{R_1} = \dfrac{0.1}{5\times1{,}000} = 0.02\ mA$ **Ans.**

EXAMPLE 10.4. Compute the input impedance, voltage gain and power gain of the amplifier circuit shown in figure 10.12. The op-amp shown is ideal.

Solution: (i) Given amplifier is an inverting amplifier

where $V_{out} = -\dfrac{R_F}{R_{in}} V_{in} = -\dfrac{1M\Omega}{1M\Omega} \times V_{in} = -V_{in}$

so $A_v = \dfrac{V_{out}}{V_{in}} = -1$ **Ans.**

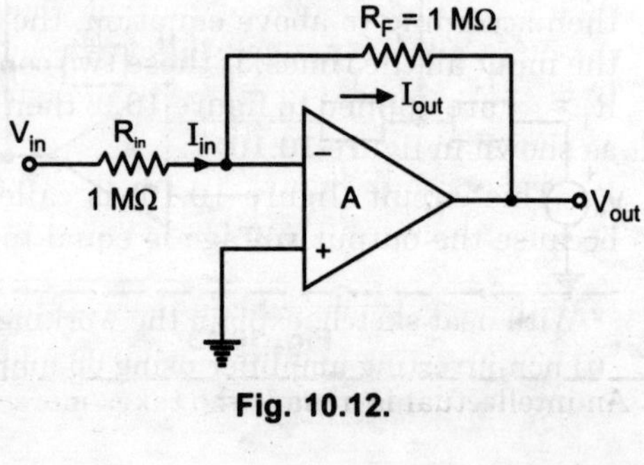

Fig. 10.12.

(ii) Because it is a unity gain inverter,

so $\qquad I_{in} = I_{out}$

or $\qquad A_{in} = \frac{I_{out}}{I_{in}} = 1$ **Ans.**

(iii) Power gain of op-amp circuit,

$$A_p = |A_v \times A_i| = |1 \times 1| = 1 \quad \textbf{Ans.}$$

EXAMPLE 10.5. Design an inverting op-amp circuit with a voltage gain of $A_v = \frac{V_{out}}{V_{in}} = -30$ and an input resistance that is largest value possible but under constraint that the largest resistance value is limited to 1 MΩ. **(RGTU, Bhopal, Sem. Exam., 2004-05)(05 marks)**

Solution: Voltage gain, $\qquad \frac{V_{out}}{V_{In}} = \frac{-R_F}{R_1} = -30$

or $\qquad R_F = 30\, R_1$

Obviously R_F will be larger resistance, being 30 times of R_1 and largest resistance value is limited to 1 MΩ.

So, $\qquad R_F = 1\ \text{M}\Omega$ **Ans.**

and $\qquad R_1 = \frac{R_F}{30} = \frac{1\times10^6}{30} = 33\ \text{k}\Omega$ **Ans.**

EXAMPLE 10.6. Design an inverting op-amp circuit with a voltage gain of $A_v = \frac{V_{out}}{V_{in}} = -8$. When the input voltage is $V_{in} = -1$ V, the maximum current in R_1 anmd R_F must be no larger than 15 μA. Determine the minimum values of R_1 and R_F.

Solution: Voltage gain, $\qquad A_v = \frac{V_{out}}{V_{in}} = -8.$

Because of the virtual ground on the inverting input, we have

Resistance, $\qquad R_1 = \frac{V_{in}}{I_1} = \frac{1\text{V}}{15\times10^{-6}\,\text{A}} = 66.66\ \text{K}\Omega$ (We use 68 kΩ as standard value)

and Resistance $\qquad R_F = -A_F\, R_1 = -(-8)\times 68 = 384\ \text{k}\Omega$ (We use 390 kΩ as standard value]

EXAMPLE 10.7. Obtain the transfer function $\frac{V_o(s)}{V_{in}(s)}$ for the circuit shown in figure 10.13. Assume the op-amp to be ideal. Y_i represents the admittance of branch i. **(GATE Examination, 1996)**

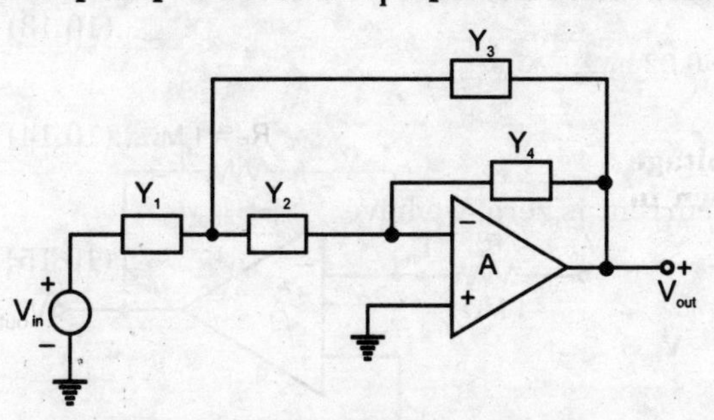

Fig. 10.13.

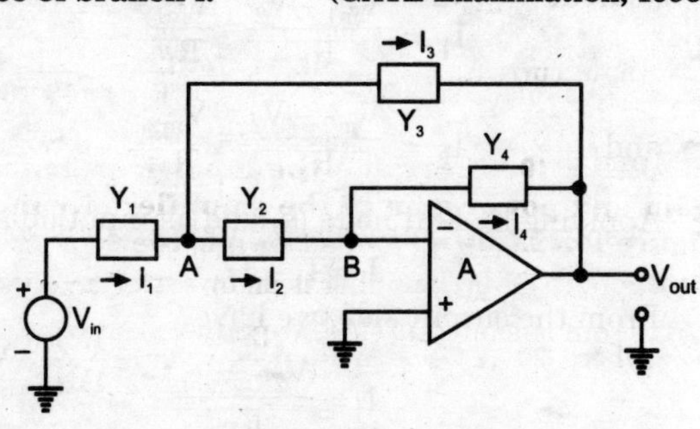

Fig. 10.14.

An intellectual is a man who takes more words than necessary to tell more than be knows.

— **Dwight D. Eisenhower**

Solution: The given circuit is redrawn with direction of flow of different currents marked there on, as shown in figure 10.14.

Now,
$$I_1 = (V_{in} - V_A)\,Y_1 \quad \text{...(i)}$$
$$I_2 = V_A Y_2 = I_4 = -V_{out} Y_4 \quad \text{(Node B being at 0 potential) ...(ii)}$$
$$I_3 = (V_A - V_{out})\,Y_3 \quad \text{...(iii)}$$

From equation (ii), we have
$$V_A = -V_{out}\frac{Y_4}{Y_2} \quad \text{...(iv)}$$

Since
$$I_1 = I_2 + I_3$$
$$(V_{in} - V_A)Y_1 = -V_{out} Y_4 + (V_A - V_{out})\,Y_3$$

or
$$V_{in} Y_1 = -V_{out} Y_4 + V_A Y_3 - V_{out} Y_3 + V_A Y_1$$
$$= -V_{out} Y_4 - V_{out} \times \frac{Y_3 Y_4}{Y_2} - V_{out} Y_3 - V_{out}\frac{Y_1 Y_4}{Y_2}$$

or
$$\frac{V_{out}}{V_{in}} = \frac{-Y_1 Y_2}{Y_2 Y_4 + Y_3 Y_4 + Y_3 Y_2 + Y_1 Y_4} = \frac{-Y_1 Y_2}{Y_2 Y_3 + Y_4 (Y_1 + Y_2 + Y_3)} \quad \textbf{Ans.}$$

10.7. SUMMER OR ADDER CIRCUIT (Important)

Since the input impedance of an op-amp is extremely large, more than one input signal can be applied to the inverting amplifier. Such circuit gives the addition of the applied signals at the output. Hence, it is called as **summer** or **adder** circuit. Figure 10.15 shows the circuit with two input signals.

Let us assume that op-amp is ideal and is producing a finite output. Because of this, open-loop gain A of op-amp becomes infinite. This means that
$$A = \infty$$

We know that
$$A = \frac{V_{out}}{V_{id}}$$

or
$$V_{id} = \frac{V_{out}}{A} = \frac{V_{out}}{\infty} = 0$$

or
$$V_1 - V_2 = 0$$

or
$$V_1 = V_2$$

But
$$V_1 = 0$$

so
$$V_2 = 0 \quad \text{...(10.12)}$$

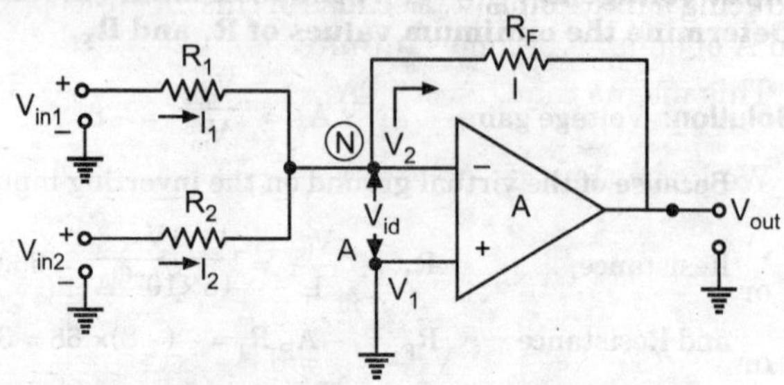

Fig. 10.15. Inverting summer

Now, from the input side, we have
$$I_1 = \frac{V_{in1} - V_2}{R_1} = \frac{V_{in1}}{R_1} \quad \text{...(10.13)}$$

and
$$I_2 = \frac{V_{in2} - V_2}{R_2} = \frac{V_{in2}}{R_2} \quad \text{...(10.14)}$$

Applying KCL at node Ⓝ and as input op-amp current is zero, we have
$$I = I_1 + I_2 \quad \text{...(10.15)}$$

From the output side, we have
$$I = \frac{V_2 - V_{out}}{R_F}$$

Adopting the right attitude can convert a negative stress into a positive one. — **Hans Selye**

But $V_2 = 0$ so, $I = -\frac{V_{out}}{R_F}$ ...(10.16)

Substituting equations (10.16), (10.13) and (10.14) in equation (10.15), we obtain

$$-\frac{V_{out}}{R_F} = \frac{V_{in1}}{R_1} + \frac{V_{in2}}{R_2}$$

or $$V_{out} = -\left[\frac{R_F}{R_1} V_{in1} + \frac{R_F}{R_2} V_{in2}\right] \quad ...(10.17)$$

> **DO YOU KNOW**
>
> Voltage follower, because of its three unique characteristics of extremely high input impedance, unity transmission gain and extremely low output impedance, is an ideal device to serve as a buffer amplifier

If three resistances are equal, $R_1 = R_2 = R_F$, then, we have

$$V_{out} = -(V_{in1} + V_{in2}) \quad ...(10.18)$$

By properly selecting R_F, R_1 and R_2, we can have weighted addition of the input signals like $aV_{in1} + bV_{in2}$, as indicated by equation (10.17). Infact in such a way, n input voltages can be added. Thus, the magnitude of the output voltage is the sum of the input voltages and hence circuit is called as summer or adder circuit.

10.8. NON-INVERTING SUMMING AMPLIFIER (RGTU, Bhopal, Sem. Exam., 2007-08)

The circuit discussed in last article is inverting summing amplifier, which can be noticed from the negative sign in equation (10.17). However, a summer that gives non-inverted sum of the input signals is known as **non-inverting summing amplifier**. Figure 10.16 shows this circuit.

Let us assume that op-amp is ideal and is producing a finite output. Because of this, open-loop gain A of op-amp becomes infinite.

This means that $A = \infty$

We know that $A = \frac{V_{out}}{V_{id}}$

or $V_{id} = \frac{V_{out}}{A} = \frac{V_{out}}{\infty} = 0$

or $V_1 - V_2 = 0$

or $V_1 = V_2$ ...(10.19)

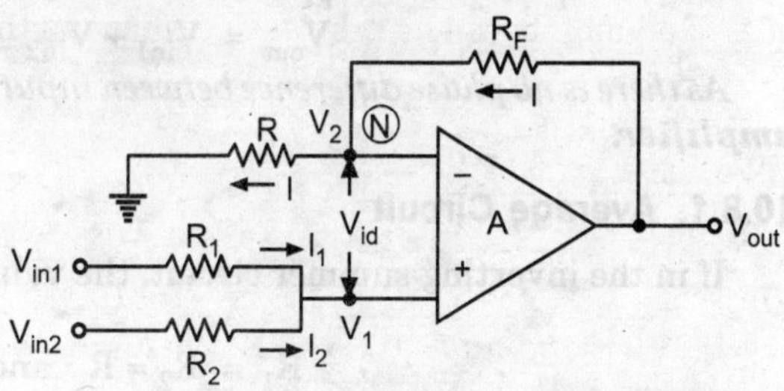

Fig. 10.16. Non-inverting summing amplifier

From the input side, we have

$$I_1 = \frac{V_{in1} - V_1}{R_1} \text{ and } I_2 = \frac{V_{in2} - V_1}{R_2}$$

But as the input current of op-amp is zero, therefore, we write

$$I_1 + I_2 = 0$$

or $$\frac{V_{in1} - V_1}{R_1} + \frac{V_{in2} - V_1}{R_2} = 0$$

or $$\frac{V_{in1}}{R_1} + \frac{V_{in2}}{R_2} = V_1\left[\frac{1}{R_1} + \frac{1}{R_2}\right]$$

or $$V_1 = \frac{(R_2 V_{in1} + R_1 V_{in2})}{(R_1 + R_2)} \quad ...(10.20)$$

Now at node Ⓝ, we have

$$I = \frac{V_2}{R} = \frac{V_1}{R} \quad ...(10.21)$$

and

$$I = \frac{V_{out} - V_2}{R_F} = \frac{V_{out} - V_1}{R_F} \quad ...(10.22)$$

Equating the two equations (10.21) and (10.22), we obtain

$$\frac{V_1}{R} = \frac{V_{out} - V_1}{R_F}$$

or

$$\frac{V_{out}}{R_F} = V_1\left[\frac{1}{R} + \frac{1}{R_F}\right]$$

or

$$V_{out} = V_1 \frac{[R + R_F]}{R} \quad ...(10.23)$$

Substituting equation (10.20) in equation (10.23), we obtain

$$V_{out} = \frac{(R_2 V_{in1} + R_1 V_{in2})[R + R_F]}{R(R_1 + R_2)}$$

i.e.,

$$V_{out} = \frac{R_2(R + R_F)}{R(R_1 + R_2)} V_{in1} + \frac{R_1(R + R_F)}{R(R_1 + R_2)} V_{in2} \quad ...(10.24)$$

Equation (10.24) shows that the output is weighted sum of the inputs.

Now, if $R_1 = R_2 = R = R_F$, we have

$$V_{out} = V_{in1} + V_{in2} \quad ...(10.25)$$

As there is no phase difference between input and output, it is called as ***non-inverting summing amplifier.***

10.8.1. Average Circuit

If in the inverting summer circuit, the values of resistance are selected as follows :

$$R_1 = R_2 = R \quad \text{and} \quad R_F = \frac{R}{2}$$

then from equation (10.17), we obtain

$$V_{out} = -\left[\frac{R/2}{R} V_{in1} + \frac{R/2}{R} V_{in2}\right] = -\frac{(V_{in1} + V_{in2})}{2} \quad ..(10.26)$$

Thus, the magnitude of the output voltage is the average of the two input voltages. Hence, circuit acts like an **averager.***

Similarly, average of n inputs can be calculated by selecting.

$$R_1 = R_2 = R_3 = ... = R_n = R \quad \text{and} \quad R_F = \frac{R}{n}.$$

EXAMPLE 10.8. Determine the output voltage for the configuration shown in figure 10.17.

Solution: From figure 10.17, it may be observed that the circuit is an inverting adder circuit.

Also, $R_1 = R_2 = R_3 = 1\ k\Omega$

and $R_F = 1\ k\Omega$

---

* If a question is asked in examination based on average circuit, then students are advised first to dscribe an inverting summer with circuit diagram and then as a special case should describe an average circuit.

Therefore, we have

$$V_{out} = -\left[\frac{R_F}{R_1}V_{in1} + \frac{R_F}{R_2}V_{in2} + \frac{R_F}{R_3}V_{in3}\right]$$

$$= -[V_{in1} + V_{in2} + V_{in3}]$$

or $V_{out} = -[2 + 1 + 4] = -7$ V **Ans.**

> **DO YOU KNOW?**
>
> If feedback resistor is made zero or R_1 is made ∞ (by keeping it open-circuited) in a non-inverting voltage amplifier circuit, voltage follower is obtained.

EXAMPLE 10.9. Design an adder circuit using an op-amp to get the output expression as $V_{out} = -(V_1 + 10\ V_2 + 100\ V_3)$ where V_1, V_2 and V_3 are the inputs. Given that R_F = 100 kΩ.

Solution: The output in figure 10.18 is given by

$$V_{out} = -R_F\left[\frac{V_1}{R_1} + \frac{V_2}{R_2} + \frac{V_3}{R_3}\right] = -\left[\frac{R_F}{R_1}V_1 + \frac{R_F}{R_2}V_2 + \frac{R_F}{R_3}V_3\right]$$

Comparing the above expression with the given expression for the output i.e., $-(V_1 + 10\ V_2 + 100\ V_3)$, we get

$$R_1 = R_F = 100\ k\Omega \quad \textbf{Ans.}$$

$$R_2 = \frac{R_F}{10} = \frac{100}{10} = 10\ k\Omega \quad \textbf{Ans.}$$

and

$$R_3 = \frac{R_F}{100} = \frac{100}{100} = 1\ k\Omega \quad \textbf{Ans.}$$

EXAMPLE 10.10. An op-amp has feedback resistor R_F = 12 kΩ and the resistance on the input sides are R_{s_1} = 12 kΩ, R_{s_2} = 2 kΩ and R_{s_3} = 3 kΩ. The corresponding inputs are V_{i1} = + 9 V, V_{i2} = – 3 V and V_{i3} = –1 V. Non-inverting terminal is grounded. Determine the output voltage.

Solution: The circuit is shown in figure 10.17, where,

R_F = 12 kΩ

R_1 = 12 kΩ

R_2 = 2 kΩ

and R_3 = 3 kΩ

V_1 = 9 V;

V_2 = – 3 V

and V_3 = – 1 V

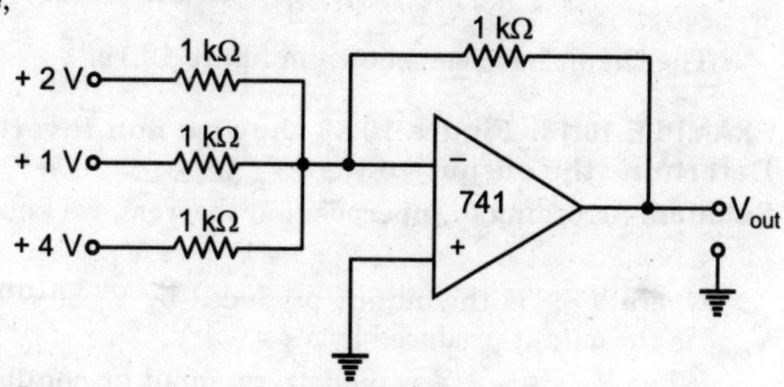

Fig. 10.17

We know that the output voltage is given by

$$V_{out} = -R_F\left[\frac{V_1}{R_1} + \frac{V_2}{R_2} + \frac{V_3}{R_3}\right]$$

$$V_{out} = -12\ k\Omega\left[\frac{9}{12\ k\Omega} + \frac{-3}{2\ k\Omega} + \frac{-1}{3\ k\Omega}\right] = -9 + 18 + 4 = 13\ V \quad \textbf{Ans.}$$

EXAMPLE 10.11. Sketch the circuit of summing amplifier using op-amp to get

$$V_{out} = -V_1 + 2\ V_2 - 3\ V_3$$

Solution : We know that the output voltage is given by

$$V_{out} = -\left[\frac{R_F}{R_1}V_1 + \frac{R_F}{R_2}V_2 + \frac{R_F}{R_3}V_3\right]$$

Comparing the above expression with the given expression for the output

i.e., $V_{out} = -V_1 + 2\ V_2 - 3\ V_3 = -[V_1 - 2\ V_2 + 3\ V_3]$

we have $\frac{R_F}{R_1} = 1, \frac{R_F}{R_2} = 2, \frac{R_F}{R_3} = 3$

Taking R_F = 6 kΩ we have

$$R_1 = R_F = 6\ k\Omega$$

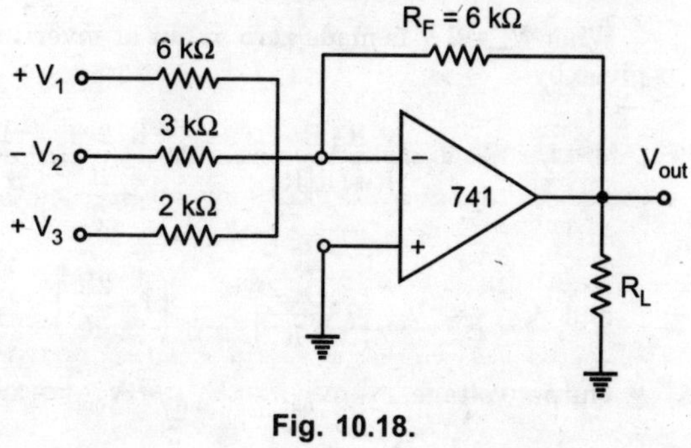

Fig. 10.18.

$$R_2 = \frac{R_F}{2} = 3\text{ k}\Omega$$

$$R_3 = \frac{R_F}{3} = 2\text{ k}\Omega$$

The circuit has been shown in figure 10.21.

EXAMPLE 10.12. Realize a circuit to obtain $V_{out} = -2V_1 + 3V_2 + 4V_3$ using an operational amplifier. Use minimum value of resistance as 10 kΩ.

Solution: We know that the output of an inverting summer or adder is given by

$$V_{out} = -\left[\frac{R_F}{R_1}V_1 + \frac{R_F}{R_2}V_2 + \frac{R_F}{R_3}V_3\right]$$

Comparing the above expression with the given expression for the output

i.e., $V_{out} = -2V_1 + 3V_2 + 4V_3$

$= -[2V_1 - 3V_2 - 4V_3]$

we have $\frac{R_F}{R_1} = 2, \ \frac{R_F}{R_2} = 3$ and $\frac{R_F}{R_3} = 4$

Resistance R_3 will be of minimum value of 10 kΩ.

Then, $R_F = 4R_3 = 4 \times 10 = 40\text{ k}\Omega$ **Ans.**

$$R_2 = \frac{R_F}{3} = \frac{40}{2} = 13.33\text{ k}\Omega \quad \textbf{Ans.}$$

$$R_1 = \frac{R_F}{2} = \frac{40}{2} = 20\text{ k}\Omega \quad \textbf{Ans.}$$

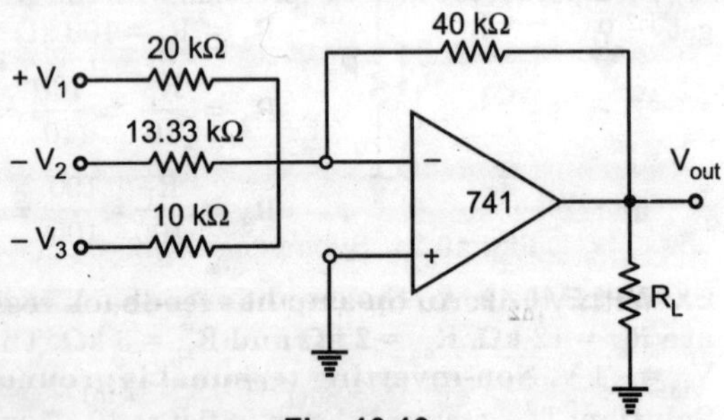

Fig. 10.19.

The circuit has been shown in figure 10.19.

EXAMPLE 10.13. Figure 10.20 shows a non-inverting op-amp summer with $V_1 = -2$ V and $V_2 = -$V. Determine the output voltage V_{out}. **(GATE Examination, 1994)**

Solution: According to superposition theorem, we know that

$$V_{out} = V_{out_1} + V_{out_2}$$

where V_{out_1} is the output produced by V_1 of + 2 V and V_{out_2} is the output produced by $V_2 = -1$ V.

When $V_{out_2} = -1$ V is made zero, input at non-inverting input terminal will be

$$V_{in_1} = V_1 \times \frac{R\|R}{R + (R\|R)} = 2 \times \frac{R/2}{R + \frac{R}{2}} = \frac{2}{3}\text{ V}$$

$$V_{out_1} = V_{in_1}\left(1 + \frac{R_F}{R}\right) = \frac{2}{3}\left(1 + \frac{2R}{R}\right) = 2\text{ V}$$

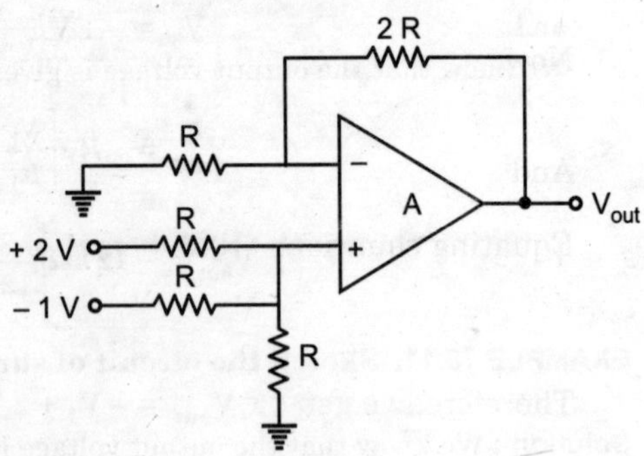

Fig. 10.20.

When $V_1 = 2$ V is made zero, input at inverting terminal is given by

$$V_{in_1} = V_2 \times \frac{R\|R}{R + (R\|R)} = -1 \times \frac{R/2}{R + \frac{R}{2}} = \frac{-1}{3}\ V$$

and $V_{out_2} = V_{in_2}\left(1 + \frac{R_F}{R}\right) = \frac{-1}{3}\left(1 + \frac{2R}{R}\right) = -1\text{ V}$

Output voltage, $V_{out} = V_{out_1} + V_{out_2} = 2 + (-1) = 1\text{ V}$ **Ans.**

10.9. SUBTRACTOR OR DIFFERENCE AMPLIFIER

Similar to the summer circuit, the subtraction of two input voltages is possible with the help of op-amp circuit, called as subtractor or difference amplifier circuit. The circuit diagram has been shown in figure 10.21. To find the relation between the inputs and output, let us use superposition principle. Let V_{out1} be the output, with input V_{in1} acting, assuming V_{in2} to be zero. And V_{out2} be the output, with input V_{out2} acting, assuming V_{in1} to be zero.

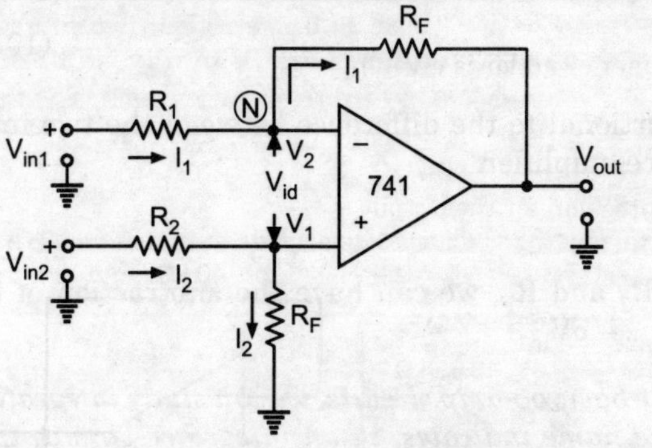

Fig. 10.21. Subtractor circuit

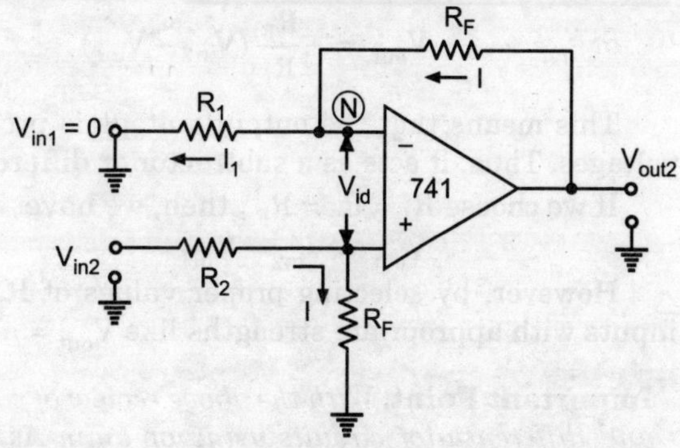

Fig. 10.22.

With V_{in2} zero, the circuit acts as an inverting amplifier. Hence, we can write

$$V_{out1} = -\frac{R_F}{R_1} V_{in1} \quad ...(10.27)$$

While with V_{in1} as zero, the circuit reduces to as shown in figure 10.22.
Because of virtual ground, we have, $V_1 = V_2$
Applying voltage divider rule to the input V_{in2} loop, we have

$$V_1 = \frac{R_F}{R_2 + R_F} V_{in2} \quad ...(10.28)$$

Now,

$$I = \frac{V_2}{R_1} = \frac{V_1}{R_1} \quad ...(10.29)$$

And

$$I = \frac{V_{out2} - V_2}{R_F} = \frac{V_{out2} - V_1}{R_F} \quad ...(10.30)$$

Equating equations (10.29) and (10.30), we obtain

$$\frac{V_1}{R_1} = \frac{V_{out2} - V_1}{R_F}$$

Therefore, we get

$$V_{out2} = \left[\frac{R_1 + R_F}{R_1}\right] V_1$$

or

$$V_{out2} = \left[1 + \frac{R_F}{R_1}\right] V_1 \quad ...(10.31)$$

Substituting V_1 from equation (10.28) in equation (10.31), we obtain

$$V_{out2} = \left[1 + \frac{R_F}{R_1}\right]\left[\frac{R_F}{R_2 + R_F}\right] V_{in2} \quad ...(10.32)$$

Hence, using superposition principle, we have

DO YOU KNOW?

The main purpose of the differential amplifier stage is to provide high gain to the difference-mode signal and cancel the common-mode signal.

$$V_{out} = V_{out1} + V_{out2} = -\frac{R_F}{R_1}V_{in1} + \left[1+\frac{R_F}{R_1}\right]\left[\frac{R_F}{R_2+R_F}\right]V_{in2} \quad ...(10.33)$$

Now, if the resistances are selected such that $R_1 = R_2$, then, we have

$$V_{out} = -\frac{R_F}{R_1}V_{in1} + \left[1+\frac{R_F}{R_1}\right]\left[\frac{R_F}{R_2+R_F}\right]V_{in2} = -\frac{R_F}{R_1}V_{in1} + \frac{R_F}{R_1}V_{in2}$$

or
$$V_{out} = +\frac{R_F}{R_1}(V_{in2} - V_{in1}) \quad ...(10.34)$$

This means that the output voltage is proportional to the difference between the two input voltages. Thus, it acts as a subtractor or difference amplifier.

If we choose $R_1 = R_2 = R_F$, then, we have

$$V_{out} = V_{in2} - V_{in1} \quad ...(10.35)$$

However, by selecting proper values of R_1, R_2 and R_F, we can have the subtraction of two inputs with appropriate strengths like $V_{out} = aV_{in2} - bV_{in1}$.

Important Point: *With the above review of some basic op-amp circuits, we can study integrator and differentiator circuits using op-amp. As the name indicates, in an integrator circuit, the output voltage is the integration of the input voltage. Whereas in the differentiator circuit, the output voltage is the differentiation of the input signal. The integrator circuit can be obtained without using active devices like op-amp, transistors etc. Such a circuit is called as* ***passive integrator.*** *While an integrator using an active device like op-amp is called as* ***active integrator.***

10.10. AN ACTIVE OP-AMP INTEGRATOR (RGTU, Bhopal, Sem. Exam., 2006-07)

Now, let us consider an active op-amp integrator circuit as shown in figure 10.23.

Let us assume that the op-amp is ideal and is producing a finite output. Because of this, the open-loop gain A of the op-amp becomes infinite. This means that

$$A = \infty$$

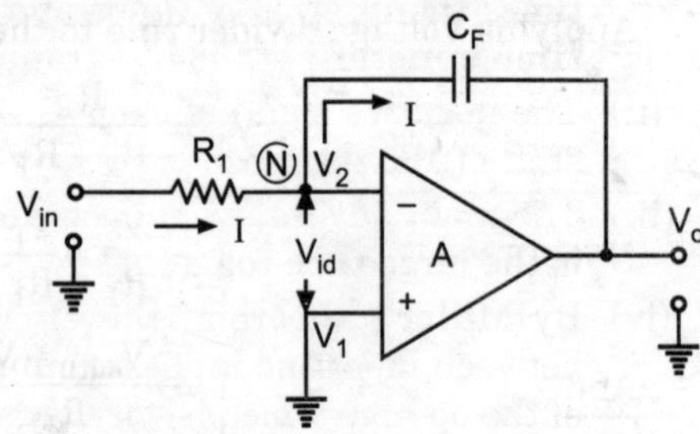

Fig. 10.23. An op-amp integrator

We know that
$$A = \frac{V_{out}}{V_{id}}$$

$$V_{id} = \frac{V_{out}}{A}$$

or
$$V_{id} = \frac{V_{out}}{\infty} \equiv 0$$

or
$$V_1 - V_2 = 0$$

or
$$V_1 = V_2$$

Now, since $V_1 = 0$

so $V_2 = 0$

Since, input current of op-amp is zero, the entire current I flowing through R_1, also flows through C_F, as shown in figure 10.23.

From input side, we can write,

$$I = \frac{V_{in} - V_2}{R_1} = \frac{V_{in}}{R_1} \quad ...(10.36)$$

DO YOU KNOW?

Integrators are preferred over differentiators for analog computers as the gain of integrator decreases with increasing frequency and hence signal to noise ratio of integrator is higher than that of differentiator.

From output side, we can write,

$$I = C_F \frac{d(V_2 - V_{out})}{R_1}$$

or

$$I = -C_F \frac{dV_{out}}{dt} \quad ...(10.37)$$

Equating two equations (10.36) and (10.37), we obtain

$$\frac{V_{in}}{R_1} = -C_F \frac{dV_{out}}{dt} \quad ...(10.38)$$

Integrating both sides, we obtain

$$\int_0^t \frac{V_{in}}{R_1} dt = -C_F \int \frac{dV_{out}}{dt} . dt$$

i.e.

$$\int_0^t \frac{V_{in}}{R_1} dt = -C_F V_{out} \quad ...(10.39)$$

or

$$V_{out} = -\frac{1}{R_1 C_F} \int_0^t V_{in} \, dt + V_{out}(0) \quad ...(10.40)$$

where $V_{out}(0)$ is the constant for integration, indicating the initial output voltage.

Few important points about integrator

(*i*) Equation (10.40) shows that the output is $-1/R_1C_F$ times the integral of input and R_1C_F is called as **time constant** of the integrator.

(ii) The negative sign indicates that there is a phase shift of 180° between input and output.

(iii) The main advantage of such an active integrator is the large time constant.

(iv) By Miller's theorem, the effective capacitance between inverting input terminal and the ground becomes $C_F(1 - A_v)$ where A_v is the gain of the op-amp which is very large. Due to such large effective capacitance, time constant is very large and thus a perfect integration results due to such circuit.

(v) Sometimes, a resistance $R_{comp} = R_1$ is connected to the non-inverting terminal to provide the bias compensation. This has been shown in figure 10.24.

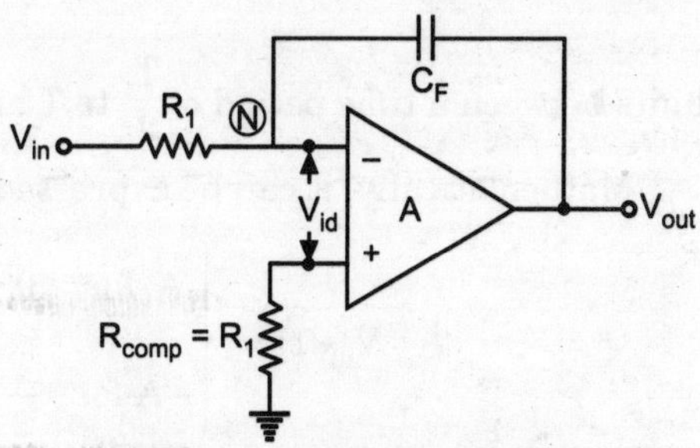

Fig. 10.24. Integrator with bias compensation

Important Point: *As the input current of op-amp is zero, the node Ⓝ is still can be treated at ground potential in this circuit. Hence, the above analysis is equally applicable to the integrator circuit with bias compensation. And the output is the perfect integration of the input.*

10.10.1. Input and Output Waveforms for Integrator

Let us observe the output waveforms for an integrator when we apply various input signals. For simplicity , let us assume that the time constant $R_1C_F = 1$ and the initial voltage $V_{out}(0) = 0$ V.

(i) Step input signal

Let the input waveform is of step type, with a magnitude of A units as shown in figure 10.25. Mathematically, the step input can be expressed as under :

$$V_{in}(t) = \begin{cases} A & \text{for } t \geq 0 \\ 0 & \text{for } t < 0 \end{cases} \quad ...(10.41)$$

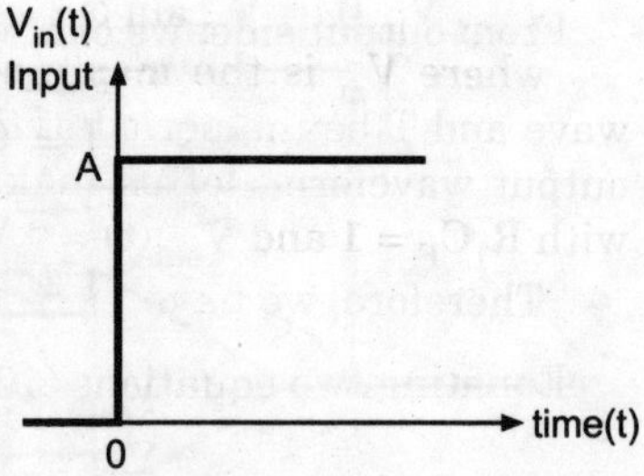

Fig. 10.25. Step input signal

From equation (10.40), with $R_1C_F = 1$ and $V_{out}(0) = 0$, we can write,

$$V_{out}(t) = -\int_0^t V_{in}(t)\,dt \quad ...(10.42)$$

$$V_{out}(t) = -\int_0^t A\,dt = -A\int_0^t dt = -A[t]_0^t$$

or $\quad V_{out}(t) = -A\,t$

Thus, output waveform is a straight line with a slope of – A where A is magnitude of the step input. The output waveform has been shown in figure 10.26.

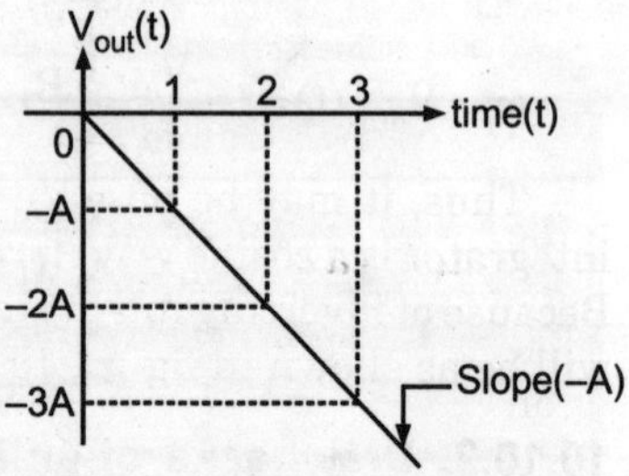

Fig. 10.26. Output waveform for step input

(ii) Square wave input signal

Let the input waveform is a square wave as shown in figure 10.27. It may be observed that the square wave is made up of steps i.e. a step of A between time period of 0 to $\frac{T}{2}$ while a step of – A units between a time period of $\frac{T}{2}$ to T and so on.

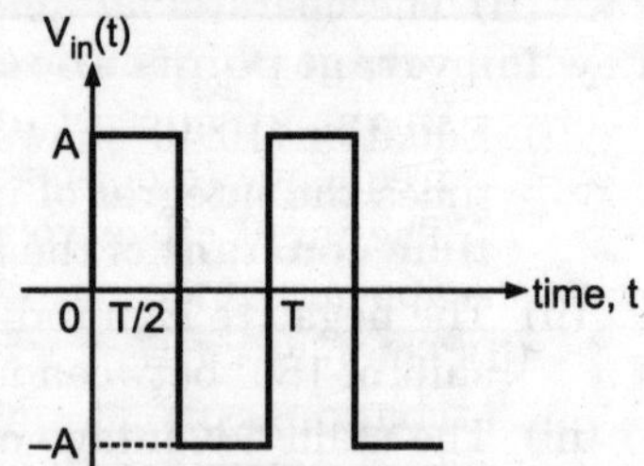

Fig. 10.27. Square wave input signal

Mathematically, it can be expressed as,

$$V_{in}(t) = \begin{cases} A, \text{for} & 0 < t < \frac{T}{2} \\ -A \text{ for} & \frac{T}{2} < t < T \end{cases} \quad ...(10.43)$$

This is the expression for the input signal for one period.

As discussed earlier, the output for step input is a straight line with a slope of – A. Therefore, for the period 0 to $\frac{T}{2}$, output will be a straight line with slope – A. From $t = \frac{T}{2}$ till t = T, the slope of the straight line will become – (– A) i.e. + A.

Hence, the output can be expressed mathematically for one period as under :

$$V_{out}(t) = \begin{cases} -At \text{ for} & 0 < t < \frac{T}{2} \\ +At \text{ for} & \frac{T}{2} < t < T \end{cases} \quad ...(10.44)$$

The output waveform has been shown in figure 10.28.

DO YOU KNOW ?

The op-amp integrator and differentiator are useful for signal waveshaping.

(iii) Sine wave input signal

Let the input waveform is purely sinusoidal with a frequency of ω rad/sec. Mathematically, it can be expressed as under :

$$V_{in}(t) = V_m \sin \omega t \qquad ...(10.45)$$

where V_m is the maximum amplitude of the sine wave and T be the period of the waveform. To find the output waveform, let us make use of equation (10.40) with $R_1C_F = 1$ and $V_{out}(t) = 0$ V.

Therefore, we have

$$V_{out}(t) = -\int V_{in}\, dt$$

or $$V_{out}(t) = -\int V_m \sin \omega t\, dt$$

or $$V_{out}(t) = -V_m\left[\frac{1}{\omega}(-\cos \omega t)\right]$$

or $$V_{out}(t) = -\frac{V_m}{\omega}(-\cos \omega t) \qquad ...(10.46)$$

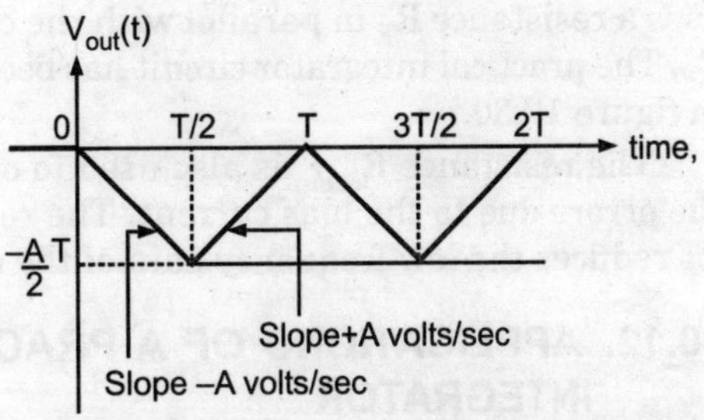

Fig. 10.28. Output waveform for square

Thus, it may be observed that the output of an integrator is a cosine waveform for a sine wave input. Because of inverting integrator, the output waveform will be as shown in figure 10.29.

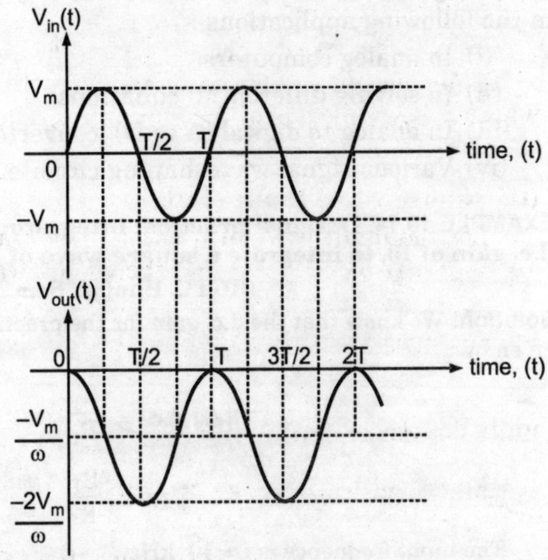

Fig. 10.29. Sine wave input and cosine output

10.10.2. Drawbacks of an ideal Integrator

(RGTU, Bhopal, Sem. Exam., 2006-07)

(i) The operational amplifier has input offset voltage (V_{ios}) and the input bias current (I_b). In the absence of input voltage or at zero frequency (d.c.), op-amp gain is very high. The input offset voltage gets amplified and appears at the output as an error voltage.

(ii) The bias current also results in a capacitor charging current and adds its effect in an output error voltage. The two components, due to high d.c. gain of op-amp cause output to ramp up or down, depending upon the polarities of offset voltage and/or bias current. After some time, output of op-amp may achieve its saturation level. Hence, there is a possibility of op-amp saturation due to such an error voltage and it is very difficult to pull op-amp out of saturation, due to such an error voltage. Thus, the output of an ideal integrator in the absence of input signal is likely to be offset towards the positive or negative saturation levels.

(iii) In the presence of the input signal also, the two components namely offset voltage and bias current, contribute an error voltage at the output. Thus, it is not possible to get a true integration of the input signal at the output. Output waveform may be distorted due to such an error voltage.

(iv) Another limitation of an ideal integrator is its bandwidth, which is very small. Hence, an ideal integrator can be used for a very small frequency range of the input only.

*Because of all these limitations, an ideal integrator is not used in practice. Some additional components are used alongwith the basic integrator circuit to reduce the effect of an error voltage, in practice. Such an integrator is called as **Practical Integrator Circuit**.*

10.11. A PRACTICAL INTEGRATOR

The limitations of an ideal integrator can be minimised in the practical integrator circuit, which uses a resistance R_F in parallel with the capacitor C_F. The practical integrator circuit has been shown in figure 10.30.

The resistance R_{comp} is also used to overcome the errors due to the bias current. The resistance R_F reduces the low frequency gain of the op-amp.

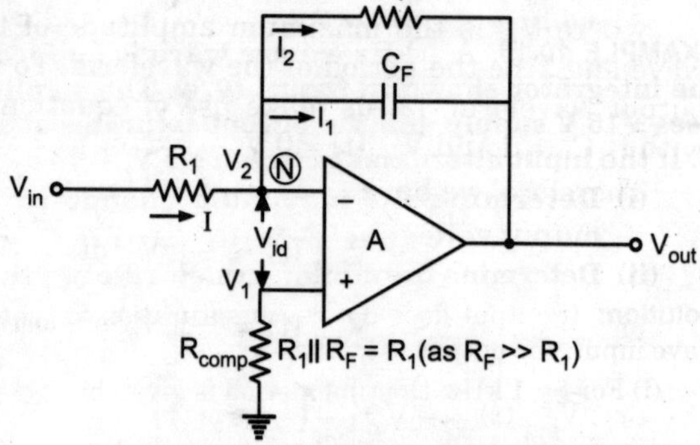

Fig. 10.30. A Practical Integrator circuit

10.12. APPLICATIONS OF A PRACTICAL INTEGRATOR

The integrator circuits are most commonly used in the following applications :

(i) In analog computers.
(ii) In solving differential equations.
(iii) In analog to digital (A to D) converters.
(iv) Various signal waveshaping circuits.

DO YOU KNOW ?

An inverting amplifier is a very versatile component and can be used for performing number of mathematical stimulation such as analog inverter, paraphase amplifier, phase shifter, adder, integrator, differentiator.

EXAMPLE 10.14. Design a practical integrator circuit with a d.c. gain of 10, to integrate a square wave of 10 kHz.

(RGTU, Bhopal, Sem. Exam., 2007-08)

Solution: We know that the d.c. gain for the practical integrator is given by

$$|A| \text{ d.c.} = \frac{R_F}{R_1}$$

or $$10 = \frac{R_F}{R_1} \quad ...(i)$$

The input frequency is f = 10 kHz.

Now, for the proper integration, $f \geq 10f_a$ where f_a is the break frequency of the practical integrator.

Therefore, $$\frac{f}{f_a} = 10$$

or $$f_a = \frac{f}{10} = \frac{10 \times 10^3}{10} = 1000 \text{ Hz} \quad ...(ii)$$

Now, for the practical integrator, we have

$$f_a = \frac{1}{2\pi R_F C_F}$$

or $$1000 = \frac{1}{2\pi R_F C_F}$$

or $$R_F C_F = 15915 \times 10^{-4}$$

Selecting $R_1 = 10\ k\Omega$, from equation (i), we get

$$R_F = 10R_1 = 10 \times 10 = 100\ k\Omega$$

Also, we have $$C_F = \frac{15915 \times 10^{-4}}{100 \times 10^3}$$

or $$C_F = 15915 \times 10^{-9} \text{ F} \sim 16 \text{ nF}$$

And $$R_{comp} = R_1 \,||\, R_f$$

or $$R_{comp} = \frac{10 \times 100}{10 + 100} = 9.09\ k\Omega$$ **Ans.**

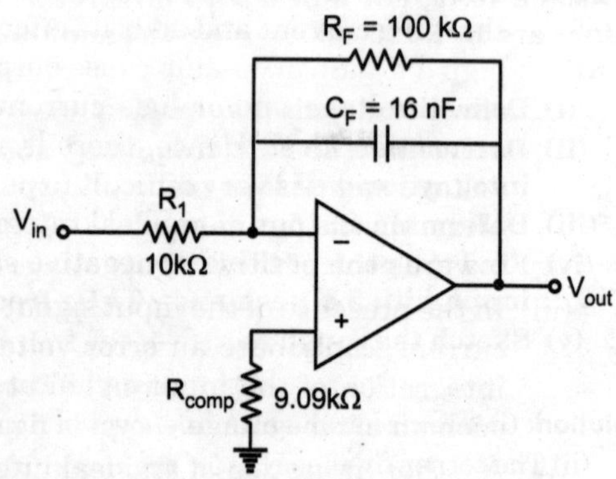

Fig. 10.31.

Hence, the practical integrator circuit wll be as shown in figure 10.31.

EXAMPLE 10.15. A 1 kHz square wave is applied to the integrator shown in figure 10.32. The amplifier uses ± 15 V supply and the output saturates at ± 14 V. If the input alternates between ± 5 V,

(i) Determine the maximum change in the output voltage

(ii) Determine the minimum slew rate required.

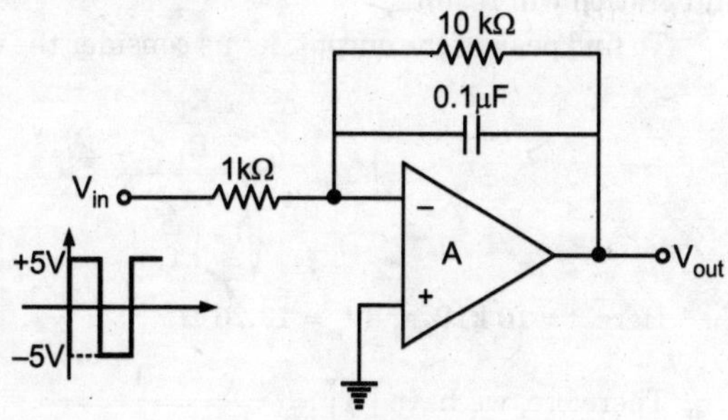

Fig. 10.32.

Solution: If output does not reach saturation for square wave input, the output is triangular.

(i) For f = 1 kHz, the time period is given by

$$T = \frac{1}{f} = \frac{1}{1 \times 10^3} = 1 \times 10^{-3} \text{ sec}$$

The change in the output will be

$$\Delta V_{out} = \frac{V_{in}\,T}{2R_1C_F} = \frac{5\text{ V} \times 1 \times 10^{-3}}{2 \times 1 \times 10^3 \times 0.1 \times 10^{-6}} = 25 \text{ V}$$

It seems that output may saturate as saturation level is ± 14 V. But 25 V change means the output will alternate between $\pm\frac{25}{2}$ i.e., ± 12.5 V. Hence, output will not be saturated and will be triangular in nature.

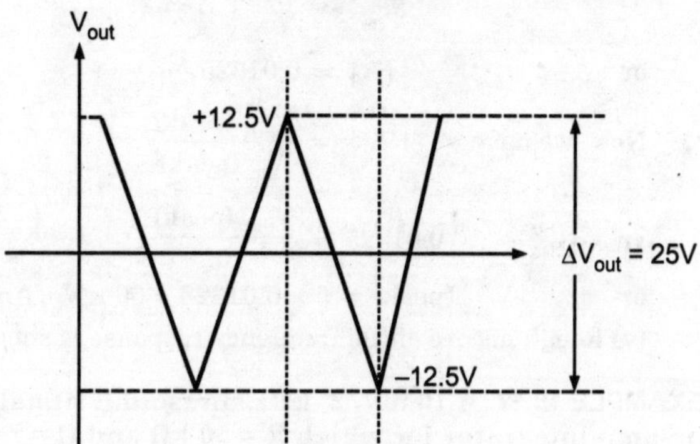

Fig. 10.33.

(ii) The expression for slew rate is given by

$$S = 2\pi f_m V_m$$

or

$$S = 2\pi \times 1 \times 10^3 \times 5 \text{ V/sec}$$
$$= 0.03141 \text{ V/μsec} \quad \textbf{Ans.}$$

This is the minimum slew rate required.

EXAMPLE 10.16. For a practical integrator, the component values are R_1 = 120 kΩ, R_F = 1.2 MΩ and the capacitor C_F = 10 nF.

(i) Draw the circuit diagram.

(ii) Determine the safe frequency above which true integration will take place.

(iii) Determine the d.c. gain.

(iv) Find the peak of the output voltage for a sine wave input with 5 V peak and 10 kHz frequency.

(v) Sketch the rough nature of the frequency response.

(GATE Examination, 1996)

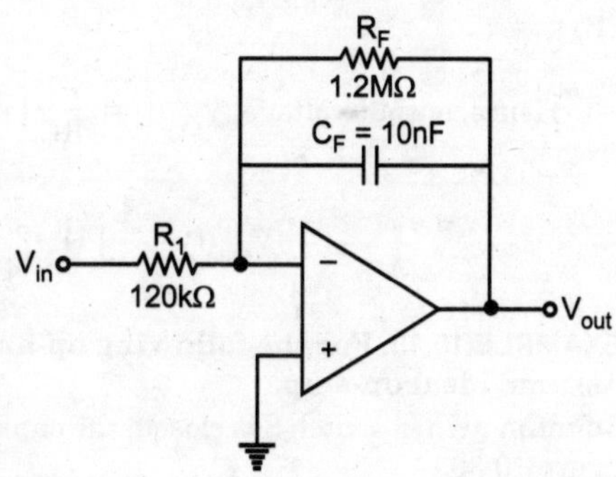

Fig. 10.34. (a)

Solution: (i) The circuit diagram is shown in figure 10.34 (a).

(ii) The corner frequency f_a for the circuit is given by

$$f_a = \frac{1}{2\pi R_F C_F} = \frac{1}{2\pi \times 1.2 \times 10^6 \times 10 \times 10^{-9}} = 13.26 \text{ Hz}$$

But, the safe frequency above which true integration will result is generally taken as 10 times the corner frequency f_a.

Therefore, Safe frequency = 10 f_a = 10 × 13.26 = 132.6 Hz **Ans.**

(iii) The d.c. gain is given by

$$|A|\text{ d.c.} = \frac{R_F}{R_1} = \frac{1.2 \times 10^6}{120 \times 10^3} = 10 \text{ i.e. } 20 \log |A| \text{ dB} = 20 \log 10 = 20 \text{ dB} \quad \textbf{Ans.}$$

(iv) The sine wave has frequency of 10 kHz which is much higher than safe frequency. Hence true integration will result.

To find peak of the output, let us consider the expression for the gain of the practical integrator circuit as

$$|A| = \frac{\frac{R_F}{R_1}}{\sqrt{1+\left(\frac{f}{f_a}\right)^2}}$$

Here, f = 10 kHz and f_a = 13.26 Hz

Therefore, we have $|A| = \dfrac{10}{\sqrt{1+\left(\dfrac{10\times10^3}{13.26}\right)^2}}$

or $|A| = 0.01325$

Now, we have $|A| = \dfrac{V_{out}\text{ (peak)}}{V_{in}\text{ (peak)}}$

or $0.01325 = \dfrac{V_{out}\text{ (peak)}}{5}$

or V_{out} (peak) = 5 × 0.01325 = 66 mV **Ans.**

(v) Rough nature of the frequency response is shown in figure 10.34 (b)

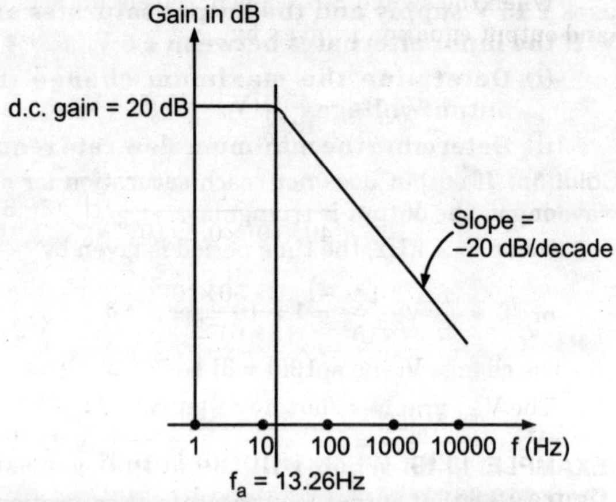

Fig. 10.34. (b)

EXAMPLE 10.17. A 10 mV, 2 kHz sinusoidal signal is applied to the inverting input terminal of an op-amp integrator for which R = 50 kΩ and C = 2 μF. Determine the output voltage.

Solution : The equation for the sinusoidal signal of rms value of 10 mV and frequency 2 kHz is given as

$$V_{in}(t) = \sqrt{2}\, V_{rms} \sin 2\pi f t = \sqrt{2} \times 10 \sin 4{,}000\, \pi\, t \text{ mV}$$

(assuming $V_{in}(t) = 0$ at $t = 0$)

$$\text{Scale factor} = -\frac{1}{CR} = -\frac{1}{2\times10^{-6}\times50\times10^{3}} = -10$$

$$\text{Hence, output voltage, } V_{out}(t) = \frac{-1}{RC}\int v_{in}(t)dt = -10\int_0^t \sqrt{2}\times10 \sin 4{,}000\pi t$$

$$V_{out}(t) = -141.42\left[\frac{-\cos 4{,}000\pi t}{4{,}000\pi}\right]_0^t = 0.0113\,(\cos 4{,}000\,\pi t - 1) \text{ mV} \quad \textbf{Ans.}$$

EXAMPLE 10.18. For the following op-amp circuit sketch V_{out} for 50 ms after the switch S is opened. Assume ideal op-amp.

Solution : When switch S is closed, the capacitor remains short-circuited, and the circuit becomes as shown in figure 10.36.

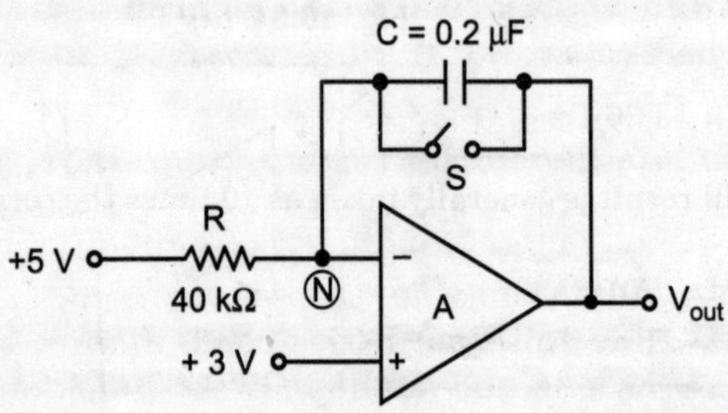

Fig. 10.35.

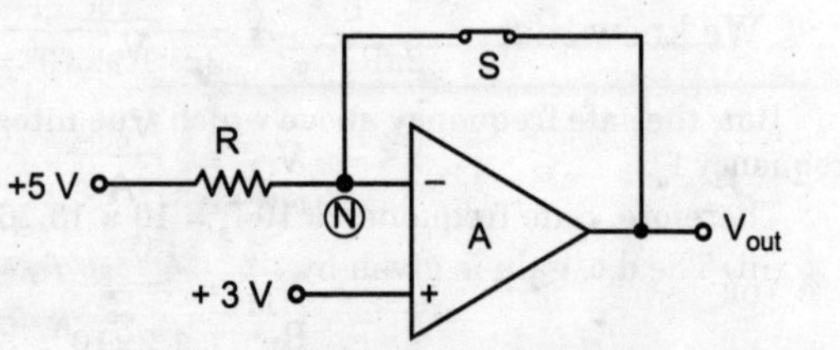

Fig. 10.36.

Since op-amp is assumed ideal, therefore, we can write

$$V_2 = V_1$$

and so output will be, $V_{out} = V_1 = V_2 = +3$ V

When the switch S is opened, the circuit acts like an integrator and output equation is given by

$$V_{out} = -\frac{1}{RC}\int_0^t (V_{in} - V_1)dt + V_{out}|_{t=0}$$

or $$V_{out} = \frac{1}{40\times10^3\times0.2\times10^{-6}}\int_0^{50ms}(5-3)dt + 3$$

or $$V_{out} = \frac{-1\times2\times50\times10^{-3}}{8\times10^{-3}} + 3$$

$$V_{out} = -12.5 + 3 = -9.5 \text{ V} \quad \textbf{Ans.}$$

The V_{out} graph is shown in figure 10.37.

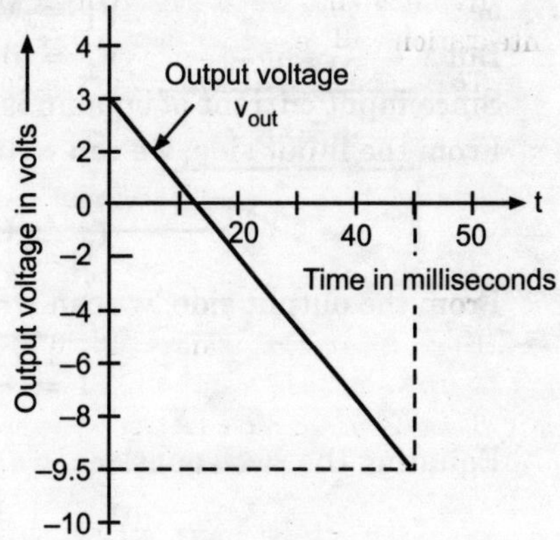

Fig. 10.37.

EXAMPLE 10.19. When will the output get saturated in figure 10.38.

Solution: The output equation is given as

$$V_{out} = -\frac{1}{RC}\int_0^t V(t)\,dt + A$$

or $$V_{out} = \frac{-1}{500\times10^3\times10\times10^{-6}}\int_0^t -0.5\,dt = [0.1t]_0^t$$

Output will get saturated when it attains the value of supply voltage, i.e., of 12 V.

Thus, $V_{out} = 0.1t = 12$ V

or time duration required for saturation of output voltage i.e.,

$$t = \frac{12}{0.1} = 120 \text{ seconds} \quad \textbf{Ans.}$$

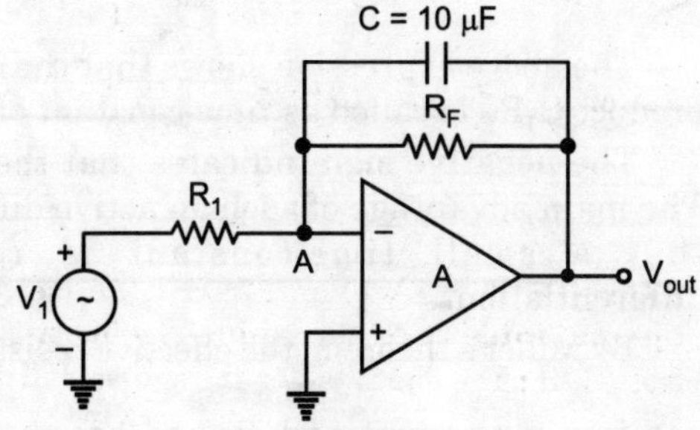

Fig. 10.38.

10.13. AN ACTIVE OP-AMP DIFFERENTIATOR **(Very Important)**

It may be observed from earlier discussion that in passive circuit, exchanging the positions of R and C, differentiator circuit can be obtained from the integrator circuit. Similarly, active differentiator circuit can be obtained by exchanging the positions of R and C in the basic active integrator circuit. The op-amp differentiator circuit has been shown in figure 10.39.

Let us assume that op-amp is ideal and is producing a finite output. Because of this, the open-loop gain A of the op-amp becomes infinite. This means that

$$A = \infty$$

We know that $$A = \frac{V_{out}}{V_{id}}$$

or $$V_{id} = \frac{V_{out}}{A}$$

or $$V_{id} = \frac{V_{out}}{\infty}$$

or $$V_{id} = 0$$

or $$V_1 - V_2 = 0$$

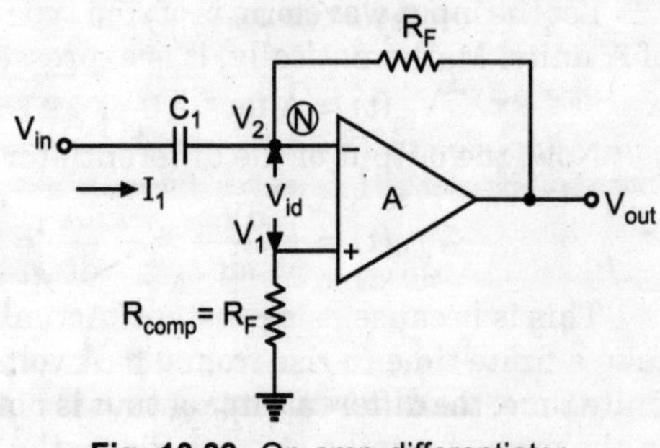

Fig. 10.39. Op-amp differentiator

or $\qquad V_1 = V_2$

But $\qquad V_1 = 0, \quad \text{so} \quad V_2 = 0$

Since input current of op-amp is zero, entire current I_1 flows through the resistance R_F.

From the input side, we can write,

$$I_1 = C_1 \frac{d(V_{in} - V_2)}{dt} = C_1 \frac{dV_{in}}{dt} \qquad ...(10.47)$$

From the output side, we can write,

$$I = \frac{(V_2 - V_{out})}{R_F} = -\frac{V_{out}}{R_F} \qquad ...(10.48)$$

Equating the two equations, we get

$$C_1 \frac{dV_{in}}{dt} = -\frac{V_{out}}{R_F} \qquad ...(10.49)$$

or

$$V_{out} = -C_1 R_F \frac{dV_{in}}{dt} \qquad ...(10.50)$$

The above expression shows that the output is C_1R_F times the differentiation of the input and product C_1R_F is called as time-constant of the differentiator.

The negative sign indicates that there is a phase shift of 180° between input and output. The main advantage of such an active differentiator is that a small time-constant is required for differentiation.

By Miller's theorem, the effective resistance between non-inverting input and ground becomes $\frac{R_F}{1 + A_v}, \frac{R_F}{A_v}$ where A_v is the gain of the op-amp which is very large. Hence, effective R_F becomes very small and hence the condition $R_FC_1 << T$ gets satisfied at all the frequencies.

In practice, a resistance $R_{comp} = R_F$ is connected to the non-inverting terminal to provide the bias compensation. This has been shown in figure 10.40.

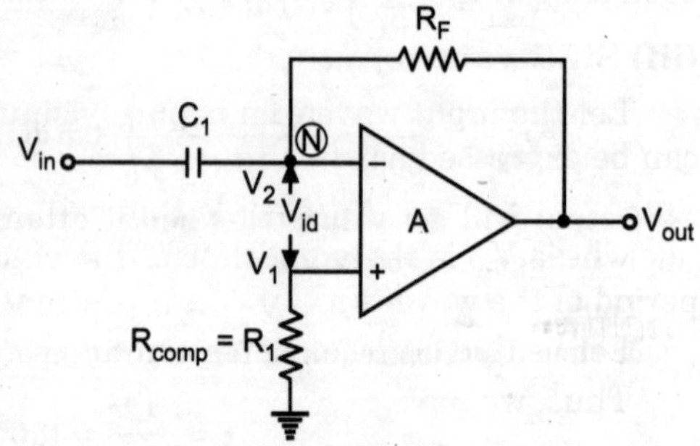

Fig. 10.40. Differentiator with bias compensation

10.13.1. Input and Output Waveforms for Differentiator

Let us study the output waveforms, for various input signals.

For simplicity, we assume that the values of R_F and C_1 are selected to have time constant $R_F C_1$ as unity.

(i) Step input signal

Let the input waveform is of step type with a magnitude of A units. Mathematically, it is expressed as,

$$V_{in}(t) = A \text{ for } t \geq 0 \qquad ...(10.51)$$

Now, the output of the differentiator will be,

$$V_{out}(t) = -\frac{dV_{in}}{dt} = -\frac{dA}{dt} = 0 \qquad ...(10.52)$$

This is because A is constant. Actually, the step input take a finite time to rise from 0 to A volts. Because of this finite time, the differentiator output is not zero but appears in the form of a spike at t = 0. Since, the circuit acts as an

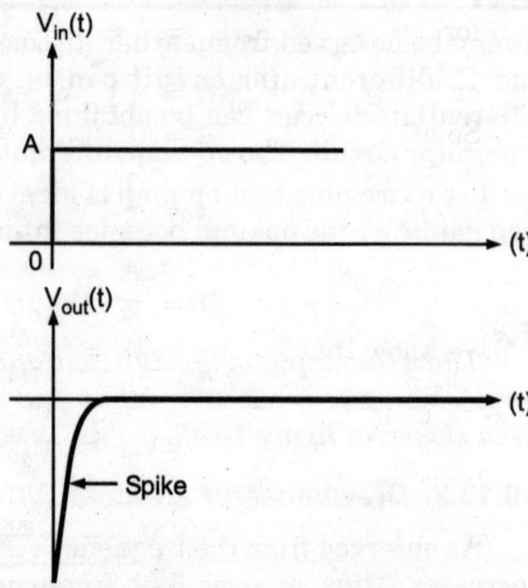

Fig. 10.41. Input and output waveforms for step input

inverting differentiator, the negative going spike or pulse appears at t = 0 and after that output remains zero. Both input and output waveforms of the differentiator with a step input, have been shown in figure 10.42.

(ii) Square wave input signal

The square wave is made of steps i.e. step of A volts from t = 0 to $t = \frac{T}{2}$, while a step of – A volts from $t = \frac{T}{2}$ to t = T and so on.

Mathematically, it can be expressed as,

$$V_{in}(t) = \begin{cases} A \text{ for } & 0 < t < \frac{T}{2} \\ -A \text{ for } & \frac{T}{2} < t < T \end{cases} \qquad ...(10.53)$$

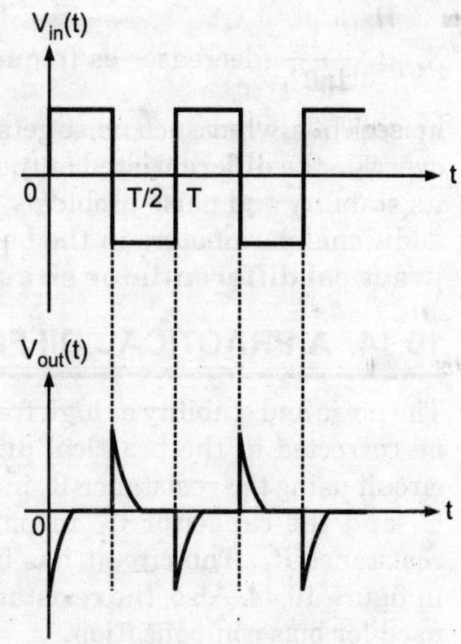

Fig. 10.42. Input and output for square wave input

The differentiator behaves similar to its behaviour to step input. For positive going pulse, the output shows negative going pulse and for negative going input, the output shows positive going pulse. Hence, the total output for the square wave input is in the form of pulses or spikes. The input and output waveforms have been shown in figure 10.42.

(iii) Sine wave input

Let the input waveform is purely sinusoidal with a frequency of ω rad/sec. Mathematically, it can be expressed as,

$$V_{in}(t) = V_m \sin \omega t \qquad ...(10.54)$$

where V_m is the amplitude of the sine wave and T is the period of the waveform.

Let us find out the expression for the output.

Thus, we have

$$V_{out}(t) = \frac{-dV_{in}(t)}{dt}$$

for $\quad R_F C_1 = 1 = -\frac{d}{dt}(V_m \sin \omega t)$

or $\quad V_{out}(t) = -V_m \omega \cos \omega t \qquad ...(10.55)$

So at $\quad t = 0, V_{out}(t) = -V_m \omega$

$$t = \frac{T}{4}, V_{out}(t) = 0$$

$$t = \frac{T}{2}, V_{out}(t) = +V_m \omega \text{ and so on.}$$

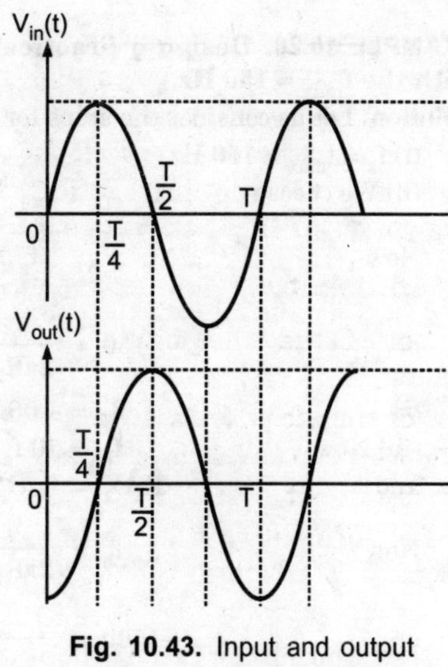

Fig. 10.43. Input and output waveforms for sine wave input

Thus, the output of the differentiator is a cosine waveform, for a sine wave input. The input and output waveform have been shown in figure 10.43.

10.13.2. Drawbacks of an Ideal Differentiator

As observed from the frequency response, the gain of the differentiator increases as frequency increases. Thus, at some high frequency, the differentiator may become unstable and break into the oscillations. There is possibility that op-amp may go into saturation. Also the input impedance

$X_{C1} = \dfrac{1}{2\pi f C_1}$ decreases as frequency increases. This makes the circuit very much sensitive to the noise. Thus, when such noise gets amplified due to high gain at high frequency, noise may completely override the differentiated output. Hence, the differentiator circuit suffers from the limitations on its stability and noise problems, at high frequencies. These problems can be corrected using some additional parameters in the basic differentiator circuit. Such a differentiator circuit is called as **practical differentiator circuit.**

10.14. A PRACTICAL DIFFERENTIATOR

The noise and stability at high frequency can be corrected in the practical differentiator circuit using the resistance R_1 in series with C_1 and the capacitor C_F in parallel with resistance R_F. The circuit has been shown in figure 10.44. Also, the resistance R_{comp} is used for bias compensation.

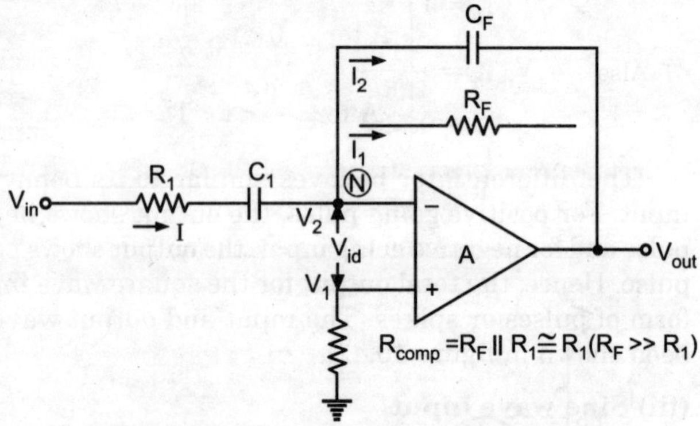

Fig. 10.44. A Practical differentiator circuit

10.15. APPLICATIONS OF PRACTICAL DIFFERENTIATOR

The practical differentiator circuits are most commonly used in :

(i) In waveshaping circuits to detect the high frequency components in the input signal.

(ii) As a rate-of-change detector in the FM demodulators.

EXAMPLE 10.20. Design a practical differentiator circuit that will differentiate an input signal with the f_{max} = 150 Hz.

Solution: Let us consider the steps for design.

(i) $f_a = f_{max} = 150$ Hz

(ii) We choose $C_1 = 1\ \mu F$

Now $$f_a = \frac{1}{2\pi R_F C_1}$$

or $$150 = \frac{1}{2\pi R_F \times 1 \times 10^{-6}}$$

or $R_f = 1.06\ k\Omega$

(iii) Now, $f_b = 10\ f_a = 10 \times 150 = 1500$ Hz

and $R_1C_1 = R_FC_F$

Now $$f_b = \frac{1}{2\pi R_1 C_1}$$

or $$1500 = \frac{1}{2\pi \times R_1 \times 1 \times 10^{-6}}$$

But $R_1 = 106.1\ \Omega$ and $R_1C_1 = R_FC_F$

or $106.1 \times 1 \times 10^{-6} = 1.06 \times 10^3\ C_F$; $C_F = 0.1\ \mu F$ **Ans.**

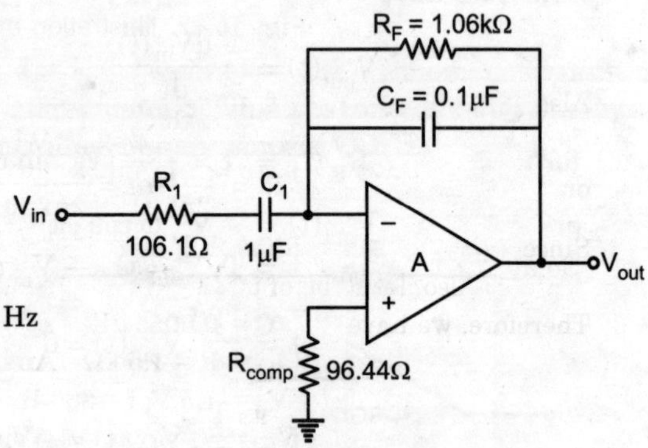

Fig. 10.45.

(iv) $$R_{comp} = R_1 \,||\, R_F = \frac{R_1 R_F}{R_1 + R_F} = \frac{106.1 \times 1.06 \times 10^3}{106.1 + 1.06 \times 10^3} = 96.44\ \Omega \quad \textbf{Ans.}$$

However, generally R_{comp} is selected equal to R_1. The designed differentiator circuit has been shown in the figure 10.45.

EXAMPLE 10.21. The input to an op-amp differentiatior circuit is a sinusoidal voltage of peak value of 10 μV and frequency 2 kHz. Determine the output voltage if R = 50 kΩ and C = 2 μF.

Solution: The equation for the sinusoidal signal of peak value of 10 μV and frequency 2 kHz is given as

$$V_{in} = V_{max} \sin 2\pi ft = 10 \sin 2 \times 2{,}000 \pi t$$

$$V_{in} = 10 \sin 4{,}000 \pi t \ \mu V$$

EXAMPLE 10.22. Design a differentiator to differentiate an input signal that varies in frequency from 10 Hz to 1 kHz. The input is a sine wave peak-to-peak amplitude 3 V at 200 Hz. Sketch the output voltage waveform.

Solution: The circuit has been shown in figure 10.46

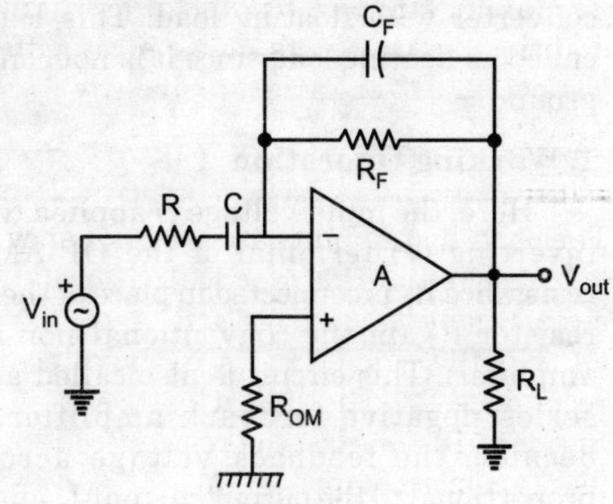

Fig. 10.46.

$$f_\alpha = 1 \text{ kHz}$$

Also $$f_\alpha = \frac{1}{2\pi RC}$$

$$C = 0.1 \ \mu F$$

$$R = \frac{1}{2\pi f_\alpha C} = \frac{1}{2\pi \times 1\times 10^3 \times 0.1\times 10^{-6}}$$

$$= 1.59 \text{ k}\Omega = 1.5 \text{ k}\Omega \text{ (say)}$$

$$f_b = 20 f_\alpha = 20 \text{ kHz}$$

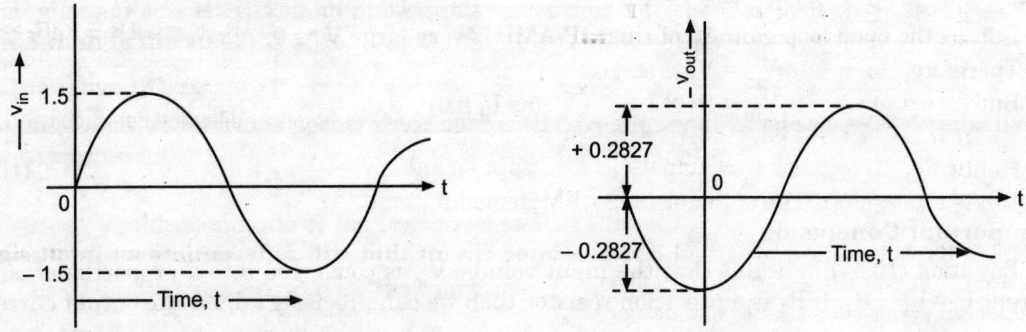

Fig. 10.47. Illustration of waveforms for example 10.22

Also, $$f_b = \frac{1}{2\pi R'C}$$

or $$R' = \frac{1}{2\pi f_b C} = \frac{1}{2\pi \times 20\times 10^3 \times 0.1\times 10^{-6}} = 79.5 \ \Omega$$

Since $$R'C = RC'$$

Also, let R′ be of 82 Ω.

Therefore, we have $$C = 0.0055 \ \mu F$$

$$R_{OM} = R = 1.5 \text{ k}\Omega$$ **Ans.**

$$V_p = 1.5 \text{ V}, f = 200 \text{ Hz}$$ ...(given)

$$V_{in} = V_p \sin \omega t = 1.5 \sin (2\pi \times 200)t = 1.5 \sin 400 \pi t$$

$$V_{out} = -RC\frac{dv_{in}}{dt} = -1.5 \times 10^3 \times 0.1 \times 10^{-6} \frac{d}{dt}(1.5 \sin 400 \pi t)$$

$$V_{out} = -0.2827 \cos 400 \pi t$$ **Ans.**

10.16. VOLTAGE TO CURRENT (V TO I) CONVERTERS (VOLTAGE CONTROLLED CURRENT SOURCE VCCS)

The voltage to current (V to I) converters may be classified into two categories, depending on the position of the load. They are as under:

(i) V to I converter with floating load

(ii) V to I converter with grounded load.

10.16.1. Voltage to Current Converter (VCCS) with Floating Load

Figure 10.48 shows a voltage to current converter with floating load. This load (R_L) is called as floating load since it is not connected to ground.

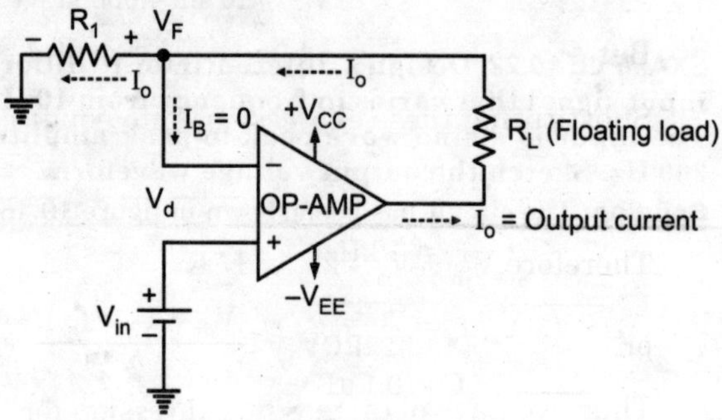

Fig. 10.48. Voltage to current converter with floating load

1. Working Operation

Here, the input voltage is applied to the non-inverting (+) terminal of the OP-AMP. Load resistance R_L is connected in place of the feedback resistor R_F (in the conventional non-inverting amplifier). This circuit is also called as current series negative feedback amplifier. This is because the feedback voltage across R_1 is proportional to the output current I_0 and appears in series with the input voltage. Apply KVL to the input loop of figure 10.48 to write,

$$V_{in} = V_d + V_F \quad \text{... (10.56)}$$

But, as the open loop gain A_V of this OP-AMP is very large $V_d \cong 0$.

Therefore, $V_{in} = V_F$

But $V_{in} = R_1 \times I_o$ (as $I_B \cong 0$)

Hence, $$I_o = \frac{V_{in}}{R_1} \quad \text{... (10.57)}$$

2. Important Conclusion

Equation (10.57) indicates that the input voltage V_{in} is converted into a proportional output current $I_o = V_{in} / R_1$. If R_1 is a precision resistor then we can precisely adjust the output current.

3. Why is the circuit called VCCS ?

Equation (10.57) shows that the output current $I_o = k\,V_{in}$ where $k = 1/R_1$ (This means that it is a constant). Therefore, OP-AMP is behaving like a current source which is controlled by the input voltage. Hence this circuit is called as a voltage controlled current source VCCS.

4. Applications

Voltage to current converter is used in applications such as low voltage dc and ac voltmeters, LED and Zener diode testers etc.

10.16.2. Voltage to Current Converter with Grounded Load

This is the other type of V to I converter, in which one terminal of the load is connected to ground. The V to I converter with grounded load is as shown in figure 10.49.

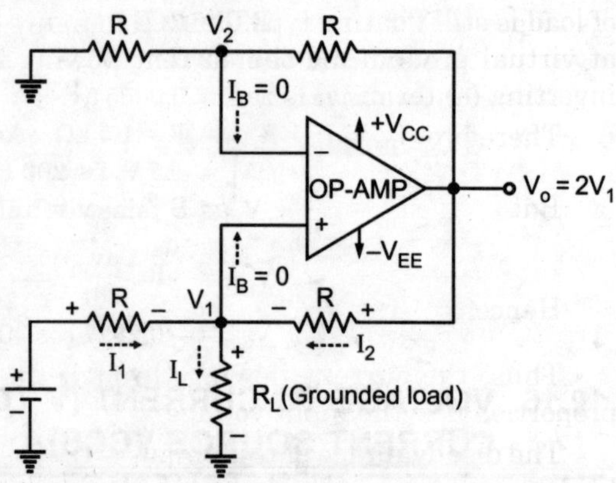

Fig. 10.49. Voltage to current converter with grounded load

Analysis of the Circuit

The analysis of the circuit shown in figure 10.49 can be done by following two steps : First step is to determine the voltage V_1 at the non-inverting (+) terminal and the second step is to establish relationship between V_1 and the load current I_L.

Applying KCL at node V_1, we can write that,

$$I_L = I_1 + I_2 \quad \text{... (10.58)}$$

But $$I_1 = \frac{V_{in} - V_1}{R} \quad \text{and} \quad I_2 = \frac{V_o - V_1}{R}$$

Substituting these expressions into equation (10.58), we get,

$$I_L = \frac{V_{in} - V_1}{R} + \frac{V_o - V_1}{R}$$

Therefore, $$V_{in} + V_o - 2V_1 = I_L R \quad \text{... (10.59)}$$

or $$V_1 = \frac{V_{in} - V_o - I_L R}{2} \quad \text{... (10.60)}$$

Thus, we have obtained the expression for V_1.

The OP-AMP in figure 10.49 is connected in the non-inverting mode. Therefore, gain of the circuit will be,

$$A_{VF} = 1 + \frac{R}{R} = 2.$$

The output voltage is given by,

$$V_o = A_{VF} \times V_1 = 2\,V_1$$

Substituting V_1 from equation (10.60), we get

$$V_o = V_{in} + V_o - I_L R$$

Therefore $$I_L R = V_{in}$$

or $$I_L = \frac{V_{in}}{R} \quad \text{... (10.61)}$$

Equation (10.61) shows that the load current is dependent on the input voltage and resistor R. Here, it may be noted that all the resistors in figure 10.49 must be equal in value.

10.16.3. V-I Converter with Inverting Amplifier

The V-I converter with input applied to the inverting terminal is shown in figure 10.50. The type of load is still floating type. Point B in figure 10.50 is at virtual ground and the current flowing into the inverting (–) terminal is zero.

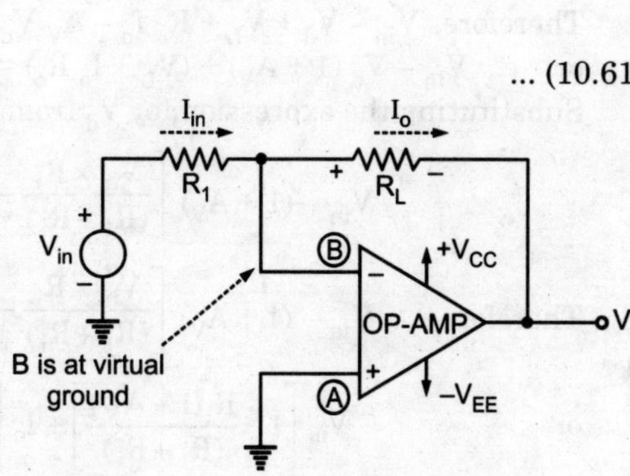

Fig. 10.50. V-I converter with inverting amplifier

Therefore $I_{in} = I_o$

But $$I_{in} = \frac{V_{in}}{R_1} \quad \text{as B is at virtual ground}$$

Hence, $$I_o = \frac{V_{in}}{R_1} \quad \text{... (10.62)}$$

Thus, the current flowing through the load is proportional to the input voltage V_{in}.

The disadvantage of this circuit is that the entire load current has to be supplied by the input source.

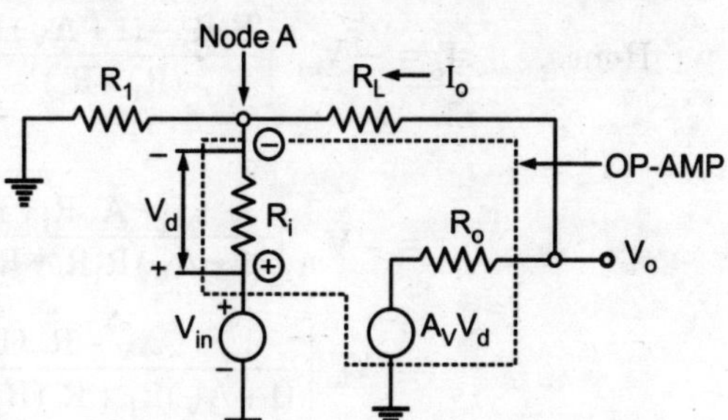

Fig. 10.51. Equivalent circuit of voltage to current converter

10.16.4. Practical OP-AMP Limitations

Till now, we have assumed that the OP-AMP is an ideal one. Now let us see the effect of practical OP-AMP on the operation of V to I converters. Let us consider the equivalent circuit shown in figure 10.51.

From the equivalent circuit, by applying KVL, we get

$$V_{in} - V_d + V_L + R_o I_o - A_V V_d = 0 \quad \text{... (10.63)}$$

At Node A, applying KCL, we get,

$$I_o = \frac{V_d}{R_i} - \frac{V_{in} - V_d}{R_1} = 0$$

Therefore, $$I_o = \frac{V_{in} - V_d}{R_1} - \frac{V_d}{R_i} \quad \text{... (10.64)}$$

or $$I_o = \frac{V_{in}}{R_1} - \frac{V_d}{R_1} - \frac{V_d}{R_i}$$

or $$I_o = \frac{V_{in}}{R_1} - V_d\left[\frac{1}{R_1} + \frac{1}{R_i}\right] = \frac{V_{in}}{R_1} - V_d\frac{(R_1 + R_i)}{R_1 R_i}$$

or $$V_d = \left[\frac{V_{in}}{R_1} - I_o\right] \times \frac{R_1 R_i}{(R_1 + R_i)} = \frac{V_{in} \times R_i}{(R_1 + R_i)} - \frac{I_o R_1 R_i}{(R_1 + R_i)} \quad \text{... (10.65)}$$

Now, let us consider equation (10.63)

$$V_{in} - V_d + V_L + R_o I_o - A_V V_d = 0$$

Let $V_L = I_o R_L$

Therefore, $V_{in} - V_d + V_L + R_o I_o - A_V V_d = 0$

or $$V_{in} - V_d (1 + A_V) + (V_L + I_o R_o) = 0 \quad \text{... (10.66)}$$

Substituting the expression for V_d from equation (10.65), we get

$$V_{in} - (1 + A_V)\left[\frac{V_{in} \times R_i}{(R_1 + R_i)} - \frac{I_o R_1 R_i}{(R_1 + R_i)}\right] + V_L + I_o R_o = 0$$

Therefore $$V_{in} - (1 + A_V)\left[\frac{V_{in} \times R_i}{(R_1 + R_i)} - \frac{I_o R_1 R_i}{(R_1 + R_i)}\right] + V_L + I_o R_o = 0$$

or $$V_{in}\left[1 - \frac{R_i(1 + A_V)}{(R_1 + R_i)}\right] + I_o\left[\frac{(1 + A_V) R_1 R_i}{(R_1 + R_i)} + R_o\right] V_L = 0$$

Hence, $$I_o = -V_{in}\left[\frac{R_1 R_i - (1 + A_V) R_i}{(R_1 + R_i)}\right] - \left[\frac{V_L}{\frac{(1 + A_V) R_1 R_i}{(R_1 + R_i)} + \frac{R_o R_1}{(R_1 + R_i)} + \frac{R_o R_i}{(R_1 + R_i)}}\right]$$

$$= -V_{in}\left[\frac{A_V R_i + R_1}{(1 + A_V) R_1 R_i + R_o R_1 + R_o R_i}\right] - \left[\frac{V_L}{(1 + A_V)(R_1 \,||\, R_i) + R_o}\right]$$

$$= V_{in}\left[\frac{A_V - R_1 / R_i}{(1 + A_V) R_1 + R_o (R_i + R_1) / R_i}\right] - \left[\frac{V_L}{(1 + A_V)(R_1 \,||\, R_i) + R_o}\right]$$

Therefore, $$I_o = \frac{V_{in}}{R_1}\left[\frac{A_V - (R_1 / R_i)}{1 + A_V + (R_o / R_i) + (R_o / R_i)}\right] - \left[\frac{V_L}{(1 + A_V)(R_1 \,||\, R_i) + R_o}\right] \quad \text{... (10.67)}$$

Now, let us consider it with the standard form of expression for I_o which states that

$$I_o = A_{VF} V_{in} - \frac{1}{R_{oF}} V_L \quad ...(10.68)$$

After comparing equations (10.67) and (10.68), we get,

$$A_{VF} = \frac{1}{R_1}\left[\frac{A_V - (R_1/R_0)}{1 + A_V + (R_o/R_o) + (R_o/R_i)}\right] \quad ... (10.69)$$

and

$$R_{oF} = (R_1 \| R_i)(1 + A_V) + R_o \quad ... (10.70)$$

Conclusion

From equation (10.70) for A_{VF}, it is obvious that as $A_V \to \infty$ we get the ideal result i.e., $A_{VF} \to (1/R_1)$ and $R_{oF} \to \infty$. But if A_V is finite then A_{VF} will exhibit some error and due to finite value of R_o, I_o will be slightly dependent on V_L.

10.16.5. Voltage to Current Converter with Grounded Load

This is the other type of V to I converter, in which one terminal of the load is connected to ground. The V to I converter with grounded load is as shown in figure 10.52 (a).

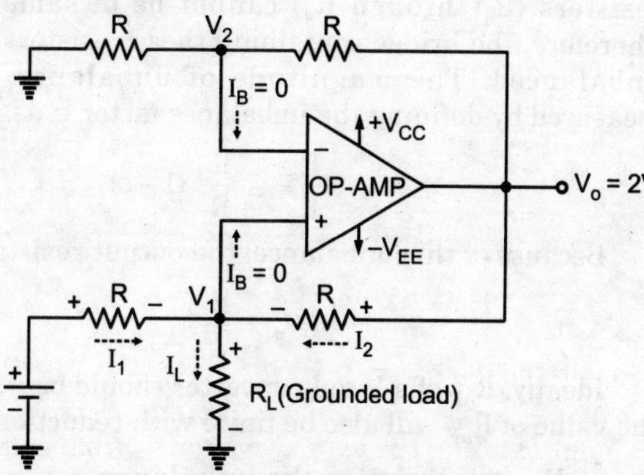

Fig. 10.52. (a) Voltage to current converter with grounded load

Analysis of the circuit

The analysis of the circuit shown in figure 10.52 (a) can be done by following two steps ; First step is to determine the voltage V_1 at the non-inverting (+) terminal and the second step is to establish relationship between V_1 and the load current I_L.

(i) First, we obtain voltage V_1.

Apply KCL at node V_1 to write :

$$I_L = I_1 + I_2 \quad ... (10.71)$$

But $I_1 = \dfrac{V_{in} - V_1}{R}$ and $I_2 = \dfrac{V_o - V_1}{R}$.

Substituting these expressions into equation (10.71), we get,

$$I_L = \frac{V_{in} - V_1}{R} + \frac{V_o - V_1}{R}$$

Therefore, $V_{in} + V_o - 2V_1 = I_L R \quad ... (10.72)$

or

$$V_1 = \frac{V_{in} + V_o - I_L R}{2} \quad ... (10.73)$$

Thus, we have obtained the expression for V_1.

(ii) Then, we establish relation between V_{in} and I_L.

The op-amp in figure 10.52 (a) is connected in the non-inverting mode. Therefore, gain of the circuit is,

$$A_{VF} = 1 + \frac{R}{R} = 2.$$

The output voltage is given by,

$$V_o = A_{VF} \times V_1 = 2\,V_1$$

Substituting V_1 from equation (10.73), we get,

$$V_o = V_{in} + V_o - I_L R$$

Therefore $\quad I_L R = V_{in}$

or
$$I_L = \frac{V_{in}}{R} \quad \text{... (10.74)}$$

Equation (10.74) shows that the load current is dependent on the input voltage and resistor R. Note that all the resistors in figure 10.52(a) must be equal in value.

10.16.6. Effect of Resistance Mismatches

In the circuit of figure 10.52 (a), all the resistances except R_L are of same value R. Let us redraw this circuit as shown in figure 10.52 (b).

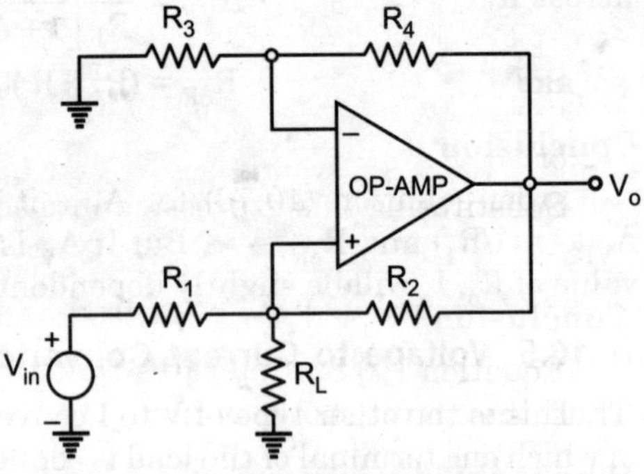

Fig. 10.52. (b)

In practice due to resistance tolerances, all the resistors (R_1 through R_4) cannot be of same value. Therefore, the bridge containing these resistors will be unbalanced. The magnitude of unbalance can be measured by defining the imbalance factor ∈ as,

$$\frac{R_4}{R_3} = \frac{R_2}{R_1}(1 - \varepsilon) \quad \text{... (10.75)}$$

Because of this unbalance, the output resistance can no more be ∞. It is given by,

$$R_{oF} = \frac{R_1}{\epsilon} \quad \text{... (10.76)}$$

Ideally, R_{oF} of a V to I converter should be ∞. However, practically, due to non-zero value of ∈, the value of R_{oF} will also be finite with reduction in the value of ∈, R_{oF} goes on increasing.

Remedies to minimize the imbalance

(i) The imbalance may be reduced by using the resistors of precise value.

(ii) Imbalance can be reduced by introducing trimmers to adjust the values of R_1, R_2, R_3 and R_4 to precisely correct values.

10.17. CURRENT TO VOLTAGE CONVERTER (I TO V) OR CURRENT CONTROLLED VOLTAGE SOURCE (CCVS)

The current to voltage (I to V) converter is as shown in figure 10.53. It may be noted that this circuit is a special case of inverting amplifier. The open loop gain A_V of the op-amp is very large. Hence, as per the concept of virtual short, we have

$$V_1 \cong V_2 \quad \text{... (10.77)}$$

Input impedance of the op-amp is very high. Therefore, the currents entering into the two input terminals is very small.

Hence, $\quad I_{B1} = I_{B2} = 0 \quad$ ... (10.78)

Gain of the inverting amplifier is given by,

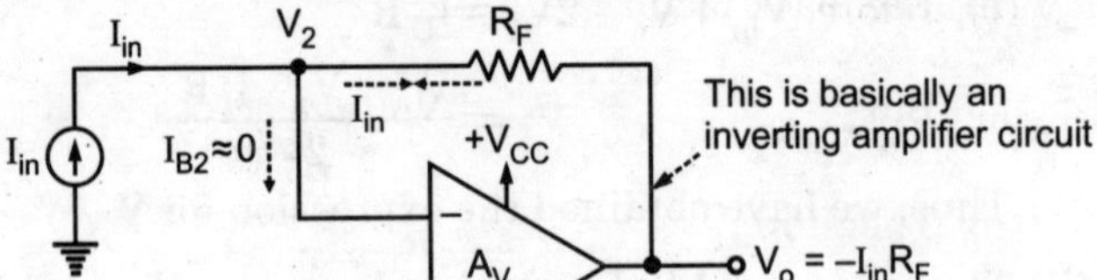

Fig. 10.53. Current to voltage converter

$$A_{VF} = -\frac{R_F}{R_1} = \frac{V_o}{V_{in}} \quad \text{...(10.79)}$$

Therefore, $$V_o = -\frac{R_F}{R_1} \times V_{in} \quad \text{... (10.80)}$$

But $V_1 \cong V_2$ and $V_1 = 0$ as the non-inverting (+) terminal is connected to ground.

Hence, $V_2 = 0$

Thus, the inverting terminal (–) also is at ground potential and the entire input voltage appears across R_1.

So, $$I_{in} = \frac{V_{in}}{R_1} \quad \text{... (10.81)}$$

or $$V_{in} = I_{in} R_1$$

Substituting this expression into equation (10.80), we get,

$$V_o = -R_F I_{in} \quad \text{... (10.82)}$$

Conclusion

Equation (10.82) indicates that the output voltage is proportional to the input current I_{in}. Thus, the circuit shown in figure 10.53 converts the input current into a proportional voltage.

Why is this Circuit called CCVS ?

Equation (10.82) shows that V_o is proportional to I_{in}. In other words, the op-amp is equivalent to a current controlled voltage source (CCVS).

10.17.1. Sensitivity of the I-V Converter

The output voltage $V_o = -R_F I_{in}$. Hence the gain of this converter is equal to $-R_F$. The magnitude of the gain i.e., R_F is also called as **sensitivity** of 1 to V converter. The amount of change in output voltage ΔV_o for a given change in the input current ΔI_{in} is decided by the sensitivity of the I to V converter. By keeping R_F variable, it is possible to vary the sensitivity as per requirements.

10.17.2. Closed Loop Parameters

(i) We have analyzed the I to V converter with the ideal op-amp assumption. Now let see how the non-idealities of op-amp affect the closed loop parameters.

(ii) **Type of Feedback :** As discussed earlier, the type of negative feedback is current shunt (or shunt) feedback.

The Loop gain T

(i) The loop gain $T = A_V \beta$

where A_V = Open loop gain and β = Feedback factor.

(ii) Feedback factor $$\beta = \frac{V_2}{A_V V_d} \quad \text{... (10.83)}$$

But V_2 = Voltage across R_i (Consider figure 10.54 (b)).

Therefore, $$V_2 = \frac{R_i}{R_i + R_F + R_o} \times A_V V_d \quad \text{... (10.84)}$$

(iii) Substituting this into equation (10.83), we get,

$$\beta = \frac{R_i \cdot A_V V_d}{(R_i + R_F + R_o) A_V V_d} = \frac{R_i}{(R_i + R_F + R_o)} \quad \text{... (10.85)}$$

(iv) Hence, the loop gain $$T = A_V \beta = \frac{A_V R_i}{(R_i + R_F + R_o)} \quad \text{... (10.86)}$$

This is the required expression for the loop gain.

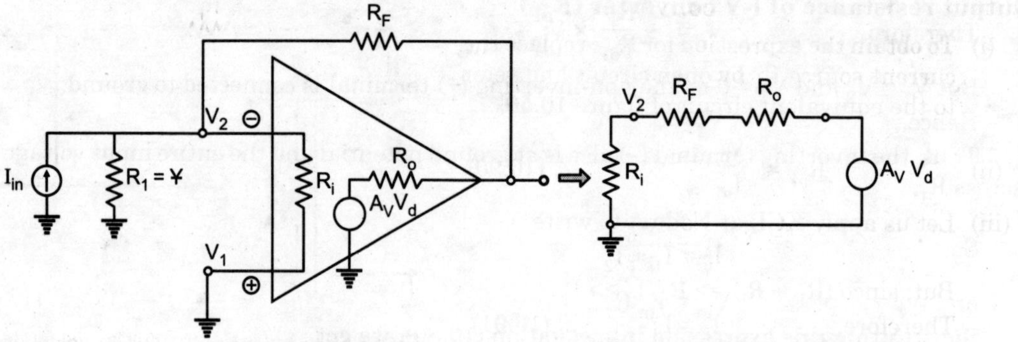

(a) Equivalent circuit (b) Simplified equivalent circuit

Fig. 10.54.

Input resistance of I-V converter (R_{iF})

To obtain the expression for R_{iF}, we consider figure 10.55.

(i) $$R_{iF} = \frac{V_2}{I_i} \quad \text{... (10.87)}$$

(ii) Let us apply KCL at Node A of figure 10.55 to write

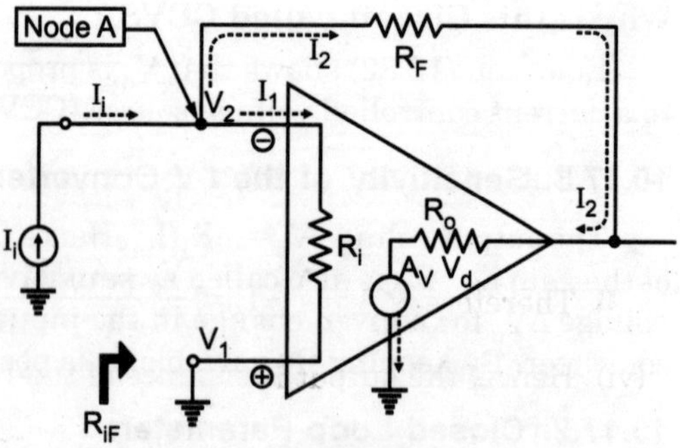

Fig. 10.55. Equivalent circuit to calculate input resistance

$$I_i = I_1 + I_2 = \frac{V_2}{R_i} + \frac{(V_2 - A_V V_d)}{(R_F + R_o)}$$

Note that $V_d = (V_1 - V_2) = 0 - V_2$ since $V_1 = 0$

Therefore, $V_d = -V_2$

Also $$I_i = \frac{V_2}{R_i} + \frac{(V_2 - A_V V_d)}{(R_F + R_o)}$$

or $$I_i = \frac{(R_F + R_o)V_2 + (V_2 + V_2 A_V)R_i}{R_i(R_F + R_o)}$$

or $$I_i = \frac{V_2[R_F + R_o + R_i(1 + A_V)]}{R_i(R_F + R_o)} \quad \text{... (10.88)}$$

(iii) Substituting this into equation (10.87), we get,

$$R_{iF} = \frac{V_2}{V_2[R_F + R_o + R_i(1 + A_V)]/R_i(R_F + R_o)} = \frac{R_i(R_{iF} + R_o)}{R_F + R_o + R_i(1 + A_V)}$$

Therefore, $$R_{iF} = \frac{R_i(R_F + R_o)}{R_F + R_o + R_i + R_i A_V}$$

(iv) Then, we divide numerator and denominator by $(R_F + R_o + R_i)$ to get

$$R_{iF} = \frac{R_i(R_F + R_o)/(R_F + R_o + R_i)}{1 + [R_i + A_V/(R_F + R_o + R_i)}$$

But $$T = \frac{R_i A_V}{(R_F + R_o + R_i)} \quad \text{and} \quad \frac{R_i(R_F + R_o)}{(R_F + R_o + R_i)} = R_i \parallel (R_F + R_o)$$

Therefore, $$R_{iF} = \frac{R_i \parallel (R_F + R_o)}{1 + T} \quad \text{... (10.89)}$$

This is the required expression.

Output resistance of I-V converter (R_{oF})

(i) To obtain the expression for R_{oF}, replace the current source I_{in} by open circuit and refer to the equivalent circuit of figure 10.56.

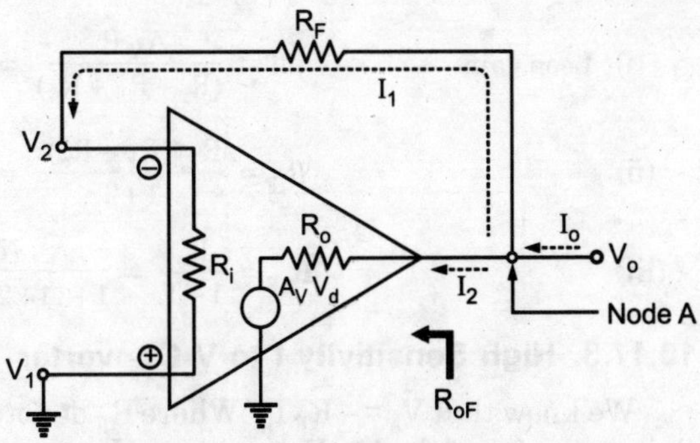

Fig. 10.56. Equivalent circuit to calculate R_{oF}

(ii) $$R_{oF} = \frac{V_o}{I_o} \quad \text{... (10.90)}$$

(iii) Let us apply KCL at Node A to write

$$I_o = I_1 + I_2$$

But, since $(R_F + R_i) >> R_o$, $I_2 >> I_1$.

Therefore, $$I_o \cong I_2 \quad \text{... (10.91)}$$

(iv) Considering figure 10.56, we get

$$I_o = I_2 = \frac{V_o - A_V V_d}{R_o} \quad \text{... (10.92)}$$

(v) But, $V_d = V_1 - V_2$ and $V_1 = 0$, hence, $V_d = -V_2$. Substituting this value into equation (10.92), we get,

$$I_o = I_2 = \frac{V_o + A_V V_2}{R_o} \quad \text{... (10.93)}$$

But $$V_2 = \frac{R_i}{R_i + R_F} V_o$$

Therefore $$I_o = \frac{V_o + A_V[R_i/(R_i + R_F)]V_o}{R_o} = V_o\left[\frac{1 + A_V R_i/(R_i + R_F)}{R_o}\right] \quad \text{... (10.94)}$$

(vi) Hence, the output resistance of the I-V converter is given by,

$$R_{oF} = \frac{V_o}{I_o} = \frac{R_o}{1 + [A_V R_i/(R_i + R_F)]} \quad \text{... (10.95)}$$

We know that $$T = \frac{A_V R_i}{R_i + R_F + R_o} \text{ hence } \frac{A_V R_i}{R_i + R_F} \cong T$$

Hence $$R_{oF} \cong \frac{R_o}{1+T} \quad \text{... (10.96)}$$

This is the required expression for R_{oF}.

Summary of analysis of I to V converter

(i) Loop gain : $T = \dfrac{A_V R_i}{(R_i + R_F + R_o)}$

(ii) Input resistance : $R_{iF} = \dfrac{R_i \| (R_F + R_o)}{1+T}$

(iii) Output resistance : $R_{oF} = \dfrac{R_o}{1+T}$

EXAMPLE 10.23. Obtain the closed loop parameters of a current to voltage converter circuit implemented using op-amp 741. Assume that the value of feedback resistance $R_F = 1.5$ MΩ.

Solution : The parameters of op-amp 741 are as follows :

$$R_i = 2 \text{ M}\Omega, A_V = 2 \times 10^5, R_o = 75\ \Omega$$

(i) Loop gain $T = \dfrac{A_V R_i}{(R_i + R_F + R_o)} = \dfrac{2\times10^5 \times 2\times10^6}{2\times10^6 + 1.5\times10^6 + 75} = 114.28 \times 10^3$ **Ans.**

(ii) $R_{iF} = \dfrac{R_i \parallel (R_F + R_o)}{1+T} = \dfrac{2\times10^6 \parallel (1.5\times10^6 \times 75)}{1+114.28\times10^3} = 7.5\ \Omega$ **Ans.**

(iii) $R_{oF} = \dfrac{R_o}{1+T} = \dfrac{75}{1+114.28\times10^3} = 656\ \text{M}\Omega$ **Ans.**

10.17.3. High Sensitivity I to V Converter

We know that $V_o = -R_F I_{in}$. Where R_F denotes the sensitivity of the I to V converter. Hence, one might feel that to increase the sensitivity, we have to increase the value of R_F to an unrealistically high value. But the high sensitivity I to V converter shown in figure 10.57 proves it to be wrong.

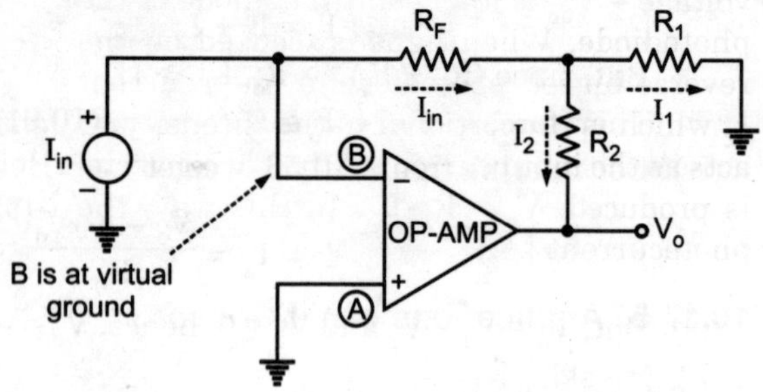

Fig. 10.57. High sensitivity I to V converter

(i) Let us apply KCL at node V_1 to write

$$I_{in} - I_1 - I_2 = 0 \quad ...(10.97)$$

(ii) But $I_{in} = \dfrac{V_B - V_1}{R_F} = \dfrac{0 - V_1}{R_F}$ *

Therefore, $I_{in} = -\dfrac{V_1}{R_F}$

And $I_1 = \dfrac{V_1}{R_1}, \quad I_2 = \dfrac{V_1 - V_0}{R_2}$

(iii) Substituting the expressions for I_{in}, I_1 and I_2 into equation (10.97), we obtain

$$-\frac{V_1}{R_F} - \frac{V_1}{R_1} - \frac{(V_1 - V_0)}{R_2} = 0 \quad ... (10.98)$$

(iv) We have, $-\dfrac{V_1}{R_F} - \dfrac{V_1}{R_1} - \dfrac{V_1}{R_2} + \dfrac{V_2}{R_2} = 0$

Therefore, $\dfrac{V_0}{R_2} = V_1\left[\dfrac{1}{R_F} + \dfrac{1}{R_1} + \dfrac{1}{R_2}\right]$

Substituting $V_1 = -R_F I_{in}$, we get

$$V_o = R_2 \times -R_F I_{in}\left[\frac{1}{R_F} + \frac{1}{R_1} + \frac{1}{R_2}\right] = -\left[1 + \frac{R_2}{R_F} + \frac{R_2}{R_1}\right] R_F I_{in}$$

Hence, $V_o = -k R_F I_{in}$

where $k = 1 + \dfrac{R_2}{R_F} + \dfrac{R_2}{R_1}$

The Sensitivity of this circuit is kR_F: As the value of k is always going to be higher than 1, $k R_F > R_F$. Hence, the circuit of figure 10.57 has a higher sensitivity than the original I to V converter. The sensitivity of the original I to V converter has been increased by a factor k.

* Since $V_B = 0$ due to virtual ground.

10.17.4. Applications of I to V Converter

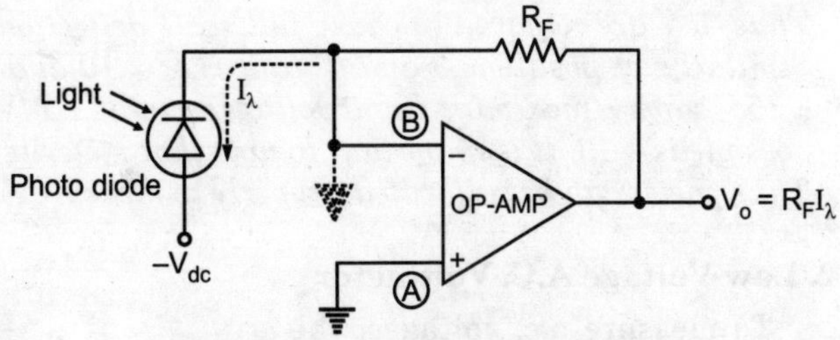

Fig. 10.58. Photodetector amplifier

One of the most important application of I to V converter is as the photodetector amplifier as shown in figure 10.58. D is a photodiode which is reverse biased. Point B i.e., the inverting terminal of the OP-AMP is at virtual ground potential and a negative voltage $- V_{dc}$ is applied at the anode of the photodiode. When light is focused on the reverse biased photodiode, a photocurrent I_2 which is proportional to the intensity of light starts flowing through it as shown. This current acts as the input current to the I-V converter. Hence the output voltage proportional to photocurrent is produced. $V_o = R_F\, I_\lambda$. In this way, the I to V converter provides amplification to the small photocurrent I_λ.

10.17.5. Applications of V-I Converter

1. Low Voltage D.C. Voltmeter

The V-I converter circuit with floating load may be modified as a low voltage d.c. voltmeter circuit. The resistance R_L is to be replaced by D'Arsonval meter movement. This is shown in the figure 10.59.

For accurate readings, it is essential to make the nulling of op-amp. This is done by offset voltage compensating network. Let R_m be the meter resistance. The effective Thevenin's equivalent of compensating network is 10Ω. Hence, when switch S is in position 1, the effective R_1 is 10 Ω + 1 kΩ ≅ 1 kΩ.

Let V_{in} = 1 V, then,

$$I_o = \frac{V_{in}}{R_1} \cong \frac{1\text{ V}}{1\text{k}\Omega} = 1\text{ mA}$$

So, current of 1 mA i.e., input of 1V will result full scale deflection, if meter has full scale deflection current of 1 mA.

When switch is in position 2 then R_1 = 5 kΩ. Thus, for full scale deflection, we have

$$1 \times 10^{-3} = \frac{V_{in}}{5 \times 10^{3}}$$

$$\therefore \qquad V_{in} = 5\text{ V}$$

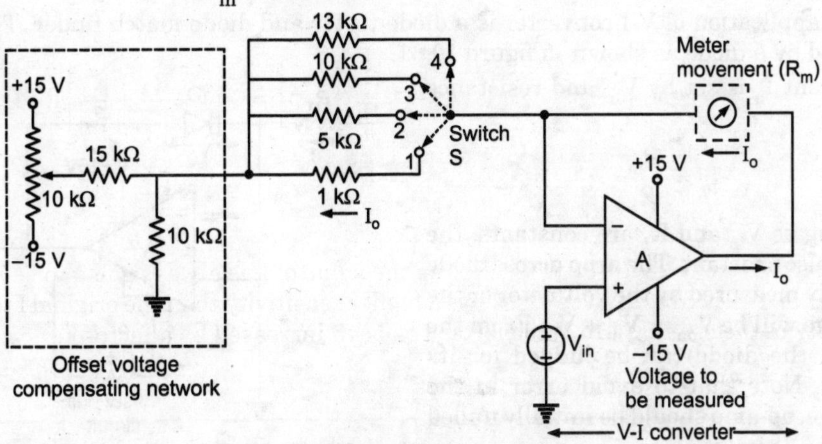

Fig. 10.59. D.C. voltmeter

Thus, 5 V are required to cause full scale deflection. Thus, range of voltmeter increases by 5. Similarly, at positions 3 and 4, ranges of × 10 and 13 are also obtained. Remember that for ± 15 V supply, maximum input voltage range is ± 14V hence with ± 15V supply, maximum range possible is × 13. It is important to note that meter resistance R_m does not affect the value of I_O. Thus, meter can be calibrated properly to measure d.c. voltages.

2. Low Voltage A.C. Voltmeter

To measure a.c. voltages, the full wave bridge rectifier and D'Arsonval meter movement are used. The combination of meter and rectifier is used in the feedback loop of a.c. voltmeter. The circuit is shown in figure 10.60.

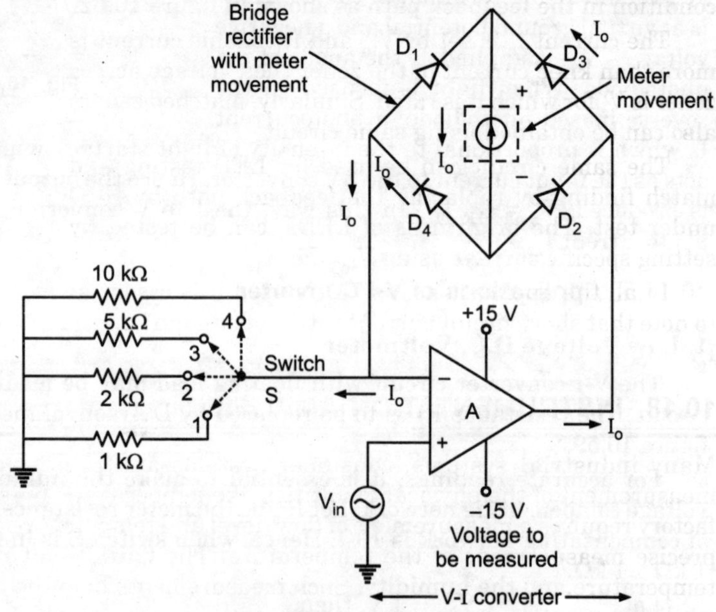

Fig. 10.60. A.C. Voltmeter

The alternating current I_o is converted to d.c. During positive half cycle of V_{in}, the diode D_3 and D_4 conduct while for negative half cycle of V_{in}, the diodes D_1 and D_2 conduct. The direction of current through meter is always same for both the half cycles of V_{in}. Thus, meter gives the average (dc) value of the rectified current. If it is necessary to calibrate the meter in terms of r.m.s values of input of the rectifies current. If it is necessary to calibrate the meter in terms of r.m.s values of input voltage then the factor 1.11 must be considered which relates r.m.s and average values for purely sinusoidal waveforms. The range can be extended using the selector switch S and various resistances as shown in figure 10.70. At position 2, × 2 can be obtained. At position 3, × 5 range can be obtained. At position 4, × 10 range can be obtained. It must be noted that due to low slew rate of 741 op-amp, the input a.c. voltage frequency should not be greater than 4 kHz. Similarly, PIV ratings of the diodes must be large than the saturation voltages of op-amp.

3. Diode Tester and Match Finder

Another application of V-I converter is a diode tester and diode match finder. The resistance R_L is replaced by a diode as shown in figure 10.61.

The current I_o is set by V_{in} and resistance R_1.

$$I_o = \frac{V_{in}}{R_1}$$

So as long as V_{in} and R_1 are constants, the current I_o is also constant. The drop across diode can be directly measured by the voltmeter or the output voltage will be $V_{out} = V_{in} + V_D$. From the value of V_D, the diode can be judged for its performance. Note that to avoid error in the output voltage, op-amp should be initially nulled using offset null circuit.

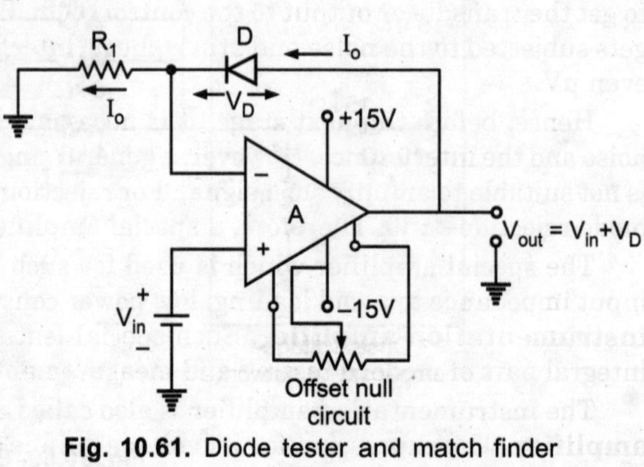

Fig. 10.61. Diode tester and match finder

Similarly, replacing diodes one after the other, the perfectly matched diodes can be found out. R_1 can be selected according to V_{in} and testing current I_o required in the circuit.

4. Zener Diode Tester

The V-I converter circuit can also be used to test zener diode. The zener diode is connected in reverse biased condition in the feedback path as shown in figure 10.62.

The current I_o is set by V_{in} and R_1. If this current is more than knee current of the zener then voltage across zener is V_z for which it is rated. Similarly, matched zeners also can be obtained using same circuit

The same circuit can be used for LED testing and match finding by replacing the feedback path by LED under test. The brightness of LEDs, can be tested by setting specific current using V_{in} and R_1.

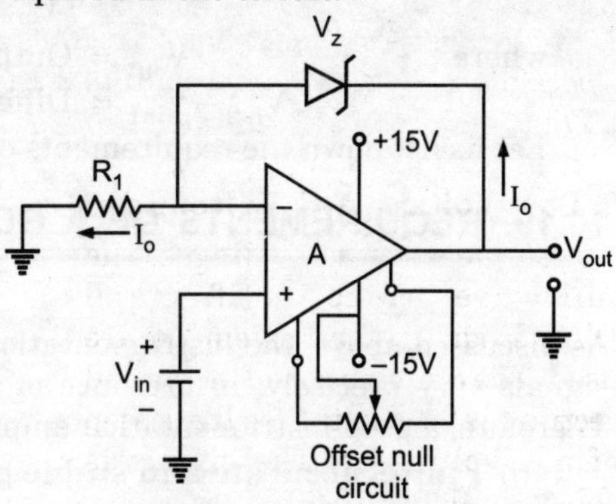

Fig. 10.62. Zener diode tester

In all the applications, as 741 is used it is essential to note that short circuit current rating of op-amp i.e., 25 mA should not be exceeded by the current I_o.

10.18. INSTRUMENTATION AMPLIFIER **(Very Important)**

Many industrial systems, consumer systems and process control systems require a precise measurement of the physical quantities like temperature, pressure, humidity, weight etc. A sugar factory requires a measurement of flow, level and temperature of juice. The plastic furnaces require precise measurement of the temperature. The dairy plant requires a precise measurement of temperature and the humidity. Such measurements help the industries to have the production of quality products.

The measurement of the physical quantities is generally carried out with the help of a device called as **transducer.** A transducer is a device which converts one form of energy into another. For example, a thermocouple converts the heat energy into an electrical energy, microphone converts the sound energy into an electrical energy, strain gauge converts pressure, force like mechanical energy into an electrical energy. Such a proportional electrical signal output from a transducer can be used further to control or operate the other parts of the system or can be used to get the display or recording of the measured physical quantity.

But, most of the transducer outputs are generally of very low level signals. Such low level signals are not sufficient to drive the next stage of the system. One more difficulty in the practical systems is that the transducer used may be mounted on pieces of equipment or structures which are remote from the control location. Long connecting wires or cables are required in such a case, to get the transducer output to the control room. Because of this, the signal which itself is low level, gets subjected to the noise and atmospheric interference. Such a signal may be as low as few mV or even μV.

Hence, before the next stage, it is necessary to amplify the level of such signal, rejecting the noise and the interference. However, a general single ended amplifier like high gain emitter amplifier is not suitable to amplify such signal. For rejection of noise, such amplifiers must have high common mode rejection ratio. Therefore, a special amplifier is used to amplify such signals.

The special amplifier which is used for such a low level amplification with high CMRR, high input impedance to avoid loading, low power consumption and some other features is known as an **instrumentation amplifier.** Such special featured instrumentation amplifiers have become an integral part of modern testing and measurement instrumentation.

The instrumentation amplifier is also called as **data amplifier** and is basically a **difference amplifier.** The expression for its voltage gain is generally of the form,

$$A = \frac{V_{out}}{V_{in2} - V_{in1}} \quad ...(10.99)$$

where V_{out} = Output of the amplifier

$V_{in2} - V_{in1}$ = Differential input which is to be amplified

Let us list down the requirements of an instrumentation amplifier.

10.19. REQUIREMENTS OF A GOOD INSTRUMENTATION AMPLIFIER

(RGTU, Bhopal, Sem. Exam. 2008-09)

As discussed above, the instrumentation amplifiers are used to amplify the low level differential signals very precisely, in presence of the large common mode noise and interference signals. Therefore, a good instrumentation amplifier has to meet the following specifications :

(i) **Finite, accurate and stable gain**

As very low level signals are required to be amplified by the instrumentation amplifiers, high and finite gain is the basic requirement. It is usually in the range of 1 to 1000. The gain has to be accurate and closed loop gain must be stable in nature.

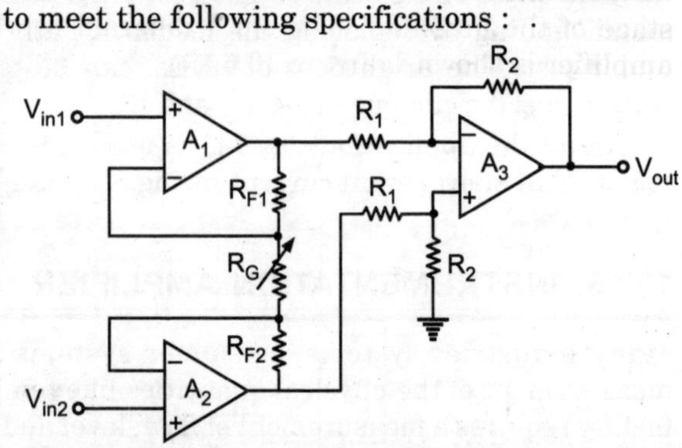

Fig. 10.63. Three op-amps instrumentation amplifier

(ii) **Easier gain adjustment**

Not only finite and stable gain is required but a variable gain over the prescribed range is also required. The gain adjustment must be easier and precise. Generally, such gain adjustment is done continuously using a potentiometer or is done digitally with the help of switches, which are JFET or MOSFET switches.

(iii) **High input impedance**

To avoid the loading of input sources, input impedance of the instrumentation amplifier must be very high (ideally infinite). The differential mode input impedance Z_{id} is the equivalent impedance between the two input terminals. The common mode input impedance Z_{ic} is the equivalent impedance between each input terminal and ground.

(iv) **Low output impedance**

Extremely low output impedance (ideally zero) to avoid the loading on the immediate stage.

(v) **High CMRR**

The output of transducer, when transmitted with long transmission lines has presence of larger common mode noise voltages. The instrumentation amplifier must amplify only the differential input, completely rejecting the common mode input component. Thus, it must have ideally infinite CMRR.

(vi) **Low power consumption**

The power consumption of an instrumentation amplifier should be as low as possible.

(vii) **Low thermal and time drifts**

The parameters of the instrumentation amplifier, should not drift with temperature or time.

(viii) **High slew rate**

The slew rate of the instrumentation amplifier must be as high as possible to provide maximum undistorted output voltage swing.

(ix) The amplifier must have differential input so that it can be amplified.

Let us discuss the various circuits which can be used as instrumentation amplifier and their advantages and drawbacks.

10.20. THREE OP-AMPS INSTRUMENTATION AMPLIFIER*

(RGTU, Bhopal, Sem. Exam; 2006-07, 2010-11)(10 marks)

Another commonly used instrumentation amplifier circuit is one using three op-amps. This circuit provides high input resistance for accurate measurement of signals from transducers. In this circuit, a non-inverting amplifier is added to each of the basic difference amplifier inputs. The circuit is shown in figure 10.64.

The op-amps A_1 and A_2 are the non-inverting amplifiers forming the input or first stage of the instrumentation amplifier. The op-amp A_3 is the normal difference amplifier forming an output stage of the amplifier. The block diagram representation of the three op-amp instrumentation amplifier is shown in figure 10.64.

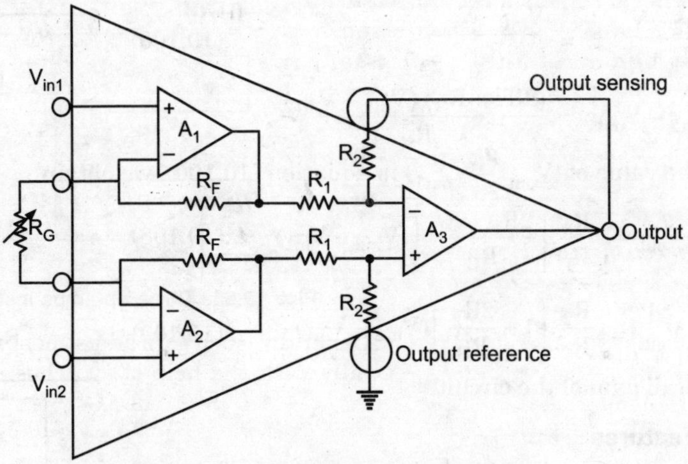

Fig. 10.64. Block diagram representation

Analysis of Three Op-amps Instrumentation Amplifier

It may be observed that the output stage is a standard basic difference amplifier. Hence, if the output of the op-amp A_1 is V_{out1} and the output of the op-amp A_2 is V_{out2}, from the previous discussion, we can write,

$$V_{out} = \frac{R_2}{R_1}(V_{out2} - V_{out1}) \qquad ...(10.100)$$

Let us find out the expression for V_{out2} and V_{out1} in terms of V_{in1}, V_{in2}, R_{F1}, R_{F2} and R_G.

Let us consider the first stage as shown in figure 10.75.

The node N_1 potential of op-amp A_1 is V_{in1}. From the realistic assumption, the potential of node N_2 is also V_{in1}. And hence potential of N_6 is also V_{in1}.

The node N_4 potential of op-amp A_2 is V_{in2}. From the realistic assumption, the potential of node N_3 is also V_{in2}. And hence potential of N_7 is also V_{in2}.

The input current of op-amp A_1 and A_2 both are zero. Hence, current I remains same through R_{F1}, R_G and R_{F2}.

Applying Ohm's law between the nodes N_5 and N_6, we obtain

---

* Describe Instrumentation amplifier. What is the most desired characteristic of it and maximum limit to that? **(RGTU, Sem. Exam., 2006-07) (10 marks)**

$$I = \frac{V_{out1} - V_{out2}}{R_{F1} + R_G + R_{F2}} \quad ...(10.101)$$

Let $$R_{F1} = R_{F2} = R_F \quad ...(10.102)$$

Therefore, $$I = \frac{V_{out1} - V_{out2}}{2R_F + R_G} \quad ...(10.103)$$

Now from the observation of nodes N_6 and N_7, we have

$$I = \frac{V_{N_6} - V_{N_7}}{R_G} = \frac{V_{in1} - V_{in2}}{R_G} \quad ...(10.104)$$

Equating the two equations (10.103) and (10.104), we get

$$\frac{V_{out1} - V_{out2}}{2R_F + R_G} = \frac{V_{in2} - V_{in1}}{R_G} \quad ...(10.105)$$

or $$\frac{V_{out2} - V_{out1}}{2R_F + R_G} = \frac{V_{in1} - V_{in2}}{R_G} \quad ...(10.106)$$

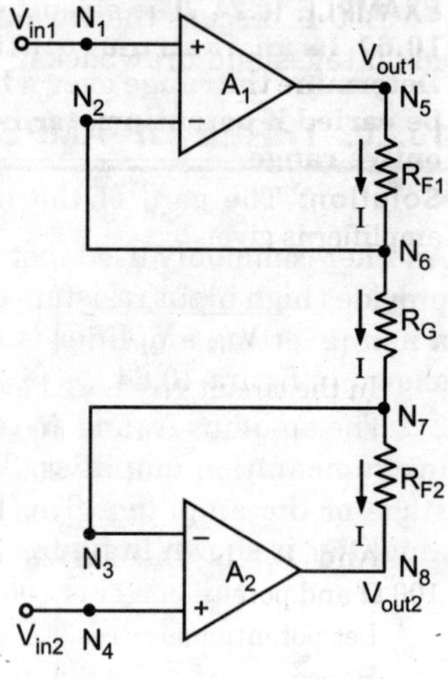

Fig. 10.65.

or $$V_{out2} - V_{out1} = \frac{(2R_F + R_G)(V_{in2} - V_{in1})}{R_G} \quad ...(10.107)$$

Substituting the value of $(V_{out2} - V_{out1})$ in equation (10.100), we obtain

$$V_{out} = \frac{R_2}{R_1}.\left[\frac{2R_F + R_G}{R_G}\right](V_{in2} - V_{in1}) \quad ...(10.108)$$

or $$V_{out} = \frac{R_2}{R_1}.\left(1 + \frac{2R_F}{R_G}\right)(V_{in2} - V_{in1}) \quad ...(10.109)$$

This is the overall gain of the circuit.

10.20.1. Salient Features

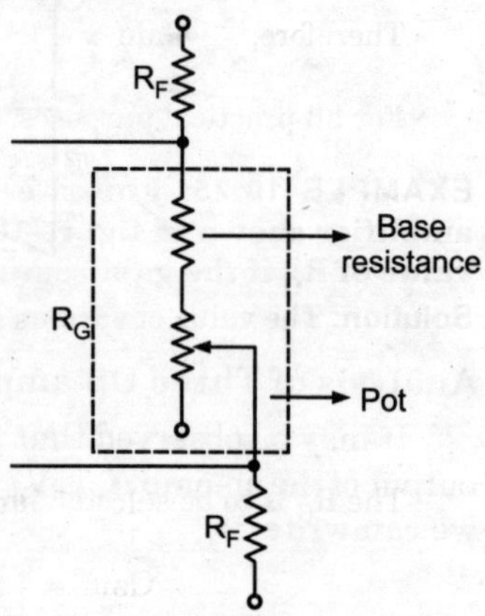

Fig. 10.66.

Following are various salient features of three op-amps instrumentation amplifier circuit :

(i) With the help of variable resistance R_G, the gain can be easily varied, without disturbing the symmetry of the circuit.

(ii) Gain depends on external resistances and hence can be adjusted accurately and made stable by selecting high quality resistances.

(iii) The input impedance depends on the input impedance of non-inverting amplifiers which is extremely high.

(iv) The output impedance is the output impedance of the op-amp A_3 which is very very low. This is as required by any instrumentation amplifier.

(v) The CMRR of the op-amp A_3 is very high and most of the common mode signal will be rejected.

(vi) By trimming one of the resistances of the output stage, CMRR can be made extremely high, as required by a good instrumentation amplifier.

Thus, the circuit satisfies all the requirements of a good instrumentation amplifier and hence very commonly used in practical applications.

Important Point : *The resistance R_G is generally implemented as the series combination of the suitable base resistance and the pot. This is shown in the figure 10.76.*

EXAMPLE 10.24. The circuit shown in figure 10.67 is an instrumentation amplifier. Determine the range over which its gain can be varied if potentiometer is varied over its entire range.

Solution: The gain of the instrumentation amplifier is given by

$$\frac{V_{out}}{V_{in2} - V_{in1}} = \left(1 + \frac{2R_F}{R_G}\right)\left(\frac{R_2}{R_1}\right)$$

In the circuit given, we have

$$R_2 = 100\ \Omega$$
$$R_1 = 200\ \Omega$$
$$R_F = 100\ k\Omega$$

And R_G is the series combination of 100 Ω and potentiometer of 100 kΩ.

Let potentiometer resistance is 0 Ω at start,

Hence, $R_G = 100 + 0 = 100\ \Omega$

Therefore,

$$\text{Gain} = \left(1 + \frac{2 \times 100 \times 10^3}{100}\right)\left(\frac{100}{200}\right) = 1000.5 \quad \textbf{Ans.}$$

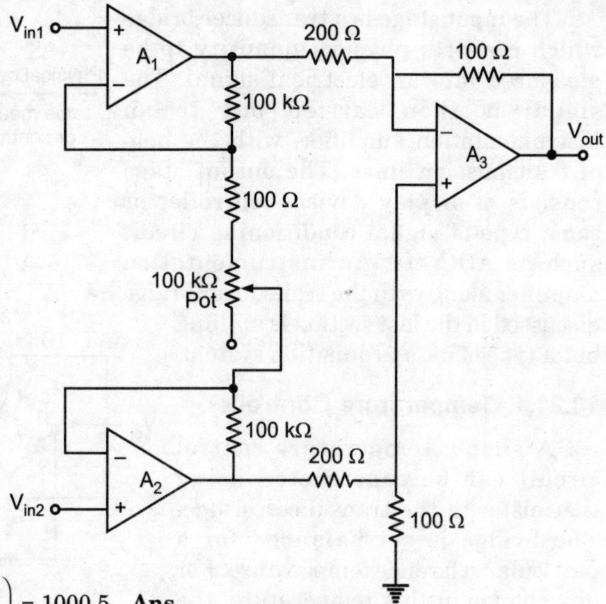

Fig. 10.67.

When the potentiometer resistance is at its maximum value, then

$$R_G = 100 + 100\ k\Omega = 100.1\ k\Omega.$$

Therefore, $$\text{Gain} = \left(1 + \frac{2 \times 100 \times 10^3}{100.1 \times 10^3}\right)\left(\frac{100}{200}\right) = 1.499 \quad \textbf{Ans.}$$

For all practical purposes, the gain can be varied from 1.5 to 1000.5.

EXAMPLE 10.25. For the instrumentation amplifier shown in figure 10.68, determine the value of R_G if the gain required is 1000.

Solution: The value of various resistances are

$$R_1 = 100\ k\Omega$$
$$R_2 = 100\ k\Omega$$
$$R_F = 470\ k\Omega$$

The R_G is to be selected for gain of 1000.

$$\text{Gain} = \left(1 + \frac{2R_F}{R_G}\right)\left(\frac{R_2}{R_1}\right)$$

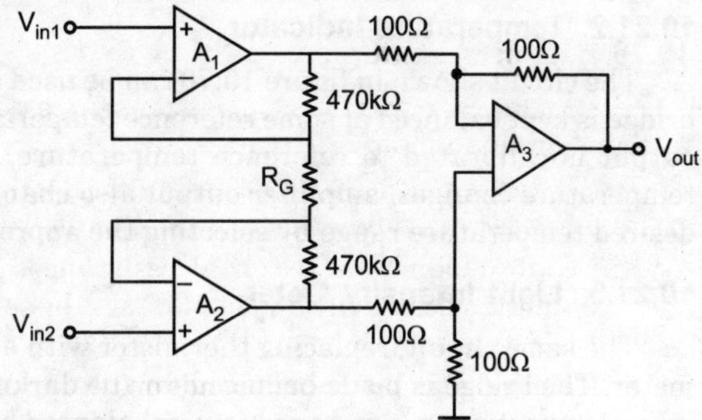

Fig. 10.68.

or $$100 = \left(1 + \frac{2 \times 470}{R_G}\right)\left(\frac{100}{100}\right)$$

or $$100 \times 1 = 1 + \frac{940}{R_G}$$

or $$\frac{940}{R_G} = 99 \quad \text{or} \quad R_G = 9.49\ \Omega.$$

10.21. APPLICATIONS OF INSTRUMENTATION AMPLIFIER

As discussed earlier, the instrumentation amplifier along with the transducer bridge can be used in many practical applications. The general form of such system can be called as 'data acquisition system' and can be represented in the block diagram form as shown in figure 10.69.

The input stage is a transducer bridge which converts physical quantity to be measured into an electrical signal. The signal is then carried out to an instrumentation amplifier, with the help of transmission lines. The output stage consists of display device, controller or some type of signal conditioning circuit such as ADC etc. An instrumentation amplifier along with the transducer bridge discussed in the last section is nothing but a type of data acquisition system.

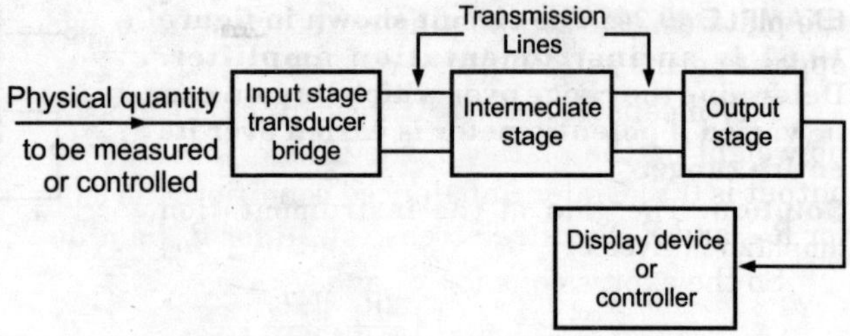

Fig. 10.69. Data acquisition system

10.21.1. Temperature Controller

A simple temperature control circuit can be constructed using thermistor in the transducer bridge. The bridge is set balanced for a particular reference temperature. For any change in this temperature, the instrumentation amplifier produces the output voltage. Now, this voltage can be used to drive the relays which in turn controls the ON-OFF of the heating unit, to control the temperature. This system is shown in figure 10.70.

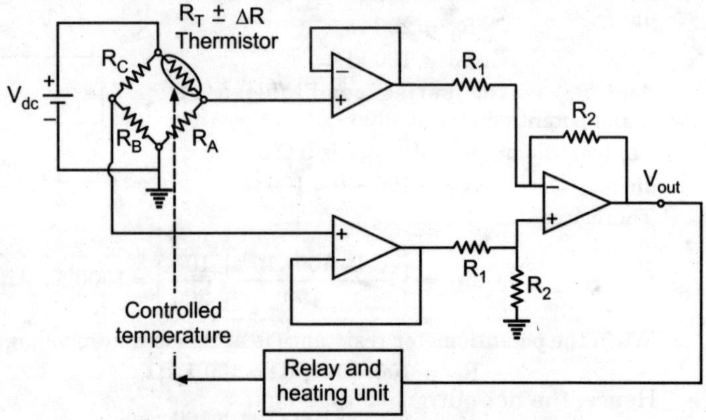

Fig. 10.70. Simple temperature controller

10.21.2. Temperature Indicator

The circuit shown in figure 10.70 can be used as a temperature indicator. As explained earlier, bridge is kept balanced at some reference temperature when V_{out} = 0 V. The meter connected at the output is calibrated to reference temperature, corresponding to this reference condition. As temperature changes, amplifier output also changes. The meter can be calibrated to indicate the desired temperature range by selecting the appropriate gain of the amplifier.

10.21.3. Light Intensity Meter

The same circuit, replacing thermistor with a photocell can be used as a simple light intensity meter. The bridge is made balanced for the darkness condition. When light falls on the photocell, its resistance changes and produces unbalanced bridge condition. This produces the output, which inturn produces the meter deflection. The meter can be calibrated in terms of lux to measure the light intensity. Such a light intensity meter is very accurate and stable.

10.21.4. Analog Weight Scale

The similar circuit, but using strain gauges in all the four arms of the bridge can be used as simple analog weight scale. The bridge is shown in figure 10.71.

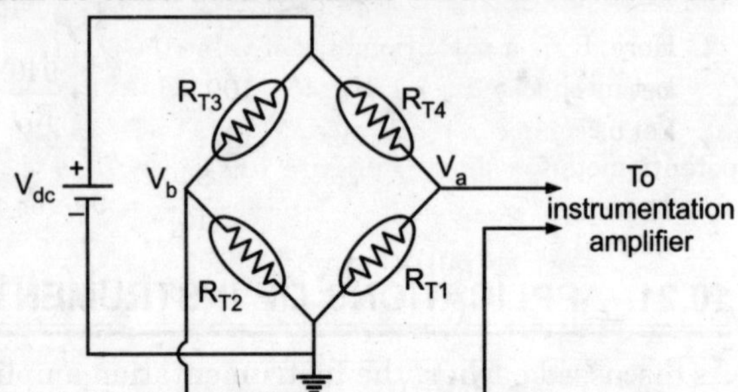

Fig. 10.71. Bridge for analog weight scale

The elements are mounted on the base of the weight platform. When the weight is placed on the platform, the strain gauges in the opposite arm elongate while the strain gauges in other two opposite arms get compressed.

This means R_{T1} and R_{T3} resistances decrease. While R_{T2} and R_{T4} resistances increase. Under no weight on the platform, the bridge is balanced and $R_{T1} = R_{T2} = R_{T3} = R_{T4} = R$ and the bridge output is 0 V. Under unbalanced condition, the change in resistance is ΔR ohms. So increase of ΔR for R_{T2} and R_{T4} while decrease of ΔR for R_{T1} and R_{T4}.

So the expressions for V_a and V_b are,

$$V_b = \frac{(R+\Delta R)\,V_{dc}}{(R-\Delta R+R+\Delta R)} \quad ...(10.110)$$

or $$V_b = \frac{(R+\Delta R)\,V_{dc}}{2R} \quad ...(10.111)$$

$$V_a = \frac{(R-\Delta R)\,V_{dc}}{(R-\Delta R+R+\Delta R)} \quad ...(10.112)$$

or $$V_a = \frac{(R-\Delta R)\,V_{dc}}{2R} \quad ...(10.113)$$

or $$V_{ab} = V_b - V_a = \frac{(R+\Delta R)\,V_{dc}}{2R} - \frac{(R-\Delta R)\,V_{dc}}{2R} \quad ...(10.114)$$

$$= \frac{2\Delta R\,V_{dc}}{2R} = \frac{\Delta R\,V_{dc}}{R} \quad ...(10.115)$$

Hence, the net output voltage is,

$$V_{out} = \frac{A\,\Delta R\,V_{dc}}{2} \quad ...(10.116)$$

Important Point: *The output voltage is proportional to the change in the resistance, which inturn depends upon the applied weight. The meter connected at the output can be calibrated in kilograms.*

EXAMPLE 10.26. Design the instrumentation amplifier to have a variable differential gain in the range 5-200. Use a 50 kΩ potentiometer.

(RGTU, Bhopal, Sem. Exam; 2004-05)

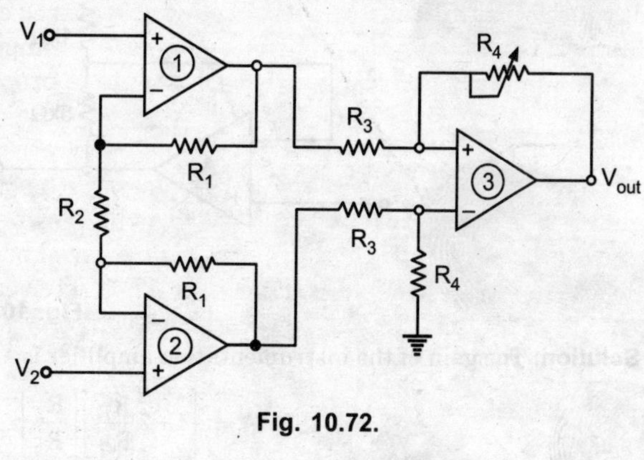

Fig. 10.72.

Solution: It may be observed that the given circuit is a three op-amp instrumentation amplifier.

Therefore, we have

$$A_F = \left[1 + \frac{2R_1}{R_2}\right] \times \frac{R_4}{R_3}$$

Here, R_4 is a potentiometer of value 0 – 50 kΩ

Let us choose R_3 = 1 kΩ

Let us connect a resistance of 1 kΩ in series with potentiometer as shown in figure 10.84.

Therefore, we have

$$\left.\frac{R_4}{R_3}\right|_{min} = \frac{1\,k}{1\,k} = 1$$

and $$\left.\frac{R_4}{R_3}\right|_{max} = \frac{(50+1)k}{1\,k} = 51$$

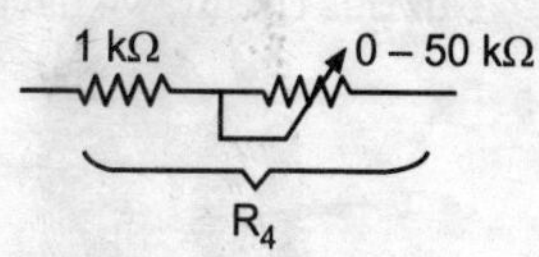

Fig. 10.73.

Since the range of gain required is 5 – 200, therefore, we have

$$5 = \left[1+\left(\frac{2R_1}{R_2}\right)\right]\times 1$$

or $$1+\frac{2R_1}{R_2} = 5$$

or $$\frac{2R_1}{R_2} = 4 \quad \text{or} \quad 2\,R_1 = 4\,R_2$$

Therefore, $$R_1 = 2\,R_2$$

Let us choose $$R_2 = 10\ \text{k}\Omega \quad \text{or} \quad R_1 = 20\ \text{k}\Omega \quad \textbf{Ans.}$$

EXAMPLE 10.27. Determine the expression for the output voltage in the circuit shown in figure 10.74.

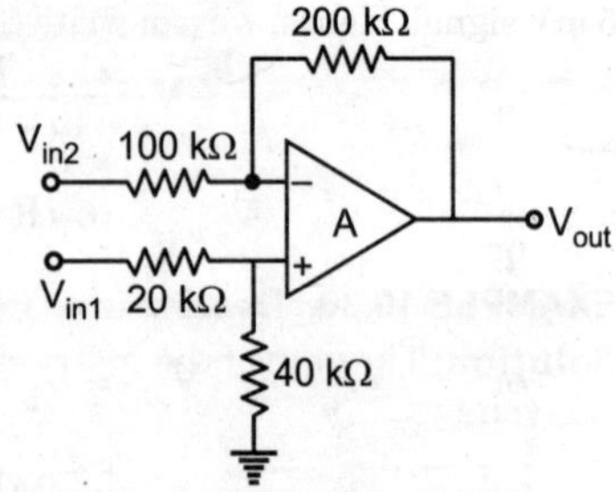

Fig. 10.74.

Solution: The circuit is a difference amplifier, hence the output voltage can be expressed as,

$$V_{out} = -\frac{200}{100}V_{in2} + \left[1+\frac{200}{100}\right]\left[\frac{400}{20+40}\right]V_{in1}$$

$$V_{out} = -2V_{in2} + [3]\left[\frac{4}{6}\right]V_{in1} = 2(V_{in1} - V_{in2}) \quad \textbf{Ans.}$$

The output voltage is proportional to the difference between the two inputs.

EXAMPLE 10.28. Calculate the gain of instrumentation amplifier shown in figure 10.75, assuming potentiometer is at maximum position.

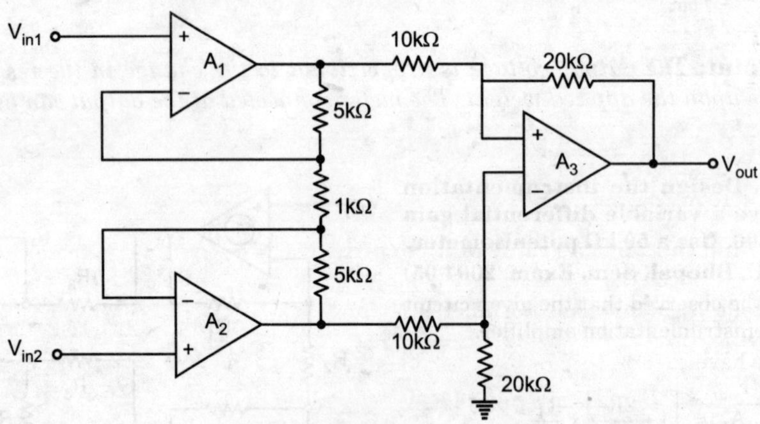

Fig. 10.75.

Solution: The gain of the instrumentation amplifier is

$$A = \left[1+2\frac{R_F}{R_G}\right]\left[\frac{R_2}{R_1}\right]$$

In this circuit, we have

$$R_F = 5\ \text{k}\Omega$$
$$R_G = 1\ \text{k}\Omega$$
$$R_1 = 10\ \text{k}\Omega$$
$$R_2 = 20\ \text{k}\Omega$$

Therefore, $$A = \left[1+\frac{2\times 5}{1}\right]\left[\frac{20}{10}\right] = 11\times 2 = 22 \quad \textbf{Ans.}$$

EXAMPLE 10.29. Determine the output voltage of the circuit shown in figure 10.76.

Solution: Let the output of op-amp A_1 is V_{out1}. The op-amp A_1 acts as an inverting summer.

Therefore,

$$V_{in1} = -\left[\frac{5}{1}\times(1\text{ mV})+\frac{5}{1}\times(5\sin\omega t)\right]$$

$$V_{in1} = -[5 + 25\sin\omega t]\text{ mV}$$

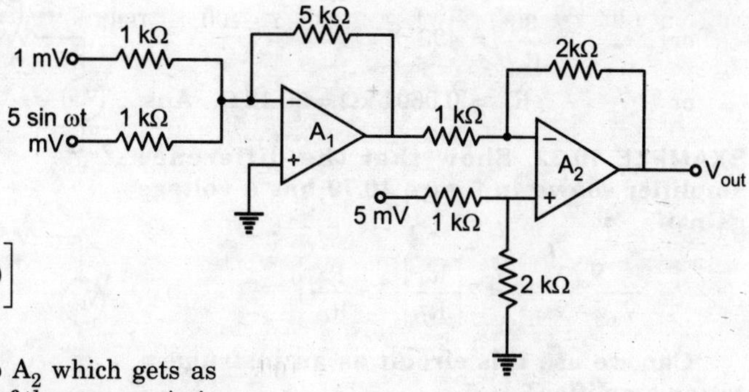

Fig. 10.76.

This V_{out1} acts as one input to op-amp A_2 which gets as difference amplifier, while the second input of the op-amp A_2 is 5 mV signal. Hence, we can write the expression for V_{out} as under:

$$V_{out} = -[5+25\sin\omega t]\left(-\frac{2}{1}\right)+\left(1+\frac{2}{1}\right)\left(\frac{2}{1+2}\right)(5)$$

or $\quad V_{out} = 10 + 50\sin\omega t + 10 = 20 + 50\sin\omega t$ mV **Ans.**

EXAMPLE 10.30. Determine the output for the circuit shown in figure 10.77.

Solution: The gain of the instrumentation amplifier is given by

$$A = \left[1+2\frac{R_F}{R_G}\right]\left[\frac{R_2}{R_1}\right]$$

Here $\quad R_F = 10\text{ k}\Omega$

$R_G = 5\text{ k}\Omega$

$R_1 = 1\text{ k}\Omega$

$R_2 = 2\text{ k}\Omega$

Therefore, $A = \left[1+\frac{2\times10}{5}\right]\left[\frac{2}{1}\right] = 10$

Also, $\quad V_{out} = A\,(V_{in2} - V_{in1})$

$= 10\,(2 - 1)$

$= 10$ mV **Ans.**

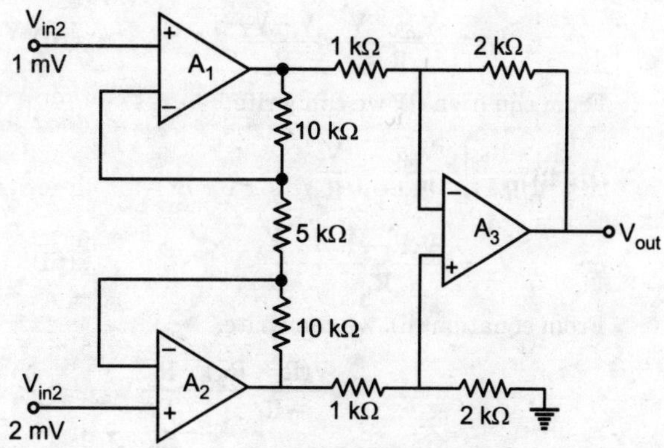

Fig. 10.77.

EXAMPLE 10.31. Find the suitable value of R_G to provide the output of 3 V, from the circuit shown in figure 10.78.

Solution: The output V_{out} = 3 V while the differential input is $V_{in2} - V_{in1} = 5 - 2 = 3$ mV.

We have

$$\text{Gain, } A = \frac{V_{out}}{V_{in2}-V_{in1}} = \frac{3}{3\times10^{-3}} = 1000$$

This is the overall gain required.

$$\text{Now, } A = \left(1+2\frac{R_F}{R_G}\right)\left(\frac{R_2}{R_1}\right)$$

Here, $\quad R_F = 15\text{ k}\Omega$,

$R_1 = 1\text{ k}\Omega$,

$R_2 = 2\text{k}\Omega$

Therefore, $1000 = \left[1+\frac{2\times15}{R_G}\right]\left[\frac{2}{1}\right]$

or $\quad 500 = 1+\frac{30}{R_G}$

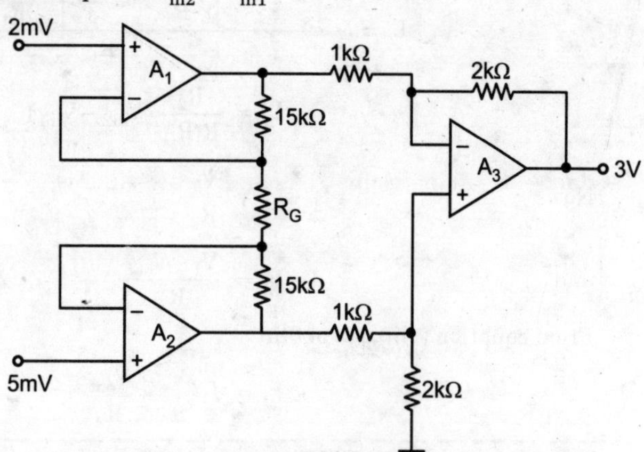

Fig. 10.78.

or $$\frac{30}{R_G} = 499$$

or $$R_G = 0.0601\ \text{k}\Omega = 60.12\ \Omega \quad \textbf{Ans.}$$

EXAMPLE 10.32. Show that the difference amplifier shown in figure 10.79 has a voltage gain of

$$\frac{V_{out}}{V_{in2} - V_{in1}} = 2\left[\frac{R_2}{R_1}\right]\left[1 + \frac{R_2}{R_G}\right]$$

Can we use this circuit as an instrumentation amplifier?

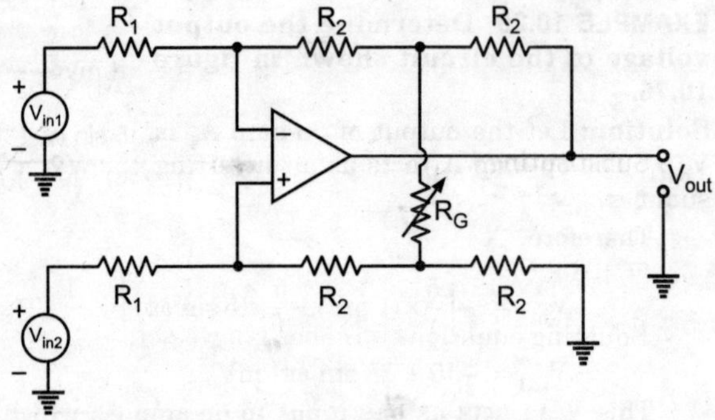

Fig. 10.79.

Solution: The various currents and voltages areas are shown in figure 10.80. The voltage inverting and non-inverting input terminals is same as V.

We can write the equations as under :

$$I = \frac{V_{in1} - V}{R_1} = \frac{V - V_x}{R_2}$$

as op-amp input current is zero ...(i)

$$I_1 = \frac{V_{in2} - V}{R_1} = \frac{V - V_Y}{R_2} \quad ...(ii)$$

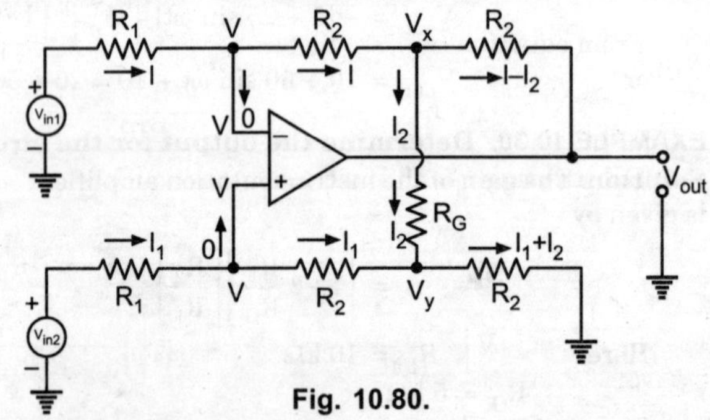

Fig. 10.80.

From equation (i), we can write,

$$V\left[\frac{R_1 + R_2}{R_1 R_2}\right] - \frac{V_{in1}}{R_1} = \frac{V_x}{R_2}$$

or $$V_x = \frac{V(R_1 + R_2)}{R_1} - \frac{R_2}{R_1} V_{in1} \quad ...(iii)$$

From equation (ii), we can write,

$$V_Y = \frac{V(R_1 + R_2)}{R_1} - \frac{R_2}{R_1} V_{in2} \quad ...(iv)$$

Now $$I_2 = \frac{V_x - V_Y}{R_G} \quad ...(v)$$

Substituting equations (iii) and (iv) in equation (v) we obtain

$$I_2 = \frac{V\left[\frac{R_1 + R_2}{R_1}\right] - \frac{R_2}{R_1} V_{in1} - V\left[\frac{R_1 + R_2}{R_1}\right] + \frac{R_2}{R_1} V_{in2}}{R_G}$$

$$I_2 = \frac{R_2}{R_1 R_G}(V_{in2} - V_{in1}) \quad ...(vi)$$

Now $$I_1 + I_2 = \frac{V_Y}{R_2} \quad ...(vii)$$

and $$I - I_2 = \frac{V_x - V_{out}}{R_2} \quad ...(viii)$$

From equation (viii), we obtain

$$I_2 = I - \frac{V_x}{R_2} + \frac{V_{out}}{R_2}$$

A little knowledge that acts is worth infinitely more than much knowledge that is idle.

— **Khalil Gibran**

$$I_2 = \frac{V_{in1}}{R_1} - \frac{V}{R_1} - \frac{V_x}{R_2} + \frac{V_{out}}{R_2}$$

Substituting V_x, we get
$$I_2 = \frac{V_{in1}}{R_1} - \frac{V}{R_1} - \frac{V(R_1+R_2)}{R_1R_2} + \frac{V_{in1}}{R_1} + \frac{V_{out}}{R_2}$$

or
$$I_2 = \frac{2}{R_1}V_{in1} + \frac{V_{out}}{R_2} - V\left[\frac{1}{R_1} + \frac{(R_1+R_2)}{R_2}\right] \quad ...(ix)$$

Equating equations (vi) and (ix), we get,

$$\frac{R_2}{R_1R_G}(V_{in2} - V_{in1}) = \frac{2}{R_1}V_{in1} + \frac{V_{out}}{R_2} - V\left[\frac{1}{R_1} + \frac{(R_1+R_2)}{R_2}\right]$$

or
$$V\left[\frac{R_1+2R_2}{R_1R_2}\right] = V_{in1}\left[\frac{2}{R_1} + \frac{R_2}{R_1R_G}\right] - \frac{R_2}{R_1R_G}V_{in2} + \frac{V_{out}}{R_2} \quad ...(x)$$

From equation (vii), we have

$$I_1 + I_2 = \frac{V(R_1+R_2)}{R_1R_2} - \frac{V_{in2}}{R_1} \quad ...(xi)$$

Substituting I_1 and I_2 we get

$$\frac{V_{in2} - V}{R_1} + \frac{R_2}{R_1R_G}(V_{in2} - V_{in1}) = \frac{V(R_1+R_2)}{R_1R_2} - \frac{V_{in2}}{R_1}$$

or
$$V\left[\frac{(R_1+R_2)}{R_1R_2} + \frac{1}{R_1}\right] = V_{in2}\left[\frac{2}{R_1} + \frac{R_2}{R_1R_G}\right] - V_{in1}\left[\frac{R_2}{R_1R_G}\right] \quad ...(xii)$$

Substituting in equation (x), in place of $V\left[\frac{R_1+2R_2}{R_1R_2}\right]$, we get

$$V_{in2}\left[\frac{2}{R_1} + \frac{R_2}{R_1R_G}\right] - \frac{V_{in1}R_2}{R_1R_G} = V_{in1}\left[\frac{2}{R_1} + \frac{R_2}{R_1R_G}\right] - \frac{R_2}{R_1R_G}V_{in2} + \frac{V_{out}}{R_2}$$

or
$$\frac{V_{out}}{R_2} = V_{in2}\left[\frac{2}{R_1} + \frac{R_2}{R_1R_G} + \frac{R_2}{R_1R_G}\right] - V_{in1}\left[\frac{2}{R_1} + \frac{R_2}{R_1R_G} + \frac{R_2}{R_1R_G}\right]$$

or
$$\frac{V_{out}}{R_2} = \left[\frac{2}{R_1} + \frac{2R_2}{R_1R_G}\right](V_{in2} - V_{in1}) \quad \text{or} \quad V_{out} = \frac{2R_2}{R_1}\left[1 + \frac{R_2}{R_G}\right]$$

Important Point: *The circuit shows that the difference amplifier gain can be varied by varying the resistance R_G. So gain adjustment is possible but gain variation is not linear with respect to the potentiometer R_G, which makes it difficult for the calibration. The linear gain control is possible by adding one more-op-amp and modifying the circuit.*

EXAMPLE 10.33. The circuit shown in figure 10.81 is a difference amplifier with linear gain control. Show that,

$$V_{out} = \frac{R_2R_G}{R_1R_3}(V_{in2} - V_{in1})$$

(GATE Exaimination, 1995)

Solution: The various currents and voltages are shown in figure 10.82.

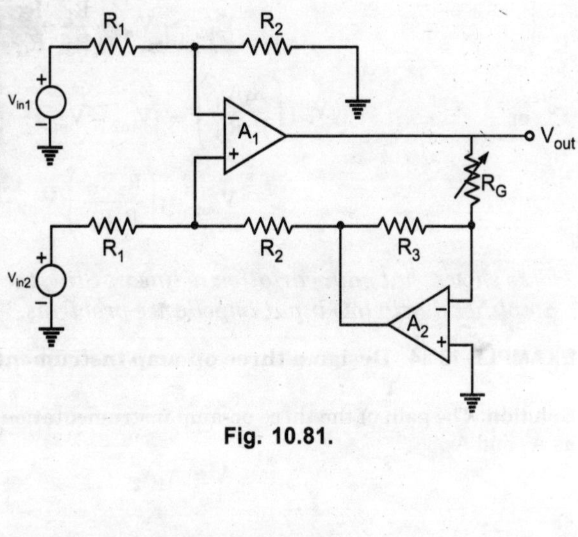

Fig. 10.81.

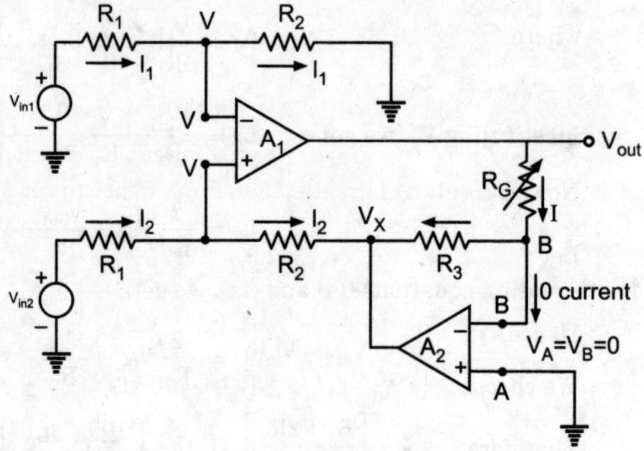

Fig. 10.82.

For input V_{in1}, we can write,

$$I_1 = \frac{V_{in1} - V}{R_1} = \frac{V}{R_2} \qquad \text{...(i)}$$

Therefore, $$\frac{V_{in1}}{R_1} = V\left[\frac{1}{R_1} + \frac{1}{R_2}\right]$$

or $$V = \frac{R_2 V_{in1}}{(R_1 + R_2)} \qquad \text{...(ii)}$$

For input V_{in2}, we can write,

$$I_2 = \frac{V_{in2} - V}{R_1} = \frac{V - V_x}{R_2} \qquad \text{...(iii)}$$

or $$\frac{V_{in2} - V}{R_1} = \frac{V - V_x}{R_2}$$

or $$\frac{V_{in2}}{R_1} = V\left[\frac{1}{R_1} + \frac{1}{R_2}\right] - \frac{V_x}{R_2}$$

or $$V_{in2} = \frac{V(R_1 + R_2)}{R_2} - \frac{R_1}{R_2} V_x \qquad \text{...(iv)}$$

Substituting V from equation (ii) in equation (iv), we obtain

$$V_{in2} = V_{in1} - \frac{R_1}{R_2} V_x \qquad \text{...(v)}$$

Now, we analyze the op-amp A_2. The point A is ground hence point B is also at virtual ground. So current I through R_G is same as current through R_3.

$$I = \frac{V_{out} - V_B}{R_G} = \frac{V_B - V_x}{R_3} \qquad \text{...(vi)}$$

But $$V_B = 0, \text{ hence, we have}$$

$$\frac{V_{out}}{R_G} = -\frac{V_x}{R_3}$$

or $$V_x = -\frac{R_3}{R_G} V_{out} \qquad \text{...(vii)}$$

Substituting in equation (v), we obtain

$$V_{in2} = V_{in1} + \frac{R_1}{R_2} \cdot \frac{R_3}{R_G} V_{out}$$

or $$V_{out}\left(\frac{R_1 R_3}{R_1 R_G}\right) = (V_{in2} - V_{in1})$$

or $$V_{out} = \left(\frac{R_2 R_G}{R_1 R_3}\right)(V_{in2} - V_{in1}) \qquad \text{...Hence Proved}$$

This shows that gain variation is linear. Still, the circuit is not preferred for an instrumentation amplifier due to low input impedance problems.

EXAMPLE 10.34. Design a three op-amp instrumentation amplifier to vary gain from 1 to 10000.
(RGTU, Bhopal, Sem. Exam., 2008-09)

Solution: The gain of the three op-amp instrumentation amplifier can be assumed to be product of two gains as A_1 and A_2.

$$A = A_1 A_2$$

where $$A_1 = \left(1+\frac{2R_F}{R_G}\right)$$

and $$A_2 = \left(\frac{R_2}{R_1}\right)$$

Now, A_1 is always greater than unity hence to get A_1A_2 as minimum gain unity, A_2 must be less than one.

Let $$A_2 = \frac{1}{5}$$

Hence, $$\frac{R_2}{R_1} = \frac{1}{5} \quad \text{...(i)}$$

We choose $$R_1 = 100\ k\Omega$$

Therefore, $$R_2 = \frac{R_1}{5} = \frac{100}{5} = 20\ k\Omega$$

Now, we design for gain A_1 which is $\left(1+\frac{2R_F}{R_G}\right)$

Let R_G be the combination of the base resistance R_{GB} and the pot of 100 kΩ say. With a pot on zero resistance position, we have

$$R_G = R_{GB} + 0 = R_{GB}, \text{ form which gain will be minimum.}$$

Therefore, $$A = A_1A_2 = 10000$$

Hence, $$10000 = \left(1+\frac{2R_F}{R_{GB}}\right)\left(\frac{1}{5}\right)$$

$$1+\frac{2R_F}{R_{GB}} = 50000 \quad \text{or} \quad \frac{2R_F}{R_{GB}} = 49999$$

or $$\frac{R_F}{R_{GB}} = 24999.5 \quad \text{..(ii)}$$

When potentiometer will be at its maximum position, gain will be minimum and

$$R_G = R_{GB} + 100\ k\Omega$$

or $$\left(1+\frac{2R_F}{(R_{GB}+100\times10^3)}\right)\left(\frac{1}{5}\right) = 1$$

$$\frac{2R_F}{R_{GB}+100\times10^3} = 5-1 = 4$$

or $$\frac{R_F}{R_{GB}+100\times10^3} = \frac{4}{2} = 2 \quad \text{...(iii)}$$

Substituting from equation (ii), we obtain

$$R_F = 24999.5 = R_{GB}$$

or $$\frac{24999.5\ R_{GB}}{R_{GB}+100\times10^3} = 2$$

or $$24999.5\ R_{GB} = 2\ (R_{GB} + 100\times10^3)$$

or $$24999.5\ R_{GB} = 2R_{GB} + 200\times10^3 = 200\times10^3$$

or $$R_{GB} = \frac{200\times10^3}{24997.5} = 8$$

This is the base resistance required in series with the pot of 100 kΩ **Ans.**

Thus, we have $R_F = 24999.5 \times 8 = 199.99\ k\Omega = 200\ k\Omega$ **Ans.**

The designed circuit is shown in figure 10.94.

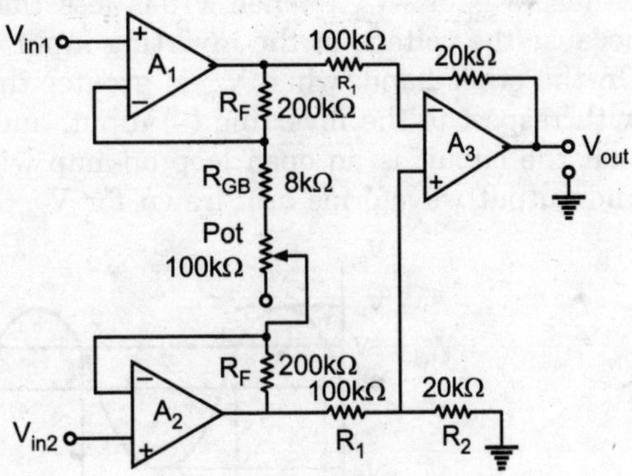

Fig. 10.83.

10.22 INTRODUCTION TO NON-LINEAR APPLICATIONS

In the previous articles, we have studied linear behaviour of op-amp circuits. The linearity of these circuits is achieved by providing negative feedback and by using linear elements in the feedback path.

An operational amplifier with positive feedback or with open loop configuration, causes it to operate primarily in saturation. This bistable behaviour of operational amplifier is highly non-linear. There are number of applications of operational amplifier in this mode such as, comparators, schmitt trigger circuits, etc. Non-linear behaviour of operational amplifier can also be achieved with negative feedback, but using non-linear elements such as diode, FET in the feedback path. In this chapter, let us study the basic circuits which use the non linear behaviour of the operational amplifier, such as comparators and schmitt triggers.

10.23 OP-AMP AS COMPARATOR

A comparator may be defined as a circuit which compares a signal voltage applied at one input of an op-amp with a known reference voltage at the other input, and produces either a high or a low output voltage, depending on which input is higher. Since comparator output has two voltage levels, either high or low, it is not linearly proportional to input voltage. Op-amp in an open loop configuration or sometimes a positive feedback configuration can be used as a comparator.

10.23.1. Non-Inverting Comparator

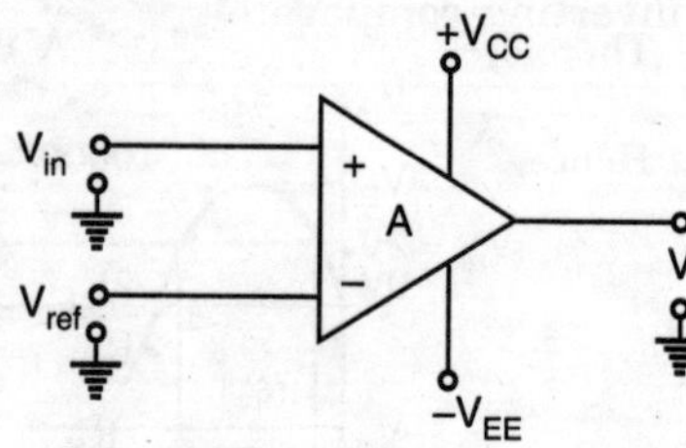

Fig. 10.84. A voltage comparator

Figure 10.84 shows an open loop op-amp which is used as a comparator. Here, a known reference voltage is applied at the inverting input of the op-amp and time-varying signal (V_{in}) is applied at the non-inverting input of the op-amp. Since, time varying signal is applied at the non-inverting input of the op-amp, the circuit is called as **non inverting comparator.**

Recall that the open loop gain of the op-amp is very large with a small input voltage, the op-amp will go into saturation, and the output voltage will be either $+V_{sat}$ or $-V_{sat}$. When V_{in} is less than V_{ref}, the output voltage V_{out} is at $-V_{sat}$ ($\cong -V_{EE}$) because the voltage at the inverting input (–) is higher than that at the non inverting (+) input. On the other hand, when V_{in} is greater than V_{ref}, the non inverting (+) input becomes positive with respect to the inverting (–) input, and V_{out} goes to $+V_{sat}$ ($\cong +V_{CC}$). Therefore, we can any that the circuit is an open loop op-amp with output $\pm V_{sat}$, as shown in figure 10.85. The input and output waveforms are drawn for $V_{ref} = +2$ V, and $V_{ref} = -2$ V.

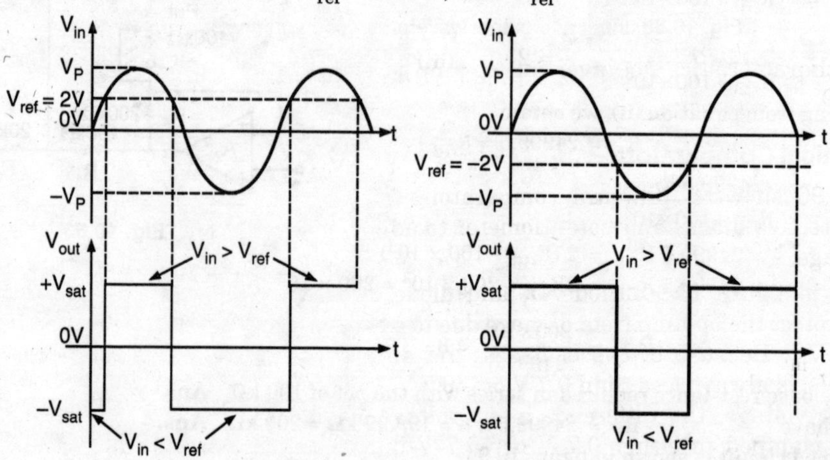

Fig. 10.85. Input and output waveforms for non-inverting comparator

The input and output relationship for comparator can be represented by transfer characteristics, as shown in figure 10.86. The switching of output voltage from $+V_{sat}$ to $-V_{sat}$ or vice-versa occurs when the input voltage crosses the reference voltage level.

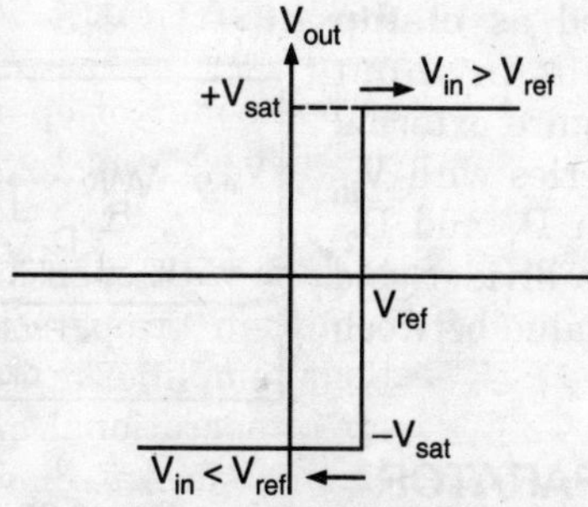

Fig. 10.86. Transfer characteristics for non-inverting comparator

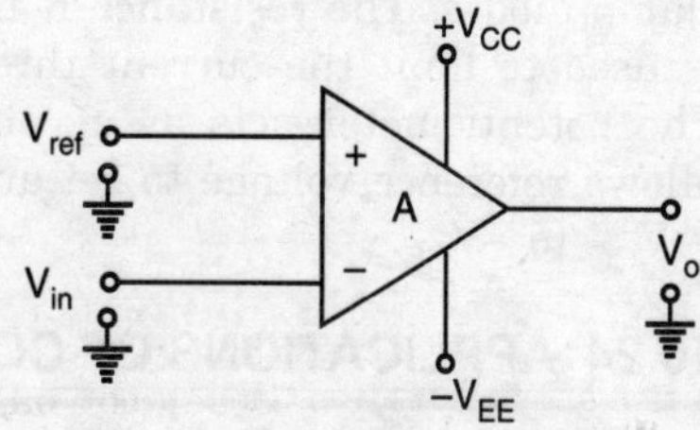

Fig. 10.87. Inverting comparator

10.23.2. Inverting Comparator

Figure 10.87 shows inverting comparator in which the reference voltage V_{ref} is applied to the non inverting (+) input and signal voltage (V_{in}) is applied to the inverting (–) input of the op-amp.

When V_{in} is less than V_{ref}, the output voltage V_{out} is at $+V_{sat}$ ($\cong +V_{CC}$) because the voltage at the inverting input (–) is less than that at the non inverting (+) input. On the other hand, when V_{in} is greater than V_{ref}, the non inverting (+) input becomes negative with respect to the inverting (–) input, and V_{out} goes to $-V_{sat}$ ($\cong -V_{EE}$). Figure 10.88 shows that input and output waveforms for inverting comparator.

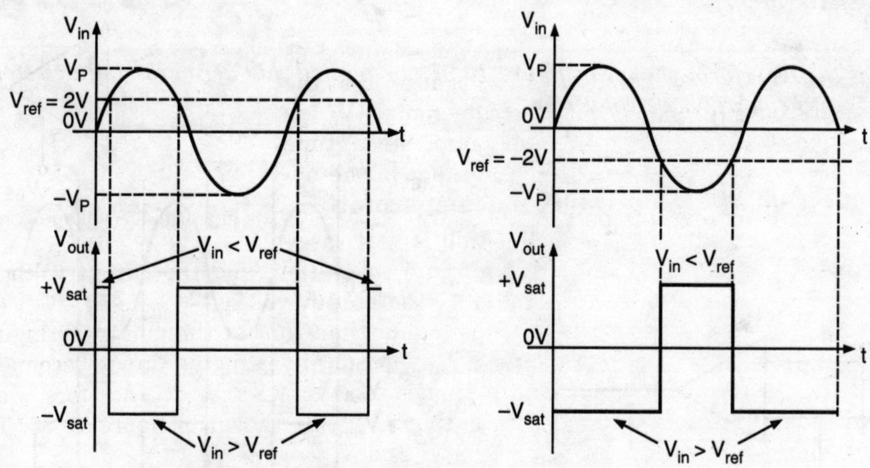

Fig. 10.88. Input and output waveforms for inverting comparator

Transfer characteristics for inverting op-amp is shown in figure 10.89

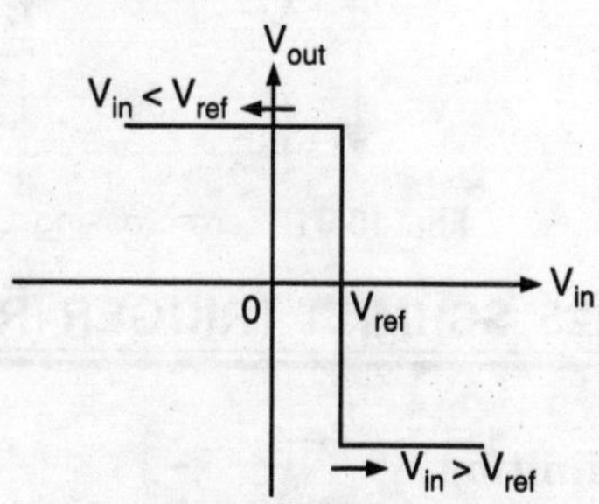

Fig. 10.89. Transfer characteristic for inverting comparator

10.23.3. Practical Comparator

Figure 10.90 shows a practical comparator circuit. It consists of protective diodes and potentiometer to adjust the reference voltage.

As shown in figure 10.90, diode D_1 and diode D_2 are connected to protect the op-amp from damage due to excessive input voltage V_{in}. Because of these diodes, the difference input voltage V_{id} is always less than 0.7 V or – 0.7 V. In case of excess input voltage, the difference input voltage V_{id} of the op-amp is clamped to either 0.7 V or –0.7 V due to the forward biasing of one of diodes.

Hence, these diodes are also called as **clamp diodes**. Some op-amps have built-in input protections circuitry, they do not require external clamp diodes. The resistance R in series with V_{in} is used to limit the current through D_1 and D_2. The potentiometer acts as a voltage divider and allows reference voltage to set any value between $+V_{CC}$ to $-V_{EE}$.

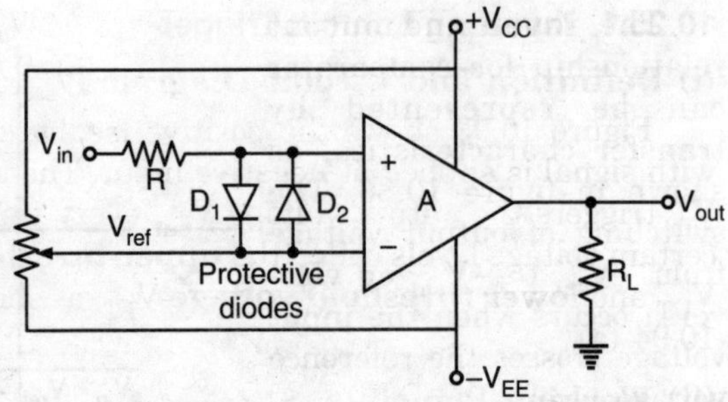

Fig. 10.90. Practical comparator circuit

10.24 APPLICATIONS OF COMPARATOR

In this article, let us discuss various applications of comparator in details.

10.24.1. Zero Crossing Detector

The basic comparator can be used as a zero crossing detector. A typical circuit for such a detector is shown in figure 10.91. It is a non inverting comparator circuit with $V_{ref} = 0$ V. During the positive half cycle, the input voltage is positive i.e., above the reference voltage. Hence, the output voltage is $-V_{sat}$. During negative half cycle, the input voltage V_{in} is negative, i.e., below the reference voltage. The output voltage is then $-V_{sat}$. Thus, the output voltage switches between $+V_{sat}$ and $-V_{sat}$ whenever the input signal crosses the zero level. This has been illustrated in figure 10.91.

Looking at the waveform shown in figure 10.91, we realize that zero crossing detector can be used as sine to square wave converter.

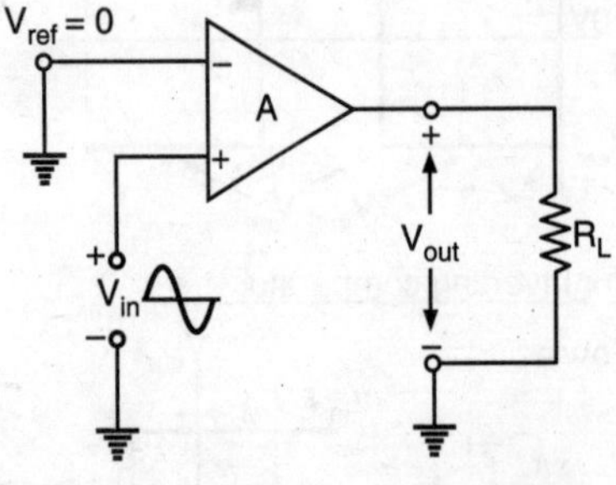

Fig. 10.91. Zero-crossing Detector

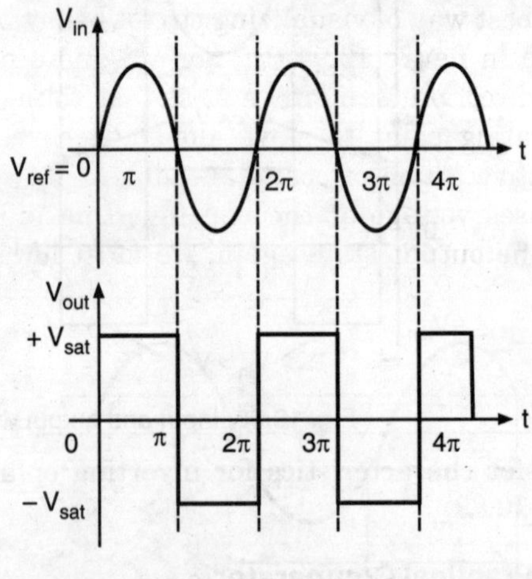

Fig. 10.92. Waveforms for zero crossing detector

10.25 SCHMITT TRIGGER (REGENERATIVE COMPARATOR)

(RGTU, Bhopal, Sem. Exam. 2010-11)(10 Marks)

Definition

Comparator circuits with positive feedback are known as schmitt triggers or regenerative comparators. They admit only two stable states $+V_{sat}$ and $-V_{sat}$. All intermediate states are unstable because the circuit has natural tendency to saturate. Depending on the side at which we apply the external input, a schmitt trigger can be of the inverting or the non-inverting type.

10.25.1. Inverting Schmitt Trigger

(i) Definition and Circuit Diagram

Figure 10.93 shows the positive feedback comparator with signal is applied at negative input. The input voltage V_{in} triggers the output voltage V_{out} every time it exceeds certain voltage levels called the **upper threshold voltage** V_{UT} and **lower threshold voltage** V_{LT}, as shown in figure 10.94 (a).

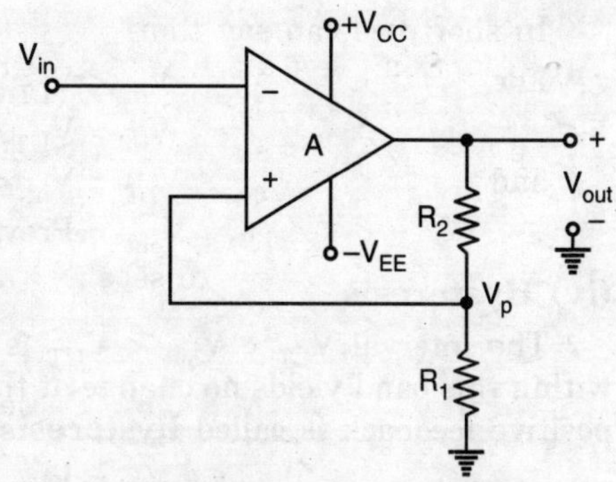

Fig. 10.93. Positive feedback comparator

(ii) Working Principle

As shown in the figure 10.93, these threshold voltages are derived using potential divider $R_1 - R_2$. The voltage across R_1 is fedback to the non inverting (positive) input of the op-amp. The voltage across R_1 depends on the value and polarity of the output voltage V_{out}. When $V_{out} = +V_{sat}$, the voltage across R_1 is called as the upper threshold voltage V_{UT} and is given as

$$V_{UT} = \frac{(+V_{sat}) \times R_1}{R_1 + R_2} \qquad ...(10.117)$$

On the other hand, when $V_{out} = -V_{sat}$, the voltage across R_1 is called as the lower threshold voltage V_{LT} and is given as

$$V_{LT} = \frac{(-V_{sat}) \times R_1}{R_1 + R_2} \qquad ...(10.118)$$

(iii) Transfer Characteristics

The best way of visualizing circuit behaviour is by driving the voltage transfer characteristics, as shown in figure 10.94 (b). For V_{in}, more negative than V_{LT}, the comparator saturates in the positive direction **(see curve a)** so that voltage across R_1 is V_{UT}. Gradually, increasing V_{in} causes the operating point to move along the curve **(b)** until V_{in} reaches V_{UT}. At this point, by the regenerative action of positive feedback, V_{out} changes from $+V_{sat}$ to $-V_{sat}$ **(see curve c)**. This in turn causes voltage at the non inverting input (V_p) to change from V_{UT} to V_{LT}. If we wish to change the output state again, we must lower V_{in} all the way down to V_{TL} **(see curve d).**

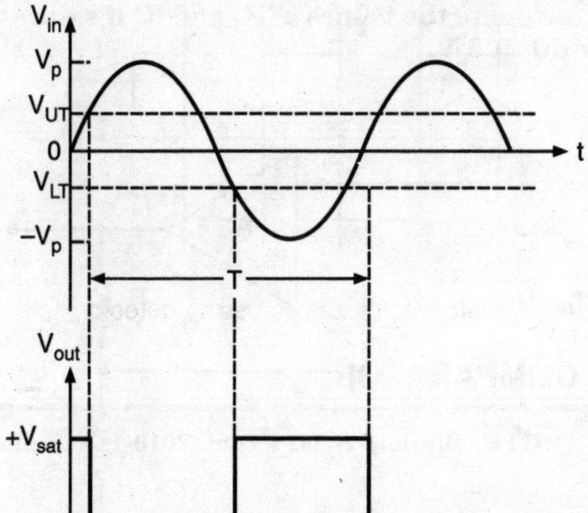

Fig. 10.94 (a) Input and output waveforms of Schmitt trigger

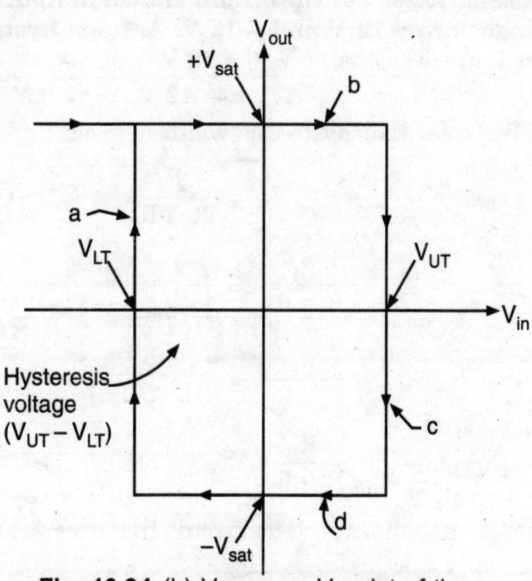

Fig. 10.94 (b) V_o versus V_{in} plot of the hysteresis voltages

In short, we can say that,

For $V_{in} < V_{LT}$, $V_{out} = +V_{sat}$ and $V_p = V_{UT}$,

$V_{in} > V_{UT}$, $V_{out} = -V_{sat}$ and $V_p = V_{LT}$,

and $V_{LT} < V_{in} < V_{UT}$

V_{out} = Previous state and V_p = Previous threshold voltage

(iv) Hysteresis

The interval $V_{LT} < V_{out} < V_{UT}$ is called **dead band** or **dead zone** since variation of V_{out} within this band yields no change at the output. This dead band condition in the comparator with positive feedback is called **hysteresis.** The hysteresis width is defined as

$$V_H = V_{UT} - V_{LT}$$

or $$V_H = \frac{+V_{sat} \times R_1}{R_1 + R_2} - \frac{-V_{sat} \times R_1}{R_1 + R_2}$$

or $$V_H = \frac{R_1}{R_1 + R_2}[+V_{sat} - (-V_{sat})] \quad ...(10.119)$$

The hysteresis width can be varied by changing the ratio $\frac{R_1}{R_2}$.

EXAMPLE 10.35. In circuit of figure 10.93, R_1 = 120 Ω and R_2 = 51 kΩ. Determine the threshold voltages if supply voltage are + 15 V and – 15 V.

Solution: Given that $+V_{sat} = +V_{CC} = +15$ V

$-V_{sat} = -V_{EE} = -15$ V

$R_1 = 120\ \Omega,\ R_2 = 51\ k\Omega$

$V_{in} = 1$ V

We know that V_{UT} and V_{LT} are given as under :

$$V_{UT} = \frac{(+V_{sat}) \times R_1}{R_1 + R_2} = \frac{15 \times 120}{120 + 51000} = 35.2 \text{ mV} \quad \textbf{Ans.}$$

$$V_{ULT} = \frac{(-V_{sat}) \times R_1}{R_1 + R_2} = \frac{-15 \times 120}{120 + 51000} = -35.2 \text{ mV} \quad \textbf{Ans.}$$

EXAMPLE 10.36. For the circuit shown in figure 10.95, evaluate the values of R_1 and R_2 if saturation voltages are + 12 V and – 12 V. Assume hystresis width = 6 V.

Solution: Given that $+V_{sat} = 12$ V

$-V_{sat} = -12$ V, $V_H = 6$ V

We know that hysteresis width is given by

$$V_H = \frac{R_1}{R_1 + R_2}[(+V_{sat}) - (-V_{sat})]$$

or $$\frac{R_1}{R_1 + R_2} = \frac{V_H}{[+V_{sat} - (-V_{sat})]}$$

or $$\frac{R_1}{R_1 + R_2} = \frac{6}{[12 - (-12)]}$$

or $$\frac{R_1}{R_1 + R_2} = 0.25$$

or $$R_1 = 0.25\ R_1 + 0.25\ R_2$$

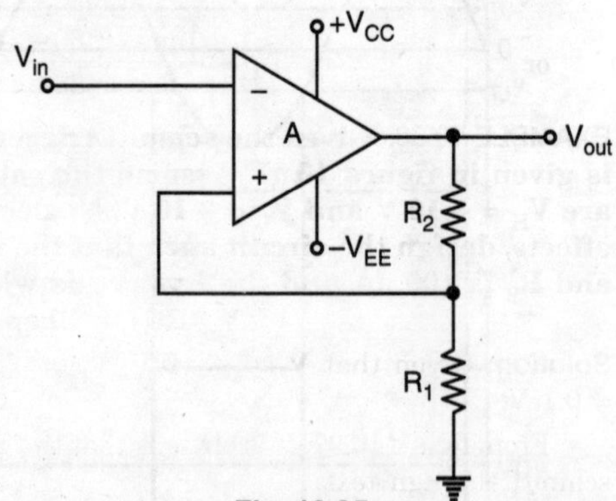

Fig. 10.95.

or $\quad 0.75\ R_1 = 0.25\ R_2$

or $\quad \dfrac{R_1}{R_2} = 0.3333$

Let us take $\quad R_1 = 10\ k\Omega$, then we have

$$R_2 = \frac{10000}{0.3333} = 30\ k\Omega \quad \textbf{Ans.}$$

EXAMPLE 10.37. A schmitt trigger with hysteresis width $V_H = 2$ V and lower threshold level $V_{LT} = -1.5$ V converts a 1 kHz sine wave of amplitude 5 V_{p-p} into a square wave. Calculate the time duration of the negative and positive portion of the output waveform.

(GATE Examination, 1998)

Solution: To get the clear idea of the problem, let us first draw the input and output waveforms as shown in figure 10.96.

Given that $\quad V_{p-p} = 5\ V,\ V_{LT} = -1.5\ V$

$V_H = 2\ V,\ f = 1\ kHz$

$V_H = V_{UT} + |V_{LT}|$

or $\quad 2 = V_{UT} + 1.5$

or $\quad V_{UT} = 0.5\ V$

Also, we have

$$V_m = \frac{V_{p-p}}{2} = \frac{5}{2} = 2.5\ V$$

In figure 10.96, the angle θ can be calculated as under:

$$-1.5\ V = V_m \sin(\pi + \theta)$$

$$= -V_m \sin\theta = -2.5 \sin\theta$$

or $\quad \sin\theta = 0.6$

or $\quad \theta = 36.87$

Time period for 1 cycle can be given as under :

$$T = \frac{1}{f} = \frac{1}{1000} = 1\ ms$$

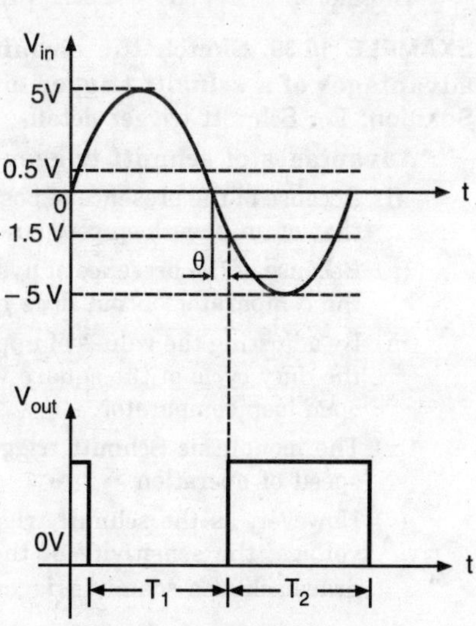

Fig. 10.96.

Period T_1 exists for angle 0° to 216.87° (180° + 36.87°) and period T_2 exists of angle 216.87° to 360°. we have

$$T_1 = \frac{1\ ms \times 216.87}{360} = 0.602\ ms$$

$$T_2 = T - T_1 = 0.398\ ms$$

or $\quad T_2 = \dfrac{1\ ms \times (360 - 216.87)}{360} = 0.398\ ms \quad$ **Ans.**

EXAMPLE 10.38. Given the schmitt trigger, whose circuit diagram is given in figure 10.97. Assume the saturated output voltages are $V_H = +10$ V and $V_L = -10$ V. Neglecting input bias current effects, design the circuit such that the maximum current in R_1 and R_2 is 100 μA and the hysteresis width is 0.1 V.

(RGTU, Bhopal, Sem. Exam; 2007-08)

Solution: Given that $V_H = +10$ V, $V_L = -10$ V, $I_{max} = 100$ μA, $V_{HV} = 0.1$ V.

From figure 10.97, it is obvious that the given circuit is an inverting schmitt through text.

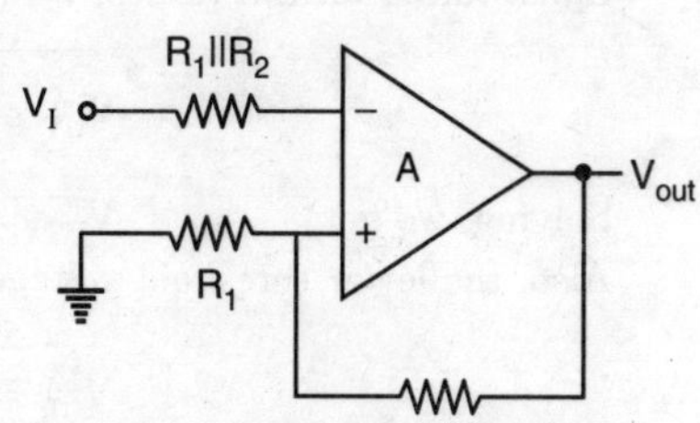

Fig. 10.97. Circuit for example 10.38

Therefore, we have, $V_{HV} = \dfrac{R_2}{(R_1+R_2)} \times 2\,V_{sat}$

Substituting all the values, we obtain

$$0.1 = \frac{R_2}{R_1+R_2} \times 2 \times 10 \qquad \text{(Since } V_{sat} = V_H = 10 \text{ V)}$$

or $\dfrac{R_2}{R_1+R_2} = 5 \times 10^{-3}$

Assuming $R_2 = 1\text{ k}\Omega$, we have

$R_1 = 199\text{ k}\Omega$

Therefore, $R_1 || R_2 = 1\text{ k}\Omega || 199\text{ k}\Omega = 995\text{ k}\Omega$ **Ans.**

EXAMPLE 10.39. Sketch the circuit and characteristics of basic Schmitt trigger. What are the advantages of a schmitt trigger circuit in analog electronics? Explain its working.

Solution: For Schmitt trigger details, please go through text.

Advantages of schmitt trigger over the conventional comparators

(i) Because of the presence of positive feedback, the speed of response of a schmitt trigger is higher than that of an open loop comparator.

(ii) Because of the presence of hysteresis, the problem of relay chattering is eli ninated. This means that the comparator output does not change in the hysteresis interval.

(iii) By adjusting the values of upper trigger level V_{UT} and lower trigger level V_{LT}, it is possible to adjust the duty cycle of the square wave output voltage of the schmitt trigger. This is not possible in the open loop comparator.

(iv) The monolithic Schmitt triggers are generally uncompensated op-amps. Hence, they have higher speed of operation.

(v) However, as the schmitt trigger does not respond in its hysteresis range to any change in input voltage, the sensitivity of the schmitt trigger decreases with increase in hysteresis. This is the drawback of a schmitt trigger.

EXAMPLE 10.40. Determine the threshold voltages and the hysteresis voltage of the Schmitt's trigger circuit shown in figure 10.98.

(RGTU, Bhopal, Sem. Exam; 2007-08)

Solution: Given that $R_1 = 80\text{ k}\Omega$, $R_2 = 20\text{ k}\Omega$, $V_{CC} = \pm 12.5$ V

The given circuit is an inverting Schmitt trigger. The upper threshold voltage is given by,

$$V_{UT} = \frac{R_2}{R_1+R_2} \times V_{sat}$$

Substituting all the values, we obtain

$$V_{UT} = \frac{20\text{ k}\Omega}{80\text{ k}\Omega + 20\text{ k}\Omega} \times 12.5$$

Solving, we get $V_{UT} = 2.5$ V **Ans.**

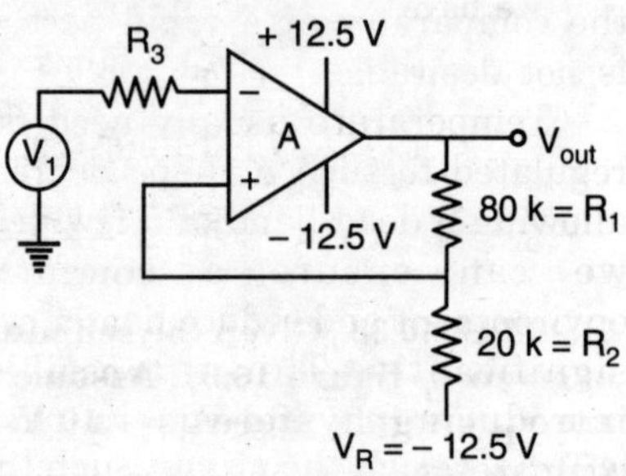

Fig. 10.98.

Also, the lower threshold voltage is given by

$$V_{LT} = \frac{R_2}{R_1+R_2} \times (-V_{sat}) = \frac{20\text{ k}}{20\text{ k} + 80\text{ k}} \times (-12.5) = -2.5 \text{ V} \quad \textbf{Ans.}$$

We know that hysteresis voltage is given by,

$$V_{HV} = \frac{R_2}{R_1+R_2} \times 2\,V_{sat} = \frac{20\text{ k}}{20\text{ k} + 80\text{ k}} \times 2 \times 12.5 = 5 \text{ V} \quad \textbf{Ans.}$$

10.25.2. Schmitt Trigger Applications

We have already seen one important application of schmitt trigger as sine to square wave converter. It can be used to eliminate **comparator chatter** in **signal** shaping and in ON/OFF control. It is a building block of relaxation oscillators.

(i) Schmitt Triggers for Eliminating Comparator Chatter

Chattering may be defined as production of multiple output transistion as the input signal swings through the threshold region of a comparator. This happens due to the fact that ac noise is present in the practical circuits. Figure 10.99 shows input signal with ac noise and how comparator output chatters. Even if noise is very less, it takes a very small noise spike to cause chatter due to high comparator gains.

This problem can be solved by using schmitt trigger circuits. This is because they exhibit hysteresis. In case of hysteresis, as soon as the input signal crosses the present threshold level once, the output changes its state and activates the other threshold level, so that the input signal must swing back to the new threshold in order to make the output of the circuit to change its state again. Let us consider figure 10.99. By making hysteresis width greater than the maximum peak amplitude of noise a dead zone is created such that noise within this zone no longer causes multiple output transitions.

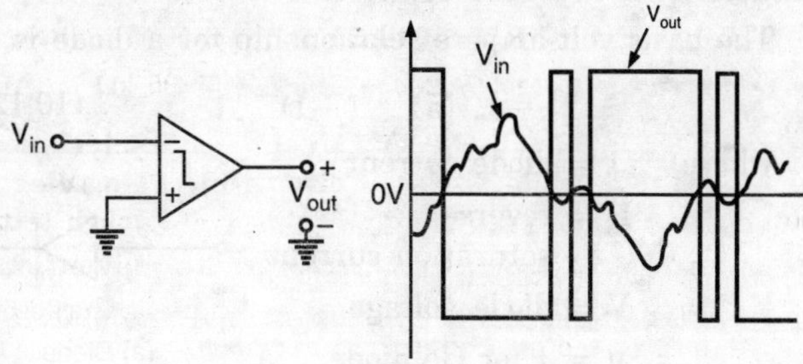

Fig. 10.99.

(ii) Schmitt Triggers in ON/OFF controllers

In ON/OFF controller, such as temperature controller, when temperature is below set point, heater is made ON and when it is above setpoint heater is made OFF. Now, let us consider that temperature is just below the set point, and heater is ON. As soon as the temperature reaches the set point, the comparator which is controlling the heater makes it OFF. The smallest temperature drop following the heater OFF will make comparator to go in its activate state and switch on the heater. As a result, the heater will be cycled ON and OFF by the comparator at a rapid pace. This is not desirable.

Temperature usually need not be regulated to such a shape degree. By allowing a dead band of a few degrees we can ensure a comfortable environment and reduce heater cycling significantly. This is possible by introducing hysteresis in the the comparator.

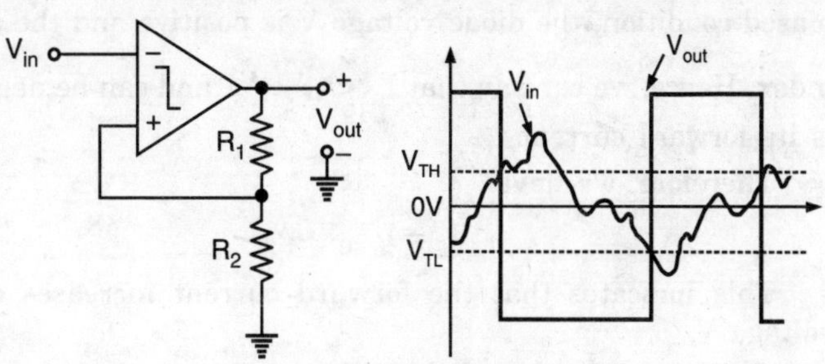

Fig. 10.100.

10.26 LOG AND ANTILOG OP-AMP AMPLIFIERS

Uptill now, we have discussed many linear and non-linear applications of op-amp. One more important non-linear application of op-amp is logarithmic and antilogarithmic amplifiers. Use of a diode or a BJT in the feedback path, forms the basis of such amplifiers. These amplifiers are commonly called as Log and Antilog amplifiers. These amplifiers have several areas of application. The mathematical operations may require ln (x), log (x) or sinh (x) calculations, such calculations can all be performed with Log amplifiers. To have direct dB display on digital instruments. Such log amplifiers are used in such instruments to calculate dB value. With the help of basic logarithmic function, Log amplifiers can be used to perform many other mathematical operations like division, multiplication, square root calculation, power extraction, true rms conversion etc. Before studying the Log and Antilog amplifier circuits, let us revise the diode current equation which forms the basic of the analysis of such amplifiers.

10.27 FUNDAMENTALS OF LOG AMPLIFIERS **(Important)**

Let us consider a diode as shown in figure 10.101 (a) and the corresponding volt-ampere (V–I) characteristics in figure 10.101 (b).

The basic volt-ampere relationship for a diode is given as,

$$I = I_o \left(e^{V/\eta V_T} - 1\right) \quad \text{...(10.120)}$$

where I = diode current

I_o = reverse saturation current

V = diode voltage

η = 1 for Ge diode, 2 for Si diode

$V_T = \dfrac{kT}{q}$ = voltage equivalent of temperature

and k = Boltzmann's constant = 8.62×10^5 eV/°K

T = temperature in °K

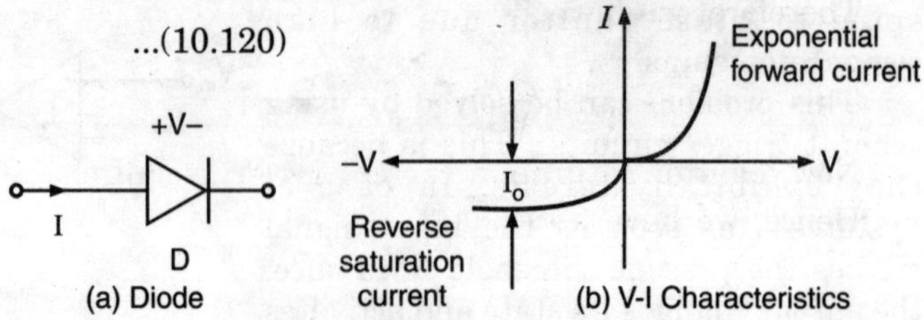

Fig. 10.101.

It can be observed that at room temperature of 27°C, i.e., T = 300°K, the voltage equivalent of temperature is,

$$V_T = 0.02586 \text{ V} \cong 26 \text{ mV at } 300°\text{K}$$

The value is frequently used in analysing the diode circuits.

It can be seen from the V – I characteristics shown in figure 10.101(b) that for a forward biased condition, the diode voltage V is positive and the exponential term $e^{V/\eta V_T}$ has a positive index. Hence, we can say that I << $e^{V/\eta V_T}$ and can be neglected. The corresponding diode current is its forward current I_f.

Therefore, we have

$$I_f = I_o \, e^{V/\eta V_T} \quad \text{...(10.121)}$$

This indicates that the forward current increases exponentially with respect to the bias voltage V.

Taking natural logarithm of both sides, we have

$$\ln [I_f] = \ln [I_o \, e^{V/\eta V_T}] \quad \text{...(10.122)}$$

or $$\ln [I_f] = \ln [I_o] + \ln [e^{V/\eta V_T}] \quad \text{...(10.123)}$$

or $$\ln [I_f] = \ln [I_o] + \frac{V}{\eta V_T} \quad \text{...(10.124)}$$

or $$V = \eta V_T [\ln (I_f) - \ln (I_o)] \quad \text{...(10.125)}$$

The above equation (10.125) is the basic equation used in Log and Antilog amplifiers.

Now, equation (10.125) can be expressed as,

$$V = \eta V_T \ln\left[\frac{I_f}{I_o}\right] \quad ...(10.126)$$

Similar to the diode, the equation for the collector current in BJT can be written as,

$$I_C = \alpha\, I_s\, (e^{V_{BE}/V_T} - 1) \quad ...(10.127)$$

where I_C = collector current

I_s = emitter saturation current

V_{BE} = base emitter voltage

$\alpha \cong 1$ (constant)

Therefore, we have

$$\frac{I_C}{I_s} = e^{V_{BE}/V_T} - 1 \quad ...(10.128)$$

Now, emitter saturation current is very small of the order of 10^{-13} A.

Hence, we have

$$e^{V_{BE}/V_T} = \frac{I_C}{I_s} + 1 = \frac{I_C}{I_s} \quad ...(10.129)$$

As I_s is very small, we have $1 << (I_C/I_s)$.

Therefore,
$$\ln\,(e^{V_{BE}/V_T}) = \ln\left(\frac{I_C}{I_s}\right) \quad ...(10.130)$$

or
$$\frac{V_{BE}}{V_T} = \ln\left(\frac{I_C}{I_s}\right) \quad ...(10.131)$$

or
$$V_{BE} = V_T \ln\left(\frac{I_C}{I_s}\right) \quad ...(10.132)$$

This equation is similar to the equation (10.126) and forms the basis of Log and Antilog amplifier circuits where BJT is used instead of a diode.

In equation (10.127), V_{BE} represents voltage across transistor, similar to the V in equation (10.126), representing voltage across diode.

Now, voltage across BJT is actually $V_{CB} + V_{BE}$. But in op-amp circuits, when BJT is used in the feedback path as a diode, V_{CB} voltage becomes zero. Hence V is replaced by V_{BE} while writing the equation for the BJT. For BJT used as a diode, $V_{CB} = 0$.

10.28 BASIC LOGARITHMIC AMPLIFIER

With the background of diode and BJT current equations, let us study the basic logarithmic amplifier circuit using op-amp. The fundamental Log amplifier is formed by placing a diode or a transistor in the negative feedback path of the op-amp.

10.28.1. Basic Log Amplifier Using Diode

The circuit diagram of basic Log amplifier using a diode is shown in figure 10.102.

The diode D is used in the negative feedback path. The node A is grounded hence node B is at virtual ground. Hence, $V_2 = 0$.

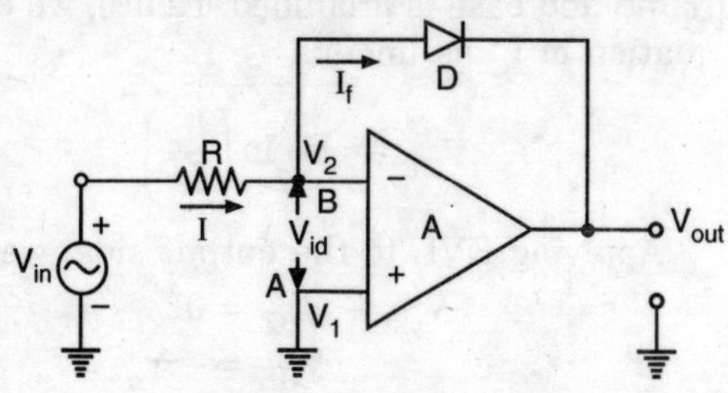

Fig. 10.102. Basic log amplifier using diode

Therefore, we have

$$I = \frac{V_{in} - V_2}{R} = \frac{V_{in}}{R} \qquad ...(10.133)$$

As the op-amp input current is zero, we have

$$I = I_f = \text{diode current} \qquad ...(10.134)$$

Now, I_f is the current through diode and voltage across diode is $V_B - V_o$ i.e., $- V_o$. Hence, using equation (10.126), we can write

$$- V_{out} = \eta V_T \ln\left[\frac{I_f}{I_o}\right] \qquad ...(10.135)$$

Substituting $I_f = I = \frac{V_{in}}{R}$, we have

$$V_{out} = - \eta V_T \ln\left[\frac{V_{in}}{RI_o}\right] \qquad ...(10.136)$$

As I_o R is constant d.c. voltage, let us denote it is V_{ref}.
Therefore, we have

$$V_{out} = -\eta V_T \ln\left[\frac{V_{in}}{V_{ref}}\right] \qquad ...(10.137)$$

where $V_{ref} = I_o R$

Important Point: Thus, the output voltage V_{out} is a function of logarithm of the input voltage V_{in}. The circuit gives the logarithm to base e, i.e., natural logarithm. But the same circuit can be used to find out logarithm values to base 10 by the proper scaling as,

$$\log_{10} X = 0.4343 \ln (X) \qquad ...(10.138)$$

10.28.2. Basic Log Amplifier Using Transistor

The basic log amplifier can be obtained by using a transistor as a diode in the negative feedback path of a diode in the negative feedback path of an op-amp, as shown in figure 10.103.

The node B is at virtual ground hence $V_2 = 0$.

Therefore, we have

$$I = \frac{V_{in} - V_2}{R} = \frac{V_{in}}{R} \qquad ...(10.139)$$

As the op-amp input current is zero, we have

$$I = I_C = \text{collector current} \qquad ...(10.140)$$

The voltage $V_{CB} = 0$ as the collector is at virtual ground and base is grounded. Hence, we can write the equation of I_C as under :

$$V_{BE} = V_T \ln\left(\frac{I_C}{Is}\right) \qquad ...(10.141)$$

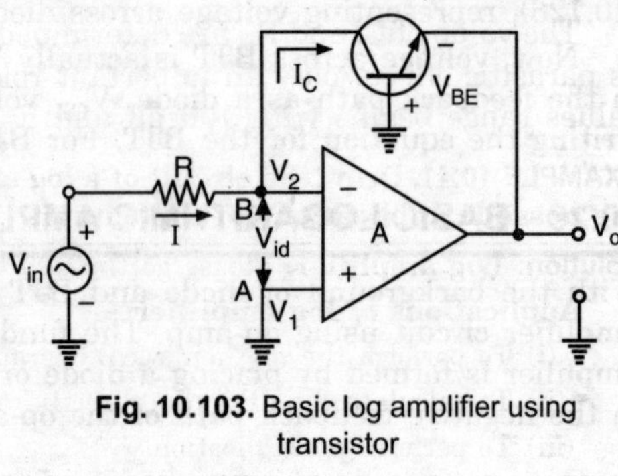

Fig. 10.103. Basic log amplifier using transistor

Applying KVL to the output side, we obtain

$$V_{out} + V_{BE} = 0 \qquad ...(10.142)$$

or

$$V_{BE} = - V_o \qquad ...(10.143)$$

and

$$I_C = I = \frac{V_{in}}{R}$$

Substituting the above value of I_C in equation (10.141), we have

$$-V_{out} = V_T \ln\left(\frac{V_{in}}{RI_s}\right) \quad ...(10.144)$$

Let $V_{ref} = R\,I_s$,

Therefore, $$V_{out} = -V_T \ln\left[\frac{V_{in}}{V_{ref}}\right] \quad ...(10.145)$$

The equation is similar to the equation (10.137), which gives the output proportional to the logarithm of the input voltage V_{in}.

10.28.3. Drawbacks of Basic Circuit

The reverse saturation current I_o for the diode changes with temperature. In fact, it doubles for every ten degree celsius rise in the temperature. Similarly the emitter saturation current varies significantly from one transistor to other and also with the temperature. Hence, it is very difficult to set the term V_{ref} for the circuit. The term V_T which is $\frac{kT}{q}$ also changes with temperature which appears in both the equations. Thus, temperature affects the performance and accuracy of the basic logarithmic amplifier circuit. Hence it is must to provide some sort of temperature compensation to reduce the errors.

10.28.4. Stability Considerations

Due to the presence of an active element in the feedback path, the circuit shows the tendency to oscillate. The circuit is compensated by means of an emitter resistance R_E and a capacitor C_f across the negative feedback element. This arrangement lowers the gain as the frequency rises and avoids the possibility of the oscillations. Hence, such a compensation is called as frequency compensation. The typical frequency compensation network is shown in figure 10.104.

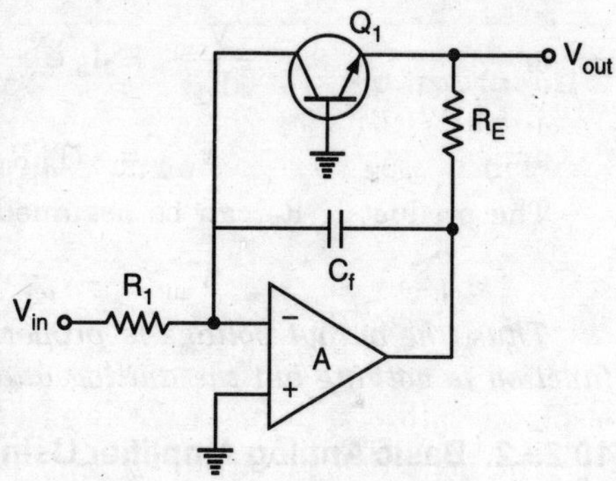

Fig. 10.104. Log amplifier with frequency compensation

The value of C_f and R_E are determined experimentally as parasitic effects play an important role. Generally, C_f values range begins with 100 pF and R_E with 1 kΩ.

EXAMPLE 10.41. Draw the circuit of a log amplifier using op-amps and explain its operation. Enlist the possible applications of this circuit.

Solution: Log amplifier: Please go through text.

Applications of log amplifier:

(i) To perform the mathematical operations such as I_n (x) or log (x) or sinh (x).

(ii) To calculate the dB value directly.

(iii) To perform multiplication.

(iv) To perform division.

(v) True rms conversion.

(vi) Power extraction.

(vii) Square root calculation.

10.29 ANTILOG AMPLIFIER*

The Log amplifier can easily be tuned around to provide the antilog or exponential function which is called as **Antilog amplifier.** The basic antilog amplifier can be obtained by using a diode or a transistor.

10.29.1. Basic Antilog Amplifier Using Diode

The circuit diagram of the basic antilog amplifier using diode is shown in figure 10.105.

The position of diode and resistance are exchanged as compared to log amplifier circuit. The node A is grounded and hence node B is at virtual ground. Hence, $V_2 = 0$.

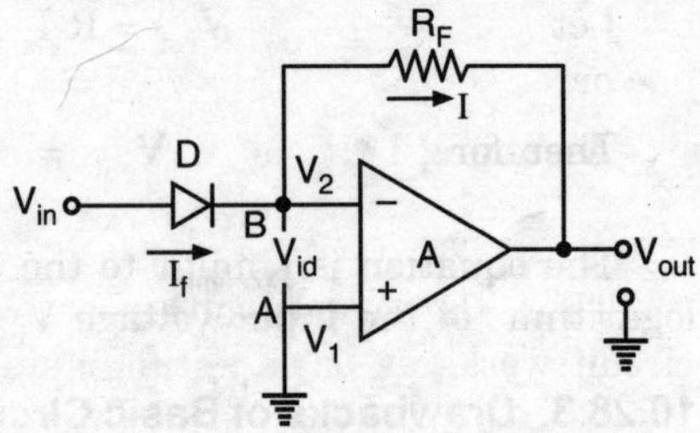

Fig. 10.105. Basic antilog amplifier using diode

Now, the current flowing through the diode is I_f and the voltage across diode is V_{in} itself, as B is the virtual ground. Hence from the diode current equation we can write,

$$I_f = I_o \, e^{V_{in}/\eta V_T} \quad \text{...(10.146)}$$

As op-amp input current is zero, the current I must be same as I_f. Therefore, we have

$$I = I_f = \frac{V_2 - V_{out}}{R_F} = \frac{-V_{out}}{R_F} \quad \text{...(10.147)}$$

or $$\frac{-V_{out}}{R_F} = I_o \, e^{V_{in}/\eta V_T}$$

or $$V_{out} = -(I_o R_F) \, e^{V_{in}/\eta V_T} \quad \text{...(10.148)}$$

The product $I_o R_F$ can be assumed to be V_{ref}, so we obtain

$$V_{out} = -V_{ref} \, e^{V_{in}/\eta V_T} \quad \text{...(10.149)}$$

Thus, the output voltage is proportional to the exponential function of V_{in}. The exponential function is nothing but the antilog and thus circuit works as an Antilog amplifier.

10.29.2. Basic Antilog Amplifier Using Transistor

The same circuit providing antilog of the input can be obtained by using a transistor instead of a diode. This is shown in figure 10.106.

The node B is at virtual ground hence $V_2 = 0$. Thus, both collector and base of the transistor are at ground potential and $V_{CB} = 0$. Hence, the voltage across the transistor is V_{BE} and we can write the expression for its collector current as under :

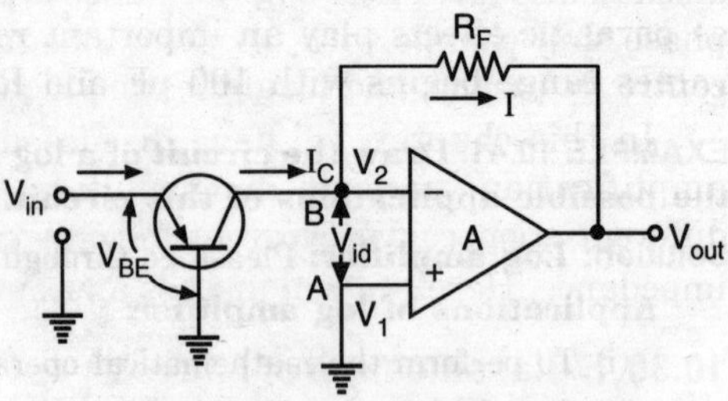

Fig. 10.106. Basic antilog amplifier using transistor

$$I_C = I_s \, e^{V_{BE}/V_T} \quad \text{...(10.150)}$$

But as seen from figure 10.106, $V_{BE} = V_{in}$

Therefore, $$I_C = I_s \, e^{V_{in}/V_T} \quad \text{...(10.151)}$$

Now, the current I_C and current I are same as op-amp input current is zero.

---

* Draw the circuit of an antilog amplifier using op-amp and explain its operation.

(RGTU, Bhopal, Sem. Exam; 2004-05)(10 marks)

Therefore, we have $$I = I_C = \frac{V_2 - V_{out}}{R_F} = \frac{-V_{out}}{R_F} \qquad ...(10.152)$$

or $$\frac{-V_{out}}{R_F} = I_s\, e^{V_{in}/V_T}$$

or $$V_{out} = -I_s\, R_F\, e^{V_{in}/V_T} \qquad ...(10.153)$$

Assuming $I_s\, R_F$ as V_{ref}, we can write

$$V_{out} = -V_{ref}\, e^{V_{in}/\eta V_T} \qquad ...(10.154)$$

Thus, the output voltage is proportional to the exponential of V_{in} i.e., antilog of V_{in}. Thus, circuit works as basic antilog amplifier.

Important Point: In both the above circuits, it can be seen that the terms I_o, I_s and V_T are present in the output equation. All these are the function of temperature. Hence, as temperature changes, these parameters also change and cause serious errors at the output. So, the basic antilog circuits also face the same limitations as that of basic Log amplifier circuits. And hence temperature compensation is must for the Antilog amplifier circuits as well.

10.30. INTRODUCTION TO ACTIVE FILTERS

An electric filter is a network designed to attenuate certain frequencies but pass others without attenuation. A filter circuit, therefore, possesses at least one pass band–a band of frequencies in which the output is approximately equal to the input (i.e., attenuation is zero) and an attenuation band in which output is zero (i.e., attenuation is infinite). The frequencies that separate the different pass and attenuation bands are called the cut-off frequencies.

Filters may be of any type such as electrical, mechanical, pneumatic, hydraulic, acoustical etc., but the most commonly used filters are of the electrical type.

Electrical filters are used in practically all circuits which require separation of signals according to their frequencies. Applications include (but are certainly not limited to) noise rejection and signal separation in industrial and measurement circuits, feedback of phase and amplitude control in servo-loops, smoothing of digitally generated analog (D-A) signals, audio-signal shaping and sound enhancement, channel separation, and signal enhancement in communication circuits.

Filter are specified, analyzed, and designed somewhat differently than the circuits we have discussed so far. Filter's performance is specified in terms of frequency response, how the gain and phase shift change with frequency. Angular frequency–3 dB frequency, bands, ripple, roll off rate, centre frequency, and magnification factor are among the terms unique to filters.

In this chapter, we have discussed how op-amp circuits are employed for providing dc/ac amplification, perform mathematical operations like summing, averaging, integration and differentiation, convert current-to-voltage and voltage-to-current signals, and provide very high input impedance. Here, we wiil discuss another important field of application using op-amps : filters.

10.30.1 Classification of Filters

Basically, a filter may be defined as a frequency selective circuit. It is designed to pass a specific band of frequencies and block or attenuate signals of frequencies which are lying outside this band. The filters may be classified into two most basic categories namely the digital filters and the analog filters. The digital filters work entirely in the digital domain. Digital filters use the digital data as the input signal. We shall not discuss the digital filters in this chapter. Figure 10.107 shows the classification of filters.

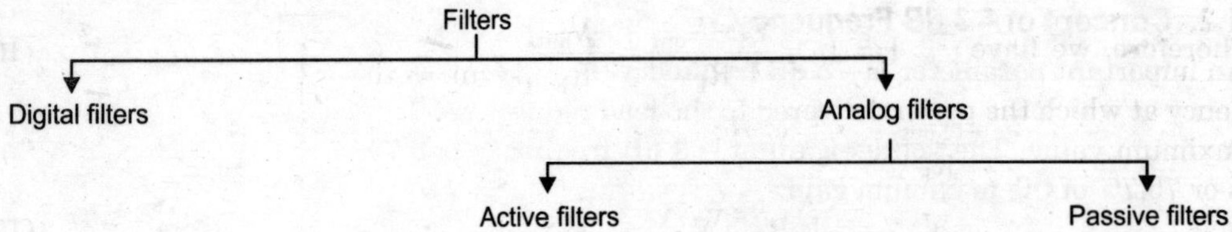

Fig. 10.107. Classification of filters

The other type of filters is the analog filters which may again be classified into following two categories

(i) active filters and

(ii) passive filters

In previous semester, we have studied about the passive filters i.e., the filters using passive components such as R, L and C. The passive filters can be made more effective by using an active device such as a transistor, FET or OP-AMP alongwith the passive components. Such filters are then known as active filters. The analog filters are designed to process analog signals. AF and RF are the audio and radio frequency ranges respectively in which the filters operate. The operating frequency range of a filter depends on the type of components used. For example, the R-C filters can operate only in the audio frequency (AF) range whereas L-C filters or crystal filters are used in the radio frequency (RF) range.

10.31 FEW DEFINITIONS RELATED TO FILTERS

The procedure to design, analyze and specify filters is different from the other circuits. The filter performance is specified in terms of the frequency response, gain and phase shift variations.

The important aspects related to the filters can be listed as under:

(i) Angular frequency

(ii) Critical frequency

(iii) 3 dB frequency

(iv) Bands, ripple and roll off rates.

(v) Centre frequency and Quality factor (Q).

DO YOU KNOW ?

A graph of the magnitude of the gain versus frequency is called a frequency response plot.

10.31.1. Concept of Frequency Response

(i) Definition

The frequency response of a filter may be defined as a graph of gain (and phase) versus frequency. The gain and phase are plotted on the Y-axis while frequency is plotted on the X-axis. The frequency axis is a logarithmic axis. This means that we plot (log f) on the X-axis in order to cover a large frequency range in a small space.

(ii) dB Expression

The gain plotted on the Y-axis is generally scaled in decibel (dB). We know that

$$dB = 10 \log_{10}\left(\frac{P_{out}}{P_{in}}\right)$$

or

$$dB = 20 \log_{10}\left(\frac{V_{out}}{V_{in}}\right)$$

10.31.2. Concept of – 3 dB Frequency

An important parameter is – 3 dB frequency ($f_{-3\text{ dB}}$). This is the frequency at which the power delivered to the load reduces to 50% of its maximum value. The voltage gain at – 3 dB frequency is 0.707 times or 70.7% of the maximum gain.

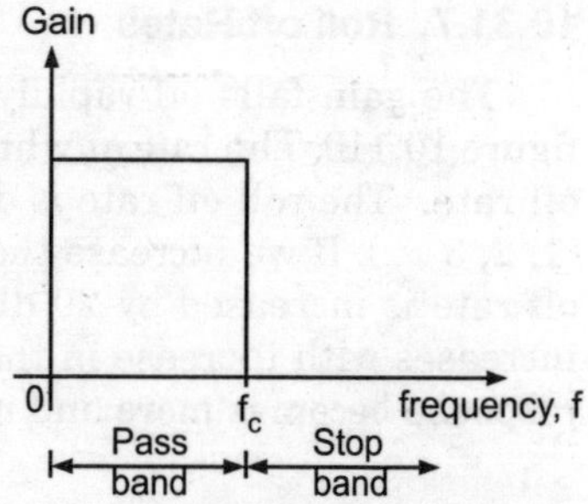

Fig. 10.108. Frequency response of an ideal low pass filter

10.31.3. Angular Velocity

Sometimes, the frequency is expressed as angular velocity ω in radians/second. We already know that the relation between ω and f is as under :

$$\omega = 2\pi f \text{ radians/sec.}$$

10.31.4. Phase Plot and Critical Frequency

The phase shift between output and input keeps changing with change in frequency. At critical frequency f_c, this phase shift will be a specific multiple of 45°, i.e., 45°, 90°, 135°. For many filters, the – 3 dB frequency and the critical frequency f_c are one and the same.

DO YOU KNOW ?

Because at a given frequency, the gain has a specific magnitude as well as a phase angle, the gain of a filter is said to be complex number.

10.31.5. Normalized Frequency Response

The graph of log $\left(\frac{f}{f_o}\right)$ versus gain ($A - A_o$) dB is called as the normalized frequency response. The critical frequency of the normalized frequency response is $\left(\frac{f_o}{f_o}\right) = 1$. The gain axis is adjusted to set the passband gain A_o to 0 dB. This is done by adjusting A_o from each point on the vertical axis.

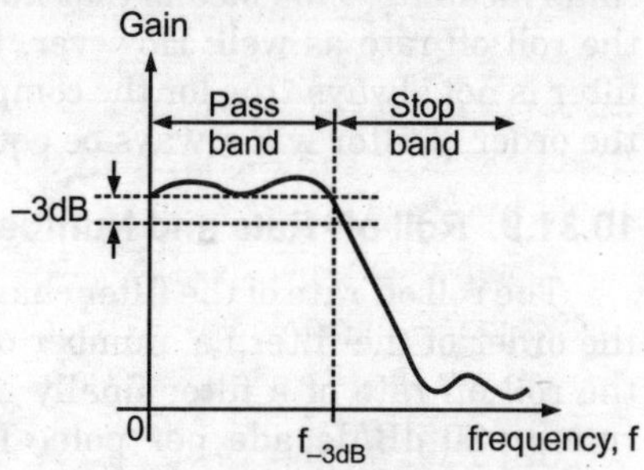

Fig. 10.109. Practical low pass filter response

10.31.6. Characteristics and Terminology

The frequency response graphs discussed so far are of a low pass filter. A low pass filter passes all the frequencies below its cut-off frequency and stops all the frequencies which are above the critical frequency. Figure 10.109 shows the frequency response of an ideal low pass filter. The critical frequency forms the boundary between the two bands called as pass band and stop band of the filter.

(i) Pass band

It is the band or range of frequencies which are allowed to pass through to the output by the filter without any attenuation.

(ii) Stop band

It is the band or range of frequencies which are not allowed to pass through the output by the filter.

(iii) Practical Frequency Response

Practical frequency response is different from the ideal one. The ideal response can not be obtained practically. Figure 10.110 shows the practical frequency response of a low pass filter. The – 3 dB frequency $f_{-3\text{ dB}}$ forms the boundary between the pass band and stop band. For some filters, gain varies up and down in the pass band as well as attenuation band. The amount of pass band ripple allowable in the pass band is an important parameter.

It is the weak man who urges compromise, never the strong man. — **Elbert Hubbard**

10.31.7. Roll off Rates

The gain falls off rapidly in the stop band as shown in figure 10.110. The rate at which it falls off is called as the roll off rate. The roll off rate is decided by the order of the filer (1, 2, 3 ....). If we increase the filter order by 1, then the roll off rate is increased by 20 dB/decade. The filter complexity increases with increase in the filter order but the frequency response becomes more and more sharp.

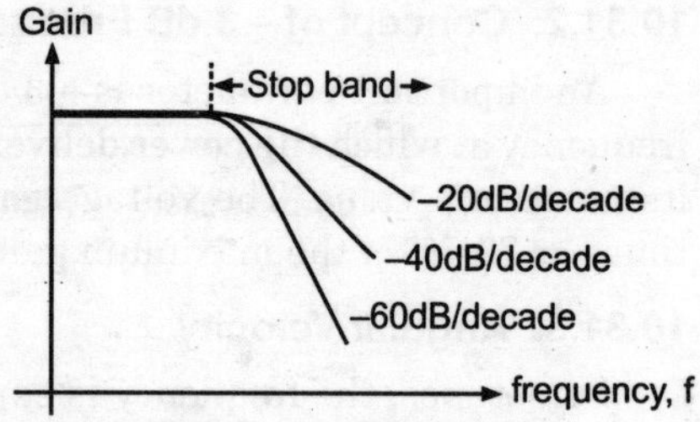

Fig. 10.110. Roll off rate

Table 10.2.

S.No.	Filter order	Roll off Rate
1.	1.	20 dB/decade
2.	2.	40 dB/decade
3.	3.	60 dB/decade
4.	4.	80 dB/decade

10.31.8. Poles and Order

For the general purpose high pass and low pass filters, the term *pole* and *order* will have the same meaning. This means that number of poles will be equal to the filter order. This also tells us the roll off rate as well. However, this relation of equality between number of poles and order of filter is not always true for the complex filters. However, for all the filters described in this chapter, the order of filter will always be equal to the number of poles.

10.31.9. Roll off Rate and Number of Poles (Order)

The roll off rate of the filter characteristics is dependent on the order of the filter i.e. number of poles. It is observed that the roll off rate of a filter finally approaches 6 dB/active per role or 20 dB/decade per pole. The number of poles and corresponding roll off rates are listed in table 10.3.

> **DO YOU KNOW?**
>
> With increasing the order of a filter, the actual stopband response of the filter approaches its ideal stopband characteristics.

Table 10.3. Relation between roll off rate and number of poles

Order of Filter	Number of Poles	Roll off Rate
1.	1	20 dB/decade
2.	2	40 dB/decade
3.	3	60 dB/decade
4.	4	80 dB/decade

The higher order filters offer higher attenuation at frequencies beyond the break frequency. The higher order filters are used when the transition band is to be as narrow as possible. The higher order filters are more complex and expensive. For typical applications, generally, the filter order is in the range of two to six.

10.32 TYPES OF FILTERS BASED UPON FREQUENCY RESPONSE

Irrespective of its realization, a filter always fall under one of the five types shown in figure 10.5. In other words, we can say that the filter can also be classified based on their frequency response characteristics as shown in figure 10.111.

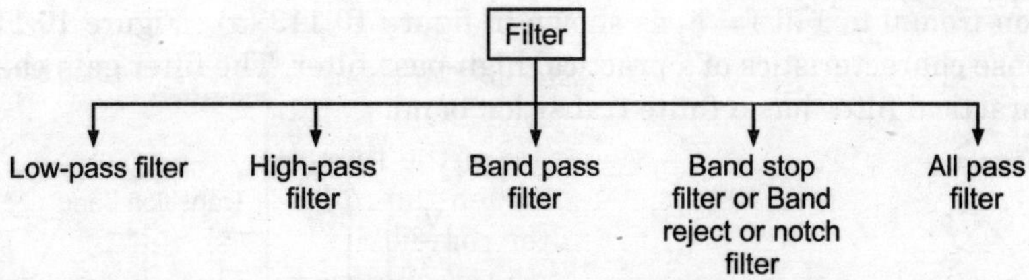

Fig. 10.111. Filter types

10.33 FREQUENCY RESPONSE OF VARIOUS FILTERS

Frequency response characteristics of a filter is the graph of filter gain on Y-axis versus frequency on X-axis, where filter gain is defined as :

$$\text{Filter gain} = \left|\frac{V_{out}}{V_{in}}\right| \qquad ...(10.155)$$

Figure 10.112 shows the ideal and practical characteristics of a low-pass filter.

10.33.1. Frequency Response Characteristics of a Low-pass Filter (LPF)

The frequency response characteristics of an ideal low-pass filter is shown in figure 10.112(a). It shows that a low-pass filter has a constant gain from 0 Hz to a high cut-off frequncy f_C. The frequencies between 0 and f_C are known as passband frequencies whereas the frequencies beyond f_C are known as the stopband frequencies. At $f = f_C$, the filter gain makes a sudden transition to zero. Therefore, all the frequencies beyond f_C are completely attenuated.

Figure 10.112 (b) shows the frequency response of a practical low-pass filter. It shows that the filter gain does not change suddenly at $f = f_C$. Instead as f increases, the gain reduces gradually. At $f = f_C$, the gain is down by 3 dB and after f_C, it reduces at a higher rate as shown in figure 10.112(b).

However, by using special design techniques, precision component values and high speed op-amps, it is possible to design a practical filter having the frequency characteristics very close to the ideal characteristics.

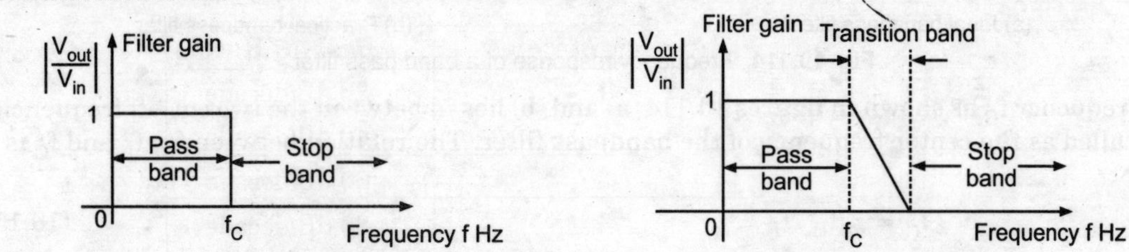

(a) Ideal low-pass filter (b) Practical low-pass filter

Fig. 10.112 Frequency response of a low-pass filter

10.33.2. Frequency Response of a High-pass Filter (HPF)

Figure 10.113 (a) shows the frequency response characteristics of an ideal high-pass filter. Its stopband extends from $f = 0$ to $f = f_C$ where f_C is the cut-off frequency. The passband will be for all frequencies above f_C.

The gain of an ideal high-pass filter is 0 over its stopband and constant (equal to 1) over its passband. The gain makes a

> **DO YOU KNOW ?**
>
> A wide bandpass filter has a wide passband between two cut-off frequencies f_H and f_L where $f_H >> f_L$ and two stopbands at $0 < f < f_L$ and $f > f_H$.

sudden transition from 0 to 1 at $f = f_C$ as shown in figure 10.113 (a). Figure 10.113 (b) shows the frequency response characteristics of a practical high-pass filter. The filter gain changes gradually. Therefore, the practical filter has a finite transition band.

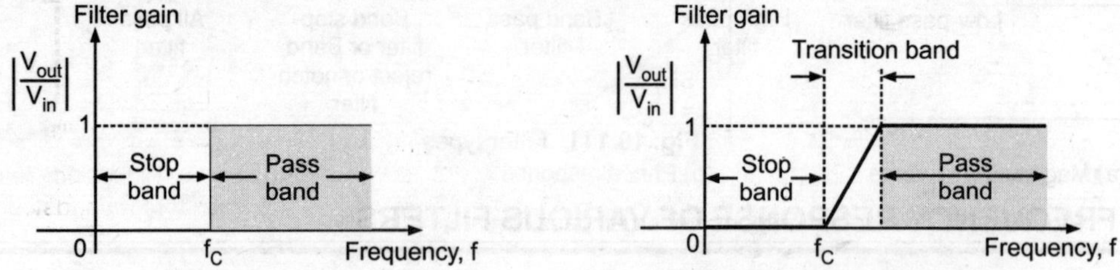

(a) Ideal high-pass filter (b) Practical high-pass filter

Fig. 10.113. Frequency response of a high-pass filter

10.33.3. Frequency Response of a Bandpass Filter (BPF) **(Important)**

Figure 10.114(a) shows the frequency response characteristics of an ideal bandpass filter. Its passband extends between the two cut-off frequencies f_L and f_H with $f_H > f_L$. The frequencies outside this passband lie in the stopband or attenuation band.

The gain of an ideal bandpass filter is 0 over the stopband and constant (equal to 1) over its passband. The gain will make sudden transitions from 0 to 1 at $f = f_L$ and from 1 to 0 at $f = f_H$ as shown in figure 10.114(a). Figure 10.114(b) shows the frequency response of a practical bandpass filter. The gain does not change suddenly at $f = f_L$ and $f = f_H$ instead it change gradually. At the cut-off frequencies, f_L and f_H, the gain is down by 3 dB with respect to the passband gain.

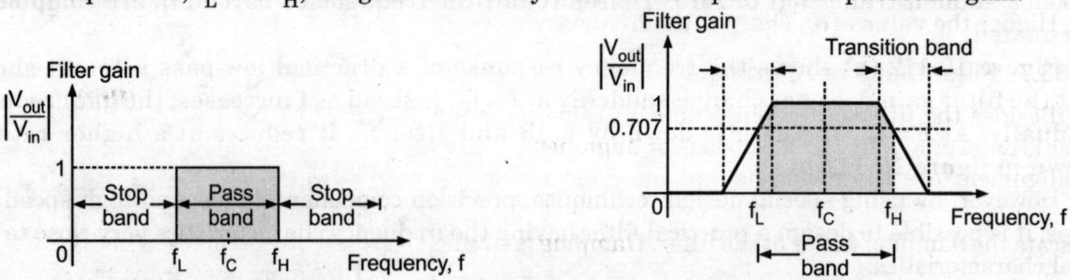

(a) Ideal bandpass filter (b) Practical bandpass filter

Fig. 10.114. Frequency response of a band pass filter

Frequency f_C as shown in figures 10.114 (a) and (b) lies in between the two cut-off frequencies. It is called as the center frequency of the bandpass filter. The relation between f_H, f_L and f_C is as under :

$$f_C = \sqrt{f_H \cdot f_L} \quad \text{...(10.156)}$$

10.33.4. Frequency Response of an All Pass Filter

An all pass filter is characterised by | H | = 1 for all the frequencies and $\measuredangle H = -t_o \omega$, where t_o is a constant, as shown in figures 10.115 (a) and (b), where H is filter gain. The all pass filter passes all the input signals without any attenuation. But, it delays each signal proportional to its frequency. Hence, it is also called as a delay filter.

DO YOU KNOW ?

Notch filter is a narrow band-stop filter. It notches out or blocks frequencies near centre frequency f_c, also called a notch-out frequency.

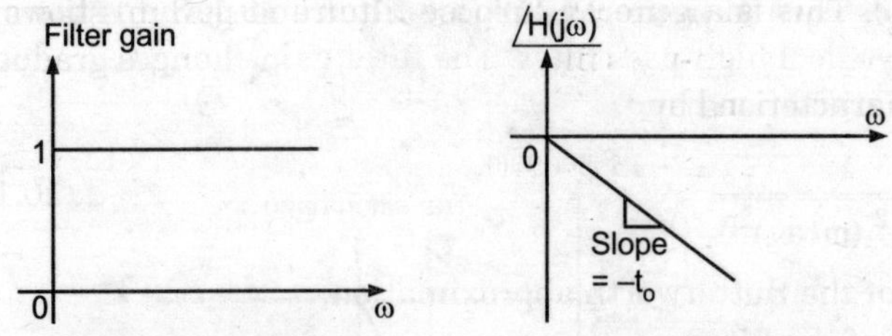

(a) Magnitude response (b) Phase response

Fig. 10.115. Frequency response of an all pass filter

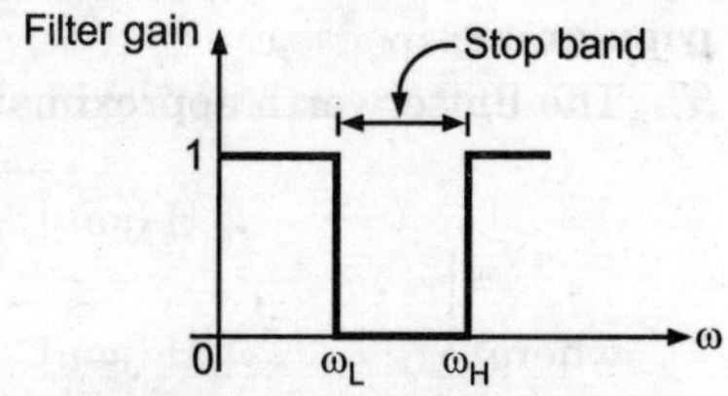

Fig. 10.116. Frequency response (idealized) of a bandstop filter

10.33.5. Bandstop Filter

A bandstop or band reject filter is complementary to the bandpass filter. It blocks out the frequency components within the stopband as shown in figure 10.116.

10.33.6. Quality Factor (Q) of Filter

Figure 10.117 shows the practical frequency response of a bandpass filter. Instead of specifying the roll off rates, Q and bandwidth are specified. The relation between centre frequency f_c, Q and bandwidth is a as under :

$$Q = \frac{f_C}{B}$$

Higher the value of Q, sharper the frequency response.

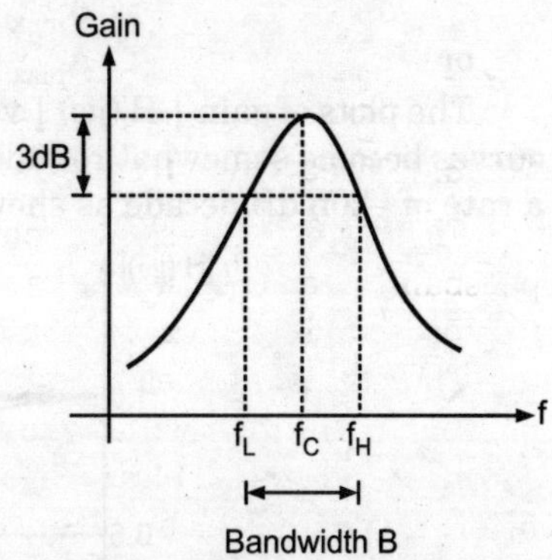

Fig. 10.117. Practical response of a bandpass filter

10.33.7. Filter Class or Alignment

Besides the filter order, the shape of transition band of a filter is determined by the filter class or alignment. Filter class or alignment are the terms having same meaning and they indicate the damping factor of the filter. Damping factor is $\left(\frac{1}{Q}\right)$ and Q is the quality factor.

> **DO YOU KNOW ?**
>
> Filter behaviour is usually described by frequency response plot. The gain magnitude is usually expressed in dB. Critical frequency occurs when the gain drops by 3 dB below the pass band gain. The critical frequency is the dividing line between the stop band and the passband.

The damping factor is denoted by α

Therefore, $$\alpha = \frac{1}{Q}$$

The alignment is responsible in determining the shape of the transition band and sometimes the passband and stopband as well.

The popular filter classes or alignments may be listed as under :

(i) Butterworth alignment.
(ii) Chebyshev alignment
(iii) Bessel alignment
(iv) Eliptic alignment.

In this section, let us compare these alignments for same critical frequency and identical DC gains.

1. The Butterworth Alignment

The Butterworth alignment exhibits moderate amplitude and phase response. It exhibits the fastest roll off of any monotonic i.e., single slope filter. For this filter, the 3 dB down frequency is equal to the critical frequency ($f_{3\,dB} = f_C$). This is a general purpose filter and perhaps the most popular one.

The Butterworth approximation is characterized by,

$$|H(j\omega)| = \frac{1}{\sqrt{1+\epsilon^2 (\omega/\omega_c)^{2n}}} \quad ...(10.157)$$

where $|H(j\omega)|$ = Gain of the Butterworth approximation.

n = Order of the filter

ω_c = Cut-off frequency and ϵ = Constant

Here, ϵ decides the maximum passband variations because

$$A_{max} = A(\omega_c) = 20 \log_{10} (1+\epsilon^2)^{1/2}$$

or

$$A_{max} = 10 \log_{10} (1+\epsilon^2) \quad ...(10.158)$$

The plots of gain $|H(j\omega)|$ versus ω are maximally flat at $\omega = 0$ as shown in figure 10.118. These curves become somewhat rounded near ω_c and they roll off into the attenuation band (stopband) at a rate of –20n dB/decade as shown in figure 10.118.

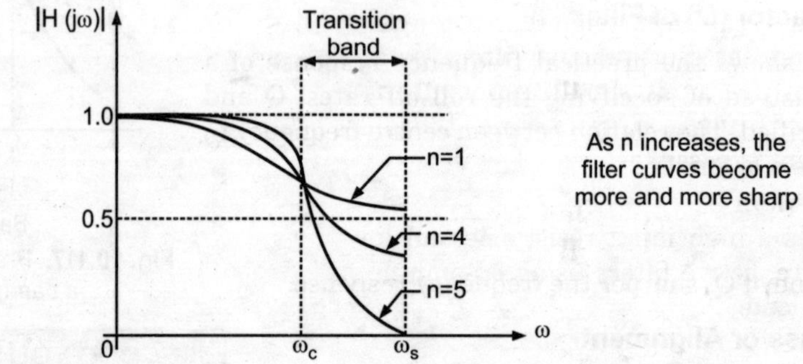

Fig. 10.118. Butterworth low pass response for various values of n

Hence we get more and more sharp curves with increase in the value of n or as n increases, the filter response becomes more and more close to the brickwall filter response. The expression for the gain of a second order (n = 2) assuming $\epsilon = 1$ is given by,

$$|H(j\omega)| = \frac{1}{\sqrt{1+(\omega/\omega_c)^4}} \quad ...(10.159)$$

Similarly, it is possible for us to obtain the expressions for the gains of filter having higher order.

Salient Features of Butterworth approximation

(i) As seen from figure 10.118, the plots of gain against frequency with Butterworth approximation are maximally flat.

(ii) This means that the ripple produced in the passband and attenuation band is minimum.

(*iii*) So when the flatness of frequency response is more important, we have to use the Butterworth approximation.

> **DO YOU KNOW ?**
>
> The Butterworth filter offers optimum flatness. Increase in damping coefficient causes considerable passband attenuation and slow initial roll off, but assures good transient, and time delay response. Decrease in damping coefficient (Chebyshev) allows much faster initial roll off, but introduces passband ripple and poor-transient response.

10.33.8. Realization of Practical Filters

The first step towards design of practical filters is to decide the specifications of the filter.

Specifications of a Filter

The filter design is based on the following specifications:

(i) Break frequency.

(ii) Maximum allowable ripple in the passband.

(iii) Attenuation levels in the transition and stop band.

(iv) Phase response and associated time delays.

From the specifications, one has to decide the order of the filter and the alignment.

10.34 PERFORMANCE COMPARISON OF ACTIVE AND PASSIVE FILTERS*

10.34.1. Salient Features (Advantages) of Active Filters

(RGTU, Bhopal, Sem. Exam; 2004-05) (05 marks)

The active filters consisting of an op-amp and passive components such as resistors, capacitors and inductors have following advantages over the passive filters :

1. Flexibility in gain and frequency adjustment

In passive filters, the input signal is attenuated while passing through the filter. But in active filters, as an op-amp can provide a gain, the input signal is not attenuated. In addition, the active filter is easy to tune, therefore, easy frequency adjustment is possible.

DO YOU KNOW ?

Active filters are preferred over passive filters because they provide gain and frequency adjustment flexibility. They pose no loading problem because of high input impedance and low output impedance and they are comparatively less expensive.

2. No loading problem

The op-amp has a high input resistance and low output resistance. Therefore, active filters using op-amp do not load the input source or load.

3. Low cost

Due to the easy availability of a variety of cheaper op-amps and due to the absence of inductors in the active filters, the active filters are more economical than the passive filters.

The active filters are used in almost all the fields of Electronics such as communication, signal processing, radio, telephone, radar, space satellites and biomedical equipments.

4. No insertion loss

The active filters do not exhibit any insertion loss. Whereas the passive filters always show some loss of signal in their pass band.

5. Pass band gain

The active filter can be designed to provide some pass band gain. This is not possible for the passive filters.

6. Interstage isolation and control of impedance

Active filters also allow for interestage isolation an control of input and output impedance.

7. Small component size

The components required for the active filters are of smaller size as compared to those used for the passive filters.

8. Use of inductors can be avoided

It is possible to avoid the use of inductors in the complex active filters. This reduces the size of the filters to a great extent.

9. Using the active filter technology, the rapid design of stable economical filters is possible.

---

* Enumerate advantages of active filters w.r.t. the passive filters. Why and how many sections are required for a composite passive filter?.

10.34.2. Drawbacks of Active Filters

(i) Active filters require dc power supply for their operation, while passive filters do not.
(ii) Active filters are limited in their frequency range. This is because an op-amp has a finite gain bandwidth product. The passive filters do not have this limitation.
(iii) Active filters can not handle large amounts of power. Passive filter can handle hundreds of watts.

Important Point: The active filters are suitable for low or moderate frequencies (upon 1 MHz with typical devices) and they cannot handle large power.

10.35 CONCEPT OF TRANSFER FUNCTION H(s)

The behaviour of any circuit is uniquely characterized by its **transfer function H (s).** The transfer function relates the output of a circuit to its input. If the input and output voltages are used then the transfer function is voltage transfer function (VTF). Similarly, we can have the current transfer function that relates the input and output currents.

Once the transfer function is known, we can obtain the response v_o (t) to the input v_i(t)

$$H(s) = \frac{V_{out}(s)}{V_{in}(s)} \quad ...(10.160)$$

We can obtain output v_{out}(t) as :

$$v_{out}(t) = £^{-1}\{H(s)\, V_{in}(s)\} \quad ...(10.161)$$

Where $£^{-1}$ represents the inverse Laplace transfer V_{in}(s) is the Laplace transform of the input v_{in}(t).

Transfer function is a rational of s as shown below :

$$H(s) = \frac{a_m s^m + a_{m-1}s^{m-1} + ... + a_1 s + a_0}{b_n s^n + b_{n-1}s^{n-1} + ... + b_1 s + b_0} \quad ...(10.162)$$

The degree of denominator polynomial is n and it denotes the order of the filter (first order, second order etc.) H(s) can be represented in terms of the roots of numerator and denominator polynomials as :

$$H(s) = H_o \frac{(s - z_1)(s - z_2) ... (s - z_m)}{(s - p_1)(s - p_2) ... (s - p_n)} \quad ...(10.163)$$

where H_o = Scaling factor

The zeros of H(s) are located at $s = z_1, s = z_2 ..., s = z_m$ and its poles are located at $s = p_1, s = p_2 ...., s = p_n$.

The roots of H(s) can be plotted in the complex s plane where the X-axis represents the real part of s and Y-axis represents its imaginary part. Just by looking at the pole-zero plot of the circuit, we can obtain a lot of important information about the circuit.

DO YOU KNOW?

Active filters using OTA's have many advantages. Filter parameters, such as mid-band gain, critical frequency or type of filter can be controlled by adjusting g_m of the OTA.

10.35.1. Transfer Function and Stability

The pole zero plot of a circuit can tell us about whether the circuit is stable or unstable. The circuit will be stable if all the poles of transfer function lie in the left half of the s-plane as shown in figure 10.119, i.e. for s < 0.

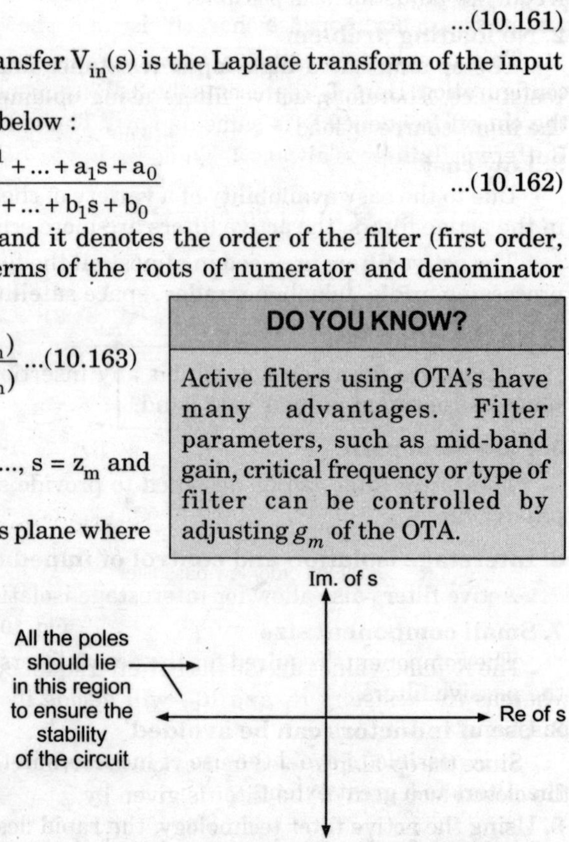

Fig. 10.119. Transfer function and stability

10.36 INTRODUCTION TO LOW PASS FILTER (LPF)

Let us begin the detailed discussion from the first order low pass filter. As discussed earlier, a low-pass filter has a constant gain form 0 Hz to a high cut-off frequency f_c. Therefore, the bandwidth is also f_c. At high cut-off frequency f_c, the gain is reduced by 3 dB and for $f > f_c$, it decreaes with the increase in input frequency. The frequencies between 0 and f_c are known as passband frequencies whereas the range of frequencies beyond f_c are attenuated and are therefore known as stop band frequencies.

As far as frequency response of a low-pass filter is concerned, an ideal filter has no loss in the passband and infinite loss (or attenuation) in the stop band. Although ideal filter response is not practically possible, however, it is possible to achieve a practical response that approximates the ideal response by employing special design techniques, precision components and high spped op-amps such as LM318 or ICL 8017. The general second order filter is discussed after that. Then we shall discuss the two techniques for designing specific types of second order low pass filters. Finally, cascading of first and second order filters to product filters of whatever order is necessary, has been elaborated.

10.37. THEORY AND ANALYSIS OF FIRST ORDER LOW-PASS FILTER (LPF)

1. Circuit diagram and frequency response

As a matter of fact, a first order low-pass filter is also known as one pole low-pass Butterworth filter. Figure 10.120 (a) shows the circuit diagram of a first order low pass filter (LPF) using op-amp and its frequency response characteristics is shown in figure 10.120 (b).

This circuit uses a R-C network for filtering and the op-amp as used in the non-inverting configuration. Here, f_H represents the 3 dB frequency ($f_{3\,dB}$). Since, this is the Butterworth alignment, the cut-off frequency f_C is same as $f_{3\,dB}$. The value of damping factor for the first order low pass Butterworth filter is always 1. Thus, we have

$$Q = 1. \left(\text{Because } Q = \frac{1}{\alpha}\right)$$

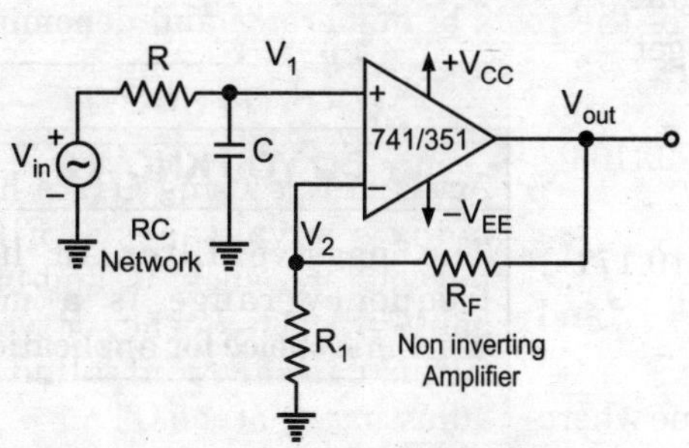

(a) First order low-pass filter

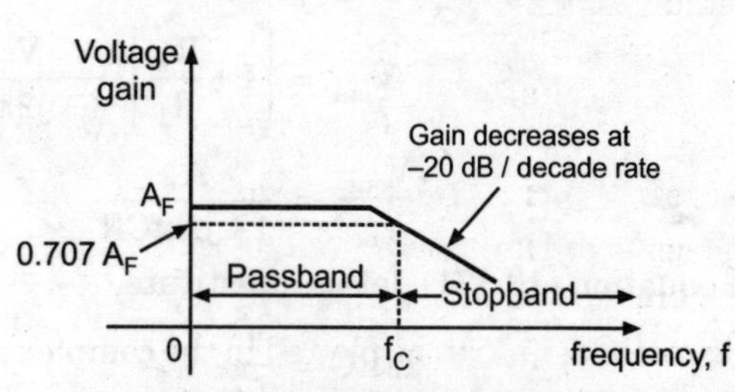

(b) Frequency response characteristics

Fig. 10.120.

The R and C values decide the cut-off frequency of the filter whereas the resistors R_1 and R_F will decide its gain in the passband.

Since the op-amp is used in the non-inverting configuration, the closed loop gain of the filter is given by,

$$A_{max} = A_o = A_F = 1 + \frac{R_F}{R_1} \quad \text{...(10.164)}$$

It is also called as the passband gain and denoted by A_{max} or A_o.

> **DO YOU KNOW ?**
>
> The active filters have their limitation too. High frequency response is limited by the gain-bandwidth (GBW) product and slew rate of the op-amp. Moreover, the high frequency active filters are move expensive than the passive filters.

2. Analysis of the filter circuit

Let us consider figure 10.120 (a). We have to obtain the expression for V_1 which is the voltage across capacitor C. The capacitor C and resistor R form a voltage divider network across the input voltage V_{in}. Therefore, voltage at the non-inverting terminal of op-amp is given by,

$$V_1 = \frac{-jX_C}{R - jX_C} V_{in} \quad ...(10.165)$$

But, the reactance of the capacitor is given by,

$$X_C = \frac{1}{2\pi fC} \quad ...(10.166)$$

Substituting the above value of X_C in equation (10.165), we obtain

$$V_1 = \frac{-j\left[\dfrac{1}{2\pi fC}\right] \cdot V_{in}}{R - j\left[\dfrac{1}{2\pi fC}\right]} = \frac{-jV_{in}}{2\pi fRC - j} \quad ...(10.167)$$

Dividing numerator and denominator by $-j$, we obtain

$$V_1 = \frac{V_{in}}{1 - \dfrac{2\pi fRC}{j}} \quad ...(10.168)$$

But $\frac{1}{j} = -j$, therefore, equation (10.168) becomes,

$$V_1 = \frac{V_{in}}{(1 + j2\pi fRC)} \quad ...(10.169)$$

Thus, we have obtained the expression for V_1.

The filter output voltage V_{out} is given by,

$$V_{out} = A_F \times V_1 \quad ...(10.170)$$

where A_F = Closed loop voltage gain

Substituting equations (10.164) and (10.167), we get

$$V_{out} = \left[1 + \frac{R_F}{R_1}\right] \cdot \frac{V_{in}}{1 + j2\pi fCR}$$

or

$$\frac{V_{out}}{V_{in}} = \frac{A_F}{1 + j2\pi fCR} \quad ...(10.171)$$

> **DO YOU KNOW ?**
>
> The passive filter in high frequency range is a more economic choice for applications.

In equation (10.171), let us substitute,

$$2\pi RC = \frac{1}{f_C}$$

where f_C = Cut-off frequency $= \dfrac{1}{2\,\pi RC}$ ...(10.172)

Therefore,

$$\frac{V_{out}}{V_{in}} = \frac{A_{VF}}{1 + j\left[\dfrac{f}{f_C}\right]} \quad ...(10.173)$$

This is the expression for the gain of the low-pass filter (LPF) as a function of the operating frequency f.

The gain magnitude and phase angle equations of the low-pass filter can be obtained by converting equation (10.173) into its equivalent polar form as under :

$$\frac{V_{out}}{V_{in}} = \left|\frac{V_{out}}{V_{in}}\right| \angle\theta$$

or
$$\frac{V_{out}}{V_{in}} = \frac{A_F}{\sqrt{1+\left(\frac{f}{f_C}\right)^2}} \angle -\tan^{-1}\left(\frac{f}{f_C}\right) \quad ...(10.174)$$

The first term in equation (10.174) represents the magnitude of the filter gain.

3. Verification of the operation

The operation of the first order low-pass filter (LPF) can be verified from equation (10.174) as under:

(i) At very low frequency $f < f_C$, we have

$$\left|\frac{V_{out}}{V_{in}}\right| \cong A_F = A_o \text{ or } A_{max}, \text{ i.e., constant}$$

Thus, at very low frequencies, the filter gain is constant. This has been shown in figure 10.120 (b).

As frequency increases, $\left(\frac{f}{f_C}\right)$ ratio increases and filter gain decreases.

(ii) At $f = f_C$, we have

$$\left|\frac{V_{out}}{V_{in}}\right| = \frac{A_F}{[1+1]^{1/2}} = \frac{A_F}{\sqrt{2}} = 0.707\, A_{V'F} \quad ...(10.175)$$

This is shown in figure 10.120 (b). Thus, at $f = f_C$, the filter gain reduces by 3 dB, as the dB value of 0.707 is – 3 dB.

(iii At very high frequencies, i.e. for $f > f_C$, we have

$$\left|\frac{V_{out}}{V_{in}}\right| < A_F \quad ...(10.178)$$

The filter gain will keep decreasing with increase in frequency. This decrease takes place at a constant rate of – 20 dB/decade as shown in figure 10.120 (b). Here, decade means 10 times and the gain is expressed in dB. The negative sign indicates that the filter gain is decreasing.

4. Design Procedure (Design steps)

The designing procedure for a first order low-pass filter is as under :

(i) We choose the cut-off frequency f_C.

(ii) We select a value of C usually between 0.001 and 0.1 μF.

(iii) We calculate the value of R using,

$$R = \frac{1}{2\pi f_c C}$$

(iv) Finally, we select the values of R_1 and R_F depending on the desired value of passband gain A_F as,

$$A_F = 1 + \frac{R_F}{R_1}$$

10.37.1. Frequency Scaling

Once the filter is designed, sometimes, it is essential to change the value of cut-off frequency f_c. The method used to change the original cut-off frequency f_c to a new cut-off frequency f_{c1}, is called as frequency scaling. To achieve such a frequency scaling, the standard value capacitor C is selected first. The required cut-off frequency can be achieved by calculating the corresponding value of resistance R. However, to achieve frequency scaling, a potentiometer is used. Hence, the resistance R is usually a potentiometer with which required cut-off frequency f_c can be adjusted and changed later on if required.

EXAMPLE 10.42. For a first order Butterworth low-pass filter, calculate the cut-off frequency if R = 10 kΩ and C = 0.001 μF. Also calculate the passband voltage gain if R_1 = 10 kΩ and R_F = 100 kΩ.

Solution: (i) We know that the cut-off frequency for a first order Butterworth low-pass filter (LPF) is given by

$$f_c = \frac{1}{2\pi RC} = \frac{1}{2\pi \times 10\times10^3 \times 0.001\times10^{-6}} = 15.915 \text{ kHz} \quad \textbf{Ans.}$$

(ii) Also, the passband voltage gain is given by

$$A_F \text{ or } A_o = 1+\frac{R_F}{R_1} = 1+\frac{100 \text{ k}\Omega}{10 \text{ k}\Omega} = 11 \quad \textbf{Ans.}$$

EXAMPLE 10.43. Design a first order low-pass filter for the following specifications :

(i) Pass band voltage gain = 2

(ii) cut-off frequency f_c = 10 kHz.

Solution: The design procedure of a first order low-pass filter is as under :

(a) We choose the value of R_F and from that calculate R_1.

(b) We choose the capacitance C, usually between 0.001 and 0.1 μF.

(i) The voltage gain in the passband is given to be equal to 2. Let us choose R_F = 10 kΩ.

Therefore, $$A_F = 1+\frac{R_F}{R_1}$$

or $$\frac{R_F}{R_1} = 1$$

or $$R_F = R_1 = 10 \text{ k}\Omega.$$

(ii) Let us choose $$C = 0.001 \ \mu F$$

But $$f_c = \frac{1}{2\pi RC}$$

or $$R = \frac{1}{2\pi f_c C} = \frac{1}{2\pi \times 10\times10^3 \times 0.001\times10^{-6}} = 15.9 \text{ k}\Omega. \quad \textbf{Ans.}$$

DO YOU KNOW ?

A first order low pass filter has a – 20 dB decade roll-off rate. The roll-off rate increases with the order of the filter. An nth order filter has a roll-off rate of – 20 n dB/decade.

EXAMPLE 10.44. Design a low-pass filter having a cut-off frequency of 2 kHz with a passband gain of 2.5.

Solution: Given that cut-off frequency, f_c = 2 kHz

Let the capacitor C be of 0.01 μF value. Then, we have

$$R = \frac{1}{2\pi f_c C} = \frac{1}{2\pi \times 2\times10^3 \times 1\times10^{-8}} = 7.95 \text{ k}\Omega \text{ (practical value 8.2 k}\Omega)$$

Since the passband gain is 2.5, so, we have

$$A_F = 1+\frac{R_F}{R_1} = 2.5 \text{ or } R_F \text{ should be equal to } 1.5 R_1$$

Since $$R_F \,||\, R_1 = R$$

or $$\frac{R_F R_1}{R_F + R_1} = R \text{ or } \frac{1.5 R_1 R_1}{1.5R_1 + R_1} = 8.2 \text{ k}\Omega$$

or $$R_1 = \frac{2.5}{1.5}\times 8.2 = 13.66 \text{ k}\Omega \text{ (say 15 k}\Omega) \quad \text{Ans.}$$

and $$R_F = 13.66 \times 1.5 = 20 \text{ k}\Omega \quad \textbf{Ans.}$$

Voltage $$V_A = I_2Z_2 + V_B = I_2Z_2 + 1 \quad ...(10.179)$$

10.38. A SECOND ORDER LOW-PASS BUTTERWORTH FILTER*

1. Circuit diagram and frequency response

As a matter of fact, a second-order filter can be obtained by coupling two first-order active filters, but this would not be economical because it needs two op-amps. In practice, the second order low-pass filter can be obtained simply by inserting an additional RC network into the first order low-pass filter as shown in figure 10.121 (a). Here, R_2C_2 is this additional RC network. The frequency response characteristics of the second order low-pass Butterworth filter is shown in figure 10.121(b). It shows that the response in the stopband rolls off at a rate of – 40 dB/decade. This rate is double the rate of the first order filter. This means that second order filters have a sharper frequency response characteristics. Second-order filters are important because higher-order filters are designed using them. The circuit illustrated in figure 10.121(a) is also called a two-pole low-pass filter (LPF) because it has two bypass circuits.

DO YOU KNOW ?

A second order filter consists of two RC pairs and has a roll-off rate of – 40 dB/decade.

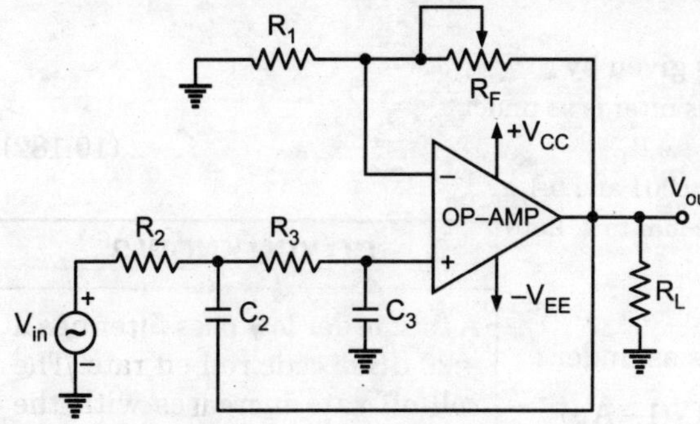

(a) Second order low-pass Butterworth filter

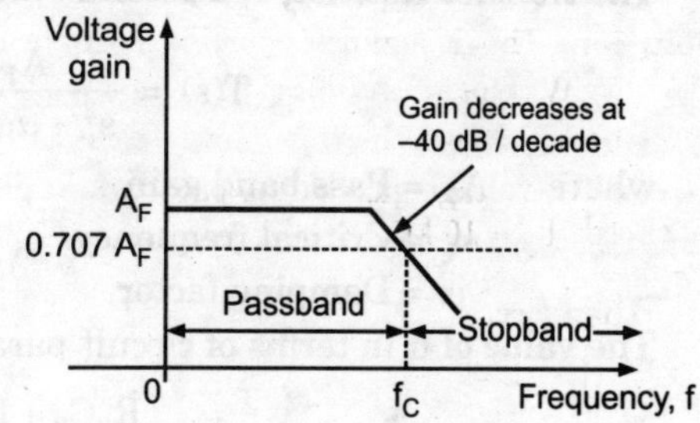

(b) Frequency response

Fig. 10.121.

2. Voltage Gain Magnitude

The resistors R_F and R_1 will decide the gain of the filter. The gain can be made variable by keeping R_F variable. The cut-off frequency f_C is determined by R_2, C_2, R_3 and C_3 as under :

$$f_C = \frac{1}{2\pi\sqrt{R_2R_3C_2C_3}} \quad \text{...(10.180)}$$

The voltage gain magnitude is given by,

$$\left|\frac{V_{out}}{V_{in}}\right| = \frac{A_F}{\sqrt{1+\left(\frac{f}{f_C}\right)^4}} \quad \text{...(10.181)}$$

where $A_F = 1 + \frac{R_F}{R_1}$ = Passband gain of the filter

f = Frequency of the input signal (Hz)

f_C = Cut-off frequency

The second order filters are important because higher order filters can be designed using them.

3. Designing Procedure

The designing procedure for a second order Butterworth low-pass filter is as under :

* Draw the circuit of a second order low pass filter (LPF) and obtain its voltage transfer function. **(RGTU, Bhopal, Sem. Exam; 2007-08)(10 marks)**

(i) We choose the cut-off frequency f_C.

(ii) For simplicity, we assume $R_2 = R_3 = R$ and $C_2 = C_3 = C$. Then, choose C in the range of 0.001 μF to 0.1 μF.

(iii) We calculate the value of R as,

$$R = \frac{1}{2\pi f_H C}$$

(iv) Due to equal resistors and capacitors ($R_2 = R_3$ and $C_2 = C_3$), the passband gain

$$A_F = \left(1 + \frac{R_F}{R_1}\right) \text{ has to be equal to 1.586}$$

Therefore, $R_F = 0.586\ R_1$.

A gain of 1.586 is essential to obtain the Butterworth response.

Now, we choose $R_1 \leq 100\ k\Omega$ and calculate R_F.

4. Transfer Function

The transfer function of a second order filter is given by

$$T(s) = \frac{A_F\ \omega_c^2}{s^2 + \alpha\omega_c s + \omega_c^2} \quad ...(10.182)$$

where A_F = Pass band gain

ω_c = Critical frequency

α = Damping factor.

The value of α in terms of circuit parameters is as under :

$$\alpha = \frac{R_2C_2 + R_1C_2 + R_1C_1(1 - A_F)}{\sqrt{R_1R_2C_1C_2}}$$

For $R_1 = R_2 = R$ and $C_1 = C_2 = C$, we have

$$\alpha = \frac{RC + RC + RC\,(1 - A_F)}{RC}$$

or

$$\alpha = \frac{2RC + RC\,(1 - A_F)}{RC} = \frac{RC\,[2 + 1 - A_F]}{RC}$$

or

$$\alpha = 3 - A_F \quad ...(10.183)$$

EXAMPLE 10.45. Design a second order low-pass filter for a cut-off frequency of 1 kHz and passband gain of 1.587.

Solution: To design the second order low-pass filter, we follow the steps given below :

(i) Given that cut-off frequency $f_c = 1$ kHz.

(ii) Let $C_2 = C_3 = 0.005$ μF.

(iii) Then, we have

$$R_2 = R_3 = \frac{1}{2\pi f_c C} = \frac{1}{2\pi \times 1 \times 10^3 \times 0.005 \times 10^{-6}}$$

or $R_2 = R_3 = 31.83\ k\Omega$.

(iv) Since $R_F = 0.586\ R_1$

Let $R_1 = 33\ k\Omega$

then $R_F = 19.338\ k\Omega$.

Therefore, the component values for a second order low-pass filter are as under :

$R_1 = 33\ k\Omega$, $R_2 = R_3 = 31.83\ k\Omega$

$R_F = 19.338\ k\Omega$, $C_2 = C_3 = 0.005$ μF. **Ans.**

EXAMPLE 10.46. For a second order low-pass filter, calculate the cut-off frequency and pass band voltage gain if the component values are R_1 = 12 kΩ, R_F = 7 kΩ, $R_2 = R_3$ = 33 kΩ, $C_1 = C_2$ = 0.002 µF.

Solution: (i) Cut-off frequency (f_c) is given by

$$f_c = \frac{1}{2\pi\sqrt{R_2R_3C_2C_3}}$$

But since $R_2 = R_3 = R$ and $C_2 = C_3 = C$, we obtain

$$f_c = \frac{1}{2\pi RC} = \frac{1}{2\pi\times 33\times 10^3\times 0.002\times 10^{-6}} = 2.411 \text{ KHz} \quad \textbf{Ans.}$$

(ii) Passband voltage gain (A_{VF}) is given by

$$A_F = 1+\frac{R_F}{R_1} = 1+\frac{7}{12} = 1.583 \quad \textbf{Ans.}$$

EXAMPLE 10.47. Design a second-order low-pass filter at a high cut-off frequency of 2 kHz using Op-amp.

Solution: Given that : High cut-off frequency, f_c = 2 KHz

Let $C_3 = C_2 = 0.033\ \mu F$

Then $R_3 = R_2 = \dfrac{1}{2\pi f_c C} = \dfrac{1}{2\pi\times 2\times 10^3\times 0.033\times 10^{-6}} = 2.4\ k\Omega$ (say 2.5 kΩ)

However, to minimize the effect of offset current (which may be quite important in a low pass filter), the resistance at the inverting terminal must be equal to the resistance at the non-inverting terminal.

So, $$2R = \frac{R_F R_1}{R_F + R_1} = \frac{0.586 R_1^2}{1.586 R_1} \qquad \because R_F = 0.586\ R_1$$

or $$2\times 2.4\ k\Omega = \frac{0.586 R_1}{1.586}$$

or $R_1 = 13\ k\Omega$

R_1 may be taken of a value of 5 kΩ and then $R_F = 0.586 \times 15 = 8.79\ k\Omega$

R_F may be taken as a pot of 10 kΩ. **Ans.**

EXAMPLE 10.48. Design a second-order low-pass filter at a high cut-off frequency of 1.0 kHz.

Solution: Given that : High cut-off frequency, f_c = 1 kHz

Let $C_3 = C_2 = 0.0047\ \mu F$

Then, $$R_3 = R_2 = \frac{1}{2\pi f_c C}$$

$$R_3 = \frac{1}{2\pi\times 1\times 10^3\times 0.0047\times 10^{-6}} = 33.86\ k\Omega = \text{(say 33 k}\Omega)$$

Let R_1 be equal to 30 kΩ.

So, $R_F = 0.586\ R_1 = 0.586 \times 30 = 17.58\ k\Omega$ (say 20 kΩ)

Thus, required components will be as under :

$R_3 = R_2 = 33\ k\Omega$

$C_3 = C_2 = 0.0047$ mF

$R_1 = 30\ k\Omega$

and $R_F = 20\ k\Omega$ pot. **Ans.**

10.39 INTRODUCTION TO 555 IC TIMER CIRCUIT

One of the most versatile linear ICs is the 555 timer which was first introduced in early 1970 by Signetic Corporation giving the name as SE/NE 555 timer. The 555 is a monolithic timing circuit that can produce accurate and highly stable time delays or oscillations. Like general-purpose op-amps, it is very much reliable, easy to use and cheaper in cost. It has a variety of applications including monostable and astable multivibrators, dc-dc converters, digital logic probes, waveform generators, analog frequency meters and tachometers, temperature measurement and control devices, voltage regulators etc. The timer basically operates in one of two modes either as a monostable (one-shot) multivibrator or as an astable (free-running) multivibrator.

The SE 555 is designed for the operating temperature range from – 55°C to 125°C while the NE 555 operates over a temperature range of 0°C to 70°C. The important features of the 555 timer are as under:

It operates from a wide range of power supplied (+ 5 V to + 18 V) supply voltage, sinking or sourcing 200 mA of load current; proper selection of only a few external components allows timing intervals of several minutes or frequencies as high a several hundred kHz; it has a high current output; the output can drive TTL; it has a temperature stability of 50 parts per million (ppm) per degree Celsius change in temperature, or equivalently 0.005%/°C, it has an adjustable duty cycle; the maximum power dissipation per package is 600 mW and its trigger and reset inputs are logic compatible.

DO YOU KNOW ?

A timer 555 can be used in such applications as waveform generators, digital logic probes, infrared transmitters, burglar alarms, toxic gas alarms and electric eyes.

10.40 PIN CONFIGURATION

The 555 IC is available as an 8-pin metal can, an 8-pin mini DIP (dual-in-line package) or a 14-pin DIP, as illustrated in figure 10.122.

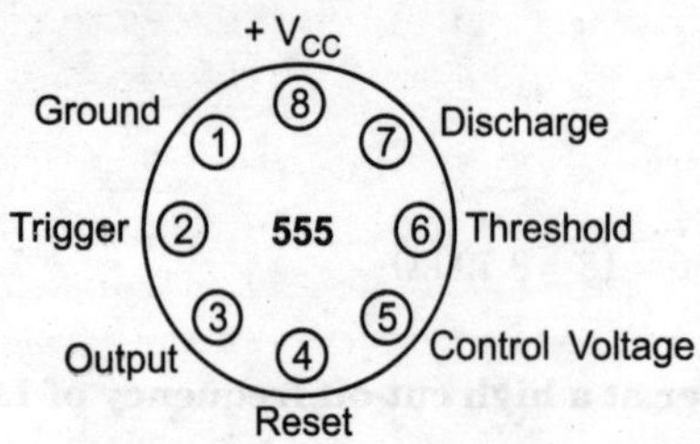

(a) Top view of metal can package

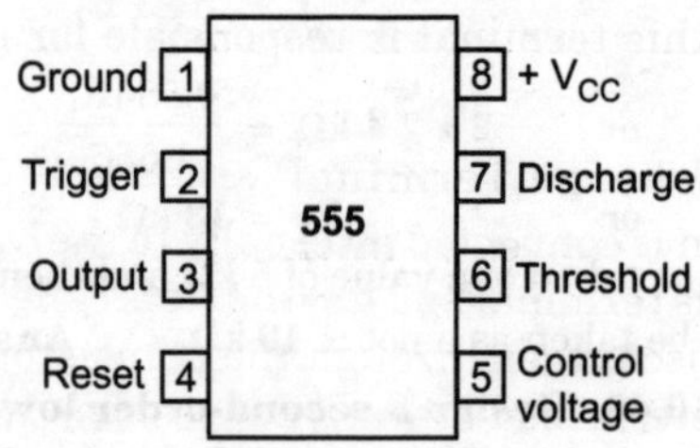

(b) 8-Pin DIP

Fig. 10.122.

This IC consists of 23 transistors, 2 diodes and 16 resistors,. The explanation of terminals coming out of the timer IC is as under :

The pin number used in the following discussion refer to the 8-pin DIP and 8-pin metal can packages.

Pin 1 : Grounded terminal

All the voltages are measured with respect to this terminal.

Pin 2 : Trigger Terminal

This pin is an inverting input to a comparator that is responsible for transition of flip-flop from set to reset. The output of the timer depends on the amplitude of the external trigger pulse applied to this pin.

Pin 3 : Output Terminal

Output of the timer is available at this pin. There are two ways in which a load can be connected to the output terminal either between pin 3 and ground pin(pin 1) or between pin 3 and supply pin (pin 8). The load connected between pin 3 and supply pin

DO YOU KNOW ?

For a timer 555, when the load is connected between the output terminal (pin 3) and grounded terminal (pin 1), it is called grounded load. This is also sometimes called normally off load.

is called the normally on load and that connected between pin 3 and ground pin is called the normally off load.

Pin 4 : Reset terminal

To disable or reset the timer, a negative pulse is applied to this pin due to which, it is referred to as reset terminal. When this pin is not to be used for reset purpose, it should be connected to + V_{CC} to avoid any possibility of false triggering.

Pin 5 : Control Voltage Terminal

The function of this terminal is to control the threshold and trigger levels. Thus, either the external voltage or a pot connected to this pin determines the pulse width of the output waveform. The external voltage applied to this pin can also be used to modulate the output waveform. When this pin is not used, it should be connected to ground through a 0.01 μF to avoid any noise problem.

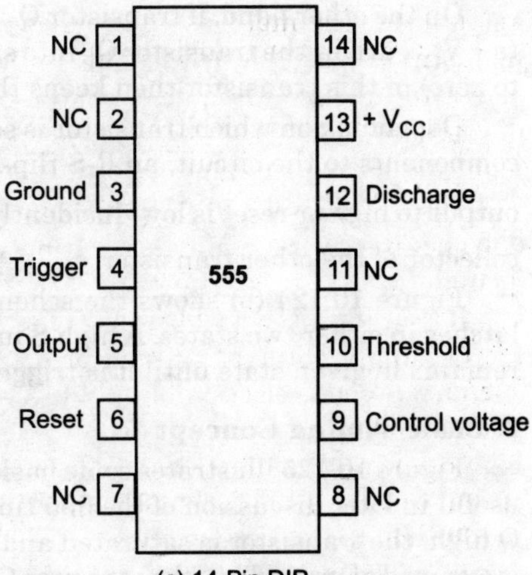

(c) 14-Pin DIP

Fig. 10.123. 555 Timer Pin Configuration

DO YOU KNOW ?

For a timer 555, the normally control terminal is connected to ground through a 0.01 μF bypass capacitor so as to prevent noise coupled onto this pin from causing false triggering.

Pin 6 : Threshold Terminal

This is the non-inverting input terminal of comparator 1, which compares the voltage applied to this terminal with a reference voltage of + $\frac{2}{3}$ V_{CC}. The amplitude of voltage applied to this terminal is responsible for the set state of flip-flop.

Pin 7 : Discharge Terminal

This pin is connected internally to the collector of transistor and mostly a capacitor is connected between this terminal and ground. It is called discharge terminal because when transistor saturates, capacitor discharges through the transistor. When the transistor is cut-off, the capacitor charges at a rate determined by external resistor and capacitor.

Pin 8 : Supply Terminal

A supply voltage of + 5 V to +18 V is applied to this terminal with respect to ground (pin 1).

10.41 555 TIMER BASICS

The 555 timer combines a relaxation oscillator, two comparators, an R-S flip-flop, and a discharge capacitor.

1. R-S Flip-Flop

A pair of cross-coupled transistors is shown in figure 10.124 (a). Each collector drives the opposite base through resistance R_B. In such circuits, one transistor is saturated while the other is cut-off. For instance, if transistor Q_1 is saturated, its collector voltage is almost zero.

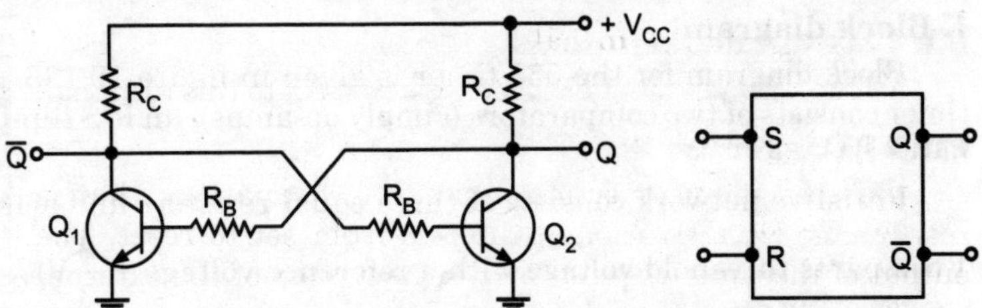

(a) Part of an R-S flip-flop

(b) Symbol

Fig. 10.124. Illustration of R-S flip-flop

Hence, there is no base drive for transistor Q_2 and it goes into cut-off and its collector voltage approaches + V_{CC}. This high voltage produces enough base current to keep transistor Q_1 in saturation.

On the other hand, if transistor Q_1 is cut-off, its collector voltage, which is approximately equal to + V_{CC}, drives the transistor Q_2 into saturation. The low collector voltage (which is approximately to zero) of this transistor then keeps the transistor Q_1 in cut-off.

Depending on which transistor is saturated, the Q output is either low or high. By adding more components to the circuit, an R-S flip-flop is obtained. R-S flip-flop is a circuit that can set the Q output to high or reset is low. Incidently, a complementry (opposite) output $\overline{Q}$ is available from the collector of the other transistor.

Figure 10.124 (b) shows the schematic symbol for an R-S flip-flop of any design. The circuit latches in either two states. A high S input sets Q to high; a high R input resets Q to low. Output Q remains in given state until it is triggered into the opposite sate.

2. Basic Timing Concept

Figure 10.125 illustrates some basic ideas that will prove useful in later discussion of the 555 timer. Assuming output Q high, the transistor is saturated and the capacitor voltage is clamped at ground i.e., the capacitor C is shorted and cannot charge. The non-inverting input voltage of the comparator is referred to as the threshold voltage while the inverting input voltage is referred to as the control voltage. With R-S flip-flop set, the saturated transistor holds the threshold voltage at zero. However, the control voltage is fixed at $\frac{2}{3}V_{CC}$ (i.e., at 10V) because of the voltage divider.

Let us consider that a high voltage is applied to the R input. This resets the flip-flop. R-S Output Q goes low and the transistor is cut-off. Capacitor is now free to charge. As the capacitor C charges, the threshold voltage rises. Eventually, the threshold voltage becomes slightly greater than (+ 10 V). The output of the comparator then goes high, forcing the R-S flip-flop to set. The high Q output saturates the transistor, and this quickly discharges the capacitor. The two waveforms are depicted in figure 10.125 (b). An exponential rise is across the capacitor C, and a positive going pulse appears at the output $\overline{Q}$. Thus, capacitor voltage V_C is exponential while the output is rectangular, as illustrated in figure 10.125 (b).

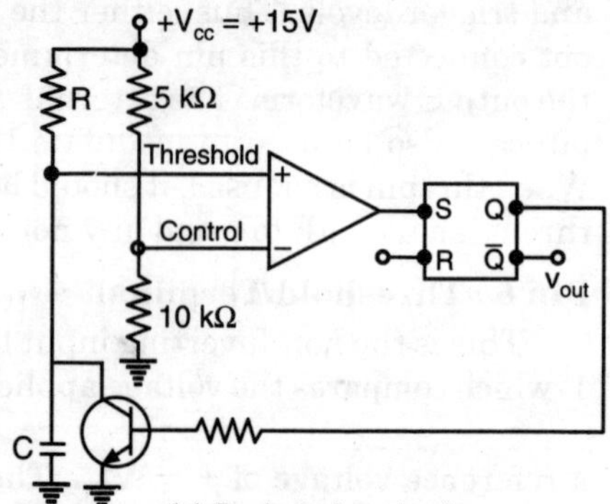

(a) Basic timing circuit

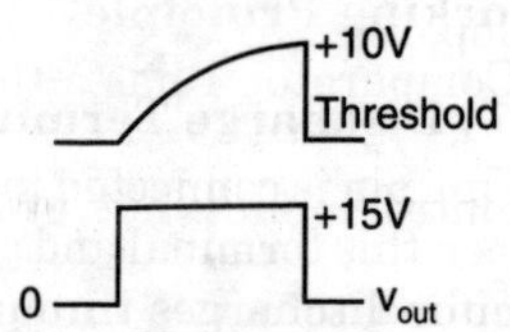

(b) Capacitor voltage and output waveform

Fig. 10.125

10.42 DETAILS OF 555 TIMER IC UNIT

1. Block diagram

Block diagram for the 555 timer is given in figure 10.126. As shown in block diagram, a 555 timer consists of two comparators (simply op-amps), an R-S flip-flop, two transistors and a resistive network.

Resistive network consists of three equal resistors and acts as a voltage divider. Comparator 1 compares threshold voltage with a reference voltage + $\frac{2}{3}V_{CC}$ volts. Comparator 2 compares the trigger voltage with a reference voltage + $\frac{1}{3}V_{CC}$ volts. Output of both the comparators is supplied to the flip-flop. Flip-flop assumes its state according to the output of the two comparators.

The happiness of every country depends upon the character of its people rather than the form of its government. — **Thomas C. Haliburton**

One of the two transistors is a discharge transistor of which collector is connected to pin 7. This transistor saturates or cuts-off according to the output state of the flip-flop. The saturated transistor provides a discharge path to a capacitor connected externally. Base of another transistor is connected to a reset terminal. A pulse applied to this terminal resets the whole timer irrespective of any input.

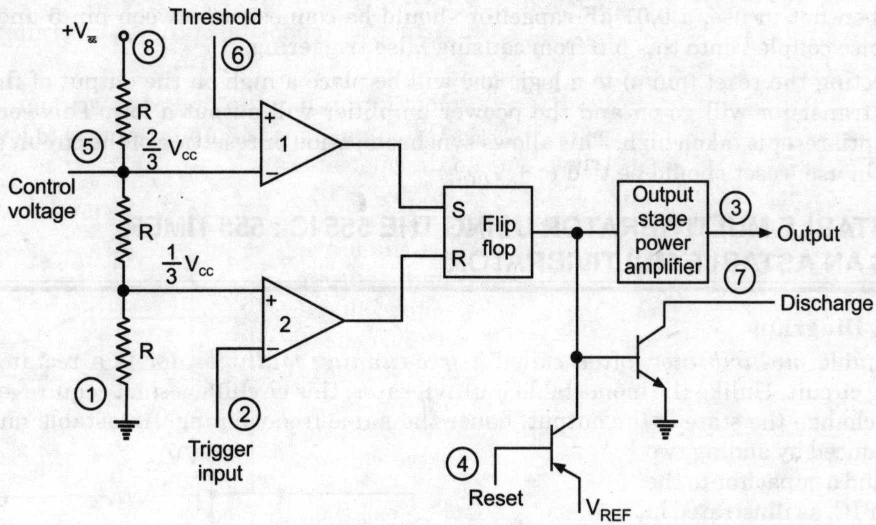

Fig. 10.126. Details of 555 timer IC Unit

2. Working Principle

Comparator 1 has a threshold input (pin 6) and a control input (pin 5). In most applications, the control input is not used, so that the control voltage equals $+\frac{2}{3}V_{CC}$. Output of this comparator is applied to set (S) input of the flip-flop. Whenever the threshold voltage exceeds the control voltage, comparator 1 will set the flip-flop and its output is high. A high output from the flip-flop saturates the discharge transistor and dischages the capacitor connected externally to pin 7. The complementary signal out of the flip-flop goes to pin 3, the output. The output available at pin 3 is low. These conditions will prevail until comparator 2 triggers the flip-flop. Even if the voltage at the threshold input falls below $+\frac{2}{3}V_{CC}$ i.e., comparator 1 cannot cause the flip-flop to change again. It means that the comparator 1 can only force the flip-flop's output high.

> **DO YOU KNOW ?**
>
> Although the timer 555 has been used in a wide variety of often unique applications, it is very hard on its power supply lines, requiring quite a bit of current, and injecting many noise transients.

To change the output of flip-flop to low, the voltage at the trigger input must fall below $+\frac{1}{3}V_{CC}$. When this occurs, comparator 2 triggers the flip-flop, forcing its output low. The low output from the flip-flop turns the discharge transistor off and forces the power amplifier to output a high. These conditions will continue, independent of the voltage on the trigger input. Comparator 2 can only cause the flip flop to output low.

From the above discussion, it is concluded that for having low output from the timer 555, the voltage on the threshold input must exceed the control voltage or $+\frac{2}{3}V_{CC}$. This also turns the discharge transistor on. To force the output from the timer high, the voltage on the trigger input must drop below $+\frac{1}{3}V_{CC}$. This also turns the discharge transistor off.

A voltage may be applied to the control input to change the levels at which the switching occurs. When not in use, a 0.01 μF capacitor should be connected between pin 5 and ground to prevent noise coupled onto this pin from causing false triggering.

Connecting the reset (pin 4) to a logic low will be place a high on the output of flip-flop. The discharge transistor will go on and the poower amplifier will output a low. This condition will continue until reset is taken high. This allows synchronisation or resetting of the circuit's operation. When not in use, reset should be tied to $+V_{CC}$.

10.43 ASTABLE MULTIVIBRATOR USING THE 555 IC : 555 TIMER AS AN ASTABLE MULTIVIBRATOR

1. Circuit Diagram

An astable *multivibrator*, often called a *free-running multivibrator* is a rectangular wave generating circuit. Unlike the monostable multivibrator, this circuit does not require any external trigger to change the state of the output, hence the name free-running. An astable multivibrator can be produced by adding two resistors and a capacitor to the basic timer IC, as illustrated in figure 10.127. The timing during which the output is either high or low is determined by the externally connected two resistors and a capacitor. The details of the astable multivibrator circuit are given below.

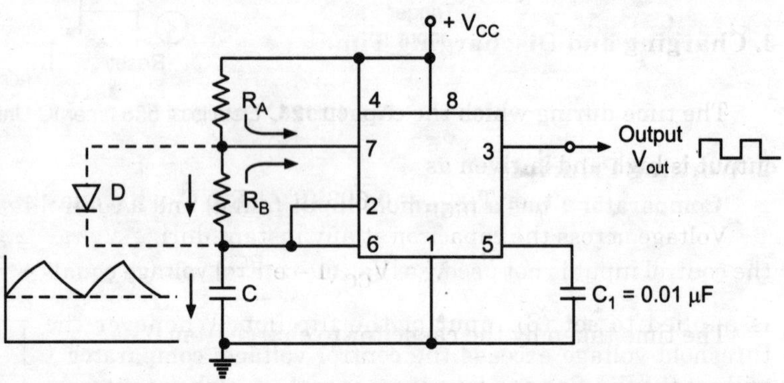

Fig. 10.127. Circuit of the Timer 555 as an astable Multivibrator

Pin 1 is grounded; pins 4 and 8 are shorted and then tied to supply $+V_{CC}$, output (V_{out}) is taken form pin 3; pin 2 and 6 are shorted and connected to ground through a capacitor C, pin 7 is connected to supply $+V_{CC}$ through a resistor R_A; and between pins 6 and 7, a resistor R_B is connected. At pin 5, either a bypass capacitor of 0.01 μF is connected or modulation input is applied.

2. Astable Operation

For explaining the operation of the timer 555 as an astable multivibrator, necessary internal circuity with external connections are shown in figure 10.128(a).

In figure 10.128 (a), when Q is low or output v_{out} is high, the discharging transistor is cut-off and the capacitor C begins charging toward V_{CC} through resistances R_A and R_B. Because of this, the charging time constant is $(R_A + R_B)C$. Eventually, the threshold voltage exceeds $+\frac{2}{3}V_{CC}$, the comparator 1 has a high output and triggers the flip-flop, so that its Q is high and the timer output is low. With Q high, the discharge transistor saturates and pin 7 grounds so that the capacitor C discharges through resistance R_B with a discharging time constant R_B C. With the discharging of capacitor, trigger voltage at inverting input of comparator 2 decreases. When it drops below $\frac{1}{3}V_{CC}$, the output of comparator 2 goes high and this resets the flip-flop so that Q is low and the timer output is high. This proves the auto-transition in output from low to high and then to low as, illustrated in figure 10.128 (b). Thus, the cycle repeats.

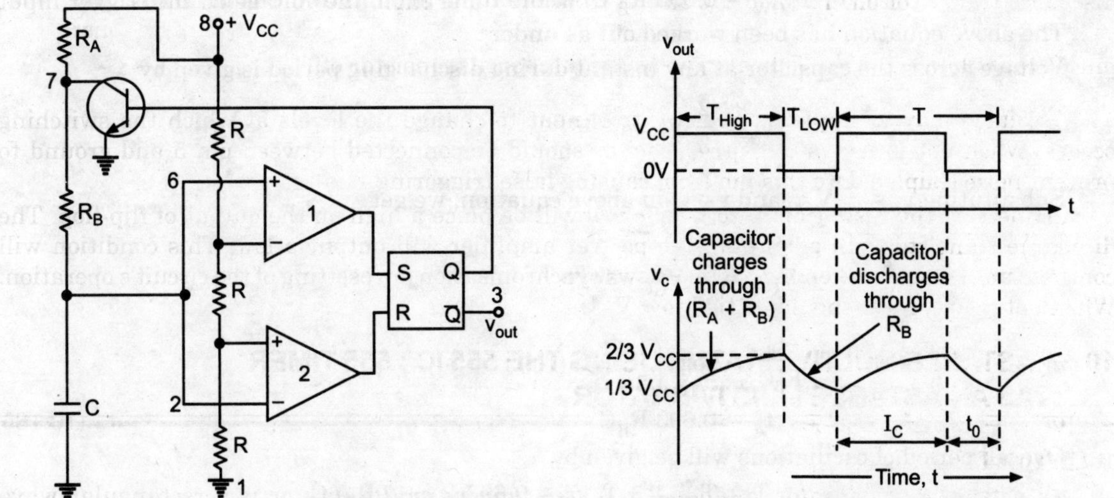

(a) Internal circuitry with external connections (b) Capacitor and output voltage waveforms

Fig. 10.128. Astable Operation

3. Charging and Discharging Time

The time during which the capacitor C charges from $\frac{1}{3}V_{CC}$ to $\frac{2}{3}V_{CC}$ is equal to the time, the output is high and is given as

$$t_c \text{ or } T_{HIGH} = 0.693\,(R_A + R_B)\,C, \text{ which has been proved below.}$$

Voltage across the capacitor at any instant during charging period is given as,

$$v_c = V_{CC}\,(1 - e^{-t/RC})$$

The time taken by the capacitor to charge from 0 to $+\frac{1}{3}V_{CC}$ is

$$\frac{1}{3}V_{CC} = V_{CC}\,(1 - e^{-t/RC})$$

or $$e^{\frac{-t}{RC}} = 2/3$$

or $$t = RC \log_e \frac{3}{2} = 0.405\,RC$$

The time taken by the capacitor to charge from 0 to $+\frac{2}{3}V_{CC}$ will be

$$t_2 = RC \log_e 3 = 1.0986\,RC$$

Therefore, the time taken by the capacitor to charge from $+\frac{1}{3}V_{CC}$ to $+\frac{2}{3}V_{CC}$ will be given by

$$t_c = (t_2 - t_1) = (1.0986 - 0.405)RC = 0.693\,RC$$

Substituting $R = (R_A + R_B)$ in above equation, we obtain

$$T_{HIGH} = t_c = 0.693\,(R_A + R_B)C \qquad ...(10.184)$$

where R_A and R_B are in ohms and C in farads.

The time during which the capacitor discharges from $+\frac{2}{3}V_{CC}$ to $+\frac{1}{3}V_{CC}$ is equal to the time the output is low and is given as

t_d or $T_{LOW} = 0.693\ R_B\ C$, where R_B is in ohms and C is in farads.

The above equation has been worked out as under :

Voltage across the capacitor at any instant during discharging period is given by

$$v_c = \frac{2}{3} V_{CC}\ e^{-t/R_B C}$$

Substituting $v_c = \frac{1}{3} V_{CC}$ and $t = t_d$ in above equation, we get

$$\frac{1}{3} V_{CC} = \frac{2}{3} V_{CC}\ e^{-t_d/R_B C}$$

or
$$\frac{-t_d}{R_B C} = \log_e\left(\frac{1}{2}\right) = -\ 0.693$$

or
$$t_d = 0.693\ R_B C \qquad ...(10.185)$$

Overall period of oscillations will be given by

$$T = T_{HIGH} + T_{LOW} = 0.693(R_A + 2\ R_B)\ C \qquad ...(10.186)$$

The frequency of oscillations being the reciprocal of the overall period of oscillations T is given by

$$f = \frac{1}{T} = \frac{1}{0.693(R_A + 2R_B)C} = \frac{1.44}{(R_A + 2R_B)C} \qquad ...(10.187)$$

Equation (10.187) indicates that the frequency of oscillation f is independent of the collector supply voltage + V_{CC}.

4. Duty Cycle

The term duty cycle is used in conjunction with the astable multivibrator.

The duty cycle, the ratio of the time t_c during which the ouput is high to the total time period T is given as

% duty cycle,
$$D = \frac{t_c}{T} \times 100 = \frac{R_A + R_B}{R_A + 2R_B} \times 100 \qquad ...(10.188)$$

From the above equation, it is obvious that square wave (50% duty cycle) output cannot be obtained unless R_A is made zero. However, there is a danger in shorting resistance R_A to zero. With $R_A = 0\ \Omega$, terminal 7 is directly connected to + V_{CC}. During the discharging of capacitor through R_B and transistor, an extra current will be supplied to the transistor from V_{CC} through a short between pin 7 and + V_{CC}. It may damage the transistor and hence the timer.

5. Square-wave oscillator

However, a symmetrical square wave can be obtained if a diode is connected across resistor R_B, as illsutrated in dotted lines in figure 10.127. The capacitor C charges through R_A and diode D to approximately + $\frac{2}{3}$ V_{CC} and discharges through resistor R_B and terminal 7 (transistor) until the capacitor voltage drops to $\frac{1}{3}$ V_{CC}. Then, the cycle is repeated. To obtain a square wave output, R_A must be a combination of a fixed resistor R and a pot as shown in figure 10.130, so that the pot can be adjusted to give the exact square wave.

6. Free-Running Ramp Generator

The free running ramp generator can be obtained if we replace resistors R_A and R_B in the circuit 10.127, with a current mirror to provide charging and discharging of capacitor C at a constant rate.

An astable multivibrator configured performing the function of free-running ramp generator is depicted in figure 10.127.

For all practical purposes, the time period of the ramp waveform is equal to the charging time (the discharging time of the capacitor is relatively negligible in comparison to its charging time) and is given by

$$T = \frac{V_{CC} \times C}{3I_C} \quad ...(10.189)$$

where $I_C = \frac{V_{CC} - V_{BE}}{R}$, constant current in amperes and C is in farads.

Thus free-running frequency of the ramp generator may be given by

$$f_0 = \frac{1}{T} = \frac{3I_C}{V_{CC} \times C} \quad ...(10.190)$$

The output waveform of the generator is shown in figure 10.129 (b).

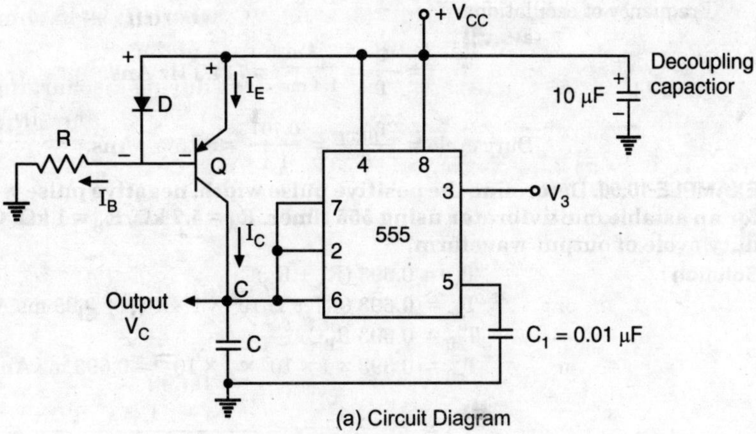

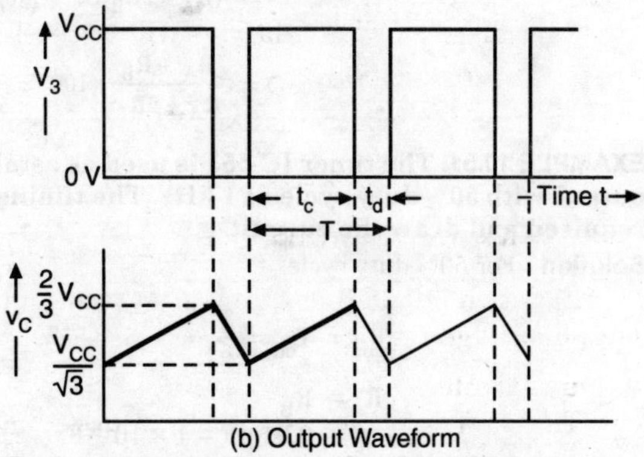

Fig. 10.129. Free-running ramp generator

7. Special Note

Although the timer 555 has been used in a wide variety of often unique applications it is very hard on its power supply lines, requiring quite a bit of current, and injecting many noise transients. This noise will often be coupled into adjacent ICs falsely triggering them. The 7555 is a CMOS version of the 555. Its quiescent current require-ments are considerably lower than that of 555, and the 7555 does not contaminate the power supply lines. It is pin compatible with the 555. Hence, this CMOS version of the 555 should be the first choice when a 555 timer IC is to be used.

EXAMPLE 10.49. Determine the frequency and duty cycle for the 555 astable multivibrator output for C = 0.01 μF, R_A = 2 kΩ and R_B = 100 kΩ.

Solution : Charging period,

$$T_{HIGH} = 0.693\,(R_A + R_B)C$$
$$T_{HIGH} = 0.693(2 + 100) \times 10^3 \times 0.01 \times 10^{-6}$$
$$T_{HIGH} = 0.707 \text{ ms}$$

Discharging period,

$$T_{LOW} = 0.693\, R_B C$$
$$C = 0.693 \times 100 \times 10^3 \times 0.01 \times 10^{-6}$$
$$C = 0.693 \text{ ms}$$

Overall period of oscillations,

$$T = T_{HIGH} + T_{LOW}$$
$$= 0.707 + 0.693 = 1.4 \text{ ms}$$

Frequency of oscillations,

$$f = \frac{1}{T} = \frac{1}{1.4 \text{ ms}} = 714.3 \text{ Hz } \textbf{Ans.}$$

$$\text{Duty cycle} = \frac{T_{HIGH}}{T} = \frac{0.707}{1.4} = 50.5\% \quad \textbf{Ans.}$$

EXAMPLE 10.50. Determine the positive pulse width, negative pulse width and free running frequency for an astable multivibrator using 555 timer. $R_A = 4.7$ kΩ, $R_B = 1$ kΩ, $C = 1$ μF, $C_1 = 0.01$ μF. What is the duty cycle of output waveform?

Solution :

$$T_{on} = 0.693\,(R_A + R_B)C$$

or $$T_{on} = 0.693\,(4.7 + 1)\,10^3 \times 1 \times 10^{-6} = 3.95 \text{ ms } \textbf{Ans.}$$

$$T_{off} = 0.693\,R_B C$$

or $$T_{off} = 0.693 \times 1 \times 10^3 \times 1 \times 10^{-6} = 0.693 \text{ ms } \textbf{Ans.}$$

$$f = \frac{1.4}{(R_A + 2R_B)C} = \frac{1.4}{(4.7 + 2 \times 1)\,10^3 \times 1 \times 10^{-6}} = 208.9 \text{ Hz} \quad \textbf{Ans.}$$

$$D = \frac{R_A + R_B}{R_A + 2R_B} \times 100 = \frac{4.7 + 1}{4.7 + 2 \times 1} \times 100 = 85\% \quad \textbf{Ans.}$$

EXAMPLE 10.51. The timer IC-555 is used as astable multivibrator. It is desired to have square-wave output with 50% duty cycle of 1 kHz. The timing capacitor is of 0.01 μF. Find the values of resistors required and draw the curcuit.

Solution : For 50% duty cycle

$$T_{on} = T_{off} = \frac{T}{2}$$

$$R_A = R_B$$

$$C = 0.01 \text{ μF} = 1 \times 10^{-8} \text{ F}$$

$$f = 1 \text{ kHz} = 1{,}000 \text{ Hz}$$

Frequency is given by equation

$$f = \frac{1.44}{(R_A + R_B)\,C} = \frac{1.44}{2R_A C}$$

or $$R_B = R_A = \frac{1.44}{2 \times f \times C}$$

$$= \frac{1.44}{2 \times 1{,}000 \times 1 \times 10^{-8}}$$

$$= 72 \text{ kΩ (68 standard value) } \textbf{Ans.}$$

Circuit diagram is shown in figure 10.130.

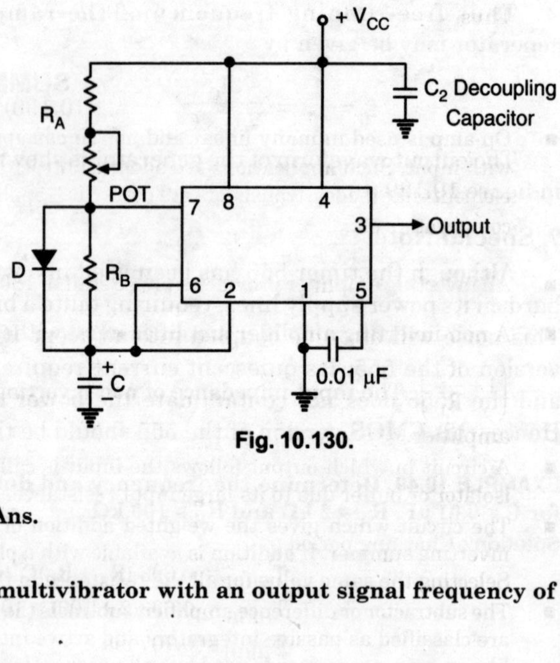

Fig. 10.130.

EXAMPLE 10.52. Design a 555 timer as an astable multivibrator with an output signal frequency of 700 Hz and 50% duty cycle.

Solution : For 50% duty cycle $T_{on} = T_{off}$ and $R_A = R_B$

and frequency is given by equation $f = \dfrac{1.44}{(R_A + R_B)C}$

Substituting f = 700 Hz; $R_B = R_A$ and C = 0.01 μF (assumed value) in above equation, we have

$$700 = \frac{1.44}{2\,R_A \times 0.01 \times 10^{-6}}$$

or $$R_A = R_B = \frac{1.44}{700 \times 2 \times 0.01 \times 10^{-6}} = 103 \text{ kΩ (100 kΩ standard value)}$$

Thus $C = 0.01\ \Omega F$

$$R_A = R_B = 100\ k\Omega \quad \textbf{Ans.}$$

EXAMPLE 10.53. Design a 555 timer as an astable multivibrator with an output signal frequency of 800 Hz and 60 percent duty cycle.

Solution: Percent duty cycle, $D = \frac{R_A + R_B}{R_A + 2R_B} \times 100 = 60$

or $R_A + R_B = 0.6\ R_A + 1.2\ R_B$

or $R_B = 2\ R_A$

Frequency is given by $f = \frac{1.44}{(R_A + 2R_B)C}$

Substituting f = 800Hz, $R_B = 2R_A$ and $C = 0.01\ \mu F$ (assumed value in above equation), we have

$$800 = \frac{1.44}{5R_A \times 0.01 \times 10^{-6}}$$

or $R_A = \frac{1.44}{800 \times 5 \times 0.01 \times 10^{-6}} = 36{,}000\ \Omega = 36\ k\Omega$

and $R_B = 2R_A = 2 \times 36 = 72\ k\Omega$

Thus, $C = 0.01\ \mu F$, $R_A = 26\ k\Omega$ and $R_B = 72\ k\Omega$ **Ans.**

SUMMARY

- **Op-amp is used in many linear and non linear applications. In linear applications output varies linearly with input. Such applications are adder, subtractor, voltage follower etc. In non-linear circuits, non-linear elements like diodes, transistors are used. Such applications are precision rectifiers, log and antilog amplifiers, comparators etc.**
- **An inverting amplifier provides a phase shift of 180° between input and output and its voltage gain is $-\frac{R_F}{R_1}$.**
- **A non-inverting amplifier amplifies input without introducing any phase shift and its voltage gain is $\left(1 + \frac{R_F}{R_1}\right)$. The input impedance of non inverting amplifier is extremely large compared to inverting amplifier.**
- **A circuit in which output follows the input is called as voltage follower. Its gain is one. It is used as an isolator or buffer due to its large input resistance.**
- **The circuit which gives the weighted addition of the applied inputs, with 180° phase shift is called as inverting summer. If addition is available with a phase shift, the circuit is called as non-inverting summer. Selecting the same value for all the resistance in the summer circuit, average circuit can be obtained.**
- **The subtractor or difference amplifier, subtracts the input signals with appropriate strengths. The integrators are classified as passive integrators and active integrators. The passive is one in which no active element like op-amp, transistor is used but all passive elements like resistances are capacitors are used. An active integrator uses an op-amp. The passive integrator has the limitations of large time constant, less gain and large values of R and C requirement. The main advantage of an active integrator is its large time constant, which helps in perfect integration.**
- **Op-amp is used in many linear and nonlinear applications. In linear applications output varies linearly with input. Such applications are adder, subtractor, voltage follower etc. In non linear circuits, nonlinear elements like diodes, transistors are used. Such applications are precision rectifiers, log, antilog amplifiers and comparators etc.**
- **An inverting amplifier provides a phase shift of 180° between input and output and its voltage gain is $-\frac{R_F}{R_1}$.**

- A non-inverting amplifier amplifies input without introducing any phase shift and its voltage gain is $\left(1+\frac{R_F}{R_1}\right)$. The input impedance of non inverting amplifier is extremely large compared to inverting amplifier.
- A circuit in which output follows the input is called as voltage follower. Its gain is one. It is used as an isolator or buffer due to its large input resistance.
- The circuit which gives the weighted addition of the applied inputs with 180° phase shift is called as inverting summer. If addition is available with a phase shift, the circuit is called as non-inverting summer. Selecting the same value for all the resistance in the summer circuit, average circuit can be obtained.
- The subtractor or difference amplifier subtracts the input signals with appropriate strengths. The integrators are classified as passive integrators and active integrators. The passive is one in which no active element like op-amp, transistor is used but all passive elements like resistances are capacitors are used. An active integrator uses an op-amp. The passive integrator has the limitations of large time constant, less gain and large values of R and C requirement. The main advantage of an active integrator is its large time constant, which helps in perfect integration.
- The frequency response of an ideal integrator shows that the gain reduces from its d.c. level value to zero, at a rate of 20 dB/decade.
- In ideal integrator, input offset voltage and input bias current causes error in the output, due to its high gain. As bandwidth is small, integration range is very small. Hence practical integrator is preferred.
- The frequency response of a practical integrator shows that d.c. gain reduces. Also the bandwidth of a practical integrator is large.
- The basic integrator can be used to obtain the circuits like summing integrators and the difference integrator.
- Similar to the integrator, the differentiator is also classified as passive and the active differentiator. The gain of an ideal differentiator increases with frequency at a rate of 20 dB/decade. Due to this, differentiator become unstable at high frequencies and may break into oscillations. The input impedance decreases with increase in the frequency, which makes the ideal differentiator very sensitive to noise. These disadvantages are eliminated in the practical differentiator. The gain of the practical differentiator decreases after certain frequency at a rate of 20 dB/decade and thus ensures the stability.
- The integrators are used in wave shaping circuits, analog computers and analog to digital converters. The differentiators are used in rate-of-change detector in FM demodulators and in wave shaping circuits.
- The basic differentiator can be used to obtain summing differentaitor.
- The integrators are preferred over differentiators because :
  (i) gain of integrator decreases with increase in the frequency.
  (ii) the input impedance is constant
  (iii) small bandwidth hence not sensitive to noise
  (iv) easy to introduce initial conditions.
- The analog computers use the integrators. Analog computers is a device which is used to do certain linear mathematical operation. The potentiometer, summer, summing integrator, scale changer etc. are the building blocks of an analog computer. The analog computers are used to perform the following operations :
  (i) Implementing given equation
  (ii) Solving differential equation
  (iii) Solving simultaneous equation
  (iv) Implementing given transfer function.
- The analog computers use the integrators.
- The voltage to current (V to I) converters may be classified into two categories, depending on the position of the load. They are as under:
  (i) V to I converter with floating load
  (ii) V to I converter with grounded load.
- Voltage to current converter is used in applications such as low voltage dc and ac voltmeters, LED and Zener diode testers etc.
- One of the most important application of I to V converter is as the photodetector amplifier
- The V-I converter circuit with floating load may be modified as a low voltage d.c. voltmeter circuit. The resistance R_L is to be replaced by D'Arsonval meter movement.
- Another application of V-I converter is a diode tester and diode match finder.
- The V-I converter circuit can also be used to test zener diode. The zener diode is connected in reverse biased condition in the feedback path.

- Many industrial systems, consumer systems and process control systems require a precise measurement of the physical quantities like temperature, pressure, humidity, weight etc. A sugar factory requires a measurement of flow, level and temperature of juice. The plastic furnaces require precise measurement of the temperature.
- The measurement of the physical quantities is generally carried out with the help of a device called as **transducer.**
- The instrumentation amplifiers are used to amplify the low level differential signals very precisely, in presence of the large common mode noise and interference signals.
- In the active filter synthesis, it is necessary to simulate frequency dependent elements such as inductances. This is achieved by using the **impedance converters**. Infact, the impedance converters are active RC circuits that are specially designed for such simulations.
- One of the most popular configuration is the *generalized impedance converter.*
- This converter can simulate inductances as well as a frequency-dependent resistances.
- Basically, inductor is a bulky component which often increases the size of RLC filters. To overcome this problem, we can simulate the inductor using a GIC converter. Hence, one of the popular GIC applications is to realize an inductorless filter starting from passive RLC prototype. The inductance in such a filter is replaced by the synthetic inductance implemented using the GIC.
- Non linear behaviour of operational amplifier can be achieved by connecting it in
  - Positive feedback configuration
  - Open loop configuration
  - Negative feedback configuration with non linear elements such as diode, FET in the feedback path.
- The operational amplifier can be used as a comparator, schmitt trigger in the non linear behaviour.
- A comparator is a circuit which compares a signal voltage applied at one input of an op-amp with a known reference voltage at the other input, and produce either a high or low output voltage, depending on which input is higher.
- There are two types of comparators.
  - Non inverting comparator (signal is applied at non inverting input of op-amp).
  - Inverting comparator (signal is applied at inverting input of op-amp).
- In practical comparator circuits protective diodes are used to protect circuit from excess input voltage.
- Comparator output voltages can be made independent of power supply by connecting voltage limiter components such as zener in the diodes feedback path.
- It is also possible to limit the voltage swing of the output voltage V_0 to either to a positive direction or to a negative direction only. This is achieved by connecting a combination of zener diodes and rectifier diode in the feedback path.
- The comparator performance parameter are :
  - Accuracy
  - Logic threshold
  - Strobe function
  - Response time
  - Positive output level
  - Negative output level
  - Strobe current
  - Strobe release time
  - Saturation voltage
- Limiting factors to use op-amp as a comparator :
  - Slew rate
  - Output voltage compatibility
  - Switching time
- Use of special comparator IC is a better option because of its faster switching. Built-in noise immunity, output compatibility, strobing and latching characteristics.
- The KM 311 and 399 are the two most popular comparator ICs.
- Most common comparator applications are
  - Zero crossing detector
  - Level detector
  - Window detector
  - Duty cycle controller
  - Pulse generator
- Comparator circuits with positive feedback are known as schmitt triggers or regenerative comparators.
- There are two types of schmitt triggers :

- Inverting schmitt trigger (signal is applied at negative input)
- Non inverting schmitt trigger (signal is applied at positive input)

■ The difference between upper threshold (V_{UT}) and lower threshold (V_{LT}) voltage is known as hysteresis. It is also called as dead band or dead zone.

■ Major application of schmitt triggers are :
  - Sine to square wave converter
  - Eliminating comparator scatter
  - ON/OFF controller

■ Using positive feedback in op-amp as a comparator reduces it response time and exhibits hysteresis.

■ Rectifier circuits cannot rectify voltage below $V_{D(ON)} = 0.7$ V, the cut-in voltage of the diode.

■ To achieve precision rectification we need a circuit that keeps V_o equal to V_i for $V_i > 0$ V. This can be achieved by circuits called as precision rectifiers.

■ Precision rectifiers consists of inverting or non inverting amplifier with diode or diodes in the feedback loop of op-amp.

■ To obtain V_{rms} at the output of the full wave rectifier we need to further amplify V_{avg} by the amount 1.11.

■ The important non linear applications of an op-amp are log and antilog amplifiers. The base of log and antilog amplifiers is the V – I characteristics of diode and logarithmic equations of transistor.

■ The basic log amplifier uses a diode or a transistor in the negative feedback path of an op-amp. But the output of such circuit contains temperature dependent terms like V_T which causes error. Hence temperature compensation must be provided to log amplifiers.

■ The errors due to temperature dependent terms get eliminated due to temperature compensated circuits. But still the errors due to bias currents, offset voltages are still present these can be minimized by properly selecting the specifications of op-amp and using FET input op-amp.

■ Due to presence of active element in the feedback path log circuit show tendency to oscillate. Such a tendency is compensated by using emitter resistance and feedback capacitor across negative feedback element. Such a compensation is called as frequency compensation.

■ The log amplifiers can easily be turned around to provide the antilog or exponential function, which are called as antilog amplifiers. The basic antilog circuit uses a diode or transistor. This circuit also faces the same problem as that of basic log circuit. Hence temperature compensated antilog are used.

■ Log and antilog circuit can be used to obtain log ratio amplifier, analog multiplier and analog divider circuits.

■ Log ratio amplifier gives output which is proportional to the logarithm of the input voltage ratio.

SHORT QUESTIONS WITH ANSWER

Q.1. What do you mean by Closed Loop? Explain.

Ans. It is a term referring to the properties of an op-amp or other amplifier circuit with feedback (in contrast to open loop).

Q.2. What is a Controlled Source?

Ans. It is a source in which the voltage or current is controlled by another voltage or current elsewhere in the circuit. It is also known as a dependent source.

Q.3. What do you mean by an Inverting Amplifier?

Ans. It is an amplifier in which the output voltage has the opposite polarity as the input voltage.

Q.4. What do you mean by a Non inverting Amplifier?

Ans. It is an amplifier in which the output voltage has the same polarity as the input voltage.

Q.5. What do you mean by an Inverting Input?

Ans. It is the input terminal of op-amp for which the output voltage will change in the cpposite direction.

Q.6. Define Loop Gain.

Ans. It is defined as the product of all gain factors around a loop in a feedback circuit.

Q.7. What is a Non inverting input?

Ans. It is the input terminal of op-amp for which the output voltage changes in the same direction.

Q.8. Define Transconductance.

Ans. It is the ratio of output current to input voltage in a circuit.

Q.9. Define Transresistance.

Ans. It is defined as the ratio of output voltage to input current.

Q.10. What is meant by VCCS?

Ans. It represents voltage-controlled current source (VCCS).

Q.11. What do you mean by VCVS ?

Ans. It stands for voltage-controlled voltage source (VCVS).

Q.12. Define closed-loop voltage gain.

Ans. It is defined as the voltage gain of an amplifier circuit with a closed feedback path between the output and the amplifier's input.

Q.13. What is a linear amplifer?

Ans. Any type of amplifier producing an output which is a faithful reproduction of the input with virtually no distortion is called a linear amplifier.

Q.14. Define virtual ground.

Ans. It is the situation in which the inverting input of an operational amplifier is at ground potential even though it is not connected correctly to ground..

Q.15. What is a voltage follower?

Ans. It is an operational amplifier circuit whose output is exactly the same as its input. It is frequently known as a buffer since it isolates a load from a given signal source.

Q.16. What are the advantages of a voltage follower circuit?

Ans. (i) Gain of voltage follower is very close to unity.

(ii) It has high input impedance.

(iii) The circuit may be used as a buffer to connect high impedance signal source to a low impedance load which may even be capacitive.

Q.17. Name the type of feedback used if an external component is connected between the output terminal and inverting input.

Ans. It is known as the negative feedback.

Q.18. What is an inverting amplifier?

Ans. The input signal is applied to the inverting terminal. The output voltage is fed back to the input by the $R/-R_i$ network. The non inverting input terminal is grounded. The gain of the inverting amplifier is given by

$$V_o/V_1 = -R/R_i$$

Q.19. What are the advantages of using a voltage follower amplifier ?

Ans. Voltage follower has three unique characteristic viz extremely high input impedance, extremely low output impedance and unity transmission gain and is therefore, an ideal circuit device for use as a buffer amplifier.

Q.20. In what way is the voltage follower a special case of the non-inverting amplifier ?

Ans. If feedback resistor is made zero or R_1 is made ∞ (by keeping it open-circuited) in a non-inverting amplifier circuit, voltage follower is obtained.

Q.21. What are the applications of an inverting amplifier ?

Ans. Inverting amplifier is a very versatile component and can be used for performing number of mathematical stimulation such as analog inverter, paraphase amplifier, phase shifter, adder, integrator, differentiator.

Q.22. Are the characteristics of dual op-amp differential amplifier similar to those of non-inverting amplifier ?

Ans. No, the characteristics of a dual op-amp differential amplifier are identical to those of a non-inverting amplifier.

Q.23. What is a Differentiator?

Ans. It is a circuit whose output voltage is proportional to the slope of input voltage.

Q.24. What do you mean by an Instrumentation Amplifier?

Ans. It is the description of a composite differential amplifier system possessing high-quality gain and common-mode rejection specifications.

Q.25. What do you mean by CCCS?

Ans. It represents a current-controlled current source.

Q.26. What are CCVS ?

Ans. It represents a current-controlled voltage source.

Q.27. What do you mean by an Integrator?

Ans. It is a circuit whose output voltage is proportional to the area under the curve of the input voltage.

Q.28. What do you mean by an AC Integrator?

Ans. It is a circuit that integrates time-varying portion of the input signal.

Q.29. What do you mean by a Summing Circuit?

Ans. It is an electronic circuit whose output voltage is proportional to the sum of two or more input voltages.

Q.30. What is a difference amplifier? Explain.

Ans. It is defined as an operational amplifier circuit whose output voltage is proportional to the voltage difference of its two inputs. It is also called a differential amplifier.

Q.31. What is an Instrumentation amplifier?

Ans. It is a precision difference amplifier whose gain may be varied by a single resistor.

Q.32. What is a current-to-voltage converter?

Ans. It is a network in which the input signal in the form of a current is changed to a proportional output voltage.

Q.33. What are the properties of an instrumentation amplifier?

Ans.
(a) High gain accuracy
(b) High CMRR
(c) High gain stability with low temperature coefficient
(d) Low DC offset
(e) Low output impedance.

Q.34. What is a differential amplifier ?

Ans. Differential amplifier is a combination of inverting and non-inverting amplifiers and amplifies the voltage difference between two input lines either of which is grounded.

Q.35. What is the drawback of dual op-amp differential amplifier ?

Ans. The possible reduction in the amplitude of the output signal is the drawback of dual op-amp differential amplifier.

Q.36. How does negative feedback affect the performance of an inverting amplifier ?

Ans. Negative feedback decrease the voltage gain, non-linear distortion, noise, output impedance and total output offset voltage and increases the input impedance and bandwidth.

Q.37. Give the example of linear circuits.

Ans. Adder, subtractor, differentiator, integrator, voltage-to-current converter, current-to-voltage converter, instrumentation amplifier fall under the category of linear circuits.

Q.38. What is an adder or summing amplifier ?

Ans. Adder or summing amplifier is a circuit that provides an output voltage proportional to or equal to the algebraic sum of two or more input voltages multiplied by a constant gain factor.

Q.39. What is an integrator ?

Ans. An integrator is a circuit that performs a mathematical operation called integration.

Q.40. What are the applications of integrators ?

Ans. Integrators are widely used in ramp or sweep generators, filters, analog computers etc.

Q.41. What do you mean by Anti logarithmic Amplifier?

Ans. An amplifier whose output voltage is proportional to the antilogarithm ('inverse logarithm') of the input voltage.

Q.42. What do you mean by a Logarithmic Amplifier?

Ans. It is an amplifier whose output voltage is proportional to the logarithm of the input voltage.

REVIEW QUESTIONS

1. Explain the working of op-amp inverting amplifier. Derive the expression for its voltage gain.
2. Explain the working of op-amp non-inverting amplifier. Derive the expression for its voltage gain.
3. Compare inverting and non-inverting amplifier.
4. Explain the use of voltage follower circuit.
5. Explain about inverting VCVS.
6. Explain about non-inverting VCVS.
7. What do you mean by a buffer amplifier? Explain.
8. Write few important applications of op-amp.
9. Write various parameters that should be considered for dc and ac applications.
10. Explain the working of inverting summer with suitable example.
11. Explain the working of non-inverting summer with neat circuit diagram.
12. Explain how the average circuit can be derived from the summer.
13. Show that the output of the subtractor is proportional to the difference between the two input voltages.
14. What is passive integrator ? Which neat circuit diagram explain the working of the passive integrator.
15. What is an active integrator ? With neat circuit diagram explain the working of an active integrator.
16. What are the advantages of an active integrator over a passive integrator ?
17. Explain the draw the output waveform of the ideal integrator circuit when the input is
    (i) Sine wave
    (ii) Step input
    (iii) Square wave.
18. Derive the frequency response of an ideal active integrator. At what frequency op-amp gain is 0 dB ?
19. Explain the various errors in an ideal integrator circuit. How these errors are minimised ?
20. Explain the practical integrator circuit. Explain its advantages.
21. Derive the frequency response of a practical integrator.
22. Explain the working of integrator with Run, Set and Hold more.
23. Explain the application of practical integrator.
24. Design a practical integrator circuit with a d.c. gain of 20, to integrate a square wave of 25 kHz.
25. Explain the working of a summing integrator.
26. Explain how the integrator of the difference of currents an be obtained using basic integrator circuit.
27. What is passive differentiator circuit ? Draw and explain.
28. Explain the working of an ideal active differentiator.
29. Derive the frequency response of an ideal active differentiator.
30. What are the limitations of an ideal active differentaitor.
31. Explain the working of practical differentiator.
32. Derive the frequency response of a practical differentiator.
33. Explain the applications of a practical differentiator.
34. What is an analog computer? Explain the various building blocks of an analog computer.
35. Why integrators are preferred over the differentiators ?
36. Draw and explain the double integrator circuit.
37. From R_1 and R_F in the lossy integrator so that the peak gain is 20 dB and the gain is 3 dB down from its peak when $\omega = 10000$ rad/sec. Use $C_F = 0.01\ \mu F$.
38. Design the practical differentiator circuit to differentiate a sine wave signal of 2 sin (2π 300 t). Draw the output waveform.
39. In an inverting adder circuit, the input voltage are 0.3 V, 0.5 V, 0.1 V while $R_1 = R_2 = R_3 = 1\ k\Omega$. If $R_F = 10$ kΩ, calculate the output voltage. (**Ans.** 10 + 50 sin ωt mV))

40. Calculate the output voltage for the circuit shown in the figure 10.131 if the inputs are

V_{in1} = (50 mv) sin (1000 t) and

V_{in2} = (10 mv) sin (3000 t)

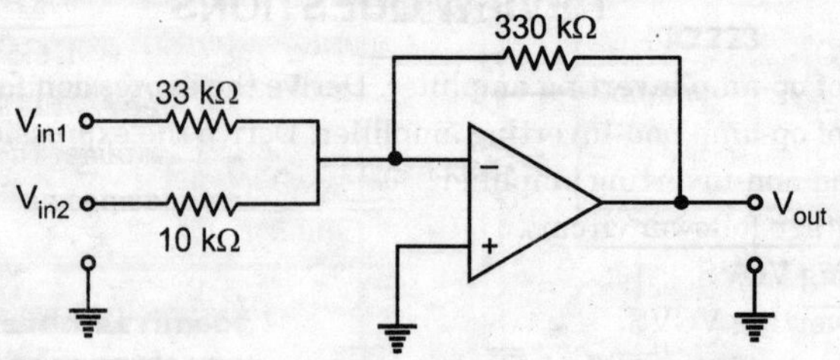

Fig. 10.131.

41. Design a practical integrator circuit to properly process input sinusoidal waveform upto 1 kHz. The input amplitude of signal is 10 mV.
42. Explain how an op-amp can be used as comparator.
43. Draw and explain inverting and non-inverting comparator circuits.
44. Explain practical comparator circuit with the help of neat diagram.
45. How is voltage limiting implemented in comparator circuits ?
46. List the important characteristics of comparators.
47. What is the necessity of separate comparators ICs ?
48. Draw and explain simplified circuit diagram of LM 311 comparator.
49. Draw and explain simplified circuit diagram of IC 339.
50. Explain op-amp as a schmitt trigger.
51. Design schmitt trigger having upper and lower thresholds of 120 mV input to this circuit is 1 V peak to peak triangular wave of 100 Hz. Draw the hysteresis loop.
52. Discuss schmitt trigger applications :
    (a) Eliminating Comparator Chatter Circuit
    (b) ON/OFF Controller
53. State the advantages of using positive feedback in op-amp as a comparator.
54. Draw the circuit diagram and explain the basic Log amplifier using
    (a) Diode (b) Transistor
55. What do you understand by precision rectifiers ? How do they differ from conventional rectifiers?
56. Draw and explain half wave precision rectifier circuit.
57. With help of neat diagram and waveform, explain the full wave precision rectifier circuit.

UNIT V

Regulated power supplies, Series and shunt regulators, current limiting circuit, Introduction to IC voltage regulators, fixed and adjustable switching regulators, SMPS, UPS.

Above syllabus has been covered in following chapter:

Chapter 11

VOLTAGE REGULATORS

Syllabus

Regulated power supplies, Series and shunt regulators, current limiting circuit, Introduction to IC voltage regulators, fixed and adjustable switching regulators, SMPS, UPS.

Chapter Outline

- Introduction
- Block diagram of Regulate D power supply
- Factors Responsible for the Variation in Load Voltage
- Performance Parameters for Overall Power Supply
- Basic Voltage Regulators
- Broad Classification of Voltage Regulators
- Shunt Regulator using Op-amp
- Series Regulator Using Op-amp
- Salient Features of IC Voltage Regulators
- General Purpose Linear IC 723 Regulator
- Protection Circuits in Regulators
- Three Terminal Fixed Voltage Regulators
- Three terminal Adjustable Regulator : LM 317
- Comparisons
- Limitations of Linear Voltage Regulators
- Basic Switching Regulator
- Block Diagram of Switched Mode Power Supply (SMPS)
- Types of Switching Regulators
- Applications of SMPS
- Uninterruptable Power Supplies (UPS)

11.1. INTRODUCTION

A **voltage regulator** is that kind of circuit which keeps the output voltage constant inspite of changes in the load current or input voltage. Its input is unregulated pulsating d.c. voltage obtained from filter and rectifier. Its output is constant d.c. voltage which is almost ripple free. The basic block diagram of voltage regulator has been shown in figure 11.1.

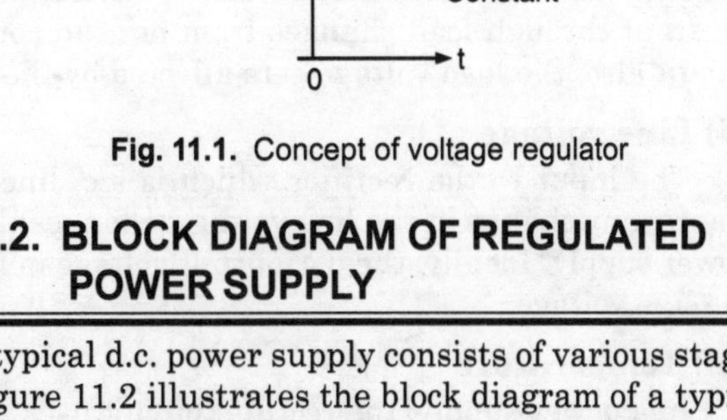

Fig. 11.1. Concept of voltage regulator

11.2. BLOCK DIAGRAM OF REGULATED POWER SUPPLY

A typical d.c. power supply consists of various stages. Figure 11.2 illustrates the block diagram of a typical d.c. power supply consisting of various circuits. The nature of voltages at various points has also been shown in figure 11.2.

The a.c. voltage (230 V, 50 Hz) is connected to the primary of the transformer. The transformer steps down the a.c. voltage, to the level required for the desired d.c. output. Thus, with suitable turns ratio, we get desired a.c. secondary voltage. The rectifier circuit converts this a.c. voltage into a **pulsating d.c. voltage. A** pulsating d.c. voltage means a unidirectional voltage containing large varying component called **ripple** in it. The filter circuit is used after a rectifier circuit, which reduces the ripple content in the pulsating d.c. and tries to make it smoother. Still then the filter output contains some ripple. This voltage is called **unregulated d.c. voltage.** A circuit used after the filter is a regulator circuit which not only makes the d.c. voltage smooth and almost ripple free but it also keeps the d.c. output voltage constant though input d.c. voltage varies under certain conditions. It keeps the output voltage constant under variable load conditions, as well. The output of a regulator is called **d.c. supply,** to which the load can be connected. Now a days, complete regulator circuits are available in the integrated circuit (IC) form.

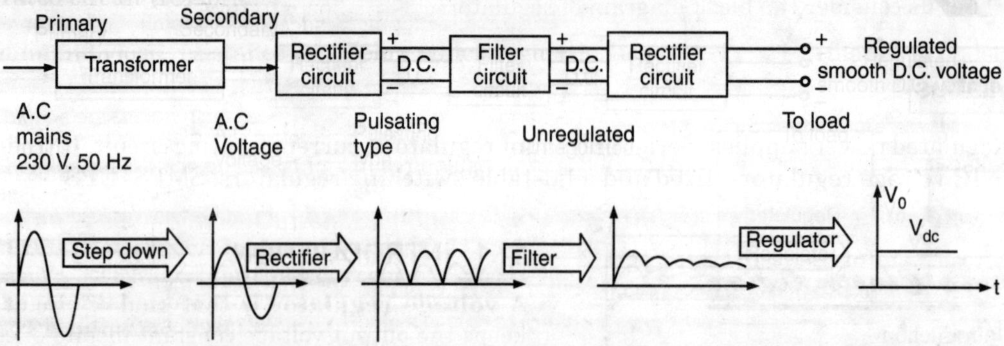

Fig. 11.2. Block diagram of regulated power supply with waveforms

> **Important Point :** Thus, a voltage regulator circuit is the one which is designed to keep the output voltage of a power supply nearly constant, under varying input voltage conditions and varying load conditions.

11.3. FACTORS RESPONSIBLE FOR THE VARIATION IN LOAD VOLTAGE

The various factors which affect the load voltage in a power supply may be listed as under :

(i) Load current (I_L)

An ideal power supply maintains a contant voltage at its output terminals, in spite of changes in load current. But, practically in a power supply without regulator the load voltage decreases as the load current I_L increases. For a practical power supply with regulator load voltage must be constant through load changes from no load condition to full load condition. But with regulator circuit also, the load voltage gets affected by the load current in a power supply.

(ii) Line voltage

The input to the rectifier which is a.c. line voltage decides the level of the output voltage. Hence, any change in the line voltage, changes the load voltage and affects the performance of the power supply. Ideally, the d.c. output voltage must remain constant irrespective of any changes in the line voltage.

(iii) Temperature

In a power supply, the rectifier circuit uses the P-N-junction diodes. The diode characteristics are temperature sensitive. the other semiconductor devices used in power supplies have their characteristics, temperature dependent. Hence, the temperature is an important factor responsible for the changes in the load voltage.

The voltage regulator circuit in a power supply, has to consider these factors and provide constant d.c. output voltage.

11.4. PERFORMANCE PARAMETERS FOR OVERALL POWER SUPPLY

The performance of the overall power supply is judged by specifying some parameters, based on the factors discussed above. Let us see detail the various power supply performance parameters.

11.4.1. Load Regulation

1. Definition

The load regulation may be defined as the change in the regulated output voltage when the load current is changed from minimum (no load) to maximum (full load).

2. Mathematical Expression

Let us consider the block diagram of regulator circuit shown in figure 11.3.

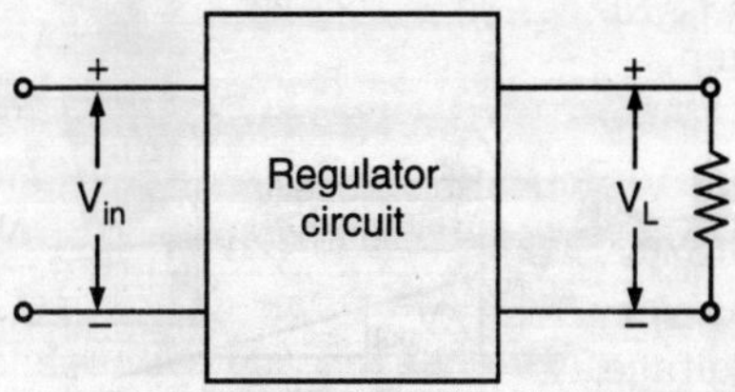

Fig. 11.3. Block diagram of regulator

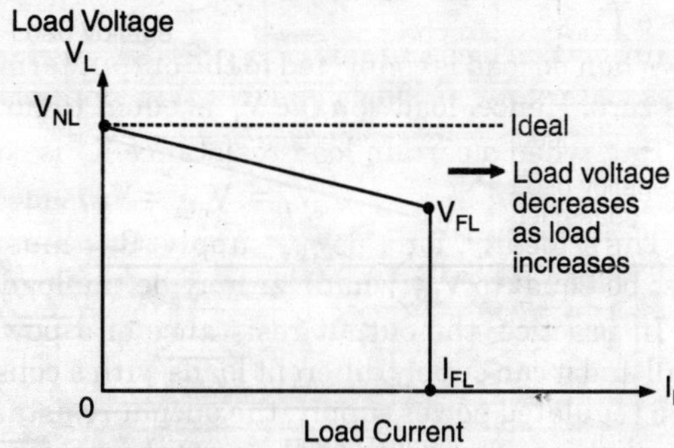

Fig. 11.4. Regulation characteristics

The load regulation is denoted as LR and mathematically expressed as under:

$$LR = V_{NL} - V_{FL}$$

where V_{NL} = load voltage with no load current

V_{FL} = load voltage with full load current

The load regulation is often expressed as percentage by dividing the LR by full load voltage and multiplying result by 100.

This means that,

$$\% \, LR = \frac{V_{NL} - V_{FL}}{V_{FL}} \times 100$$

3. Regulation Characteristics

The graph of load current against load voltage is called regulation characteristics of a power supply. The ideal value of load regulation is zero. Less the regulation, better is the performance of regulator. The regulation characteristics has been shown in figure 11.4.

11.4.2. Line Regulation

1. Definition

The input to the unregulated power supply i.e. rectifier circuit is 230 V a.c. supply. This line voltage may change, under the different load conditions. This affects the output voltage of rectifier which is V_{in} for a regulator circuit. Hence, the characteristics which gives source effect on regulator performance is defined. The line regulation is also called as source regulation and is denoted as SR.

2. Mathematical Expression

The SR may be defined as the change in the regulated load voltage for a specified range of line voltage, typically 230 V ± 10 %. Mathematically, it may be expressed as under :

$$SR = V_{HL} - V_{LL}$$

where V_{HL} = load voltage with high line voltage

V_{LL} = load voltage with low line voltage

The percentage source regulation is defined as under:

$$\% \text{ SR} = \frac{\text{SR}}{V_{nom}} \times 100$$

where V_{nom} = nominal load voltage

11.4.3. Output Resistance

To understand the effect of output resistance on the load voltage, let us consider the representation of power supply in terms of no load voltage V_{NL} and output resistance R_{out} as shown in figure 11.5.

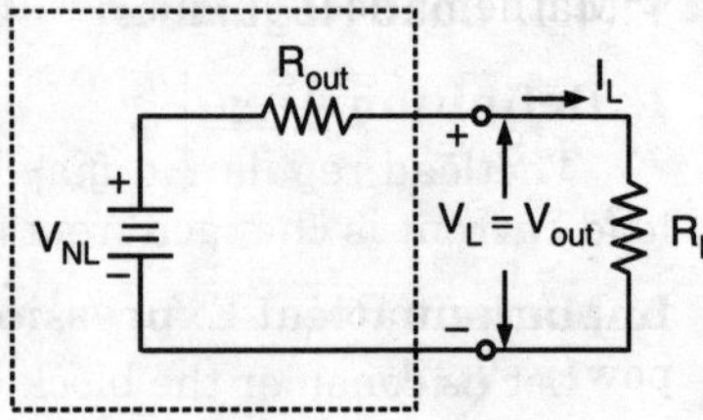

Fig. 11.5. Concept of R_{out}

When no load is connected to the output terminals, the current I_L is zero. Hence, load voltage V_L is equal to no load voltage V_{NL}.

But, when a certain load resistance R_L is connected then,

$$V_L = V_{out} = V_{NL} - I_L R_{out}$$

Thus, ideally, for a power supply, R_{out} must be zero and V_{out} must be equal to V_{NL} whatever may be the load current drawn.

In practice, the output resistance of a power supply is very small and it can supply different loads with a constant load voltage. For a regulated power supply, the output resistance is in the range of milliohms. The value of R_{out} may be obtained by finding the slope of a load regulation characteristics. We consider the load regulation characteristics as shown in figure 11.6. The slope of this characteristics is the output resistance.

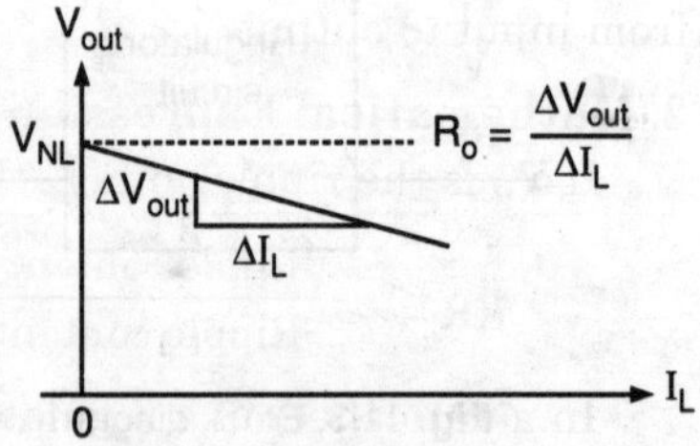

Fig. 11.6. Load regulation characteristics

Hence mathematically, the output resistance may be expressed as under :

$$R_{out} = \left. \frac{\Delta V_{out}}{\Delta I_L} \right|_{V_{in} \text{ and temperature constant}}$$

11.4.4. Voltage Stability Factor (S_V)

Another way of specifying the dependence of output voltage on line voltage is defining voltage stability factor denoted as S_V. It is the percentage change in the output voltage which occurs per volt change in the input line voltage, with load current and the temperature are assumed constant. Mathematically, it may be expressed as under :

$$S_V = \left. \frac{\Delta V_{out}}{\Delta V_{in}} \right|_{I_L \text{ and temperature constant}}$$

Important Point: Smaller the value of this factor, better is the performance of power supply.

11.4.5. Temperature Stability Factor (S_T)

1. Definition

As discussed earlier, the semiconductor devices having temperature dependent characteristics cause change in the output voltage as temperature changes. Thus, the temperature stability of power supply will be determined by temperature coefficients of various temperature sensitive semiconductor devices. By selecting low temperature coefficient devices, the output voltage of a power supply may be fairly independent of temperature.

2. Particular Example

For example, zener diodes having breakdown voltages in the range of 5 to 8 V have very low temperature coefficient and hence are always preferred and used in the power supply circuits.

3. Mathematical Expression

Mathematically, Temperature stability factor may be expressed as under :

$$S_T = \left.\frac{\Delta V_{out}}{\Delta T}\right|_{V_{in} \text{ and } I_L \text{ constant}}$$

Important Point: The value of this factor must be as small as possible, ideally zero, for a power supply.

11.4.6. Ripple Rejection

1. Definition

The output of a rectifier and filter circuit consists of ripples. The ripple rejection may be defined as a factor that indicates how effectively the regulator circuit rejects the ripples and attenuates it from input to output.

2. Mathematical Expression

If V_R is the ripple voltage then the ripple rejection factor may be defined as under :

$$RR = \frac{\text{Ripple content in output}}{\text{Ripple content in input}} = \frac{V_{R(out)}}{V_{R(in)}}$$

In a data sheet of a regulator, this factor is expressed in units decibels (dB) as under :

$$RR' = 20 \text{ Log}_{10} \text{ RR dB}$$

Important Point: As $V_{R(out)}$ is always less than $V_{R(in)}$, RR' i.e. RR in dB is always negative when defined as $V_{R(out)}/V_{R(in)}$. In practice, the values of all the performance parameters must be as small as possible. Smaller the values of these parameters, better is the performance of the power supply circuit.

11.5. BASIC VOLTAGE REGULATORS

1. Elements

The basic voltage regulator in its simplest form consists of,

(i) Voltage reference, V

(ii) Error amplifier

(iii) Feedback network

(iv) Active series or shunt control element

2. Brief Description

The voltage reference generates a voltage level which is applied to the comparator circuit, which is generally error amplifier. The second input to the error amplifier is obtained through feedback network. Generally, using the potential divider, the feedback signal is derived by sampling the output voltage. The error amplifier converts the difference between the output sample and the reference voltage into an error signal. This error signal, in turn, controls the active element of the regulator circuit, in order to compensate the change in the output voltage. Such an active element is generally a transistor. Thus the output voltage of the regulator is maintained constant.

11.6. BROAD CLASSIFICATION OF VOLTAGE REGULATORS

There are two types of voltage regulators available as follows :

(i) Shunt voltage regulator

(ii) Series voltage regulator

Each type provides a constant d.c. output voltage which is regulated. Let us discuss the functional block diagram and working principle of each of these two types of regulator circuits.

11.6.1. Shunt Voltage Regulator

1. Definition

The heart of any voltage regulator circuit is a control element. If, such a control element is connected in shunt with the load, the regulator circuit is called shunt voltage regulator.

2. Block Diagram

Figure 11.7 shows the block diagram of a shunt voltage regulator circuit.

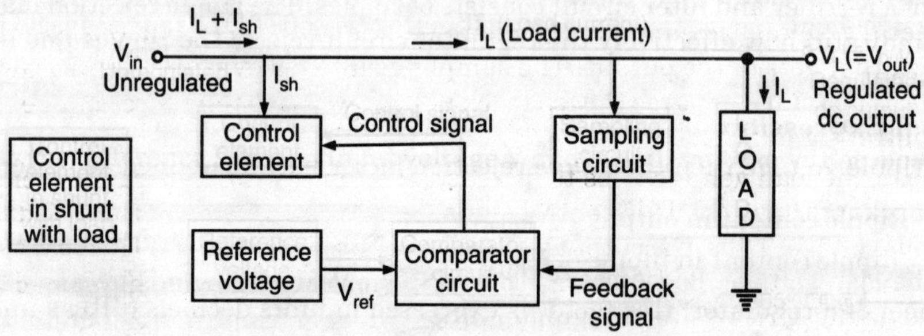

Fig. 11.7. Block diagram of shunt voltage regulator

3. Working Operation

The unregulated input voltage V_{in}, tries to provide the load current. But, part of the current is taken by the control element, to maintain the constant voltage across the load. If there is any change in the load voltage, the sampling circuit provides a feedback signal to the comparator circuit. The comparator circuit compares the feedback signal with the reference voltage and generates a control signal which decides the amount of current required to be shunted to keep the load voltage constant. For example, if load voltage increases then comparator circuit decides the control signal based on the feedback information, which draws increased shunt current I_{sh} value. Because of this, the load current I_L decreases and hence the load voltage decreases to its normal. Thus, control element maintains the constant output voltage by shunting the current, hence the regulator circuit is called as voltage shunt regulator circuit.

As observed from the block diagram, only part of the load current required to be diverted, passes through the control element. **Thus, the control element is low current, high voltage rating component.** The efficiency depends on the load current I_L. Hence, shunt regulators are not preferred for varying load conditions.

11.6.2. Series Voltage Regulator

1. Definition

If in a voltage regulator circuit, the control element is connected in series with the load, the circuit is called as series voltage regulator circuit.

2. Block Diagram

Figure 11.8 shows the block diagram of a series voltage regulator circuit.

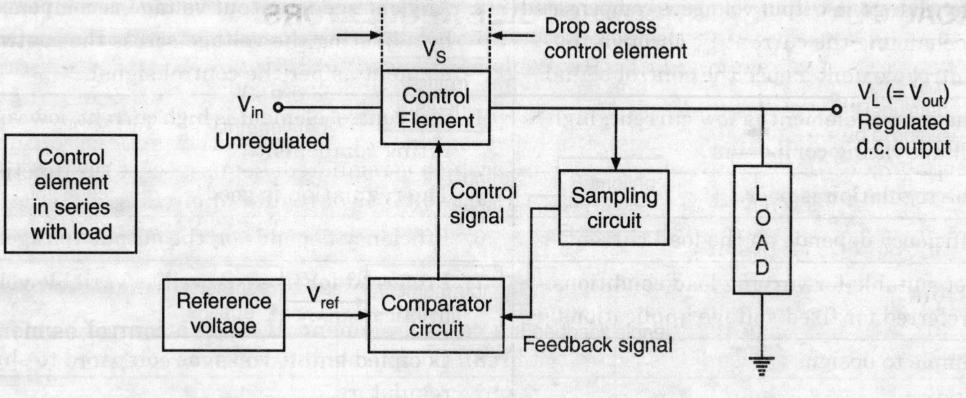

Fig. 11.8. Block diagram of series voltage regulator

3. Working Operation

The unregulated d.c. voltage is the input to the circuit. The control element, controls the amount of the input voltage, that gets to the output. The sampling circuit provides the necessary feedback signal. The comparator circuit compares the feedback voltage with the reference voltage to generate the appropriate control signal.

As an example, if the load voltage tries to increase, the comparator generates a control signal based on the feedback information. This control signal causes the control element to decrease the amount of the output voltage. Thus, the output voltage is maintained constant. Thus, control element which regulates the load voltage, based on the control signal is in series with the load and hence the circuit is called as the series voltage regulator circuit.

In series regulators, the entire current passes through the control element and hence **control element is high current, low voltage rating component.** As input current and load current are same, the efficiency depends on output voltage. It provides good regulation than shunt regulators. It can be used for fixed voltage as well as variable voltage requirements. To compensate for the drop across the control element, input voltage V_{in} must be at least 2 to 3 V more than output voltage.

11.6.3. Performance Comparison of Shunt and Series Regulators

TABLE 11.1.

<table>
<tr><th>S.No.</th><th>Performance of shunt regulator</th><th>Performance of series regulator</th></tr>
<tr><td>1.</td><td>The control element is in parallel with the load.
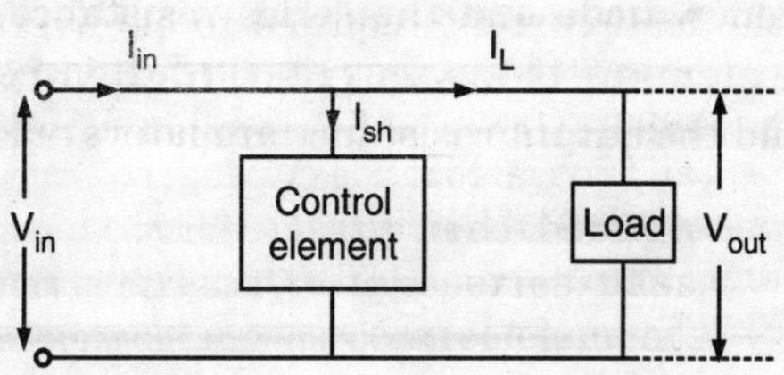
</td><td>1. The control element is in series with the load.
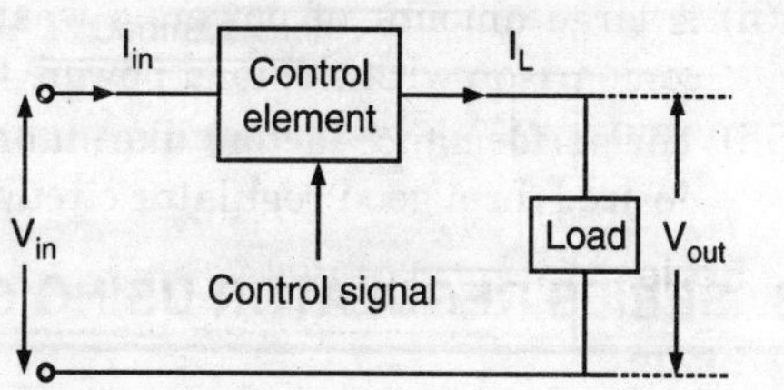
</td></tr>
<tr><td>2.</td><td>Only small current passes through the control element, which is required to be diverted to keep output constant.</td><td>2. The entire load current I_L always passes through the control element.</td></tr>
</table>

3.	Any change in output voltage is compensated by changing the current I_{sh} through the control element as per the control signal.	3. Any change in output voltage is compensated by adjusting the voltage across the control element as per the control signal.
4.	The control element is low current, high voltage rating component.	4. The control element is high current, low voltage rating component.
5.	The regulation is poor.	5. The regulation is good.
6.	Efficiency depends on the load current.	6. Efficiency depends on the output voltage.
7.	Not suitable for varying load conditions. Preferred for fixed voltage applications.	7. Preferred for fixed as well as variable voltage applications.
8.	Simple to design.	8. Complicated to design as compared to shunt regulators.
9.	Examples are : Zener shunt regulator, transistorised shunt regulator etc.	9. Examples are : Series feedback type regulator, series regulator with preregulator and foldback limiting etc.

11.7. SHUNT REGULATOR USING OP-AMP

1. Definition and Schematic Diagram

An op-amp can be used as a voltage comparator. The circuit diagram is shown in figure 11.9. The potential divider formed by the resistances R_1 and R_2, provides the part of the output voltage as a feedback. The op-amp compares this with the reference voltage generated by zener diode. The op-amp provides the control (signal) current to shunt control element which is the transistor Q. The current through the resistance R is thus controlled to drop a voltage across R so that the output voltage is maintained constant.

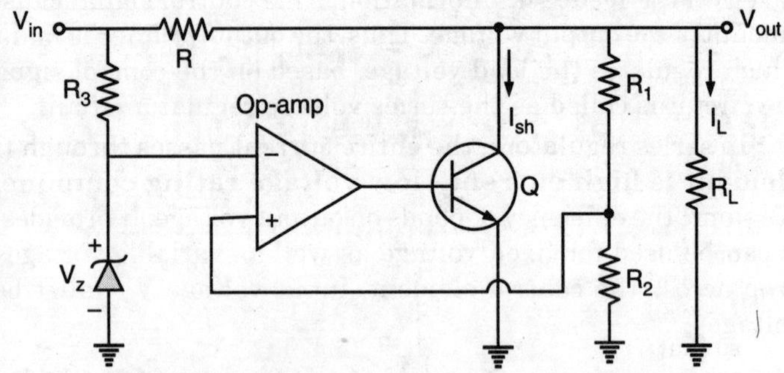

Fig. 11.9. Shunt regulator using op-amp

2. Limitations of Shunt Regulator Using op-amp

The shunt regulator is generally suffered with the following limitations:

(i) The maximum load current that can be supplied is limited.

(ii) A large amount of power is wasted in the zener diode and the series resistance R, in comparison with the load power.

(iii) The performance factors like stability factor and the output resistance are not very low, as desired for a good regulator circuit.

11.8. SERIES REGULATOR USING OP-AMP

1. Definition and Schematic Diagram

The series regulator circuit using op-amp has been shown in figure 11.10. The op-amp is used as a comparator. It compares the part of the output voltage obtained from the potential divider circuit as a feedback with the reference voltage generated by the zener. The output of op-amp drives the series pass transistor Q_1. If there is any change in the output voltage, the control signal from op-amp, controls the conduction of the transistor Q_1. Thus, the output voltage is maintained at constant level.

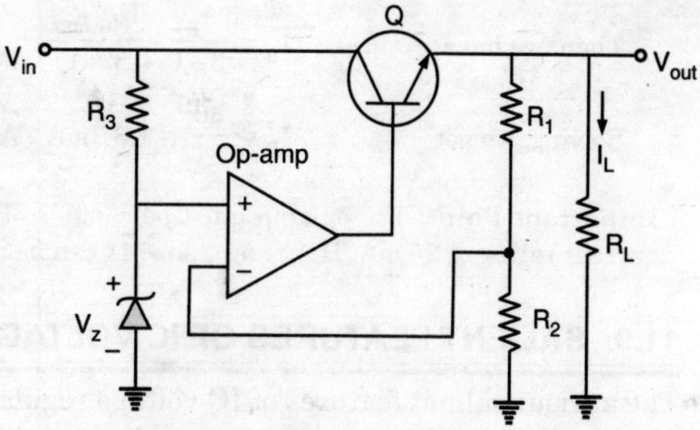

Fig. 11.10. Series regulator using op-amp

2. Expression for Output Voltage

The output voltage is maintained constant at a value which is given by

$$V_{out} = \left(1+\frac{R_1}{R_2}\right)V_Z \quad ...(11.1)$$

This version of the series regulator with op-amp is used in IC 723.

EXAMPLE 11.1. Design an op-amp series voltage regulator to meet the following specifications: V_{in} = (18 ± 3) volts, V_{out} = 9 volts at I_{out} = 10 to 50 mA, Zener available V_Z = 5.6 V, $P_{Z\,max}$ = 0.5 W.

Solution: The required circuit has been shown in figure 11.11.

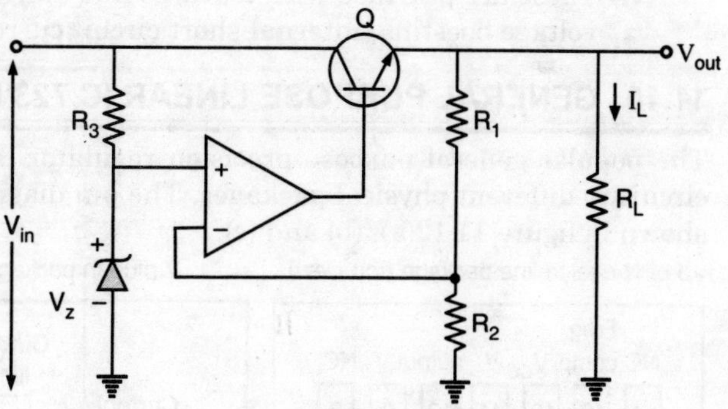

Fig. 11.11.

We know that the output voltage is given by

$$V_{out} = \left(1+\frac{R_1}{R_2}\right)V_Z$$

Substituting values, we get

$$9 = \left(1+\frac{R_1}{R_2}\right)5.6$$

or $$1+\frac{R_1}{R_2} = 1.6071$$

or $$\frac{R_1}{R_2} = 0.6071$$

Let $R_1 = 10\ k\Omega$

so that $R_2 = 16.47\ k\Omega$

Minimum zener current is (say) 5 mA, which is same as current through R_3.

Therefore, we have

$$I_{Z\,min} = \frac{V_{in(min)} - V_Z}{R_3}$$

Substituting the values, we obtain

$$50 \times 10^{-3} = \frac{[18-3]-5.6}{R_3}$$

Solving, we obtain $R_3 = 1.88\ k\Omega$

For $V_{max} = 18 + 3 = 21$ V, we have

$$I_Z = \frac{21-5.6}{1.88\times10^3} = 8.19\ mA$$

Also, $P_D = I_Z \times V_Z = 8.19 \times 10^{-3} \times 5.6 = 0.045$ W

Thus, $P_D < P_{Z\,max}$

Hence, the elements values for the designed circuit are as follows :

$R_1 = 10\ k\Omega$

$R_2 = 16.47\ k\Omega$

and, $R_3 = 1.88\ k\Omega$

Let us assume β of the series transistor as 30.

Then, we have $$I_B = \frac{I_E}{\beta+1} = \frac{I_{out\ max}}{\beta+1}$$

Solving, we get $$I_B = \frac{50}{31} = 1.61 \text{ mA} \quad \textbf{Ans.}$$

Important Point: The op-amp must be capable of supplying this current. The op-amp 741 has output current rating of 25 mA. Hence op-amp 741 can be used.

11.9. SALIENT FEATURES OF IC VOLTAGE REGULATORS

The various salient features of IC voltage regulators may be listed as under :

(i) Easy to use.
(ii) It greatly simplifies power supply design.
(iii) Due to mass production, low in cost.
(iv) IC voltage regulators are versatile.
(v) Conveniently used for load regulation.
(vi) These are provided with features like built in protection, programmable output, current/voltage boosting, internal short circuit current limiting etc.

11.10. GENERAL PURPOSE LINEAR IC 723 REGULATOR

The popular general purpose precision regulator is IC 723. It is a monolithic linear integrated circuit in different physical packages. The pin diagram alongwith the various packages has been shown in figure 11.12(a), (b) and (c).

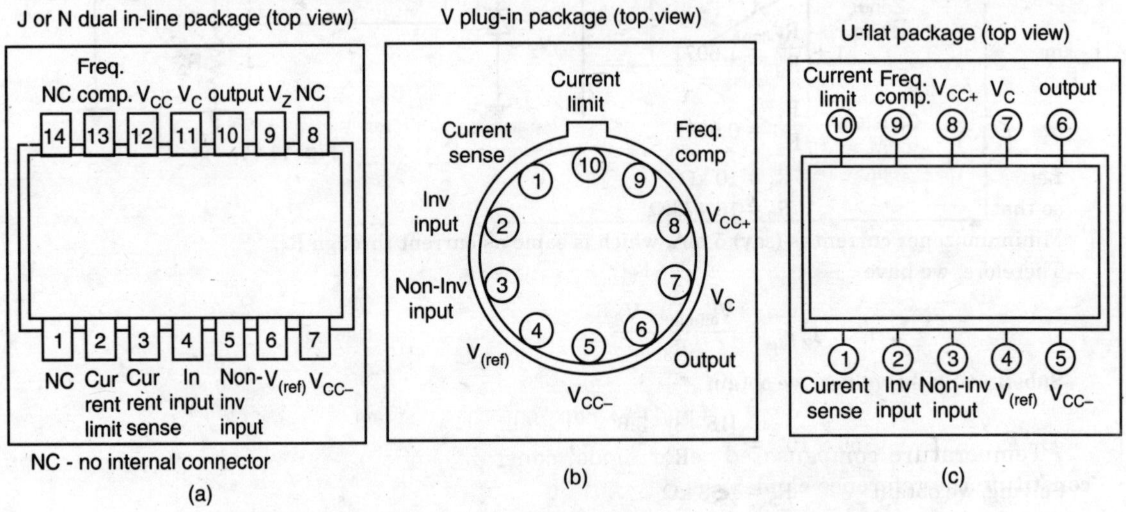

Fig. 11.12. Available packaging for μA 723 precision voltage regulator

11.10.1. Salient Features of IC 723

(i) It works as voltage regulator at output voltage ranging from 2 to 37 volts at currents upto 150 mA.
(ii) It can be used at load currents greater than 150 mA with the use of suitable NPN or PNP external pass transistors.
(iii) Input and output short-circuit protection is provided.
(iv) It has good line and load regulation (0.03%)
(v) Wide variety of applications of series, shunt, switching and floating regulator.

(vi) Low temperature drift and high ripple rejection.
(vii) Low standby current drain.
(viii) Small size, lower cost.
(ix) Relative ease with which power supply can be designed.
(x) It provides a choice of supply voltage.

11.10.2. Internal Architecture of IC 723

1. Functional block diagram

The functional block diagram of IC 723 can be divided into four major blocks as under :

(i) Temperature compensated voltage reference source, which is zener diode.
(ii) An op-amp circuit used as an error amplifier.
(iii) A series pass transistor capable of a 150 mA output current.
(iv) Transistor used to limit output current.

2. Functioning

The functioning of the above blocks can be explained with the help of a simplified functional block diagram of IC 723 as shown in figure 11.13.

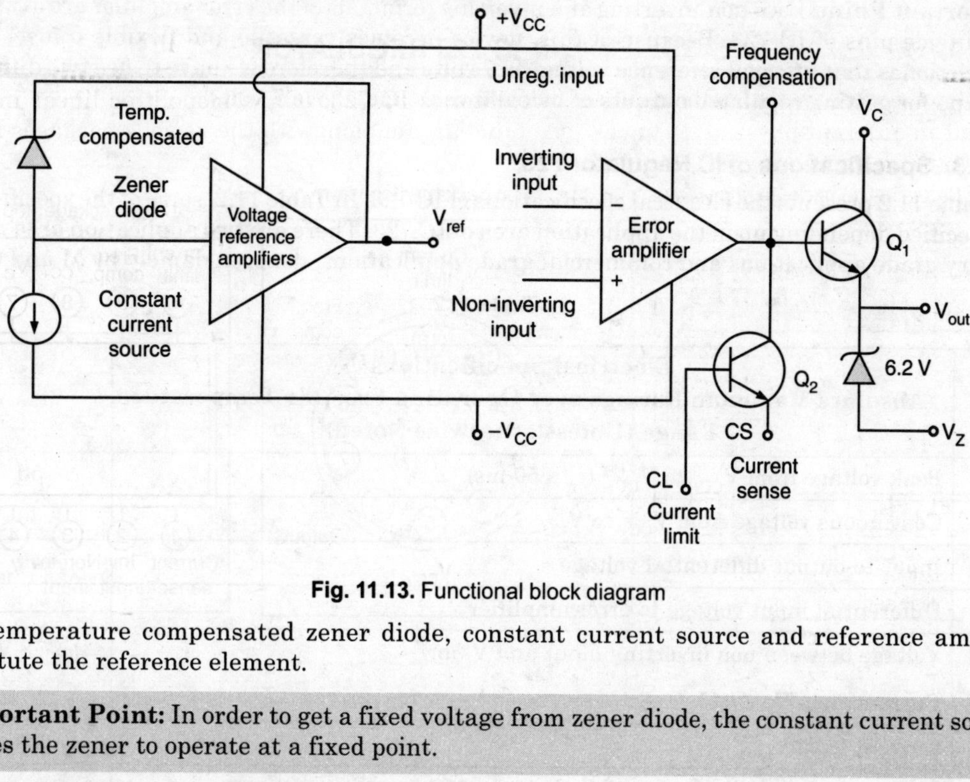

Fig. 11.13. Functional block diagram

Temperature compensated zener diode, constant current source and reference amplifier constitute the reference element.

> **Important Point:** In order to get a fixed voltage from zener diode, the constant current source forces the zener to operate at a fixed point.

Output voltage is compared with this temperature compensated reference potential of the order of 7 volts. For this, V_{ref} is connected to the non-inverting input of the error amplifier.

This error amplifier is high gain differential amplifier. It's inverting input is connected to the either whole regulated output voltage or part of that from outside. For later case, a potential divider of two scaling resistors is used. Scaling resistors help in getting multiplied reference voltage or scaled up reference voltage.

Error amplifier controls the series pass transistor Q_1, which acts as variable resistor. The series pass transistor is a small power transistor having about 800 mW dissipation. The unregulated power supply source (< 36 V d.c.) is connected to collector of series pass transistor.

Transistor Q_2 acts as current limiter in case of short circuit condition. It senses drop across R_{sc} placed in series with regulated output voltage externally.

The frequency compensation terminal controls the frequency response of the error amplifier. The required roll-off is obtained by connecting a small capacitor of 100 pF between frequency compensation and inverting input terminals.

The internal structure can be represented in a more simplified form as shown in figure 11.14.

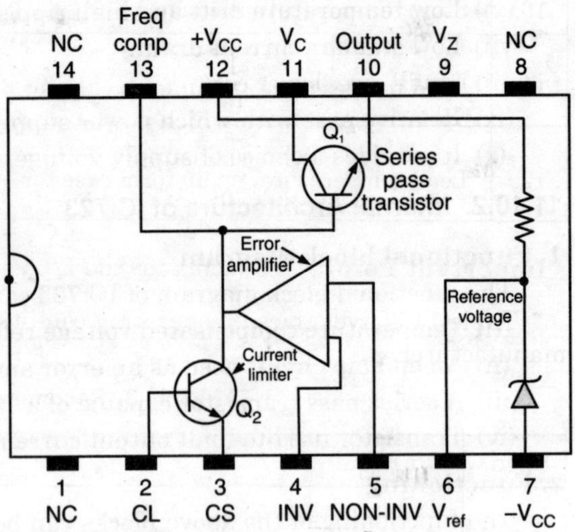

Fig. 11.14. Simplified internal structure of IC 723

Important Point: Both non-inverting and inverting terminals of the error amplifier are available on outside pins of IC 723. Because of this, device becomes versatile and flexible to use. Only restriction is that internal reference voltage is 7 volts and therefore we have to use two different circuits for getting regulated outputs of below 7 volts and above 7 volts.

11.10.3. Specifications of IC Regulator 723

Table 11.2 presents the electrical specifications of IC 723. In Table 11.2, some of the specifications are specified depending upon the application area of IC 723. There are two application area namely military grade applications and commercial grade applications, denoted namely by M and C.

Table 11.2.

S.No.	Electrical Specifications Absolute Maximum Ratings over Operating Free-Air Temperature Range (Unless otherwise Noted)	
1.	Peak voltage from V_{CC+} to V_{CC-} ($t_w \le 50$ ms)	50 V
2.	Continuous voltage from V_{CC+} to V_{CC-}	40 V
3.	Input-to-output differential voltage	40 V
4.	Differential input voltage to error amplifier	± 5 V
5.	Voltage between non inverting input and V_{CC-}	8 V
6.	Current from V_Z	25 mA
7.	Current from $V_{(ref)}$	15 mA
8.	Continuous total dissipation at (or below) 25°C free-air temperature J or N package L package (see Note 1) U package	 1000 mW 800 mW 675 mW
9.	Operating free-air temperature range: µA 723M Circuits µA 723 C Circuits	– 55°C to 125°C 0°C to 150°C

10.	Storage temperature range	– 65°C to 150°C
11.	Lead temperature $\frac{1}{16}$ in form case for 60s, J, L, or U package	300°C
12.	Lead temperature $\frac{1}{16}$ in form case for 10s, N package	260°C

Important Point: This rating for the L package requires a heat sink.

Table 11.3 gives the safe operating conditions of IC 723. These conditions are provided by the manufacturer.

TABLE 11.3. Recommended Operating Conditions

S.No.	Parameter	Min	Max	Unit
1.	Input voltage, V_{in}	9.5	40	V
2.	Output voltage, V_{out}	2	37	V
3.	Input-to-output differential voltage $V_C - V_{out}$	3	38	V
4.	Output current, I_{out}		150	mA

11.10.4. Applications of IC 723

The various regulator circuits as per the requirement can be achieved using IC 723. Some of them are discussed below:

11.10.4.1. Basic Low Voltage Regulator (V_{out} = 2 to 7 volts)

The resistor, R_{sc} is connected between CL and CS pins. The current limit transistor remains non-conductive unless drops across R_{sc} is 0.6 V (equal to V_{BE} drop). The value of R_{sc} can be determined by following expression:

$$R_{sc} = \frac{V_{sense}}{I_{limit}} = \frac{0.6}{I_{limit}} \quad ...(11.2)$$

I_{limit} can be selected as 1.2 to 1.5 times the maximum load circuit. Potential divider made up of R_1 and R_2 is connected between V_{ref} and non-inverting terminals.

$$V_{non\text{-}inverting} = V_{ref} \times \frac{R_2}{R_1 + R_2} \quad ...(11.3)$$

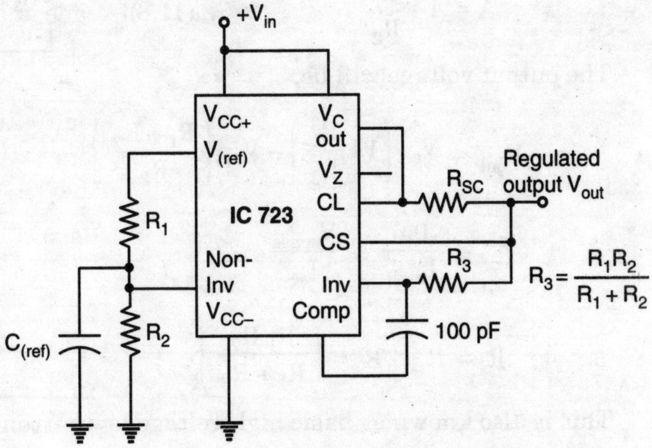

Fig. 11.15. Basic low voltage regulator

As the series pass transistor is working as emitter follower, therefore, we have

$$V_{out} = V_{ref} \times \frac{R_2}{R_1 + R_2} \quad ...(11.4)$$

Also, R_1 and R_2 can be chosen between 1 kΩ to 10 kΩ i.e.,

$$R_3 = R_1 \parallel R_2$$

or

$$R_3 = \frac{R_2}{R_1 + R_2} \quad ...(11.5)$$

Maximum load current can be 150 mA.

11.10.4.2. Low Voltage High Current Regulator

Output voltage from + 2 to + 7 V and load current can be more than 150 mA. For this, one transistor is connected externally, shown as Q_1 in figure 11.16.

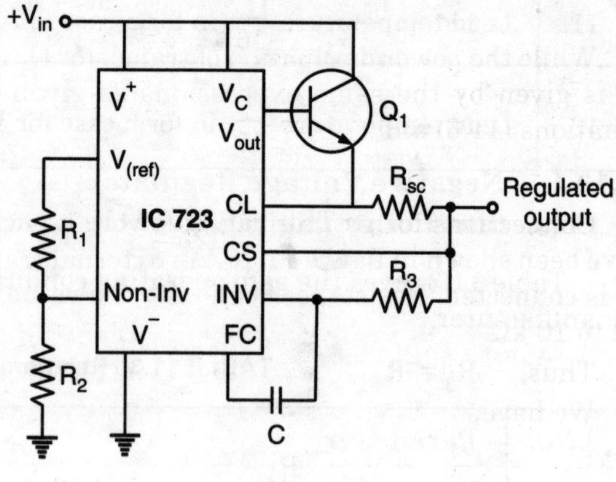

Fig. 11.16.

The functional expressions are similar to that of basic low voltage regulator circuit.

This means that, $V_{out} = V_{ref} \times \dfrac{R_2}{R_1 + R_2}$

$$R_{sc} = \frac{0.6}{I_{limit}}$$

Power dissipation of transistor =

$$= [V_{in(max)} - V_{out(min)}] \times I_{L(max)} \quad ...(11.6)$$

Power dissipation of IC =

$$[V_{in(max)} - V_{out(min)}] \times \frac{I_{L(max)}}{h_{fe(min)\ of\ Q_1}} \quad ...(11.7)$$

11.10.4.3. Basic Positive High Voltage Regulator

For this type, output voltage varies from + 7 V to + 37 V and $I_L \le 150$ mA.

The non-inverting terminal connected to V_{ref} through R_3. Because of this arrangement, the error amplifier acts as non-inverting amplifier.

The gain is given by,

$$A = 1 + \frac{R_1}{R_2} \quad ...(11.8)$$

The output voltage will be,

$$V_{out} = V_{ref}\left(1 + \frac{R_1}{R_2}\right) = V_{ref}\left(\frac{R_1 + R_2}{R_2}\right)$$

$$R_{sc} = \frac{0.6}{I_{limit}} = \frac{V_{sense}}{I_{sc}}$$

$$R_3 = R_1 \,||\, R_2 = \left(\frac{R_1 R_2}{R_1 + R_2}\right)$$

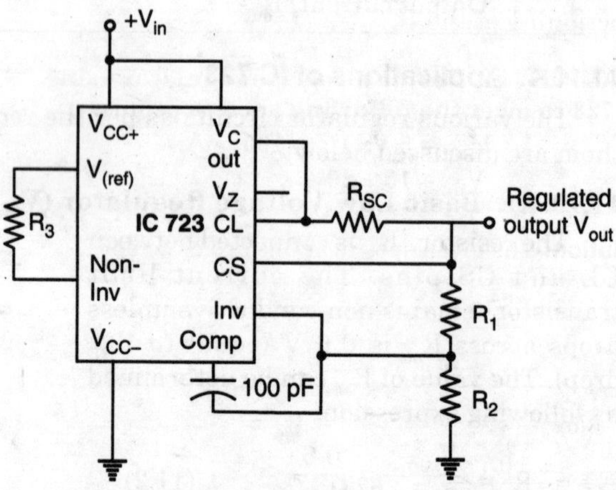

Fig. 11.17. Basic high voltage regulator

This is also known as basic high voltage low current regulator.

11.10.4.4. Positive High Voltage High Current Regulator

For this type, output voltage from + 7 V to + 37 V and load current $I_L > 150$ mA. For this, an external transistor Q_1 is connected, as shown in figure 11.18.

For this, different expressions are similar to basic high voltage regulator and reproduced for the convenience i.e.,

$$V_{out} = V_{ref}\left(\frac{R_1 + R_2}{R_2}\right) \quad ...(11.9)$$

$$R_{sc} = \frac{0.6}{I_{limit}} = \frac{V_{sense}}{I_{sc}}, \; R_3 = \frac{R_1 R_2}{R_1 + R_2}$$

While the power dissipation of transistor Q_1 and the IC is given by the same expressions as given by the equations (11.6) and (11.7).

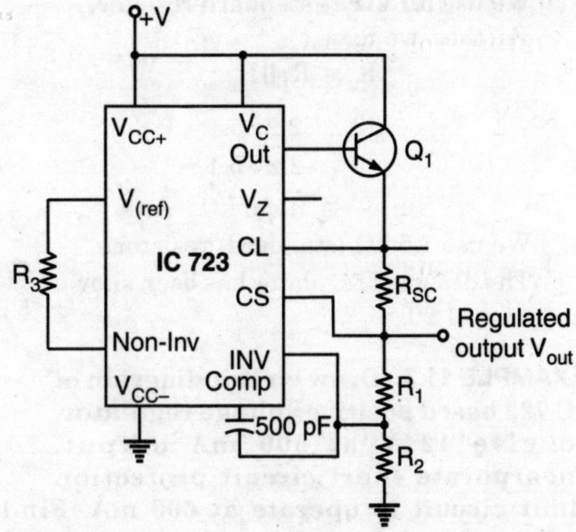

Fig. 11.18. Positive high voltage regulator

11.10.4.5. Negative Voltage Regulator

Connections for getting negative voltage regulator have been shown in figure 11.19. An external transistor Q_1 is connected. Resistances can take the value from 1 kΩ to 10 kΩ.

Thus, $R_3 = R_4$ ...(11.10)

We have,

$$V_{out} = \left[\frac{V_{ref}}{2} + \frac{R_1 + R_2}{R_1}\right] \quad ...(11.11)$$

If magnitude of $-V_{in}$ is less than 9 V, we connect V_{CC+} and V_C to a positive supply such that V_{CC+} to V_{CC-} is greater than 9 V, for proper functioning of the IC.

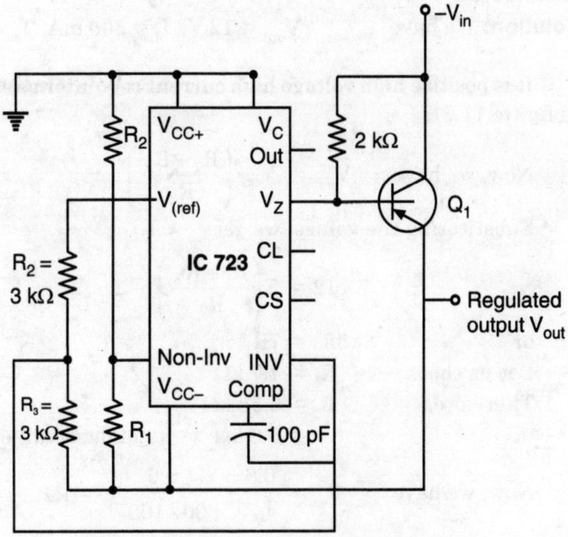

Fig. 11.19. Negative voltage regulator

EXAMPLE 11.2. Design a voltage regulator using IC 723 to meet the following specifications:

$V_{out} = 5$ V; $I_{out} = 100$ mA.

$V_{in} = 15 \pm 20\%$

$I_{sc} = 150$ mA; $V_{sense} = 0.7$ V.

Solution: The given specifications are as under :

$V_{out} =$ 5 V; $I_{out} = 100$ mA.

$V_{in} = 15 \pm 20\%$

$I_{sc} = 150$ mA; $V_{sense} = 0.7$ V.

Now, $R_{sc} =$

$$\frac{V_{sense}}{I_{sc}} = \frac{0.7}{150 \times 10^{-3}} = 4.67 \; \Omega.$$

Neglecting input bias current of an error amplifier, we can write,

$$R_1 = \frac{V_{ref} - V_{out}}{I_D}$$

where I_D = potential divider current = 1 mA

and $V_{ref} = 7.15$ V for IC 723

Therefore, $R_1 = \frac{7.15 - 5}{1 \times 10^{-3}} = 2.15$ kΩ

We use 2.2 kΩ as standard resistance.

Now, we have $V_{out} = V_{ref} \cdot \frac{R_2}{R_1 + R_2}$

Substituting values, we get $5 = 7.15 \cdot \frac{R_2}{(2.2 + R_2)}$

or $2.2 + R_2 = 1.43 \, R_2$

or $0.43 \, R_2 = 5.11$ kΩ

We use 5.1 kΩ as standard resistor.

Further, we have

$$R_3 = R_1 || R_2$$

$$= \frac{2.2 \times 5.1}{2.2 + 5.1}$$

$$= 1.536 \text{ k}\Omega \quad \textbf{Ans.}$$

We use 1.5 kΩ (standard) resistor.

The designed regulator has been shown in figure 11.20.

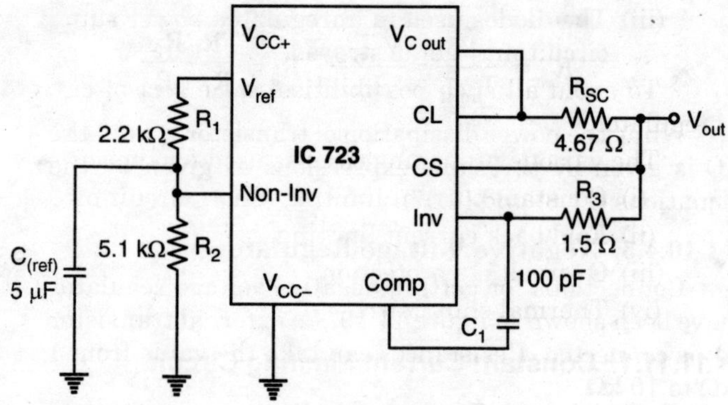

Fig. 11.20.

EXAMPLE 11.3. Draw circuit diagram of IC 723 based positive voltage regulator, to give 12 V at 500 mA output. Incorporate short circuit protection limit circuit to operate at 600 mA. Find all resistor values, calculate their wattage and specify the type of resistors.

Solution: We have $V_{out} = 12$ V, $I_L = 500$ mA, $I_{sc} = 600$ mA

It is positive high voltage high current regulator as shown in figure 11.21.

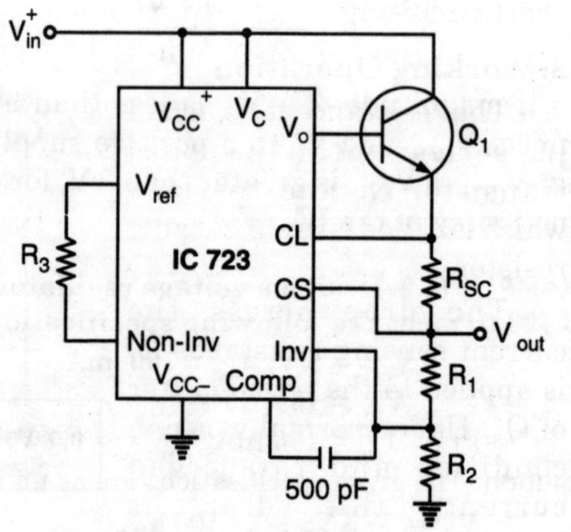

Fig. 11.21.

Now, we have $\quad V_{out} = V_{ref}\left(\frac{R_1 + R_2}{R_2}\right)$

Substituting the values, we get

$$12 = 7\left(\frac{R_1 + R_2}{R_2}\right)$$

or $\quad 5R_2 = 7R_1$

Let us choose $\quad R_1 = 4.7$ kΩ

Therefore, $\quad R_2 = 6.58$ kΩ

$\cong 6.8$ kΩ (standard value)

Now, we have $\quad R_{sc} = \frac{0.6}{I_{sc}} = \frac{0.6}{600 \times 100^{-3}} = 1\Omega$

$$R_3 = R_1 || R_2 = \frac{R_1 R_2}{R_1 + R_2} = 2.7 \text{ k}\Omega$$

Power wattage $\quad P_{sc} = (I_{sc})^2 R_{sc} = (600 \times 10^{-3})^2 \times 1 = 0.36 \text{ W} = 360 \text{ mW}$

All the resistor type can be of metal film resistors. For the power rating of R_1 and R_2, let us assume the input current to the inverting terminals is zero.

Therefore, we have $\quad I = \frac{V_{out}}{R_1 + R_2} = s\frac{12}{(4.7 + 6.8) \times 10^3} = 1.043$ mA

or $\quad P_1 = I^2R_1 = (1.043 \times 10^{-3})^2 \times 4.7 \times 10^3 = 5.112$ mW

and $\quad P_2 = I^2R_2 = (1.043 \times 10^{-3})^2 \times 6.8 \times 10^3 = 7.387$ mW $\quad$ **Ans.**

Hence, both R_1 and R_2 may be selected safely of 1/16th watt power rating.

11.11. PROTECTION CIRCUITS IN REGULATORS (Very Important)

The series regulator using transistors has no short circuit protection. If the load terminals are shorted accidently then,

(i) Large amount of load current will flow.

(ii) The pass transistor Q will get destroyed.

(iii) The diodes used in unregulated power supply, supplying the input voltage V to regulator circuit may get destroyed.

To avoid all such possibilities, some sort of current limiting must be provided to the series regulator.

The various protecting circuits used for the series voltage regulators may be listed as under :

(i) Constant current limiting (Short circuit protection).
(ii) Fold back current limiting.
(iii) Overvoltage protection.
(iv) Thermal shut down.

11.11.1. Constant Current Limiting Circuit

1. Definition and Schematic Diagram

A block of series regulation using simple current limiting circuit has been shown in figure 11.22. It is also called as **short circuit protection circuit** since it provides protection against short circuiting.

2. Working Operation

The resistance R_4 is added in series with the pass transistor Q_1 and the load which is called current sensing resistor.

The drop across the current sensing resistance R_4 is applied to the base-emitter of Q_3. Under normal working condition and rated load current, this drop is insufficient to turn on the transistor Q_3 and hence pass transistor Q_1 continues to supply the rated load current.

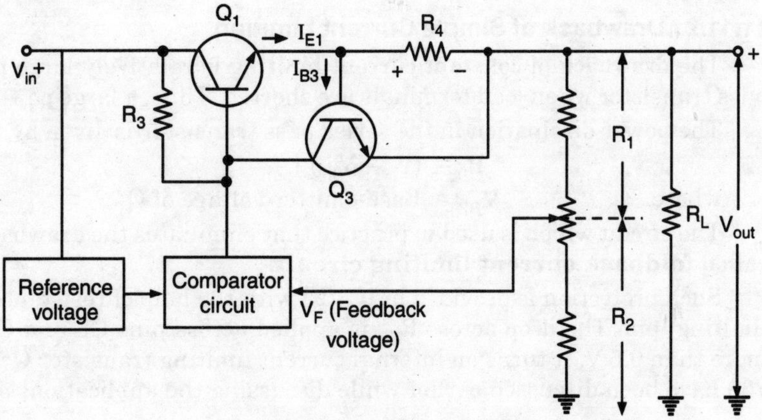

Fig. 11.22. Current limiting circuit

When current increases due to over loading or short circuit conditions then drop across R_4 increases more than 0.6 V. The 0.6 V is sufficient to turn on the transistor Q_3. The collector current of Q_3 flows through R_3 and decreases the base voltage of Q_1. This decreases the load voltage. This decrease in output voltage prevents the large load current.

The value of R_4 can be selected to adjust the rated current of the circuit. As drop across R_4 should be less than 0.6 V under normal working condition, for rated current of 1 A, R_4 can be selected as 0.7 Ω. For a rated current of 7 A, R_4 must be selected as 0.1Ω and so on.

Let us study the variation of load current against load voltage with simple current limiting.

When $R_L = \infty$ i.e. output terminals open, the output voltage is $V_{out} = V_{reg.}$ The load current is zero.

When load increases, the load resistance decreases and the load current increases. The load current can increase till the drop across R_4 is not equal to 0.6 V. This load resistance at which drop across R_4 is 0.6 V (say) R_{min}. When load

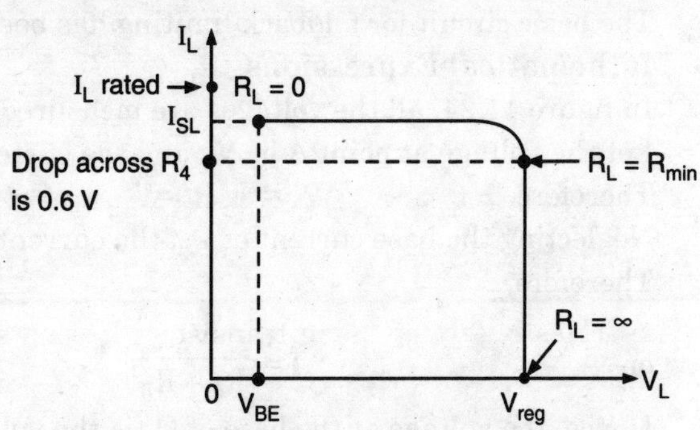

Fig. 11.23. Load current against load voltage

current further increases and R_L becomes zero (i.e., short circuit), the load voltage decreases and the load current gets maintained at a value less than rated current. After $R_L = R_{min}$, the regulation is lost and load voltage decreases as R_L decreases. This has been shown in figure 11.23.

3. Mathematical Expressions

Let I_{SL} = load current when load terminals are shorted.

In such a case, drop across R_4 is equal to V_{BE} to turn on the transistor Q_3.

Therefore, we have

$$V_{BE} = I_{SL} R_4$$

or
$$I_{SL} = \frac{V_{BE}}{R_4} \quad ...(11.12)$$

The minimum load resistance R_{min} below which regulation is lost may be calculated as under :

$$R_{min} = \frac{V_{reg}}{I_{SL}} \quad ...(11.13)$$

The practical value of R_{min} will be slightly greater or less than this.

11.11.2. Drawback of Simple Current Limiting

The drawback of constant current limiting is relatively large power dissipation in the series pass transistor when load terminals are shorted. Thus, a large power rating transistor is required.

The power dissipation in the series pass transistor is given by,

$$P_D = (V_{in} - V_{BE}) I_{SL} \quad ...(11.14)$$

where V_{BE} = Base-emitter voltage of Q_1

The circuit which is used in practice that eliminates the drawback of simple current limiting is called **foldback current limiting circuit**.

Such protection is provided in IC 723 with the help of resistance R_{SC} connected to CL (current limiting) pin. The drop across R_{SC} is applied across pins CL and CS. When this voltage becomes more than 0.6 V, it turns on internal current limiting transistor Q_3. The calculations of R_{SC} for IC 723 have been discussed earlier while discussing the applications of IC 723.

11.11.3. Foldback Current Limiting

1. Definition

The drawback of constant current limiting is relatively large power dissipation in the series pass transistor when the load terminals are shorted. Thus, a large power rating transistor is required. The foldback limiting technique allows us to provide the essential load current at rated voltage but reducing the short circuit current. Hence, the series pass transistor gets utilised efficiently.

2. Schematic Diagram

The basic circuit for foldback limiting has been shown in figure 11.24.

3. Mathematical Expressions

In figure 11.24, all the voltages are measured with respect to a common point.

Let the voltage at point A be V_A and the current flowing through R_4 is almost I_L.

Therefore,
$$V_A = I_L R_4 = V_{out} \quad ...(11.15)$$

Neglecting the base current of Q_3, the current flowing through R_5 and R_6 is same as I.

Therefore,

or
$$I = \frac{V_A}{R_5 + R_6} \quad ...(11.16)$$

Hence, the voltage at the base of Q_3 is the voltage across R_6.

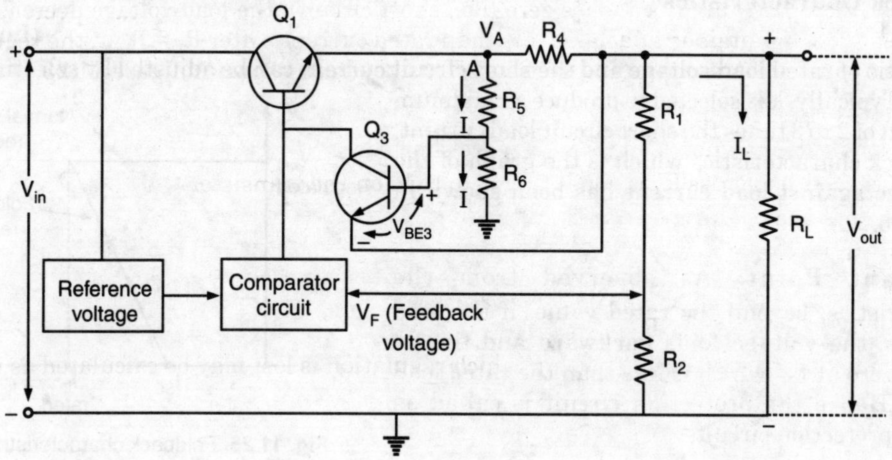

Fig. 11.24. Foldback current limiting

Therefore, $$V_{B3} = \frac{V_A}{R_5 + R_6} \cdot R_6 = (I_L R_4 + V_{out}) \left(\frac{R_6}{R_5 + R_6}\right) \quad ...(11.17)$$

Let $$k = \frac{R_6}{R_5 + R_6}$$

So that, we have $$V_{B3} = k(I_L R_4 + V_{out}) \quad ...(11.18)$$

The voltage at the emitter of Q_3 is given by

$$V_{E3} = V_{out} \quad ...(11.19)$$

or $$V_{BE3} = V_{B3} - V_{E3} = k(I_L R_4 + V_{out}) - V_{out}$$

or $$V_{BE3} = kI_L R_4 + (k - 1) V_{out}$$

or $$kI_1R_4 = V_{BE3} - (k - 10 V_{out}$$

or $$I_L = \frac{V_{BE3} + (1 - k)V_{out}}{kR_4} \quad ...(11.20)$$

Thus, if the output terminals are shorted, the output voltage V_{out} reduces to zero. Hence, we get from the equation (11.20).

$$I_{SC} = I_L = \frac{V_{BE3}}{k R_4} \quad ...(11.21)$$

The rated current may be written as under :

$$I_L = I_{SC} + \frac{(1 - k)V_{out}}{k R_4} \quad ...(11.22)$$

where I_L = Rated load current

I_{SC} = Short circuit current

The rated load current I_L is also called I_{knee} known as **knee current.**

It may be observed from the equation (11.22) that the rated load current is more than the short circuit current.

Important Point: Thus when the output terminals are shorted, the current decreases rather than increasing. Therefore, the series pass transistor Q_1 gets protected automatically.

4. Foldback Characteristics

(Important)

The ratio of rated load voltage and the short circuit current can be adjusted by selecting proper value of k. Typically, k is selected to produce a maximum load current of 2 to 3 times the short circuit load current. The foldback characteristics which is the graph of the load voltage against load current has been shown in figure 11.25.

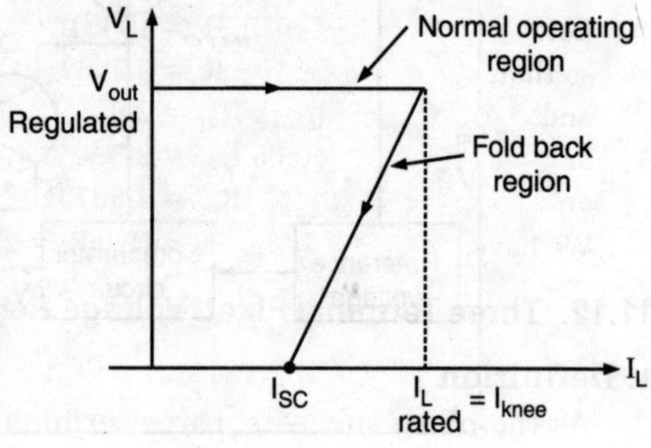

Fig. 11.25. Foldback characteristics

Important Point: As observed from the characteristics, beyond the rated value, if the load increases, the voltage folds backward and finally becomes zero at I_{SC} which is less than the rated load current. Hence the protection circuit is called as foldback protection circuit.

The foldback current limiting circuit design with IC 723 has been explained in the example below:

EXAMPLE 11.4. Design a regulator using IC 723, to have V_{out} = 6 V and the load current of 1 A. Use the foldback protection to get I_{sc} of 250 mA. (GATE Examination, 1996)

Solution: It is a low voltage less than 7 V and high current more than 150 mA, regulator. The circuit has been shown in figure 11.26.

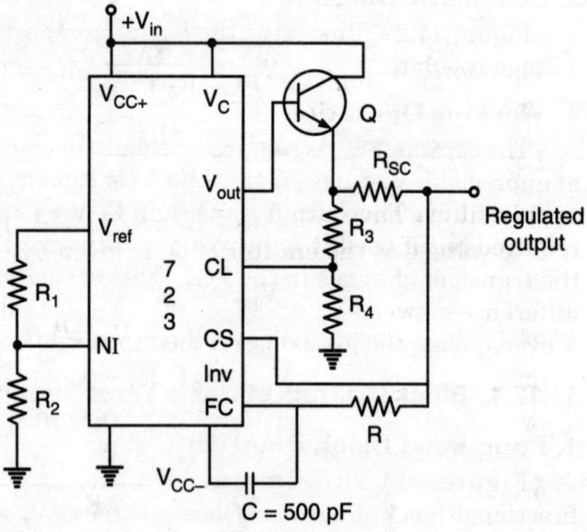

Fig. 11.26.

Now $\quad V_{out} = 6 \text{ V and } V_{ref} = 7 \text{ V}$

$$V_{out} = V_{ref}\left[\frac{R_2}{R_1 + R_2}\right]$$

or $$6 = \frac{7\,R_2}{R_1 + R_2}$$

or $$6R_1 + 6R_2 = 7\,R_2$$

or $$R_2 = 6\,R_1$$

Let $$R_1 = 2.7 \text{ k}\Omega$$

and $$R_2 = 16.2 \text{ k}\Omega$$

Now, let us calculate foldback circuit components R_3, R_4 and R_{SC}.

We have $$I_{SC} = \frac{V_{BE}}{k\,R_{SC}}$$

or $$250 \times 10^{-3} = \frac{0.7}{k\,R_{SC}}$$

or $$k\,R_{SC} = 2.8 \qquad \text{...(i)}$$

While I_L is given by, $$I_L = I_{SC} + \frac{(1-k)V_{out}}{k\,R_{SC}}$$

Substituting (i), we get

$$1 = 250 \times 10^{-3} + \frac{(1-k)\times 6}{2.8}$$

or $$(1-k)6 = 2.1$$

or $$1 - k = 0.35$$

or $$k = 0.65$$

Now, $k = \frac{R_4}{R_3 + R_4}$

or $R_3 = 0.35\ R_4$

Let $R_4 = 10\ k\Omega$

so that, $R_3 = 3.5\ k\Omega$

and $k\ R_{SC} = 2.8$

or $0.65\ R_{SC} = 2.8$

or $R_{SC} = 4.307\ \Omega$

We have, $R = R_1 \| R_2 = 2.31\ k\Omega$. **Ans.**

11.12. Three Terminal Fixed Voltage Regulators

1. Definition

As the name suggests, three terminal voltage regulators have three terminals namely input which is unregulated (V_{in}), regulated output (V_{out}) and common or a ground terminal. These regulators do not require any feedback connections.

2. Schematic Diagram

Figure 11.27 illustrates the basic three terminal voltage regulator.

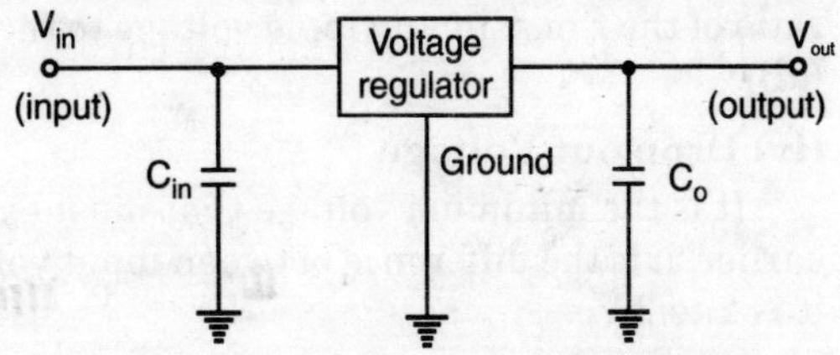

Fig. 11.27. Basic three terminal voltage

3. Working Operation

The capacitor C_{in} is required if regulator is located at appreciable distance, more than 5 cm from a power supply filter. The output capacitor C_o may not be required but if used it improves the transient response of the regulator i.e. regulator response to the transient changes in the load. This capacitor also reduces the noise present at the output. The difference between V_{in} and V_{out} ($V_{in} - V_{out}$) is called as drop out voltage and it must be typically 2.0 V even during the low point on the input ripple voltage, for the proper functioning of the regulator.

11.12.1. Block Diagram of Basic Three Terminal IC Regulator

1. Functional Block Diagram

Figure 11.28 illustrates the functional block diagram of basic three terminal IC regulator.

2. Working Operation

This is basically a series voltage regulator circuit. A part of output voltage is taken with the help of potential divider formed by R_1 and R_2. This is compared with reference voltage, V_{ref}, internally generated with the help of zener diode. After comparison, a control signal is generated which is applied through protective circuit to the series pass transistor working as control element. This element works as a variable resistance. The control signal adjusts the control element in such a way that output voltage remains constant.

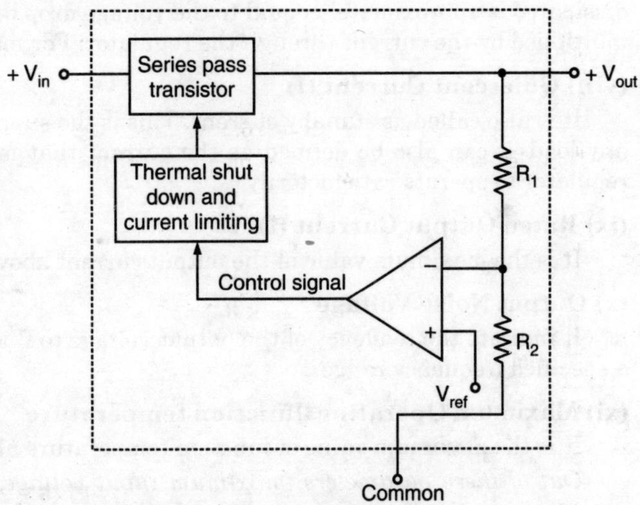

Fig. 11.28. Functional block diagram of a typical three terminal IC voltage regulator

Thermal shut down means that the chip will automatically turn itself off if the internal temperature exceeds, typically, 175°C. The current limiting circuit will protect the chip from excessive load current. Because of the thermal shut down and current limiting, the IC voltage regulator chip is almost indestructible.

11.12.2. Data Sheet Specifications of IC Linear Regulators

A number of performance parameters are generally specified for the IC voltage regulators by the manufacturers. These are as under:

(i) Line Regulation

It is defined as the change in the output voltage for a given change in the input voltage. It is expressed as a percentage of output voltage or in millivolts.

(ii) Load Regulation

As defined earlier, it is the change in output voltage over a given range of load currents i.e. from full load to no load. It is usually expressed in millivolts or as a percentage of output voltage.

(iii) Ripple Rejection

It indicates regulator's ability to reject ripple voltage present in the input. It is defined as the ratio of the r.m.s. input ripple voltage to the r.m.s. output ripple voltage. It is expressed in decibels (dB).

(iv) Drop out Voltage

It is the minimum voltage that must exist between input and output terminals. As mentioned earlier, it is the difference between input voltage V and output voltage V. For most of the regulators it is 2 to 3 V.

(v) Output Resistance (R)

It is the rate of change of output voltage with respect to the output current. It should be as small as possible.

(vi) Maximum Input Voltage ($V_{in\,max}$)

This is the maximum input voltage that can be applied to the regulator safely.

(vii) Maximum Power Dissipation (P_{Dmax})

This is the maximum power which regulator can dissipate without damage. The actual power dissipated is approximately equal to the voltage drop between the input and the output terminals multiplied by the current through the regulator. For many regulators it is internally limited.

(viii) Quiescent Current (I)

It is also called as standby current. This is the supply current drawn by the regulator without any load. it can also be defined as the current that must flow from the ground terminal of the regulator to operate satisfactorily.

(ix) Rated Output Current (I)

It is the maximum value of the output current above which current limiting occurs.

(x) Output Noise Voltage

It indicate the tendency of the output voltage to fluctuate above its prescribed d.c. value, over a specified frequency range.

(xi) Maximum Operating Junction temperature

It is the maximum value of junction temperature above which thermal shutdown occurs.

Out of these parameters, maximum input voltage, maximum power dissipation, maximum operating junction temperature etc. are called as absolute maximum ratings of IC regulator, while other ratings are called as electrical characteristics of IC regulator.

11.12.3. IC Series of Three Terminal Fixed Voltage Regulators

1. Definition

The popular IC series of three terminal regulators is μA 78XX and μA 79XX. The series μA 78XX is the series of three terminal positive voltage regulators while μA 79XX is the series of three terminal negative voltage regulators. The last two digits denoted as XX, indicate the output voltage of the IC.

Such series is available with seven voltage options indicated in Table 11.4.

Table 11.4.

S.No.	Device type	Output voltage	Device type	Output voltage
1.	7805	5.0 V	7905	–5.0 V
2.	7806	6.0 V	7906	–6.0 V
3.	7808	8.0 V	7908	–8.0 V
4.	7812	12.0 V	7912	–12.0 V
5.	7815	15.0 V	7915	–15.0 V
6.	7818	15.0 V	7918	–18.0 V
7.	7824	24.0 V	7924	–24.0 V

2. Packages

The 79XX series voltage regulators are available with same seven options as 78XX series as indicated in table 11.4. In addition, two extra voltages –2 V and –5.2 V are also available with ICs 7902 and 7905.2 respectively.

These ICs are provided with adequate heat sinking and can deliver output currents more than 1 A. These ICs do not require external components These are provided with internal protection, overload and short circuit protection.

The two series are available in various versions like low power and high power versions. The low power versions are available in plastic or metal packages, like small signal transistors. The higher power versions are packaged in TO–3 type metal cans or in TO–220 type moulded plastic packages like power transistors.

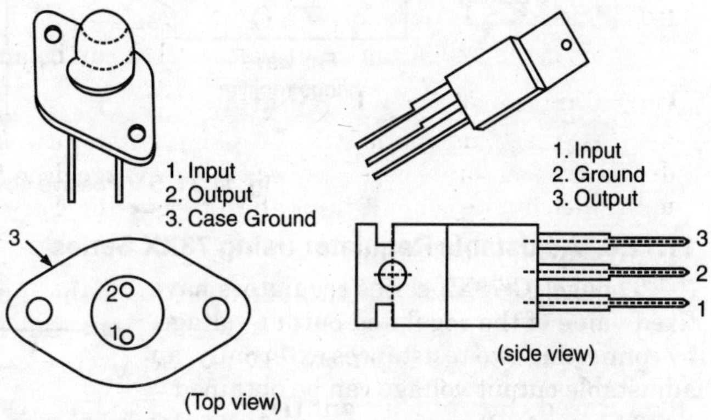

(a) TO-3 type metal package (b) TO-220 type plastic package

Fig. 11.29.

These have been shown in figure 11.29(a) and (b).

11.12.4. Typical Connection of IC 7805 Regulator

Figure 11.30 shows the typical connection of μA 7805 regulator The maximum input voltage which can be applied to IC 7805 is 35 V. The maximum dropout voltage is 2.5 V. The output of 7805 regulator is 5 V fixed. Hence minimum input voltage required is 5 + 2.5 = 7.5. So, input range of IC 7805 regulator is 7.5 V to 35 V to get fixed 5 V as regulated output. This has been illustrated in figure 11.30.

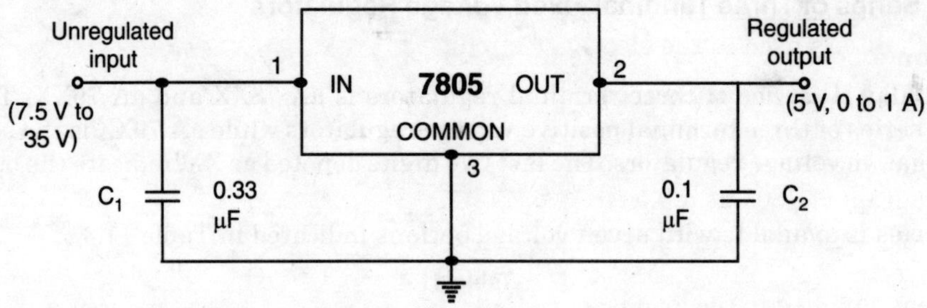

Fig. 11.30. Typical circuit connection of µA 7805 regulator

11.12.5. Positive 5V Power Supply using IC 7805

A 5 V output voltage supply system using full wave bridge rectifier, capacitor filter and IC regulator 7805 has been shown in figure 11.31. The a.c. line voltage is 230 V which is stepped down to 15 V using a transformer. A full wave rectifier alongwith the capacitor voltage provides the unregulated voltage input to IC 7805 regulator. This input contains a.c. ripple of few volts. The IC 7805 regulator provides the regulated output of 5 V.

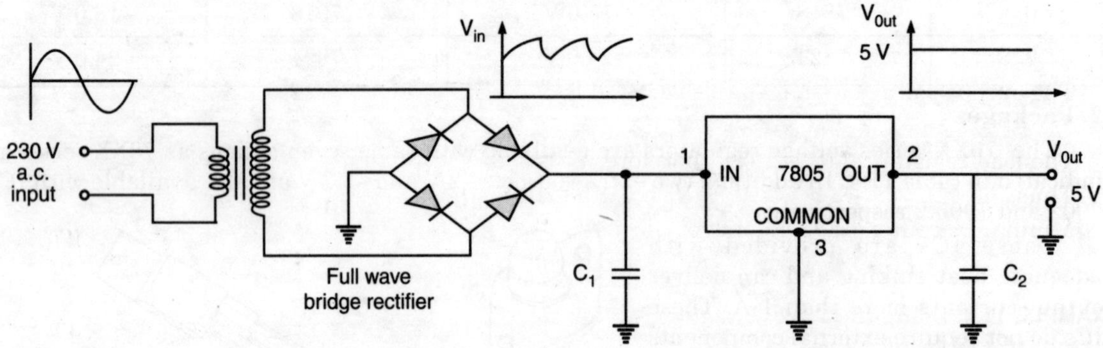

Fig. 11.31. 5 V Positive power supply

11.12.6. Adjustable Regulator using 78XX Series

Though IC 78XX series regulators have fixed value of the regulated output voltage, by connecting two resistances externally, an adjustable output voltage can be obtained.

The typical connection of 78XX IC regulator to obtain variable output voltage has been shown in figure 11.32.

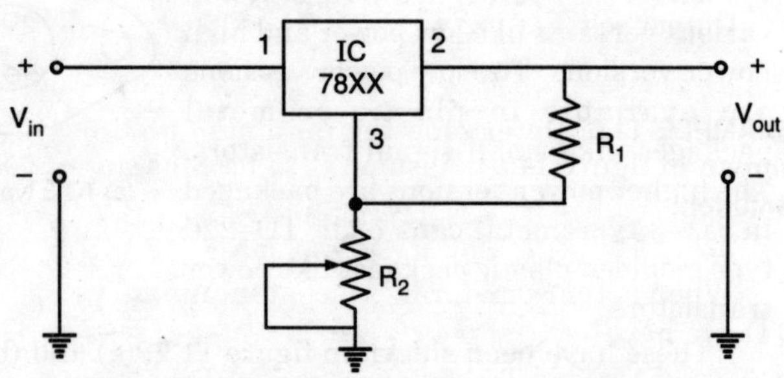

Fig. 11.32. Adjustable regulator using IC 78XX

$$V_{out} = V_{reg}\left[\frac{R_2}{R_1 + R_2}\right]$$

where

V_{reg} = Regulated fixed voltage of IC

By varying R_2, variable output voltage can be obtained.

11.12.7. Applications of IC 78XX and 79XX **(Very Important)**

These ICs are regulator ICs and are basically used to provide constant d.c. voltage to various components in complex electronic circuits. The IC 7805 is typically used to provide constant 5 V supply to the digital circuits. The IC 7812 and 7912 are used to provide dual supply of ± 12 V to operational amplifiers used in the electronic circuits.

The 7805 as a 0.5 A Current Source

The 7800 regulators can also be employed as current sources. A typical connection diagram of 7805 as a 0.5 A current source is depicted in figure 11.33.

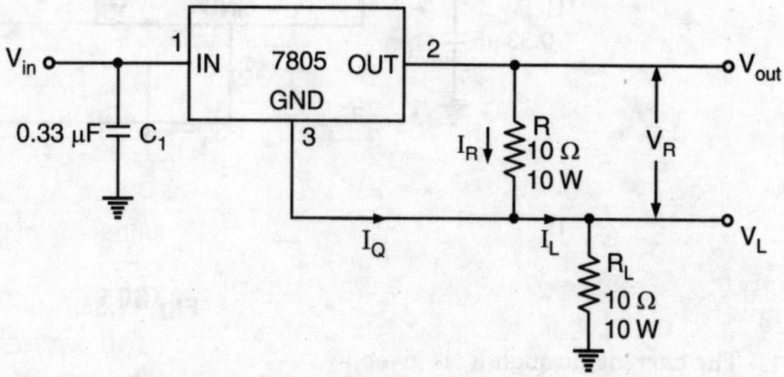

Fig. 11.33. The 7805 as a 0.5 A current source

The current supplied to the load is given by

$$I_L = \frac{V_R}{R} + I_Q$$

where I_Q is quiescent current in amperes (4.3 mA typically for the 7805 IC)

In figure 11.33, we have

$$V_R = V_{23} = 5\text{ V and } R = 10\ \Omega$$

So, $$I_L = \frac{5}{10} = 0.5$$

The output voltage with respect to ground is

$$V_{out} = V_R + V_L$$

The load resistance, $R_L = 10\ \Omega$, therefore, $V_L = 5$ V

Thus, $$V_{out} = V_R + V_L = 5 + 5 = 10\text{ V}$$

Minimum input voltage required,

$$V_{in} = V_{out} + \text{ drop out voltage } 10 + 2 = 12\text{ V} \quad \textbf{Ans.}$$

EXAMPLE 11.5. Using the 7812 voltage regulator, design a current source that will deliver a 0.5 A current to a 25 Ω, 10 W load.

Solution: According to figure 11.33, the resistance required R will be

$$R = \frac{V_R}{I_L} = \frac{12}{0.5} = 24\ \Omega \qquad \text{(neglecting quiescent current } I_Q)$$

Output voltage, $V_{out} = V_R + V_L = V_R + I_L R_L = 12 + 0.5 \times 25 = 24.5$ V

Input voltage $= V_{out} + 2 = 24.5 + 2 = 26.5$ **Ans.**

EXAMPLE 11.6. Determine the minimum and maximum output voltage of the voltage regulator shown in figure 11.34. Assume I_Q to be constant at 10 mA.

Solution: $V_{REG} = 15$ V

$I_Q = 10$ mA

When potentiometer R_2 is set at minimum i.e., at 0 Ω, we have

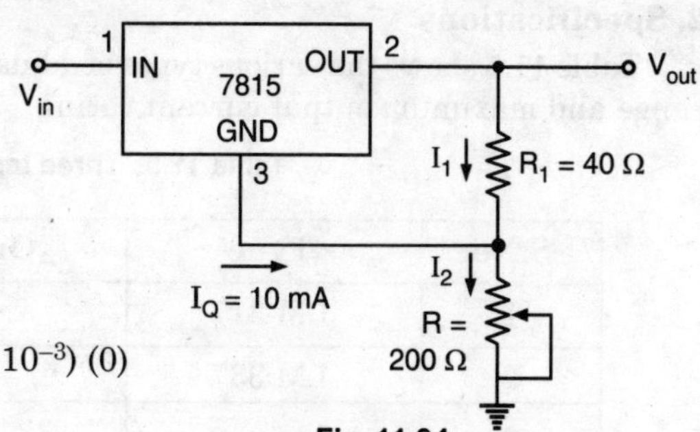

Fig. 11.34.

Output voltage, $$V_{out} = \left(1 + \frac{R_2}{R_1}\right) V_{REG} + I_Q R_2$$

$$V_{out} = \left(1 + \frac{0}{40}\right) \times 15 + (10 \times 10^{-3})\ (0)$$

$$V_{out} = 15\text{ V} \quad \textbf{Ans.}$$

EXAMPLE 11.7. Using 7805, design a current source to deliver 0.2 A current to a 22Ω, 10 W load.

Solution: Figure 11.35 shows the required circuit diagram.

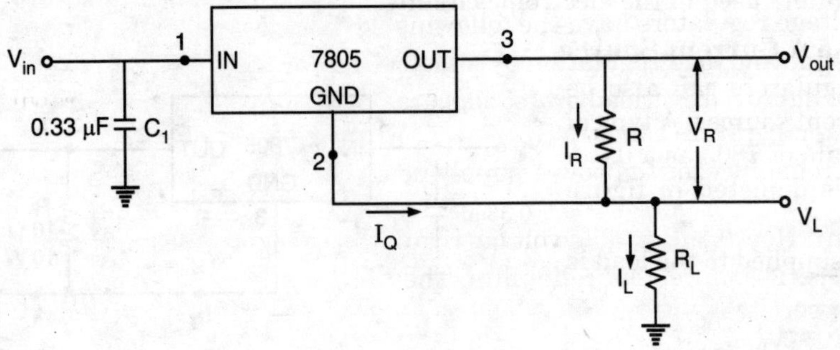

Fig. 11.35.

The current through R_L is given by,

$$I_L = I_R + I_Q = \frac{V_R}{R} + I_Q$$

Now $\quad I_Q = 4.2$ mA for 7805 IC.

and $\quad V_R = 5$ V for 7805 IC.

Therefore, we have $\quad I_L = \frac{5}{R} + 4.2 \times 10^{-3}$ $\quad$ (Here, we have not neglected I_Q)

or $\quad 0.2 = \frac{5}{R} + 4.2 \times 10^{-3}$

or $\quad R = 25.33\ \Omega$

Now $\quad V_{out} = V_R + R_L = 5 + I_L R_L = 5 + 0.2 \times 22 = 9.4$ V

Therefore, we have $\quad V_{in} = V_{out}$ + drop out voltage = 9.4 + 2 = 11.4 V $\quad$ **Ans.**

11.13. THREE TERMINAL ADJUSTABLE REGULATOR : LM 317

1. Definition

Besides fixed voltage regulators, IC voltage regulators are available which allow the adjustment of the output voltage. The output voltage can be adjusted from 1.2 V to as high as 57 V with the help of such regulators. In such regulator ICs, the common terminal plays the role of control input and hence called as ADJUSTMENT (ADJ) terminal. The LM 317 series is the most commonly used three terminal adjustable regulators. These devices are available in a variety of packages which can be easily mounted and handled. The power rating of such regulators is 1.5 A. The maximum input voltage of LM 317 is 40 V.

2. Specifications

Table 11.5 shows the various types of adjustable regulators and the corresponding output voltage range and maximum output current rating.

Table 11.5. Three terminal adjustable regulators

S.No.	Type	Output voltage	Output current
1.	LM 317	+ 1.2 to 57 V	1.5 A
2.	LM 337	– 1.2 to – 47 V	1.5 A
3.	LM 358	+ 1.2 to 32 V	5 A
4.	LM 350	+ 1.2 to 32 V	3 A
5.	LM 396	+ 1.25 to 15 V	10 A

3. Salient Features

Adjustable voltage regulators have the following advantages over fixed voltage regulators :

(i) Improved line and load regulation by a factor of 10 or more

(ii) Because of improved overload protection, greater load current can be drawn over the given operating temperature range.

(iii) Improved reliability for the power supply using these regulators.

Important Point: Hence, adjustable voltage regulators have become more popular because of their versatility, performance, and reliability. The LM 317 is the most widely used adjustable voltage regulator series.

11.13.1. Connection Diagram of LM 317 Regulator

1. Definition and Schematic Diagram

The typical connection diagram of LM 317 regulator has been shown in figure 11.36. The LM 317 requires only two external resistances R_1 and R_2, to set the required output voltage. Internally, it develops a reference voltage of 1.25 V between OUT and ADJ terminals, which is denoted as V_{REF}. This voltage is impressed across the resistance R_1. Hence, for constant V_{REF} and R_1, the current I_{R1} is also constant. Hence, the resistance R_1 which sets the current I_{R1} is called **current set resistor or program resistor.**

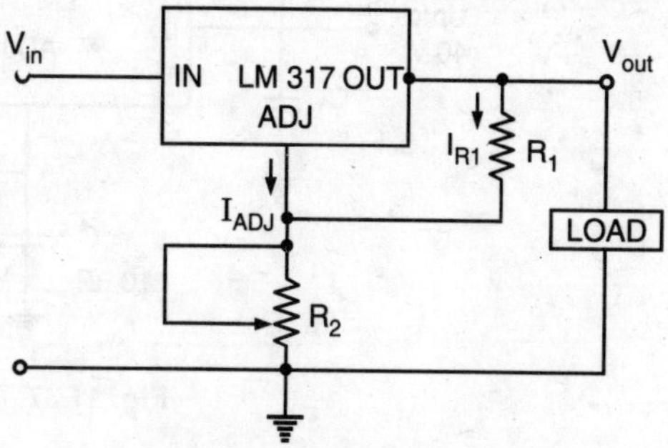

Fig. 11.36. The Schematic diagram of LM 317 regulator

2. Working Operation

The current I_{ADJ} flows out from the ADJ terminal and is similar to the quiescent current I_Q in the fixed voltage regulators. This current, along with I_{R1}, flow through the resistance R_2. The design of LM 317 is such that the current I_{ADJ} is very small and remains constant with line and load changes. The maximum value of I_{ADJ} is 100 μA which is very small.

Applying KVL, we can write

$$V_{out} = V_{R1} + V_{R2}$$

or $$V_{out} = I_{R1} R_1 + (I_{R1} + I_{ADJ}) R_2 \quad ...(11.23)$$

Now $$I_{R1} = \frac{V_{REF}}{R_1} = \frac{1.25}{R_1} \quad ...(11.24)$$

Therefore, we have

or $$V_{out} = 1.25 + \frac{1.25}{R_1} . R_2 + I_{ADJ} R_2$$

or $$V_{out} = 1.25 \frac{(R_1 + R_2)}{R_1} + I_{ADJ} R_2 \quad ...(11.25)$$

The current I_{ADJ} is very small and hence the drop $I_{ADJ} R_2$ is also very small and can be neglected.

Therefore, $$V_{out} = 1.25 \left[1 + \frac{R_2}{R_1}\right] \quad ...(11.26)$$

Thus, the output voltage is a function of R_1 and R_2 Keeping R_1 fixed and varying R_2, the output voltage can be adjusted. The program resistor R_1 is generally 240 Ω.

Important Point: Generally, no capacitors are required. But, if LM 317 is located far from the power supply filter then the capacitors C_1 and C_2 are required. The input capacitor C_1 is generally 0.1μF disc or 1μF tantalum while the output capacitor C_2 is in the range of 1 to 1000 μF. To get high ripple rejection, the capacitor C_3 is used. To avoid the discharging of the capacitors through low current points into the regulator, the protecting diodes are used. Figure 11.37 shows the functional diagram of LM 317 alongwith the capacitors and the protecting diodes.

Important Point: The diodes are essential if output voltage is higher than 25 V .

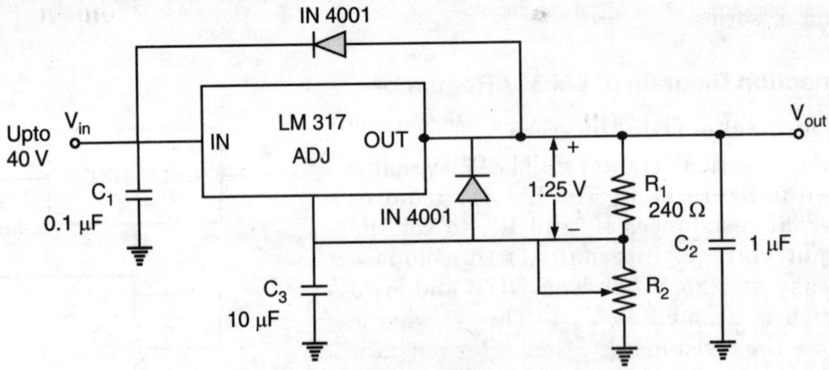

Fig. 11.37. Functional diagram of LM 317

11.13.2. Practical Regulator using LM 317

The practical regulator using LM 317 has been shown in figure 11.38. Filter capacitors are necessary only if LM 317 is situated far away from the filter capacitor of unregulated input power supply. If LM 317 is far away from input power supply, then, a 0.1 μF disc type of 1 μF-tantalum capacitor should be used at the inpuit of LM 317 as shown in figure 11.38. The ouput capacitor C_o is optional. We can connect C_o to improve the transient response, C_o should be in the range of 1 to 1000 μF.

The adjustment terminal is bypassed with a capacitor C_2 as shown in figure 11.38. This will improve the ripple rejection capacity to a great extent. With $C_2 = 10$ μF, the ripple rejection ratio as high as 80 dB is obtainable at any output level. When the filter capacitors are used, it is necessary to use the protective diodes as shown in figure 11.38. These diodes do not allow the filter capacitors to discharge through the low current points of the regulator. These diodes are required only for high output voltages (above 25 V) and for higher values of output capacitance (25 μF and above)

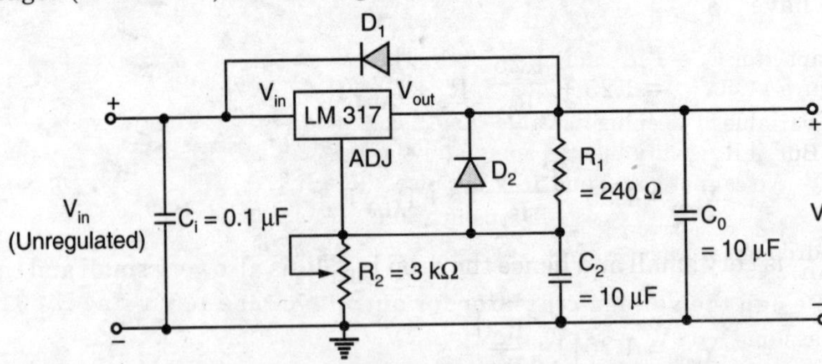

Fig. 11.38. Practical LM 317 circuit with capacitors and protective diodes

11.13.3. Adjustable Negative Voltage Regulators (LM 337 Series)

The LM 337 series of adjustable voltage regulators provides an adjustable negative output voltage. This series is a complement to the LM 317 series devices. The LM 337 devices are available in the same voltage and current ratings as LM 317 series. This is shown in table 11.6. Different types of packages available for LM 337 series have been shown in figure 11.39.

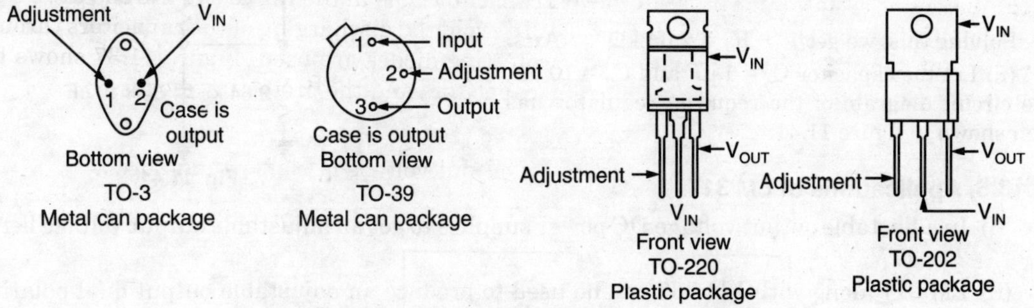

Fig. 11.39. Different standard packages for LM 337 series

Table 11.6 Adjustable negative voltage regulators

S.No.	Device	Output voltage (volts)	Output current (Amp)	Maximum V_{in}
1.	LM 337	– 1.2 to – 37	1.5	40
2.	LM 337 H	– 1.2 to – 37	0.5	40
3.	LM 337 HV	– 1.2 to – 47	1.5	50
4.	LM 337 HVH	– 1.2 to – 47	0.5	50
5.	LM 337 LZ	– 1.2 to – 37	0.10	40
6.	LM 337 M	– 1.2 to – 37	0.5	40

11.13.4. Examples on Adjustable Voltage Regulators

EXAMPLE 11.8. Calculate the component values for LM 317 to get output voltage of 12 Volts, with input of 20 Volts. If the output is made variable, calculate the range of output voltage.

Solution: Given : $V_{in} = 20\text{ V}, \quad V_o = 12\text{ V}$

(i) The output voltage $V_0 = 1.25\left[1+\frac{R_2}{R_1}\right]$

Let us assume $R_1 = 240\ \Omega$

Therefore $12 = 1.25\left[\frac{240+R_2}{240}\right]$

or $R_2 = 2064\ \Omega$ or $2.064\text{ k}\Omega$ **Ans.**

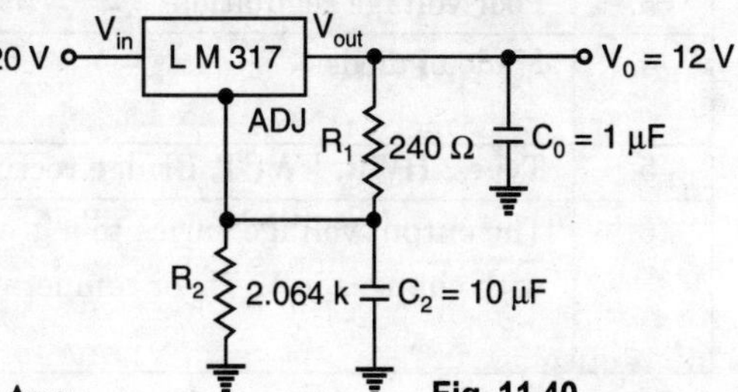

Fig. 11.40

(ii) Let the capacitor $C_0 = 1\ \mu\text{F}$ and $C_2 = 10\ \mu\text{F}$. The circuit diagram of the required regulator circuit has been shown in figure 11.40.

If R_2 is made variable by keeping its value same, i.e., 2064 Ω, then the output voltage will vary in the range of 1.25 V to 12 V. But, if R_2 is variable and adequately larger than 2064 Ω, then we can vary the output voltage from 1.25 V to 18V. We cannot go beyond 18 V because

$$V_{o\,(max)} = V_{in} - \text{Drop out voltage}$$

and drop out voltage = 2 V

EXAMPLE 11.9. Design the voltage regulator for output voltage 15 V using LM 317.

Solution: Let us assume $V_{in} = 20\text{ V}$ and $V_o = 15\text{ V}$

(i) The output voltage $V_o = 1.25\left[1+\frac{R_2}{R_1}\right]$

We assume $R_1 = 240\ \Omega$

Therefore, $15 = 1.25\left[1 + \frac{R_2}{240}\right]$

Solving this, we get $R_2 = 2.64\ k\Omega$ **Ans.**

(ii) Let the capacitor $C_o = 1\ \mu F$ and $C_2 = 10\ \mu F$. The circuit diagram of the required regulator has been shown in figure 11.41.

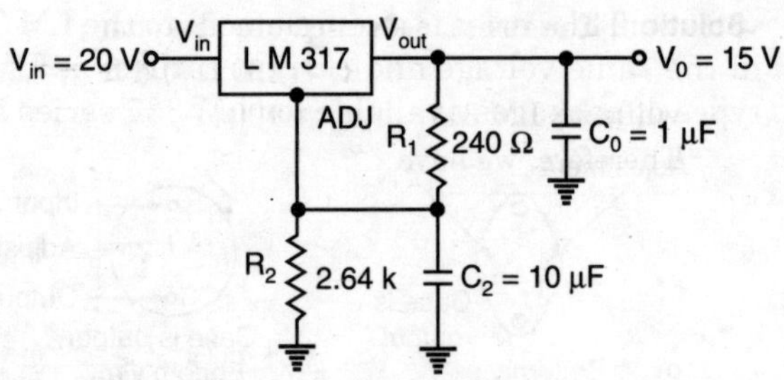

Fig. 11.41.

11.13.5. Applications of LM 317

(i) In adjustable output voltage DC power supplies to get an adjustable output voltage between 2 to 37 V.

(ii) LM 317 along with LM 337 can be used to produce an adjustable output dual polarity (±) regulated dc power supply.

11.14. PERFORMANCE COMPARISONS

11.14.1. Performance Comparison of Rectifier and Regulator

S.No.	Rectifier	Regulator
1.	Its function is to convert ac input to dc output	Its function is to maintain the output voltage constant
2.	Output voltage waveform is pulsating dc 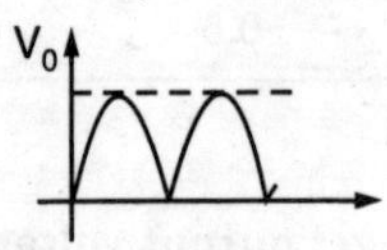	Output voltage waveform is pure dc 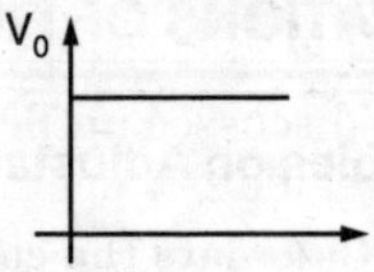
3.	Poor voltage regulation	Very good voltage regulation
4.	Made of diodes	Contains devices such as transistor, zener diode
5.	Types: HWR, FWCT, Bridge rectifiers	Types: Series or shunt regulators
6.	The output voltage varies to a great extent with changes in V_{in}, I_L or temperature.	The output voltage remains constant irrespective of changes in V_{in}, I_L and temperature

11.14.2. Performance Comparison of 78XX and LM 317 Regulators

S.No.	78XX Regulators	LM 317 Regulator
1.	78XX is a fixed positive voltage regulator series	LM 317 is an adjustable positive voltage regulator series
2.	The voltage available in this series are 5, 6, 8, 12, 15, 18 and 24 Volts	Output voltage can be adjusted between 2 to 27 Volts
3.	Maximum output current is 1 A	Maximum output current is 1.5 A

EXAMPLE 11.10. Determine the regulated output voltage for the LM 317 voltage regulator shown in figure 11.42.

Solution: The resistances used are as under :

$$R_1 = 220\ \Omega \text{ and } R_2 = 1.5\ k\Omega$$

while for LM 317, $I_{ADJ} = 100\ \mu A$

Therefore, we have

$$V_{out} = 1.25\left[1+\frac{R_2}{R_1}\right] + I_{ADJ} R_2$$

or,
$$V_{out} = 1.25\left(1+\frac{1.5\times10^3}{220}\right) + 100 \times 10^{-6} \times 1.5 \times 10^3$$

or,
$$V_{out} = 9.92 \text{ V} \quad \textbf{Ans.}$$

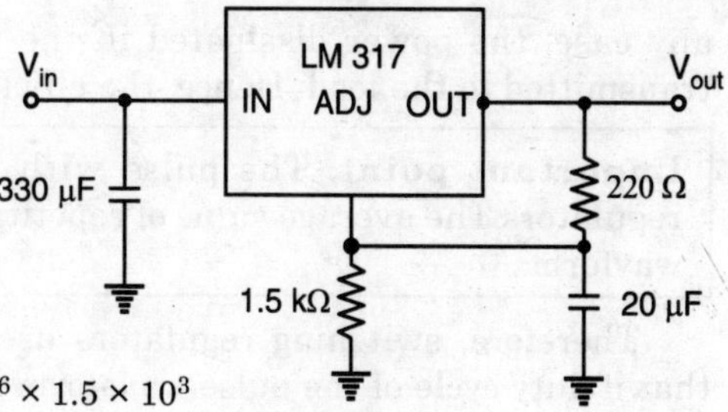

Fig. 11.42.

EXAMPLE 11.11. Determine the range in which output voltage can be varied with the help of LM 317 regulator using R_1 = 820 Ω and R_2 as 10 kΩ potentiometer.

Solution: For LM 317, the current I_{ADJ} = 100 μA, we have

When R_2 is minimum i.e. R_2 = 0 then

$$V_{out} = 1.25\left[1+\frac{R_2}{R_1}\right] + I_{ADJ} R_2 = 1.25 \text{ V}$$

When R_2 is maximum, i.e. R_2 = 10 kΩ, then,

$$V_{out} = 1.25\left(1+\frac{10\times10^3}{820}\right) + 100 \times 10^{-6}\ 10 \times 10^3 = 17.49 \text{ V} \quad \textbf{Ans.}$$

Thus, the output can be varied in the range 1.25 V to 17.49 V

11.15. LIMITATIONS OF LINEAR VOLTAGE REGULATORS

The regulators discussed uptill now are called as linear regulators because the series pass transistor used in such regulators operates in the region. But such linear regulated power supplies have following limitations :

(i) The required input step down transformer is bulky and expensive.
(ii) Because of low line frequency (50 Hz), large values of filter capacitors are required.
(iii) The efficiency is very low.
(iv) Input must be greater than the output voltage.
(v) As large is the difference between input and output voltage, more is the power dissipation in in the series pass transistor.
(vi) For higher input voltages, efficiency decreases.
(vii) The need for dual supply, is not economical and feasible to achieve with the help of linear regulators.

The switching regulators overcome all these limitations.

11.16. BASIC SWITCHING REGULATOR

1. Definition and Important Aspects

The operating principle of switching regulators is completely different than that of linear regulators. The switching regulator are also called as switched mode regulators. Such a switching regulator requires an external transistor and a choke. The series pass transistor in such a regulator is used as a controlled switch and is operated in cut-off region or saturation region. Hence, the power transmitted across such a transistor is in the form of discrete pulses rather than a steady flow of current. When the transistor is operated in the cut-off region, there is no current and dissipates no power While when it is operated in the saturation region, a negligible voltage drop appears across it and hence dissipates very small power, providing maximum current to load. In any case, the power dissipated in the transistor is very small. Almost the entire power gets transmitted to the load. Hence, the efficiency of the switching regulators is always very high.

Important point: The pulse with modulation is the basic principle of the switching regulators.The average value of repetitive pulse waveform is proportional to the area under the wavform.

Therefore, switching regulators use the fact that if duty cycle of the pulse waveform is varied, the average value of the voltage also changes proportionally.

Important Point: The duty cycle of the pulse waveform is the ratio of the ON time t_{on} to the period T of the pulse waveform.

Mathematically, it can be expressed as under:

$$\text{duty cycle } \delta = \frac{t_{on}}{t_{on} + t_{off}}$$

or $$\delta = \frac{t_{on}}{T} = t_{on} f$$

when t_{on} = on time of pulse

t_{off} = off time of pulse

$$T = \text{time period} = t_{on} + t_{off} = \frac{1}{f}$$

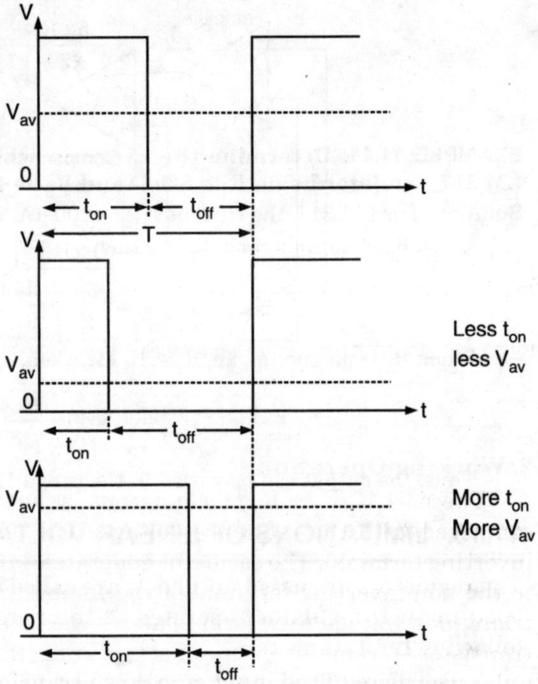

Fig. 11.43. Pulse width modulation

2. Waveforms

This basic pulse width modulation has been shown in the figure 11.43.

3. Major Elements

The basic switching regulator consists of four major components :

(a) Voltage source V_{in}

(b) Switching transistor

(c) Pulse generator, V_{pulse}

(d) Filter F_1

These blocks are connected together as shown in figure 11.44, to obtain the switching regulator.

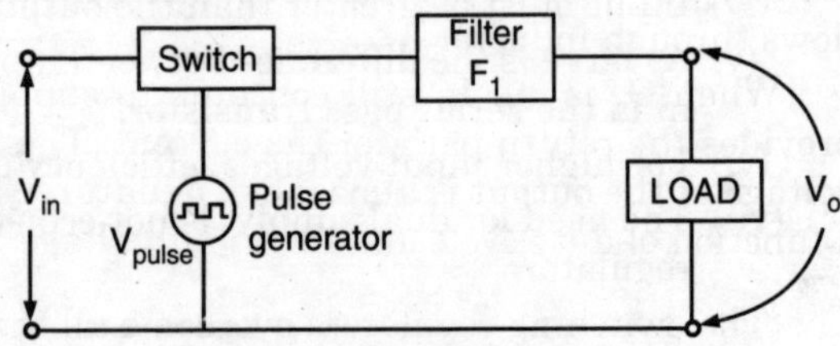

Fig. 11.44. Basic switching regulator

A voltage source V_{in} is a d.c. supply which is a battery, unregulated or regulated voltage.

4. Basic Requirements

It has to satisfy the following requirements :

(i) It has to supply required power and the losses associated with the regulator.

(ii) It must be high to satisfy the minimum requirements of the regulator.

(iii) It must be large to supply sufficient dynamic range of line and load changes.

Important Point: The switch is generally a transistor. The pulse generator output makes it ON and OFF. The pulse generator produces a required pulse waveform. The most effective range of pulse waveform frequency is 20 kHz. The typical operating frequency range is 10 to 50 kHz. The filter F_1 may be RC, RL or RLC Most commonly used filter is RLC. It converts the pulse waveforms obtained from the switch into a d.c. output voltage.

11.17. BLOCK DIAGRAM OF SWITCHED MODE POWER SUPPLY (SMPS)

1. Functional Block Diagram

Figure 11.45 shows the functional block diagram of basic switching voltage regulator, which uses transistor Q_1 as a switch.

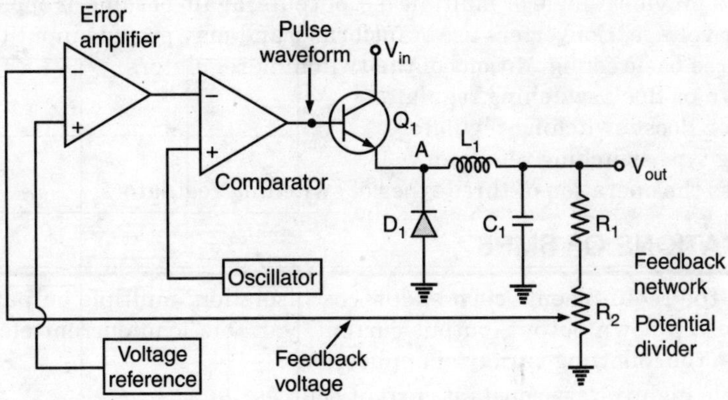

Fig. 11.45. Functional block diagram of switching regulator

2. Working Operation

The part $R_2/R_1 + R_2$ of the output is feedback to the inverting input of error amplifier. It is compared with the reference voltage. The difference is amplified and given to the comparator inverting terminal. The oscillator generates a triangular waveform at a fixed frequency. It is applied to the non-inverting terminal of the comparator. The output of the comparator is high when the triangular voltage waveform is above the level of the error amplifier output Because of this, the transistor Q_1 remains in cut-off state. Thus, the output of the comparator is nothing but a required pulse waveform. The period of this pulse waveform is same as that of oscillator output say T. The duty cycle is denoted as $\delta = t_{on}/T$ or t_{on} f as discussed earlier. This duty cycle is controlled by the difference between the feedback voltage and the reference voltage. When Q_1 is on in saturation state, $V_{CE\,(sat)}$ for Q_1 is zero. Hence, entire input voltage V_{in} appears at point A. Thus, the current flows through inductor L_1.

When Q_1 is off, L_1 still continue to supply current through itself to the load. The diode D_1 provides the return path for the current. The capacitor C_1 acts to smooth out the voltage and the voltage at the output is almost d.c. in nature. The output voltage V_{out} of the switching regulator is a function of duty cycle and the input voltage V_{in} Mathematically, it is expressed as under :

$$V_{out} = \frac{t_{on}}{T} V_{in} = d\,V_{in} \qquad ...(11.34)$$

Thus when T is constant, output is proportional to t_{on} This method is called as **pulse width modulation (PWM).** When t_{on} is constant, the output is inversely proportional to the period T i.e. proportional to frequency of the pulse waveform. This method is **called frequency modulation**.

Important Point: The PWM technique is commonly used though it is more complex as it is most suitable for the high current applications.

Important Point: A high switching frequency allows small values of L_1 and C_1 and thus reduces size, cost and weight. It also reduces the ripple at the output. But, the efficiency decreases and electrical noise increases. On the other hand, low switching frequency improve efficiency and reduce noise but require large filtering components. As a result of this, the range of operating frequency to get maximum efficiency, is 10 to 50 kHz.

11.18. TYPES OF SWITCHING REGULATORS

The term **switched mode regulator** is used to describe a circuit which takes d.c. input (unregulated) and provides a single d.c. output of the same or opposite polarity, and of a lower or higher voltage. The switched mode regulators use an inductor and there is no input to output isolation.

The term **switched mode converter** is used to describe a circuit which takes d.c. inpuit (unregulated) and provides single or multiple d.c. outputs, again of same or opposite polarity and of a lower or higher voltage. Converters use transformer and may provide input to output isolation.

There are three basic configurations of the switching regulators.

(i) Step down or Buck switching regulator
(ii) Step up or Boost switching regulator
(iii) Inverting type switching regulator.

Let us discuss the operation of three types of switching regulators.

11.19. APPLICATIONS OF SMPS

Depending upon the requirements such as low cost, isolation, multiple outputs, step up output, inverted output, step down output, output current, variable load current etc., various types of SMPS are used in the following variety of applications:

(i) Adjustable high voltage constant current sources.
(ii) Battery powered systems.
(iii) Telecommunication circuits.
(iv) Personal computers.
(v) Printers.
(vi) Video games.
(vii) Motor and industrial control systems.
(viii) Automotive applications.

11.19.1. Performance Comparison of SMPS and Linear Regulators

S.No.	Linear Regulator	Switched Mode Regulator
1.	Series pass transistor used acts in active region.	Series pass transistor used acts either in cut-off region or saturation region.
2.	Series pass transistor does not act as a switch.	Series pass transistor acts as a switch.
3.	Power transmitted across a transistor is in linear fashion.	Power transmitted across a transistor is in the form of discrete pulses.
4.	Step down transformer is bulky and expensive.	Small size transformer is the feature.
5.	Weight is high.	It is light in weight.
6.	Cost is low.	Cost is high.
7.	Simple to design	Complex to design.
8.	Efficiency is low.	Efficiency is very high.
9.	Ripples are less at the output.	Ripples are more at the output.
10.	Due to low line frequency, large filter capacitors are required.	High switching frequency allows small values of choke and capacitor.
11.	Power handling capacity is low.	Power handling capacity is high.
12.	No switching losses.	Switching losses are high
13.	Response to load variations is fast.	Response to load variations is slow.

11.19.2. Advantages and Disadvantages of SMPS

Advantages

(i) Low power dissipation in the series pass transistor as it operates as a switch and not in the active region.

(ii) High efficiency (upto 95%) due to reduced power dissipation in the transistor.

(iii) Small size : This is due to the smaller size of L and C at high operating frequencies and need of smaller heat sink of the series pass transistor.

(iv) Higher power handling capacity.

Disadvantages

(i) Increased switching loss in the series pass transistor due to high frequency switching.

(ii) Radio frequency interference (RFI) to the neighbouring electronic circuits.

(iii) There is no isolation between input and output.

(iv) The load requires separate protection circuitry.

(v) The transient response is slow as compared to the linear power supplies.

(vi) Ripple content in the output is higher than that for a linear power supply.

(vii) Load regulation is poor as compare to the linear regulators.

11.19.3. Specifications of SMPS

Some of the important specifications of SMPS are as under :

1. DC output voltage Say +12V or + 5V or ± 12 V etc.
2. Multiple outputs or single output.
3. Maximum load current.
4. Ripple content in the output voltage waveform.
5. Switching frequency.
6. Output power.
7. Type of Regulator : Buck, Boost etc.
8. Isolation provided or not.
9. Battery operated or mains operated.
10. Magnitude of RFI/EMI interferences.
11. Protections such as over voltage protection, overcurrent protection etc.
12. Load regulation.
13. Power efficiency.

11.20. UNINTERRUPTABLE POWER SUPPLIES (UPS)

Need of UPS

Certain application areas such as personal computers, computer work stations, medical equipments, intensive care units (ICU) need the continuous supply of high quality sinusoidal votlage. For such loads, the user cannot depend solely on the sinusoidal voltage available at mains supply. This is due to frequent outage, poor quality of voltage waveform, fluctuations in mains voltage etc. Hence, an alternative system to mains supply has been developed called uninterruptable power **supplies (UPS)**. The UPS systems have become an integral part of our day to day life. We see these systems along with the computer, or with expensive instruments in the hospitals etc. The UPS is used for supplying power to such critical loads. The UPS system not only supplies an uninterrupted power to these critical loads, but, it takes care of the following things as well:

(i) It regulates the load voltage when the mains voltage fluctuates.

(ii) It suppresses the line transient (spikes) that otherwise would have been easily coupled to the load.

(iii) It supplies a pure, harmonic free signal to the load.

11.20.1. Types of UPS

Depending on the arrangement of the basic blocks of the UPS system, they can be classified into following configurations :

(i) On line or inverter preferred UPS system.
(ii) Off line or line preferred UPS system.
(iii) Line interactive UPS.
(iv) In the following subarticles, these configurations have been discussed in details.

11.20.2. On Line UPS

The configuration of an on line UPS system has been shown in figure 11.46.

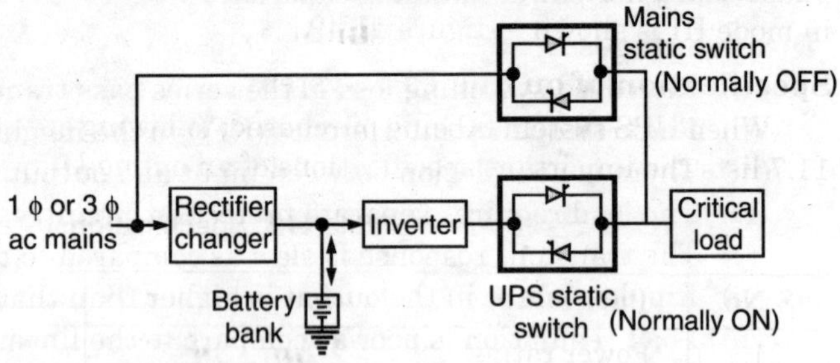

Fig. 11.46. ON line (inverter preferred) UPS system.

Working Operation of an on line UPS

In the on line UPS system, the load is always connected to the inverter through the UPS static switch. The UPS static switch in figure 11.46 is a Normally ON switch. It turns off only when the UPS system fails. In that case, the Mains static switch is turned on to connect the ac mains directly to the load. This switch is a Normally OFF switch and used only when UPS is to be bypassed. A static switch consists of a pair of antiparallel SCRs. Therefore, conduction through a static switch is bidirectional. Static switches are preferred over conventional mechanical switches because they can be activated electrically. Earthing and sparking does not take place while operating the static switches.

The operations can be divided into two modes :

(i) Mode I (Mains ON)

When the ac mains is on, the rectifier circuit will supply the power to the inverter as well as to the battery. So, it acts as a rectifier cum charger. The inverter output is connected to the load via UPS static switch. Battery will be charged in this mode.

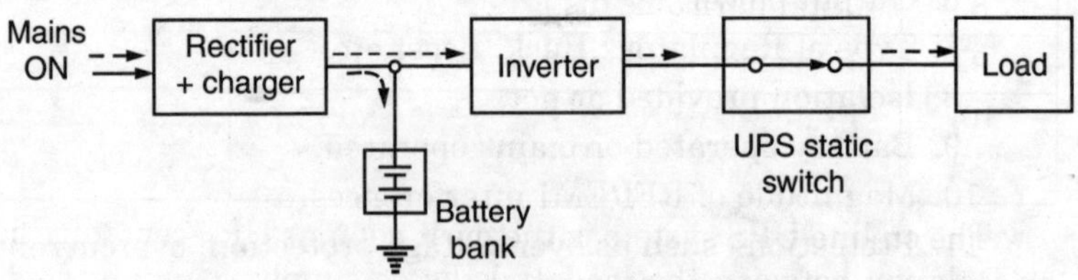

Fig. 11.47. Equivalent circuit for mode I (Mains ON)

The equivalent circuit for this mode is shown in figure 11.47.

(ii) Mode II (Mains OFF)

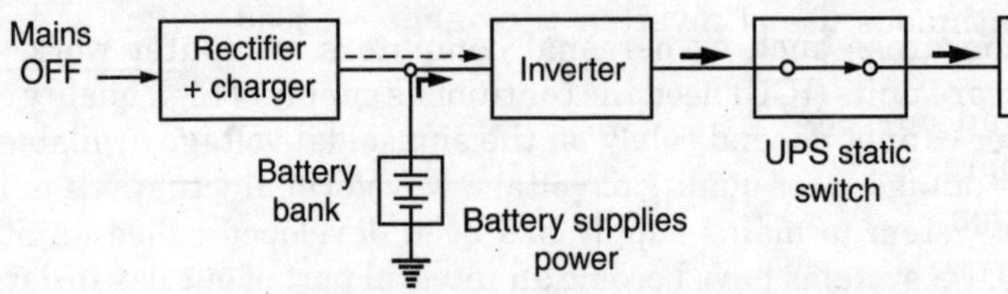

Fig. 11.48. Equivalent circuit for mode II (Mains OFF)

Under the condition of mains outage, the rectifier output is zero. Hence, the battery bank now supplies power to the inverter. The inverter then drives the load through the UPS static swtich. This, transfer from rectifier to battery bank is instantaneous and during the transfer period, there is no power interruption. After the restoration of the mains, the rectifier again starts supplying power to the inverter and will also charge the battery bank. The equivalent circuit and power flow in mode II is shown in figure 11.48.

(iii) When UPS fails

Due to some reason, if UPS fails, then the normally off mains static switch is turned on to connect the mains to the load bypassing the UPS system. The UPS static switch automatically turns off and disconnects the load from the UPS. The transfer of load from UPS to mains takes place within 1/4 cycle period with no phase change. The equivalent circuit and power flow in mode III is shown in figure 11.49.

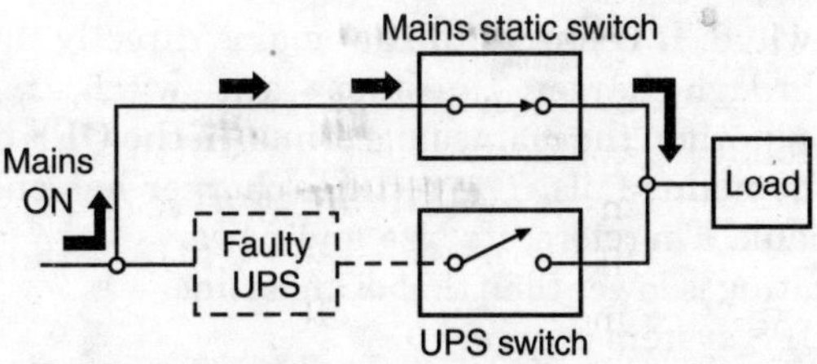

Fig. 11.49. Equivalent circuit for mode III

Specification of on line UPS system

When UPS system is being purchased, following specifications should be carefully studied. Table 11.7 lists the important specifications of an online UPS system

Table 11.7. Specifications of an online UPS

S.No.	Parameter	Expected value
1.	Power rating	1 KVA, 2 KVA, 5 KVA, etc.
2.	Output voltage	230 V ±0.5%
3.	Output frequency	50 Hz ± 2Hz
4.	Input voltage	190 to 260 Volts
5.	Nature of output voltage waveform	Sinusoidal
6.	Power factor	> 0.8 lagging
7.	Backup time	30 minutes to 4 hours
8.	Total harmonic distortion	< 5%
9.	Efficiency of the inverter	> 85%
10.	Protection circuits	(a) Over voltage and under voltage cutout (b) Over current protection

The on line UPS system is the most popular UPS configuration of all. The on line UPS provides an isolation between the load and the ac supply. This is its advantage over the rest of the UPS configurations.

Features of on line UPS system

Some of the important features are as under :

(i) It provides isolation between ac source and load

(ii) Due to the continuous use of inverter, the quality of load voltage is better than that of mains.

(iii) All voltage and current spikes are suppressed

(iv) Better regulation

11.20.3. Off Line UPS

The block diagram for the off line or line preferred UPS has been shown in figure 11.50. If we observe carefully, this diagram is same as figure 11.46, i.e., on line UPS. The only difference is that the Mains static switch here is a normally on

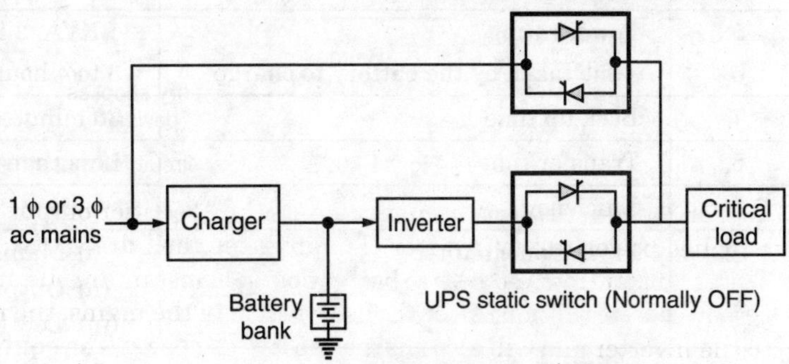

Fig. 11.50. OFF line (line preferred) UPS system

switch. It connects the ac mains directly to the load when the mains is on. The battery charges through charger. The other static switch, i.e., UPS static switch is normally off switch. It is closed only when the mains fails. Thus, in the OFF line UPS, the inverter comes into the circuit only when the mains fails. The rectifier/charger has only one function to perform, i.e., to charge the battery bank. Therefore, its size and power rating is lower than that of an on line UPS system charger.

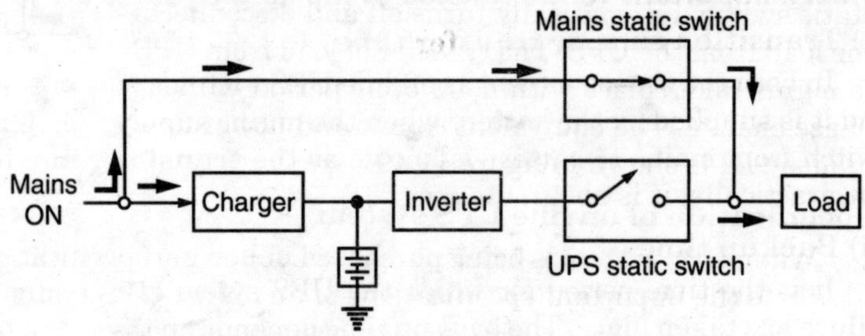

Fig. 11.51. Equivalent circuit for Mode I (Mains ON)

Working Operation

Mode I (Mains ON)

When the mains is ON, the power is supplied to the load via the normally ON mains static switch. The battery is charged through the charger. The equivalent circuit and power flow for this mode is shown in figure 11.51.

Mode II (Mains OFF)

When the mains is off, the mains static switch gets open circuited and UPS static switch get closed automatically. Battery charging will stop and battery starts supplying power through inverter as shown in figure 11.52. The total time taken to sense the power failure and make a changeover from mains to UPS is about 5 msec.

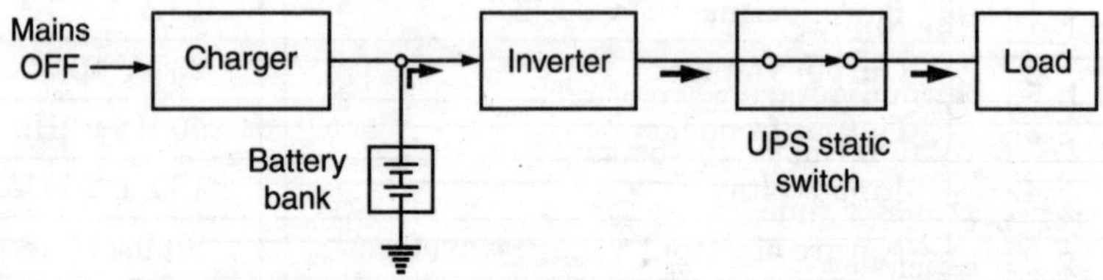

Fig. 11.52. Equivalent circuit Mode II (Mains OFF)

Specifications of off line UPS

The important specifications for the off line UPS are as listed in table 11.8.

Table 11.8. Specifications of off line UPS

S.No.	Parameter	Expected value
1.	Output voltage	(i) Mains on : 230 ±5% Volts (ii) UPS on : 230 ± 0.5% Volts
2.	Output voltage waveform	Sinusoidal
3.	Output frequency	(i) Mains on : 50 Hz ± 3% (ii) UPS on : 50 Hz ± 2 Hz
4.	Input voltage (Battery voltage)	24 V, 48 V, 72 V etc. with AH capacity.
5.	Power rating	1 KVA, 2 KVA etc.
6.	Time taken by the battery to charge	3 to 4 hours
7.	Back up time	30 minutes or more
8.	Transfer time	Less than 5 msec.
9.	Efficiency	> 85%
10.	Protection circuits	(i) Mains over voltage (ii) Output over current (iii) Low battery

The off line UPS system uses the UPS sparingly only when the mains is OFF, therefore, it is more economical as compared to the on line UPS system. However, as the UPS is not being used continuously all the advantages of using the inverter are lost. This sytem is therefore less reliable than the online UPS.

11.20.4. Important Terms Related to the UPS System

(i) Transition time or transfer time

In both the on line as well as off line UPS systems, power is supplied by the mains when it is ON and it is supplied by the battery when the mains supply fails. The time taken by the UPS system to switch from mains to battery is known as the **transfer time**. Ideally transfer time should be zero but practically, it is about 4 to 5 msec.

(ii) Backup time

It is the time period for which the UPS sytem can supply power to the load, after the mains failure has taken place. The back up time depends on the capacity of the battery and the efficiency of the inverter.

11.20.5. Performance Comparison of UPS Systems

S.No	Parameter	On line	OFF line
1.	Output voltage waveform	Sinusoidal	Quasi square
2.	Total harmonic distortion	Low	High
3.	Transfer time	0	< 5 ms
4.	Efficiency	Low	High
5.	Cost	Costliest	Less costly
6.	Applications	Critical loads	General purpose loads
7.	Isolation between the load and mains	Full isolation	No isolation
8.	Size and Weight	Highest	Low
9.	Rectifier/charger power rating	Highest	Lowest
10.	Operation	Continuous	Only in absence of mains
11.	Reliability	Very reliable	Not so reliable

11.20.6. Applications of UPS

Now a days, UPS are being used in almost all the fields. Some of the important application areas are as under :

(i) As power supply for computers
(ii) For supplying a continuous power to medical equipments
(iii) In intensive care units
(iv) For EPABX
(v) Flight simulation

EXAMPLE 11.12. (i) For the given network shown in figure 11.53 determine the range of R_L and I_L that will result in VRL being maintained at 10 V. (ii) Determine the maximum wattage rating of diode. **(RGTU, Bhopal, Sem. Exam. 2006-07)**

Solution: (i) Range of R_L and I_L :

Assume $I_{Z(min)} = 5$ mA

(i) Value of I_R

$$I_R = \frac{V_i - V_z}{R_S} = \frac{50 - 10}{1k} = 40 \text{ mA}$$

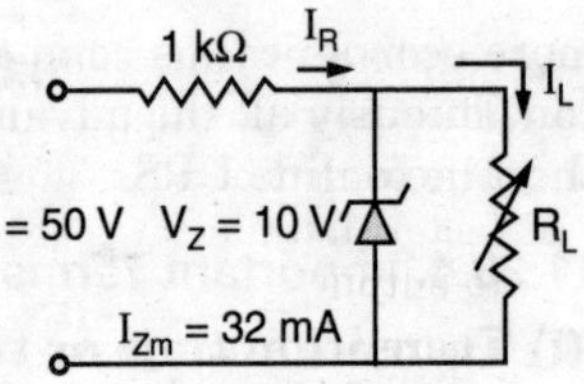

Fig. 11.53.

(ii) Value of $I_{L(min)}$

$$I_R = I_{L(min)} + I_{Z(min)}$$

Therefore $40 = I_{L(min)} + 32$ $\quad I_{L(Min)} = 8$ mA

(iii) Value of $I_{L(max)}$

$$I_R = I_{L(max)} + I_{Z(min)}$$

Hence, $40 = IL(max) + 5$ $\quad I_{L(max)} = 35$ mA

Therefore, Range of I_L = 8 mA to 35 mA **Ans.**

(iv) Value of R_L

$$R_{L(max)} = \frac{V_Z}{I_{L(min)}} = \frac{10}{8 \text{ mA}} = 1.25 \text{ k}\Omega$$

$$R_{L(min)} = \frac{V_Z}{I_{L(max)}} = \frac{10}{35 \text{ mA}} = 285.7 \ \Omega$$

Therefore, Range of R_L = 285.7 Ω to 1.25 kΩ **Ans.**

Part (ii): Maximum wattage rating of diode

$$P_{zm} = I_{Z(max)} \times V_Z = 32 \times 10^{-3} \times 10 = 0.32 \text{ W} \quad \textbf{Ans.}$$

SUMMARY

- A **voltage regulator** is that kind of circuit which keeps the output voltage constant inspite of changes in the load current or input voltage. Its input is unregulated pulsating d.c. voltage obtained from filter and rectifier. Its output is constant d.c. voltage which is almost ripple free.
- The a.c. voltage (230 V, 50 Hz) is connected to the primary of the transformer. The transformer steps down the a.c. voltage, to the level required for the desired d.c. output. Thus, with suitable turns ratio, we get desired a.c. secondary voltage. The rectifier circuit converts this a.c. voltage into a **pulsating d.c. voltage.** A pulsating d.c. voltage means a unidirectional voltage containing large varying component called **ripple** in it. The filter circuit is used after a rectifier circuit, which reduces the ripple content in the pulsating d.c. and tries to make it smoother. Still then the filter output contains some ripple. This voltage is called **unregulated d.c. voltage.**
- A circuit used after the filter is a regulator circuit which not only makes the d.c. voltage smooth and almost ripple free but it also keeps the d.c. output voltage constant through input d.c. voltage varies under certain conditions. It keeps the output voltage constant under variable load conditions, as well. The output of a regulator is called d.c. supply, to which the load can be connected. Now a days, complete regulator circuits are available in the integrated circuit (IC) form.
- A voltage regulator circuit is the one which is designed to keep the output voltage of a power supply nearly constant, under varying input voltage conditions and varying load conditions.
- An ideal power supply maintains a constant voltage at its output terminals, in spite of changes in load current. But, practically in a power supply without regulator the load voltage decreases as the load current I_L increases. For a practical power supply with regulator load voltage must be constant through load changes from no load condition to full load condition. But, with regulator circuit also, the load voltage gets affected by the load current in a power supply.
- The input to the rectifier which is a.c. line voltage decides the level of the output voltage. Hence, any change in the line voltage, changes the load voltage and affects the performance of the power supply. Ideally, the d.c. output voltage must remain constant irrespective of any changes in the line voltage.
- In a power supply, the rectifier circuit uses the p-n-junction diodes. The diode characteristics are temperature sensitive. the other semiconductor devices used in power supplies have their characteristics, temperature dependent. Hence, the temperature is an important factor responsible for the changes in the load voltage.
- The load regulation may be defined as the change in the regulated output voltage when the load current is changed from minimum (no load) to maximum (full load).
- The input to the unregulated power supply i.e. rectifier circuit is 230 V a.c. supply. This line voltage may change, under the different load conditions. This affects the output voltage of rectifier which is V_{in} for a

regulator circuit. Hence, the characteristics which gives source effect on regulator performance is defined. The line regulation is also called as source regulation and is denoted as SR.

- The output of a rectifier and filter circuit consists of ripples. The ripple rejection may be defined as a factor that indicates how effectively the regulator circuit rejects the ripples and attenuates it from input to output.
- There are two types of voltage regulators available as follows :
  (i) Shunt voltage regulator
  (ii) Series voltage regulator

  Each type provides a constant d.c. output voltage which is regulated.
- The shunt regulator is generally suffered with the following limitations:
  (i) The maximum load current that can be supplied is limited.
  (ii) A large amount of power is wasted in the zener diode and the series resistance R, in comparison with the load power.
  (iii) The performance factors like stability factor and the output resistance are not very low, as desired for a good regulator circuit.
- The various salient features of IC voltage regulators may be listed as under :
  (i) Easy to use.
  (ii) It greatly simplifies power supply design.
  (iii) Due to mass production, low in cost.
  (iv) IC voltage regulators are versatile.
  (v) Conveniently used for load regulation.
  (vi) These are provided with features like built in protection, programmable output, current/voltage boosting, internal short circuit current limiting etc.
- The series regulator using transistors has no short circuit protection. If the load terminals are shorted accidently then,
  (i) Large amount of load current will flow.
  (ii) The pass transistor Q will get destroyed.
  (iii) The diodes used in unregulated power supply, supplying the input voltage V to regulator circuit may get destroyed.
- The various protecting circuits used for the series voltage regulators may be listed as under :
  (i) Constant current limiting (Short circuit protection).
  (ii) Fold back current limiting.
  (iii) Over voltage protection.
  (iv) Thermal shutdown.
- The drawback of constant current limiting is relatively large power dissipation in the series pass transistor when the load terminals are shorted. Thus, a large power rating transistor is required. The foldback limiting technique allows us to provide the essential load current at rated voltage but reducing the short circuit current. Hence, the series pass transistor gets utilised efficiently.
- As the name suggests, three terminal voltage regulators have three terminals namely input which is unregulated (V_{in}), regulated output (V_{out}) and common or a ground terminal. These regulators do not require any feedback connections.
- Adjustable voltage regulators have the following advantages over fixed voltage regulators :
  (i) Improved line and load regulation by a factor of 10 or more
  (ii) Because of improved overload protection, greater load current can be drawn over the given operating temperature range.
  (iii) Improved reliability for the power supply using these regulators.

SHORT QUESTIONS WITH ANSWERS

Q.1. What do you mean by shunt regulator?

Ans. It is a voltage regulator circuit in which a transistor is placed in parallel with a power source to vary its resistance automatically, which, in turn, controls the output voltage or current.

Q.2. What is a voltage regulator?

Ans. It is a circuit that maintains, its output voltage at predetermined value independent of changes in the load current or input voltage as long as the input is greater than the predetermined output voltage.

Q.3. What is a Switching Regulator?

Ans. It is a voltage regulator utilizing switching circuits with varying duty cycle to achieve regulation, which results in very high efficiency.

Q.4. What do you mean by Ripple?

Ans. It is the term describing the AC component appearing at the output of a DC power supply.

Q.5. What is the need for voltage regulation?

Ans. Unregulated power that is supplied gives impulse DC i.e., floating DC. This has to be regulated i.e., made constant before feeding them in to electronic circuits. This is the need of voltage regulation.

Q.6. What are the limitations of linear IC voltage regulators?

Ans. (a) There is no short circuit protection

(b) The output voltage whether positive or negative is fixed.

Q.7. How will you classify voltage regulators?

Ans. (i) Series regulators

(ii) Switching regulators

Q.8. What are the essential features of a general purpose voltage regulator?

Ans. (i) The regulated output voltage is fixed at a value as specified by the manufacturer.

(ii) The irregulated input voltage must be at least 2V more than the regulated output voltage.

(iii) The load current may vary from 0 to rated maximum output current.

(iv)The 1C has a temperature sensor which turns off the IC when it becomes too hot.

Q.9. What is meant by line regulation and load regulation?

Ans. Line regulation is defined as the percentage change in the output voltage for a change in input voltage. Load regulation is defined as the change in output voltage for a change is load current.

Q.10. Mention the major functional blocks of switching regulators.

Ans. Voltage source, Switch, Pulse generator and Filter.

Q.11. Name any two IC voltage regulators

Ans. LM 340 and LM 320.

Q.12. Define the term ripple factor.

Ans. Ripple factor is defined as the ratio of AC component voltage to the DC component in voltage.

Q.13. Differentiate between positive and negative regulators.

Ans. A regulator which gives a positive regulated voltage is called positive regulator. A regulator which gives a negative regulated voltage is called a negative regulator.

Q.14. What is the order of the voltage drop across the current limiting resistor is an IC regulator when the limiting action occurs?

Ans. 15mV, for $5\text{mA} < I_o < 1.5$ A.

Q.15. What are the advantages and disadvantages of switching regulators?

Ans. Advantages:

(i) It has high efficiency

(ii) It produces a high output ripple

(iii) The response to load transistors is not always as fast as linear regulators.

(iv) Switching regulators throw current transients back into the unregulated supply which are some what larger than the maximum load current.

Disadvantages:

(i) It produces a high output ripple.

(ii) It is a very complex circuit.

(iii) The response to load transistors is not always as fast as linear regulators.

Q.16. What are the two types of regulators? How do they differ in their operation?

Ans. The two types of regulators are series and switching. Series regulators are those that use a transistor in its linear portion and switching regulators are those that use a transistor as a high frequency ON/OFF switch

Q.17. Why are series regulators called linear regulators?

Ans. As series regulators use transistors in its linear portion they are called series or linear regulators.

Q.18. Name the important parts of series regulator power supply using discrete components.

Ans. (i) Reference voltage

(ii) Error amplifier

(iii) Series pass transistor

(iv) Feedback network

Q.19. What is the function of series pass transistor?

Ans. It is used as an emitter follower and therefore provides sufficient current gain to drive the load.

Q.20. What is meant by current limit protection?

Ans. In order to protect the circuit under short circuit conditions, a resistor is used in the circuit. The value of $I_{limit} = \frac{V_{sence}}{R_{SC}} = \frac{0.5\text{ V}}{R_{SC}}$ is their limiting current.

Q.21. What is meant by current fold back?

Ans. In current limiting techniques the load current is maintained at a present value and when overload condition occurs, the output voltage V_0 drops to zero. However if the load is short circuit, maximum current flows through the regulator. To protect the regulator current fold back techniques is used.

Q.22. What is the function of voltage regulator?

Ans. It is to regulate the unregulated power supply giving impure DC to a signal of constant DC voltage.

Q.23. Define ripple rejection.

Ans. The IC regulator not only keeps the output voltage constant but also reduces the amount of ripple voltage. It is usually expressed in decibels.

Q.24. What are two types of regulators and differentiate them?

Ans. (i) Series voltage regulator

(ii) Switching voltage regulator.

In series regulator a power transistor is connected in series between the DC input and the load. The output voltage is controlled by the continuous voltage drop across series pass transistor. But switching regulator operates the power transistor as a high frequency on/off switch, so the power transistor does not conduct current continuously.

Q.25. What is the advantage of a switched capacitor filter?

Ans. (i) Low system cost.

(ii) Low external component count.

(iii) High accuracy.

(iv) Excellent temperature stability.

Q.26. What is an ordinary power supply?

Ans. The combination of a transformer, a rectifier and a filter constitutes an ordinary dc power supply, also called an unregulated power supply.

Q.27. What is a regulated power supply?

Ans. A regulated power supply is an electronic circuit designed to provide constant dc voltage of predetermined value across load terminals, which is independent of variations in load current, ac mains voltage and temperature.

Q.28. Why unregulated power supply is not good enough for many of the applications in electronics?

Ans. Since the output voltage available from an unregulated power suply varies with the variations in load current, ac mains voltage and temperature and variations in dc output voltage may cause inaccurate of erratic operation or malfunctioning of many electronic circuits. Unregulated power supply is not considered suitable for many of the applications in electronics.

Q.29. What is a regulated power supply and what are its parts?

Ans. A transformer, a rectifier circuit, a filter circuit, a voltage regulator along with some current limiting arrangement constitute a regulated voltage supply.

Q.30. What do you mean by the parameter 'load regulation'?

Ans. The load regulation is the change in regulated output voltage when the load current changes from minimum to maximum value i.e.,

$$\text{Load regulation} = \text{No-load voltage, } V_{NL} - \text{Full-load voltage } V_{FL}.$$

$$\% \text{ LR} = \frac{V_{NL} - V_{FL}}{V_{FL}} \times 100$$

Q.31. Why pass transistor is so called?

Ans. Collector and emitter terminals of a transistor connected in a transistor series voltage regulator are in series with the load and whole load current pass through it, so it is called a pass transistor.

Q.32. Why emitter follower regulator is better than Zener regulator?

Ans. The advantage of an emitter follower regulator is that the changes in Zener current are reduced by a factor β and thus the effect of Zener effect is largely reduced and much more stabilised output is obtained.

Q.33. What is the need of foldback current limiting?

Ans. A problem with the simple current limiting, provided for protection against over-load/short-circuit, is that a large amount of power is dissipated in the series pass resistor while the regulator remains short-circuited. The foldback current limiting is the solution of this problem.

Q.34. What is meant by 'dropout' voltage of an IC voltage regulator?

Ans. The drop-out voltage, typically 2 V, is the minimum amount of voltage across the input-output terminals that is required to be maintained if the IC is to operate as a regulator.

Q.35. What is the function of error amplifier in IC voltage regulator?

Ans. An error amplifier is essentially an op-amp with extremely high open-loop voltage gain. Because of high gain of an op-amp, the error voltage between the two input terminals approaches zero.

Q.36. What is the categorisations of linear voltage regulators?

Ans. Linear.

Q.37. What are the limitations of linear voltage regulators?

Ans. The main drawback of linear voltage regulator is the power dissipation in the pass transistor which is operated in its linear mode. Other drawbacks of linear voltage regulators are that regulated power supplies using these regulators require a step down transformer (most expensive and bulky component of the power supply) and large fixed filter capacitors to reduce the ripple.

Q.38. The demand for on-line UPS system is less than for off-line UPS system. Why?

Ans. The demand for on-line UPS system is less than for off-line UPS system because the former is costlier one.

Q.39. Why are electronic generators in great demand?

Ans. There is acute shortage of power in the country and during summer, power-cuts of 1 to 4 hours duration are common in cities and villages. Every home dreams of buying a generator during that time, and in comparison to kerosene operated generators, the electronic features.

REVIEW QUESTIONS

1. Explain the working principle of a regulated power supply.
2. Describe the operation of a regulated power supply. Give the standard representation of IC voltage regulator
3. Explain the working of a series regulated power supply.

4. Draw the circuit diagram of an electronic voltage regulator using a zener diode for voltage reference and a series pass transistor, and explain its operation.
5. What is the function of voltage regulator?
6. What are fixed output voltage regulators?
7. Explain three terminal IC voltage regulator using block diagram.
8. What are adjustable voltage regulators?
9. How does an IC LM 340 series incorporates thermal shutdown and current limiting?
10. What are the various types of IC regulators? Explain the operation of any IC voltage regulator.
11. Explain, with a neat block diagram of the internal blocks, the working of an IC voltage regulator.
12. Discuss how a three-terminal voltage regulator IC be employed to realize an adjustable voltage output.
13. List and explain the characteristics of three-terminal IC regulator. Explain the protection used in 78XX.
14. What do you mean by current sensing technique in IC 723?
15. Discuss the IC 723 as
    (i) Low voltage regulator.
    (ii) High voltage regulator.
16. What is function of voltage regulator? Explain current limiting features of IC 723 regulator.
17. (a) What is voltage regulator?
    (b) Design a voltage regulator using IC 723 for output voltage of 5V.
18. Explain the current fold back characteristics of 723 voltage regulator.
19. Describe where current boosting is done in a line voltage regulator IC 723.
20. What are dual voltage supplies. Give their advantages.
21. Draw the block diagram of SMPS and explain its working.
22. What are the advantages of switching mode dc voltage regulator over a linear mode one. Give the basic block diagrams of the two types and explain their operation. What is the purpose of energy storage elements in "switching mode" one. **(U.P.S.C. I.E.S. Examination. 2000)**
23. Explain the operation of a switching regulator using the relevant block diagram.
24. Discuss the limitations of linear voltage regulators. Explain switch-mode power supplies.
25. Give the circuit and required waveforms of a switched-mode power supply and explain its operation.
26. What is SMPS? Compare it with linear power supply of switched mode.
27. What is the principle of switch mode power supplies? Discuss its advantages and disadvantages.
28. What is CVT? How does it differ from an ordinary transformer? Explain its operation. Why the use of CVT is necessary in computers? Compare CVT and stabilizer on the basis of input/output voltage, maintenance, life, efficiency etc.
29. What is UPS? Why the UPS system is becoming more and more popular now-a-days? Compare on-line UPS, off-line UPS and electronic generator.
30. Write short notes in the following:
    (i) Regulated power supply.
    (ii) Zener controlled transistor voltage regulator.
    (iii) Three terminal regulator ICs.
    (iv) Dual voltage supply.
    (v) SMPS.
    (vi) CVT.
    (vii) UPS system.
    (viii) On-line UPS and Off-line UPS.
31. Explain the following:
    (i) What is regulated power supply?
    (ii) How will you classify voltage regulators?
    (iii) What is a voltage reference? Why is it needed?
    (iv) What is a three-terminal regulator?
    (v) How can a three-terminal voltage regulator IC circuit be modified to provide variable voltage output?
    (vi) How can dual voltage supply be obtained by using two three-terminal IC voltage regulators?

(vii) How does a CVT work?

(viii) Why do we use a CVT and not a voltage stabilizer for computer applications?

(ix) Why electronic generators are in high demand?

31. Explain the following:
32. What are the two types of static UPS systems ?
33. State the applications of UPS.
34. What are the advantages of UPS ?
35. Define transfer time and back up time.
36. Define ON line and OFF line UPS.
37. Which UPS is most popularly used ? Why ?
38. State the applications of UPS.
39. What is the need of UPS ?
40. State advantages and disadvantages of ON line UPS.
41. What are the merits and demerits of OFF line UPS ?
42. Explain the need of a UPS system and state its types.
43. With the help of block diagram explain the operation of an on line UPS. Why is it called as inverter preferred UPS ?
44. Compare different types of UPS systems.
45. Write a short note on : Line interactive UPS.
46. In connection with UPS systems explain the following terms :
    (a) Transfer time (b) Back up time (c) Static switches.
47. Explain the necessity of uninterrupted power supplies.
48. Explain short-break UPS.
49. Explain No-break UPS.
50. With the help of block diagram explain the operation of an on line UPS. Why is it called as inverter preferred UPS ?
51. State the important specifications of ON line UPS.
52. State the important specifications of OFF line UPS.
53. What is SMPS ?
54. How is SMPS different from linear power supply ?
55. What are the requirements of a power supply ?
56. State the types of power supplies.
57. Explain the principle of SMPS.
58. Give the classification of SMPS.
59. State advantags and disadvantages of SMPS.
60. What are the applications of SMPS?
61. State the specification of SMPS.
62. Why is the efficiency of SMPS high?

NUMERICAL PROBLEMS

1. Design a current source of 1 A using IC 7812 voltage regulator to a 25 Ω, 10 W load.
2. Using the 7805C voltage regulator, design a current source that would deliver a 0.5 A current to the 48 Ω, 10 W load. **[Ans.** $R = 10\ \Omega$, $V_{out} = 29$ V]
   $V_{in} = V_0$ + drop out voltage = 31 V]
3. Determine the regulated output voltage in the circuit of figure 11.23 with $R_1 = 540\ \Omega$ and $R_2 = 2.4\ k\Omega$ **[Ans.** $V_{out} = 6.8$ volts]
4. Determine the minimum and maximum output voltage of the voltage regulator shown in figure 11.27. Assume I_Q of value 20 mA (instead of 10 mA). **[Ans.** 15 V; 94 V]
5. In figure 11.40, determine the regulated output voltage if $R_1 = R_2 = 2\ k\ \Omega$ and $V_{REF} = 2V$ **[Ans.** 4V]
6. Determine the duty cycle of the pulses at the base of transistor Q in figure 11.40, if $V_{REF} = 2.5$ V, $R_1 = 6\ k\Omega$, $R_2 = 2\ k\Omega$ and regulated input voltage is 20 V.

□□□

MODEL QUESTION PAPERS

MODEL QUESTION PAPERS

MODEL QUESTION PAPER – 1

Note : *Attempt any five questions. All questions carry equal marks. Assume any data missing.*

1. (*a*) Draw the piece-wise linear volt-ampere characteristic of a *p-n* diode. What is the circuit model for the ON state and the OFF state ? Explain the term 'Potential Energy Barrier' in connection with *P-N* junction. Hence discuss forward bias and reverse bias.

   (*b*) Give brief answer to each of the following :

   (*i*) Tunneling in diodes

   (*ii*) Zener breakdown

   (*iii*) Diffusion capacitance

   (*iv*) Junction capacitance

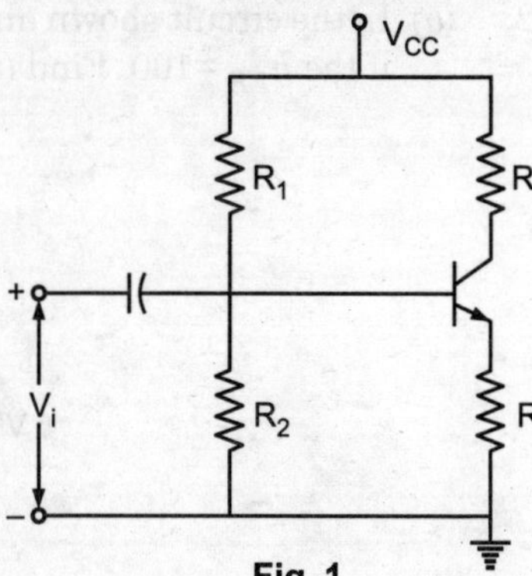

Fig. 1.

2. (*a*) Find the value of I_C, I_B and V_{CE} from the circuit given ahead if $\beta = 55$, $V_{CC} = 22.5$, $R_C = 5.6$ K, $R_e = 1$ K, $R_1 = 90$ K and $R_2 = 10$ K.

   (*b*) Define the Pinch-off voltage V_p. Sketch the depletion region before and after pinch off. How does FET behaves in the region before and after pinch off ?

3. (*a*) Calculate the voltage gain, current gain, input resistance and output resistance of a common emitter amplifier. State their typical order.

   (*b*) Discuss the doping levels for emitter base and collector regions and their physical size in BJT.

4. (*a*) What is the relationship between the transfer gain with feedback A_f and that without feedback A ? Explain the three fundamental assumptions which are made in order that the above expression be satisfied exactly.

   (*b*) Sketch the circuit of phase shift oscillator. Explain its working. Determine its frequency of oscillation.

5. (*a*) Draw the circuit of monostable multivibrator. Explain its working.

   (*b*) Draw and explain Darlington emitter follower circuit. Why the input impedance is high than that of a single stage emitter follower ?

6. Explain the following with the aid of diagram :

   (*i*) differentiator using OP-AMP

   (*ii*) log amplifier using OP-AMP

   (*iii*) 555 timer in astable mode

7. (*a*) Explain V-I characteristics of SCR. What is meant by Gate ON and OFF times ?

   (*b*) Discuss turn-off and turn OFF mechanism in SCR.

8. Write short notes on any *two* of the following :

   (*i*) Varactor diode

   (*ii*) h-parameter

   (*iii*) Efficiency of power amplifiers

   (*iv*) Schmitt trigger

   (*v*) UJT

MODEL QUESTION PAPER – 2

Note : *Attempt any five questions. All questions carry equal marks. Assume any data missing.*

1. (*a*) Distinguish between Zener breakdown and Avalanche breakdown. Draw the characteristics if a Zener diode and explain Zener breakdown.
   (*b*) Consider a circuit consisting of a diode *D*, a resistance *R* and a signal source V_i in series. Define (*i*) static characteristics (*ii*) dynamic characteristic (*iii*) transfer characteristic (*iv*) what is the correlation between (*ii*) and (*iii*) ?
2. (*a*) In the circuit shown ahead, determine whether or not the siliocn transistor is in saturation. if the h_{EE} =100. Find (*i*) I_B and I_C (*ii*) repeat with the 2 K emitter resistance added.

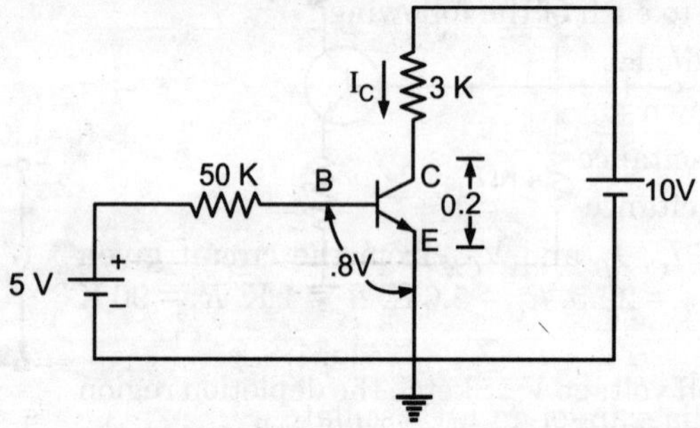

Fig. 1.

   (*b*) Explain the working of enhancement model in a MOSFET with circuit symbol and characteristics.
3. (*a*) Draw a self-bias circuit. Explain qualitatively why such a circuit is an improvement on the fixed bias as far as stability is concerned. How is load line drawn for a self bias circuit ?
   (*b*) Define *h*-parameter of a transistor. Find the input and output resistance in CE configuration.
4. (*a*) Draw a circuit of voltage series feedback amplifier. Derive the expression for the following :
   (*i*) Voltage gain with feedback A_{vf}
   (*ii*) Output resistance with feedback R_{of}
   (*b*) Sketch the circuit of Weinbridge oscillator. Explain its working. Determine the frequency of operation.
5. (*a*) Draw the circuit of bistable multivibrator. Explain its working with the help of waveforms.
   (*b*) Draw boostrapped Darlington circuit and explain how this increases the input resistance.
6. Explain the following with the aid of diagram :
   (*i*) Integrator using OP-AMP
   (*ii*) Instrumentation amplifier using OP-AMP
   (*iii*) 555 timer in monostable mode
7. (*a*) Explain series and parallel connection of SCR.
   (*b*) Explain V-I characteristics of SCR in detail.
8. Write short notes on any *two* of the following :
   (*i*) Varactor diode
   (*ii*) UJT
   (*iii*) Power amplifier
   (*iv*) Schmitt trigger

MODEL QUESTION PAPER – 3

Note : *Attempt any five questions. Assume missing data suitably..*

1. (*a*) Define the following : 8

   (*i*) Cut-in (knee) voltage (*ii*) depletion layer

   (*iii*) reverse saturation current (I_{CBO}).

   (*b*) For the circuit shown in fig. 1, draw the loadline. Determine the operating point and show it on the load line. Assume V_{BE} = 0.7 volts. 12

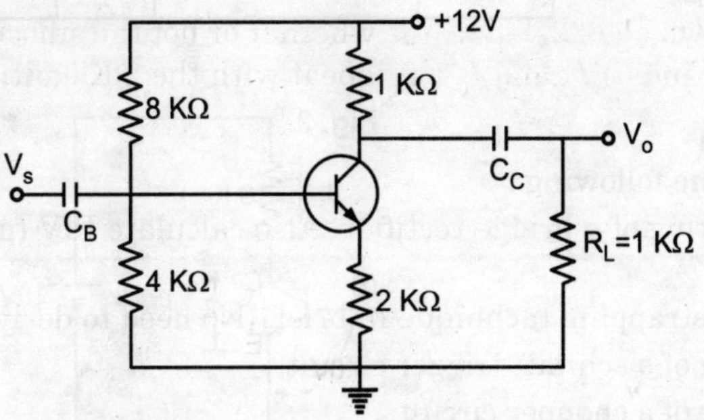

Fig. 1.

2. (*a*) Draw the circuit diagram of an L-C oscillator and explain its working. 10

   (*b*) Give classification of power amplifiers. What is P_{Dmax} rating ? Explain. 10

3. (*a*) Define the following : 8

   (*i*) CMRR

   (*ii*) Offset voltages

   (*iii*) Slew rate

   (*b*) Draw circuit diagram of a difference amplifier using OP-AMP and derive an expression for its gain. 12

4. (*a*) Explain V-I characteristics of an SCR. 10

   (*b*) Discuss the working of a simple inverter. 10

5. (*a*) Explain the working of a clamper circuit with input and output wave forms. 10

   (*b*) Draw the circuit diagram of a monostable multivibrator using 555 IC and explain its working. 10

6. (*a*) Show that the maximum conversion efficiency of a class-B amplifier system is 78.5%. 10

   (*b*) Draw circuit digram of an emitter follower. Give its low frequency *h*-parameter model. Analyse its feedback mechanism (no derivation desired). 10

7. (*a*) Derive an expression for resonant frequency of a tank (tuning) circuit of BJT amplifiers. 8

   (*b*) Workout the following quantites for the circuit shown in figure 2.

   (*i*) a.c. emitter current

   (*ii*) a.c. voltage at emitter, base and collector

   (*iii*) a.c. voltage gain

   Assume h_{ie} = 250 Ω, h_{fe} = 50.

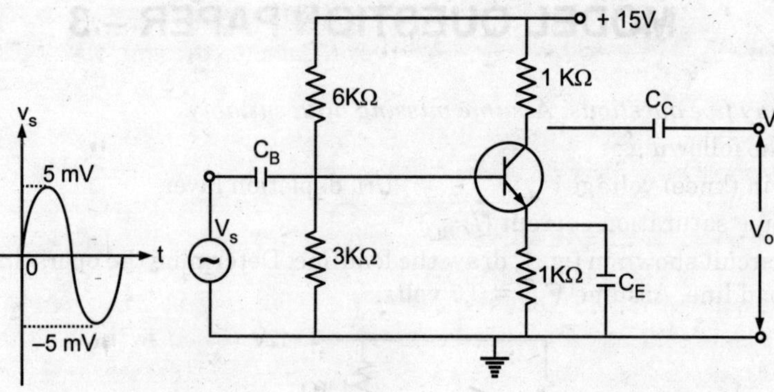

Fig. 2.

8. Attempt any two of the following :

   (*i*) Explain the working of a bridge rectifier. Also calculate PIV (peak inverse voltage) for the same.

   (*ii*) Discuss the Bootstrapping tachnique in brief. (No need to derive expressions).

   (*iii*) Give the working of a schmitt trigger circuit.

   (*iv*) Give the working of a chopper circuit.

INDEX

INDEX

H

J

L

M

N

O

P

Win Prizes

ATTENTION READERS

Win Prizes

We request for your frank assessment regarding various aspects of the book as given below :

ELECTRONICS DEVICES AND CIRCUITS

(For RGTU)

By: Dr. SANJAY SHARMA

Please Fill up the Shet in Neat and Clean Writing

(i) What topics of your syllabus, important for your examination point of view, are not covered in this book ?

....

....

....

....

(ii) What are the chapters wherein the treatment of the subject matter is not systematic or organised or updated ?

....

....

....

....

(iii) Have you come across misprints/mistakes/factual inaccuracies in the book? Please specify the chapters and the page numbers.

....

....

....

....

(iv) What other books on the same subject have you found/heard better than the present book? Please specify in terms of price and quality

....

....

....

....

(v) Further suggestions and comments for the improvement of this book.

....

....

....

....

(PS: If need be please attach a separate sheet to write your views)

OTHER DETAILS

(i) Who recommended you this book ? (Please tick (✔) in the box near the option relevant to you.

☐ Teacher ☐ Friends ☐ Bookseller

....

....

....

....

(ii) Name of the recommending teacher, his designation and address :

....

....

....

....

(iii) Name and address of the bookseller you purchased the book from :

....

....

....

....

(iv) Name and address of your institution (Please mention the University or Board as the case may be)

....

....

....

....

(v) Your name and complete postal address :

....

....

....

....

(vi) Course you are enrolled in :

....

....

....

....

The best assessment will be awarded each month: The award will be in the form of our publications as decided by the Editorial Board.

Please mail the filled up coupon at your earliest to :

S.K. KATARIA & SONS®
4885/109, Prakash Mahal, Dr. Subhash Bhargav Lane,
Opposite Delhi Medical Association, Daryaganj, New Delhi-110002 (INDIA)
Phone: +91-11-23243489, 23269324, 43551243; Telefax: +91-11-23243489
e-mail: katariabooks@yahoo.com; katariabook@gmail.com
Website : www.skkatariaandsons.com; For ebooks: www.katsonbooks.com